REACTION MECHANISMS IN ORGANIC CHEMISTRY

REACTION MECHANISMS IN ORGANIC CHEMISTRY

Florin Badea

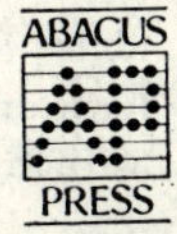

ABACUS PRESS
Tunbridge Wells, Kent

Revised, up-dated English version of the third edition of
MECANISME DE REACTIE IN CHIMIA ORGANICĂ
published in Romanian by Editura Ştiinţifică, Bucharest, in 1974

ABACUS PRESS,
Abacus House, Speldhurst Road, Tunbridge Wells, Kent, England

British Library Cataloguing in Publication Data

Badea, Florin
Reaction mechanisms in organic chemistry.
Bibl. — Index.
ISBN 0-85626-002-9
1. Title 2. Constantinescu, Take V
3. Hammel, John
547'.1'39 QD258
Chemistry, Organic
Chemical reactions

Printed in Romania

Contents

Part 1.
CHEMICAL STRUCTURE

1 The symmetry of organic compounds

2 The chemical bond

Part 2
CHEMICAL REACTIVITY

3 Reactions of organic compounds

4 Elementary reactions. Reaction intermediates

Part 3

HETEROLYTIC REACTIONS

(A) SUBSTITUTION REACTIONS (ALIPHATIC AND AROMATIC)

5 Aliphatic nucleophilic substitution

(A) Methods for the investigation of the mechanism

(B) Reactivity in nucleophilic substitution

6 Aliphatic electrophilic substitution

7 Electrophilic aromatic substitution

8 Aromatic nucleophilic substitution

(B) ELIMINATION AND ADDITION REACTIONS

9 Elimination reactions

10 Addition reactions

Part 4
HOMOLYTIC AND PERICYCLIC REACTIONS

11 Homolytic reactions

12 Thermal and photochemical transformations of unsaturated compounds

Preface

Ever since organic synthesis was discovered, the attempt to prepare an unknown compound by a known method or by an easily conceivable route led often to the discovery of a new reaction, affording a product completely different from the expected one. The baffled organic chemist, faced with such a puzzle, began to exert his mind and to use his skill in order to find out the mysterious laws which govern bond breaking and bond formation in organic reactions. Thus the interest in organic mechanisms arose. However, the first theories relied more on fantasy and imagination rather than on solid facts.

The beginning of accurate studies of organic mechanisms was marked by the use of the methods of physical chemistry, beside the classical methods of organic chemistry (isolation, purification, physical and chemical identification and analysis). The new theories, based mainly on electronic structure, as well as on the positive identification and isolation of certain intermediates, have largely contributed to the interest in the reaction mechanisms. Quantum chemistry, even if applied with certain approximations, enabled in its turn the development of the modern theories of organic reactions. The first part of this book is a simple outline of the present concepts of organic structure and reactivity, which the advanced reader may prefer to skip.

The later developments were based mainly on controversies over the interpretation of experimental results. The reader must be aware of the fact that no known experimental method can afford a detailed and precise description of the way in which the bonds change during a chemical reaction. Actually, the chemist has to take into account a variety of experimental data, obtained by many different techniques, in order to be able to reach a conclusion concerning the reaction mechanism. The correctness of this conclusion depends often on the degree of precision of the experimental techniques used. Clearly, different interpretations of the same data are possible in such a case and more than one reaction mechanism may be found to account for the chemical results. The material dealt with in the present book is not free from such ambiguities. In some instances, alternative explanations are therefore presented, together with their experimental support, in order to enable the reader to reach his own conclusion.

It is not surprising that in his account of the story an author sometimes shows some partiality; the reader himself may also have his own preferences and his weaknesses. The present author is no exception to this rule. A great deal of emphasis was thus given to the contributions of Romanian chemists, and especially of the School of the late Professor Costin D. Nenitzescu, an oustanding personality to whom the author owes so much.

I gratefully acknowledge my indebtness to Dr. T. Constantinescu, who checked the accuracy of the English version and the chemical text. I also acknowledge my gratitude to Professors E. Ciorănescu, M. Avram, I. Necșoiu and A. Balaban for their encouragement and valuable advice during the progress of this manuscript. I am also indebted to many friends in chemistry who have contributed to the improvement of the Romanian editions.

FLORIN BADEA

PART 1
Chemical structure

In any chemical reaction, the structures of the reacting molecules are continuously changing until the reaction products are formed. The process implies a new arrangement of the atoms and bonds and may be described by changes in the geometry and in the electron distribution of the molecules involved. The object of any study of reaction mechanism is to draw up a complete picture of these transformations and to devise a reaction scheme in agreement with the observed reaction energetics and with the molecular geometry of the products. The research worker in this field is thus permanently faced with structural problems, with either geometrical or electronic aspects. It is not by chance that two of the most renowned books on reaction mechanisms [1,2] have stressed in their title the importance of chemical structure for the understanding of reaction mechanisms.

It is assumed that the reader is familiar with the basic principles of chemical structure and in this part only those items of structural organic chemistry that will be necessary in the later chapters will be referred to.

[1] C. K. Ingold, *Structure and Mechanism in Organic Chemistry,* Cornell University Press, Ithaca, New York, 1953; 2nd ed., 1969.

[2] E. S. Gould, *Mechanism and Structure in Organic Chemistry,* Rinehart and Winston, New York, 1959.

1
The symmetry of organic compounds

1.1 Symmetry and chemistry

Symmetry is a general geometrical property of the world in which we live. The Earth, except for some irregularities at the surface, can be regarded as a sphere, having the highest known symmetry. The same is true for the Sun, the Moon, the stars. The human body itself has a plane of symmetry. Man's tools, as well as other objects produced by his imagination and skill, do not lack the pleasant and artistic aspect conferred by symmetry. This general property is also found in the atoms and molecules with which a chemist is concerned.

A chemical compound can be regarded, in a simple manner, as a collection of points (the atoms) and lines (the bonds) connected in a certain way, which make up a geometrical framework. It is easily seen that some regularly built organic molecules, such as benzene, cubane [1], prismane [2], or tetrahedrane [3], possess a high symmetry. However, there are only a few molecules with such regular bond frameworks. Moreover, these compounds, except benzene which is indeed very common, may be regarded merely as chemical curiosities devised by the ingenuity, skill and application of the research worker in organic chemistry, trying to build up all conceivable geometrical structures. An advanced student should find no

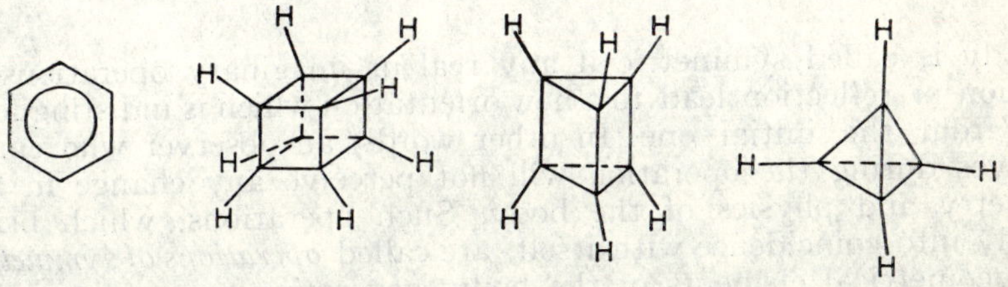

1 2 3 4

[1] P. E. Eaton and J. W. Cole, Jr., *J. Amer. Chem. Soc.*, **86**, 3157 (1964); E. B. Fleisher, *ibid.*, **86**, 3889 (1964).

[2] W. E. Wilzbach and L. Kaplan, *J. Amer. Chem. Soc.*, **87**, 4004 (1965).

[3] L. B. Rodewald and H. Lee, *J. Amer. Chem. Soc.*, **95**, 623 (1973); **95**, 3084 (1973).

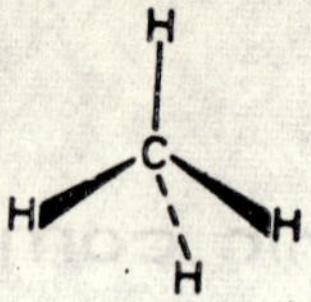

5 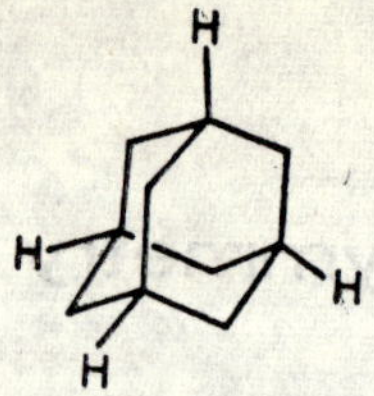 6

difficulty in perceiving that tetrahedrane (**4**), methane (**5**) and even adamantane (**6**)[4] are geometrically related through their similar symmetry. The purpose of this chapter is to present a classification of chemical compounds, particularly organic species, through their symmetry properties.

One may wonder whether the knowledge of the molecular symmetry is of any use for the study of organic reaction mechanisms. This question will be duly answered in the last sections of this chapter. For the moment, let us mention that a research worker is frequently surprised to find that molecules made up of different kinds and/or numbers of atoms display similar physical and chemical properties. He may even notice that a slight modification in the structure of his compounds can dramatically change the course of the reaction. Often, the answer to such chemical puzzles is found in the symmetry of the compounds or of the reaction intermediates.

1.2 Operations and elements of symmetry

A body is called symmetric if any real or imaginary operations of rotation or reflection lead to a new orientation which is indistinguishable from the initial one. In other words, an observer who closes his eyes during the operation will not perceive any change in the geometry and physics of the body. Such operations, which bring a body into coincidence with itself, are called *operations of symmetry*. The geometrical elements of the body, generating symmetry operations, are called *symmetry elements*; they may be a point, an axis or a plane, with respect to which the symmetry operations are performed.

1.2.1 *The rotation operation*

The rotation operation is performed with respect to a symmetry axis of the body. The order of the rotation axis is given by the ratio between 2π (a complete rotation about the axis to bring the body

[4] P. von R. Schleyer and R. C. Ford, *Chem. Rev.*, **64**, 262 (1964).

into the initial orientation) and θ, the actual rotation performed in order to obtain a new equivalent representation:

$$n = \frac{2\pi}{\theta} \tag{1.1}$$

where n is always an integer. The axis thus defined is a n-fold rotation axis, C_n.

Let us consider an organic molecule with a set of rotation axes, for example benzene (Figure 1.1). The rotation axis of highest order is perpendicular to the molecular plane and passes through the centre of the molecule. (For a better understanding, the six carbon atoms will be labelled, as shown in Figure 1.1*a*.) A simple rotation of 60° about the axis generates the representation *b* of the molecule, physically indistinguishable from *a*, although different as the labels show. This axis is then a sixfold rotation axis, C_6.

The presence of this axis implies a set of other rotation operations. Let us perform with *b* another rotation C_6 leading to representation *c*. This representation is thus obtained from *a* by two successive C_6 rotations, but could also result directly by a rotation by 120° (that is a C_3 rotation). Using the algebraic formalism, this conclusion can be expressed as follows:

$$C_6 \cdot C_6 = C_6^2 = C_3 \tag{1.2}$$

Representation *d* results in its turn either by three successive C_6 rotations or directly by a twofold rotation about the same axis. Figure 1.1 indicates all the rotations implied by the highest rotation axis of the benzene molecule, namely:

$$C_6;\ C_6^2 = C_3;\ C_6^3 = C_2;\ C_6^4 = C_3^{-1};\ C_6^5 = C_6^{-1};\ C_6^6 = C_1 = E \tag{1.3}$$

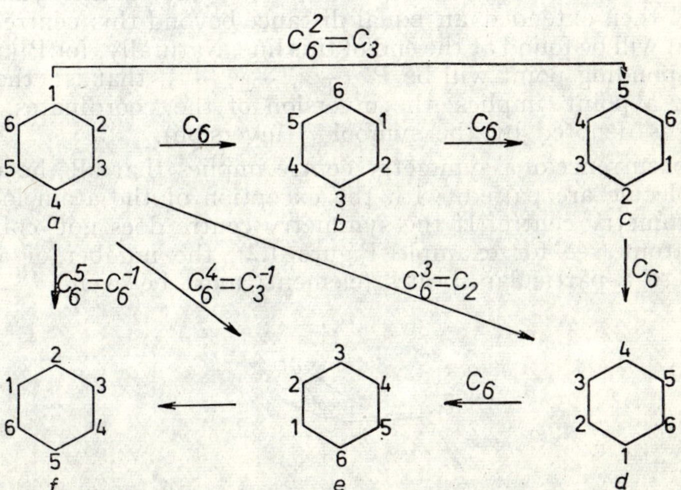

Figure 1.1 Possible rotations of the benzene molecule about the C_6 rotation axis.

Relations (1.2) and (1.3) can be generalized as follows:

(a) The product of two or more symmetry operations is defined by the successive performance of the individual operations.

(b) A n-fold rotation axis, C_n, implies the existence of all powers of the rotation operation, *i. e.* C_n, C_n^2, C_n^3, ..., C_n^{n-1}, $C_n^n = C_1$.

(c) The last rotation of the set, $C_n^n = C_1$, brings the molecule into the initial orientation. The same result could be obtained without performing any operation at all (leaving the molecule in the initial position). This case corresponds to the so-called identity operation, denoted by E.

(d) Sincc n is an integer, the powers of C_n could be simplified as follows:

$$C_n^k = C_p\,,\ \text{if}\ p = n/k\ \text{is an integer.} \tag{1.4}$$

If n/k is not an integer, the following simplification is used:

$$C_n^k = C_n^{k-n},\ \text{with}\ k - n\ \text{negative or positive.} \tag{1.5}$$

The simultaneous occurrence of negative and positive powers of C_n requires the definition of the positive sense of the rotation, which is usually a clockwise rotation. Clearly,

$$C_n^{k-n} = C_p^{-1},\ \text{if}\ p = n/k - n\ \text{is an integer.} \tag{1.6}$$

1.2.2 *The inversion operation*

A body admits a symmetry centre if any point (P) of the body is reflected in this centre into another point (P′), equivalent to it. In other words, if a line is drawn from a point to the symmetry centre and then extended an equal distance beyond this centre, a similar point will be found at the end of this line. Naturally, for P (x,y,z), the corresponding point will be P′ $(-x, -y, -z)$, that is: the reflection in a point implies the inversion of the coordinates. This operation is denoted by the symbol i (inversion).

The occurrence of a symmetry centre implies that all the atoms of the molecule are paired, with the exception of the atom located in the symmetry centre. If the symmetry centre does not coincide with an atom (see for example Figure 1.2), the number of atoms belonging to a particular species (element) must be even.

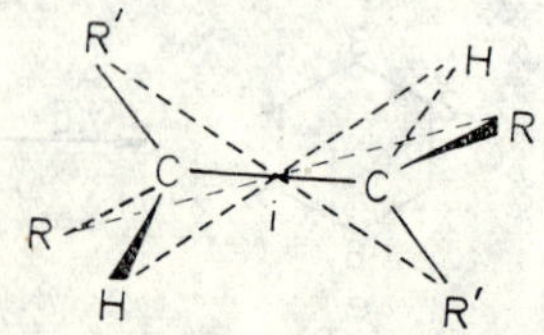

Figure 1.2 Molecule with an inversion centre.

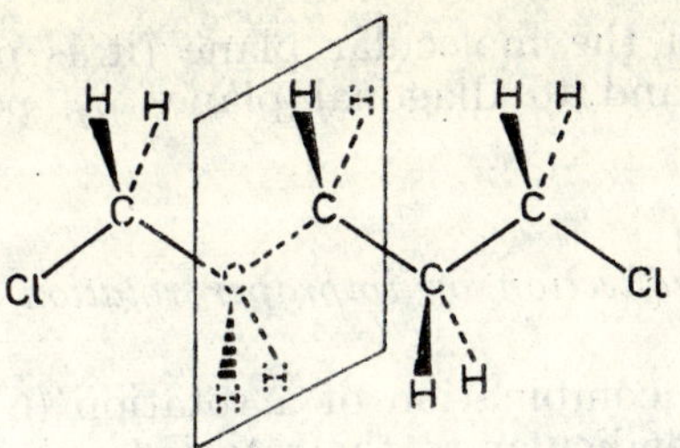

Figure 1.3 A symmetry plane divides a molecule in two halves, which are related to each other as an object to its mirror image.

If we repeat the inversion operation (i^2), we come across the initial representation. An additional operation (i^3) will correspond to a single inversion. We can therefore write:

$$i^n = E, \text{ if } n \text{ is even, and}$$
$$i^n = i, \text{ if } n \text{ is odd.} \tag{1.7}$$

1.2.3 *The reflection in a plane of symmetry*

This symmetry operation is denoted by σ (from German: Spiegel = mirror). The existence of a plane of symmetry implies that the molecule is made up of two halves, one of them being the mirror image of the other. Any point situated in one half of the body must be reflected in an equivalent, indistinguishable point in the other half. As it is easily seen from Figure 1.3:

$$\sigma^n = E, \; n \text{ even, or}$$
$$\sigma^n = \sigma, \; n \text{ odd.} \tag{1.8}$$

An organic molecule, such as benzene, may have several reflection planes (Figure 1.4). In such cases, we refer to a horizontal plane,

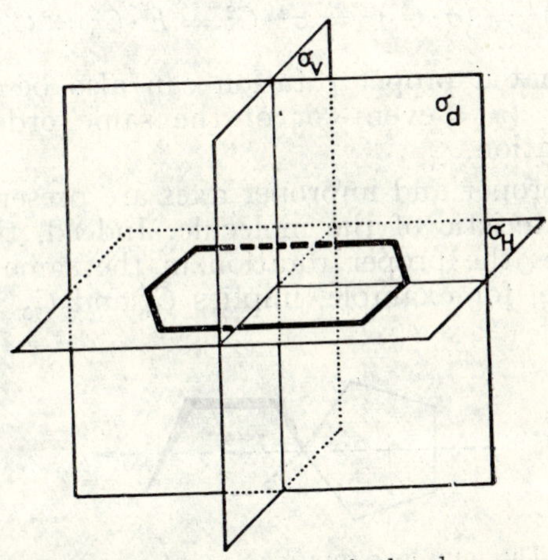

Figure 1.4 Some of the symmetry planes of the benzene molecule; σ_H is the molecular plane; σ_v (only one of the three σ_vs is represented) divides two opposite bonds; σ_d (only one represented) passes through two opposite carbon atoms.

σ_H, coinciding with the molecular plane (it is usually unique*), to vertical planes σ_v and to diagonal planes σ_d, perpendicular to the horizontal plane.

1.2.4 *The rotation-reflection or improper rotation*

This operation is a combination of a rotation (C_n) with a reflection (σ) in a plane perpendicular to the rotation axis. Any of these component operations, taken alone, may not constitute an element of symmetry for the body. However, taken together, they are able to generate a new representation of the body, equivalent to the original one. The symbol of the rotation-reflection operation is S_n, where n is the order of the improper axis.

$$S_n = \sigma C_n \tag{1.9}$$

The understanding of this symmetry operation is of utmost importance for the organic chemist as the absence of the element is related to optical activity (see for example Section 1.8).

Let us consider the spirane shown in Figure 1.5. It is easily seen that the axis passing through the central atom (shared by the two cyclopentane rings) is a C_2 axis. However, if the molecule is rotated only by 90° about this axis and then the non-equivalent representation which results is reflected in a plane perpendicular to the axis and passing through the spirane atom (N.B. this is not a symmetry plane of the molecule), an equivalent representation is also obtained. Using this example we can make the following generalizations:

(a) An improper rotation does not exclude a proper rotation. Indeed, if we repeat an improper rotation:

$$(S_n)^2 = (\sigma \cdot C_n)^2 = \sigma^2 \cdot C_n^2 = E \cdot C_n^2 = C_{n/2} \tag{1.10}$$

It is clear that a proper rotation can also be involved, either of a lower order ($n =$ even) or of the same order ($n =$ odd) as the improper rotation.

(b) If only proper and improper axes are present, the improper rotation is characteristic of the molecule. Indeed, the improper rotation axis implies the proper rotation in the same manner as the C_6 axis of benzene, for example, implies C_3 and C_2 axes (coinciding with C_6).

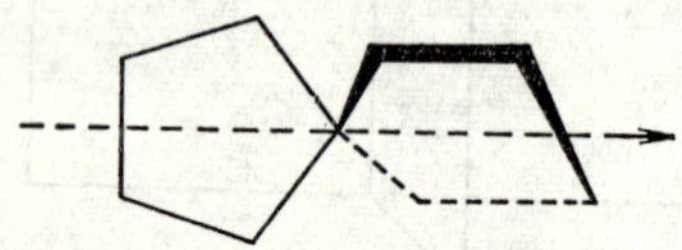

Figure 1.5 The improper rotation, S_4, of a spirane molecule.

* An obvious exception is cubane (**2**), which has three σ_Hs (see p. 19).

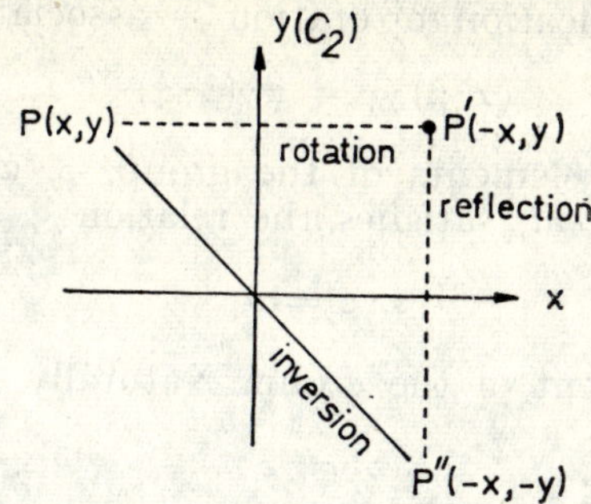

Figure 1.6 The identity between a twofold improper rotation and an inversion around a symmetry centre.

The presence of other symmetry elements in a molecule may imply the presence of an improper axis. Thus, each molecule has the identity element, E. Since

$$C_1 = E, \tag{1.11}$$

we can write:

$$\sigma = E \cdot \sigma = C_1 \cdot \sigma = S_1 \tag{1.12}$$

The occurrence of a onefold improper axis confers a higher order of symmetry than a onefold proper rotation axis (bodies possessing only the element $C_1 = E$ are actually asymmetric, as no symmetry operation, except the identity operation, can be performed on them). The same is true for higher orders of the improper axis, $S_n > C_n$.

It can be easily demonstrated (Figure 1.6) that an inversion operation, i, implies a twofold improper rotation:

$$i = S_2 \tag{1.13}$$

The inversion in a point is therefore superior to a rotation or to a reflection in a mirror plane ($S_2 > S_1$).

1.2.5 *Group theory*[5–7]

A collection of elements (either specified or, more importantly, unspecified) is called a *group* if it satisfies the following four axioms:

(a) A multiplication operation in the group is defined in such a manner that:

$$\alpha \cdot \beta = \gamma \tag{1.14}$$

α, β, γ being any element of the group.

[5] D. S. Schonland, *Molecular Symmetry*, D. van Nostrand, London (1965).

[6] L. H. Hall, *Group Theory and Symmetry in Chemistry*, McGraw-Hill, New York (1969).

[7] H. H. Jaffé and M. Orchin, *Symmetry in Chemistry*, J. Wiley, New York (1965).

(b) The multiplication operation is associative, that is:

$$(\alpha \cdot \beta) \cdot \gamma = \alpha \cdot (\beta \cdot \gamma) \tag{1.15}$$

(c) One of the elements of the group, ε, called the *identical element* or *unity element,* satisfies the relation:

$$\alpha \cdot \varepsilon = \alpha \tag{1.16}$$

where α is any element of the group. Naturally,

$$\varepsilon^n = \varepsilon \tag{1.17}$$

for any value of n.

(d) For any element of the group, α, there is another element, β, called the *inverse element,* so that:

$$\alpha \cdot \beta = \varepsilon \tag{1.18}$$

The inverse element is sometimes denoted by α^{-1}.

The molecular symmetry elements constitute a so-called *point group* *. Indeed, all the symmetry operations defined for a molecular species satisfy the above-mentioned axioms:

(a) The operation defined in a molecular point group is the multiplication of symmetry operations, as indicated in Section 1.2.1. The product of two symmetry operations is the result of the successive performance of these operations. It can be easily demonstrated that such a product is equivalent to another operation in the group (see for example the C_6 set of elements found in the benzene group).

(b) The second axiom is also satisfied, as can be proved from the examples already given (the C_6 group).

(c) The identity operation corresponds, in a molecular point group, to the identity, E. For mathematical reasons, this operation, which is actually no operation at all, is also taken into consideration.

(d) For any element of the molecular point group, there is an inverse element, defined by the sense (positive or negative) of the corresponding symmetry operation. Thus:

$$C_n \cdot C_n^{-1} = E \tag{1.19}$$

$$i \cdot i^{-1} = i^2 = E \tag{1.20}$$

$$\sigma^n \cdot \sigma^n = \sigma^{2n} = E \quad (n \text{ odd}) \tag{1.21}$$

$$S_n \cdot S_n^{-1} = \sigma^2(C_n \cdot C_n^{-1}) = E \cdot E = E \tag{1.22}$$

It is not difficult to see that the inversion i and the reflection σ correspond also to their inverse element.

* When all symmetry operations are performed on a molecule, at least one point, situated on the symmetry elements, does not change its place.

1.3. Molecular point groups

Any molecule can be assigned to a molecular point group, depending on its characteristic set of symmetry elements. The notation of molecular symmetry groups, based on the $C - \sigma - i$ symbols of symmetry elements, largely used in chemistry, is called the Schönfliess notation, and will be adopted throughout this book *. For a rapid identification, it is convenient[6] to distinguish three types of molecular point groups.

1.3.1 *Type 1: molecules with low symmetry*

This type includes molecules possessing either no symmetry elements (except the identity element E) or only one characteristic element: a rotation (C_n) or a reflection in a centre (i) or in a mirror plane (σ).

Group C_1. This group has no symmetry element, except the identity, E. We may consider therefore that it has a onefold proper rotation axis, since $C_1 = E$.

This group includes all molecules possessing one asymmetric atom (C, N, P, Si, etc.). It will be shown later that only in this case the term asymmetric is correctly used.

Groups C_n have only a n-fold proper rotation axis ($n = 2$, 3, 4, etc.). It can be shown that molecules belonging to these groups, even if symmetric, possess a non-superimposable mirror image. Such molecules are called *dissymmetric* (see Section 1.8). The number of elements in the group is n: C_n, $C_n^2 ... C_n^{n-1}$, $C_n^n = C_1 = E$.

Group C_s possesses only two elements, E and σ. Many planar molecules, such as the non-symmetric aromatic compounds (Figure 1.7) belong to this molecular point group.

Group C_i has also only two elements, E and i.

It must be remembered that the last two groups (C_s and C_i, respectively) imply also an uncharacteristic improper rotation (S_1 and S_2, respectively, see p. 25).

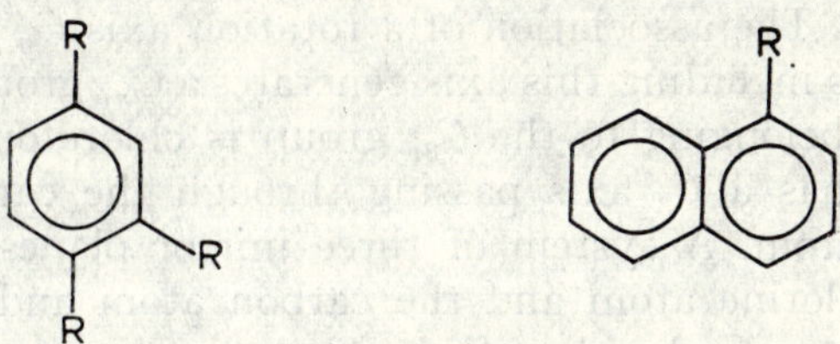

Figure 1.7 The point group C_s.

* In crystallography a different notation is used.

1.3.2 *Type 2: molecules of intermediate symmetry*

Characteristic for this type of symmetry groups is the presence of a rotation axis, C_n, associated with other symmetry elements (σ, C_2, i). The lowest symmetry groups are the S_n groups (n even), the highest ones are the D_{nh} and D_{nd} groups.

Groups S_n. The presence of an improper rotation axis is usually associated with other symmetry elements, such as proper rotations, symmetry centre or mirror planes. Let us consider, for example, the spirane from Figure 1.5, possessing an S_4 axis. This improper rotation will generate other elements by its successive repetition, that is:

$$S_4 = \sigma C_4; \quad (S_4)^2 = (\sigma C_4)^2 = C_2; \quad (S_4)^3 = (\sigma C_4)^3 = \sigma C_4^3 = S_4^3;$$
$$(S_4)^4 = (\sigma C_4)^4 = E$$

Thus, the S_4 group implies two improper rotations* (S_4, $S_4^3 = S_4^{-1}$), one proper rotation (C_2) of lower order than the improper rotation, and the identity, that is four elements in all.

Let us now assume that a molecule possesses an improper rotation axis S_3 (n odd). That will imply the following elements:

$$S_3 = C_3; \ (S_3)^2 = (\sigma C_3)^2 = C_3^2; \ (S_3)^3 = (\sigma C_3)^3 = \sigma; \ (S_3)^4 = (\sigma C_3)^4 = C_3;$$
$$(S_3)^5 = (\sigma C_3)^5 = \sigma C_3^2 = S_3^{-1}; \quad (S_3)^6 = (\sigma C_3)^6 = E$$

The above equations show that the S_3 group involves two proper rotations, two improper rotations, a reflection and the identity, that is six operations in all. This number can be obtained by multiplying by two the order of $S(2n = 2 \times 3)$.

Comparing the two examples we shall note that the S_n group with n even possesses a lower symmetry (only n elements in the group) than the S_n group with n odd ($2n$ elements). Moreover, the improper rotation S_n is characteristic if n is even (the improper rotation has a higher order than the proper rotation), but is not characteristic if n is odd (proper and improper rotation axes of the same order). Actually, S_n groups include only molecules whith n even (see groups C_{nh}).

Groups C_{nv}. The association of a rotation axis C_n with n vertical reflection planes including this axis generates a C_{nv} group. An example of a molecule belonging to the C_{3v} group is chloroform (Figure 1.8). This molecule has a C_3 axis passing through the carbon atom and the hydrogen atom. A system of three mirror planes, each passing through one chlorine atom and the carbon atom and including the rotation axis, can also be identified. The symmetry elements of the

* The reader will note that the symbol $(S_n)^p$ means the p-fold repetition of an S_n operation, whereas S_n^p means σC_n^p.

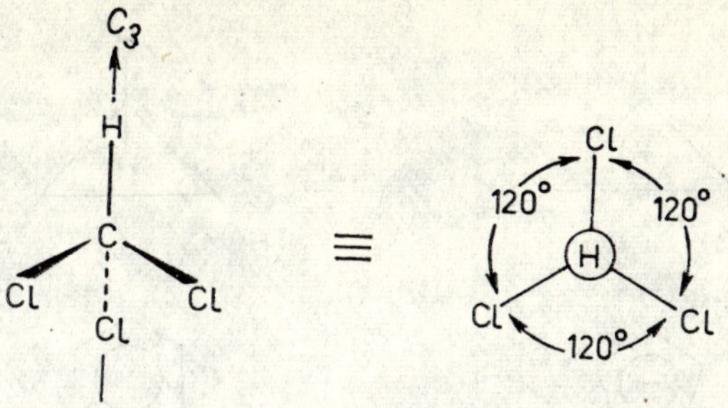

Figure 1.8 The symmetry of chloroform.

group are: E, C_3, $C_3^2 = C_3^{-1}$, σ_v, σ_v', σ_v'', that is $2n$ elements. Obviously, the C_{nv} group is more symmetric than the S_n group (n even).

Groups C_{nh}. A rotation axis, C_n, perpendicular to a horizontal mirror plane, σ_H, gives rise to a C_{nh} group. A few such molecules can be seen in Figure 1.9. For the C_{3h} group, the elements of symmetry are:

$$C_3;\ C_3^{-1};\ \sigma;\ \sigma C_3 = S_3;\ \sigma C_3^{-1} = S_3^{-1};\ E.$$

It must be noted that when n is odd the C_{nh} group is identical with the S_n group. That is the reason why in the S_n groups are included only molecules with n even.

In the identification of this group one must be careful to exclude molecules with other mirror planes than the horizontal molecular plane. For example, both planar boric acid (Figure 1.9) and methenium ion (CH_3^+) have a C_3 axis perpendicular to the molecular mirror plane. However, no other reflection planes can be identified in boric acid, whereas in methenium ion other reflection planes, perpendicular to the molecular plane, are also present. Boric acid belongs obviously to the C_{3h} group, while for methenium ion we must find a higher symmetry group.

Groups D_n. A D_n group is generated by a high-order rotation axis, C_n ($n \geqslant 2$), and n C_2 axes perpendicular to it. Actually, from the chemical point of view (see p. 47), these groups must be considered of low symmetry. However, there are only a few molecular

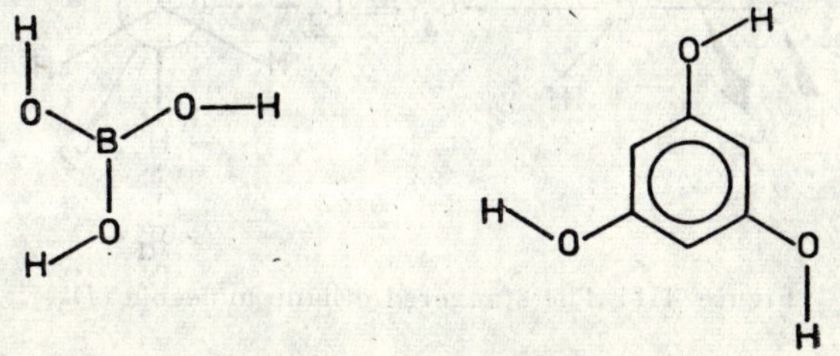

Figure 1.9 Molecules with C_{3h} symmetry.

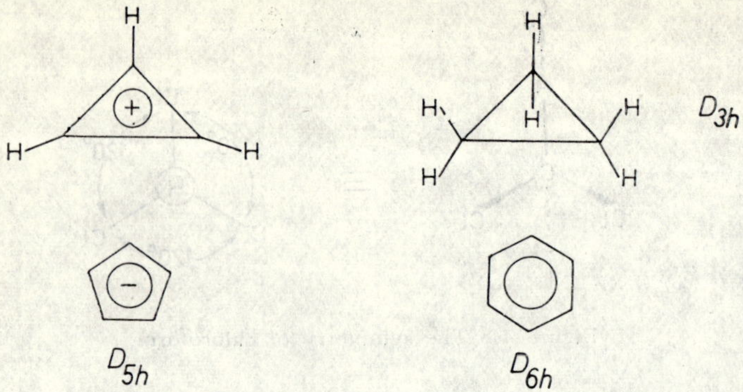

Figure 1.10 Molecules with D_{nh} symmetry.

species which belong to these groups*. Usually, D_n elements are associated either with σ_H or σ_d mirror planes.

Groups D_{nh}. To these groups belong all cyclic neutral or ionic polyene systems (Figure 1.10). For the benzene molecule (D_{6h}), the C_6 axis perpendicular to the molecular plane (σ_H) will also coincide with improper rotation axes ($2S_6$, $2S_3$, $S_2 = i$), in addition to the proper rotations associated with C_6 (C_6^{-1}, $2C_3$, C_2). Six C_2 axes are contained in the molecular plane, each of them being also found in a vertical plane or a diagonal plane. The total number of elements in the D_{6h} group is $4n = 4 \times 6 = 24$.

Groups D_{nd}. The D_{nd} groups are generated by the association of the D_n elements with n diagonal reflection planes. An example (Figure 1.11) is the staggered ethane molecule. The C_3 axis, passing through the C—C bond, coincides also with an S_6 axis. A number of three diagonal planes, intersecting at the molecular axis, pass through two opposite hydrogen atoms. The three C_2 axes intersect at the centre of the C—C bond, each of them bisecting a dihedral angle H—C(1)—C(2)—H. It can be shown that these groups possess also a number of $4n$ elements.

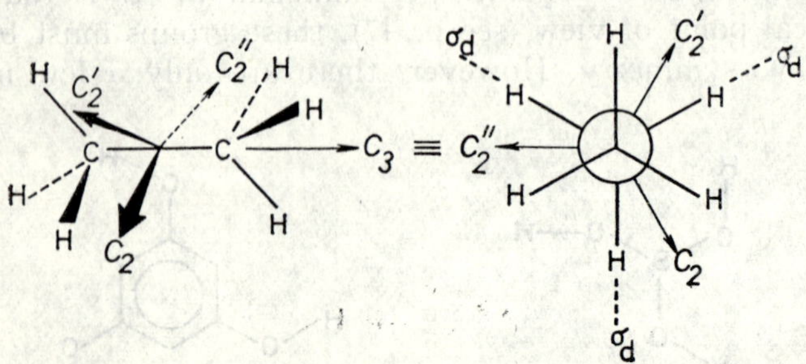

Figure 1.11 The staggered ethane molecule (D_{3d}).

* See p. 46 for some examples.

1.3.3 *Type 3: molecules of high symmetry*

This type of symmetry groups includes linear molecules or molecules related through their geometry to regular geometric bodies, such as the tetrahedron (T), the octahedron (O) or the icosahedron (I).

Linear molecules. The molecular axis of a linear molecule contains a C_∞ rotation axis. Indeed, any small rotation about this axis will bring the molecule into coincidence with itself, hence:

$$n = \frac{360°}{\sim 0°} \approx \infty$$

At the same time, a system of infinite mirror planes will intersect at the molecular axis. Linear molecules with a symmetry centre, such as H—H, Cl—Cl, O=C=O, etc., belong to the $D_{\infty h}$ group, the centre of these molecules being also the site of intersection of an infinite number of C_2 axes, all contained in a σ_H reflection plane. Molecules devoid of a symmetry centre (H—Cl, C=O, D—C≡C—H, etc.) are ascribed to the $C_{\infty v}$ group.

Group T_d. The molecule of methane (Figure 1.12) has a system of four C_3 axes, each of them passing through one hydrogen atom and the carbon atom. This system of four C_3 rotation axes is characteristic to the tetrahedron group (T) and will be sufficient for identification. The other elements of the T_d group are easily found if we inscribe the molecule in a cube (Figure 1.12). It can be seen that each of the x, y and z coordinate axes contain a C_2 proper rotation and an S_4 improper rotation axis. A number of six diagonal reflection planes bisect the six different H—C—H angles. Thus, the T_d group has 24 elements.

Group O_h. To this group belong the octahedral complexes as well as cubane (see p. 19). Using the latter as an example, we can identify a characteristic system of three C_4 rotation axes.

Group I_d. Molecules belonging to this group are rarely encountered in organic chemistry. We shall mention, as an example, the

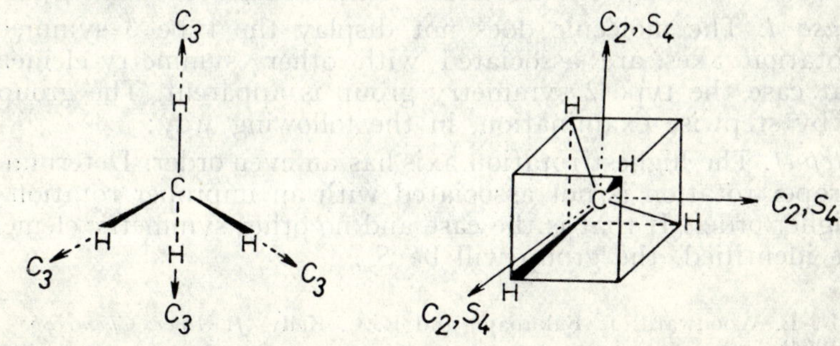

Figure 1.12 Methane (T_d).

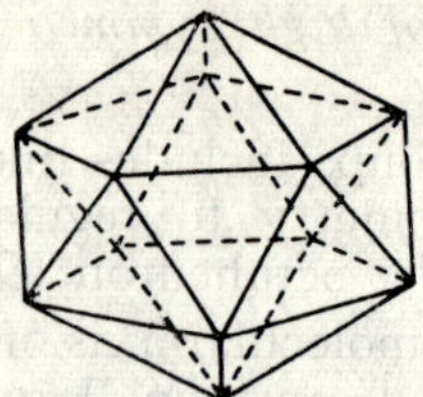

Figure 1.13 The icosahedron (I_h). A C_5 rotation axis passes through each of the 12 vertices.

icosahedrane (Figure 1.13). It can be seen that characteristic to I groups is the presence of twelve C_5 rotation axes. Dodecahedrane [8,9] belongs to the same group.

1.4 The rapid identification of molecular point groups

In the literature [5–7], several procedures are recommended for the rapid identification of the symmetry point groups. The following procedure, based on the three types of symmetry groups[6], seems very convenient. Indeed, a simple inspection of the molecular structure is very often sufficient to decide whether the molecules belong to the high-symmetry type or to the low-symmetry type.

Case 1. The molecule is linear. In that case only two groups can be envisaged: $D_{\infty h}$ or $C_{\infty v}$. The decision is easily made after ascertaining whether the molecule has or has not a centre of symmetry.

Case 2. The molecule has a regular geometry, suggesting type 3 symmetry group. In that case we must identify more than one high-order rotation axes $C_n (n \geqslant 3)$. The T_d group will be indicated by four C_3 axes, the O_h by three C_4 axes and the I_d by twelve C_5 axes.

Case 3. The low-symmetry type of group is indicated by the presence of only one symetry element (C_n, σ or i). If the molecule has no symmetry elements, it belongs to the C_1 group. If it has only one rotation axis C_n, the C_n group must be considered. The presence of only one reflection plane corresponds to the C_s group, and the identification of only one inversion centre indicates the C_i group.

Case 4. The molecule does not display the type 3 symmetry, but rotation axes are associated with other symmetry elements. In that case the type 2 symmetry group is apparent. The group is found by stepwise examination, in the following way:

Step 1. The highest rotation axis has an even order. Determine if this proper rotation is not associated with an improper rotation S_n of a higher order. If that is the case and no other symmetry elements can be identified, the group will be S_n.

[8] R. B. Woodward, T. Fukunagu and R. C. Kelly, *J. Amer. Chem. Soc.*, **86**, 3162 (1964).

[9] H. P. Schultz, *J. Org. Chem.*, **30**, 1361 (1965).

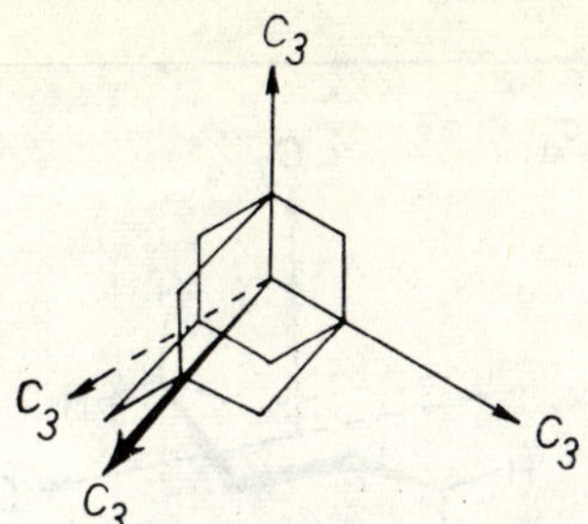

Figure 1.14 The T_d symmetry of adamantane.

Step 2. The rotation axis is also associated with mirror planes This may indicate the C_{nv} group. To make sure, identify the n vertical mirror planes.

Step 3. The rotation axis C_n is perpendicular to the molecular plane. If that plane contains no C_2 axes, the C_{nh} group must be ascribed to the molecule.

Step 4. If the rotation axis C_n is perpendicular to the molecular plane and this plane contains n C_2 axes, the molecule belongs to the D_{nh} group.

Step 5. The molecule is not planar, but the C_n axis is associated with nC_2 axes perpendicular to it. The D_n elements are present here. The D_{nh} group being excluded (the molecule is not planar), look for diagonal planes, which indicate the D_{nd} group.

1.5 Examples of molecular group identification

Example 1. Find the molecular group of the acetylene molecule.

The molecule is linear and will correspond to Case 1 examined above. Since the molecule has a symmetry centre located in the middle of the $C{\equiv}C$ bond, it belongs to the $D_{\infty h}$ group.

Example 2. Find the molecular point group of adamantane.

The geometry of the molecule (Figure 1.14) suggests type 3 symmetry. A number of four C_3 axes passing through the bridgehead CH groups can be identified, suggesting thus the T group. A careful examination reveals also the presence of the other elements of the T_d group. Thus, methane and adamantane belong to the same symmetry group.

Example 3. Identify the molecular point group of the eclipsed ethane molecule (Figure 1.15).

Figure 1.15 The eclipsed ethane (D_{3h}).

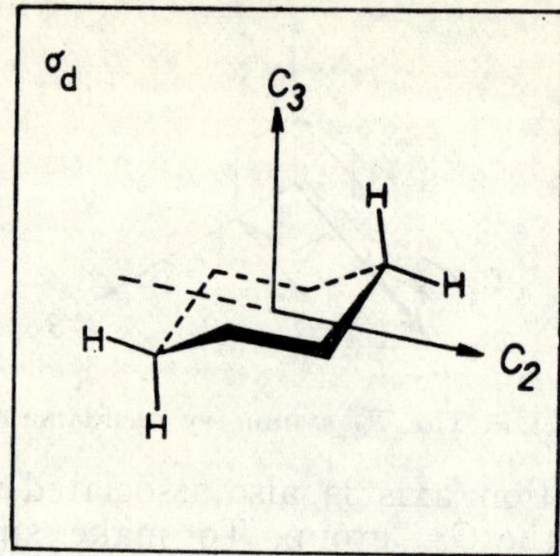

Figure 1.16 Cyclohexane (chair, D_{3d}). Only one C_2 axis represented.

The molecule has a C_3 rotation axis passing through the molecular axis. A mirror plane passes through the middle of that axis and is perpendicular to it. This plane contains three C_2 axes, which indicates the D_{nh} goup. This example is significant in that it shows that D_{nh} groups are not always associated with planar molecules.

Example 4. Find the molecular point group of the chair form of cyclohexane (Figure 1.16).

A C_3 rotation axis, passing through the centre of the molecule, is easily identified. Intersecting at this axis, there is a system of three diagonal planes passing through two opposite carbon atoms. This indicates either C_{3v} or D_{3d} groups. One should now look for C_2 axes. These are actually found, passing through the middle of two opposite C--C bonds. Cyclohexane belongs therefore to the D_{3d} group.

1.6 The representations of symmetry elements

1.6.1 *Symmetry elements as operators*

A symmetry element indicates the operation which can be performed on any point of the molecular frame in order to obtain a point equivalent to the former. Thus, any symmetry element is an operator. In order to define it better, let us consider a point P from the three-dimensional space, whose position is defined by the vector **p**. The action of a symmetry element R will bring the point P into the point P′, defined by the vector **p**′. Using the mathematical formalism, we may express this by the equation:

$$\mathrm{R}\mathbf{p} = \mathbf{p}' \tag{1.23}$$

Now, **p**′ will have the same scalar value as the vector **p**, the two vectors differing only in their direction. We may wonder what relation there is between the new direction and the operator R.

In order to find such a relation it is useful to define the vector **p** through three *unitary vectors*[5], called also *base vectors*, oriented along the x, y, z coordinate axes.

$$\mathbf{p} = x_1\mathbf{e}_1 + x_2\mathbf{e}_2 + x_3\mathbf{e}_3 \tag{1.24}$$

The action of the operator R will also change the base vectors $\mathbf{e}_1$, $\mathbf{e}_2$, $\mathbf{e}_3$ into the vectors $\mathrm{R}\mathbf{e}_1$, $\mathrm{R}\mathbf{e}_2$ and $\mathrm{R}\mathbf{e}_3$. These new vectors result as a linear combination of the initial vectors, that is:

$$\mathrm{R}\mathbf{e}_1 = r_{11}\mathbf{e}_1 + r_{12}\mathbf{e}_2 + r_{13}\mathbf{e}_3 \tag{1.25 a}$$

$$\mathrm{R}\mathbf{e}_2 = r_{21}\mathbf{e}_1 + r_{22}\mathbf{e}_2 + r_{23}\mathbf{e}_3 \tag{1.25 b}$$

$$\mathrm{R}\mathbf{e}_3 = r_{31}\mathbf{e}_1 + r_{32}\mathbf{e}_2 + r_{33}\mathbf{e}_3 \tag{1.25 c}$$

the terms r_{ij} being scalars and depending on the particular symmetry operation.

Relation (1.23) can be thus rewritten as follows:

$$\mathrm{R}\mathbf{p} = \mathbf{p}' = x_1'\mathbf{e}_1 + x_2'\mathbf{e}_2 + x_3'\mathbf{e}_3 \tag{1.26}$$

in which:

$$x_1' = r_{11}x_1 + r_{12}x_2 + r_{13}x_3 \tag{1.27 a}$$

$$x_2' = r_{21}x_1 + r_{22}x_2 + r_{23}x_3 \tag{1.27 b}$$

$$x_3' = r_{31}x_1 + r_{32}x_2 + r_{33}x_3 \tag{1.27 c}$$

Relations (1.27) can be written in turn as a matrix product:

$$\begin{pmatrix} x_1' \\ x_2' \\ x_3' \end{pmatrix} = \begin{pmatrix} r_{11} & r_{12} & r_{13} \\ r_{21} & r_{22} & r_{23} \\ r_{31} & r_{32} & r_{33} \end{pmatrix} \begin{pmatrix} x_1 \\ x_2 \\ x_3 \end{pmatrix}$$

The square matrix is called the *matrix representation* of the symmetry element R in the three-dimensional space. Of course, a matrix representation can be thus defined for any set of base vectors $\mathbf{e}_1, \mathbf{e}_2, \dots, \mathbf{e}_n$, determining a n-dimensional space. The three-dimensional physical space allows, however, the setting up of representations for the most common symmetry elements. Let us first consider the identity element E. This operator will "change" any point into itself, that is:

$$x_1' = 1 \cdot x_1 + 0 \cdot x_2 + 0 \cdot x_3$$

$$x_2' = 0 \cdot x_1 + 1 \cdot x_2 + 0 \cdot x_3$$

$$x_3' = 0 \cdot x_1 + 0 \cdot x_2 + 1 \cdot x_3$$

The matrix representation of the identity will thus be the *unitary matrix* (1.28):

$$E = \begin{pmatrix} 1 & 0 & 0 \\ 0 & 1 & 0 \\ 0 & 0 & 1 \end{pmatrix} \tag{1.28}$$

Let us further consider other symmetry elements. We may choose our coordinate system in such a manner that for a certain $\mathbf{Re}_i$ vector:

$$\mathbf{Re}_i = r_{i1}\mathbf{e}_1 + \ldots + r_{ii}\mathbf{e}_i + \ldots + r_{in}\mathbf{e}_n$$

all terms $r_{ij}\mathbf{e}_j$ $(i \neq j)$ vanish except for $i = j$, that is:

$$\mathbf{Re}_i = 0 \cdot \mathbf{e}_1 + \ldots + 1 \cdot \mathbf{e}_i + \ldots + 0 \cdot \mathbf{e}_n$$

It is convenient to choose this vector as the $\mathbf{Re}_1$ vector. Therefore, we shall determine the matrix representation of any symmetry element which does not affect the $\mathbf{e}_1$ vector. Considering a rotation by $\alpha°$ about the z axis (containing the $\mathbf{e}_1$ vector), it can be easily shown (Figure 1.17) that:

$$\mathbf{Re}_1 = \mathbf{e}_1$$
$$\mathbf{Re}_2 = \cos\alpha\, \mathbf{e}_2 + \sin\alpha\, \mathbf{e}_3$$
$$\mathbf{Re}_3 = -\sin\alpha\, \mathbf{e}_2 + \cos\alpha\, \mathbf{e}_3$$

which gives the matrix representation:

$$\mathbf{C}_\alpha = \begin{vmatrix} 1 & 0 & 0 \\ 0 & \cos\alpha & \sin\alpha \\ 0 & -\sin\alpha & \cos\alpha \end{vmatrix} \tag{1.29}$$

For a reflection in the $z-y$ plane, the operator σ_{xy} will not change the $\mathbf{e}_1$ and $\mathbf{e}_2$ vectors, but will reverse the $\mathbf{e}_3$ vector:

$$\sigma_{xy} = \begin{vmatrix} 1 & 0 & 0 \\ 0 & 1 & 0 \\ 0 & 0 & -1 \end{vmatrix} \tag{1.30}$$

For each symmetry element, one may thus find a great number of representations, differing in their numerical value but expressing the same kind of information regarding the effect of the considered operator. The form of the matrices depends upon the particular base vectors chosen.

In Section 1.6.2 we shall try to classify and characterise all possible representations of the symmetry elements.

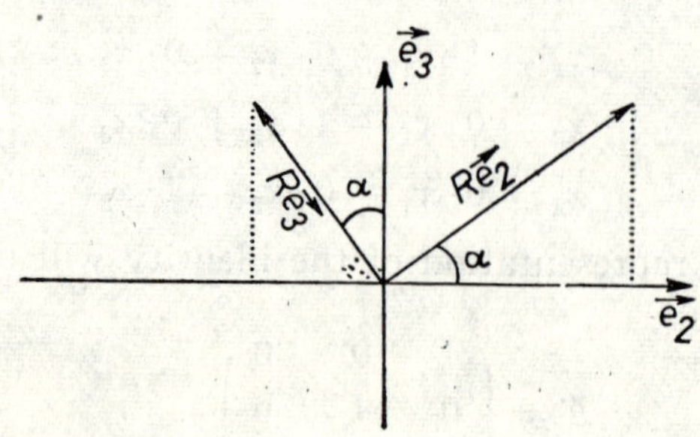

Figure 1.17 The representation of a rotation axis in the two-dimensional space.

1.6.2 *Reducible and irreducible representations*

The group theory can also be applied to the matrix representations of the symmetry elements. The operation defined in the group is the matrix multiplication, the product

$$\mathbf{A} \cdot \mathbf{B} = \mathbf{C} \tag{1.31}$$

being defined by the relation:

$$c_{kl} = \sum a_{kj}\, b_{jl} \tag{1.32}$$

in which c_{kl} is the element of the kth row and lth column of **C**. Equation (1.32) shows that each element of the product matrix is the sum of products between the kth row elements of **A** and the lth column of **B**. The unit element in the group is the unitary matrix. To each matrix **A** corresponds an inverse matrix $\mathbf{A}^{-1}$, which satisfies the equation:

$$\mathbf{A} \cdot \mathbf{A}^{-1} = \mathbf{A}^{-1} \cdot \mathbf{A} = \mathbf{E} \tag{1.33}$$

Two matrices **A** and **B** will be related by a similarity transformation if there is a matrix **Q** so that:

$$\mathbf{Q}\mathbf{A}\mathbf{Q}^{-1} = \mathbf{B} \tag{1.34}$$

It can be shown that matrix **B** is an equivalent representation of **A**, the transformation matrix **Q** defining the new set of vectors. Indeed, let us consider the following representations **A**:

$$\mathbf{A} = \begin{vmatrix} m & 1 \\ 1 & m \end{vmatrix} \tag{1.35}$$

in the two-dimensional space of the base vectors $\mathbf{e}_1$ and $\mathbf{e}_2$. Let us now change these vectors into:

$$\mathbf{e}_1' = (1/\sqrt{2})\, \mathbf{e}_1 + (1/\sqrt{2})\, \mathbf{e}_2$$

$$\mathbf{e}_2' = (1/\sqrt{2})\, \mathbf{e}_1 - (1/\sqrt{2})\, \mathbf{e}_2$$

which gives the transformation matrix:

$$\mathbf{Q} = \begin{pmatrix} 1/\sqrt{2} & 1/\sqrt{2} \\ 1/\sqrt{2} & -1/\sqrt{2} \end{pmatrix} \tag{1.36}$$

It can be easily shown that **Q** is identical to its inverse.

Performing the similarity transformation defined by (1.34) we obtain:

$$\mathbf{B} = \begin{pmatrix} 1/\sqrt{2} & 1/\sqrt{2} \\ 1/\sqrt{2} & -1/\sqrt{2} \end{pmatrix} \begin{pmatrix} m & 1 \\ 1 & m \end{pmatrix} \begin{pmatrix} 1/\sqrt{2} & 1/\sqrt{2} \\ 1/\sqrt{2} & -1/\sqrt{2} \end{pmatrix} =$$

$$\begin{pmatrix} m/\sqrt{2} + 1/\sqrt{2} & m/\sqrt{2} + 1/\sqrt{2} \\ m/\sqrt{2} - 1/\sqrt{2} & 1/\sqrt{2} - m/\sqrt{2} \end{pmatrix} \begin{pmatrix} 1/\sqrt{2} & 1/\sqrt{2} \\ 1/\sqrt{2} & -1/\sqrt{2} \end{pmatrix} = \begin{pmatrix} m+1 & 0 \\ 0 & m-1 \end{pmatrix}$$

The following conclusions may be drawn by comparing the two equivalent representations **A** and **B**.

a. The two matrices, even if different, have one common feature, namely the same trace *:

$$\mathrm{Tr}\,\mathbf{A} = m + m = 2m$$

$$\mathrm{Tr}\,\mathbf{B} = (m + 1) + (m - 1) = 2m$$

In other words, during a similarity transformation the trace is invariant. The representations of a given operator can be thus characterised by their trace.

b. If m is variable, it must be noted that both matrices have the same roots. A similarity transformation does not change the roots of a matrix.

c. Matrix **B** is completely diagonalised, that is it has all elements, except the diagonal elements, zero. We shall say that matrix **A** (or **B**) is a reducible representation of the two-dimensional space, and this representation will be denoted by Γ.

One may also notice that the reducible representation can be considered as a sum of two representations, Γ_1 and Γ_2, each of them being a first-order matrix. Naturally, these two representations will be irreducible representations. In general, a representation in an n-dimensional space can be reduced to a sum of irreducible representations:

$$\Gamma = \Gamma_1 + \Gamma_2 + \ldots + \mathrm{a}_i\Gamma_i + \ldots + \Gamma_k \qquad (1.37)$$

The number of the irreducible representations (k) is identical with the dimension n of the space if each different irreducible representation occurs only once in the reducible representation ($\mathrm{a}_i = 1$).

We have thus found a criterion for the classification of matrix representations, namely according to their irreducible representations. As our interest lies in point group representations, let us mention that each element R of the point group may transform a representation Γ_i into another representation $\mathrm{R}\Gamma_i$, that is:

$$\mathrm{R}\Gamma = \mathrm{R}\Gamma_1 + \mathrm{R}\Gamma_2 + \ldots + \mathrm{a}_i\mathrm{R}\Gamma_i + \ldots + \mathrm{R}\Gamma_n \qquad (1.38)$$

In discussing a certain molecular point group we have to find out all its irreducible representations and then to determine for any irreducible representation the effect produced by each symmetry element. In dealing with such problems we shall refer to the *character* χ of the irreducible representation, which is defined as the numerical value of the trace of the irreducible representation.

* The trace is defined as the sum of the diagonal elements, a_{kk}.

1.6.3 *Character tables*

We have established so far that we can characterise a reducible representation of a molecular point group by its irreducible representations (or their characters, respectively) and by the effect produced upon them by the symmetry elements. Let us consider again the representation **A** (1.35) in the two-dimensional space of the point group C_2, whose elements are:

$$E = \begin{pmatrix} 1 & 0 \\ 0 & 1 \end{pmatrix}; \quad \mathbf{C}_{180°} = \begin{pmatrix} \cos 180° & \sin 180° \\ -\sin 180° & \cos 180° \end{pmatrix} = \begin{pmatrix} -1 & 0 \\ 0 & -1 \end{pmatrix}$$

The effect produced by these elements (**EA** and $\mathbf{C}_{180°}$ **A**, respectively) will be different. Indeed:

$$\mathbf{EA} = \begin{pmatrix} 1 & 0 \\ 0 & 1 \end{pmatrix}\begin{pmatrix} m & 1 \\ 1 & m \end{pmatrix} = \begin{pmatrix} m & 1 \\ 1 & m \end{pmatrix} = \mathbf{A}$$

$$\mathbf{C}_{180°}\mathbf{A} = \begin{pmatrix} -1 & 0 \\ 0 & -1 \end{pmatrix}\begin{pmatrix} m & 1 \\ 1 & m \end{pmatrix} = \begin{pmatrix} -m & -1 \\ -1 & -m \end{pmatrix} = -\mathbf{A}$$

E leaves **A** unchanged; we shall say that **A** is symmetric with respect to **E**. $\mathbf{C}_{180°}$ leaves unchanged the numerical value of the matrix, but reverses its sign: we shall say that **A** is antisymmetric with respect to this operation. We shall denote the two different effects by $+1$ and -1, respectively.

The same must be true for the effect produced by a symmetry element upon the characters of the irreducible representations.

Let us now consider the three-dimensional point group C_3, with the elements **E**, $\mathbf{C}_3$ and $\mathbf{C}_3^{-1}$. The representation of **E** was given in (1.28). From (1.29), with $\alpha = 120°$ and $\alpha = -120° = 240°$, respectively, it is found that:

$$\mathbf{E} = \begin{pmatrix} 1 & 0 & 0 \\ 0 & 1 & 0 \\ 0 & 0 & 1 \end{pmatrix}; \quad \mathbf{C}_3 = \begin{pmatrix} 1 & 0 & 0 \\ 0 & -1/2 & \sqrt{3}/2 \\ 0 & \sqrt{3}/2 & -1/2 \end{pmatrix}; \quad \mathbf{C}_3^{-1} = \begin{pmatrix} 1 & 0 & 0 \\ 0 & -1/2 & -\sqrt{3}/2 \\ 0 & -\sqrt{3}/2 & -1/2 \end{pmatrix}$$

Examination of the three representations of the elements of point group C_3 leads to the following conclusions:

a. The group has only two irreducible representations (this conclusion is borne out by the examination of the last two representations).

b. The C_3 and C_3^{-1} have in Γ_2 the same character, $\chi = -1$. We shall say that Γ_2 is doubly degenerated and that C_3 and C_3^{-1} fall into the same class of symmetry elements. Thus, the number of the irreducible representations is given by the number of the classes of symmetry elements.

c. The characters of the representation Γ_1 are not changed by the application of all symmetry elements, being $+1$ throughout,

This is the result of the way we have chosen the coordinate system (see p. 36). The same must be true for any point group, irrespective of its dimension. Γ_1 will have throughout the character $+1$. We shall call Γ_1 a symmetric irreducible representation. Very often, after Mulliken [10], this representation is denoted by A.

d. The characters of Γ_2 are $+2$ in E and -1 in C_3 or C_3^{-1}. This representation is usually termed E *.

The above conclusion can be tabulated as in Table 1.1.

TABLE 1.1

Character tables for the point group C_3

C_3	E	$2C_3$
A	$+1$	$+1$
E	$+2$	-1

Such tables are called *character tables* and contain, as will be seen later (see Chapter 2), all the information concerning the irreducible representations of a point group.

In setting up the character tables of point groups having a higher dimension, the following rules ** are useful:

If a point group G has the dimension n (n elements), the dimensions n_i of the irreducible representations must satisfy the relation:

$$\sum_i n_i^2 = n \qquad (1.39)$$

Equation (39) shows that n determines unequivocally the number of all irreducible representations, regardless of the number of times (a_i) a distinct representation might occur in the reducible representation. The actual number of the distinct irreducible representations is given by the number of classes of symmetry elements.

In order to determine rapidly the number of classes, one may use the following rules: E is always in a class by itself; C_n^k and C_n^{-k} will be in the same class of symmetry operations if there are either n vertical mirror planes or n C_2 rotation axes perpendicular to them. The same rule holds for improper rotations; in C_{nv} groups, the reflections σ_v belong to the same class. In D_{nh} groups, the σ_v and σ_d mirror planes are in separate classes; σ_H and i are always found in a class by themselves.

[10] R. S. Mulliken, *Phys. Rev.*, **43**, 279 (1933).

* After Mulliken [10], the symbols used for irreducible representations are A_1, A_2 etc., if the character of the unidimensional representation Γ is $+1$; if not, it is denoted by B_1, B_2 etc. The symbol E (not to be confused with the identity) is reserved for bidimensional representations.

** See Reference 5, Section 6.1.

Example 5. Determine the number of irreducible representations for point group C_3.

Equation (1.39) indicates three irreducible representations of first-order dimension. This is quite clear for the E representation (1.28), but is not apparent for the representations of the other two elements. It is however possible to find a transformation matrix $\mathbf{Q}$:

$$\mathbf{Q} = \begin{pmatrix} 1 & 0 & 0 \\ 0 & -i/\sqrt{2} & 1/\sqrt{2} \\ 0 & i/\sqrt{2} & 1/\sqrt{2} \end{pmatrix}; \quad \mathbf{Q}^{-1} = \begin{pmatrix} 1 & 0 & 0 \\ 0 & i/\sqrt{2} & -i/\sqrt{2} \\ 0 & 1/\sqrt{2} & 1/\sqrt{2} \end{pmatrix}$$

By applying equation (1.34) we find the equivalent representation:

$$\mathbf{C}_3 = \begin{pmatrix} 1 & 0 & 0 \\ 0 & \varepsilon^* & 0 \\ 0 & 0 & \varepsilon \end{pmatrix}; \quad \mathbf{C}_3^{-1} = \begin{pmatrix} 1 & 0 & 0 \\ 0 & \varepsilon & 0 \\ 0 & 0 & \varepsilon^* \end{pmatrix}$$

in which each Γ_i has the dimension one, and $\varepsilon = e^{-ia} = \cos a - i \sin a$ and $\varepsilon^* = \cos a + i \sin a$. The character table of point group C_3 can also be written:

C_3	E	C_3	C_3^{-1}
Γ_1	$+1$	$+1$	$+1$
Γ_2	$+1$	ε^*	ε
Γ_3	$+1$	ε	ε^*

It is seen that the **E** representation from Table 1.2 results from the sum of the corresponding characters of Γ_2 and Γ_3 (for E: $+1 + (+1) = +2$; for C_3 or C_3^{-1}: $\varepsilon + \varepsilon^* = \cos a - i \sin a + \cos a + i \sin a = 2 \cos a = 2 \cos 120° = -1$). Usually, only condensed character tables, such as Table 1.1, are given.

Example 6. Determine the number of irreducible representations in the point group C_{3v}.

a. The six elements of this group fall into three classes: E, $2C_3$ and $3\sigma_v$.

b. The characters in the representation Γ_1 (A_1) will be $+1$ throughout (see p. 40).

c. The characters $E\Gamma_i$ will be $+1$ for the unidimensional representations (each different representation is symmetric with respect to the identity operation), $+2$ for the bidimensional representations (see above) or $+3$ for triply-degenerated representations (occurring only in the T and O point groups).

d. The other characters tabulated in a character table are found using equation (1.40):

$$\sum_R \chi_j(R)\, \chi_k(R) = \delta_{jk} n \tag{1.40}$$

where δ_{jk} is the Kronecker delta which can have only two values, $+1$ for $j = k$ and 0 for $j \neq k$, respectively; j and k are the indices of the two representations, $\chi_j(R)$ and $\chi_k(R)$ are the characters of jth and kth representations, respectively; n is the dimension of group G.

Example 7. Determine the character table of point group C_{3v}.

We have seen from Example 6 that this group has three classes of symmetry elements. There will be only three representations. Equation (1.39) tells us that one representations must be doubly-degenerated. From (1.40) we can write:

$$\chi_2(E)\,\chi_2(E) + \chi_2(C_3)\,\chi_2(C_3) + \chi_2(\sigma)\,\chi_2(\sigma) = 3^*$$

$$\chi_1(E)\,\chi_2(E) + \chi_1(C_3)\,\chi_2(C_3) + \chi_1(\sigma)\,\chi_2(\sigma) = 0^*$$

which gives: $\chi_2(C_3) = -1$ and $\chi_2(\sigma) = -1$

In a similar way we find: $\chi_3(C_3) = -1$ and $\chi_3(\sigma) = 0$. The character table will be:

C_{3v}	E	$2C_3$	$3\sigma_v$
A	$+1$	$+1$	$+1$
B	$+1$	-1	-1
E	$+2$	-1	0

1.7 Applications of group theory and symmetry to chemistry

Group theory and symmetry have many and various applications in chemistry. However, it is not possible either to treat them or to review the most important ones without neglecting the major subject of this book. Here we shall therefore restrict our discussions to some molecular properties of interest in studies of organic mechanisms. The application of the group theory in molecular orbital theory will be presented in the next chapter. For a complete account of the applications related to chemistry, the reader is referred to the literature devoted to group theory[5–7].

1.7.1 *Structure and symmetry*

As each story has its beginning and its end, the story of an organic reaction must begin and end somewhere. Very often, the first chapter and the last chapter of a novel have a common element, as is always true for a chemical story. Indeed, the story must begin with the structure of the reactants and usually ends (if it ever ends) with a detailed description of the structure of the reaction products.

* Only the characteristic elements of each class of symmetry operations have been considered. This procedure is justified by the fact that all $\chi_i(R)$ are identical if the symmetry operations fall into the same class.

The structure of a chemical compound is defined by the number and kind of atoms of the molecule, by the chemical bonds involved, as well as by the molecular geometry. The geometry of the molecular frame is the province of *Stereochemistry*, but a study of organic mechanism is incomplete without a thorough knowledge of the molecular geometry of reactants, intermediates and reaction products.

Ascribing a point group to a molecular structure implies the precise description of the molecular geometry. Indeed, the elements of a point group, as has been shown in the preceding section, provide a mathematical relation between all the points of the chemical structure. According to the result of certain symmetry operations, the positions in a molecule can be divided into *equivalent positions* and *special positions*.

Equivalent positions cannot be distinguished, one from another, by applying symmetry operations. These are positions which are moved during a symmetry operation. Conversely, a special position is not moved by any of the symmetry operations to which the molecule is subjected. It is situated at the intersection of all symmetry elements of the molecule. Thus, benzene has no special position, all symmetry elements of the molecule intersecting at a non-material point, the centre of the molecule. Monochlorobenzene (group C_{2v}) has four special positions: the chlorine atom, carbons 1 and 4 and the hydrogen bonded to carbon 4; all these material points lie on the C_2 axis, which is found at the intersection of the two mirror planes of the group.

The ascertainment of the equivalent positions, as well as of the special positions, is of utmost importance in the prediction of both chemical reactivity and physical properties of the molecule. Let us take the example of aromatic substitution, on one hand, and n.m.r. spectra, on the other. The D_{6h} group of benzene molecule tells us that all the positions on the aromatic ring are equivalent. The reactivity of one position will be one-sixth of the observed reactivity. The molecular point group indicates also that all hydrogen atoms are equivalent. A single n.m.r. peak will be observed. If we refer to chloro-benzene, the molecular point group C_{2v} indicates non-equivalence for certain positions. Considering the C_2 axis (or the mirror plane perpendicular to the molecular plane) we see that C-2 and C-6, as well as C-3 and C-5, respectively, are equivalent, while C-1 and C-4 are special positions. This is the reason why the positions in a monosubstituted benzene derivative are designated as *ortho*, *meta* and *para*. In n.m.r. spectroscopy, three different peaks will be observed for the hydrogen atoms (to a first approximation).

These examples are instructive in that they show a direct relation between the symmetry of a molecule and its physical and chemical properties. It is thus possible to infer from properties the correct geometry of the molecule and, reciprocally, one may anticipate certain properties from the known molecular structure. In this respect the i.r. spectra are of special interest.

The absorption of energy in the infrared region is due to transitions between vibrational levels of the molecule [11] (see also p. 87). A vibrating *non-linear* molecule will have $3n-6$ degrees of vibrational freedom (n is the number of constituent atoms) and hence $3n-6$ fundamental frequencies. Even for simple organic molecules, this number is very large. However, the selection rules indicate that some of the expected frequencies will appear in the Raman, some in the infrared spectrum and some in neither. Now, in order to apply the general selection rules to a particular structure one needs the symmetry group and the character table *. Thus, the observed i.r. or Raman spectra may indicate or confirm a particular geometry of the molecule, in accordance with the corresponding molecular point group. Unfortunately, only a few organic molecules are suitable for this theoretical treatement. Usually, the research chemist applies the i.r. spectroscopy to structural problems in an empirical manner.

1.7.2 *Molecular rotations about single bonds*

We have assumed so far that molecules are rigid bodies. Actually, this is not true. Rotations about molecular axes (the bonds within the molecule) are possible, and such rotations can occur more rapidly than the actual phenomenon under observation. For example, the rate of the magnetic resonance of the hydrogen nucleus limits the investigative value of n.m.r. [12] only to structures with a lifetime longer than $10^{-2}-10^{-4}$ s. For short-lived conformers, a resultant geometry (symmetry, respectively) will be observed.

Let us examine, for instance, the molecule of cyclohexane. In order to realise the tetrahedral bond angle and to minimise repulsions between non-bonded atoms (see p. 176), the molecule adopts the geometry shown in Figure 1.16. However, there are at least two such possible geometries, which are interconvertible by simple rotations about C—C bonds. If they equillibrate more slowly than the time-scale of the property measured, two different kinds of hydrogen atoms will be observed, as the D_{3d} symmetry of the molecule (see p. 34) indicates. (In a conformer, six atoms point up and down — *axial* hydrogens, and the other six point outward — *equatorial* hydrogens.) This is actually observed in the low-temperature spectra [13], the rotation being more affected than the proton resonance by lowering the temperature. At room temperature, however, a single peak is recorded. At this temperature the equilibrium (1.41) is

[11] M. Avram and G. D. Mateescu, *Infrared Spectroscopy*, Ed. Tehnică, Bucharest (1966); Wiley-Interscience (1972).

* See, for instance, Reference 5, Chapter 8.

[12] L. M. Jackman and S. Sternhell, *Applications of Nuclear Magnetic Resonance Spectroscopy in Organic Chemistry*, Pergamon Press (1969).

[13] F. A. L. Anet and A. J. Brown, *J. Amer. Chem. Soc.*, **89**, 760 (1967) and references cited therein.

$$\text{(chair, } H_a \text{ axial)} \rightleftharpoons \text{(chair, } H_e \text{ equatorial)} \quad (1.41)$$

faster than the hydrogen resonance, each hydrogen being thus viewed almost simultaneously both in equatorial and in axial position.

This example shows that, depending on the time-scale of the property measured, we may observe either the actual symmetry of the rigid structure or an average geometry of the equilibrating species.

A different situation is encountered in the case of i.r. spectra. Here, due to the nearly instantaneous trasitions between vibrational levels, the time-scale is very small. Consequently, all possible geometries assumed by a given molecular species can be identified.

1.7.3 *Melting point and solubility*

The melting point is the temperature at which the lattice forces in the crystal are compensated by the kinetic motion of the molecules. A high symmetry of the molecular species determines a large lattice force in the crystal. The solubility of a crystal is also determined by the symmetry of the component molecules. The higher its symmetry, the higher the melting point and the lower the solubility, respectively.

Both properties are of interest in studies of organic reaction mechanisms. Indeed, a symmetric reaction product will be easy to isolate by crystallisation. If several reaction products can be separated from the reaction mixture, the more symmetric one will have a higher melting point.

The following example is related to the subject of organic mechanism. von Schleyer and his co-workers have studied the aluminium chloride equilibration of bicyclo- and polycyclo-hydrocarbons. In the equilibration of *endo*-tetrahydrobicyclopentadiene at room temperature, only the transformation into the more stable *exo*-isomer has been observed. At higher temperature, however, a multitude of $C_{10}H_{16}$ isomers is formed, a high melting product (m.p. ~ 270°C) being easily separated in nearly pure form from the complex reaction mixture. This indicated a highly symmetric $C_{10}H_{16}$ hydrocarbon, and thus the easiest route to adamantane was discovered [4,14].

The symmetry properties, as related to the solubility and the melting point, are of great interest in the applied chemistry. Let us consider the example of xylenes. Assuming a rapid rotation about

[14] P. von R. Schleyer and M. M. Donaldson, *J. Amer. Chem. Soc.*, 82, 4645 (1960).

the C—CH_3 bonds, the *para* isomer will have a higher symmetry (D_{2h}) than the other two isomers (C_{2v}). Actually, *p*-xylene is isolated from the mixture of xylenes by freezing. An efficient separation based on their boiling points is almost impossible [15].

1.8 Symmetry and optical activity

All physical properties of a molecule are affected by symmetry. One of the most important for the study of organic reaction mechanism is optical rotation.

From the very beginning of the research on optical activity of organic compounds, Pasteur [16] realised that this property is associated with a characteristic structural feature, namely the molecular *dissymmetry*. Later on, van't Hoff and Le Bel related the phenomenon to the *asymmetric* carbon atom. The label of asymmetric attached to optically active organic molecules has been preferred, until recently, in most of the chemical literature. However, the term is not entirely correct in that it suggests a complete lack of symmetry. Molecules like *trans*-1,2-dichlorocyclopropane (**7**), hexahydrophenalene (**8**), possess one symmetry element (C_2) or more (C_3, C_3^{-1}), respectively, although having non-superimposable mirror images and being thus optically active. Moreover, if in ethylene (point group D_{2h}) we destroy the molecular plane, without affecting the C_2 axes (**9**), the molecular species which result will preserve the D_2 symmetry. It can be seen that this species, although having a higher symmetry than **7** and **8**, is optically active. (Such twisted ethylenes can be obtained either by exciting the normal molecules or by substituting the hydrogen atoms with bulky groups [17].) This kind of optically active molecules, which are not entirely devoid of symmetry, are called *dissymmetric*.

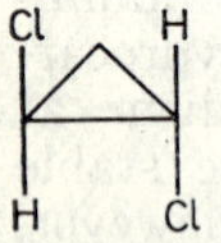

7(C_2) 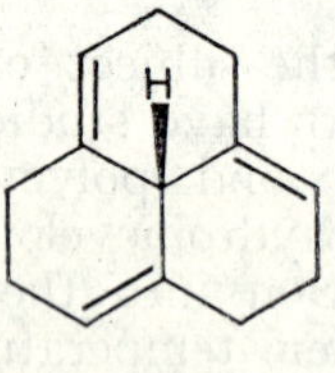 **8**(C_3) 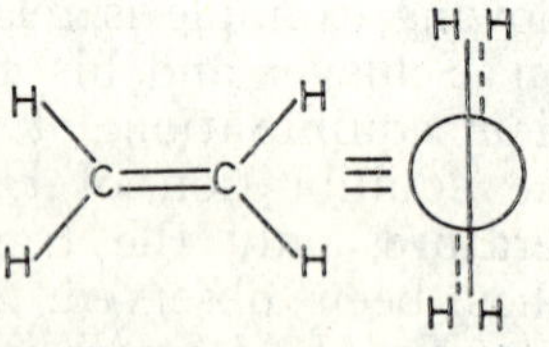 **9**(D_{2h})  **9**(D_2)

It was concluded long ago that the presence of a symmetry plane or centre is necessary for the lack of optical activity. However, the spirane shown in Figure 1.5 has neither symmetry plane nor symmetry centre, and is nevertheless optically inactive. It was

[15] R. E. Kirk and D. F. Othmer, *Encyclopedia of Chemical Technology*, vol. 15, p. 189, The Interscience Encyclopedia, Inc., New York (1956).

[16] For a history of the subject see, for example, E. Eliel, *Stereochemistry*, Chapter 2, McGraw-Hill, New York (1962).

[17] H. O. Kalinowski and H. Kessler, *Topics in Stereochemistry*, 7, 295 (1972).

thus realised that the necessary and sufficient condition for optical activity is *the absence of any improper rotation axis*, $S_n(n = 1, 2, ...)$. Obviously, this rule includes the older one referring to a symmetry plane ($\sigma = S_1$) or a symmetry centre ($i = S_2$). It shows also that optically active molecules may have a certain degree of symmetry. Recently [18–20], the more general term of *chiral* * has been accepted to designate molecules which can exist in two non-superimposable enantiomeric forms. The correspondence between all these terms and the symmetry groups is shown below:

	Asymmetric	Dissymmetric	Symmetric
Point group	C_1	$C_2 ... C_n$; D_n	S_n; C_{nh}; C_{nv}; D_{nh}; D_{nd}; T_d; O_h
	Chiral	Chiral	Achiral

The term asymmetric can be used therefore in connection with the C_1 group.

1.9 Nomenclature of stereoisomers [21]

In studies devoted to organic mechanisms, one should be able to indicate the molecular geometry in a complete, unambiguous, and concise manner. For achiral molecules (see above) this is simply done by indicating the corresponding molecular point group. However, even in such situations, and especially in the case of chiral molecules, this procedure is not sufficient. Indeed, the designation of the C_2 symmetry for compound **7**, for example, does not tell us at all which of the two possible enantiomers we are referring to. The description of the *configuration* of the carbon atoms** is clearly necessary. There are several systems of indicating the absolute configuration, the most logical one being the system R—S † proposed by Cahn, Ingold and Prelog some years ago [18, 19], and now generally accepted [21, 22]. This system is based on the identification of the *chiral elements* of the molecule and the *sequence rule*, which specifices the order in which the ligands about a chiral element must be viewed.

18 R. S. Cahn, C. K. Ingold and V. Prelog, *Experientia*, **12**, 81 (1956).
19 R. S. Cahn, C. K. Ingold and V. Prelog, *Angew. Chem. Intern. Ed. Engl.*, **5**, 385 (1966).
20 K. R. Hanson, *J. Amer. Chem. Soc.*, **88**, 2731 (1966).
* In Greek *chiral* means: of or pertaining to the hand.
21 IUPAC, Tentative Rules for the Nomenclature of Organic Chemistry. Section E. Fundamental Stereochemistry. *J. Org. Chem.*, **35**, 2849 (1970).
** Configuration is the spatial arrangement of the *ligands* (=substituents, groups) about a centre, axis or plane. Configurative isomers have the same molecular parameters (bond length, bond angles and torsional angles about molecular axes). They differ only in their spatial arrangement.
† From the Latin: *R*ectus = right, *S*inister = left.
22 W. Döpke, *Zeitschr. für Chem.*, **13**, 445 (1973).

1.9.1 *Chiral elements*

A chiral element is a geometric element about which the chirality of the molecule is generated. As in the case of symmetry elements, chiral elements can be a *centre*, an *axis* or a *plane*.

A *chiral centre* can be located in a material point of the molecule (an atom such as C, N^+, P^+, etc.) or in a non-material point of the molecular frame, as for the symmetry centre. Formally, a symmetry centre is transformed into a chiral centre by substituting successively the identical ligands with different ligands.

Cahn, Ingold and Prelog [19] showed that there are four possible different chiral arrangements about a chiral centre *C* (not to be confused with the carbon atom), namely *C*abcd (four different ligands), *C*aabb and *C*abbb (only two different ligands), and even *C*aaaa (identical ligands).

The arangement *C*abcd, which is the most important for organic reaction mechanisms, corresponds to the asymmetric carbon atom (see above), but also to such molecules as the chiral adamantane **10** *, with a non-material chiral centre, or to the hetero derivatives **11**—**13**. It must be noted that a lone pair of electrons can also be regarded as a ligand.

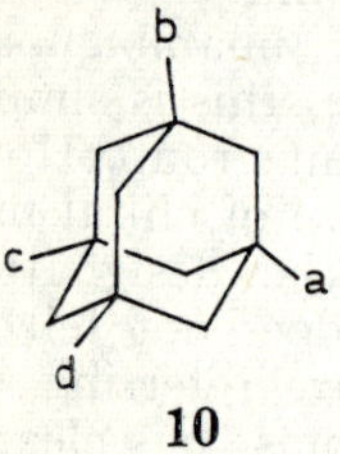

10 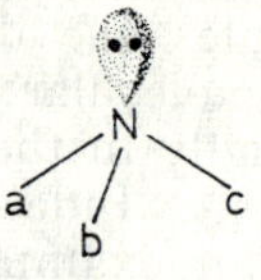 **11** 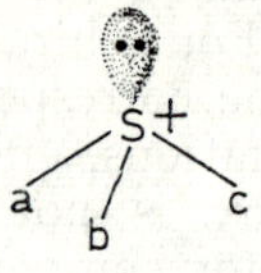 **12** 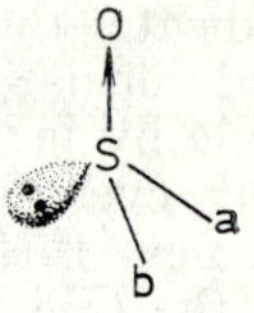 **13**

The arrangement *C*abbb is encountered in derivatives having a C_3 axis, such as **8**.

A *chiral axis* is a molecular axis about which the chiral arrangement is built up. Examples of molecules possessing chiral axes are: chiral allenes, **14**, spiranes † with even number of atoms in the

* This example shows that methane and adamantane, which have the same symmetry, have also similar properties.

† Other spiranes, like spirobihydantoin **15**, have a chiral centre, located in the spiranic atom [19]. The arrangement is *C*aabb.

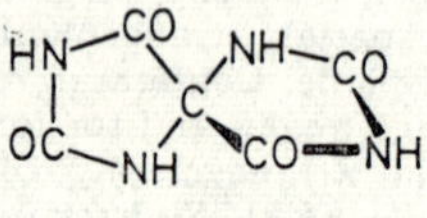

15

rings [23], **16**, and biphenyls substituted in *ortho* with bulky groups [24], **17** (atropisomerism).

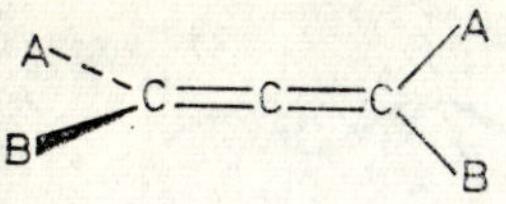

14 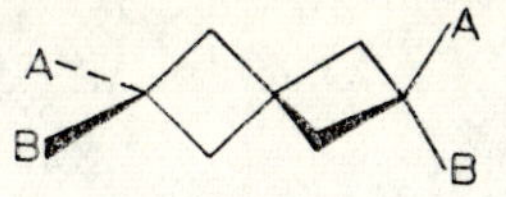 **16** 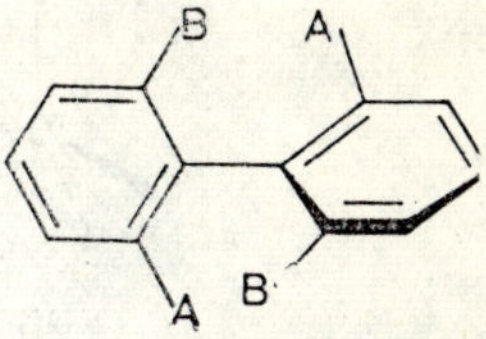 **17**

Chiral plane. Let us compare benzene with the paracyclophane [25] **18**. In **18**, the σ_H symmetry plane of the benzene molecule has been destroyed by bridging the *para* positions. The molecule is still achiral, σ_v and σ_d being present. If we remove these symmetry elements by introducing a substituent in ring A (**19**), the new molecule will be chiral:

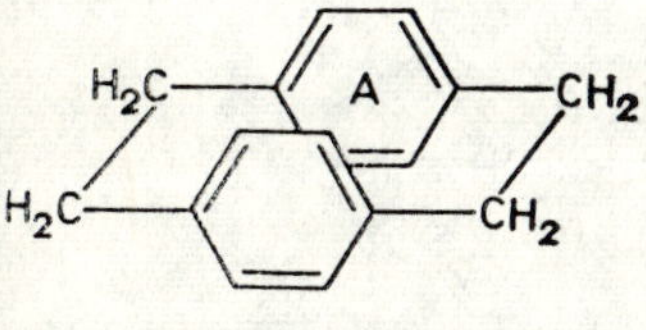

18 (achiral) 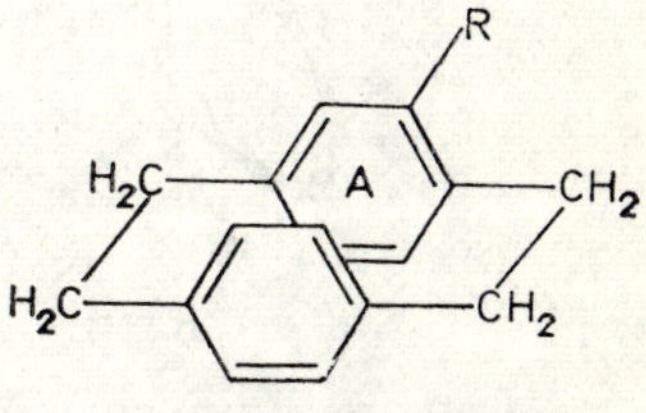 **19** (chiral)

The chirality of **19** results from the dissymmetric arrangement of the substituents about ring A. We say that A is a chiral plane.

1.9.2 *The sequence rules*

In a tetrahedral molecule which contains an asymmetric carbon atom, Cabcd, we may arrange the ligands according to the sequence: a > > b > c > d, a having the highest *priority* and d the lowest. If we view the molecule from the side remote from d, the other ligands

[23] H. J. Backer and H. B. J. Shurink, *Rec. Trav. Chim.*, **50**, 921 (1931); S. E. Jensen and W. J. Pope, *J. Soc. Chem. Ind. (London)*, **51**, 316 (1932); see also Reference 19 for various chirality elements in spiranes.

[24] For the first known example of atropisomerism see C. H. Christie and J. Kenner, *J. Chem. Soc.*, **121**, 614 (1922); for a modern discussion of the implied chirality see D. Gust and K. Mislow, *J. Amer. Chem. Soc.*, **95**, 1535 (1973), and references cited therein.

[25] D. J. Cram and J. M. Cram, *Accounts Chem. Res.*, **4**, 204 (1971).

may be arranged in clockwise (**20**) or counterclockwise (**21**) order. The *R* configuration is assigned to **20** and *S* to **21**.

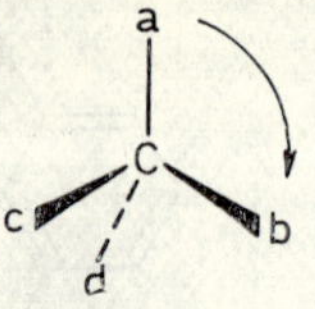

20

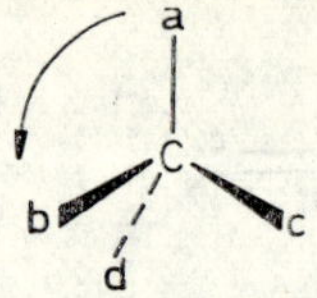

21

The sequence rules define the priority of the ligands a, b, c, d, and thus the order in which they must be viewed. Actually, there is only one rule; the other ones can be considered as subrules. This rule states: *the ligands are arranged in order of decreasing atomic number of the atom directly attached to the chiral element.* For example in CBrClFH, the sequence will be: Br > Cl > F > H. Viewing the molecule from the side remote from H (lowest priority), the possible configurations are:

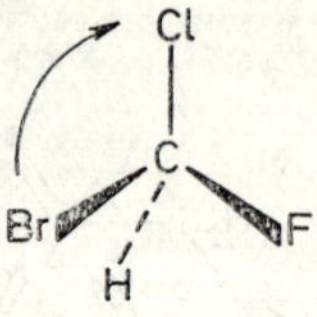

R

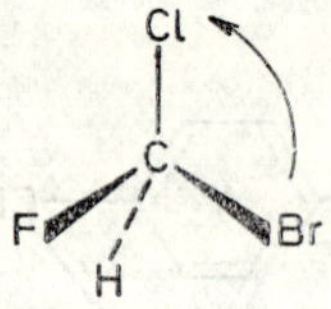

S

Let us now consider the chiral molecule $CH_3CH_2CHClCH_3$. The atomic number of the atoms directly attached to the asymmetric carbon decreases in the order Cl > C > H. There are two carbon atoms bonded to the asymmetric atom and the rule does not tell us in which order the CH_3 and CH_3CH_2 groups must be placed. Similar ambiguities are encountered in many other examples, such as: $CH_2{=}CH{-}CHCl{-}CH_2{-}CH_3$, CHDClBr etc. The subrules of the sequence rule clarify such situations.

Subrule 1. If two or more similar atoms are attached to the same chiral element, the atom substituted with atoms of higher atomic number will have a higher priority. Thus, in CH_3 the carbon atom is substituted only with hydrogen atoms, in C_2H_5 the carbon directly attached to the asymmetric centre is substituted with two hydrogens (low priority) and a carbon atom (higher priority). The sequence will be $C_2H_5 > CH_3$.

Subrule 2. When a ligand is attached to the chiral element by a multiple bond, the former is expanded at its unsaturated atoms by *replica atoms*, before the other rules are applied. The valences of these replica atoms are satisfied with *phantom atoms* of zero atomic number and mass. For instance, $-CH{=}CH_2$ will be expanded into:

$$\begin{array}{l} -\mathrm{CH}-\mathrm{CH_2}-(\mathrm{C}) \\ \quad\ | \\ \quad (\mathrm{C}) \end{array}$$

which has a higher priority than $CH_2—CH_3$. Applying the same subrule, the CHO ligand will precede CH_2OH, the phenyl group * will come before cyclohexyl etc.

Subrule 3. Higher mass number precedes lower. In CHDClBr, for example, D precedes H.

Subrule 4. If the chiral centre is located at a heteroatom possessing a lone pair of electrons, the lone pair will have the lowest priority. Thus, in ethylmethylamine † the hydrogen atom precedes the lone pair and the molecule must be viewed from the side remote from the lone pair.

Subrule 5. If the ligands differ only in their geometry, the following conventions should be used: *trans* precedes *cis*, *RR* (or *SS*) precedes *RS* (or *SR*), *R* has a higher priority than *S*. The last convention needs an exemplification. Actually, a centre substituted with two structurally identical ligands, differing only in their configuration, will not confer chirality to the molecule (Figure 1.18). However, two different arrangements about this centre are possible, which are designated by the symbols *r* and *s* (with the same meaning as *R* and *S*).

We have presented up to now only the subrules necessary to define the configuration of a chiral centre. Other subrules have been established for molecules with chiral axis or chiral plane [19,21]. As

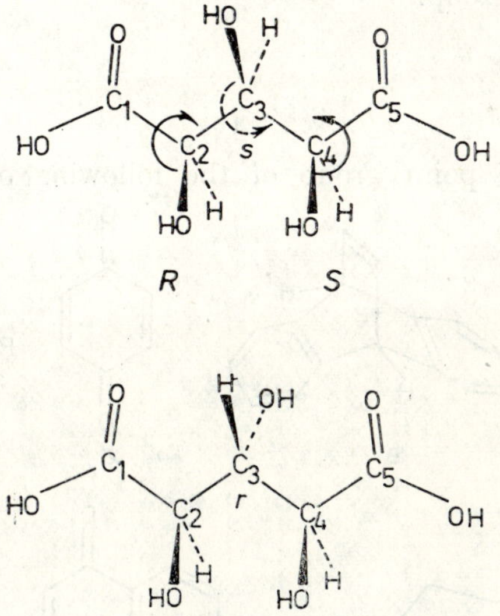

Figure 1.18 The two possible rearrangements of the ligands about the C_3 atom in trihydroxyglutaric acid. The two molecules, possessing a symmetry plane, are achiral (the mirror plane contains C_3HOH, the achirality of the molecule not being changed by the configuration of this carbon atom).

* Kekulé structures should be considered for aromatic ligands.

† Ethylmethylamine is configurationally not stable. However, we shall apply the $R—S$ convention with no regard to the stability of a given configuration.

we are especially interested in reactions at a chiral centre, the above rules will be sufficient.

Finally, we must note that the change of the configurational symbol (*R* or *S*) in a reaction does not necessarily imply a change of the configuration at the reaction site, as it might be simply the result of a change in the priority of ligands. The following reactions of sulphoxide **22**[26] illustrate the point:

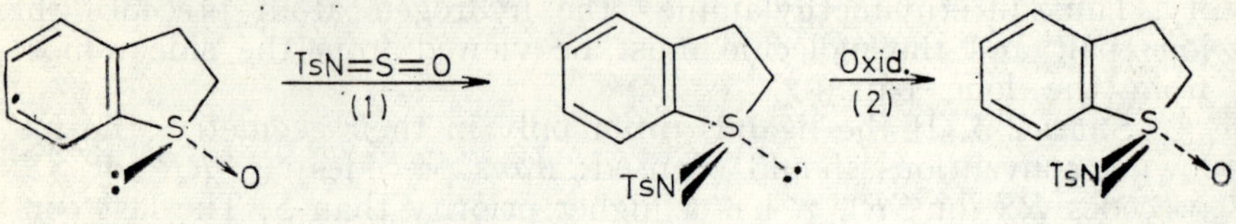

TsN=S=O (1) Oxid. (2)

S-**22** *R* *S*

In the first reaction (1), the change of the configurational symbol is determined indeed by a change in the arrangement of ligands (an inversion), since substituting O for NTs does not modify the priority sequence. In the second reaction, the configurational symbol is reversed, even though the three remaining ligands do not change their initial arrangement around the chiral centre (retention of configuration). However, this time the priority sequence is changed by the oxidation of the sulphur atom, the oxygen taking precedence over the TsN group.

1.10 Problems

(1) Identify the point group of the following organic molecules:

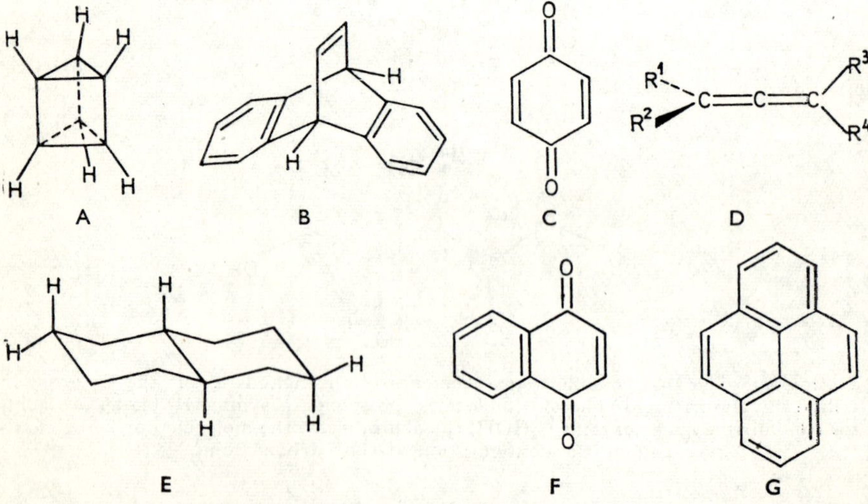

[26] F. G. Yamagishi, D. R. Rayner, E. T. Zwicker and D. J. Cram, *J. Amer. Chem. Soc.*, **95**, 1916 (1973).

(2) Discuss the symmetry group of the following organic reaction intermediates:

$$\overset{+}{C}R^1R^2R^3; \quad :\overset{-}{C}R^1R^2R^3; \quad R^1{-}\dot{\underset{\cdot}{C}}{-}R^2; \quad :C\langle^{R^1}_{R^2}$$

A B C D

if: (a) $R^1{=}R^2{=}R^3{=}H$; (b) $R^1{\neq}R^2{\neq}R^3$; (c) $R^1{=}R^2{=}R^3{=}CH_3$.

(3) What modifications undergoes the symmetry group of an allene D, with $R^1{=}R^2{=}R^3{=}R^4$ (see Problem 1), if: (a) $R^1{\neq}R^2{=}$ $= R^3 = R^4$; (b) $R^1 = R^2 \neq R^3 = R^4$; (c) $R^1 = R^3 \neq R^2 = R^4$.

(4) Indicate the symmetry groups of biphenyl, as depending upon the value of the torsional angle τ. (Indicate the significative values of this angle).

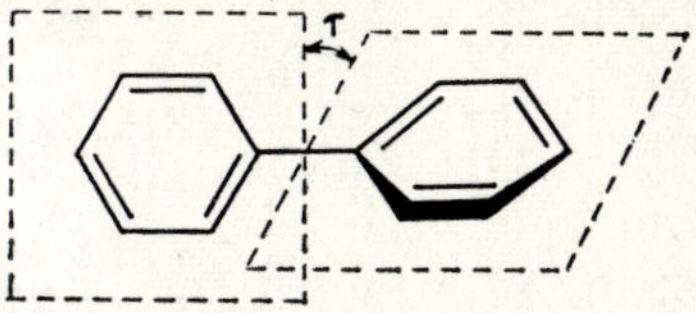

(5) Indicate for what values of n a point group D_{nh} will also possess the symmetry element i.

(6) Consider the following compound:

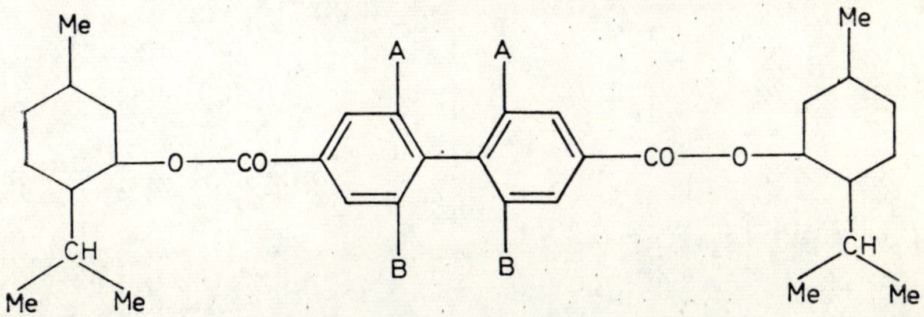

Indicate all the chirality elements if $A \neq B$. Show what happens if $A = B$ and the two alcohol moieties are enantiomeric (see also K. Mislow, *Trans. N.Y. Acad. Sci.* (2), **19**, 298 (1957)).

2
The chemical bond

Chemical structure can be described essentially in two ways: either by a geometrical model or by a quantum-mechanical model. The first treatment has been applied in the preceding chapter, affording a complete description of the geometry of compounds. However, theoretical chemistry is interested both in chemical structure and in the reflection of the structural features in the chemical reactivity. Even if the geometry of a chemical compound may determine its reactivity, the mere description of it cannot give any definite answer regarding its reactivity.

In order to understand the relation between chemical structure and chemical reactivity, it is necessary to have a more complete picture of the molecule, which includes not only its geometry, but the electronic structure as well. The quantum mechanical model, containing all these data, will be dealt with in this chapter.

Most organic reactions are processes during which chemical bonds are being broken and formed. A theory of organic reactions is first of all a theory of the covalent bond. The electronic theory [1,2] has formulated a qualitative explanation of this bond, the details of which will not be given here. A quantitative interpretation of the covalent bond has been formulated by quantum mechanics. In view of the great importance of quantum chemistry in theoretical organic chemistry, we shall briefly outline its main ideas [3,4].

2.1 Wave functions. The hamiltonian operator

The simplest atom, the hydrogen atom, is made up of a negatively charged particle, the electron, which moves in the field of a positive nucleus. In classical mechanics the problem of the particle motion amounts only to describing its trajectory and impulse.

[1] G. N. Lewis, *J. Amer. Chem. Soc.*, **38**, 762 (1916); W. Kossel, *Ann. Physik*, **49**, 229 (1916).

[2] T. M. Lowry, *J. Chem. Soc.*, 822 (1923).

[3] M. J. S. Dewar, *The Molecular Orbital Theory of Organic Chemistry*, McGraw-Hill, New York (1969).

[4] (a) A. Streitwieser, Jr., *Molecular Orbital Theory for Organic Chemists*, J. Wiley, New York (1961); (b) E. Heilbronner and H. Bock, *Das HMO-Modell und seine Anwendung*, Verlag Chemie (1968).

Unlike classical mechanics, wave mechanics is not interested in the trajectory of the particle, but in the probability (dP) of finding the electron in a volume element dv of the space around the nucleus. The probability being positive, the proportionality factor will be squared, so that

$$\mathrm{d}P = \psi^2 \mathrm{d}v. \tag{2.1}$$

The quantity ψ is a function of the position of the particle and of time. For stationary systems, it varies only with the position of the electron with respect to the nucleus. By analogy with other wave phenomena, ψ may be thought of as an amplitude, namely the electronic wave amplitude. The function ψ is usually called *orbital wave function* or, shortly, *orbital*. The function ψ^2 (or, for the complex domain, $\psi\psi^*$, where ψ^* is the complex conjugated function) represents the probability of finding the electron in a certain point or, simply, the electron density.

With the condition that the electron is found in the space around the nucleus, the probability becomes:

$$\int \psi^2 \mathrm{d}v = 1 \tag{2.2}$$

which represents the normalisation condition of the wave function [5].

The function ψ is the solution of an equation which, for the hydrogen atom, was first formulated by Schrödinger [6]:

$$-\frac{h^2}{8\pi^2 m}\left(\frac{\partial^2\psi}{\partial x^2}+\frac{\partial^2\psi}{\partial y^2}+\frac{\partial^2\psi}{\partial z^2}\right)+V\psi = E\psi \tag{2.3}$$

where E is the total energy of the system, V is the potential energy, m is the mass of the electron, h is Planck's constant and x, y, z are the coordinates of the electron.

Equation (2.3) is a differential equation and therefore may have an infinite number of solutions. Physically acceptable solutions can be obtained only for certain discrete values of E, called *eigenvalues*, which in turn are determined by an integral number n, the principal quantum number.

$$E = \frac{n^2 h^2}{8\pi^2 m r^2} \tag{2.4}$$

r = distance of the electron from the nucleus
n = 1, 2 ...

For every value of the quantum number n there is a certain energy E_n and a wave function ψ_n.

The Schrödinger equation (2.3) can be written in a simpler form, using operators, which indicate the mathematical operations to be applied to a function. We notice thus that two operations are

[5] For complex functions, the normalisation condition is $\int\psi\psi^*\,\mathrm{d}v = 1$.
[6] E. Schrödinger, *Ann. Physik*, **79**, 361 (1926).

to be applied in the left-hand side of equation (2.3). The first is multiplication of function ψ by the potential energy V; we shall call this operator the *potential operator*. The second, more complex operation is the differentiation of the function ψ. The corresponding operator, called the *laplacian operator*, is given by:

$$\nabla^2 = \frac{\partial^2}{\partial x^2} + \frac{\partial^2}{\partial y^2} + \frac{\partial^2}{\partial z^2}. \tag{2.5}$$

With this notation, equation (2.3) becomes:

$$-\frac{h^2}{8\pi^2 m}\nabla^2\psi + V\psi = E\psi. \tag{2.6}$$

By analogy with the potential operator V, the following operator, denoted by T, is called the *kinetic operator*:

$$T = -\frac{h^2}{8\pi^2 m}\nabla^2. \tag{2.7}$$

Hence, equation (2.6) becomes:

$$(T + V)\psi = E\psi. \tag{2.8}$$

The sum of the kinetic and potential operators $(T + V)$ constitutes a new operator called the *hamiltonian*. Denoting this operator by H, the Schrödinger equation can be written in a general form:

$$H\psi = E\psi. \tag{2.9}$$

Hamiltonians H can be set up for other atoms or molecules as well. The problem is to find wave functions ψ and eigenvalues E to correspond to such hamiltonians. Owing to mathematical difficulties, complete solutions of the wave functions have been determined only in a few cases. Usually, approximate solutions are sought, corresponding to inexplicit or approximately formulated hamiltonians. For the time being, this is the method of interest in organic chemistry.

2.2 The hydrogen atom

Even in the case of the hydrogen atom, solving the wave equation involves considerable mathematical difficulties. In order to obtain physically acceptable solutions, in addition to the principal quantum number n two more quantum numbers must be introduced: the angular momentum quantum number l and the magnetic quantum number m. The angular momentum quantum number l may have the integral values 0, 1, ... $n - 1$ and it determines the symmetry of the wave function ψ. The orientation of the wave function is determined by the magnetic quantum number m, which may have all integral values between $-l$ and $+l$, including zero. Solving the wave equation in this case leads to several wave functions ψ (orbitals). Usually, these orbitals are indicated by a number, representing the value of n, and a letter corresponding to the value of l (s for $l = 0$, p for $l = 1$, d for $l = 2$ etc.), *e.g.* ψ_{1s}, ψ_{2p}. The energy corresponding to an orbital is indicated by the quantum numbers n and l.

The solutions obtained for the hydrogen atom are shown in Table 2.1.

TABLE 2.1.

Orbital wave functions ψ (orbitals) for the hydrogen atom

Principal quantum number	*Angular momentum quantum number*	*Magnetic quantum number*	*Expression of ψ*
$n = 1$	$l = 0$	$m = 0$	$\psi_{1s} = \frac{1}{\sqrt{\pi a^3}} e^{-\frac{r}{a}}$
$n = 2$	$l = 0$	$m = 0$	$\psi_{2s} = \frac{1}{4\sqrt{2\pi a^3}}\left(2 - \frac{r}{a}\right) e^{-\frac{r}{2a}}$
	$l = 1$	$m = -1$	$\psi_{2p} = \frac{1}{4\sqrt{2\pi a^5}} r e^{-\frac{r}{2a}} \cos\theta$
		$m = 0$	$\psi_{2p} = \frac{1}{4\sqrt{2\pi a^5}} r e^{-\frac{r}{2a}} \sin\theta\cos\varphi$
		$m = +1$	$\psi_{2p} = \frac{1}{4\sqrt{2\pi a^5}} r e^{-\frac{r}{2a}} \sin\theta\sin\varphi$

In the expressions written in the table, the distance of the electron from the nucleus, r, and the angles θ and φ of polar coordinates, determine the position of the electron. The constant a has the value of the radius in Bohr's atomic model.

The energy of the orbitals increases downwards in the table. In the ground state of the hydrogen atom, only the lowest-energy orbital ψ_{1s} is occupied.

When the hydrogen atom is excited by absorption of energy, the electron from the orbital ψ_{1s} moves up to a higher orbital, which is not occupied in the ground state. The return to the ground state is accompanied by release of energy, corresponding to the difference in energy between the two levels.

The examinations of the expressions from Table 2.1 shows first that the function ψ_{1s} is always positive. The function ψ_{2s} is zero for $r = 2a$, admitting thus a nodal surface. The region in which

$r < 2a$ corresponds to positive values of ψ_{2s}, and the region $r > 2a$ to negative values.

Similarly, the functions ψ_{2p}, which depend on θ and φ, will admit a nodal surface, separating the positive region from the negative region of ψ.

The orbitals are customarily represented pictorially in bidimensional plots, as loci of points of equal electron density. Such representations may be seen in Figure 2.1, where the contour lines close

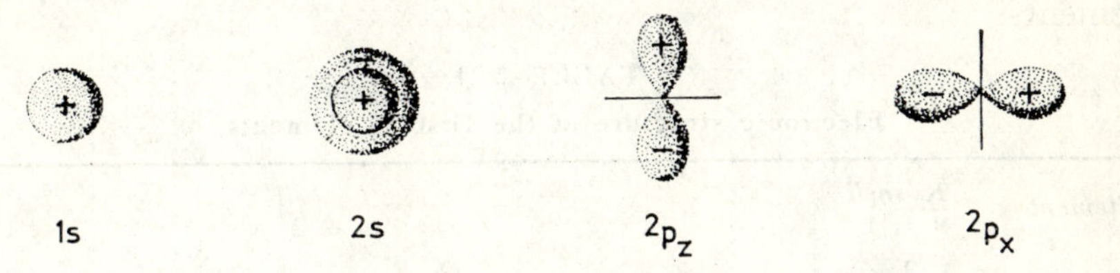

Figure 2.1 Atomic orbitals.

only regions of high electron density. This delimitation, which does not fully correspond to reality, results from the exponential nature of the functions ψ.

2.3 Atomic orbitals of polyelectronic atoms

Solution of wave equations for polyelectronic atoms is tremendously difficult. The wave equations will have the same general form:

$$H^\circ\psi^\circ = T^\circ\psi^\circ + V^\circ\psi^\circ = E^\circ\psi^\circ, \tag{2.10}$$

where H° is the hamiltonian of the polyelectronic system; T° and V° are the kinetic and, respectively, potential operator of the system; ψ° is the wave function.

The kinetic term can be developed relatively easily, but the potential term involves considerable difficulties. If, for instance, the helium atom is considered, with the two electrons a and b, the potential term must comprise three distinct terms given by: attraction between nucleus and electron a $(-\,2\,e^2/r_a)$, attraction between nucleus and electron b $(-\,2\,e^2/r_b)$ and repulsion between the two electrons $(-\,e^2/r_{ab})$. It can be demonstrated that if the repulsion between the two electrons is neglected, the problem amounts to a problem of one electron, the wave function being:

$$\psi^\circ = \varphi_b\varphi_b, \tag{2.11}$$

where φ_a and φ_b represent wave functions of one electron.

It can be shown that the atomic orbitals of polyelectronic elements do not differ essentially from those of the hydrogen atom. The main difference lies in the fact that orbitals which are vacant in the hydrogen atom are gradually occupied with electrons in the heavier atoms. In occupying these orbitals the following rules are observed: an atomic orbital can accommodate two electrons at the most and these must have opposite spin (Pauli exclusion principle); as the nuclear charge is increased, the electrons are placed by pairs into orbitals of successively higher energy; in the case of orbitals of equal energy level (the same n and l), but differing in their orientation, m, each orbital is first occupied by only one electron and only afterwards the second electron is housed (Hund's rule). As an example, Table 2.2 presents the electronic structures of the first ten elements.

TABLE 2.2

Electronic structure of the first 10 elements

Element	*K shell* $n = 1$		*L shell* $n = 2$		
	1s	2s	2p_x	2p_y	2p_z
H	↑				
He	↑↓				
Li	↑↓	↑			
Be	↑↓	↑↓			
B	↑↓	↑↓	↑		
C	↑↓	↑↓	↑	↑	
N	↑↓	↑↓	↑	↑	↑
O	↑↓	↑↓	↑↓	↑	↑
F	↑↓	↑↓	↑↓	↑↓	↑
Ne	↑↓	↑↓	↑↓	↑↓	↑↓

2.4 Molecular orbitals. The covalent bond

The electronic theory regards the covalent bond as consisting of two electrons shared by two nuclei. The complete quantum mechanical treatment of such a system is much more difficult compared to the preceding case, that of the helium atom, since we must consider the attraction between electrons and nuclei, the repulsion between the two electrons, and the repulsion between the two nuclei. If certain approximations are made, as in the case of polyelectronic elements, the problem may be reduced to the problem of one electron. There are two main methods for treating the covalent bond: *the molecular orbital method* and *the valence-bond method*. We shall introduce here the molecular orbital method, which is largely used in organic chemistry[4]. The valence-bond method, difficult to use in calculations, will be only briefly outlined. In fact, the latter approach, in its qualitative form (resonance theory), is rather well known.

2.4.1 *Molecular orbital method. σ bond*

Let us consider first the most simple molecule, the hydrogen ion-molecule, H_2^+, in which the bond between the two hydrogen nuclei, *a* and *b*, is realised by only one electron. The wave function ψ is a solution of the equation:

$$-\frac{h^2}{8\pi^2 m}\nabla^2\psi + \left(-\frac{e^2}{r_a} - \frac{e^2}{r_b}\right)\psi = E\psi, \tag{2.12}$$

where r_a and r_b are the distances between the electron and the two nuclei, respectively.

Let us assume that the electron is very close to nucleus *a* and sufficiently far from nucleus *b*, so that the term $-e^2/r_b$ is negligible as compared to $-e^2/r_a$. Equation (2.12) becomes then:

$$-\frac{h^2}{8\pi^2 m}\nabla^2\psi + \left(-\frac{e^2}{r_a}\right)\psi = E\psi. \tag{2.13}$$

This equation is similar to the Schrödinger equation of the hydrogen atom. We may therefore infer that function ψ is proportional to a function φ_a of one electron (for nucleus *a*). In a similar way we find that function ψ is proportional to a function φ_b for nucleus *b*, that is:

$$\psi = c_1\varphi_a + c_2\varphi_b \tag{2.14}$$

or, otherwise expressed, molecular orbital ψ is a linear combination of atomic orbitals φ_a and φ_b. It can be shown that equation (2.14) is also valid for the hydrogen molecule. In both cases the problem amounts to the matter of finding the coefficients c_1 and c_2 and calculating the energy of the corresponding orbitals.

When energies are calculated we must also consider the fact that the wave function ψ should satisfy a relation of the form:

$$H\psi = \varepsilon\psi$$

or

$$(H - \varepsilon)\psi = 0. \tag{2.15}$$

Introducing in equation (2.15) the value of ψ obtained from equation (2.14) we find:

$$c_1(H - \varepsilon)\varphi_a + c_2(H - \varepsilon)\varphi_b = 0, \tag{2.16}$$

in which the coefficients c_1 and c_2 are unknown. In order to obtain a set of two equations we multiply by φ_a every term of equation (2.16), integrate over all space and then repeat the same operations with φ_b:

$$c_1\int \varphi_a(H - \varepsilon)\,\varphi_a\,dv + c_2\int \varphi_a(H - \varepsilon)\,\varphi_b\,dv = 0 \tag{2.17}$$

$$c_1\int \varphi_b(H - \varepsilon)\,\varphi_a\,dv + c_2\int \varphi_b(H - \varepsilon)\,\varphi_b\,dv = 0. \tag{2.18}$$

The form of the two equations can be simplified using the following notations:

$$\alpha_i = \int \varphi_i H \varphi_i \, dv \tag{2.19}$$

$$\beta_{ij} = \int \varphi_i H \varphi_j \, dv \tag{2.20}$$

$$S_{ij} = \int \varphi_i \varphi_j \, dv. \tag{2.21}$$

Integral α_i is termed the *Coulomb integral* and has the physical significance of the energy of an electron in an atomic orbital φ_i of an atom i. Integral β_{ij} is called the *bond integral* or *resonance integral* and has the significance of the energy of interaction between two orbitals φ_i and φ_j. The last integral S_{ij} is called the *overlap integral* and is a measure of the overlap between atomic orbitals φ_i and φ_j. Using these notations, equations (2.17) and (2.18) become:

$$c_1(\alpha_a - \varepsilon) + c_2(\beta_{ab} - \varepsilon S_{ab}) = 0 \tag{2.22}$$

$$c_1(\beta_{ba} - \varepsilon S_{ba}) + c_2(\alpha_b - \varepsilon) = 0. \tag{2.23}$$

The equations have solutions (other than the banal $c_1 = c_2 = 0$) only if the following determinant is zero:

$$\begin{vmatrix} \alpha_a - \varepsilon & \beta_{ab} - \varepsilon S_{ab} \\ \beta_{ba} - \varepsilon S_{ba} & \alpha_b - \varepsilon \end{vmatrix} = 0. \tag{2.24}$$

This determinant is called the *secular determinant* and the corresponding equation, *secular equation*. In the case of the hydrogen molecule, the secular equation can be further simplified, since the nuclei are identical. Hence, $\alpha_a = \alpha_b = \alpha$, $\beta_{ab} = \beta_{ba} = \beta$ and $S_{ab} = S_{ba} = S$. The equation has two solutions for the energy ε : ε_1 and ε_2, which are functions of the integrals discussed above. Since these integrals vary with internuclear distance, we can plot in a diagram the variation of ε_1 and ε_2 as a function of the distance r between the two nuclei (Figure 2.2).

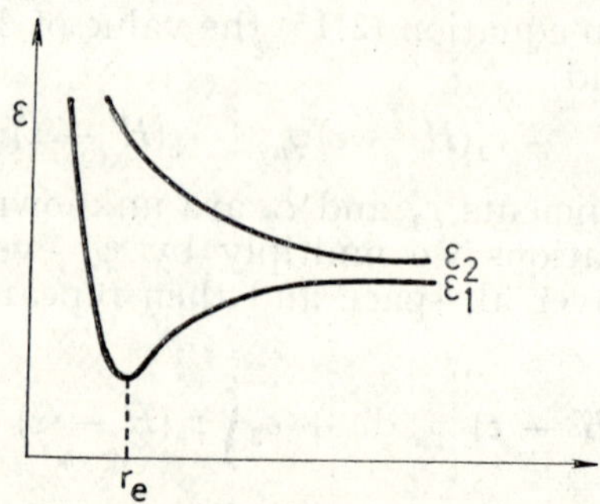

Figure 2.2 The variation of the MO's energy with the internuclear distance, r, in a diatomic molecule.

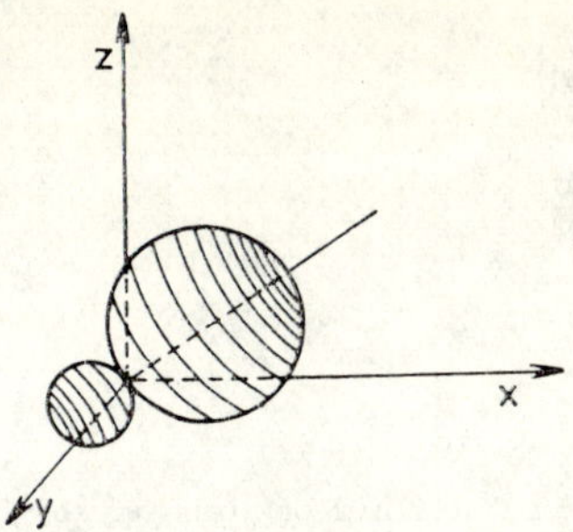

Figure 2.3 Polar representation of the tetrahedral hybridisation.

We notice that ε_1 has a minimum for a certain value of $r(r_e)$. The value of ε_2 increases continuously as the internuclear distance r decreases. The energy ε_1 corresponds to a *bonding orbital* ψ_1, which in ground state will be occupied by two electrons (of opposite spins). The orbital ψ_1 is also called the σ orbital and the corresponding bond, the σ bond. The energy ε_2 corresponds to an orbital ψ_2, vacant in ground state, and called *antibonding orbital* (σ^*). In excited molecules, an electron from a σ orbital may rise to an antibonding orbital σ^*.

If the value of the energy of the bonding orbital is known, the coefficients c_1 and c_2 can be calculated. Such an example will be presented later (p. 76).

All the considerations made so far, concerning the covalent bond of the hydrogen molecule, hold also for other two-electron bonds. However, in the latter cases we shall have to consider the nature of the atomic orbitals involved. Obviously, the hydrogen atom can use only 1*s* orbitals. The electronic structure of the carbon atom is $1s^2\ 2s^2\ 2p^2$ (Table 2.2). We find however that the tetracovalent carbon atom is an excited atom, in which an electron from the 2*s* orbital was promoted to a $2p_z$ orbital, which led to four equivalent orbitals. Since orbitals ψ_{2p} can be regarded as vectors, the situation is equivalent to a linear combination, known as *hybridisation*, of the four initial orbitals. Usually, by such a linear combination the orientation properties in the direction of the bond are improved, the atomic orbitals realising a better overlap. An example is shown in Figure 2.3. Calculations show that this type of linear combination of three *p* orbitals with an *s* orbital (sp^3 hybridisation), leading to four equivalent hybrid orbitals, with minimum energy, corresponds to a tetrahedral orientation.

In hydrocarbons, a tetrahedral carbon atom can develop only two types of two-electron bonds: $C_{sp^2} - H_{1s}$ and $C_{sp^3} - C_{sp^3}$.

a b

Figure 2.4 The overlap of hybrid sp^3 orbitals with a. a hydrogen 1*s* orbital; b. another hybrid sp^3 orbital.

Figure 2.5 Hybrid orbitals: a. sp^2; b. sp.

The secular equation for the two bonds will be similar to equation (2.24), in which in the place of integrals α_i and β_{ij} we must introduce $\alpha_{C_{sp^3}}$, α_H, β_{C-H} and β_{C-C}.

In the unsaturated carbon compounds, σ bonds can be formed by other hybrid orbitals of carbon. The latter result from the linear combination of one or two p orbitals, respectively, with an s orbital (sp and sp^2 hybridisation, respectively). The orientation of the sp^2 orbitals is trigonal, and that of sp orbitals digonal (Figure 2.5).

In these types of hybridisation, sp^2 and sp, the carbon atom still possesses one or two $2p$ orbitals, respectively. In the former case, the $2p$ orbital is perpendicular to the plane of the $\sigma_{C_{sp^2}}$ bonds, while in the latter the two $2p$ orbitals are mutually orthogonal, the intersection line of their nodal planes corresponding to the common axis of the two $\sigma_{C_{sp}}$ bonds.

2.4.2 *Molecular orbital method. π bond*

The p orbitals of C_{sp^2} or C_{sp} atoms can overlap with other p orbitals, leading to other types of two-electron bonds; p orbitals may overlap in several ways, as shown in Figure 2.6.

We may note that diagrams *a* represent symmetric orientations, while diagrams *b* antisymmetric orientations with respect to a plane passing between the two nuclei. The symmetric arrangements corre-

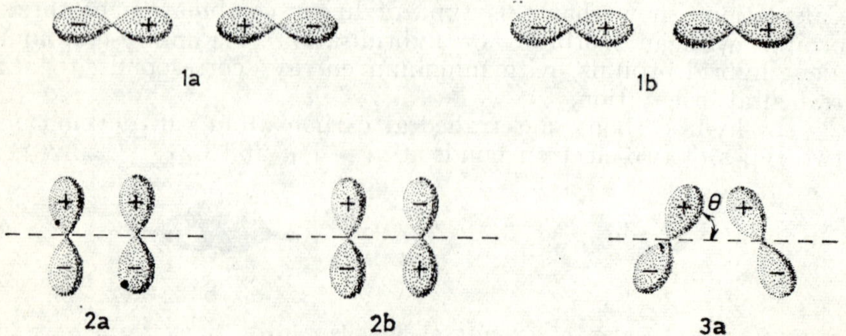

Figure 2.6 Possible ways in which two p orbitals may overlap.

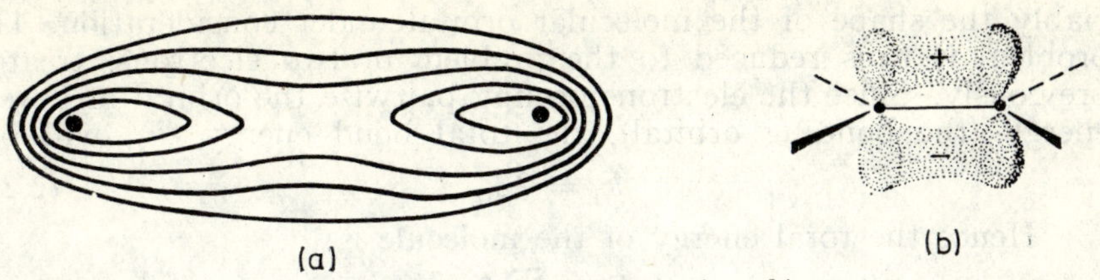

Figure 2.7 Molecular orbitals: a) σ; b) π.

spond to *bonding orbitals*, and the antisymmetric ones to *antibonding orbitals*. The orientations 1*a* and 1*b* (Figure 2.6) are not possible in the case of two adjacent nuclei. However, when the nuclei are far enough from each other, this orientation may correspond to a σ-type overlapping ($p_\sigma - p_\sigma$ bond), which will be discussed later.

The symmetric arrangement 2*a* determines another type of bond, referred to as π bond (ψ_π orbital). Unlike the σ orbital, the π orbital has a nodal plane, passing through the two nuclei (Figure 2.7).

The quantum mechanical treatment of this molecular orbital does not differ from the treatment of σ orbitals. However, the integrals (2.19) — (2.21) will have a different meaning, as they refer to $\varphi_{C_{2p}}$ atomic orbitals.

In the orientation 3*a*, the corresponding bond is partly σ and partly π. The predominance of any of these will depend on the angle θ (Figure 2.6).

2.5 The theoretical study of molecules

2.5.1 *Hückel approximation (LCAO-HMO method)*

Organic molecules, even simple ones, are built up of a great number of nuclei and bonds. The solution of such complicated systems requires certain additional approximations.

In the LCAO method (Linear Combination of Atomic Orbitals) the molecule is divided into two regions, one comprising the nuclei, the other the bonds. σ bonds are then separated from π bonds. If the molecule contains only σ bonds, they are treated individually, assuming that each one must satisfy an equation of the form:

$$-\frac{h^2}{8\pi^2 m}\nabla^2\psi_i + V\psi_i = \varepsilon_i\psi_i. \tag{2.25}$$

In this equation, V represents the average interaction between an electron and the rest of the molecule (including the second bonding electron). Obviously, V is difficult to make explicit. We shall then introduce a simplifying assumption, considering that the potential contributions of more distant regions of the molecule are negligible or, in other words, that neighbouring orbitals do not influence appreciably the shape of the molecular orbital under consideration. The problem is thus reduced to the problem of two electrons, treated previously. Since the electrons occupy pairwise the orbital of lowest energy (the bonding orbital), the total bond energy, E_i, will be:

$$E_i = 2\varepsilon_{1i}. \tag{2.26}$$

Hence the total energy of the molecule is:

$$E = \sum 2\varepsilon_{1i} \tag{2.27}$$

a value which should coincide with the *atomisation energy* of the molecule (the energy required to dissociate the molecule into free atoms).

When the molecule contains both σ and π bonds, usually only the system of π electrons is treated, assuming that σ bonds have little contribution to the total electronic energy of the molecule. As in the previous case, the molecule is divided into regions of nuclei. It is possible that certain regions comprise only two neighbouring nuclei. These regions will correspond to *localised bonds*. Other regions may extend over several nuclei, corresponding to *delocalised bonds*.

There are several methods of quantum mechanical treatment for such systems, among which the most simple is *Hückel's method* (HMO, *Hückel Molecular Orbital*). The method is based on the following approximations:

(a) For all systems containing only carbon nuclei, all Coulomb integrals are equal, as they refer to the same type of nucleus. This approximation is fair enough if there is no neighbouring interaction to modify the potential associated with the nucleus

$$\alpha_i = \int \varphi_i H \varphi_i \, dv \cong \alpha. \tag{2.28}$$

(b) For a given system, all bond integrals, β_{ij}, between adjacent nuclei are equal, while bond integrals between distant nuclei are negligible:

$$\begin{aligned} \beta_{ij} &= \int \varphi_i H \varphi_j \, dv \cong \beta && i \text{ and } j \text{ — two neighbouring nuclei} \\ \beta_{ij} &= 0 && i \text{ and } j \text{ — distant nuclei} \end{aligned} \tag{2.29}$$

The bond integral depends on the internuclear distance. For systems with practically constant internuclear distances (*e.g.* the benzene molecule), this approximation is fair enough. In the case of systems comprising bonds with various lengths (*e.g.* butadiene, see later), the approximation is less justified.

(c) In a more drastic approximation, it is assumed that all overlap integrals S_{ij} are negligible.

$$S_{ij} = \int \varphi_i \varphi_j \, dv \cong 0 \text{ for } i \neq j. \tag{2.30}$$

Although overlap integrals may differ significantly from zero [7], Hückel's method gives, rather surprisingly, satisfactory results. If

[7] See for instance Reference 4, p. 19.

some of the conclusions reached by this method do not agree with the results of more refined methods or of experiments, this is due more to the fact that electronic repulsions are neglected, than to other approximations [8]. However, the former approximation is inherent to any molecular orbital method.

2.5.2 *Localised* π *bonds*

We shall consider, as an example, the molecule of ethylene. Introducing the simplifications proposed by Hückel, the secular determinant (24) corresponding to a two-$2p$-electron system becomes:

$$\begin{vmatrix} \alpha - \varepsilon & \beta \\ \beta & \alpha - \varepsilon \end{vmatrix} = \begin{vmatrix} m & 1 \\ 1 & m \end{vmatrix} = 0, \tag{2.31}$$

with the notation

$$m = \frac{\alpha - \varepsilon}{\beta}. \tag{2.32}$$

The secular equation will be:

$$m^2 - 1 = 0, \tag{2.33}$$

with the solution: $m = \pm 1$.

From the notation used before we obtain:

$$\varepsilon_1 = \alpha + \beta \tag{2.34}$$

$$\varepsilon_2 = \alpha - \beta. \tag{2.35}$$

Since β integrals have negative values, ε_1 will have less energy than ε_2, corresponding thus to a *bonding orbital.* In the ground state, this π bonding orbital will be occupied by the two electrons. The energy $\varepsilon_2 = \alpha - \beta$ corresponds to the *antibonding orbital* which is vacant in the ground state (Figure 2.8). When the molecule is

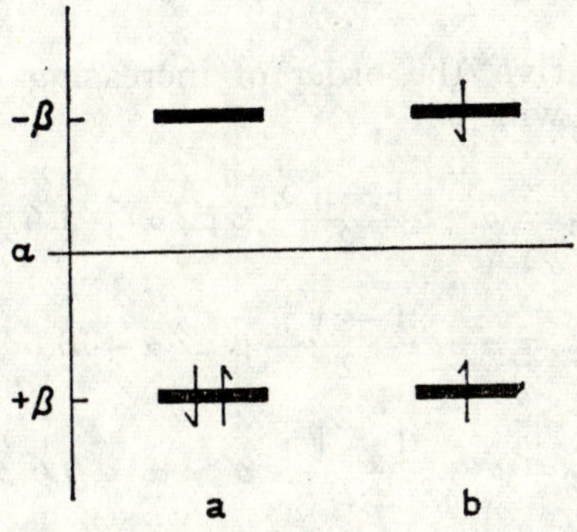

Figure 2.8 π-energy levels in the molecule of ethylene: a) ground state; b) excited state.

[8] See for instance: (a) H. H. Jaffé and M. Orchin, *Symmetry in Chemistry*, John Wiley, New York (1965); (b) Reference 4, Chapter 3; (c) D. S. Schonland, *Molecular Symmetry*, D. van Nostrand, London (1965).

excited, an electron from the bonding orbital will be promoted to the antibonding orbital. The energy absorbed in this transition is:

$$\Delta E = \varepsilon_1 - \varepsilon_2 = 2\beta. \tag{2.36}$$

2.5.3 *Delocalised* π *bonds*

We shall consider the molecule of butadiene as an example. Applying the principle of linear combination of atomic orbitals, we find the molecular orbitals of butadiene:

$$\psi_j = c_{1j}\varphi_1 + c_{2j}\varphi_2 + c_{3j}\varphi_3 + c_{4j}\varphi_4. \tag{2.37}$$

These molecular orbitals must satisfy the relation (2.15), that is:

$$(H - \varepsilon)\,\psi = 0.$$

By combining relation (2.15) with (2.37) and following the procedure used for ethylene, the secular determinant is obtained:

$$\begin{vmatrix} m & 1 & 0 & 0 \\ 1 & m & 1 & 0 \\ 0 & 1 & m & 1 \\ 0 & 0 & 1 & m \end{vmatrix} = 0. \tag{2.38}$$

Expansion of this determinant leads to the equation:

$$m^4 - 3\,m^2 + 1 = 0,$$

which admits the following solutions:

$$m_1,\ m_2 = \pm\frac{1+\sqrt{5}}{2} \quad \text{and} \quad m_3,\ m_4 = \pm\frac{1-\sqrt{5}}{2}.$$

Since β is negative, the order of increasing energy of the four butadiene orbitals will be:

$$\varepsilon_1 = \alpha + \frac{1+\sqrt{5}}{2}\,\beta \simeq \alpha + 1.6\,\beta, \tag{2.39a}$$

$$\varepsilon_2 = \alpha - \frac{1-\sqrt{5}}{2}\,\beta \simeq \alpha + 0.6\,\beta, \tag{2.39b}$$

$$\varepsilon_3 = \alpha + \frac{1-\sqrt{5}}{2}\,\beta \simeq \alpha - 0.6\,\beta, \tag{2.39c}$$

$$\varepsilon_4 = \alpha - \frac{1+\sqrt{5}}{2}\,\beta \simeq \alpha - 1.6\,\beta. \tag{2.39d}$$

In the ground state, only the first two orbitals, of energy ε_1 and ε_2, will be occupied. The energies ε_3 and ε_4 correspond to anti-

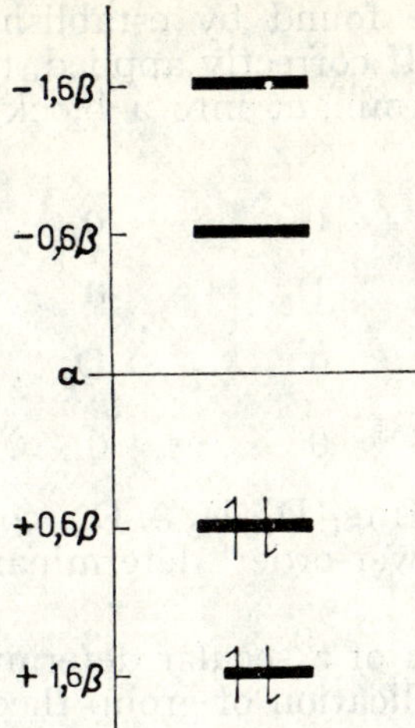

Figure 2.9 π-energy levels in the molecule of butadiene.

bonding orbitals. All these energy levels are represented in Figure 2.9.

With the above values of ε_1 and ε_2, the total energy of the π-electron system will be:

$$E = 2\varepsilon_1 + 2\varepsilon_2 = 4\alpha + 4.4\,\beta.$$

It was shown earlier in this chapter that the energy of the π level of ethylene is $2\alpha + 2\beta$. In the absence of any interaction between the four p_z orbitals of butadiene, the energy of the π system would have been $2(2\alpha + 2\beta)$.

The energy difference:

$$\Delta E = 4\alpha + 4.4\,\beta - 2\,(2\alpha + 2\beta) = 0.4\,\beta$$

is called the *delocalisation energy*.

In the general case of a π system with n electrons there will be n molecular orbitals, the energy of which can be determined similarly from the secular equation. All solutions of the secular equation will have the form:

$$\varepsilon_j = \alpha + m_j\beta, \tag{2.40}$$

where m_j is a coefficient characterising the molecular orbital j.

In the case of electronic systems larger than those considered so far, the problem is complicated by the fact that secular equations of higher order have to be solved (*e.g.* in the case of benzene, of the 6th order). The form of the secular determinant can be simplified considerably by using symmetry species and operations.

Thus, the orbitals represented in the secular determinant are replaced by *symmetry orbitals*, resulting from appropriate linear combinations of the equivalent atomic orbitals of the molecule [9].

[9] D. S. Schonland, *Molecular Symmetry*, D. van Nostrand, London (1965).

Symmetry orbitals are found by establishing the molecular point group (see Chapter 1). If correctly applied, the method further transforms the secular determinant into a block determinant having the same roots.

$$\begin{vmatrix} D_1 & 0 & 0 \ldots\ldots 0 \\ 0 & D_2 & 0 \ldots\ldots 0 \\ 0 & 0 & D_3 \ldots\ldots 0 \\ 0 & 0 & 0 \ldots\ldots D_n \end{vmatrix}.$$

The diagonal elements, $D_1, D_2 \ldots D_n$, corresponding to the symmetry orbitals, are lower-order determinants, which are easy to solve.

The transformation of a secular determinant into a block determinant implies the application of group theory in transforming a reducible representation into the corresponding irreducible representations. The following steps in finding out the lower-order determinants $D_1, D_2 \ldots D_n$ have been suggested by Streitwieser [4].

Step 1: The symmetry group of the molecule is established by inspecting the molecular frame. However, it is not necessary to use all the elements of the corresponding point group, in order to find out the symmetry orbitals. Only those elements which do not bring all the atoms into coincidence with themselves are picked up, keeping in mind that they must form a subgroup of the whole group. Thus, the reflection in the molecular symmetry plane, σ_H, and the other elements implied by it (S_n), are not considered, the atoms being either not moved by it, or being moved in a similar way as by rotations. As an example let us consider the cyclobutadiene molecule with symmetry D_{4h}. Neglecting the reflection in the molecular plane and the other operations involved by it, the D_2 group can be picked up. The consultation of character tables facilitates the choice.

Step 2: In order to find out the character of the reducible representation (R) (see p. 37), each symmetry element of the subgroup is applied to every atomic orbital, which leads to the equivalent orbitals. The corresponding character will be given by the number of untransformed orbitals obtained by the operation applied. For example, the character for (**E**)Γ will be equal to the number of atoms of the molecule. The character table gives the characters of each irreducible representation of the group. The number of times (a_j) a particular irreducible representation occurs in the reducible representation is given by equation (2.41):

$$a_j = \frac{1}{h} \Sigma \chi_{(R)} \chi_j \tag{2.41}$$

where $\chi_{(R)}$ is the character of the reducible representation as determined above, χ_j the characters of the irreducible representations taken from the character table and h the number of elements in the group.

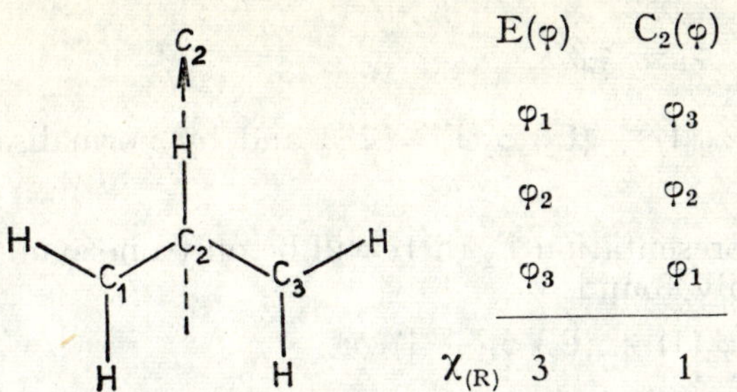

Figure 2.10 Symmetry and labelling in the allyl system.

Step 3: The symmetry orbitals for each irreducible representation are simply constructed by multiplying each atomic orbital by the symmetry element and the corresponding characters of the irreducible representation. Finally, the symmetry orbitals are obtained by normalisation. One should be careful that the number of symmetry orbitals in each representation agrees with the number calculated from (2.41).

Step 4: The symmetry orbitals of each Γ_i irreducible representation are then used to construct the determinants D_i of the block determinant, in a similar way as described on p. 61—62.

This procedure will be exemplified below.

Example 1. Allyl system (Figure 2.10).

Step 1: The point group picked up is C_2. There are only two irreducible representations in the group:

C_2	E	C_2
Γ_1	+1	+1
Γ_2	+1	−1

Step 2: Figure 2.10 shows that $\chi_E = 3$ and $\chi_{C_2} = 1$. Consulting the character table for this group it results:

$$a_1 = \frac{1}{2}(3\cdot 1 + 1\cdot 1) = 2 \text{ and}$$

$$a_2 = \frac{1}{2}(3\cdot 1 - 1\cdot 1) = 1.$$

Thus, in the first irreducible representation there will be two symmetry orbitals and in the second only one.

Step 3: The symmetry orbitals of representation Γ_1 are obtained by applying the symmetry operations of the group to each atomic orbital and multiplying by the corresponding character:

$\varphi_a' = (\mathbf{E})\,\varphi_1(1) + (\mathbf{C}_2)\,\varphi_1(1) = \varphi_1 + \varphi_3$, or by normalisation

$$\varphi_a = \frac{1}{\sqrt{2}}(\varphi_1 + \varphi_3)$$

$\varphi_b' = (\mathbf{E})\,\varphi_2(1) + (\mathbf{C}_2)\,\varphi_2(1) = 2\varphi_2$ and by normalisation

$\varphi_b = \varphi_2.$

In the representation Γ_2 there will be only one symmetry orbital, which is readily found:

$\varphi_c' = (\mathbf{E})\,\varphi_1(1) + (\mathbf{C}_2)\,\varphi_1(-1)$ or,

$$\varphi_c = \frac{1}{\sqrt{2}}(\varphi_1 - \varphi_3).$$

Step 4: The two determinants corresponding to the block determinant are found in a similar way as the secular determinants (see p. 61), substituting atomic orbitals by symmetry orbitals. Thus, D_1 will be a second-order determinant. The first element in the first row will be given by:

$$\int \varphi_a(H - \varepsilon)\,\varphi_a\,dv = \int \frac{1}{2}(\varphi_1 + \varphi_3)\,(H - \varepsilon)\,(\varphi_1 + \varphi_3)dv.$$

Substituting the integrals with the parameters defined on p. 62 and using Figure 2.10, the integral is transformed into:

$$\frac{1}{2}(2\alpha - 2\varepsilon) = \alpha - \varepsilon.$$

In a similar way are then found the other elements. The resultant determinant will be:

$$D_1 = \begin{vmatrix} \alpha - \varepsilon & \sqrt{2}\beta \\ \sqrt{2}\beta & \alpha - \varepsilon \end{vmatrix}$$

with the roots: $\varepsilon_{1,3} = \alpha \pm 1.41\,\beta$.

Determinant D_2 will have a single element. Using the same procedure:

$$D_2 = |\,\alpha - \varepsilon\,|$$

with the root $\varepsilon_2 = \alpha$.

Example 2. Cyclobutadiene.

Step 1: As has been shown above, D_2 group can be chosen. The character table is given below:

$\mathbf{D}_2$	E	C_2	C_2'	C_2''
Γ_1	1	1	1	1
Γ_2	1	1	−1	−1
Γ_3	1	−1	−1	1
Γ_4	1	−1	1	−1

Figure 2.11 shows the choice of the rotation axes and the labelling of the atoms in the molecule.

Step 2 : D_2 group has four representations, and cyclobutadiene has only four atoms in the molecular frame. Without applying equation (2.41), it can be inferred in this simple case that each representation will occur only once.

Step 3 : The symmetry orbital in Γ_1 will be:

$$\varphi_a' = (\mathbf{E})\,\varphi_1(1) + (\mathbf{C}_2)\,\varphi_1(1) + (\mathbf{C}_2')\,\varphi_1(1) \text{ or:}$$

$$\varphi_a = \frac{1}{2}(\varphi_1 + \varphi_2 + \varphi_3 + \varphi_4). \tag{2.42}$$

In a similar way we find for Γ_2:

$$\varphi_b' = (\mathbf{E})\varphi_1(1) + (\mathbf{C}_2)\varphi_1(1) + (\mathbf{C}_2')\varphi_1(-1) + (\mathbf{C}_2'')\varphi_1(-1) \text{ or:}$$

$$\varphi_b = \frac{1}{2}(\varphi_1 - \varphi_2 + \varphi_3 - \varphi_4). \tag{2.43}$$

The other symmetry orbitals have the form:

$$\varphi_c = \frac{1}{2}(\varphi_1 + \varphi_2 - \varphi_3 - \varphi_4) \tag{2.44}$$

$$\varphi_d = \frac{1}{2}(\varphi_1 - \varphi_2 - \varphi_3 + \varphi_4). \tag{2.45}$$

Step 4 : The block determinant will be made of 4 unidimensional determinants having the elements of the form $\int \varphi_i (H - \varepsilon)\varphi_i \, dv$, which can be calculated from equations (2.42)—(2.45) and by inspecting Figure 2.11. The resulted block determinant (2.46) readily affords the four roots.

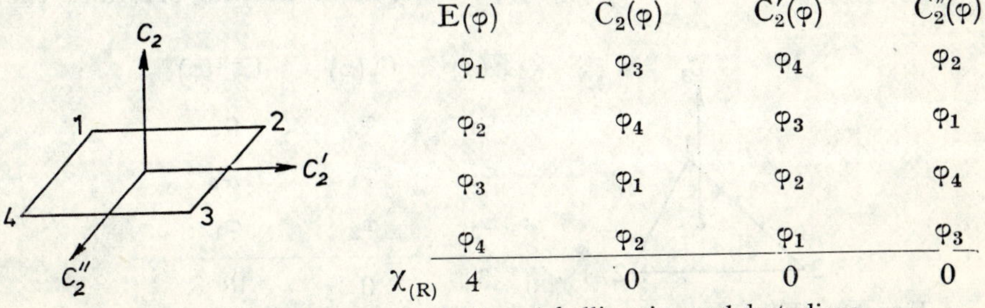

	$E(\varphi)$	$C_2(\varphi)$	$C_2'(\varphi)$	$C_2''(\varphi)$
	φ_1	φ_3	φ_4	φ_2
	φ_2	φ_4	φ_3	φ_1
	φ_3	φ_1	φ_2	φ_4
	φ_4	φ_2	φ_1	φ_3
$\chi_{(R)}$	4	0	0	0

Figure 2.11 Symmetry and atom labelling in cyclobutadiene.

$$\begin{vmatrix} \alpha + 2\beta - \varepsilon & 0 & 0 & 0 \\ 0 & \alpha - 2\beta - \varepsilon & 0 & 0 \\ 0 & 0 & \alpha - \varepsilon & 0 \\ 0 & 0 & 0 & \alpha - \varepsilon \end{vmatrix} . \qquad (2.46)$$

Example 3. Cyclopropenyl system (Figure 2.12).

Step 1: The symmetry of group C_3 is sufficient for diagonalisation. The character table of the group has been given before (see p. 40). The group has two representations, one of which is doubly degenerated.

Step 2: From Figure 2.12 and the character table one calculates:

$$a_1 = \frac{1}{3}(3 \cdot 1 + 0 \cdot 1 + 0 \cdot 1) = 1$$

$$a_2 = \frac{1}{3}[3 \cdot 2 + 0 \cdot (-1) + 0 \cdot (-1)] = 2.$$

Step 3: In representation A, only one symmetry orbital will appear, namely:

$$\varphi_a = \frac{1}{\sqrt{3}}(\varphi_1 + \varphi_2 + \varphi_3).$$

In representation E the doubly degenerated symmetry orbital will be:

$$\varphi_b = \frac{1}{\sqrt{6}}(2\varphi_1 - \varphi_2 - \varphi_3).$$

In this example we find two elements, C_3 and C_3^{-1}, which fall into the same class. The C_3 group will therefore have a doubly degenerated representation. In this case the character tables are used in the same way as for the previous examples, considering however that the elements C_3 and C_3^{-1} have the same character and therefore appear only once in the table (see p. 42).

Step 4: In order to set up the equations (2.17) and (2.18), which involve however symmetry orbitals instead of atomic orbitals, the

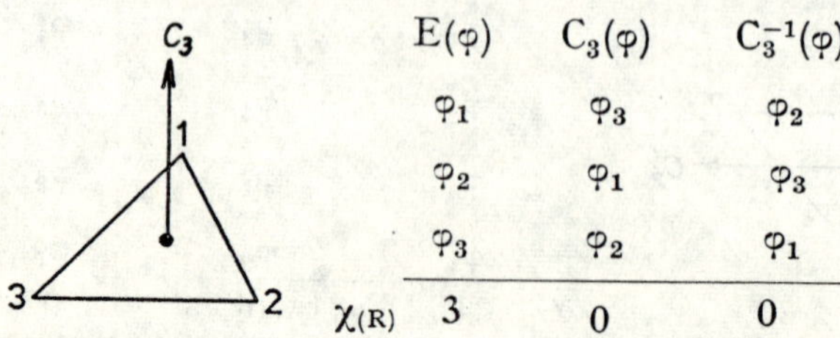

	$E(\varphi)$	$C_3(\varphi)$	$C_3^{-1}(\varphi)$
	φ_1	φ_3	φ_2
	φ_2	φ_1	φ_3
	φ_3	φ_2	φ_1
$\chi(R)$	3	0	0

Figure 2.12 Symmetry and atom labelling in the cyclopropenyl system.

following simplification will be used. The complete form of the expanded integrals $\int \varphi_b(H - \varepsilon)\varphi_b \, dv$ is replaced by the product $\varphi_b\varphi_b$. Expanding this product leads to an expression involving atomic orbitals φ_i, φ_j. The product $\varphi_i\varphi_j$ will represent α if $i = j$, or β if i and j are adjacent atoms, or will vanish if i and j are far apart in the molecule (see also p. 66). Thus:

$$\varphi_b\varphi_b = \frac{1}{\sqrt{6}}(2\varphi_1 - \varphi_2 - \varphi_3)\frac{1}{\sqrt{6}}(2\varphi_1 - \varphi_2 - \varphi_3) =$$

$$= \frac{1}{6}(4\varphi_1\varphi_1 - 2\varphi_1\varphi_2 - 2\varphi_1\varphi_3 - 2\varphi_1\varphi_2 + \varphi_2\varphi_2 + \varphi_2\varphi_3 - 2\varphi_3\varphi_1 +$$

$$+ \varphi_2\varphi_3 + \varphi_3\varphi_3) = \frac{1}{6}(4\alpha - 2\beta - 2\beta - 2\beta + \alpha + \beta - 2\beta + \beta + \alpha) =$$

$$= \frac{1}{6}(6\alpha - 6\beta) = \alpha - \beta.$$

Therefrom:

$$D_2 = \begin{vmatrix} \alpha - \beta - \varepsilon & 0 \\ 0 & \alpha - \beta - \varepsilon \end{vmatrix}$$

This simplification can also be applied in the preceding examples.

For the rapid determination of the coefficients m_j of equation (2.40), the classification into *alternant* hydrocarbons (AH) and *non-alternant* hydrocarbons proves to be very useful.

According to this classification, the carbon atoms of an extended π-system are arbitrarily divided into *starred* and *unstarred atoms*. In an alternant hydrocarbon, atoms of a certain type, *e.g.* starred, are surrounded only by atoms of opposite type (unstarred). A few examples are shown in Figure 2.13.

In the case of alternant hydrocarbons, the coefficients m_j appear in pairs: $+m_j$ and $-m_j$. The odd alternant systems possess also coefficients $m_j = 0$, for which $\varepsilon = \alpha$. The orbitals corresponding to these energies are termed *non-bonding orbitals*. Non-bonding orbitals can be occupied or vacant in the ground state, according to the nature of the system: cation (vacant orbital), free radical (orbital occupied by one electron) or anion (orbital occupied by two electrons).

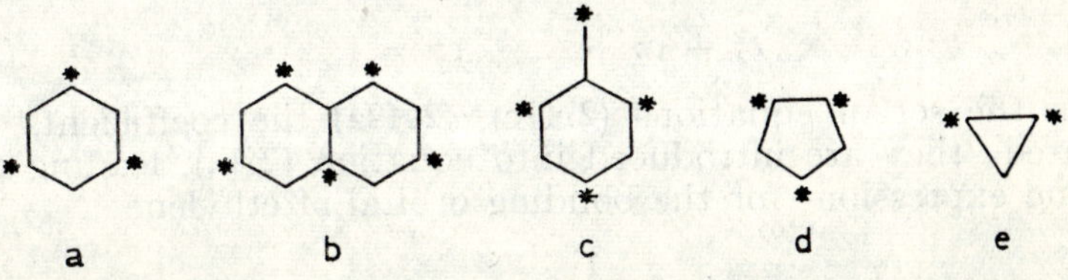

Figure 2.13 Alternant and non-alternant hydrocarbons: a, b) even alternant hydrocarbons; c) odd alternant radical; d, e) non-alternant hydrocarbons.

Since the maximum value of β is 3 (a carbon atom belonging to a π-system cannot be bonded to more than three other atoms), the problem of finding the m_j coefficients of alternant systems is greatly simplified, as the positive roots $+m_j$, once found, lead us immediately to the negative roots $-m_j$.

2.5.4 *LCAO coefficients*

The energies calculated in the preceding section may be introduced into the equations (2.22) and (2.23) which served to set up the secular determinant. This should lead to a set of n equations with n unknowns. Actually, only $n-1$ equations are obtained, the nth one being a combination of the other $n-1$. We shall illustrate the procedure in the case of ethylene, the equations corresponding to the secular determinant being:

$$c_1(\alpha - \varepsilon) + c_2\beta = 0 \tag{2.47a}$$

$$c_1\beta + c_2(\alpha - \varepsilon) = 0. \tag{2.47b}$$

Introducing the value of ε for the bonding orbital of ethylene, $(\alpha + \beta)$, equations (2.47) become:

$$-c_1 + c_2 = 0 \tag{2.47c}$$

$$c_1 - c_2 = 0.$$

They are equivalent and amount to $c_1 = c_2$.

In order to obtain the second equation, the normalisation conditions are used:

$$\int \psi^2 dv = 1 = \int (c_1\varphi_1 + c_2\varphi_2)^2\, dv =$$

$$= c_1^2 \int \varphi_1^2\, dv + c_2^2 \int \varphi_2^2\, dv + 2\, c_1 c_2 \int \varphi_1\varphi_2\, dv. \tag{2.48}$$

With the normalisation condition for atomic orbitals φ and using the Hückel approximation, $S_{ij} = \int \varphi_i\varphi_j\, dv = 0$, equation (2.48) is reduced to:

$$c_1^2 + c_2^2 = 1 \tag{2.49a}$$

or, for the more general case of an n-electron system:

$$c_1^2 + c_2^2 + \dots + c_n^2 = 1. \tag{2.49b}$$

From the set of equations (2.47c), (2.49a) the coefficients c are calculated; they are introduced into equation (2.14), leading to the following expression for the bonding orbital of ethylene:

$$\psi_1 = \frac{1}{\sqrt{2}}(\varphi_1 + \varphi_2), \tag{2.50a}$$

Figure 2.14 Bonding orbital and antibonding orbital in ethylene.

while for the antibonding orbital:

$$\psi_2 = \frac{1}{\sqrt{2}}(\varphi_1 - \varphi_2). \tag{2.50b}$$

These two orbitals are represented pictorially in Figure 2.14.

Examination of these relations shows that the molecular orbital ψ_1 has only one nodal plane, which is the nodal plane of the initial $2p$ atomic orbitals, while the antibonding orbital presents another nodal plane, passing between the two nuclei, as illustrated in Figure 2.14.

Application of the same method in the case of the four orbitals of butadiene (the energies of which are calculated as shown in Section 2.5.3) leads to the following expressions:

$$\psi_1 = 0.37\varphi_1 + 0.60\varphi_2 + 0.60\varphi_3 + 0.37\varphi_4 \tag{2.51a}$$

$$\psi_2 = 0.60\varphi_1 + 0.37\varphi_2 - 0.37\varphi_3 - 0.60\varphi_4 \tag{2.51b}$$

$$\psi_3 = 0.60\varphi_1 - 0.37\varphi_2 - 0.37\varphi_3 + 0.60\varphi_4 \tag{2.51c}$$

$$\psi_4 = 0.37\varphi_1 - 0.60\varphi_2 + 0.60\varphi_3 - 0.37\varphi_4. \tag{2.51d}$$

The number of the nodal planes corresponding to these orbitals may be deduced from the relations (2.51). The occupied low orbital ψ_1 has only the nodal plane of the four nuclei. This orbital strengthens the bond between the four nuclei. The occupied high orbital ψ_2 has one more nodal plane, between nuclei 2 and 3, strengthening thus only the C_1—C_2 and C_3—C_4 bonds. The lower antibonding orbital ψ_3 has a nodal plane between atoms 1 and 2 and a second nodal plane between atoms 3 and 4, contributing thus only to the bond between atoms 2 and 3. The higher orbital ψ_4 is antibonding in all the region of the four nuclei (Figure 2.15).

The example of butadiene is instructive since it shows that the geometry of the ground state may differ sometimes radically from that of excited states. The ground state is best represented by structure A. If butadiene is excited, an electron from the orbital ψ_2 may move up to the orbital ψ_3. As the latter contributes to the bond between atoms C_2 and C_3, the geometry of the excited state is best approximated by structure B:

$CH_2{=}CH{-}CH{=}CH_2$	$CH_2{-}CH{=}CH{-}CH_2$
A	B

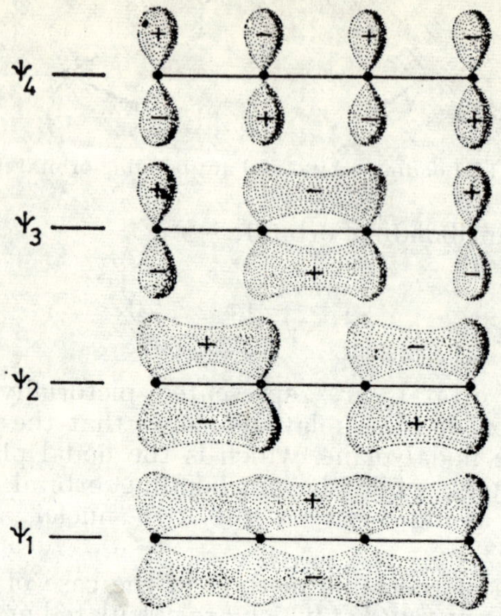

Figure 2.15 π-molecular orbitals of butadiene.

An interesting example is offered by the pentadienyl cation, which has a non-bonding orbital with $m_j = 0$. The coefficients of this orbital are calculated from the set of equations:

$$\begin{aligned}
&c_1(\alpha - \varepsilon) + c_2\beta = 0 \\
&c_1\beta + c_2(\alpha - \varepsilon) + c_3\beta = 0 \\
&c_2\beta + c_3(\alpha - \varepsilon) + c_4\beta = 0 \\
&c_3\beta + c_4(\alpha - \varepsilon) + c_5\beta = 0 \\
&c_4\beta + c_5(\alpha - \varepsilon) = 0.
\end{aligned} \tag{2.52}$$

Substituting $\varepsilon = \alpha$, we obtain:

$$c_2 = c_4 = 0,$$

$$c_1 = - c_3 = c_5 = \frac{1}{\sqrt{3}},$$

that is the coefficients of even atoms are zero, while coefficients of odd atoms adjacent to an even atom cancel each other. This result shows that the coefficients of the non-bonding orbital of odd alternant systems can be easily determined without applying to the secular determinant. The odd atoms of the molecule are marked with a star and an arbitrary value a is attributed to one of them. Knowing that unstarred atoms have coefficients zero and starred atoms adjacent to an unstarred position have coefficients which cancel each other, the coefficients of the other starred positions can be expressed as a function of a. The value of a is then calculated from equation (2.49).

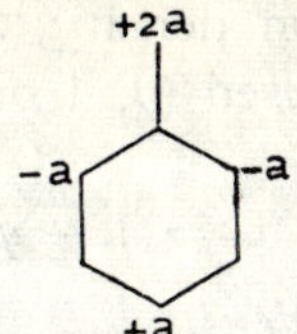

Figure 2.16 Coefficients of non-bonding orbital of the benzyl system.

We shall next consider as an example the benzyl radical (Figure 2.16). If the coefficient a is ascribed to the *para* position of this system, we find the values $-a$ for the coefficients of *ortho* positions and $+2a$ for the coefficient of the benzylic carbon. With the relation (2.49) we obtain:

$$(2a)^2 + 3a^2 = 7a^2 = 1; \quad \text{or} \quad a = \frac{1}{\sqrt{7}}\cdot$$

2.5.5 *The significance of LCAO coefficients*

The relation (2.40) was set up for the energy of a molecular orbital j. However, the same energy is also given by the relation:

$$\varepsilon_j = \int \psi_j H \psi_j \, dv \tag{2.53}$$

or, developing:

$$\varepsilon_j = \int (c_{1j}\varphi_1 + \ldots + c_{nj}\varphi_n) \; H \; (c_{1j}\varphi_1 + \ldots + c_{nj}\varphi_n) \, dv =$$

$$= c_{1j}^2 \int \varphi_1 \, H \, \varphi_1 \, dv + \ldots + c_{nj}^2 \int \varphi_n \; H \; \varphi_n \, dv + 2 \, \Sigma \, c_{tj} \, c_{sj} \int \varphi_t \, H \, \varphi_s \, dv. \tag{2.54}$$

The first integrals in the relation (2.54) are Coulomb integrals α, which have the same value, as we admitted previously. The last integrals represent bonding integrals β, which for more distant atoms t and s vanish, while for neighbouring atoms (denoted, for instance, l and m) are equal. Equation (2.54) is thus reduced to:

$$\varepsilon_j = (c_{1j}^2 + c_{2j}^2 + \ldots + c_{nj}^2)\alpha + 2\Sigma c_{lj} c_{mj} \beta = \alpha + 2\Sigma c_{lj} c_{mj} \beta \tag{2.55}$$

where coefficients l and m refer to two adjacent atoms of the molecule.

By comparing equation (2.55) with (2.40) we find that:

$$m_j = 2\Sigma c_{lj} c_{mj}. \tag{2.56}$$

The terms c_{ij}^2 represent the electron density associated with the orbital j of an atom i. The total electron density of the atom i will be:

$$q_i = \sum_j n_j \, c_{ij}^2, \tag{2.57}$$

where q_i is the total electron density, and n_j indicates the number of times that orbital ψ_j is used (0, 1 or 2).

The quantity ξ_i:

$$\xi_i = 1 - q_i, \tag{2.58}$$

is referred to as the *charge density*.

Using the coefficients of butadiene we find the values:

$$q_1 = 2(0.37)^2 + 2(0.60)^2 = 1$$
$$q_2 = 2(0.60)^2 + 2(0.37)^2 = 1,$$

and

$$\xi_1 = \xi_2 = 0.$$

It can be demonstrated that the charge density is zero for even alternant hydrocarbons and for odd alternant radicals. In the case of odd alternant cations and anions it is easily demonstrated that the charge density depends only on the coefficients of the non-bonding orbital. Indeed, equation (2.58) can be written in the form:

$$\xi = \sum_j^n c_{ij}^2 - \sum_{j=1}^{j=\frac{n-1}{2}+1} n_j c_{2ij}. \tag{2.59}$$

The first summation is made over all orbitals ψ_j. The second summation is made only over occupied orbitals. However, since $n_j = 2$ for all occupied orbitals except the non-bonding orbital, and since the coefficients of an alternant system occur in pairs, $\pm\, c_{ij}$, equation (2.59) can also be written as:

$$\xi = \sum_{j=1}^n c_{ij}^2 - \sum_{j \neq N} c_{ij}^2 - n_N c_{iN}^2$$

or

$$\xi = c_{iN}^2 - n_N c_{iN}^2, \tag{2.60}$$

where subscript N refers to the non-bonding orbital. In the case of the odd alternant cation ($n_N = 0$) the charge density will be, according to relation (2.60), c_{iN}^2, for the radical ($n_N = 1$) it vanishes, and for the anion ($n_N = 2$) it has the value $-c_{iN}^2$.

The coefficients of the non-bonding orbital are calculated as indicated in Section 2.5.4.

The sum:

$$p_{lm} = \sum_j n_j c_{lj} c_{mj}, \tag{2.61}$$

is called *bond order*, for two adjacent atoms l and m; n_j has the same significance as in relation (2.57). The summation is made over all occupied orbitals. In the case of butadiene, the following values are obtained:

$$p_{12} = 2(0.37)\ (0.60) + 2(0.60)\ (0.37) = 0.894,$$
$$p_{23} = 2(0.60)\ (0.60) + 2(0.37)\ (0.37) = 0.447.$$

A high value of p_{lm} indicates a pronounced double-bond character. The values calculated above, for instance, show that the 1,2 bond of butadiene has a more pronounced double-bond character than the 2,3 bond.

2.5.6 *Heteroatomic systems*

Polyene systems in which one or several carbon atoms are replaced by heteroatoms (X) can be treated similarly, appropriate values being ascribed to the parameters α and β. It is convenient to express these parameters as functions of the standard parameters α_0 and β_0:

$$\alpha_X = \alpha_0 + h_X\beta_0, \tag{2.62}$$

$$\beta_{CX} = k_{CX}\beta_0. \tag{2.63}$$

The standard parameters are defined for the molecule of benzene. The Coulomb integral α_X must have more negative values than the integral α_0, for elements more electronegative than carbon, and less negative values for more electropositive elements. Since the bond integral β_0 is negative, the additional parameter h_X will be either positive or negative depending on the electronegativity of the heteroatom. The heteroatom may contribute to the extended π-system either one electron (*e. g.* the oxygen of carbonyl compounds, nitrogen of pyridine) or two electrons (the oxygen of phenolate anion, nitrogen of pyrrole or aniline, etc.). The value of h_X depends on the number of electrons contributed by the heteroatom X, as well as on the charge of the latter.

The value of integral β_{CX} varies with the length of the C—X bond. For systems with bond lengths close to the C—C distance of benzene (1.39 Å), the integral β_{CX} will approach the value of the standard integral β_0 ($k_{CX} = 1$). For longer bonds, the additional parameter k_{CX} will be less than unity, and for shorter bonds greater than unity.

Relations (2.62) and (2.63) can be used also for carbon systems, as corrections to the too drastic Hückel approximations. These corrections are useful for systems in which the carbon-carbon bonds are not exactly equal (*e.g.* the molecule of butadiene) or for systems where substituents with strong inductive effects modify the value of the Coulomb integral α. Table 2.3 presents a few values of the parameters h_X and k_{CX}.

2.6 **Other calculation methods**

2.6.1 *Wheland's method*

Wheland's method differs from the HMO method by the fact that overlap integrals S_{ij}, for pairs of adjacent atoms, are considered to be different from zero and therefore must be introduced in the calcula-

TABLE 2.3

Values of parameters h_X and k_{CX} (according to Streitwieser *).

Element	*Bond*	*Example*	h_X	k_{CX}
C	CC	(benzene)	(1.39 Å) $h_C = 0.0$	$k_{CC} = 1$
	C=C	$H_2C=CH_2$	(1.34 Å) $h_C = 0.0$	$k_{C=C} = 1.1$
	C–C	$CH_2=CH-CH=CH_2$	(1.47 Å) $h_C = 0.0$	$k_{C-C} = 0.9$
B	CB		$h_B = -1.0$	$k_{CB} = 0.7$
N	CN	(pyridine)	$h_{\dot{N}} = 0.5$	$k_{CN} = 1.0$
	C – N	(aniline, NH_2)	$h_{\ddot{N}} = 1.5$	$k_{C-N} = 0.8$
	$C\overset{+}{N}$	(pyridinium, N–H)	$h_{\overset{+}{N}} = 2.0$	$k_{C\overset{+}{N}} = 1.0$
O	C=O	$R_2C=O$	$h_{\dot{O}} = 1.0$	$k_{C=O} = 1.0$
	C–O	(phenolate, O^-)	$h_{\ddot{O}} = 2.0$	$k_{C-O} = 0.8$

* Reference 4a, p. 135.

tion. For the molecule of ethylene, for instance, the secular determinant will have the form:

$$\begin{vmatrix} \alpha - \varepsilon & \beta - \varepsilon S \\ \beta - \varepsilon S & \alpha - \varepsilon \end{vmatrix} = 0.$$

We may note that in this determinant, as compared to the secular determinant of the Hückel's method, the terms β are replaced by the terms $\beta - \varepsilon S$. If the overlap integral S is constant, the system can be solved by the procedure used previously. This time, in the simplified determinant:

$$\begin{vmatrix} m & 1 \\ 1 & m \end{vmatrix} = 0$$

m represents the ratio $(\alpha - \varepsilon)/(\beta - \varepsilon S)$. The roots of this determinant will be: $m = \pm 1$. With this substitution and using a new parameter γ defined by the relation:

$$\gamma = \beta - S\alpha \tag{2.64}$$

the energies of bonding orbitals (ε_1) and antibonding orbitals (ε_2) will be:

$$\varepsilon_1 = \alpha + \frac{1}{1+S}\gamma, \tag{2.65a}$$

$$\varepsilon_2 = \alpha - \frac{1}{1-S}\gamma. \tag{2.65b}$$

The general solution obtained by Wheland's method (2.66) is similar to that of Hückel's method (2.40):

$$\varepsilon = \alpha + n_j\beta. \tag{2.66}$$

Unlike the HMO method, the coefficients n_j of the parameter γ do not occur in pairs, $+n_j$ and $-n_j$ (see equations (2.65)), which obviously complicates the solution of complex systems. However, these two methods reach similar conclusions, showing that neglecting overlap integrals S_{ij}, in Hückel's method, is not such a drastic approximation as it may seem.

2.6.2 *Dewar's method (perturbation method)*

Application of HMO method to complex electronic systems involves serious computational difficulties, in spite of the approximations made. In the perturbation method, the molecular orbitals ψ_{RS} of a molecule R—S are constructed by linear combination (perturbation) of the molecular orbitals ψ_R and ψ_S of the fragments R and S which make up the molecule R—S.

$$R\cdot + S\cdot \rightarrow R-S \tag{2.67}$$

$$a\psi_R + b\psi_S = \psi_{RS}.$$

If R—S is an even alternant molecule and R and S are odd alternant radicals, the problem is simplified considerably. In this case, the non-bonding orbitals of R and S (which can be recognised without solving the secular equation) will give rise to the occupied high orbital and the vacant low orbital of the molecule R—S. These are the orbitals of interest for the interpretation of ultraviolet and visible spectra and for predicting the chemical reactivity of the molecule R—S.

Beside equation (2.67), the orbitals ψ_{RS} must also satisfy a Schrödinger relation:

$$H\,\psi_{RS} = \varepsilon_{RS}\,\psi_{RS} \tag{2.68}$$

where H is the hamiltonian describing the molecular orbitals of the molecule R—S. Using the relations (2.67) and (2.68), the determinant (2.69) can be set up:

$$\begin{vmatrix} H_{11} - \varepsilon_{RS} & H_{12} \\ H_{21} & H_{22} - \varepsilon_{RS} \end{vmatrix} = 0, \tag{2.69}$$

where

$$H_{11} = \int \psi_R\, H\psi_R\, dv; \qquad H_{22} = \int \psi_S\, H\psi_S\, dv,$$

$$H_{12} = \int \psi_R\, H\psi_S\, dv; \qquad H_{21} = \int \psi_S\, H\psi_R\, dv.$$

On the other hand:

$$\psi_R = C_1\varphi_1 + C_2\varphi_2 \ldots + C_r\varphi_r, \tag{2.70a}$$

$$\psi_S = C_1\varphi_1' + C_2\varphi_2' \ldots + C_s\varphi_s'. \tag{2.70b}$$

Assuming that the fragments R and S are linked by the atoms r and s, it can easily be demonstrated that:

$$H_{11} = H_{22} = \alpha \tag{2.71}$$

$$H_{12} = H_{21} = C_r C_s \beta, \tag{2.72}$$

where C_r and C_s are coefficients of the orbitals of atoms r and s from the non-bonding orbitals of the fragments R and S.

By substituting the hamiltonians of (2.69), the following secular equation is obtained:

$$(\alpha - \varepsilon_{RS})^2 - C_r^2 C_s^2 \beta^2 = 0 \tag{2.73}$$

or

$$\varepsilon_{RS} = \alpha \pm C_r C_s \beta. \tag{2.74}$$

If the fragments R and S are linked by several bonds (as in a ring, for instance), it is similarly demonstrated that:

$$\varepsilon_{RS} = \alpha \pm \Sigma C_r C_s \beta, \tag{2.74a}$$

in which the summation is made over all atoms r and s which link the two fragments.

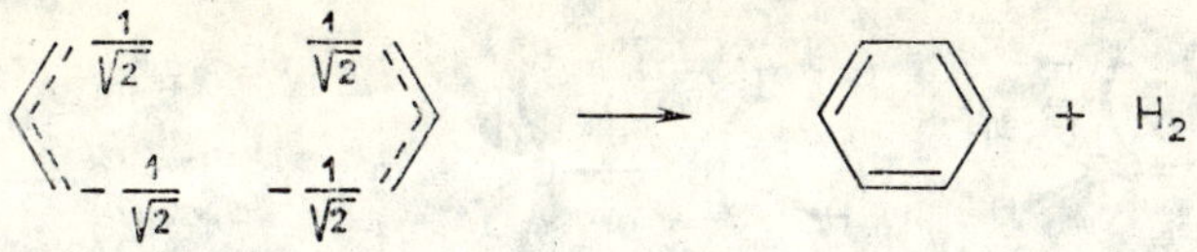

$$\frac{1}{\sqrt{2}} \quad \frac{1}{\sqrt{2}} \quad -\frac{1}{\sqrt{2}} \quad -\frac{1}{\sqrt{2}} \longrightarrow + H_2$$

We shall consider, as an example, the ring closure of two allyl radicals, leading to the benzene molecule:

Using relation (2.74a), the energies of the occupied high orbital and the vacant low orbital can be calculated:

$$\varepsilon_{RS} = \alpha \pm \left(\frac{1}{\sqrt{2}} \frac{1}{\sqrt{2}} + \frac{1}{\sqrt{2}} \frac{1}{\sqrt{2}} \right) \beta = \alpha \pm \beta.$$

This result is in perfect agreement with that obtained by solving the secular equation.

The perturbation method is largely used for predicting chemical reactivity.

2.6.3 *Spin methods*

The main disadvantage of MO methods lies in the fact that they neglect electron interactions. An improvement is made in this respect if spin functions are taken into account in setting up the molecular orbitals. However, methods based on spin function are more laborious and will not be dealt with here [10]. We shall only mention that the valence-bond (VB) treatment is based on the spin approximation. According to this method, a molecular orbital ψ is a combination of orbitals ψ_n:

$$\psi = a_1\psi_1 + a_2\psi_2 + \ldots + a_n \psi_n \tag{2.75}$$

where the functions $\psi_1 \ldots \psi_n$ correspond to various singlet spin structures which can be imagined for the system. Although the valence-bond method is the basis of the resonance theory, its quantitative application to polyelectronic systems is not practical, considering the great number of spin structures which must be taken into account.

2.7 Ground state and excited state

2.7.1 *Electronic configuration*

The electronic configuration of a system is obtained formally by introducing electrons into the available molecular orbitals. There are five kinds of molecular orbitals: bonding orbitals, σ and π, nonbonding orbitals, n, antibonding orbitals, σ^* and π^*. Their energy

[10] For an ample discussion see Reference 2, Chapter 4, Sections 10 and 11.

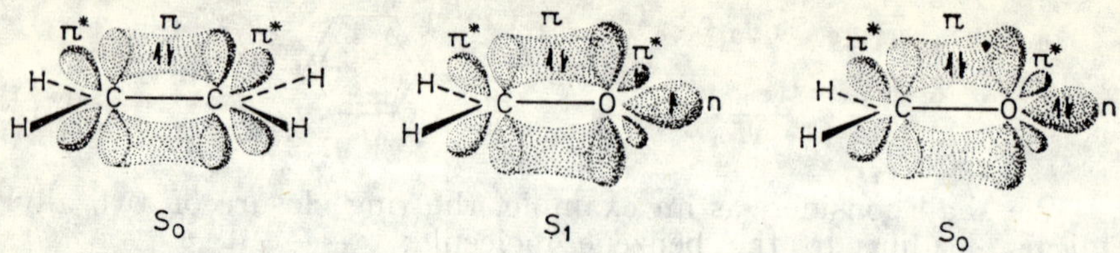

Figure 2.17 Electronic configurations of simple molecules.

levels increase in this order. The first three types, with lower energies (in equation (2.40) m_j is positive or zero), are stable molecular orbitals. The electronic configuration in which the electrons occupy stable orbitals of the molecule is called *ground state*. An *excited state* corresponds to an electronic configuration in which antibonding (unstable) orbitals are used.

The quantity $2|S| + 1$, where $|S|$ is the net spin of the molecule, is called the *multiplicity* of the electronic system. For a molecule with orbitals occupied by two electrons, according to the Pauli exclusion principle, the net spin is zero and the multiplicity is 1. Such a configuration is termed a *singlet state* S. A configuration with two singly occupied MOs has multiplicity 1 if the two electrons have opposite spins or multiplicity 3 if the spins are parallel. This latter configuration is called the *triplet state* T. In the ground state, most organic molecules are found in singlet states (S_0). In Figure 2.17 are presented the electronic configurations of a few simple molecules.

The change in the electronic configuration which occurs when an electron is transferred from an orbital to another orbital of higher or lower energy is called an *electronic transition*. The transition is accompanied by energy absorption or emission, respectively. The difference between the energy levels of the orbitals involved is given by the equation:

$$\Delta E = h\nu = h\frac{c}{\lambda} = \frac{286\,000}{\lambda} \text{ kcal mol}^{-1} \tag{2.76}$$

where h is Planck's constant, c the velocity of light and λ the wavelength of the radiation.

The electronic transitions require energies of the order 36—300 kcal mol^{-1}. Using relation (2.76) we calculate that the wavelength of the radiation absorbed or released is in the range 1000—8000 Å.

The electronic levels are not the only energy levels found in a molecule, as we shall see later.

2.7.2 *Vibrational levels. Zero-point vibrational energy*

We assumed in the foregoing that in a covalent bond the nuclei occupy certain fixed positions. Actually, this is not true, since nuclei vibrate and rotate around equilibrium positions corresponding to bond distances. Informations concerning various rotational and

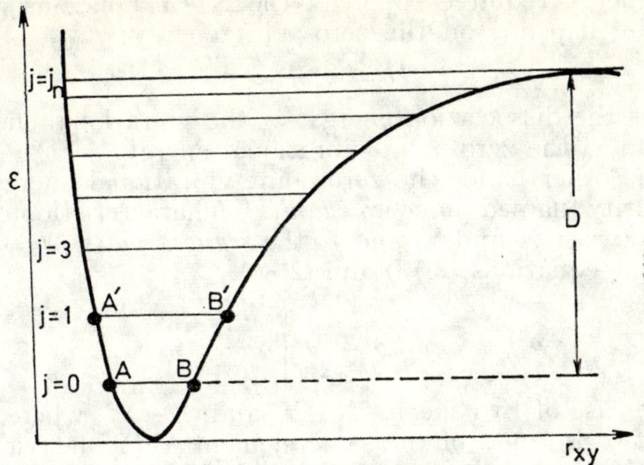

Figure 2.18 Vibrational levels in a diatomic molecule.

vibrational levels may be obtained from the study of microwave and infrared spectra [11]. From the point of view of chemical reactions, we are interested especially in the vibrational levels.

Let us consider a simple diatomic molecule X—Y, the energy of which is a function of the interatomic distance r_{XY}. If the potential energy of the system is plotted against interatomic distance, as in Figure 2.18, the minimum corresponds to the equilibrium distance; the limits (A and B) of the vibration are situated on the two branches of the potential energy curve.

The amplitude of the vibration is greater for higher vibrational energy levels. The energy of a vibrational level is also quantised:

$$\varepsilon_j = h\nu\left(j + \frac{1}{2}\right), \tag{2.77}$$

where ν is the frequency of the vibration and j the vibrational quantum number, $j = 0, 1, 2,$ etc.

When a certain critical vibrational level j_n is reached (see Figure 2.18), the bond between the two nuclei is broken. From relation (2.77) is may be seen that the vibrational energy is never zero. Even at the absolute zero temperature, where all the molecules are found in the vibrational level $j = 0$, there is still a vibrational energy:

$$\varepsilon_0 = h\nu/2 \tag{2.78}$$

called the *zero-point vibrational energy*. The majority of molecules are found at this level even at room temperature. The transition from a lower to a higher vibrational level may simply occur in the course of the thermal motion of the molecule.

[11] For details, see M. Avram and G. Mateescu, *Infrared Spectroscopy*, Wiley-Interscience, New York (1972).

The energy required to break the X—Y bond, as results from Figure 2.18, depends on the zero-point energy:

$$D = \varepsilon_n - \varepsilon_0, \tag{2.79}$$

where D is the dissociation energy, ε_n the vibrational energy for the level j_n and ε_0 the zero-point vibrational energy.

On the other hand, the zero-point vibrational energy is related to a quantity named *force constant*, f (characteristic for the type of bond between X and Y) and to the *reduced mass* $\mu = m_x m_y / m_x + m_y$ by the equations (2.78) and (2.80).

$$\nu = \frac{1}{2\pi}\sqrt{\frac{f}{\mu}}. \tag{2.80}$$

In the case of two species X—Y′ and X—Y″, where Y′ and Y″ designate two isotopes of the same element Y (Y″ heavier than Y′), the respective force constants will be identical, since the two bonds are of the same type, but the reduced masses will be different. According to equation (2.80), the heavier isotope Y″ will have a lower frequency ν and a lower zero-point vibrational energy. On the other hand, for the two species under consideration the higher vibrational levels must be the same (the species which result by bond breaking will be electronically identical). From equation (2.79) it follows therefore that the X—Y′ bond is easier to break than the X—Y″ bond. This conclusion is of great significance for the evaluation of the kinetic isotope effect, an important tool in the study of reaction mechanisms [12].

2.7.3 *Excited states*

Some electronic transitions occur so rapidly (10^{15} s), as compared to the motion of nuclei, that we may fairly assume that the nuclei do not change their positions during a transition (the Frank-Condon principle). In accordance with this principle, several situations may arise, as shown in Figure 2.19. In the first case (A), the equilibrium

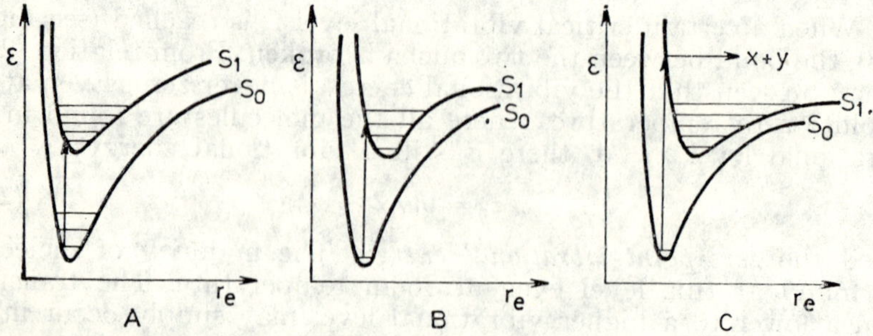

Figure 2.19. Potential energy plots for ground states and excited states.

12 See for example W. H. Saunders, Jr., in *Survey of Progress in Chemistry*, vol. 3, Academic Press, New York (1966), p. 109.

distance r_e is the same for the ground and for the excited state; the electronic transition may occur also between the vibrational levels $j = 0$ of the two states (0—0 transition). The more frequent case (B) corresponds to an equilibrium distance greater for the excited state than for the ground state. Electronic transitions from the lower vibrational level ($j = 0$) of the ground state will lead to higher vibrational levels of the excited state. In the case of such a transition (C), the molecule may dissociate.

In the course of an electronic transition, the net spin of the molecule may be conserved (singlet-singlet, triplet-triplet transitions) or may be reserved (singlet-triplet, triplet-singlet transitions). Only the transitions in which the spin is conserved have high frequencies (10^6—$10^{15}s^{-1}$, allowed transitions). Transitions with reversal of spin have much lower frequencies (10^{-1}-10^4s^{-1}). Hence we could assume that excited triplet states (T_1, T_2, etc.) are less populated than the corresponding singlet states (S_1, S_2, etc.). Actually, triplet-states are easily populated, but in an indirect manner [13, 14]. Indeed, the energy levels of excited triplet states are lower than those of singlet states (see Figure 2.20). This fact, established by quantum mechanical calculations, can be easily understood if we admit that the two electrons, occupying different orbitals, will interact more strongly in the singlet state than in the triplet state. Since the repulsion is stronger if the electrons are singlets, the excited states S_n will have higher energy than the excited triplet states T_n. By losing a part of its

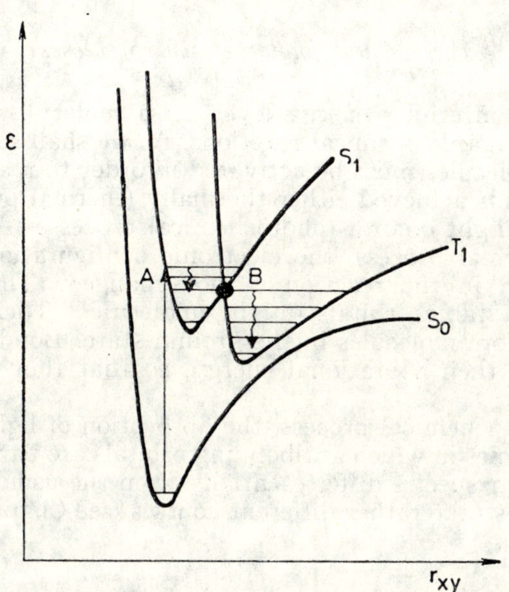

Figure 2.20 Intersystem crossing $S_1 \rightarrow T_1$.

[13] N. J. Turro, *Molecular Photochemistry*, W. A. Benjamin, New York (1965).
[14] J. G. Calvert and J. N. Pitts, Jr., *Photochemistry*, J. Wiley, New York (1966).

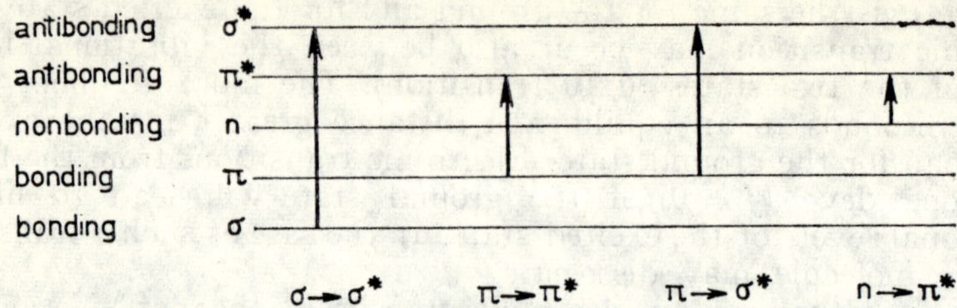

Figure 2.21 Various types of electronic transitions.

vibrational energy in collisions with other molecules, a molecule in a higher vibrational level of the singlet state S_1 may thus reach a point (B in Figure 2.20) corresponding to the end of a vibration in the state S_1 and the beginning of a vibration in the state T_1. Since the level $j = 0$ in the triplet state T_1 is lower than that in the singlet state S_1, the spin inversion occurs very easily (10^{-6}-10^{-9} s^{-1}), competing thus with the singlet-singlet ($S_1 \rightarrow S_0$) transitions. Such a process is called *intersystem crossing.*

Electronic transitions may also differ by the nature of the orbitals involved. There are thus transitions $\sigma \rightarrow \sigma^*$, $\pi \rightarrow \pi^*$, $n \rightarrow \pi^*$ etc. As a consequence of the energy levels of the 5 types of orbitals, as shown in Figure 2.21, the energetically most favoured transitions are $n \rightarrow \pi^*$ and $\pi \rightarrow \pi^*$, while $\sigma \rightarrow \sigma^*$ transitions are rarely achieved. Each one of these transitions can be singlet-singlet or singlet-triplet.

2.7.4 *Thermal processes and photochemical processes*

Information concerning energy levels in a molecule is essential for the study of organic chemical reactions. As we shall see later in more detail, the molecules must be activated in order to react. Practically, the activation is achieved either thermally (thermal processes), or by absorption of light quanta (photochemical processes).

In a thermal process, the electronic configuration of the molecules involved in the reaction is not changed. The electrons remain in the stable orbitals of the molecules. Thermal processes involve therefore molecules in the ground state. Bonds are activated by increasing their vibrational energy, so that they can be broken more easily.

In a photochemical process, the absorption of light quanta leads to excited species in which antibonding orbitals are partially occupied. Such molecular species differ esentially from the usual molecules and their reactions take rather different courses (see Chapters 11 and 12).

2.8 Problems

(1) Calculate the delocalisation energies (DE) for the allyl cation, free radical and anion. What conclusions can be drawn from these data? (see also Example 1, p.71).

(2) Calculate the total π-energy for the cyclopropenyl system I (see Example 3, p.74).

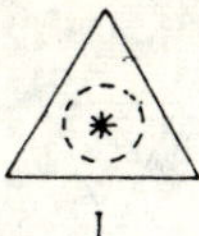

I

Calculate the delocalisation energies for the cation, the free radical and the anion. Compare the results with the values obtained for the allyl system in Problem 1. What conclusions can be drawn?

(3) Indicate the distribution of the four π electrons in the molecular orbitals of cyclobutadiene (see Example 2, p.72). What conclusions can be drawn from these results?

(4) Applying the character tables of group C_3 and D_2 (see p.40; 72), calculate the energies of the MOs of the following compounds:

(5) Calculate the energies of the π orbitals of benzene and the delocalisation energy of this system. Use C_2 (p.71) and C_6 point groups.

The character table for C_6 is given below:

$\mathbf{C_6}$	E	$2C_6$	$2C_3$	C_2
A	+1	+1	+1	+1
B	+1	−1	+1	−1
E_1	+2	+1	−1	−2
E_2	+2	+1	−1	−2

(6) Calculate π-orbital energies and delocalisation energies for hexa-1,3,5-triene. Compare the results with those obtained in the preceding problem.

(7) Calculate the π energies corresponding to the acetylene oxide II, assuming $h_O = 2$ and $k_{C-O} = 0.8$.

II

Compare the results with those obtained for the cyclopropenyl carbanion.

(8) Calculate the π-energies for the boron compound III, assuming $h_B = -1$ and $k_{CB} = 0.7$.

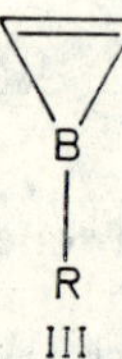

Compare the results with those of Problem 2.

(9) The secular determinant of the pentadienyl cation may be obtained by leaving out the first row and the first column of the secular determinant of benzene or of hexa-1,3,5-triene. What is the significance of this fact?

(10) Calculate the coefficients of the non-bonding orbital for the following systems:

IV V VI

(11) Calculate the charge density at C_α of the benzyl, benzhydryl and triphenylmethyl cations. Discuss the results.

(12) What is the type of hybridisation of the central atom in the molecules of allene (VII) and of carbon dioxide (VIII)?

$$\underset{\text{VII}}{H_2C = C = CH_2} \qquad \underset{\text{VIII}}{O = C = O}$$

(13) What kind of electronic transitions may occur in the molecule of formaldehyde? Write the electronic configuration for each excited species.

(14) Is the Frank-Condon principle obeyed or not in the course of a triplet-singlet ($T_1 \rightarrow S_0$) transition?

(15) The energy differences between the $j = 0$ levels of the singlet (DE_S) and of the triplet (DE_T) excited states are 72 kcal mol^{-1} and 69 kcal mol^{-1}, respectively, for a molecule A, and 120 kcal mol^{-1} and 60 kcal mol^{-1}, respectively, for a molecule B. Which molecule will undergo more readily the intersystem crossing $S_1 \rightarrow T_1$?

(16) How can the empirical value of the bond integral β be calculated using spectrophotochemical data? What kind of electronic transitions must be considered in this case?

PART 2
Chemical reactivity

3
Reactions of organic compounds

3.1 Equilibrium reactions. Thermodynamic quantities

Thermodynamics is the science which deals with energy changes during physical or chemical processes at equilibrium or tending to reach equilibrium. The laws of thermodynamics are outside the scope of this book. However, as we are interested in chemical reactions, we cannot leave out entirely the question of chemical equilibrium and certain aspects of chemical thermodynamics.

Chemical equilibrium can be studied from the point of view of statistics or of classical thermodynamics. In the statistical approach, the state of equilibrium is defined as a state of maximum probability. Let us consider a very simple case of chemical equilibrium, namely an isomerisation:

$$A \rightleftarrows B$$

In order to find the state of maximum probability of the system we must consider the various energy levels of each species, A and B, and the distribution of the molecules between these levels. The number of molecules in the ε_{iA} and ε_{iB} levels will be given by the Boltzmann equation:

$$N_{iA} = \Lambda g_{iA}\, e^{-\varepsilon_{iA}/\mathbf{k}T} \qquad (3.1)$$

$$N_{iB} = \Lambda g_{iB}\, e^{-\varepsilon_{iB}/\mathbf{k}T}, \qquad (3.2)$$

where Λ is a constant of proportionality, g_i is a factor of degeneracy, **k** is Boltzmann's constant and T is the absolute temperature.

The constant of proportionality Λ depends only on the number and type of atoms of the molecule and not on their arrangement or the energy level of the molecule. Since the two species A and B are isomeric, they will have the same constant of proportionality, Λ.

The total number of molecules A and B will be given by a summation over all ε_i levels:

$$N_A = \Lambda\, \Sigma g_{iA}\, e^{-\varepsilon_{iA}/\mathbf{k}T}, \qquad (3.3a)$$

$$N_B = \Lambda\, \Sigma g_{iB}\, e^{-\varepsilon_{iB}/\mathbf{k}T}. \qquad (3.3b)$$

The equilibrium constant for the system formed by molecules A and B is:

$$K = \frac{N_B}{N_A} = \frac{\Sigma g_{iB}\, e^{-\varepsilon_{iB}/\mathbf{k}T}}{\Sigma g_{iA}\, e^{-\varepsilon_{iA}/\mathbf{k}T}}. \qquad (3.4)$$

This expression can be written in another way, tc demonstrate the difference between ground energy levels:

$$K = e^{-(\varepsilon_{0B}-\varepsilon_{0A})/kT} \frac{\Sigma g_{iB}\, e^{-(\varepsilon_i - \varepsilon_{0B})/kT}}{\Sigma g_{iA}\, e^{-(\varepsilon_{iA}-\varepsilon_{0A})/kT}} \tag{3.4a}$$

The following sum is called the *partition function:*

$$Q = \Sigma g_i\, e^{-(\varepsilon_i - \varepsilon_0)/kT} \tag{3.5}$$

Using this function and denoting $\varepsilon_{0B} - \varepsilon_{0A}$ by $\Delta\varepsilon_0$, equation (3.4 a) becomes:

$$K = e^{-\Delta\varepsilon_0/kT} \frac{Q_B}{Q_A}. \tag{3.6}$$

The equilibrium constant depends therefore on two factors. The first, $e^{-\Delta\varepsilon_0/kT}$, indicates that the equilibrium will shift towards the isomer with lower ground-energy level ε_0. The second factor, the ratio of the partition functions, suggests that the equilibrium is displaced in favour of the isomer with a larger partition function.

Equation (3.5) shows that the partition function increases with the number of energy levels (since it is a summation over these levels) and with decreasing differences between these levels and the ground level ε_0.

Figure 3.1 illustrates a system consisting of molecules A and B, where $\varepsilon_{0A} < \varepsilon_{0B}$ and $\varepsilon_{iA} - \varepsilon_{0A} > \varepsilon_{iB} - \varepsilon_{0B}$, irrespective of the i level. If only ground levels are considered, the equilibrium is expected to be displaced towards A; the partition function, on the other hand, would favour molecules B. Clearly, the separate analysis of these factors cannot decide the actual equilibrium trend. The problem is solved only by calculating all factors in equation (3.6). Unfortunately,

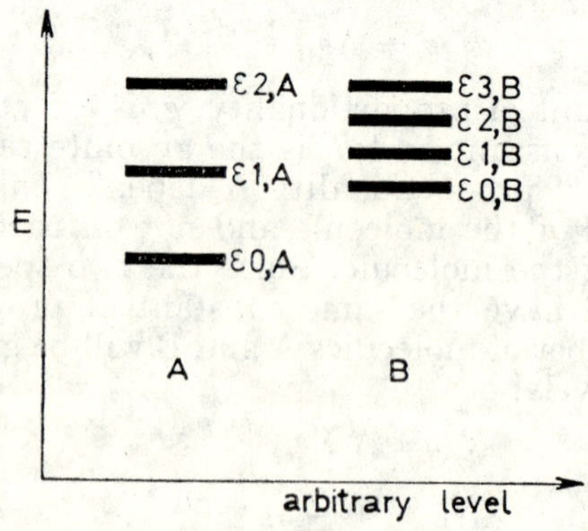

Figure 3.1 Energy levels of two isomeric molecules A and B, at equilibrium.

the computation of partition functions for organic systems is exceedingly difficult. Equation (3.6) is nevertheless useful since it indicates the possibility of displacing the equilibrium in either direction. At lower temperatures the energy factor will predominate, so that the equilibrium mixture will contain only a small amount of the unstable isomer. If the temperature is increased, the equilibrium will be displaced in favour of the unstable isomer (the exponential factors tend to 1).

The thermodynamic treatment of equilibrium starts from the useful work carried out by the system or expended on the system. It is obvious that a system which proceeds toward equilibrium will perform useful work, while a system which is forced away from equilibrium will have to consume work. If these considerations are applied to systems of constant pressure and temperature (as it is the case for most chemical experiments), then the useful work (W) is equal to the difference in free energy, ΔG.

At equilibrium, useful work is zero; hence:

$$\Delta G = -W = 0 \tag{3.7}$$

Free energy is a fundamental quantity in thermodynamics, defined as:

$$\Delta G = \Delta H - T\Delta S, \tag{3.8}$$

where ΔH is the change in enthalpy of reaction or heat of reaction, and ΔS the change in entropy.

The difference in free energy depends also on the nature of the system. For a system containing several components, it is:

$$\Delta G = n_A \left(\frac{\partial G}{\partial n_A}\right)_{n_B,\, n_C \dots} + n_B \left(\frac{\partial G}{\partial n_B}\right)_{n_A,\, n_C \dots} + \dots \tag{3.9a}$$

where n_A, n_B ... are the number of moles of components A, B ... With the notation $\left(\frac{\partial G}{\partial n_i}\right)_{n_A \dots} = \Delta \bar{G}_i$, equation (3.9a) becomes:

$$\Delta G = n_A\, \Delta \bar{G}_A + n_B\, \Delta \bar{G}_B + \dots + n_i\, \Delta \bar{G}_i. \tag{3.9b}$$

In the above relation, $\Delta \bar{G}_i$ is the difference in partial molal free energy and has the significance of the increase of free energy when component i is added, in concentrations tending to zero.

The change in partial molal free energy depends on concentration. For dilute solutions, the following equation applies:

$$\Delta \bar{G}_i = \Delta \bar{G}_{0,i} + RT \ln c_i, \tag{3.10}$$

where c_i is the concentration of compound i.

For an equilibrium reaction of the general form:

$$a\mathrm{A} + b\mathrm{B} \dots \rightleftarrows r\mathrm{R} + s\mathrm{S} \dots$$

the free energy of the system will be given by equation (3.9b), where $n_A = -a$, $n_B = -b \dots n_R = r$, $n_S = s \dots$:

$$\Delta G = -a\Delta \bar{G}_A - b\Delta \bar{H}_B - \dots + r\Delta \bar{G}_R + s\Delta \bar{G}_S + \dots$$

Combining this equation with equation (3.10) leads to:

$$\Delta G = - a\Delta \bar{G}_{\circ,\mathrm{A}} - b\Delta \bar{G}_{\circ,\mathrm{B}} - \ldots + r\Delta \bar{G}_{\circ,\mathrm{R}} + \ldots + RT \ln \frac{C_\mathrm{R}^r C_\mathrm{S}^s \ldots}{C_\mathrm{A}^a C_\mathrm{B}^b \ldots}$$

or

$$\Delta G = \Delta G^\circ + RT \ln K, \tag{3.11}$$

where ΔG° is the change in standard free energy of reaction.

At equilibrium $\Delta G = 0$ and

$$\Delta G^\circ = - RT \ln K. \tag{3.12}$$

From this equation, the equilibrium constant can be expressed as:

$$K = \mathrm{e}^{-\Delta G^\circ/RT} = \mathrm{e}^{-\Delta H^\circ/RT}\, \mathrm{e}^{\Delta S^\circ/R}\,. \tag{3.13}$$

The thermodynamic and the statistical approach lead therefore to similar conclusions. The equilibrium constant is determined by the interplay of two terms, one of energy (ΔH°) and the other of probability (ΔS°). While the partition functions are difficult to determine, standard thermodynamic quantities can be measured and calculated by relatively simple methods [1].

The equilibrium constant can be calculated as a function of these quantities or, conversely, the equilibrium constant can be measured by current chemical methods, then using this experimental value, the change in the standard free energy of reaction is calculated.

The difference in standard free energy of reaction ΔG°, should not be mistaken for the change in total energy of reaction ΔG. The former is used for the determination of the equilibrium constant, while the latter allows to decide whether the process may or may not proceed spontaneously. Indeed, according to the second principle of thermodynamics, the reactions which proceed spontaneously toward equilibrium must have $\Delta G < 0$. Such processes are thermodynamically possible. The reactions for which $\Delta G > 0$ cannot take place spontaneously and are thermodynamically impossible. The change in free energy of reaction is calculated from equation (3.8), the changes in entropy and enthalpy being obtained by conventional methods [1]. Equation (3.14), which expresses the dependence of free energy on temperature, shows that by raising the temperature a thermodynamically impossible process may become possible. Whether such a reaction is actually observed is a matter of the rate with which it may proceed.

$$\frac{\partial \dfrac{\Delta G}{T}}{\partial T} = - \frac{\Delta H}{T^2}\,, \tag{3.14}$$

[1] G. J. Janz, *Estimation of Thermodynamic Properties of Organic Compounds*, Academic Press, New York (1958).

3.2 Chemical kinetics

It is a well-known fact that chemical reactions proceed much more rapidly in concentrated than in dilute solutions. There are certain relations between the concentrations of reactants and the rate of their consumption; to set up the exact form of these relations is the main task of chemical kinetics [2].

3.2.1 *Reaction rate*

The rate of a reaction is defined by the change in the concentration of a reactant or product per time unit. It is assumed that the reaction rate is proportional to the concentration of the reactants and that it is not influenced by the presence of other molecules (*e.g.* reaction products, solvent) [3].

For a reaction of the type:

$$\alpha A + \beta B + ... = \text{products},$$

the reaction rate can be expressed as:

$$-\frac{d[A]}{dt} = -\frac{d[B]}{dt} = ... = \frac{d[\text{products}]}{dt} = k[A]^{\alpha}[B]^{\beta} ... \qquad (3.15)$$

where k is the rate constant.

The *order of reaction* is the sum of the powers of all concentration factors appearing in the rate equation. There are thus reactions and, correspondingly, rate constants of zero-order, first-order, second-order, etc.

Let us denote by a, b, etc. the initial concentrations of the reactants and by x the concentration of the reaction product at the time t. With this notation, the general equation (3.15) can be integrated, affording the expression of the rate constant. Table 3.1 presents several types of rate constants more frequently encountered in organic chemical kinetics.

The first objective of a kinetic study is to find out the reaction order and the rate constant. Equally important is to compare the rate of the reaction under investigation with rates of similar reactions. This comparison may reveal significant relationships between structure and chemical reactivity. Reaction rates, however, are not constant in time and therefore cannot be compared as such. Quantities which can be compared are the *rate constants*. Rate constants,

[2] For a detailed treatment see: (a) A. A. Frost and R. G. Pearson, *Kinetics and Mechanism*, 2nd ed., J. Wiley (1961); (b) S. W. Benson, *The Foundation of Chemical Kinetics*, McGraw-Hill, New York (1960); (c) J. E. Leffler and E. Grunwald, *Rates and Equilibria of Organic Reactions*, J. Wiley, New York (1963).

[3] Actually, solvent molecules or other species involved in the reaction may have a considerable influence on the reaction rate. Their effect, however, is included in the rate constant.

unlike reaction rates, *are not experimental quantities and their value will depend on the method by which they are calculated.* Comparison of two rate constants calculated in different manners could be utterly misleading[4].

The reaction rates and the rate constants in equation (3.15) and in the relations shown in Table 3.1 are expressed as functions

TABLE 3.1

Integrated expressions of rate constants for the main reaction types

Type of reaction	Reaction order	Expression of the rate constant*)	Units of the rate constant
$A \xrightarrow{k_0}$ products	zero	$k_0 = \frac{x}{t}$	mol $l^{-1}s^{-1}$
$A \xrightarrow{k_1}$ products	first	$k_1 = \frac{2.303}{t} \log \frac{a}{a-x}$	s^{-1}
$A + B \xrightarrow{k_2}$ products	second	$k_2 = \frac{2.303}{(a-b)t} \log \frac{b}{a} \frac{a-x}{b-x}$	l $mol^{-1}s^{-1}$
$2A \xrightarrow{k_2}$ products	second	$k_2 = \frac{1}{t}\left(\frac{1}{a-x} - \frac{1}{a}\right)$	l $mol^{-1}s^{-1}$

*) Concentrations are expressed in mol l^{-1}.

of concentrations, which is not exactly correct. Instead of concentrations, *activities* of the reactants should be used, $a_i = \gamma_i c_i$, where γ_i is the activity coefficient. Since it is not possible in every case to determine the activity coefficients, kinetic measurements must be carried out using concentrations lower than 10^{-3} mol l^{-1}, where $a_i \approx c_i$. Practically, owing to inherent analytical errors and the too slow rates at such concentrations, a compromise is reached, determinations being usually performed at concentration of *ca.* 10^{-2} mol l^{-1}.

3.2.2 *Rate-determining step. Product-determining step*

Most organic reactions proceed through several elementary steps.

Let us consider a simple reaction, with the equation:

$$A + B \rightarrow P$$

Let us assume that the rate equation corresponds to a first-order reaction:

$$\text{rate} = k_1[A] \qquad (3.16)$$

[4] E. D. Hughes, C. K. Ingold and A. J. Parker, *J. Chem. Soc.*, 4400 (1960).

If the concentrations of A and B are of the same order of magnitude, the rate equation (3.16) indicates that the reaction takes place in steps, namely:

$$\mathrm{A} \xrightarrow{k_1} \mathrm{I}; \; v_1 = k_1[\mathrm{A}] \tag{3.17a}$$

$$\mathrm{I} + \mathrm{B} \xrightarrow{k_2} \mathrm{P}; \; v_2 = k_2[\mathrm{I}][\mathrm{B}]. \tag{3.17b}$$

Equation (3.16) shows that $v_1 \ll v_2$, that is the intermediate I reacts further with B as soon as it is formed. Consequently, the rate determined experimentally is the rate of the step (3.17a). This slow elementary reaction is called the *rate-determining* (or *rate-controlling*) *step*.

A kinetic study supplies data concerning only the rate-determining step; it cannot tell us anything about the subsequent elementary processes.

Information about subsequent steps can be obtained by varying the working conditions. Let us consider the following mechanism:

$$A + \mathrm{C} \xrightarrow{k_1} \mathrm{I} \tag{3.18a}$$

$$\mathrm{I} + \mathrm{B} \xrightarrow{k_2} \mathrm{P} + \mathrm{C}, \tag{3.18b}$$

where C is a catalyst.

If $k_1 \ll k_2$, the reaction rate observed will be:

$$\frac{\mathrm{d[P]}}{\mathrm{d}t} = k_1 [\mathrm{A}][\mathrm{C}]. \tag{3.19a}$$

In the rate-determining step the intermediate I is therefore formed. Let us assume now that catalyst C is replaced by a more efficient catalyst C′, other conditions being kept constant. The rate constant for the disappearance of I, k_2, will certainly remain the same, while k_1 may change considerably, even exceeding the value of k_2. In the latter case, the kinetic equation will be:

$$\frac{\mathrm{d[P]}}{\mathrm{d}t} = k_3[\mathrm{A}]\,[\mathrm{B}]\,[\mathrm{C}']. \tag{3.19b}$$

An example of such a reaction is the acid-catalysed halogenation of ketones [5,6], a reaction which will be examined in Chapter 6.

Two important conclusions can be drawn from the facts presented above. In the first place, if the reaction mechanism (3.18) and the rate equation (3.19b) are examined, it is obvious that the rate-determining step must be preceded by a rapid step, since tri-

[5] A. Lapworth, *J. Chem. Soc.*, **85**, 30 (1904).
[6] L. P. Hammett and L. Zucker, *J. Amer. Chem. Soc.*, **61**, 2785, 2791 (1939).

molecular collisions are rather rare events. Therefore, the steps preceding the rate-controlling step are reflected in the rate equation. A second conclusion follows from the comparison of equations (3.19a) and (3.19b) which refer to the same reaction, namely that the *order of a reaction* or, more precisely, *the form of the rate equation, depends on the reaction conditions*. The cases examined so far ($k_2 \gg k_1$, or $k_1 \gg k_2$) are only limit situations for the process (3.18). The two consecutive reactions can, however, occur with comparable velocity and then the corresponding rate equations are much more complex and more difficult to verify. It is the task of the experimental chemist to choose convenient reaction conditions under which the rate equation has a simple form, easy to check experimentally and to interpret theoretically.

The reaction products are formed in the last step of the process, called the *product-determining* step. Most often, this step does not coincide with the rate-determining step. Information about the former is obtained by other means (see, for instance, p.231).

3.2.3 *Transition state theory*

The rate constant k depends on temperature. Arrhenius[7] has shown that this dependence can be expressed as an exponential relation:

$$k = Ae^{-\frac{E}{RT}} \tag{3.20}$$

where A is a frequency factor, E — the activation energy of the reaction, R — the gas constant and T — the absolute temperature.

The full significance of equation (3.20) was realised much later. Marcelin[8] was the first to show that the activation energy of the reaction is analogous to a high "pass" separating two potential energy "valleys", which the system crosses when it proceeds from the initial to the final state. Lewis[9], a little later, showed that the frequency factor A corresponds to the number of collisions calculated by the kinetic theory of gases.

The "pass" through the potential mountain was called later the *transition state* (or *activated complex*).

The variation of the potential energy of the system can be treated by simple quantum mechanical methods [10]. Let us consider for instance a displacement reaction, in which a reactant A replaces a group (or atom) C from the molecule B—C:

$$\mathrm{A + B{-}C \longrightarrow A{-}B + C} \tag{3.21}$$

[7] S. Arrhenius, *Z. physik. Chem.*, **4**, 226 (1889).
[8] A. Marcelin, *Ann. Physik*, **3**, 158 (1915).
[9] W. C. Mac Lewis, *J. Chem. Soc.*, **113**, 471 (1918).
[10] K. K. Laidler and J. C. Polanyi, *Progress in Reaction Kinetics*, vol. 3 (G. Porter, editor), Pergamon Press, Oxford (1965), p. 1.

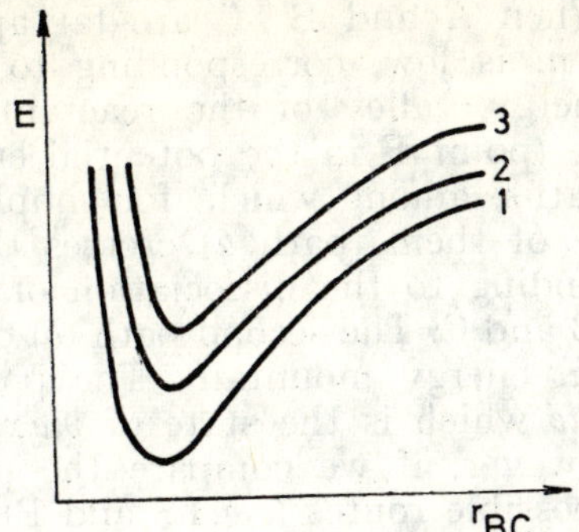

Figure 3.2 Potential energy variation of a diatomic molecule B—C as a function of the internuclear distance r_{BC} and of the distance from a molecule A.

The energy of this system is a function of two variables, the distances r_{AB} and r_{BC}. The energy of a diatomic molecule depends, as we showed previously, on the internuclear distance (see p. 62). The variation of the energy of molecule B—C as a function of distance r_{BC}, when A is still very far away, can be represented by the curve 1 on Figure 3.2 (which is similar to the curve of Figure 2.2). When A approaches the molecule B—C, the energy of the system increases (curves 2, 3, Figure 3.2) and this increase will depend on the angle from which molecule B—C is attacked by A. Quantum mechanical calculations show that the energy of the system is minimum when atom A, B and C are collinear.

The dependence of the energy of the system (3.21) on the two variables r_{AB} and r_{BC} should be represented diagramatically by a three-dimensional plot $E = f(r_{BC}r_{AB})$. Instead of this, we shall draw a more simple contour diagram of the type used for geographical maps (Figure 3.3). Each curve on Figure 3.3 is a locus of points

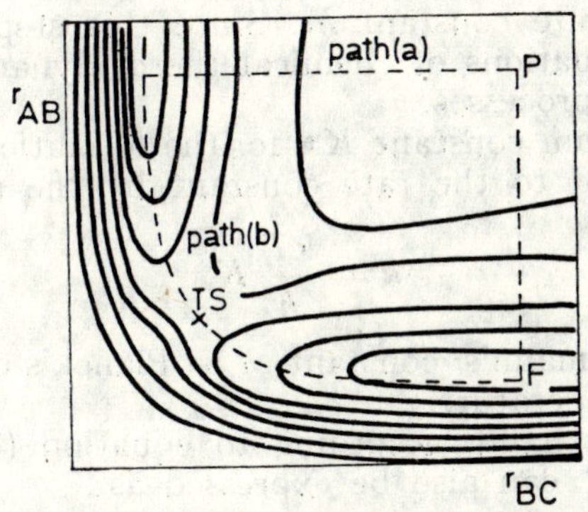

Figure 3.3 Potential energy plot for the system A + B—C as a function of the internuclear distances r_{AB} and r_{BC}; I = initial point (substrate B—C and reagent A at infinite distance); F = final point (reaction products A—B and C); P = a point corresponding to the dissociated molecules A + B + C.

of equal energy. When A and B—C are far apart (r_{AB} large), the energy of the system is low, corresponding to the initial point I in the potential energy valley of the reactants. The system can reach the final state (point F in the potential energy valley of products) by several paths, among which, for simplicity, we shall consider only two. One of them (path *a*) crosses the potential energy plateau *P*, corresponding to the dissociation of the molecule B—C into the fragments B and C. The second path (*b*) crosses a pass (point TS) in the potential energy mountain. The point TS corresponds to the *transition state* which is the state of *maximum energy* in the case of path *b*. However, if we compare the points of maximum energy for various possible routes (*e.g.*TS and P), it appears clearly that TS is the lowest of them. Obviously, the actual energy path adopted by the system when passing from valley I to valley F will be the more convenient path *b*.

The energy expended by the system throughout path *b* is lower, because the energy released when the new bond A—B is formed is partly used for breaking the bond B—C. This compensation operates better if the three atoms are situated on the same straight line. In the transition state, the three atoms A, B and C are not independent from each other, since the bond B—C is not completely broken and the new bond is not entirely formed. An interesting feature of the transition state theory is the fact that the transition state is treated like a molecular species, differing from the other molecules by its higher energy content. It is also assumed that this species is in equilibrium with the initial state:

$$A + B - C \underset{K^{\ddagger}}{\overset{k,\ \text{slow}}{\rightleftarrows}} \underset{\text{transition state}}{A \dots B \dots C} \xrightarrow{\text{rapid}} A - B + C$$

The over-all rate, as shown previously, will be determined by the rate of the slowest step, that is the formation of the activated complex (rate constant *k*). Since this step is reversible, we can apply the equations of chemical thermodynamics and statistics for equilibrium processes.

The equilibrium constant $K^{\ddagger}$ for the formation of the activated complex is related to the rate constant by the equation:

$$k = \frac{\mathbf{k}T}{h} K^{\ddagger}, \tag{3.22}$$

where **k** is Boltzmann's constant, *h* is Planck's constant, and *T* is the absolute temperature.

On the other hand, according to equation (3.13), the equilibrium constant $K^{\ddagger}$ can also be expressed as:

$$K^{\ddagger} = e^{-\frac{\Delta G^{\ddagger}}{RT}}, \tag{3.23}$$

where $\Delta G^{\ddagger}$ is the free energy of activation.

Replacing in equation (3.22), we obtain:

$$k = \frac{\mathbf{k}T}{h} e^{-\frac{\Delta G^{\ddagger}}{RT}} = \frac{\mathbf{k}T}{h} e^{-\frac{\Delta H^{\ddagger}}{RT}} e^{\frac{\Delta S^{\ddagger}}{R}}, \qquad (3.24)$$

where $\Delta H^{\ddagger}$ and $\Delta S^{\ddagger}$ represent the enthalpy and the entropy of activation, respectively.

The comparison of relations (3.20) and (3.24) shows that the enthalpy of activation is related to the energy of activation by the equation:

$$\Delta H^{\ddagger} = E - RT \qquad (3.25)$$

while the entropy of activation and the frequency factor are related by the equation:

$$\Delta S^{\ddagger} = 2.303R \log \frac{A}{T} - 2.303 \log \frac{\mathbf{k}}{h} \qquad (3.26a)$$

Introducing the values of the fundamental constants into equation (3.26a), the latter becomes:

$$\Delta S^{\ddagger} = 4.567 \log \frac{A}{T} - 49.203 \qquad (3.26b)$$

If a diagram of the type shown in Figure 3.3 is sectioned along the path adopted by the system (path *b*), a plot of potential energy *versus* the reaction coordinate is obtained.

In this diagram the transition state corresponds to an energy maximum which the molecules have to acquire in order to be able to react.

The reaction rate, as shown by equation (3.24), depends on the magnitude of the enthalpy of activation, namely it increases with the decrease of the latter, which is also obvious from Figure 3.4. However, the reaction rate depends also on the entropy factor. The larger the entropy, the greater the reaction rate.

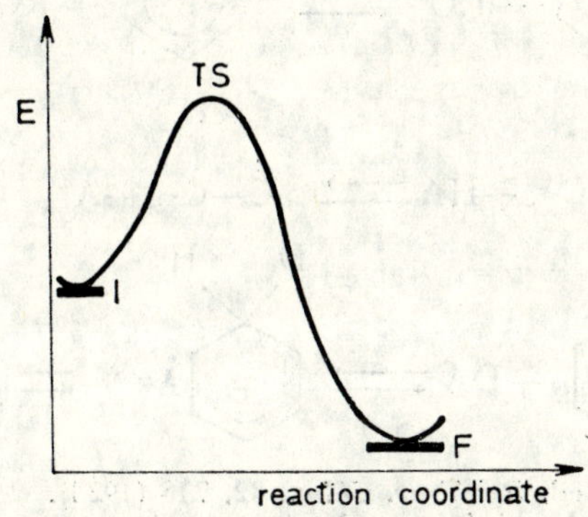

Figure 3.4 Potential energy variation *versus* a reaction coordinate.

3.3 Acids and bases

3.3.1 *Definition*

According to Brönsted's theory [11], an *acid* is a chemical species capable of releasing a proton, while a *base* is a chemical species which can accept a proton. The acid which gives up the proton to the base is itself transformed to a proton acceptor, the *conjugate base* of the acid, while the base becomes an acid after accepting the proton, the *conjugate acid* of the base. The *acid-base* equilibrium can be represented schematically:

$$\underset{\text{Acid}}{AH} + \underset{\text{Base}}{B} \rightleftharpoons \underset{\text{Conjugate base}}{A^-} + \underset{\text{Conjugate acid}}{BH^+} \qquad (3.27)$$

In equation (3.27), the charges merely symbolise that as a result of the proton-transfer, the negative charge of the species increases or decreases, respectively, by a unit. As compared to the classical definitions of acids and bases [12], Brönsted's formulation presents the advantage of being more comprehensive, so that it can be extended to electrically neutral species in any solvent. The theory stipulated that if the acid is strong, the conjugate base will be weak and, conversely, a weak acid will give rise to a strong conjugate base.

With Brönsted's theory of acids and bases it has been possible to understand the mechanisms of the reactions catalysed by acids and by bases. According to this theory, any reaction involving transfer of a proton is regarded as an acid-base equilibrium, for instance:

$$CH_3{-}CO{-}CH_2{-}COOR + RO^- \rightleftharpoons [CH_3CO\ddot{C}HCOOR]^- + ROH$$

$$CH_3NO_2 + HO^- \rightleftharpoons \bar{\ddot{C}}H_2NO_2 + HOH$$

$$C_5H_6 \text{ (cyclopentadiene)} + RO^- \rightleftharpoons C_5H_5^- \text{ (cyclopentadienyl anion)} + ROH$$

$$>C{=}C< + HA \rightleftharpoons >\overset{+}{C}{-}\underset{|}{\overset{H}{C}}< + A^-$$

$$C_6H_5H + DA \rightleftharpoons [C_6H_5(D)(H)]^+ A^- \rightleftharpoons C_6H_5D + HA$$

[11] J. N. Brönsted, *Rec. Trav. Chim.*, **42**, 718 (1923).

[12] For historical and further details on this subject see, for instance, R. P. Bell, *The Proton in Chemistry*, Methuen, London (1959).

In a basic solvent (SH), an equilibrium is reached between the acid and the solvent:

$$AH + SH \overset{K_A}{\rightleftarrows} A^- + SH_2^+ \tag{3.27a}$$

Applying the law of mass action to this equilibrium, an acidity constant, K_A, can be defined:

$$K_A = \frac{[A^-][SH_2^+]}{[AH]} \cdot \frac{\gamma_{A^-}\gamma_{SH_2^+}}{\gamma_{AH}}, \tag{3.28}$$

where γ_{A^-}, $\gamma_{SH_2^+}$ and γ_{AH} are activity coefficients.

The acidity constant is a measure of the strength of the acid in the respective solvent SH. Frequently, instead of acidity constants, pK_A values are used, defined in the same way as for pH: pK_A = = − log K_A. If the acid is stronger than the lyonium ion (the conjugate acid of the solvent), then the conjugate base of the former, A^-, will be a weaker base than the solvent SH; practically, the equilibrium (3.27a) will lie completely to the right. In general, the strongest acid which can exist in a certain medium is the lyonium ion. The solvent exerts therefore a levelling effect on strong acids. In aqueous solutions, the strongest acid is the *hydronium ion* H_3O^+. When a less basic solvent is used, strong acids can be differentiated, as the equilibrium (3.27a) is shifted to the left. Thus, in water, the nitric, hydrochloric, hydrobromic, sulphuric and perchloric acids have apparently the same strength, being almost completely dissociated. In acetic acid, which is weaker base than water, the relative acidities of HNO_3, HCl, HBr, H_2SO_4 and $HClO_4$ are in the ratios 1 : 9 : 30 : 160 : 400 [13a].

In a solvent with acid properties, the following equilibrium occurs between a dissolved base and the solvent:

$$B^- + SH \overset{K_B}{\rightleftarrows} BH + S^- \tag{3.29}$$

The equilibrium constant

$$K_B = \frac{[BH][S^-]}{[B^-]} \cdot \frac{\gamma_{BH}\gamma_{S^-}}{\gamma_{B^-}} \tag{3.30}$$

is called the *basicity constant* and measures the strength of the base in the solvent SH.

[13] (a) R. P. Bell, *Acids and Bases*, Methuen, London (1956), p. 33; (b) R. F. Cookson, *Chem. Rev.*, **74**, 5 (1974).

As in the case of acids, the solvent can exert a levelling effect on the strength of bases, the strongest base which can exist in an acidic solvent being the *lyate ion* (the conjugate base of the solvent). In water, the strongest possible base is the hydroxide ion HO^-.

3.3.2 *Determination of acidity and basicity constants*

Two types of methods are generally used for the determination of acidity and basicity constants [12,13b]. One of them consists in measuring the concentrations of conjugate acid-base pairs which appear in the equilibria of the type (3.27a) or (3.29), in dilute solutions, followed by computation of the respective constants (activities being neglected). If the species are ionic, their concentrations can be determined by measuring the conductivities of the solutions or by the electrometric method [12]. Exact results can be obtained by means of spectrophotometry if the two conjugate species have sufficiently different spectra (this happens, for instance, in the case of the indicators). The method is limited, however, to acids with pK less than 20. Solvents frequently used in these determinations are water, alcohols and mixtures of water and various organic solvents such as sulpholane [14], dimethylsulphoxide [15], etc.

According to the second method [16,17], the acid is treated, in an adequate solvent, with the conjugate base of another acid of known acidity constant; then are measured the concentrations of the two conjugate acid-base pairs involved in the equilibrium:

$$HA + B^- \rightleftharpoons HB + A^- \tag{3.31}$$

The equilibrium constant

$$K = \frac{[HB]\,[A^-]}{[B^-]\,[HA]} \tag{3.32}$$

is thus a measure of the difference between the acidity constants of the two acids, allowing the calculation of the unknown constant. The method assumes that acidity constants are the same in various media (solvents) and that the corresponding activity coefficients of the two systems are proportional. Using as reference acid 9-phenylfluorene, the pK_A of which is independent of the solvent used [16,17], Streitwieser and his co-workers [18] have determined the pK_A values for a great number of hydrocarbons (see Table 3.2).

[14] C. H. Langford and R. L. Burwell, *J. Amer. Chem. Soc.*, **82**, 1503 (1960).
[15] K. Bowden and R. Stewart, *Tetrahedron*, **21**, 261 (1965).
[16] J. B. Conant and G. W. Wheland, *J. Amer. Chem. Soc.*, **54**, 1212 (1932).
[17] W. K. McEwen, *J. Amer. Chem. Soc.*, **58**, 1124 (1936).
[18] A. Streitwieser, Jr., J. H. Hammons, E. Ciuffarin and J. I. Brauman, *J. Amer. Chem. Soc.*, **89**, 59 (1967).

TABLE 3.2
pK_A values for some organic compounds[19]

Organic compound	pK_A	*Organic compound*	pK_A
$CH(CN)_3$	−5	(Fluoradene)	11
$CH(NO_2)_3$	∼0		
$CH(SO_2CH_3)_3$	∼0		
$CH_2(NO_2)_2$	4		
Carboxylic acids	5	Cyclopentadiene	14−15
$CH(COCH_3)_3$	6	CH_3-OH	16
		CH_3-CH_2-OH	18
(fluorenylidene=CH−)$_3$CH	6.2	9-Phenylfluorene	18.5
		Indene	20
$CH_3-CH_2-NO_2$	8.5	Fluorene	22.8
CH_3-NO_2	11	CH_3-CN	25
		Ph_3CH	25
		Ph_2CH_2	33.1

3.3.3 *Acidity functions*

In the study of reactions catalysed by acids or bases it is important to define the acidity and the basicity, respectively, of the medium. For dilute aqueous solutions this amounts to determining the concentration of hydronium ions H_3O^+ (pH scale). However, in the case of many reactions taking place in concentrated acid solutions, it is found that the reaction rate is not proportional to the concentration of the acid. A small increase in the concentration of acid, for instance, may produce a considerable increase in rate. For these reactions it is necessary to find another acidity scale. Such a scale has been set up by Hammett[20,21] for media with dielectric constants over 50.

This acidity scale makes use of weakly basic indicators, especially substituted nitroanilines and azobenzenes. Considering the equilibrium:

$$IH^+ \rightleftharpoons I + H^+ \tag{3.33}$$

[19] A. Streitwieser, Jr. and J. H. Hammons, *Progress in Physical Organic Chemistry*, vol. 3, Interscience, New York (1965), p. 41.
[20] (a) L. P. Hammett, *Physical Organic Chemistry*, McGraw-Hill, New York (1940), p. 267; (b) M. A. Paul and F. A. Long, *Chem. Rev.*, **57**, 1 (1957).
[21] A recent summary: E. M. Arnett, *Progress in Physical Organic Chemistry*, vol. 1, Interscience, New York (1963), p. 223.

the constant of the indicator (I) is defined as:

$$K_{IH^+} = \frac{[I]\,[H^+]}{[IH^+]} \cdot \frac{\gamma_I\,\gamma_{H^+}}{\gamma_{IH^+}} = \frac{[I]}{[IH^+]}\,a_{H^+}\,\frac{\gamma_I}{\gamma_{IH^+}} \tag{3.34}$$

with the understanding that the proton is actually bonded to the solvent.

If the indicator is sufficiently basic, then the ratio $[I]/[IH^+]$ can be directly determined for dilute solutions, and from this the constant of the indicators is calculated, neglecting activity coefficients. When the concentration of the acid is increased, the concentration of the free base (I) may become too small to be measured, so that more weakly basic indicators must be used. However, in this case the direct determination of pK_{IH^+} is not possible, since activity coefficients can no longer be neglected. Using a reference indicator I_1 of known $pK_{I_1H^+}$, the $pK_{I_2H^+}$ of a more weakly basic indicator I_2 can be determined from the equation:

$$pK_{I_1H^+} - pK_{I_2H^+} = \log\frac{[I_1H^+]}{[I_1]} - \log\frac{[I_2H^+]}{[I_2]} \tag{3.35}$$

provided that the ratio γ_{IH^+}/γ_I is practically the same for the two pairs of conjugated species. This appears to be a fair approximation since the indicator differs from its conjugate acid only by a proton. If we accept this, equation (3.34) can be written:

$$K_{IH^+} = \frac{[I]}{[IH^+]}\,h_0,$$

where

$$h_0 = a_{H^+}\,\frac{\gamma_I}{\gamma_{IH^+}} = K_{IH^+}\frac{[IH^+]}{[I]}. \tag{3.36}$$

The quantity h_0 is called the *acidity function* and is independent of the nature of the indicator. Its value can easily be determined by measuring spectrophotometrically the ratio between the concentrations of the indicator and its conjugate form, the value of pK_{IH^+} being known. It is convenient to use, instead of h_0, the quantity H_0, defined as:

$$H_0 = -\log h_0 = -\log a_{H^+} + \log\frac{\gamma_{IH^+}}{\gamma_I} \tag{3.37}$$

It might be noted that for dilute solutions H_0 has the same value as pH. At high concentrations, H_0 increases more rapidly than pH (see Table 3.3). The function H_0 is a measure of the ability of the medium to transfer a proton to an electrically neutral substrate.

TABLE 3.3

Values of H_0 for H_2SO_4-water mixtures

$\%H_2SO_4$	H_0	$\%H_2SO_4$	H_0	$\%H_2SO_4$	H_0
5	+0.24	40	−2.28	80	− 6.82
10	−0.16	50	−3.23	90	− 8.17
20	−0.89	60	−4.32	95	− 8.74
30	−1.54	70	−5.54	100	−10.60

Many applications of the function H_0 in the study of reaction mechanisms are known [22]. Some of them will be discussed in the next section.

For neutral, proton-donating solvents, containing a negatively charged base (A^-), a function H_- can similarly be defined:

$$H_- = -\log a_{H^+} \frac{\gamma_{A^-}}{\gamma_{AH}} = pK_{A^-} + \log \frac{[A^-]}{[AH]} \tag{3.38}$$

if we admit, as previously, that the ratios γ_{A^-}/γ_{AH} do not vary with the nature of the indicator. The function H_- represents the ability of the medium to accept a proton from an electrically neutral acid.

Although applications of the function H_- are known [20b,23,24], they are less frequent than those of H_0.

3.3.4 *Reactions catalysed by acids and bases*

It is commonly stated that if a chemical species C, even if present in small concentrations, accelerates a chemical reaction without being consumed in the process, C is a catalyst for the reaction. Although the catalyst does not appear in the stoichiometrical equation, it takes part in the reaction, its effect being apparent in the rate constant. In the case of a reversible reaction, the catalyst has no effect on the equilibrium constant and therefore does not change the composition of the equilibrium mixture; it only increases the rate with which equilibrium is attained.

Let us consider the reaction:

$$A \xrightarrow{k} B, \tag{3.39}$$

the rate of which, in absence of a catalyst, depends on the rate constant k. Let us assume that a species (catalyst) is introduced, which

[22] F. A. Long and M. A. Paul, *Chem. Rev.*, **57**, 935 (1957).

[23] R. Stewart and J. P. O'Donnell, *J. Amer. Chem. Soc.*, **84**, 494 (1962); R. Stewart, J. P. O'Donnell, J. D. Cram and B. Rickborn, *Tetrahedron*, **18**, 917 (1964).

[24] E. C. Steiner and J. M. Gilbert, *J. Amer. Chem. Soc.*, **87**, 382 (1965).

is capable of forming with A a reactive intermediate I in a reaction faster than the direct transformation of A to B, and that the intermediate decomposes very rapidly yielding the product B. The over-all reaction rate in this case will be determined by the constant k_c for the formation of the intermediate:

$$A + nC \xrightarrow{k_c} I \tag{3.40}$$

$$I \longrightarrow B + nC$$

$$\text{rate} = k_c\,[A]\,[C]^n = k_c'[A]. \tag{3.41}$$

The rate constant k_c is termed the catalytic constant. This constant, as well as the reaction order with respect to the catalyst, is usually determined by plotting the logarithm of the observed rate constant (k_c') against the logarithm of catalyst concentration. The slope of the resulting line will afford the reaction order with respect to C, while the intercept with the ordinate will provide the value of k_c.

In reactions catalysed by acids and bases, the intermediates are formed in acid-base equilibria of the types previously discussed. For an acid-catalysed reaction, the rate equation should be:

$$-\frac{dS}{dt} = k_{H^+}[S]\,[H^+]\,\frac{\gamma_S\gamma_{H^+}}{\gamma^{\ddagger}}, \tag{3.42}$$

where S is the substrate and $\gamma^{\ddagger}$ the activity coefficient of the transition state.

The equation (3.42) cannot be verified, since activity coefficients for transition states cannot be determined. For dilute solutions, where activity coefficients are close to unity, equation (3.42) should hold. The fact that the agreement has actually been observed only in very few cases *, led to the discovery of the *general catalysis by acids and bases*. It was thus recognised that all species capable of accepting or losing protons, which are present in the reaction medium, can have a catalytic effect. The rate equation of such a reaction, buffered with AcOH—AcO⁻, in aqueous solution, can be formulated as follows:

$$-\frac{dS}{dt} = \{k_0 + k_{H_3O^+}[H_3O^+] + k_{HO^-}[HO^-] + k_{AcOH}[AcOH] + k_{AcO^-}[AcO^-]\}\,[S]. \tag{3.43}$$

In the cases when $k_{H_3O^+}$ exceeds by far the other constants, equation (3.43) becomes:

$$-\frac{dS}{dt} = k_{H_3O^+}[H_3O^+]\,[S] \tag{3.43a}$$

which corresponds to a *specific catalysis* by hydronium ions.

* See an ample discussion in Reference 12, Chapter 9.

If the rate constants corresponding to the acid species are much greater than those for the basic species, we are dealing with a case of *general catalysis by acids*, as for instance:

$$-\frac{dS}{dt} = (k_0 + k_{H_3O^+}[H_3O^+] + k_{AH}[AH] + k_{A'H}[A'H])[S] \qquad (3.43b)$$

In a similar manner we can define the *general catalysis by bases*, the *specific catalysis by hydroxide ions*, HO^-, etc.

The rate of an acid-catalysed reaction may depend either on the hydrogen ion concentration or on the acidity function h_0. Let us consider a catalytic process:

$$S + H^+ \underset{}{\overset{K_{SH^+}}{\rightleftarrows}} SH^+ \qquad (3.44a)$$

$$SH^+ + Y \xrightarrow[\text{slow}]{k} \text{products} + H^+ \qquad (3.44b)$$

The rate of product formation will be:

$$\frac{d\text{Prod.}}{dt} = k[SH^+][Y]\frac{\gamma_{SH^+}\gamma_Y}{\gamma^{\ddagger}} \qquad (3.45)$$

or, considering the equilibrium (3.44a):

$$\text{rate} = \frac{k}{K_{SH^+}}[S][H^+][Y]\frac{\gamma_S\gamma_Y\gamma_{H^+}}{\gamma^{\ddagger}}.$$

The logarithm of the observed rate constant will be:

$$\log k_{obs} = \log[H^+][Y] + \log\frac{\gamma_S\gamma_Y\gamma_{H^+}}{\gamma^{\ddagger}} + \log\frac{k}{K_{SH^+}}.$$

For dilute solutions, if Y is the solvent itself, the above equation can be simplified to:

$$\log k_{obs} = \log[H^+] + \text{const.} \qquad (3.46)$$

This shows that if the logarithm of the observed rate constant is plotted against the logarithm of the hydrogen ion concentration, a straight line is obtained with the slope equal to unity. This type of dependence is not encountered, however, in all cases of specific catalysis by the hydronium ion. When the reaction products do not result directly from the protonated species SH^+, that is:

$$S + H^+ \overset{K_{SH^+}}{\rightleftarrows} SH^+ \qquad (3.47a)$$

$$SH^+ \xrightarrow[k]{\text{slow}} I^+ \qquad (3.47b)$$

$$I^+ + Y \xrightarrow{\text{rapid}} \text{products} \tag{3.47c}$$

the rate equation becomes:

$$\frac{\text{dProd.}}{\text{d}t} = k\,[\text{SH}^+]\,\frac{\gamma_{\text{SH}^+}}{\gamma^{\ddagger}} \tag{3.48}$$

Combining with equation (3.47a):

$$\text{rate} = \frac{k}{K_{\text{SH}^+}}[\text{S}]\,\frac{\gamma_{\text{S}}}{\gamma^{\ddagger}}\,a_{\text{H}^+} = \frac{k}{K_{\text{SH}^+}}[\text{S}]\,h_0 \tag{3.49}$$

whence, by introducing the logarithm of the observed constant, we obtain:

$$\log k_{\text{obs}} = -H_0 + \text{constant} \tag{3.50}$$

The logarithm of the observed constant depends therefore linearly on the acidity function H_0, the slope of the plot being -1. There are numerous examples of such correlations [22]. As follows from the two mechanisms (3.44) and (3.47), a correlation with the acidity function H_0 indicates that the transition state of the rate-determining step involves only a proton, while a linear correlation with the hydrogen ion concentration implies, in addition to the proton, a molecule Y (of solvent).

3.3.5 *Lewis acids. Hard and soft acids*

According to the definition of Lewis [25a], an acid is a species capable of accepting two electrons from an electron-donating base. An acid-base equilibrium, according to this definition, will be represented as follows:

$$\text{A} + :\text{B} \overset{K}{\rightleftarrows} \text{A}:\text{B} \tag{3.51}$$

The following reactions may illustrate the definition:

$$H^+ + :\text{NH}_3 \rightleftarrows \overset{+}{\text{N}}\text{H}_4$$

$$\text{F}_3\text{B} + :\text{NH}_3 \rightleftarrows \text{F}_3\text{BNR}_3$$

$$\text{AlCl}_3 + :\text{X}{-}\text{R} \rightleftarrows \text{R}^+\text{X}{-}\text{AlCl}_3^- \text{ etc.}$$

It might be noted from the above examples that a Lewis base is also a Brönsted base. However, the Lewis definition of acids is much broader than that of Brönsted.

Making use of the linear free-energy relationships (see Section 3.5), the equilibrium constant of reaction (3.51) can be expressed in the form:

$$\log K = S_{\text{A}}S_{\text{B}}, \tag{3.52}$$

[25] (a) G. N. Lewis, *Valency and the Structure of Atoms and Molecules*, Reinhold, New York (1923), cf. Reference 23.

where S_A is a measure of the strength of the Lewis acid A, and S_B a measure of the strength of base B. If the parameters S_A and S_B were known for a series of acids and bases, it would be possible to describe equilibria involving Lewis acid-base adducts of the type A—B. Since such a relation was found to be valid only in very few cases, a four-parameter equation was proposed:

$$\log K = S_A S_B + \sigma_A \sigma_B, \qquad (3.53)$$

where σ_A, σ_B characterise the "*hardness*" of the two species A and B. The *hard* or *soft* character of an acid or of a base is defined by the equilibrium:

$$BH^+ + CH_3Hg(H_2O)^+ \rightleftharpoons CH_3HgB^+ + H_3O^+. \qquad (3.54)$$

By definition, a *hard base* has affinity for the proton (the reverse half-reaction (3.54)), and a *soft base* has affinity for the Lewis acid CH_3Hg^+ (the forward half-reaction (3.54)). These ideas led to the generalisation that *a hard base prefers to combine with a hard acid, and a soft base with a soft acid* [25b,c].

A few examples of bases and acids, classified as hard or soft, are shown in Table 3.4.

TABLE 3.4

Classification of acids and bases [25b, c]

Lewis acids	
Hard	Soft
H^+, Li^+, Na^+, K^+	Cu^+, Ag^+, Au^+, Hg^+
Be^{2+}, Mg^{2+}	Pd^{2+}, Cd^{2+}, Hg^{2+}
Al^{3+}, Si^{4+}	CH_3Hg^+
$BeMe_2$, BF_3, $B(OR)_2$	BH_3
$Al(CH_3)_3$, $AlCl_3$, AlH_3	RS^+, RSe^+
RSO_2^+, SO_3	I^+, Br^+, HO^+, RO^+
RCO^+, CO_2	I_2, Br_2
HX (molecules which form hydrogen bonds)	$(CN)_2C{=}C(CN)_2$
	CH_2 (carbenes)

Bases	
Hard	Soft
H_2O, HO, F^-	R_2S, RSH, RS^-
CH_3COO^-, PO_4^{3-}, SO_4^{2-}	I^-, SCN^-
Cl^-, CO_3^{2-}, NO_3^-	R_3P, R_3As, $(RO)_3P$
ROH, RO^-, R_2O	C_2H_2, C_6H_6
NH_3, R_3N	H^-, R^-

[25] (b) P. G. Pearson, *J. Chem. Educ.*, **45**, 581, 683 (1968); (c) T.-L. Ho, *Chem Rev.*, **75**, 1 (1975).

From the tabulated data, one may note that hard Lewis acids are positive ions or neutral molecules which have no unshared electrons in outer valence shells, have a small volume and low polarisability. Soft acids have large volumes, possess unshared electrons in outer valence shells and display high polarisability. Bases present similar characteristics. Obviously, there is a frontier domain between the two classes, in which the hard or soft character is not exactly defined. We may cite, among such bases: $C_6H_5NH_2$, Br^-, N_3^-, etc., and among acids: SO_2^+, NO, R_3C^+, etc.

3.4 Thermodynamic control and kinetic control

In many reactions which proceed reversibly, the nature of products depends on the reaction conditions used. Let us write a general reaction (3.55) in which a compound A gives rise, reversibly, to two products, B and C:

$$\mathrm{B} \underset{k_1}{\overset{k_{-1}, K_1}{\rightleftarrows}} \mathrm{A} \underset{k_{-2}}{\overset{k_2, K_2}{\rightleftarrows}} \mathrm{C}. \tag{3.55}$$

The ratio of the two products will be given by the equation:

$$\frac{d[B]}{d[C]} = \frac{k_1[A] - k_{-1}[B]}{k_2[A] - k_{-2}[C]} \tag{3.56a}$$

If the rate constants of the reverse reactions, k_{-1} and k_{-2}, are small as compared to k_1 and k_2, and if the system is not given enough time to attain the equilibrium state, equation (3.56a) is reduced to:

$$\frac{d[B]}{d[C]} = \frac{k_1}{k_2} \quad \text{or} \quad \frac{[B]}{[C]} = \frac{k_1}{k_2}. \tag{3.56b}$$

This relation shows that reaction products are controlled only by the ratio of the rate constants of the forward reactions. Such a process is said to be under *kinetic control*. If by some means (temperature increase, use of catalysts, etc.) the rates of all reactions are increased and the system is allowed to reach equilibrium, the ratio between B and C will depend on the overall equilibrium constant K:

$$\frac{[B]}{[C]} = K_1K_2 = K. \tag{3.56c}$$

In this case the reaction is considered to be under *thermodynamic control*.

In the general reaction (3.55), A represents either a stable molecule (substrate) capable of being converted into the two products B and C, or a reactive intermediate which appears in the course of the transformation of B to C. In the reactions in which A is a stable molecule, the type of control may change as a result of temperature changes. Usually, the kinetic control operates at lower temperatures, while at higher temperatures the reaction is controlled by the equilibrium constant K. As examples of such reactions we may cite electrophilic aromatic substitutions (A = aromatic substrate; B, C = monosubstituted products), the Diels-Alder reaction (A = diene; B, C = stereoisomeric *endo* and *exo* adducts), etc.

When A is a reactive intermediate (carbenium ion, carbanion), the nature of the reaction products is influenced especially by the nature of the reactant and to a lesser extent by the temperature used[26]. As in the previous case, the thermodynamic control is determined by the magnitude of the over-all equilibrium constant K. It is convenient to express this constant as a product of two factors, P and R[26]:

$$K = \frac{k_1}{k_{-1}} \frac{k_{-2}}{k_2} = \frac{k_1}{k_2} \frac{k_{-2}}{k_{-1}} = PR. \qquad (3.57)$$

The first factor, $P = k_1/k_2$, known as the partition factor, represents the ratio of the rate constants of the reactions by which A disappears from the system. The second factor, $R = k_{-2}/k_{-1}$, referred to as the reactivity factor, shows how much C is more reactive than B in generating the intermediate A. Relation (3.57) allows us to predict how the nature of the reaction products can be influenced by the change of the reactant or even by the change of its concentration. Let us consider the methanolysis of 7-norbornenyl tosylate (**1**-OTs), which may lead to two products, the unrearranged ether **1**-OMe and a tricyclic ether **3**-OMe. The reaction involves the transient carbonium ion **2**.

X
k_{-1}
k_1
k_2
k_{-2}
X

1-OTs (X = OTs)
1-OMe (X = OMe)

2

3-OMe
(X = OMe)

In pure methanol, toluene-*p*-sulphonic acid is formed in the course of the reaction; this strong acid protonates the ethers formed initially, allowing their equilibration *via* the common intermediate **2**. Under these conditions, the ether **1**-OMe results, which is the product of thermodynamic control. The reaction in buffered medium (where toluene-*p*-sulphonic acid is neutralised as it is formed) affords a mixture of ethers in the ratio **1**-OMe/**3**-OMe = 300. When the reaction is carried out with an excess of sodium

[26] S. Hünig, *Angew. Chem.*, **76**, 400 (1964); R. Gompper, *Angew. Chem.*, **76**, 412 (1964).

methoxide, although the mechanism remains unchanged (unimolecular), the ratio **1**-OMe/**3**-OMe decreases to about 1, the product **3**-OMe representing 51.5% of the ethers formed. The contrasting product distributions in the two kinetically controlled reactions are due to different partition factors P, as a result of the change in the species with which ion **2** reacts [27].

In buffered solution, the reactant is methanol, which is less reactive than the methoxide ion and therefore has higher selectivity. The exclusive formation of the ether **1**-OMe in the thermodynamically controlled reaction is explained by the large equilibrium constant of the isomerisation **3**-OMe $\overset{K}{\rightleftarrows}$ **1**-OMe. The value of this constant was obtained from the partition factor in buffered medium ($P = 300$) and from the rate constants of the hydrolysis of the two ethers, in aqueous acetone in presence of a strong acid ($R = 7 \times 10^6$). The overall equilibrium constant is then: $K = 3 \times 10^2 \times 7 \times 10^6 = 2.1 \times 10^9$, which shows that the equilibrium is completely shifted to ether **1**-OMe.

3.5 Linear free-energy relationships

The correlations between equilibrium constants or rate constants and structural factors are free-energy relationships [28]. The most simple and frequently used free-energy relationships are two-parameter equations. A number of equations with several parameters are also known.

3.5.1 *Compensation of enthalpies and entropies*

In many related reactions, involving moderate variations of the substrate structure or of the characteristics of the medium (solvent, etc.), it is noticed that activation enthalpies and entropies do not vary independently. A large entropy, for instance, is usually associated with a large enthalpy. Sometimes, the correlation of enthalpies and entropies approaches the precision of linear relationships. In many instances we encounter a linear relation of the type [28,29]:

$$\Delta H^{\ddagger} = \Delta H_0^{\ddagger} + \beta \Delta S^{\ddagger}, \tag{3.58}$$

where $\Delta H_0^{\ddagger}$ is the intercept for $\Delta S^{\ddagger} = 0$, and β the slope of the line with the dimension of temperature.

Introducing the expression of $\Delta H^{\ddagger}$ from the above relation into equation (3.8), we obtain:

$$\Delta G^{\ddagger} = \Delta H_0^{\ddagger} + (\beta - T)\Delta S^{\ddagger}. \tag{3.59}$$

[27] (a) S. Winstein and E. M. Kosower, *J. Amer. Chem. Soc.*, **81**, 4399 (1959); (b) D. Diaz, M. Brookhart and S. Winstein, *J. Amer. Chem. Soc.*, **88**, 3133 (1966).

[28] N.B. Chapman and J. Shorter, Ed., *Advances in Linear Free-Energy Relationships*, Plenum Press, New York, 1973.

[29] J. E. Leffler, *J. Org. Chem.*, **31**, 533 (1966); **20**, 1202 (1955).

It is seen that when $T = \beta$, $\Delta G^{\ddagger} = \Delta H_0^{\ddagger} =$ constant. It follows that at this temperature all the reactions which satisfy equation (3.58) must have the same rates. This is why β is called the *isokinetic temperature* and equation (3.58) the *isokinetic relationship*.

The consistency of an isokinetic relationship in a series of related reactions is an indication that these reactions have a common mechanism. Owing to the experimental errors in the determination of $\Delta H^{\ddagger}$ and especially of $\Delta S^{\ddagger}$, the interpretations based on the validity of an isokinetic relationship are sometimes hazardous. In general, one should avoid drawing conclusions from measurements carried out in the vicinity of the isokinetic temperature.

The isokinetic relationship indicates also that even a large variation of the enthalpy or entropy of activation is not a reliable indication for a change in mechanism. Since the changes in these two quantities can be mutually compensated, they must be considered together rather than individually.

3.5.2 *The Hammett ρσ equation*

Ever since chemical kinetics was used in the study of organic reactions, many investigators have tried to set up relationships between the magnitude of rate constants and the structure of the substrate. The attempts proved successful only after it was realised that rate constants can be correlated with thermodynamic quantities[30] (see Section 3.2).

The Hammett ρσ equation[30–34] was the first successful attempt in this direction. Hammett investigated the relations of *meta*- and *para*-disubstituted benzene derivatives. He found that the logarithms of their rate constants (or equilibrium constants) are linearly related if the substrates compared have the same substituents X and if the latter are similarly located with respect to the reaction centres Y and Z, respectively.

$$X\text{-}C_6H_4\text{-}Y + A \underset{}{\overset{k_X,\ K_X}{\rightleftharpoons}} X\text{-}C_6H_4\text{-}Y' + B$$

(reaction compared)

$$X\text{-}C_6H_4\text{-}Z + C \underset{}{\overset{k'_X,\ K'_X}{\rightleftharpoons}} X\text{-}C_6H_4\text{-}Z' + D$$

(reference reaction)

30 L. P. Hammett, *J. Chem. Educ.*, **43**, 464 (1966).
31 L. P. Hammett, *Chem. Rev.*, **17**, 125 (1935).
32 L. P. Hammett, *J. Amer. Chem. Soc.*, **59**, 96 (1937).
33 H. H. Jaffé, *Chem. Rev.*, **53**, 191 (1953).
34 D. Fărcașiu, *Stud. cerc. chim.*, **14**, 37 (1966).

The quantitative equations expressing this dependence have the form:

$$\log K_X = a + \rho \log K'_X \quad \text{or}$$

$$\log k_X = a' + \rho \log k'_X, \tag{3.60a}$$

where the subscript X refers to the substituent, a and a' are constants, and ρ is the slope of the plot, which varies with the nature of the reaction. The point corresponding to X = H is found also on the line, that is:

$$\log K_H = a + \rho \log K'_H \text{ or}$$

$$\log k_H = a' + \rho \log k'_H. \tag{3.60b}$$

Combining equation (3.60a) with equation (3.60b) we find:

$$\log \frac{K_X}{K_H} = \rho \log \frac{K'_X}{K'_H} \quad \text{or}$$

$$\log \frac{k_X}{k_H} = \rho \log \frac{k'_X}{k'_H}. \tag{3.61}$$

Relationships (3.61) show that the effects of substituents on the reaction centre of the reference reaction are proportional to the effects in the case of the reaction under comparison. It follows that the logarithm of the ratio of the reference reaction constants is a measure of the substituent effect. For a given substituent X this ratio is a constant and is denoted by σ_X. Relationships (3.61) appear thus in the form of a two-parameter equation, which is the usual form of the Hammett equation:

$$\log \frac{K_X}{K_H} \text{ or } \log \frac{k_X}{k_H} = \rho\sigma_X. \tag{3.62}$$

The proportionality constant ρ indicates the extent to which the effect of the substituent in the reference reaction is transmitted to the reaction centre in the reaction under investigation. This constant is independent of the nature of the substituent, but will vary characteristically with the reaction type.

The values of σ_X can be determined easily from the rate constants of the reference reaction, for which ρ is obviously unity. Hammett used as reference reaction the ionisation of *meta*- and *para*-substituted benzoic acids in water, at 25 °C, the equilibrium constants of which had been determined with high accuracy.

A few σ_X values are listed in Table 3.5. It may be noticed that, for a given substituent X, σ_X has different values for the two positions, *meta* and *para*. It is also found that electron-attracting substituents have positive σ values, while for electron-donating substituents σ is negative.

A few σ_X values from Table 3.5 are worth examining. It is thus noted that the OMe group has a positive σ_x for the *meta* position, and a negative one for *para*. The explanation of the discrepancy lies in the two opposed effects, $-I$ and $+E$, of this group. When the group is in the *meta* position, the inductive electron-withdrawing effect operates, while in the *para* position the mesomeric electron-donating effect predominates.

The plot of the logarithm of rate constants against σ_X values should give a straight line if the type of reaction under investigation involves one and the same mechanism. The slope of this line is the constant ρ. Table 3.6 lists ρ values for a few types of reactions. It is seen that ρ may assume positive or negative values.

TABLE 3.5
σ constants *

Substituent (X)	*meta* σ_m	*para* σ_p	Substituent (X)	*meta* σ_m	*para* σ_p
CH_3-	−0.069	−0.170	$-O^-$	−0.708	−1.00
CH_3-CH_2-	−0.07	−0.151	−OH	+0.121	−0.37
$(CH_3)_2CH-$	−0.068	−0.151	$-OCH_3$	+0.115	−0.268
$(CH_3)_3C-$	−0.10	−0.197	−F	+0.337	−0.062
C_6H_5-	+0.06	−0.01	−Cl	+0.373	+0.227
$-COCH_3$	+0.376	+0.502	−Br	+0.391	+0.232
−CN	+0.56	+0.660	−I	+0.352	+0.276
$-COO^-$	−0.1	0.0	$-NO_2$	+0.710	+0.778
−COOH	+0.35	+0.406	$-NH_2$	−0.16	−0.66
$-COOCH_3$	+0.37	+0.45			
$-CF_3$	+0.43	+0.54			

* From Reference 2(c), p. 173.

TABLE 3.6
ρ values*

Reaction	Solvent	Temperature (°C)	ρ
	Reversible reactions		
$ArCOOH + H_2O$	H_2O	25	1.000
2-Methylbenzoic acids + H_2O	50% EtOH	25	1.673
$ArOH + H_2O \rightleftharpoons ArO^- + H_3O^+$	H_2O	25	2.113
$Ar\overset{+}{N}H_3 + H_2O \rightleftharpoons ArNH_2 + H_3O^+$	H_2O	25	2.766
	20% dioxane	25	3.257
	70% dioxane	25	3.567
	Irreversible reactions		
$ArCOOH + Ph_2CN_2$	EtOH	30	0.940
$ArCOOMe + H_2O$	60% acetone	0	2.460
$Ar_2O^- + CH_2-CH_2$ (bridged by O, epoxide)	98% EtOH	70.4	−0.947
$ArO^- + EtI$	EtOH	42.5	−0.994
$ArCOCH_3 + Br_2 \longrightarrow ArCOCH_2Br$	0.5 M HCl in 75% AcOH	25	+0.458

* From Reference 2(c), p. 178.

A positive ρ indicates that the reaction centre of the substrate is more negatively charged in the transition state than in the initial state, as in the reference system **4**:

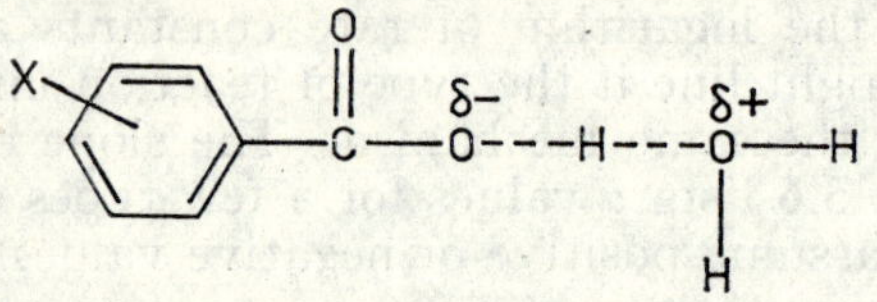

4

On the contrary, a negative ρ is associated with an increase of the positive charge in the transition state.

The $\rho\sigma$ equation is useful for demonstrating the constancy of a mechanism for a given type of reaction. On the other hand, it may offer indications regarding the nature of the transition state.

The $\rho\sigma$ equation, as shown before, was set up empirically. Subsequently, attempts were made to find a theoretical interpretation of it. Using the equations presented in Section 3.1, equation (3.62) can also be written as:

$$\Delta G_X - \Delta G_H = -2.303\ RT\ \rho\sigma = \rho(\Delta G'_X - \Delta G'_H), \qquad (3.63)$$

where subscripts have the same meaning as for the equations (3.60)—(3.62). On the other hand, using equation (3.8) we may write:

$$(\Delta H_X - \Delta H_H) - (\Delta S_X - \Delta S_H)\,T = -2.303\,RT\ \rho\sigma \qquad (3.64)$$

The substituent is assumed to affect the potential energy of the substrate, which is included in the enthalpy. The entropies (the activation entropies, respectively, when rate constants are used) are influenced especially by factors related to the medium and not by the nature of the substituent. If a Hammett equation holds for a given type of reaction, we may assume that the changes in entropy are negligible or that they are similar when the substituent is changed, so that:

$$\Delta S_X - \Delta S_H = 0 \qquad (3.65)$$

This is not the only condition to be fulfilled in the case of a linear free-energy relationship of the Hammett-equation type. From equations (3.12) and (3.8) we find:

$$\frac{\mathrm{d}\,(\ln K)}{\mathrm{d}t} = \frac{\Delta H}{RT^2} \qquad (3.66)$$

and from equation (3.6):

$$\frac{\mathrm{d}\ln K}{\mathrm{d}t} = \frac{\Delta E}{RT^2} + \frac{\mathrm{dln}\, Q_{\mathrm{fin}}/Q_{\mathrm{in}}}{\mathrm{d}t} \qquad (3.67)$$

It follows that:

$$\Delta H = \Delta E + RT^2 \frac{\mathrm{dln}\, Q_{\mathrm{fin}}/Q_{\mathrm{in}}}{\mathrm{d}t}$$

or

$$\Delta H_{\mathrm{X}} - \Delta H_{\mathrm{H}} = (\Delta E_{\mathrm{X}} - \Delta E_{\mathrm{H}}) + RT^2 \left[\left(\frac{\mathrm{dln}\, Q_{\mathrm{fin}}/Q_{\mathrm{in}}}{\mathrm{d}t}\right)_{\mathrm{X}} - \right.$$
$$\left. - \left(\frac{\mathrm{dln}\, Q_{\mathrm{fin}}/Q_{\mathrm{in}}}{\mathrm{d}t}\right)_{\mathrm{H}}\right]$$

The terms which include partition functions reflect the changes in kinetic energy, which are not directly influenced by the nature of the substituent. If a free-energy relationship is observed, these terms probably vanish.

The conditions shown above may certainly seem rather drastic and it is therefore surprising that the Hammett equation was found to apply in such a great number of cases [33, 34]. The explanation is that entropy factors and partition functions do not vary significantly if the reaction mechanism remains constant. In the case of *ortho*-substituted derivatives, where the activation entropy varies with the nature of the substituent (for steric reasons), the Hammett equation is no longer applicable.

Relationships similar to the Hammett equation have also been found in the case of aliphatic compounds [33, 34] for which it was even possible to measure the contribution of polar effects to the magnitude of the rate constant (see Taft equation, Section 3.5.4).

An important fact should be stressed here, namely that for reaction rate correlations, constants σ_{X} are referring to the substituent effect *on the transition state*. The value of σ_{X} is thus directly related to the geometry of the transition state. Since the geometry of the transition state varies from one reaction to another, the values of σ_{X} will not be perfectly constant and will depend on the type of the reference reaction.

The σ_X values listed in Table 3.5 have been determined using a certain reference reaction (the ionisation of substituted benzoic acids), where the influence of the substituent consists mainly of its inductive effect. It is therefore not surprising that in the course of time it was necessary to determine σ_{X} values on the basis of other reference reactions, for which electronic requirements at the reaction centre were different from those of the ionisation of the benzoic acids. Consequently, it is more correct to speak of a range of σ_{X} values rather than of a particular value [35, 36].

[35] H. van Beldman, P. E. Verkade and B. M. Wepster, *Rec. Trav. Chim.*, **78**, 815 (1959).
[36] G. Thirot, *Bull. Soc. chim. France*, 759 (1967).

3.5.3 *The dependence of parameter* ρ *on temperature*

Experimental data show that the parameter ρ changes with temperature. The fact should not surprise us if we remember the equations (3.8) and (3.58). It may be shown that:

$$\Delta G^{\ddagger} = \Delta H^{\ddagger}\left(1 - \frac{T}{\beta}\right) + \frac{T}{\beta}\,\Delta H_0^{\ddagger}. \tag{3.68}$$

The Hammett equation holds if the changes in free energy are proportional to the changes in enthalpy. Equation (3.68) indicates also that the sign of the proportionality factor $(1-T/\beta)$ depends on temperature. For $T < \beta$, an increase in enthalpy corresponds to an increase in activation free-energy (therefore to a decrease in rate), while for $T > \beta$, the free energy of activation is decreased. The sign of ρ should follow a similar pattern. An equation expressing the dependence of ρ on temperature may be set up by using equations (3.59) and (3.63), which lead to:

$$\frac{(\Delta G_{\mathrm{X}}^{\ddagger} - \Delta G_{\mathrm{H}}^{\ddagger})_{T_1}}{(\Delta G_{\mathrm{X}}^{\ddagger} - \Delta G_{\mathrm{H}}^{\ddagger})_{T_2}} = \frac{(\beta - T_1)\,\Delta\Delta S^{\ddagger}}{(\beta - T_2)\,\Delta\Delta S^{\ddagger}} = \frac{RT_1\,\rho_{T_1}\,\sigma_{\mathrm{X}}}{RT_2\,\rho_{T_2}\,\sigma_{\mathrm{X}}},$$

$$\text{where } \Delta\Delta S^{\ddagger} = \Delta S_{\mathrm{X}}^{\ddagger} - \Delta S_{\mathrm{H}}^{\ddagger}.$$

Since $\Delta\Delta S^{\ddagger}$ and σ_{X} are independent of temperature, it follows that

$$\frac{T_2}{T_1}\,\frac{\beta - T_1}{\beta - T_2} = \frac{\rho_{T_1}}{\rho_{T_2}}. \tag{3.69}$$

Equation (69) can also be written as:

$$\rho_T = \rho_0\left(1 - \frac{\beta}{T}\right), \tag{3.70}$$

where ρ_0 is the constant of the reaction defined at a standard temperature. Equation (3.70) shows that for $\beta = T$, $\rho = 0$, and therefore at this particular temperature no conclusion can be drawn regarding the reaction mechanism. Equation (3.70) indicates also that we can compare only those ρ values which have been determined at temperatures either higher or lower than the isokinetic temperature. Only within these domains can the change of the sign of ρ be related to changes in the distribution of charge in the transition state. In the vicinity of the isokinetic temperature, ρ offers no clues as to the reaction mechanism.

3.5.4 *The Taft* $\rho^*\sigma^*$ *equation*

The rate constant of a reaction is a complex quantity which includes medium effects and the effects of structure on the reaction centre in the transition state. For a substrate R—X, where X is the reaction centre and R the variable structural moiety, the reaction rate is influenced by the nature of R, which may supply electrons to the reaction centre or withdraw them from it (polar effect), or may prevent the access of the reactant to the reaction centre (steric effect). We could therefore write:

$$k = k_M\, k_S\, k_P, \tag{3.71}$$

where k is the observed constant, while k_M, k_S and k_P are constants reflecting the influence of medium, the steric effects and the polar effects, respectively.

Comparing the reaction rates with the rate of a reference substrate R_0X, in the hypothesis that the medium affects the transition state in the same way, we may write, using equation (3.71):

$$\log \frac{k}{k_0} = \log \frac{k_P}{k_{P_0}} + \log \frac{k_S}{k_{S_0}}. \tag{3.72}$$

The ratio k_P/k_{P_0} is thus a measure of polar effects. In analogy to the Hammett equation, equation (3.72) can be written in a more general form:

$$\log \frac{k}{k_0} = \rho^*\sigma^* + sE_S, \tag{3.73}$$

where ρ^* and σ^* have the same significance as for the Hammett equation; E_S is a constant measuring the steric effect of the substituent; and s a reaction constant which reflects the extent to which steric effects are transmitted to the reaction centre.

Equation (3.73) is a four-parameter equation. If the substituents in the reference substrate are sufficiently far from the reaction centre ($s \simeq 0$), equation (3.73) becomes a two-parameter equation. This situation corresponds to cases when a Hammett equation is verified, the substituents in the *meta* and *para* positions being located far enough from the reaction centre. In the case of aliphatic compounds or *ortho*-substituted aromatic compounds, the logarithm of the ratio of the rate constants will measure also steric effects in addition to polar effects.

It has been known for a long time [37–42] that in the reactions of aliphatic compounds steric effects may predominate over polar effects. However, when polar effects are much more decisive for the course of reaction or if these effects can be quantitatively separated from steric effects, a two-parameter equation, known as the *Taft equation* [42–44], is also applicable:

$$\log \frac{k_R}{k_0} = \rho^*\sigma_R^* \tag{3.74}$$

[37] C. K. Ingold, *J. Chem. Soc.*, 1032 (1930).
[38] F. P. Price and L. P. Hammett, *J. Amer. Chem. Soc.*, **63**, 2387 (1941).
[39] P. M. Dunbar and L. P. Hammett, *J. Amer. Chem. Soc.*, **70**, 3448 (1948).
[40] H. C. Brown, R. S. Fletcher and R. B. Johanesen, *J. Amer. Chem. Soc.*, **73**, 217 (1951).
[41] R. W. Taft, Jr., *J. Amer. Chem. Soc.*, **74**, 2729 (1952).
[42] R. W. Taft, Jr., *J. Amer. Chem. Soc.*, **75**, 4231 (1953).
[43] R. W. Taft, Jr., *J. Amer. Chem. Soc.*, **74**, 3120 (1952).
[44] R. W. Taft, Jr. and M. S. Newman, *Steric Effects in Organic Chemistry*, J. Wiley, New York (1956), p. 586; for the validity of this assumption, see however: M. Charton, *J. Amer. Chem. Soc.*, **97**, 3691 (1975).

Polar and steric effects can be separated if two reactions, A and B, can be found, for which steric requirements at the reaction centre are similar. In this case we obtain from equation (3.73):

$$\log\left(\frac{k_R}{k_0}\right)_B - \log\left(\frac{k_R}{k_0}\right)_A = (\rho_B^* - \rho_A^*)\,\sigma_R^* \qquad (3.75)$$

where subscripts A and B refer to the two reaction types.

According to a proposition of Ingold [37], Taft [42, 43, 44] used for the determination of the substituent constant, σ_R^*, the hydrolysis of esters R^1—$COOR^2$ (R^1 variable, R^2 constant) catalysed by acids (reaction A) and by bases (reaction B). The choice is justified by the fact that the intermediates of these reactions are rather similar. By definition, it is assumed that $\rho_B^* - \rho_A^* \equiv 1$. As reference substrate (k_0) are used esters of acetic acid ($\sigma_{Me} = 0$). In order to obtain σ_R^* values compatible with the σ-Hammett scale, a correction factor is introduced (1/2.48) which represents the difference between the ρ values determined for the alkaline and acid hydrolysis of ethyl benzoates, both *meta* and *para* substituted:

$$\sigma_R^* = \frac{1}{2.48}\left[\log\left(\frac{k_R}{k_0}\right)_B - \log\left(\frac{k_R}{k_0}\right)_A\right]. \qquad (3.76)$$

In Table 3.7 are listed some σ_R^* values, and in Table 3.8 some values of the parameter E_S from equation (3.73).

TABLE 3.7

Substituent constants σ_R^* †

Substituent	σ_R^*	Substituent	σ_R^*
CCl_3—	2.65	H—	0.490
F_2CH—	2.05	C_6H_5—CH=CH—	0.410
Cl_2CH—	1.94	$(C_6H_5)_2CH$—	0.405
$(CH_3)\overset{+}{N}$—CH_2—	1.90	C_6H_5—CH_2—	0.215
CH_3CO—	1.65	C_6H_5—CH(CH_3)—	0.11
C_6H_5—C≡C—	1.35	C_6H_5—CH_2—CH_2—	0.080
NC—CH_2—	1.30	CH_3—	0.000
FCH_2—	1.10	C_2H_5—	−0.100
$ClCH_2$—	1.05	$C_3H_7^n$—	−0.115
$BrCH_2$—	1.00	$C_4H_9^i$—	−0.125
CH_2=CH—	0.653	$C_4H_9^n$—	−0.130
		neo-C_5H_{11}—	−0.165
C_6H_5—	0.600	$C_3H_7^i$—	−0.190
		$C_4H_9^s$—	−0.210
		$C_4H_9^t$—	−0.300

† After Reference 2(c), p. 222; σ_R refers to R^1 in R^1COOR^2.

TABLE 3.8

Substituent steric constants E_S, at 25°C †

R^1 in R^1COOR^2	E_S	R^1 in R^1COOR^2	E_S
H–	1.24	cyclopentyl	−0.51
CH_3–	0.00	isobutyl	−0.93
C_2H_5–	−0.07	cycloheptyl	−1.10
$C_3H_7^n$ –	−0.36	C_6H_5–CH(CH_3)–	−1.19
$C_4H_9^n$ –	−0.39	$C_4H_9^t$ –	−1.54
$C_5H_{11}^n$–	−0.40	$(C_6H_5)_2CH$–	−1.76
n-octyl	−0.33	$(C_2H_5)_2CH$–	−1.98
C_6H_5–CH_2–	−0.38	$(C_2H_5)_3C$–	−3.8
C_6H_5–CH_2–CH_2–	−0.38		
$C_3H_7^i$ –	−0.47		

† After Reference 2(c), p. 228.

Two conclusions can be drawn from the values listed in Table 3.7. First, electron-withdrawing substituents have positive σ_R^* values, while electron-releasing substituents have negative σ_R^* values. In the second place, it can be noted that σ_R^* values are additive. Replacing a hydrogen atom from CH_3 by a Me group (R = Et) decreases σ_R^* by 0.1 units; replacing two hydrogen atoms by Me groups (R = Pr^i) diminishes σ_R^* by 2×0.1 units, while for the Bu^t group σ_R^* is decreased by 3×0.1 units as compared to σ_{Me}^*.

Table 3.8 shows that when the hydrogen atoms of the methyl group are gradually replaced by methyl groups, the steric effect is not proportional to the number of methyl groups.

3.5.5 *The Brown* $\rho\sigma^+$ *equation*

Brown and his co-workers[45] have found that the aromatic substitution reactions of benzene and toluene obey a linear free-energy equation:

$$\log p_f = 1.310 \log p_f/m_f + \text{const}, \tag{3.77}$$

where p_f and m_f are partial reaction rates for the *para* and *meta* positions, respectively, as compared to the rate of benzene.

It is interesting that equation (3.77) applies also for the solvolysis in aqueous acetone of *meta*- and *para*-substituted aryldimethylchloromethanes.

The most probable explanation for these results is that the electronic requirements for the two reaction types are similar:

[45] H. C. Brown, J. D. Brady, M. Graysm and W. H. Bonner, *J. Amer. Chem. Soc.*, **79**, 1897 (1957).

para-Substitution

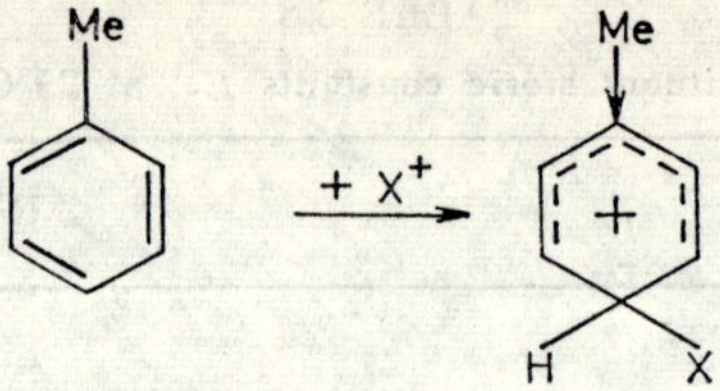

Solvolysis

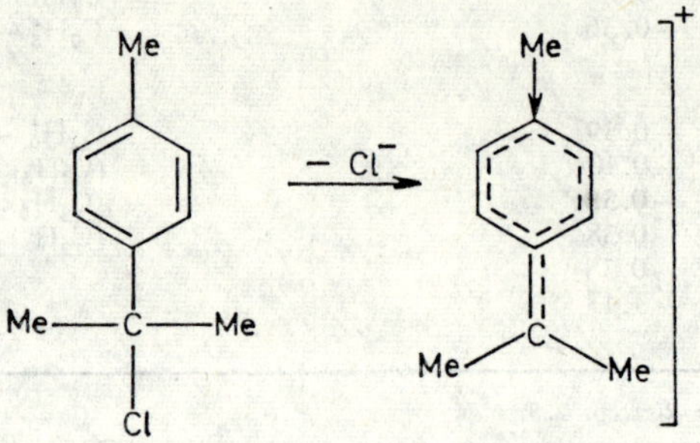

The solvolysis of aryldimethylhalogenomethanes can be used for the determination of substituent constants σ^+ applicable to the aromatic substitution [46–50]. In Table 3.9 are listed some σ^+ values, determined on the basis of this reference reaction.

TABLE 3.9

σ^+ values according to Brown [49]

Substituent	σ_m^+	σ_p^+	Substituent	σ_m^+	σ_p^+
MeO—	0.0465	−0.764	F—	0.346	−0.0714
Me—	−0.0652	−0.306	Cl—	0.391	0.112
Et—	−0.0625	−0.291	Br—	0.399	0.148
Pr^i—	−0.0589	−0.276	I—	0.353	0.132
Bu^t—	−0.0581	−0.250	NO_2—	0.662	0.777

If these values are compared with the σ values listed in Table 3.5, it is easily seen that σ_m^+ does not differ essentially from σ_m. Large differences appear in the case of σ_p^+ (with the exception of the nitro group). These facts can be accounted for if we remember that σ values are determined on the basis of ionisation constants of ben-

[46] Y. Okamoto and H. C. Brown, *J. Amer. Chem. Soc.*, **79**, 1903 (1957).
[47] H. C. Brown, Y. Okamoto and G. Ham, *J. Amer. Chem. Soc.*, **79**, 1906 (1957).
[48] Y. Okamoto and H. C. Brown, *J. Amer. Chem. Soc.*, **79**, 1909 (1957).
[49] H. C. Brown and Y. Okamoto, *J. Amer. Chem. Soc.*, **79**, 1913 (1957).
[50] Y. Okamoto and H. C. Brown, *J. Amer. Chem. Soc.*, **80**, 4979 (1958).

zoic acids, which are especially influenced by inductive effects. In the solvolysis of *meta*-substituted aryl-dimethylhalogenomethanes, only inductive effects are operative, which explains the agreement between σ_m and σ_m^+ values. In the solvolysis of *para*-substituted derivatives, which are used for the determination of σ_p^+ values, conjugative effects are predominant in supplying electrons to the reaction centre. This is why $\sigma_{p\text{-OMe}}$ and $\sigma_{p\text{-OMe}}^+$ differ widely, while $\sigma_{p\text{-NO}_2}$ and $\sigma_{p\text{-NO}_2}^+$ are practically identical. The difference between $\sigma_{p\text{-Me}}$ and $\sigma_{p\text{-Me}}^+$ suggests hyperconjugation of this group with the reaction centre.

σ^+ values are to be preferred to σ values whenever electronic requirements at the reaction centre are large.

3.5.6 *Determination of substituent constants*

Linear free-energy relationships of the $\rho\sigma$ type are useful in predicting chemical reactivity, when ρ and σ values are available. However, such predictions may be rendered difficult by the wide range of the σ values. The ideal situation would be to calculate these values starting only from theoretical assumptions concerning the known molecular structure. Such determinations have actually been carried out in some instances [51]. According to Dewar and Grisdale [52], in the calculation of σ constants one should take into account the mechanisms by which a substituent of an atom i can influence a reaction centre located at a distant atom j. These mechanisms can be:

(a) Successive polarisations of σ bonds linking the two centres (σ inductive effect). The σ inductive effect decreases exponentially with the number of bonds between the two centres i and j.

(b) Direct conjugation between substituent and reaction centre along a π-electron system (electromeric effect).

(c) Direct transmittance of the dipole field of the substituent to the reaction centre (field effect).

(d) Polarisation of the π-electron system by the electric charge of an atom adjacent to the substituent (π inductive effect).

(e) Polarisation of the π-electron system by resonance interactions with the substituent (mesomeric effect).

Taking into account these mechanisms, one can calculate by current quantum mechanical methods the charge induced by the substituent at the reaction centre [51].

Attempts are known to modify the Hammett equation, by introducing new parameters, so that σ constants would be unified. Thirot [36] proposed the following equation:

$$\sigma = \sigma^{\mathrm{H}} - r\Delta\sigma_1 - s\Delta\sigma_2, \tag{3.78}$$

[51] See, for example, E. Horowitz, *Stud. cerc. chim.*, **15**, 927 (1967).
[52] M. J. S. Dewar and P. J. Grisdale, *J. Amer. Chem. Soc.*, **84**, 3539 (1962).

where σ^H is a universal constant of substituent, calculated from the Taft σ^* values by taking hydrogen as standard group ($\sigma^H_H = 0.000$), and r, s, $\Delta\sigma_1$, $\Delta\sigma_2$ are auxiliary parameters.

Since a large number of σ^H, $\Delta\sigma_1$ and $\Delta\sigma_2$ values are known [36], equation (3.78) can be useful for the empirical determination of σ.

3.5.7 *Other types of linear free-energy relationships*

In the examples examined so far the reaction rate was correlated with structural factors, considering that solvent and steric effects are constant. However, the same type of correlations can be set up either with medium parameters if the characteristics of the substrate are kept constant, or with reactant parameters, if structural and medium factors are kept constant etc. Such correlations, which are important in the study of nucleophilic substitution, will be discussed later.

It is interesting to note that the Hammett equation was also used to correlate physical properties, such as infrared [53], visible and ultraviolet [54] spectra, nuclear magnetic resonance spectra [55–57], half-wave potentials in polarographic reductions [58] etc. Nuclear magnetic resonance spectroscopy was currently used by Dewar [59,60] to estimate the influence of field effects and of polarisation effects on the value of σ constants.

3.6 **Problems**

(1) Show that for any equilibrium reaction:

$$\mathrm{A} \underset{k_{-1}}{\overset{k_1}{\rightleftharpoons}} \mathrm{B}$$

the equation (3.79) is verified:

$$k_1 + k_{-1} = \frac{2.303}{t} \log \frac{A_0 - A_e}{A - A_e} \tag{3.79}$$

where A_0 is the initial concentration of compound A, A_e the concentration at equilibrium, and A the concentration at time t.

[53] P. R. Wells, *Chem. Rev.*, **63**, 171 (1963).
[54] C. N. R. Rao, *Ultraviolet and Visible Spectroscopy*, Butterworth, London (1961).
[55] H. S. Gutowski, D. W. McCall, B. R. McCarney and L. H. Meyer, *J. Amer. Chem. Soc.*, **74**, 4809 (1952).
[56] R. W. Taft, Jr., S. Ehrenson, I. C. Lewis and R. E. Glick, *J. Amer. Chem. Soc.*, **81**, 5352 (1959).
[57] R. W. Taft, Jr., *J. Amer. Chem. Soc.*, **79**, 1045 (1957).
[58] N. Bodor, L. Fey and A. Kövendi, *Rev. Roumaine Chim.*, **11**, 405 (1966).
[59] W. Adcock and M. J. S. Dewar, *J. Amer. Chem. Soc.*, **89**, 379 (1967).
[60] M. J. S. Dewar and Y. Takeuki, *J. Amer. Chem. Soc.*, **89**, 390 (1967).

(2) The isomerisation of n-butane to isobutane:

$$CH_3CH_2CH_2CH_3 \underset{k_{-1}}{\overset{k_1}{\rightleftharpoons}} CH_3CH(CH_3)_2$$

was studied in liquid phase under pressure, at 65°C, and in presence of 0.1 M $AlBr_3$. The rate constants for the direct and the reverse reactions were found to be: $k_1 = 0.282 \times 10^{-4}\ s^{-1}$ and $k_{-1} = 0.158 \times 10^{-4}\ s^{-1}$, respectively.

Calculate: (*a*) the equilibrium constant corresponding to the conditions specified above; (*b*) the composition of the equilibrium mixture; (*c*) the change in rates and in the equilibrium composition if the concentration of the catalyst is increased to 0.5 M; (*d*) the time necessary to attain the equilibrium for a mixture containing initially 60% isobutane; (*e*) the standard free energy of reaction; (*f*) the activation free-energy at 65°C (Boltzmann constant $\mathbf{k} = 1.380 \times 10^{-16}$ erg deg^{-1} $molecule^{-1}$, and Planck's constant $h = 6.625 \times 10^{-27}$ erg s).

(3) Triphenylcarbinol undergoes in ethanol, in the presence of sulphuric acid, the following reactions, according to Bartlett and his co-workers (P.D. Bartlett and J. D. McCollum, *J. Amer. Chem. Soc.*, **78**, 1441 (1956)):

$$Ph_3C{-}OH + H^+ \rightleftharpoons Ph_3C{-}\overset{+}{O}H_2$$

$$Ph_3C{-}\overset{+}{O}H_2 \xrightarrow{\text{slow}} Ph_3C^+ + H_2O$$

$$Ph_3C^+ + CH_3CH_2OH \longrightarrow Ph_3CH + CH_3CH{=}O + H^+$$

Considering the above mechanism, indicate the expected dependence between the observed rates and the catalysts. Considering the same mechanism explain why this reaction is observed only in a very narrow range of sulphuric acid concentration in ethanol.

(4) A negative charge at the carbon atom is destabilised by alkyl groups bonded to it. What is the explanation of the relatively high pK_a values of nitroalkanes (see Table 3.2)?

(5) Using pK_a values listed in Table 3.2, calculate the equilibrium constant and the composition of the equilibrium mixture for the reaction:

$$CH_3O^- + CH_3{-}NO_2 \rightleftharpoons CH_3OH + {}^-{:}CH_2{-}NO_2$$

assuming equal initial concentrations of methoxide ion and of nitromethane, namely 0.01 M for each of them.

(6) The acetolysis of primary tosylates I:

$$Me{-}\underset{Me}{\overset{Ph}{C}}{-}CH_2{-}X\ ;\quad X = Cl, Br, Me{-}C_6H_4{-}SO_3,\ Br{-}C_6H_4{-}SO_3$$

I

obey a linear relationship:

$$\Delta H^{\ddagger} = \Delta H_0^{\ddagger} + \beta \Delta S^{\ddagger}$$

where the slope β has a negative value. What is the significance of this apparently absurd result?

(7) Indicate the sign of the parameter ρ for the following reactions:

(a) $ArO^- + HOH \overset{K}{\rightleftharpoons} ArOH + HO^-$

(b) $Ar\overset{+}{N}H_3 + HOH \overset{K}{\rightleftharpoons} ArNH_2 + H_3O^+$

(c) $ArCOOH + N_2CPh_2 \xrightarrow{k} ArCOOCHPh_2 + N_2$

(8) If reactions (*a*) and (*b*) satisfy a Hammett equation:

$$Ar{-}X \xrightarrow{K_X} Ar{-}Y \qquad (a)$$

$$Ar{-}Y \xrightarrow{K_Y} Ar{-}X \qquad (b)$$

the reaction parameters being ρ_X and ρ_Y, respectively, calculate the ρ parameter if the equilibrium constants $K_Z = k_X/k_Y$ (Z = the substituent of the aromatic nucleus) are correlated with the σ_Z values?

(9) Considering the following mechanism:

$$R^1R^2R^3C{-}X \underset{k_{-1}}{\overset{k_1}{\rightleftharpoons}} R^1R^2R^3C^+ + X^- \qquad (a)$$

$$R^1R^2R^3C^+ \underset{Nu:^-}{\overset{k_2}{\longrightarrow}} R^1R^2R^3C{-}Nu \qquad (b)$$

in which the observed rates are correlated with σ^* values, according to a Taft equation, determine the sign of the ρ_{obs} parameter. Show in what sequence would fall the absolute values of the ρ_{obs} parameter for the following possible cases:

(i) Step (*a*) is very slow and step (*b*) very rapid ($k_2 \gg k_{-1}$; ρ_i);
(ii) idem, with $k_{-1} \gg k_2$ (ρ_{ii});
(iii) equilibrium (*a*) is rapid, while step (*b*) is slow (ρ_{iii}).

4
Elementary reactions. Reaction intermediates

In the course of organic reactions, chemical bonds in reactants may be modified, slightly or drastically, to lead to reaction products. Except for some concerted reactions, discussed in the last chapter of this book, drastic bond changes indicate the succession of some *elementary reactions*, each of them occurring with the least possible modification of the reacting species. The first object of a study of reaction mechanisms is to demonstrate the correct succession of the elementary reactions, *i.e.* the correct reaction scheme. Even for very simple reactions, one may imagine several schemes. In order to find and prove the correct one, it is necessary to plan experiments which may eventually confirm or exclude certain possibilities. In this respect, the researcher must be very familiar with the characteristics of each type of elementary reaction, with the possible intermediates involved, as well as with any change which a molecule may undergo prior to or after the reactions. This chapter will deal in a general form with such subjects. In the later chapters of the book these topics will be considered in more detail.

4.1 Classification of reagents and reactions

For any reaction, involving two or more reacting partners, it is convenient to designate one molecule as *substrate* and the other(s) as *reagent*(*s*):

$$\underset{\text{substrate}}{S} + \underset{\text{reagent(s)}}{R + \dots} \longrightarrow \underset{\text{product(s)}}{P} \tag{4.1}$$

However, the designation is arbitrary and sometimes misleading. Usually, in the case of organic reactions involving an organic molecule and an inorganic species, the decision is readily made, the organic reactant being designated as substrate. In other cases, the reagent, even if organic, may be similar in its reactivity to an inorganic reactant (for example H_3N: and R_3N:). Anyway, the substrate is always chosen to be the molecule whose bond modifications are studied in the reaction.

In the classification of elementary reactions, the chemical reactivity of the reagent is of utmost importance.

4.1.1 *Reagents*

From the point of view of the electronic theory, chemical reactions are electrical transactions [1]. More precisely, the configuration of electrons about the reacting nuclei is continuously changed starting from the reactants, passing through the transition state, until the final configuration in the products is attained. A reagent can display either affinity for the atomic nuclei (nucleophilic reagents) or for the electrons of a chemical bond (electrophilic reagents) or for both (homolytic reagents). Nucleophilic and electrophilic reagents are even-electron species (ions or neutral molecules), which are involved in polar transition states. Obviously, for a simple reaction between two reaction partners, if the reagent R (see reaction (4.1)) is a nucleophile, Nu:, the substrate S must be an electrophile, E. This shows that the classification in substrates and reagents is arbitrary:

$$\underset{\substack{\text{substrate}\\ \text{(reagent)}}}{\mathrm{E}} + \underset{\substack{\text{reagent}\\ \text{(substrate)}}}{\mathrm{Nu:}} \underset{}{\overset{k,K}{\rightleftharpoons}} \underset{\text{product(s)}}{\mathrm{P}} \qquad (4.2)$$

Homolytic reagents are odd-electron species (free radicals or free atoms). They are able to react either with even- or with odd-electron molecules, producing either odd-electron species (free radicals) or neutral even-electron species.

The reactivity of organic reagents (electrophiles, nucleophiles or free radicals) will be discussed in a later section of this chapter. However, some characteristics of these reagents must be mentioned here.

The nucleophilic reagents (nucleophiles) possess at the reaction site either a pair of unshared electrons (lone electrons) or a system of π electrons, having affinity for an electron-deficient atom, either from the same molecule or from a different molecule. Table 4.1 shows some examples of nucleophilic reagents. It can be seen that the nucleophilic character does not imply a negative charge on the reagent; neutral molecules, such as water, ammonia or their organic derivatives, manifest a reactivity similar to that of typical inorganic anions.

TABLE 4.1
Nucleophilic reagents

Type	*Examples*
Anions	F^-, Cl^-, Br^-, I^-
	CN^-, N_3^-, HO^-, RO^-, R_2N^-, NO_2^-, SCN^-, PhO^-, HS^-, RS^- etc.
Neutral molecules	
(a) polar	H_2O, R_2O, H_3N, R_3N, R_2S, pyridines etc.
(b) non-polar	alkenes, polyenes, arenes etc.
(c) electron-deficient	CX_2 (carbenes, $X = NR_2$, OR)
Cations	R_3X^+ ($X = O^+$, S^+)

[1] C. K. Ingold, *Structure and Mechanism in Organic Reactions*, Cornell University Press, Ithaca, New York, 1st ed. (1953), p. 198; 2nd ed. (1969), p. 234.

A π-electron system, as encountered in alkenes, polyenes and arenes, is nucleophilic in most of its reactions (see pp. 274, 485). This shows that the nucleophilic character is not always located at a single atom, but can be a general property of the molecule as a whole (see also ambident reagents, p. 325).

The affinity of the nucleophile Nu: towards an electrophile E (nucleophilicity) can be measured in two ways:

(a) By the equilibrium constant, K, of reaction (4.2), that is by the free-energy difference between the initial state (reactants) and the final state (products), knowing that:

$$\Delta G = -RT \ln K \text{ or } K = e^{-\frac{\Delta G}{RT}} \tag{4.3}$$

In order to evaluate the equilibrium constant we must be able to estimate the heat of reaction, ΔH, and the entropy, ΔS.

$$\Delta G = \Delta H - T\Delta S \tag{4.4}$$

For reactions in gaseous phase, the heat of reaction, ΔH, can be determined easily, if we know the structures of the reagents and products. Thus, applying the MO descriptions of the species involved, the differences in MO energies will give a good approximation to the heat of reaction (see also p. 358). However, the entropy of reaction is not directly related to the structure and cannot be as yet exactly predicted. The problem is even more difficult for the more frequent reactions occurring in solution. Both ΔH and ΔS values are affected by solvation and the solvation components cannot be evaluated. However, in considering equilibria (4.2), we are mainly interested in the variations of the equilibrium constants along a series of closely related reactions, that is less in the absolute values of ΔG and more in the variation of ΔG with structure. It is however argued [2], on theoretical grounds, that solvation and entropy changes might be neglected if the reactions are similar and the structure modifications are not too drastic.

If the electrophile in reaction (4.2) is H^+ or a protic acid, the affinity of the nucleophile Nu:, defined by the equilibrium constant, is called the *basicity*. The determination of the basicity constants is treated elsewhere in this book (see p. 108). It has been shown that the concept of basicity is not limited to the hydrogen nucleus (see p. 114). However, considering free-energy relationships (see p. 119), one would expect the basicity of a nucleophile to vary in the same manner towards different electrophiles. Actually, that was not found to be true for all electrophiles, as for example the basicity of amines with

[2] See, for example, M. J. S. Dewar, *The Molecular Orbital Theory of Organic Chemistry*, McGraw-Hill, New York (1969), Section 8.2, p. 279.

respect to Brönsted acids, or to Lewis acids such as boron derivatives[3]. The non-proportionality of the basicity of the same reagent defined with respect to different electrophiles is essentially explained by the fact that the entropy term in relation (4.4) is not proportional to structural modifications. The non-proportionality is mainly determined by steric effects and will be treated in later chapters of this book.

(b) Another way to define the affinity of a nucleophile towards an electrophile is by the free-energy barrier to be surmounted in reaction (4.2). Considering that relation (4.4) holds when thermodynamic parameters (ΔG, ΔH, ΔS) are substituted by activation parameters (see p. 104), the kinetic affinity of a nucleophile can be estimated theoretically if we know the structure of the transition state. The transition state, or activated complex, is a molecular species characterised by the highest energy adopted by the reacting molecules during the process. It cannot be either isolated or directly observed. Its structure is not exactly known, but could be inferred by minimising the energies of all conceivable structures developed by the interaction between the reacting molecules. In that way, the kinetic idefnition of the affinity of a nucleophile towards an electrophile is even more difficult to describe than in thermodynamical terms. However, the majority of practical chemical problems are concerned not so much with absolute values, as with the variation of these values along a series of related reactions. Applying the same consideration as above [2], the differences in the free energy of activation will correspond to the logarithm of the relative rate constants and could be linearly proportional to two empirical parameters, one related to the reaction type, the other to structural modifications (see free-energy relationships, p. 120). This is the usual form in which the reactivity of a nucleophile is described. One of the first known nucleophilicity scales was proposed by Swain and his co-workers [4,5]. These authors have established a linear free-energy relationship:

$$\log \frac{k_{\mathrm{Nu:}}}{k_{\mathrm{H_2O}}} = sn \tag{4.5}$$

in which $k_{\mathrm{Nu:}}$ and $k_{\mathrm{H_2O}}$ are the observed rate constants of a reaction (4.2), either with a nucleophile Nu: or with water as nucleophilic reagent, s is a reaction parameter and n defines the nucleophilicity of reagent Nu:. The reference reaction ($s = 1$) is the reaction of methyl bromide in pure water at 25 °C. Some values of the parameter n are given in Table 4.2. A large n value corresponds to a high nucleophilicity of the reagent Nu:.

[3] See, for example, H. C. Brown, *Boranes in Organic Chemistry*, Cornell University Press, Ithaca (1972), Chapter 6.

[4] C. G. Swain and C. B. Scott, *J. Amer. Chem. Soc.*, **75**, 141 (1953).

[5] C. G. Swain, R. B. Moseley and D. E. Brown, *J. Amer. Chem. Soc.*, **77**, 3731 (1955).

TABLE 4.2

The nucleophilicity of some reagents in pure water

Reagent	$n^{a)}$	$e_N{}^{b)}$
F^-	−0.70 [c]	−0.27
H_2O	0.00	0.00
$MeCOO^-$	+2.72	+0.95
Cl^-	3.04	1.24
Br^-	3.89	1.51
N_3^-	4.00	1.58
HO^-	4.20	1.65
$PhNH_2$	4.49	1.78
SCN^-	4.77	1.83
NH_3	4.80 [c]	1.84
I^-	5.03	2.06
$S_2O_3^{2-}$	6.36	2.52
CN^-	6.50 [c]	2.79
S^{2-}	7.40 [c]	3.08

a) After Swain and his co-workers[4,5] unless otherwise stated; *b)* after J.O. Edwards[7]; *c)* value extrapolated from Figure 4.2.

It can be easily seen that there is no relationship between basicity and nucleophilicity. Aniline is a very weak base ($K_b = 3.8 \times 10^{-10}$), but has about the same nucleophilicity in water as the hydroxide ion, the strongest possible base in water (see p. 108). In order to explain this apparent anomaly, we must remember that basicity is a thermodynamic quantity, whereas nucleophilicity is a kinetic quantity. The former depends on the free energy ΔG, the latter on the free energy of activation $\Delta G^{\ddagger}$. Obviously, one should not always expect these parameters to change in the same way with structural modifications.

It has been postulated by Hammond[6] that the transition state of an exothermic reaction will resemble the reactants in energy and structure, whereas in an endothermic reaction the transition state will be similar to products. The changes in free energy along a reaction coordinate for the two situations are plotted in Figure 4.1. Evidently, for the acid-base equilibrium (4.6) with AH a weak acid (for example water):

$$\text{B:} + \text{A—H} \underset{k_{\text{BH}^+}}{\overset{k_{\text{AH}}}{\rightleftharpoons}} \text{BH}^+ \qquad (4.6)$$

the case *A* in Figure 4.1 *(early transition state)* will correspond to B: = OH^- and the case *B* *(late transition state)* to B: = $PhNH_2$. However, if we consider the reaction of the same reactants with a strong electrophile (see reaction (4.2)), according to Hammond's

[6] G. S. Hammond, *J. Amer. Chem. Soc.*, **77**, 334 (1955).
[7] J. O. Edwards, *J. Amer. Chem. Soc.*, **76**, 1540 (1954); **78**, 1819 (1956).

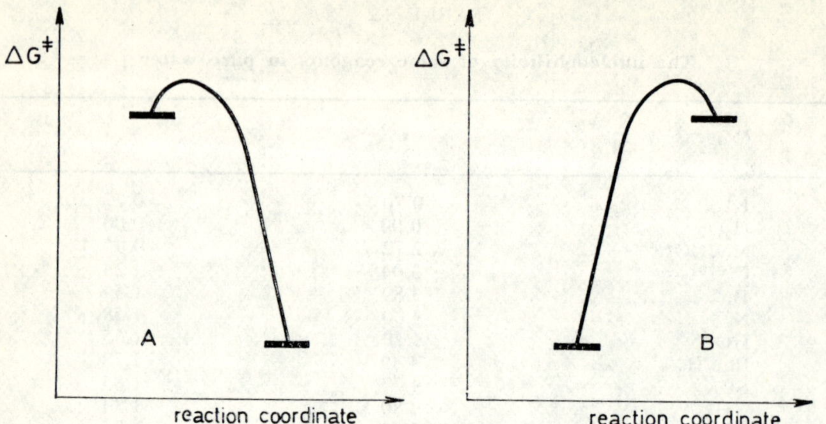

Figure 4.1 Potential energy variation *versus* a reaction coordinate: A. in an exothermic reaction (early transition state), B. in an endothermic reaction (late transition state).

postulate, both hydroxide ion and aniline will react exothermically with the electrophile, the free-energy diagram corresponding in both cases to an early transition state. Naturally, there will be no proportionality between the changes in $\Delta G^{\ddagger}$ in reaction (4.2) and the changes in ΔG in reaction (4.6). The lack of proportionality between the free-energy differences in these examples is determined by the different profile of the free-energy diagrams in the two sets of reactions. Nevertheless, if the reactions under comparison have a similar energy profile, Hammond's postulate would suggest that the change in $\Delta G^{\ddagger}$ could be a rough approximation of the change in ΔG.

Edwards [7] has established the four-parameter relationship (4.7), correlating the thermodynamic properties with nucleophilicity:

$$\log \frac{k_{\mathrm{Nu:}}}{k_{\mathrm{H_2O}}} = ae_{\mathrm{N}} + bH_{\mathrm{N}} \qquad (4.7)$$

in which H_N is a measure of the basicity of the nucleophile, e_N corresponds to its nucleophilicity, while a and b are some empirical parameters.

H_N can be determined as:

$$H_{\mathrm{N}} = \mathrm{p}K_{\mathrm{A}} + 1.74 \qquad (4.8)$$

K_A being the acidity constant of the corresponding conjugate acid in water at 25 °C, and 1.74 a correction for the $\mathrm{p}K_A$ of hydronium ion. The parameter e_N represents the standard electrode potential of the reagent in the equilibrium:

$$2\mathrm{Nu:} \rightleftharpoons \mathrm{Nu}_2^{2+} + 2\mathrm{e}^- \quad (\text{or } 2\mathrm{Nu:}^- \rightleftharpoons \mathrm{Nu}_2 + 2\mathrm{e}^-) \qquad (4.9)$$

corrected for the similar reaction of water molecules ($\varepsilon = 2.6$ v).

The nucleophilic scale based on relation (4.7) is similar to the nucleophilic scale based on relation (4.5), e_N parameters appearing to be linearly correlated with parameters n (Figure 4.2). This fact is rather surprising, if relations (4.5) and (4.7) are considered:

$$\log \frac{k_{Nu:}}{k_{H_2O}} = sn = ae_N + bH_N, \quad \text{or}$$

$$n = (a/s)\, e_N + (b/s)\, H_N \tag{4.10}$$

Relation (4.10) suggests that linearity between n and e_N should be observed either if H_N varies in a similar way with e_N or if the last term in equation (4.10) vanishes. It has been demonstrated above that the first alternative is not always true, so the last one must hold. Empirically it has been found[7] that e_N is related in its turn to the basicity of the reagent (H_N) and its polarisability (p_N):

$$e_N = cp_N + dH_N \tag{4.11}$$

where c and d are empirical constants.

When combining equations (4.10) and (4.11), the basicity terms may vanish. Relations (4.5) and (4.7) and other similar equations[8] provide therefore the same type of information. However, equations

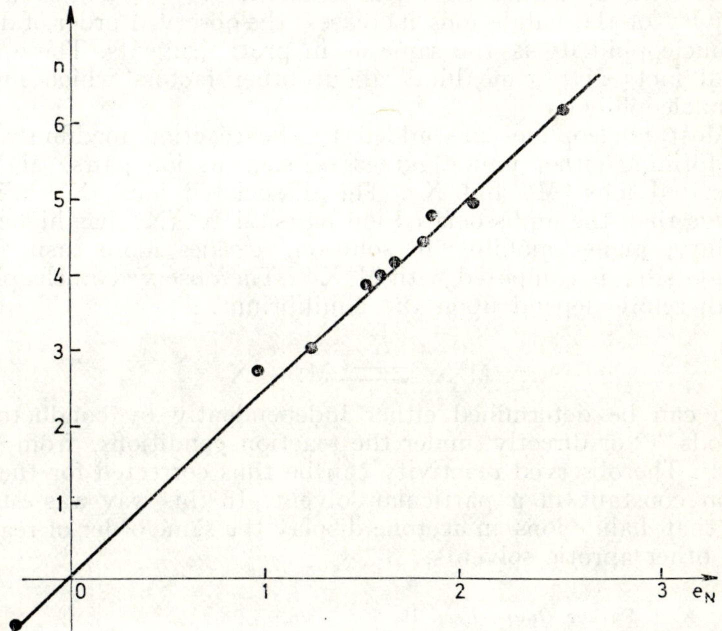

Figure 4.2 Linear correlation of the nucleophilic constants n and e_N.

[8] See for example, J. E. Leffler and E. Grunwald, *Rates and Equilibria in Organic Chemistry*, J. Wiley, New York (1963), p. 243; A. Streitwieser, Jr., *Chem. Rev.*, **56**, 571 (1956).

(4.7)—(4.11) offer a better explanation for the order of increasing nucleophilicity given in Table 4.2. Indeed, the definition of e_N shows that the higher the ability of the reagent to give up electrons, the higher its nucleophilicity in reactions (4.2). We can now understand why the sulphide ion, for example, is a better nucleophile than the hydroxide ion, even if HO^- forms stronger bonds with typical electrophiles than does the sulphide ion.

The nucleophilicity scale discussed above has been determined in water and holds only in water and other related protic solvents. According to Parker [9], a protic solvent has a reactive hydrogen, able to form strong hydrogen bonds with ionic species. The reactivity of an anion and the nucleophilicity scale could be dramatically changed by transferring the reagent from a protic into an aprotic solvent. Thus, the reactivity of the halide ions in protic solvents (such as water, methanol) decreases in the order $I^- > Br^- > Cl^- > F^-$ (see Table 4.2), but the reverse order is found in *aprotic dipolar solvents* [9–11], such as dimethylformamide (DMF, $\varepsilon_{DMF} = 37.6$), dimethylsulphoxide (DMSO, $\varepsilon_{DMSO} = 48$). The alteration of the nucleophilic scale is coupled with a high increase in reactivity, the smaller halide ions being more affected ($k_{X^-}^{DMF}/k_{X^-}^{MeOH}$ is *ca.* 10^8 for fluoride, 10^6 for chloride and *ca.* 10^4 for iodide [9]). However, in acetone, an aprotic solvent with a smaller dielectric constant ($\varepsilon_{Me_2CO} = 20$), even if the reactivity of the halide ions increases, the observed order of decreasing nucleophilicity is the same as in protic solvents. These experimental facts give some hints about other factors which influence the nucleophilicity.

Most nucleophiles are added to the reaction medium as salts MX, forming either ionic aggregates, such as ion pairs, M^+X^-, or dissociated ions, M^+ and X^-. The dissociated ions X^- are more reactive than the undissociated ion pairs, M^+X^- (X^- has higher polarisability, higher mobility in solution, accedes more easily to the reaction site, as compared with M^+X^-). The observed nucleophilicity will therefore depend upon the equilibrium:

$$M^+X^- \xrightleftharpoons{K_X} M^+ + X^- \tag{4.12}$$

which can be determined either independently by conductometric methods [12,13] or directly, under the reaction conditions, from kinetic data [14]. The observed reactivity can be thus corrected for the dissociation constant in a particular solvent. In this way was established [15] that halide ions in acetone display the same order of reactivity as in other aprotic solvents.

[9] A. J. Parker, *Quart. Rev.*, **16**, 163 (1962).
[10] B. Tchoubar, *Bull. Soc. chim. France*, 2069 (1964).
[11] F. Madaule-Aubry, *Bull. Soc. chim. France*, 1452 (1966).
[12] O. Chris and H. Elias, *Chem. Ber.*, **99**, 1 (1966).
[13] H. Elias and H. Strecker, *Chem. Ber.*, **99**, 1019 (1966).
[14] W. M. Weaver and J. D. Hutchison, *J. Amer. Chem. Soc.*, **86**, 261 (1964).
[15] S. Winstein, L. G. Savedoff, S. Smith, J. D. R. Stevens and J. S. Gall, *Tetrahedron Letters*, (9) 24 (1960).

The dissociation process (4.12) involves solvent molecules, the dissociation energy of ion pairs being largely compensated by the solvation energy of the individual ions. Weak bonds are developed between solvent molecules and ions, most of them resulting from ion-dipole interactions. The higher the polarity of the solvent (expressed by its dielectric constant), the higher the dissociation constant K_X will be[16,17]. In protic solvents, besides weak ion-dipole interactions, there are stronger hydrogen bonds between protic molecules and the anions, depending on solvent acidity (see also p. 205) and on the volume of the reagent X^-. For ions with smaller radius the hydrogen bonds will be stronger and the number of solvent molecules surrounding the ion will be greater. Hydrogen bonds and the atmosphere of solvent molecules about the reagent will screen the reaction site in a similar way as the counter-ion in an ion pair, decreasing the actual activity of the reagent. In aprotic dipolar solvents the reagents are more free to manifest their nucleophilicity and the latter will be more closely related to the basicity[9].

This explanation of nucleophilic activity can be put into a quantitative form[8,9,16,18]. In this section we shall only refer to a new parameter defined by Parker[19–21], expressing quantitatively the change in nucleophilic reactivity in passing from a protic to an aprotic solvent. This parameter was called *solvent activity coefficient* ${}^0\gamma^S$, defined by the ratio of the activity coefficient in a given solvent γ^S, and in a reference solvent (methanol) γ^0. Activity coefficients for individual ions cannot be easily determined. Indeed, when determining the activity of a salt MX, we actually observe the product of the activity coefficients $\gamma_{X^-}\gamma_{M^+}$ which cannot be split into components. In determining γ_{X^-}, Parker has used some extrathermodynamic relations the most important one based on an assumption of Grunwald, Baughman and Kohnstam[22]. According to these authors, neutral molecules or ions, having the same bulky substituents around the central atom, such as: CPh_4, $SnPh_4$, BPh_4^-, $AsPh_4^+$, must have in solution about the same activity coefficients, irrespective of the electrical charge of the central atom. By determining the solvent activity coefficients for the salt $AsPh_4^+$ BPh_4^-, and then for various other salts $AsPh_4^+X^-$, the parameter ${}^0\gamma^S_{X^-}$ can be easily calculated, as shown by the following relations:

$$\log {}^0\gamma^S_{AsPh_4^+} = \frac{1}{2} \log [{}^0\gamma^S_{AsPh_4^+}] [{}^0\gamma^S_{BPh_4}$$

$$\log {}^0\gamma^S_{X^-} = \log [{}^0\gamma^S_{AsPh_4^+}][{}^0\gamma^S_{X^-}] - \log {}^0\gamma^S_{AsPh_4^+} \qquad (4.13)$$

[16] L. D. Pettit and S. Bruckenstein, *J. Amer. Chem. Soc.*, **88**, 4783 (1966).

[17] S. Bruckenstein and L. D. Pettit, *J. Amer. Chem. Soc.*, **88**, 4790 (1966).

[18] E. S. Amis, *Solvent Effects on Reaction Rates and Mechanisms*, Academic Press, New York (1966).

[19] R. Alexander and A. J. Parker, *J. Amer. Chem. Soc.*, **89**, 5549 (1967).

[20] R. Alexander, E. C. F. Ko, A. J. Parker and T. J. Broxton, *J. Amer. Chem. Soc.*, **90**, 5049 (1968).

[21] A. J. Parker and R. Alexander, *J. Amer. Chem. Soc.*, **90**, 3313 (1968).

[22] E. Grunwald, G. Baughman and G. Kohnstam, *J. Amer. Chem. Soc.*, **82**, 5801 (1960).

In Table 4.3 are given some solvent activity coefficients determined by Parker and his co-workers [19–21]. The values included in this table and in Table 4.2 offer valuable practical information regarding nucleophilic reactivity.

TABLE 4.3

Logarithms of solvent activity coefficients, $^0\gamma^S_{Y^-}$, of some nucleophilic reagents, after Parker and his co-workers [19]

Y	log $^0\gamma_{Y^-}$						
	H_2O	$HCONH_2$	DMF[a)]	DMAc[b)]	DMSO[c)]	MeCN	HMP[d)]
$MeCOO^-$	−2.4	—	+9.5	+10.4	+6.3	+7.9	—
Cl^-	−2.0	+0.3	+6.8	+ 8.0	+5.3	+6.4	+6.7
Br^-	−1.6	+0.2	+5.2	+ 6.1	+3.4	+4.3	+5.0
N_3^-	−1.3	+0.5	+5.2	+ 6.4	+3.3	+4.8	+5.1
TsO^-	−0.6	—	+3.5	—	—	—	—
I^-	−1.0	+0.2	+2.9	+ 3.2	+1.1	+2.5	—
SCN^-	−0.7	0.0	+3.0	+ 3.4	+1.2	+2.7	+1.3

a) DMF = dimethylformamide; *b)* DMAc = dimethylacetamide; *c)* DMSO = dimethylsulphoxide; *d)* HMPT = hexamethylphosphortriamide.

Electrophilic reagents or electrophiles are even-electron reagents [1] possessing a vacant molecular orbital, located either at one atom (the reaction site) or delocalised over more atoms (ambident reagents). In their reactions they are "electron-seeking reactants", combining easily with electrons, with nucleophiles and even with electron-rich bonds (see Chapters 6 and 7). Table 4.4 lists some common electrophiles.

TABLE 4.4

Electrophilic reagents

Type	*Examples*
Cationic	H^+
	R_3C^+ (R = H, alkyl, aryl etc.)
	X^+ (X = halogen), NO_2^+, NO^+, ArN_2^+, RHg^+
Neutral	AH (Brönsted acids)
	MX_3 (Lewis acids, M = B, Al etc.)
	R_2C: (carbenes)
	Alkenes, arenes (with electron-withdrawing substituents)
	SO_3, X_2 (X = halogen)
Anionic	$Fe(CN)_6^{3-}$

Again, the electric charge is not characteristic for the reagent. Evidently, a positive charge localised at the reaction site will determine a high reactivity of the electrophile.

In order to measure the tendency of an electrophile to combine with a nucleophile, a reference reaction with a reference nucleophile must be chosen, then either the equilibrium constants or the reaction rate constants are determined. Apparently, the same kinds of difficulties are encountered in the determination of the electrophilic activity as when measuring the nucleophilicity. However, the experimental conditions are quite different for the two types of reagents. For nucleophilic reagents it is possible to choose a reference electrophile and to determine the relative nucleophilicity by performing the reference reaction under the same conditions (solvent, temperature). Moreover, some characteristics of the nucleophiles, such as solvent activity coefficients, can be determined independently of the reactions involved. However, it is impossible to do the same for the electrophilic reagents. Most of the electron-deficient species listed in Table 4.4 are generated *in situ*, as unstable primary intermediates. It is not possible to determine the concentration of such a species, nor to know if it is the single electrophile involved. However, in describing the reactivity of an electrophile we cannot ignore the process in which it is involved. Therefore, more details regarding electrophilic reactivity will be further presented in later chapters. Some quantitative relations, regarding organic electrophiles, will be mentioned later in this chapter.

Homolytic reagents are free atoms or radicals. They have an electron deficiency either localised at one atom or delocalised over more atoms. As electrophilic reagents, homolytic reagents are highly reactive species, having a tendency to combine either with electrons, or with other odd-electron species (free atoms or radicals, oxygen etc.), or with even-electron species (alkenes, arenes etc.).

4.1.2 *Classification of elementary reactions*

Organic elementary reactions are classified according to the electronic transaction implied [1]. There are two different ways in which the initial electronic configuration of reacting molecules can be changed.

In the simplest one, electrons are transferred to the organic substrate. Such processes occur in oxidation-reduction reactions, which are commonly found both in inorganic and organic chemistry. The electron transfer is usually reversible, but can sometimes be irreversible under certain reaction conditions.

An example of a reversible electron transfer is the reaction of an aromatic hydrocarbon with alkali metals:

$$\mathrm{ArH + Na \rightleftarrows ArH^- + Na^+}$$

The equilibrium is well studied both experimentally and theoretically [23]. The ions resulted are actually radical-anions and exhibit

[23] See, for example, A. Streitwieser, Jr. *Molecular Orbital Theory for Organic Chemists*, J. Wiley, New York (1961), Sections 6.5, 7.1 and 14.3, and references cited therein.

e.s.r. spectra [24]. The observed spectra are well accounted for by HMO theory.

Carbonyl compounds may add one or two electrons in a reversible process:

$$R^1R^2C{=}O + e^- \rightleftarrows \underset{\mathbf{1}}{R^1R^2\dot{C}{-}O^-} + e^- \rightleftarrows \underset{\mathbf{2}}{R^1R^2\overline{\overline{C}}{-}O^-} \qquad (4.14)$$

The formation of **1** can be followed by polarographic reduction potential. It has been found [23] that the half-wave potential for the first equilibrium correlates with the calculated MO energy differences between the carbonyl compound and the radical anion.

If reaction (4.14) is performed in a protic solvent, the dianion **2** is trapped and the reduction becomes irreversible.

$$\underset{\mathbf{2}}{R^1R^2\overline{C}{-}O^-} + SH \longrightarrow \underset{\mathbf{3}}{R^1R^2CH{-}O^-} + SH \rightleftarrows R^1R^2CHOH \qquad (4.15)$$

Even if the resulted alcohol equilibrates with its alkoxide ion, there is no base strong enough to extract the proton from **3** and lead back to **2**.

Electron-transfer reactions will be only incidentally treated in this book.

Another way of bringing about electron redistribution in molecules is by modification of the covalent bonds. Ingold [1] was the first to make a comprehensive classification of such reactions. With certain changes, it is still in use nowadays, being adopted in most reaction mechanism textbooks. Table 4.5 summarises the elementary reactions occurring with bond modification.

In Table 4.5 organic reactions are first divided according to the particular conditions under which they take place. A first group is formed by the homogeneous reactions occurring in solution. A second group is represented by the heterogeneous reactions, with large applicability in industry. Last but not least comes the group of the enzymatic reactions. In each group, the same elementary reactions are found, even if their mechanisms are quite different. This book will deal only with homogeneous reactions in solutions and some related reactions in gaseous phase.

The reactions in solution are classified according to the specific reagent involved. A first category is formed by homolytic reactions, characterised by the presence of a homolytic reagent. In the second category, the heterolytic reactions, we find involved either a nucleophilic or an electrophilic reagent. It must be noted that the same transformation of the organic substrate can be achieved in different ways.

[24] S. I. Weissman, J. Townsend, D. E. Paul and G. E. Pake, *J. Chem. Phys.*, **21**, 2227 (1953).

TABLE 4.5

Classification of organic reactions

Reaction type, examples	*Formulation*
I. Electron-transfer reactions	Redox reactions
II. Reactions with modification of the bond	
A. *Homolytic reactions*	
1. Decomposition, combination	$A-B \rightleftarrows A\cdot + B\cdot$
2. Substitution	$A\cdot + B-C \longrightarrow A-B + C\cdot$
3. Elimination	$X-A-B-Y \longrightarrow A{=}B + X\cdot + Y\cdot$
4. Addition	$A{=}B + X\cdot + Y\cdot \longrightarrow X-A-B-Y$
B. *Heterolytic reactions*	
1. Decomposition, combination	$A-B \rightleftarrows A + B{:}$
2. Substitution	
a. nucleophilic (S_N)	$Y{:} + A-X \longrightarrow Y-A + X{:}$
b. electrophilic (S_E)	$X + A-Z \longrightarrow X-A + Z$
3. Elimination (E)	$X-A-B-Y \longrightarrow A{=}B + X + Y{:}$
4. Addition	
a. nucleophilic	$Y{:} + A{=}B \longrightarrow Y-A-B{:} \longrightarrow$ products
b. electrophilic	$X + A{=}B \longrightarrow X-A-B \longrightarrow$ products
C. *Heterogeneous reactions* Decomposition, combination, substitution, elimination, additions under heterogeneous catalysis	
D. *Biochemical reactions* Decomposition, combination, substitution, elimination, biochemical additions	

The most simple reaction, involving the modification of only one bond, is the combination reaction, with its reverse — the fragmentation reaction. If the combining molecules are free atoms or free radicals, the reaction is called *colligation*, and its reverse *homolysis* [1].

$$A\cdot + B\cdot \rightleftarrows A-B \tag{4.16}$$

A bond can also be formed or broken in a heterolytic reaction, the direct reaction, called *coordination*, involving an electron acceptor (electrophile) and an electron donor (nucleophile).

$$A{:}^- + B^+ \rightleftarrows A-B \tag{4.17}$$

The indirect reaction, the *heterolysis*, has a higher energy of activation than the homolysis. It can be imagined that the heterolysis is preceded by homolysis (4.16), which is then followed by an electron transfer between the resultant free radicals:

$$A\cdot + B\cdot \longrightarrow A{:}^- + B^+ \tag{4.18}$$

Even if the actual pathway is not made up of (4.16) and (4.18), the Hess law tells us that (4.17) must be energetically disfavoured over (4.16). Homolysis will therefore occur with about the same ease in gaseous and condensed liquid phase, whereas heterolysis is observed only in solutions. In that case the solvation of the resulted polar species $A{:}^-$ and B^+ largely compensates the activation energy of the heterolysis. The same is true for all other heterolytic reactions.

A substitution reaction:

$$A + B{-}C \longrightarrow A{-}B + C \tag{4.19}$$

involves two bond modifications, the formation of the new bond A—B and the breaking of the old bond B—C. These bond modifications can be either similar to (4.16) or to (4.17), so that we may distinguish a homolytic substitution and a heterolytic substitution. However, more bond modifications imply more reaction schemes. Let us first consider the homolytic reaction. The transformation can occur either synchronously, in a single concerted step:

$$A\cdot + B{-}C \longrightarrow A \cdots B \cdots C \longrightarrow A{-}B + C\cdot \tag{4.20}$$

or in two different steps:

$$\begin{aligned} B{-}C &\longrightarrow B\cdot + C\cdot \\ A\cdot + B\cdot &\longrightarrow A{-}B \end{aligned} \tag{4.21}$$

It has been shown in the preceding chapter that the concerted reaction is favoured over the two-step process.

In the heterolytic reactions the reagent can be either a nucleophile (nucleophilic substitutions) or an electrophile (electrophilic substitutions). The heterolytic substitutions may occur concertedly, or by steps, depending on the action of solvent molecules involved in the heterolysis. Obviously, in the transition state of a concerted nucleophilic substitution **5** with large charge delocalisation, solvation will be less important than in heterolysis (4.18) which produces ionic species:

$$A{:}^- + B{-}C \longrightarrow \underset{\mathbf{5}}{\overset{\delta-}{A}{:}\cdots B \cdots \overset{\delta-}{C}} \longrightarrow A{-}B + C{:}^-$$

Such aspects will be discussed in more detail in the following chapters.

An elimination or an addition reaction involves even more bond modifications, and the number of possible reaction schemes increases correspondingly. We may distinguish at least three possible mechanisms, depending on the actual reagent: homolytic addition (or elimination), electrophilic addition (or cationic elimination), nucleophilic addition (or carbanionic elimination). The detailed reaction schemes will be duly treated in corresponding chapters.

It must be retained from the foregoing discussion that the first objective of a reaction mechanism study is to establish accurately the correct reaction type. Naturally, the implied substrate and reagent may give a hint in that direction, but not rarely this indication can be misleading. To give a single example, let us refer to reaction (4.22):

$$EtLi + EtI \longrightarrow Et{-}Et \tag{4.22}$$

Ethyl iodide undergoes easily nucleophilic substitutions. Ethyl lithium, like other organometallic compounds, manifests in many reactions nucleophilic reactivity. From the simple examination of equation (4.22) we could infer that it is a nucleophilic substitution. However, it was proved [25] that the reaction is actually homolytic, the ethyl radical being identified as a reaction intermediate. The identification of the reaction intermediate is therefore very important in establishing the reaction type. Before further discussing the reaction intermediates, some general characteristics of the two types of reaction must be briefly outlined.

Homolytic reactions occur easily both in the gaseous and in the liquid phase. In solution, the rate of a homolytic reaction is only slightly affected by the polarity of the solvent, and usual homogeneous catalysts (Brönsted acids or bases, Lewis acids) do not modify either the equation rate or the observed rate constant. However, both equation rate and reaction rate are strongly affected by very small amounts of compounds which can generate free radicals. The reactions are favoured by light and inhibited by very small amounts of substances able to combine more rapidly than the reactants with the intermediate free radical species. Polar substituents in substrate modify its reactivity only slightly. In a heterolytic reaction, on the other hand, polar substituents may dramatically change the reaction rate. Such reactions are strongly affected by the polarity of the solvent or by the presence of catalysts. They are quite insensible to radicals and not influenced by light. These characteristics permit a clear distinction between the two reaction types.

A puzzling situation results if the reaction studied manifests both characteristics of homolytic and of heterolytic processes, for example: insensibility to the reaction conditions (gaseous or liquid phase, polar or non-polar solvent), but a high susceptibility to polar substituents in the substrate, insensibility to free radicals or catalysts, high stereoselectivity and regioselectivity (which are usually observed in heterolytic reactions, but never in homolytic reactions). Such reactions are sometimes called, due to these intriguing features,

[25] H. R. Ward, R. G. Lawler and R.A. Cooper, *J. Amer. Chem. Soc.*, **91**, 746 (1969).

no mechanism reactions[26], the term suggesting the ambiguity in labelling the reaction either as homolytic or as heterolytic, on the basis of experimental data. In such reactions, the bond modifications occur in a concerted single step, through a cyclic transition state. The last chapter of this book will be devoted to these reactions.

4.2 Reaction intermediates

Most of the homogeneous organic reactions occur stepwise, cationic, anionic or free-radical species being involved as reaction intermediates. The identification of these intermediates is of utmost importance in a reaction mechanism study. It is necessary to anticipate in this chapter some aspects of their chemistry.

4.2.1 *Carbocations*

Carbocations are involved as intermediates in the nucleophilic substitution (see Chapter 5), in the aromatic electrophilic substitution, in the bimolecular electrophilic substitution at the saturated carbon atom and in the electrophilic addition. They may also appear in the course of Friedel-Crafts reactions, in the cationic polymerisation, in alkene alkylation reactions and in many other petrochemical processes. Their reactivity with respect to different nucleophiles will be treated in the corresponding chapters of this book. In this section we shall review only their fundamental properties.

(A) Nomenclature

In 1900, Gomberg[27] was the first to identify a free radical, the triphenylmethyl. Two years later Baeyer and Villiger[28] identified the coloured cationic species Ph_3C^+, proposing for such ions the name of *carbonium ions*. Soon afterwards, Gomberg[29] confirmed the discovery of Baeyer and Villiger, but pointed out that the name of carbonium ion suggested a higher valency of the central atom than the usual one, as for ammonium, sulphonium or oxonium ions. Despite this inconsistency in nomenclature, the name of carbonium ions has been preferred ever since, being generally adopted in papers, reviews and monographs[30,31] devoted to this topic. Recently, Olah[32]

[26] W. von E. Doering cited in S. J. Rhoads, *Molecular Rearrangements*, vol. 1 (P. de Mayo, editor), Interscience Publishers, New York (1963), p. 655.

[27] M. Gomberg, *Ber. dtsch. chem. Ges.*, **33**, 3150 (1900).

[28] A. Baeyer and V. Villiger, *Ber. dtsch. chem. Ges.*, **35**, 1189 (1902), namely on page 1196.

[29] M. Gomberg, *Ber. dtsch. chem. Ges.*, **35**, 2397 (1902).

[30] D. Bethel and V. Gold, *Carbonium Ions*, Academic Press, New York (1967).

[31] G. A. Olah and P. von R. Schleyer (editors), *Carbonium Ions*, Interscience, vol. 1 (1968); vol. 2 (1969); vol. 3 (1972); vol. 4 (1972).

[32] G. A. Olah, *J. Amer. Chem. Soc.*, **94**, 808 (1972).

resumed Gomberg's objection and proposed instead the generic name of *carbocations* for all cationic species with charge on carbon atoms. This name is similar to the generally adopted name [33] of *carbanions*, denoting anions with negative charge on carbon atoms. However, a large variety of cationic species of carbon derivatives are known nowadays, some of them, named *classical ions*, having a coordination number three at the positive carbon (as in triphenylmethyl cation), while others, named *non-classical ions*[34,35], have a higher coordination number, 4 or 5. For the classical ions, Olah proposed the name of *carbenium ions*, considering that they may formally result by protonation of carbenes, in a similar way as ammonium ions result from ammonia:

$$CH_2: + H^+ \longrightarrow CH_3^+$$

An alternative name, based on the protonation of alkenes, is that of *alkenium ions*. According to Olah [32], the old name of carbonium ions could still be used to designate non-classical ions, with higher coordination number at the carbon atom(s).

Olah's proposal does not seem to be generally adopted as yet. However, as we are satisfied with this new nomenclature, it shall be used throughout the book.

(B) Structure

(*a*) *Simple alkenium ions*. In the methenium ion, CH_3^+, and its simple derivatives, the central atom has an sp^2 hybridisation. The bonding orbitals of the carbenium atom will therefore lie in the same plane, the molecule possessing either a molecular plane of symmetry or a local symmetry plane, depending on the other surroundings (Figure 4.3). The unoccupied p orbital is a non-bonding orbital ($\varepsilon = \alpha$).

Carbenium ions can be directly observed and studied by physical methods. The first successfully applied methods [33] were based on the simple reaction:

$$R\cdot \longrightarrow R^+ + e^- \tag{4.23}$$

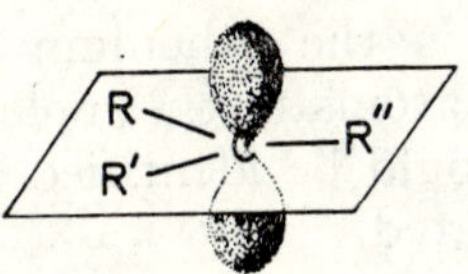

Figure 4.3 Structure of a planar carbocation.

[33] Rule C-83 IUPAC, *Pure Appl. Chem.*, **11**, 63 (1965).

[34] First mentioned by J. D. Roberts and C.C. Lee, *J. Amer. Chem. Soc.*, **75**, 5009 (1951).

[35] P. D. Bartlett, *Non-classical Ions, Reprints and Commentary*, W. A. Benjamin, New York (1965).

[36] F. H. Field and J. L. Franklin, *Electron Impact Phenomena and the Properties of Gaseous Ions*, Academic Press, New York (1957).

which occurs during the impact of an energetic particle (electron) with the radical R·. The minimum energy which an electron is required to have in order to produce the carbenium ion R^+ is related to the relative stability of the latter. It was proved in this way that primary carbenium ions are the least stable, whereas tertiary carbenium ions are the most stable among alkenium ions. Using LCAO-MO methods, Muller and Mulliken [37] demonstrated that the observed stabilisation energies can be well accounted for if it is assumed that carbenium ions have a planar structure and the vacant non-bonding orbital is stabilised by *hyperconjugation** with the neighbouring C—H bonds (Table 4.6).

TABLE 4.6

Stabilisation and formation energies (kcal mol^{-1}) of simple carbenium ions, determined by electron-impact methods [36] and calculated theoretically [37] by LCAO-MO methods

Ion	*Stabilisation energies* (kcal mol^{-1})		*Heats of formation* (kcal mol^{-1})
	Exper.	*Calc.*	
CH_3^+	0	0	262
$CH_3CH_2^+$	36	41	224
$(CH_3)_2CH^+$	66	66	190
$(CH_3)_3C^+$	84	83	166
$CH_2{=}CH{-}CH_2^+$	58	64	220

The study of gaseous ions based on electron-impact phenomena developed considerably in recent years due to the increased weight of mass-spectrometry in chemical research [38]. However, carbocations are most frequently encountered in solution. The discovery of Olah and his co-workers [31] (see also the next chapter) that simple carbenium ions are stable enough to be directly observed in *superacid* solutions, such as $FSO_3H{-}SbF_5$ [39], led to the increasing use of n.m.r. spectroscopy in this field. The only primary carbenium ion observed with certainty is the ethenium ion ($CH_3{-}CH_2^+$) [40]. For higher numbers of carbon atoms in the molecule, only secondary and tertiary carbenium ions could be identified [39]. The planarity of these ions is clearly demonstrated.

[37] N. Muller and R. S. Mulliken, *J. Amer. Chem. Soc.*, **80**, 3489 (1958).

* The overlap interaction of appropriately oriented σ bonds with π bonds is called hyperconjugation.

[38] H. M. Grubb and S. Meyerson, in *Mass Spectrometry of Organic Ions* (F. W. McLafferty, editor), Academic Press, New York (1963), p. 453.

[39] G. A. Olah, E. B. Baker, J. C. Evans, W. S. Tolgyesi, J. S. McIntire and I. J. Bastien, *J. Amer. Chem. Soc.*, **86**, 1360 (1964).

[40] G. A. Olah, J. R. DeMember, R. H. Schlosberg and Y. Halpern, *J. Amer. Chem. Soc.*, **94**, 156 (1972).

(*b*) *Delocalised alkenium ions*. Delocalised alkenium ions are extended π-systems in which the unoccupied p orbital is conjugated with double bonds or aromatic rings. As has been already shown (see Chapter 2), such systems are planar. Their stability increases with delocalisation, some of them, like the triarylmethane dyes, being stable, easily isolated molecules.

Due to their increased stabilisation, delocalised alkenium ions can be observed not only by electron-impact methods, but also by n.m.r., u.v. and i.r. spectroscopy, X-ray methods, conductometry etc.[30].

It is interesting that the simple HMO technique [41] gives a satisfactory account for the stability of these ions. Thus, the highest occupied π orbital in the allylic ion has the energy $\varepsilon = \alpha + 1.41\,\beta$, which corresponds to a delocalisation energy of 0.82 β. Considering an empirical value of 70—80 kcal mol^{-1} for β, derived from spectral data [42], the agreement with the stabilisation energy of the ion in gaseous phase is satisfactory (see Table 4.6).

Cyclic polyenic cations, such as cyclopropenylium and tropylium ions deserve special mention. Table 4.7 lists some HMO data for cyclic polyenes, including these ions. It can be seen that in the cationic species with three or seven carbon atoms in the ring, only the most stable MOs (bonding orbitals) are occupied, whereas for the corresponding free radicals or anions antibonding orbitals are partially filled with electrons. Delocalised energies (DE) are high and are comparable with the DE of benzene, the typically aromatic compound.

Using these and other data, Hückel established his famous $4n + 2$ rule [43], predicting an aromatic character for the cyclopropenylium and tropylium ions, which were unknown at that time. These planar and fully symmetric ions have been experimentally obtained later (see p. 157) and their structure, symmetry and properties have been found to be in full agreement with the theory.

(*c*) *Carbonium ions*. The most simple carbonium ion, the methonium ion CH_5^+ was first known from mass-spectrometric studies [44]. Its formation in superacid solutions has been later undoubtedly ascertained [45,46]. Careful experimental studies of Olah and his coworkers [46,47] and theoretical calculations[46–49] suggested a C_s sym-

[41] E. Heilbronner and H. Block, *Das HMO-Modell und seine Anwendung*, vol. 3, Tabellen berechneter und experimenteller Grössen, Verlag Chemie (1970).

[42] See for example Streitwieser [23], p. 208—209.

[43] E. Hückel, *Z. Physik*, **70**, 204 (1931); for historical and theoretical treatment of these systems see also: Reference 41, vol. 1 (1968), p. 115; Reference 23, p. 256; P. J. Garratt, *Aromaticity*, McGraw-Hill, London (1971).

[44] F. H. Field and M. S. B. Munson, *J. Amer. Chem. Soc.*, **87**, 3289 (1965).

[45] H. Hogeveen and C. J. Gaasbeek, *Rec. Trav. Chim.*, **87**, 319 (1968).

[46] G. A. Olah, Y. Halpern, J. Shen and Y. K. Mo, *J. Amer. Chem. Soc.*, **93**, 1251 (1971).

[47] G. A. Olah, K. Klopman and R. H. Schlosberg, *J. Amer. Chem. Soc.*, **91**, 3261 (1969).

[48] J. J. C. Mulder and J. S. Wright, *Chem. Phys. Lett.*, **5**, 445 (1970).

[49] W. A. Lathan, W. J. Hehre and J. A. Pople, *J. Amer. Chem. Soc.*, **93**, 808 (1971).

TABLE 4.7

HMO data for cyclic π-systems

Cyclic π-system	*Symmetry*	*Energies of MOs*	*Total π-electron energy*	*DE/β*
	D_{3h}	$\varepsilon_1 = \alpha + 2\beta$ $\varepsilon_2 = \varepsilon_3 = \alpha - \beta$	$2\alpha + 4\beta$	2.00
	D_{3h}		$3\alpha + 3\beta$	
	D_{3h}		$4\alpha + 2\beta$	
	D_{4h}	$\varepsilon_1 = \alpha + 2\beta$ $\varepsilon_2 = \varepsilon_3 = \alpha$ $\varepsilon_4 = \alpha - 2\beta$	$4\alpha + 4\beta$	0.00
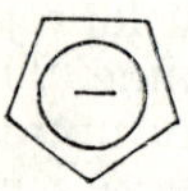	D_{5h}	$\varepsilon_1 = \alpha + 2\beta$ $\varepsilon_2 = \varepsilon_3 = \alpha + 0.61\beta$ $\varepsilon_4 = \varepsilon_5 = \alpha + 1.62\beta$	$5\alpha + 6.47\beta$	1.47
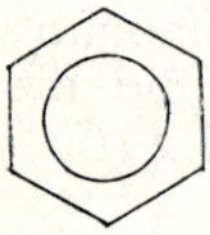	D_{6h}	$\varepsilon_1 = \alpha + 2\beta$ $\varepsilon_2 = \varepsilon_3 = \alpha + \beta$ $\varepsilon_4 = \varepsilon_5 = \alpha - \beta$ $\varepsilon_6 = \alpha - 2\beta$	$6\alpha + 8.00\beta$	2.00
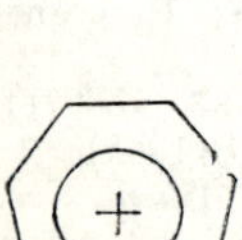	D_{7h}	$\varepsilon_1 = \alpha + 2\beta$ $\varepsilon_2 = \varepsilon_3 = \alpha + 1.24\beta$ $\varepsilon_4 = \varepsilon_5 = \alpha - 0.445\beta$ $\varepsilon_6 = \varepsilon_7 = \alpha - 1.80\beta$	$6\alpha + 8.98\beta$	2.98

metry of the ion, corresponding to three C—H bonds and a two-electron, three-centre bond, **6**:

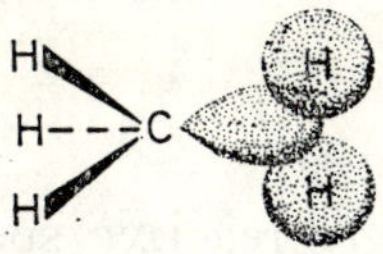

6 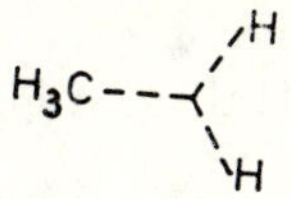 **7**

For simplicity, instead of drawing orbitals, Olah[32] proposed formula **7**, the dotted branched lines depicting the three-centre bond.

Similar in structure with the methonium ion are the H-alkonium ions in which the methyl is replaced by another alkyl **(8)**. Like the methonium ion, they may result by the protonation of a C—H bond, in superacid solutions, being thus directly observed by n.m.r. spectroscopy. If a C—C bond is protonated, a C-alkonium ion **9** is formed[32, 46].

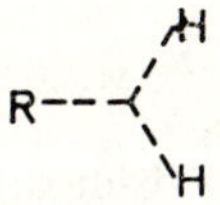

8 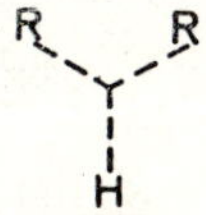 **9**

Another type of carbonium ion is represented by the ethenonium ion **10**, in which tetra-coordinated carbon atoms are found:

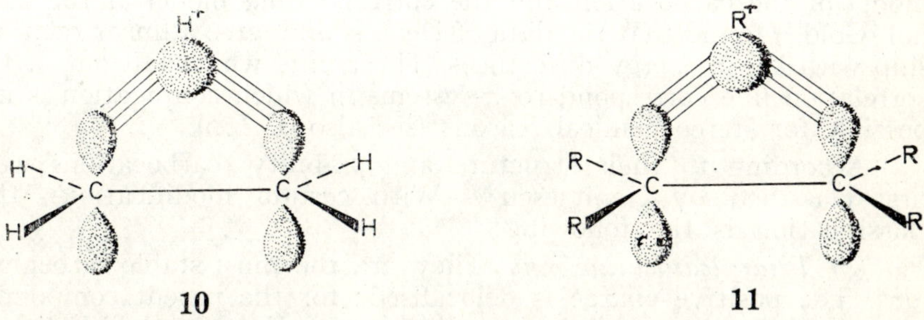

10 **11**

This and other similar ions **11** are imagined to result by the overlap interaction of the occupied π orbital of alkenes with the vacant orbital of a proton (1*s*) or of a carbenium ion (2*p*), respectively. The positive charge is delocalised over the three atoms involved in the three-centre bond. Such ions are named *bridged ions*, due to their structure. They are frequently encountered in nucleophilic substitutions, electrophilic additions, or electrophilic substitutions.

Due to the possibility of directly observing these ions by modern physical methods (see for example the norbornyl cation, p. 278), their structure is well understood at present.

(C) Stability. Classification

Carbenium and carbonium ions are higly reactive species with respect to nucleophilic reagents. If the nucleophile is water, the reaction (4.24) may be reversible:

$$R^+ + H_2O \overset{K_{R^+}}{\rightleftharpoons} R{-}OH + H^+ \qquad (4.24)$$

The negative logarithm of the equilibrium constant (pK_{R^+}) can characterise the stability of the ion R^+. If the carbocation R^+ is relatively stable, pK_{R^+} values can be directly determined by measuring equilibrium (4.24) by spectrometric methods (for example u.v. spectra [50]). Conversely, if the carbocation is unstable, the pK_{R^+} value is determined from equilibrium (4.25) involving an ion R^{1+} of known pK_{R^+} and a hydrocarbon R^2H:

$$R^{1+} + R^2H \rightleftharpoons R^1{-}H + R^{2+} \qquad (4.25)$$

Table 4.8 lists pK_{R^+} values of some representative carbocations. A negative value corresponds to an unstable cation, while a high positive value indicates that equilibrium (4.24) is shifted to the cationic species.

Wheland suggested many years ago [51] that such equilibrium constants should correlate with π-energy differences between a model of the carbocation and the corresponding model of the alcohol. Gold [52] found that the data of Deno *et al.*[50] give a linear relationship with the π-energy differences. The points which fall out of the correlation line correspond to π-systems in which conjugation is not possible for stereochemical reasons (see also p. 155).

According to their structure and stability, carbocations were first classified by Nenitzescu [53]. With certain modifications, this classification is the following:

(*a*) *Triarylcarbenium ions.* They are the most stable carbenium ions. The positive charge is delocalised; for the parent compound, triphenylcarbenium (trityl) ion, Ph_3C^+, the HMO method indicates a total π-electron energy of $18\alpha + 25.800\beta$ [41], corresponding to a DE of 1.800β. Due to their stability, the triarylcarbenium ions were the first to be discovered[28, 29]. Adequate substituents in **12**

[50] N. C. Deno, J. J. Jaruzelski and A. Schriesheim, *J. Amer. Chem. Soc.*, **77**, 3044 (1955).
[51] G. W. Wheland, *J. Chem. Phys.*, **2**, 474 (1934).
[52] V. Gold, *J. Chem. Soc.*, 3944 (1956).
[53] C. D. Nenitzescu, in Reference 31, vol. 1, p. 1.

TABLE 4.8

pK_{R^+} **values of some representative carbocations**

Carbocation	pK_{R^+}	*Carbocation*	pK_{R^+}
CH_3^+	~ -44[a]	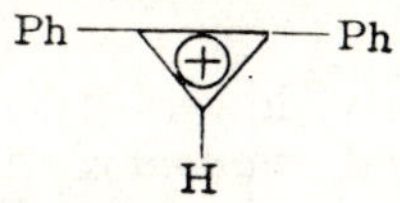 Ph—(+)—Ph, H (cyclopropenyl)	0.32[a, b]
$Ph-CH_2^+$	-22[a]	(benzo-fused (+) ring)	$+1.7$[a]
(fluorenyl, +)	-14.0[b]	Ph—(+)—Ph, Ph (cyclopropenyl)	$+2.8$[e]; $+3.1$[d]
Ph_2CH^+	-13.3[b]	((+) ring)	4.7[d]
 ((+) cyclopropenyl)	-7.4[c]	p-MeO—C_6H_4—(+)—C_6H_4—OMe-p, H	$+6.42$[f]
Ph_3C^+	-6.6[b]	(thiopyrylium, $\overset{+}{S}$)	$+8.7$[b]
(dibenzo-fused (+) ring, Ph)	-5.7[d]	$[p\text{-}Me_2NC_6H_4]_3C^+$	$+9.36$[a]
$Ph_2\overset{+}{C}$—(thienyl, S)	-5.0[b]		
(dibenzo-fused (+) ring)	-3.7[d]		

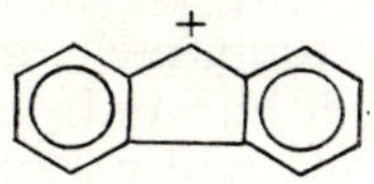

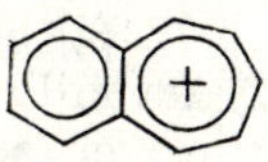

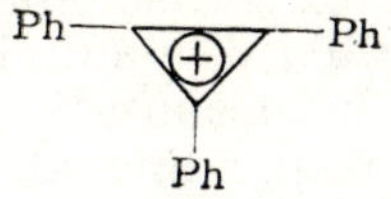

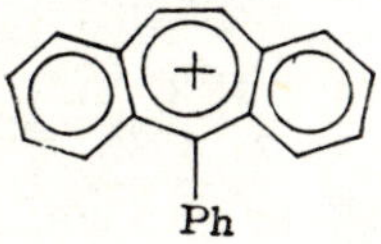

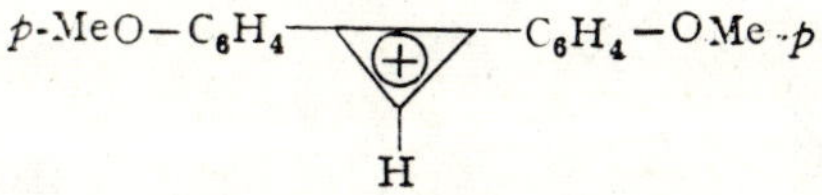

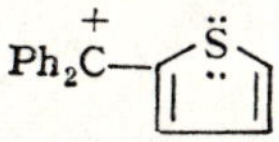

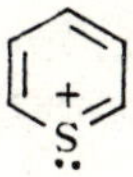

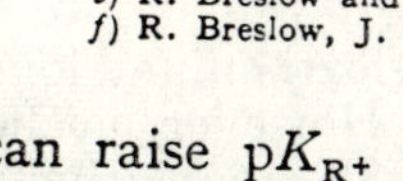

a) Reference 50.
b) R.G. Turnbo, D.L. Sullivan and R. Pettit, *J. Amer. Chem. Soc.*, **86**, 5630 (1964).
c) R. Breslow and J. T. Groves, *J. Amer. Chem. Soc.*, **92**, 984 (1970).
d) G. Berti, *J. Org. Chem.*, **22**, 230 (1957).
e) R. Breslow and H. W. Chang, *J. Amer. Chem. Soc.*, **83**, 2367 (1961).
f) R. Breslow, J. Lockhart and H. W. Chang, *J. Amer. Chem. Soc.*, **83**, 2375 (1961).

can raise pK_{R^+} values from -6.6, in the parent compound, to $+9.36$ in the triarylmethyl dyes (see Table 4.8).

The pK_{R^+} values for such ions are lower than anticipated from their π-energy. It has now been well established that these ions are non-planar (**13**), the aromatic rings determining a torsional angle τ between the plane of the three sp^2 bonds of the carbenium ion and one of the three rings:

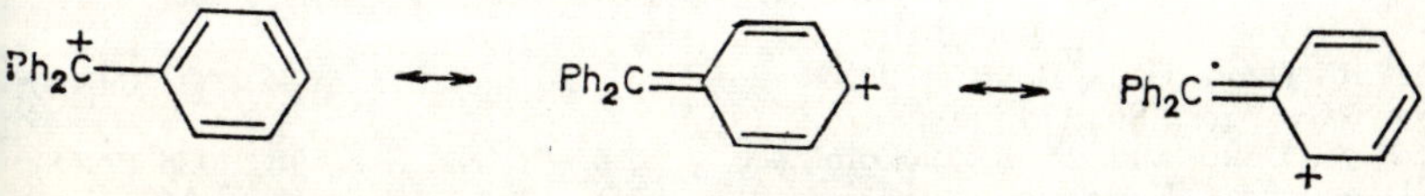

$Ph_2\overset{+}{C}$—C_6H_5 ⟷ Ph_2C=$C_6H_5^+$ ⟷ Ph_2C=$C_6H_5^+$ etc.

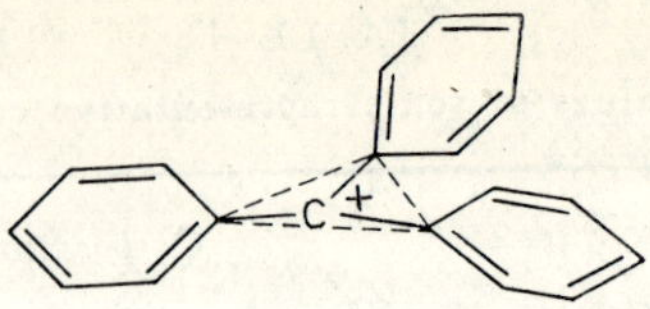

13

Non-coplanarity lowers the DE and thus the pK_{R^+} value.

(*b*) *Diaryl- and monoaryl-carbenium ions*. The total π-energy for the diphenylcarbenium ion Ph_2CH^+ is $12\alpha + 17.301\beta$, and for the benzyl ion $6\alpha + 8.721\beta$, corresponding to DEs of 1.301β and 0.721β, respectively. Actually, these systems are less stable than the triaryl system. However, as they have lower steric requirements, the stability correlates satisfactorily with the observed pK_{R^+} values.

(*c*) *Enylcarbenium ions* (allyl ions) have been directly observed in the last years by Deno[54,55]. They were generated in sulphuric acid from the corresponding alcohols or alkenes. The most stable allylic ions seem to be the cyclopentenylic ions **14**:

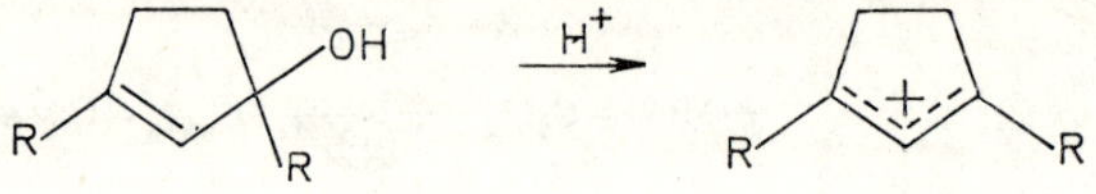

14

(*d*) *Dienylcarbenium ions* can be generated in a similar way as the enylcarbenium ions:

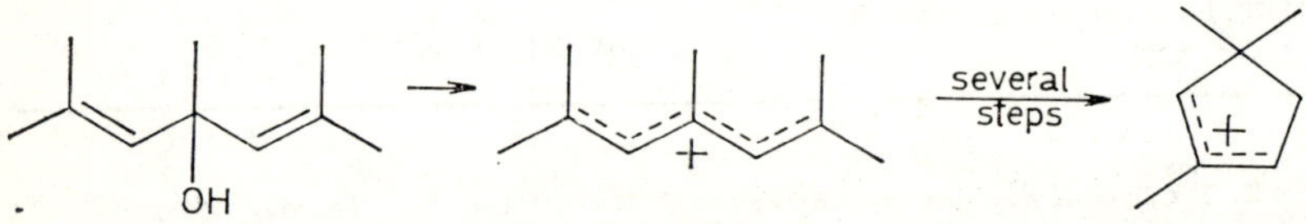

several steps (4.26)

15

Equation (4.26) shows that the acyclic dienylcarbenium ions are less stable than the cyclic enylcarbenium ions. However, cyclic carbenium ions like the cyclohexadienyl carbocations **16** are relatively stable, being involved in the aromatic substitution (see Chapter 7).

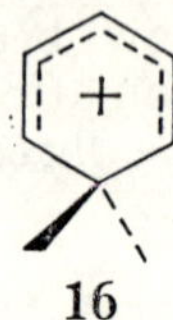

16

[54] N. C. Deno, D. B. Boyd, J. D. Hodge, C. U. Pittman, Jr. and J. O. Turner, *J. Amer. Chem. Soc.*, **86**, 1745 (1964).

[55] N. C. Deno and R. R. Lastomirsky, *J. Amer. Chem. Soc.*, **90**, 4058 (1968); N. C. Deno and C.U. Pittman, Jr., *ibid.*, **86**, 1744 (1964); N. C. Deno and C. U. Pittman, Jr., *ibid.*, **86**, 1871 (1964).

(*e*) *Alkyl and dialkylcarbenium ions* are the less stable alkenium ions. Their generation from the corresponding saturated alcohols or corresponding alkenes in sulphuric acid usually leads to a mixture of polymers and cations, the cyclic ions of type **14** being the most stable ions formed under these conditions[54]. However, in superacid solutions[39, 56], saturated alcohols may generate simple carbenium ions, corresponding only to the starting alcohol. Their formation is directly related to the H_0 values of the corresponding acid or acid mixture, respectively. The integrity of the number of carbon atoms in the carbocation, with respect to the initial substrate, can be assured by solvation[39].

(*f*) *Aromatic carbocations.* A high stability for the tropylium and cyclopropenylium ions was predicted by the $4n + 2$ rule of Hückel (see p. 151) and was later confirmed by Doering and Knox[57] who first prepared the tropylium ion, **17**, and by Breslow and Ynan[58] who succeeded to obtain the triphenylcyclopropenylium ion **18**, R = Ph. The unsubstituted cyclopropenylium ion (**18**, R = H), he parent compound of the aromatic systems with $n = 0$ in Hückel's rule, was identified only recently[59]. Table 4.8 shows that tropylium ion is more stable than trityl ion (triphenylmethyl ion), its higher stability permitting an easy preparation (see p. 161).

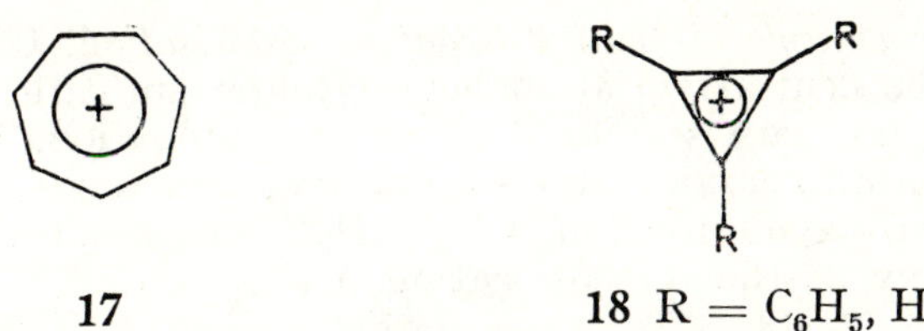

17 **18** R = C_6H_5, H

(*g*) *Homoaromatic and bicycloaromatic ions.* Winstein and Sonnenberg[60] were the first to describe a trishomocyclopropenylium ion **19** (see also p. 288), showing that such ions may possess high stability, comparable to the parent compounds. The concept of homoaromaticity was later developed by Winstein[61]. Simple HMO estimations (applying the method of α and β variation) indicate high energy stabilisations for these homoaromatic compounds.

[56] G. A. Olah, M. B. Comisarow, C. A. Cupas, C. U. Pittman, Jr., *J. Amer. Chem. Soc.*, **87**, 2997 (1967).

[57] W. von E. Doering and L. H. Knox, *J. Amer. Chem. Soc.*, **76**, 3203 (1954).

[58] R. Breslow and C. Ynan, *J. Amer. Chem. Soc.*, **80**, 5991 (1958).

[59] R. Breslow and J. T. Groves, *J. Amer. Chem. Soc.*, **92**, 984 (1970).

[60] S. Winstein and J. Sonnenberg, *J. Amer. Chem. Soc.*, **83**, 3235, 3244 (1961).

[61] S. Winstein, Special Publication nº 21, The Chemical Society, London, 1967; *Quart. Rev. (London)*, **23**, 141 (1969).

19

Goldstein[62] has shown, on theoretical grounds, that a polymethinic system **20** may have a high stability if it possesses $4n$ π-electrons; a similar system with $(4n+2)$ π-electrons will have low stability. The stable systems are called *bicycloaromatic* ions, the unstable ones, *antibicycloaromatic*. This theory gained much experimental support in recent years (see p. 170).

20

(*h*) *Carbenium ions with heteroatom stabilisation.* Carbenium ions having at the central sp^2 atom heteroatomic substituents with unshared electrons, are stabilised by $p-p$ conjugation. Measurements of the formation energies by electron-impact methods[63] have shown that the methoxymethyl ion, $CH_3OCH_2^+$, has about the same formation energy as the tertiary cation $(CH_3)_3C^+$.

$$CH_3-\ddot{\underset{\cdot\cdot}{O}}-CH_2^+ \longleftrightarrow CH_3-\overset{+}{\underset{\cdot\cdot}{O}}=CH_2$$

A special case of this type of stabilisation is that of carbonyl compounds and their salts:

$$R_2C=O \longleftrightarrow R_2\overset{+}{C}-\overset{-}{O}; \quad R_2\overset{+}{C}-OH \longleftrightarrow R_2C=\overset{+}{O}H$$

Carbonium ions (non-classical ions) is a last group of stable carbocations discovered and studied recently. Their chemistry, for the time being, cannot be discussed in a general manner, neglecting the particular reactions in which they are involved. According to Olah[32], they are classified as penta- and tetra-coordinated carbonium ions (see p. 149).

[62] M. J. Goldstein, *J. Amer. Chem. Soc.*, **89**, 6357 (1967).
[63] R. H. Martin, F. W. Lampe and R. W. Taft, Jr., *J. Amer. Chem. Soc.*, **88**, 1353 (1966).

(D) Generation

Carbocations can be generated in several ways. A detailed discussion of such reactions will be presented in separate chapters.

(*a*) *Dissociation reactions.* The formation of carbocations by unimolecular dissociation is the first step in the unimolecular nucleophilic substitution (S_N1, see Chapter 5):

$$R{-}X \rightleftharpoons R^+X^- \tag{4.27}$$

Such reactions are observed in polar solvents, or in non-polar media in presence of electrophilic catalysts, either Brönsted acids:

$$R{-}X + HA \rightleftharpoons R^+ + X^- \cdots HA \tag{4.27a}$$

or Lewis acids:

$$R{-}X + MeX_n \rightleftharpoons R^+ + MeX^-_{(n+1)} \tag{4.27b}$$

Equilibria (4.27) are easily observed if R^+ is a stable carbocation such as trityl ion (triphenylcarbenium ion). It was recognised earlier that triphenylmethyl chloride forms colourless solutions in non-polar solvents (physical solvation), but yields yellow solutions displaying electrolytic behaviour in polar solvents such as sulphur dioxide, nitromethane, etc. [29,64]. In nitromethane, trityl ion represents 0.032 %[65] of the initial concentration of triphenylmethyl chloride.

In a modern experimental technique, carbocations are generated from the corresponding halides in the presence of SbF_5. It has been shown [39] that SbF_5 is an effective solvation agent, assuring the integrity of the carbenium or carbonium ions formed.

(*b*) *Alcohol protonation.* Stable carbenium or carbonium ions can be generated by the protonation of the corresponding alcohols. Oxonium ions are formed as intermediates, which further ionise according to the general reaction (4.27). In that way, Baeyer and Villiger [28] were able to identify for the first time the trityl ion:

$$Ph_3C{-}OH + 2H_2SO_4 \rightleftharpoons Ph_3C^+ + H_3O^+ + 2HSO_4^- \tag{4.28}$$

It was later demonstrated [50, 52, 66] that equilibrium (4.28) depends upon the acidity function (H_0) of the medium.

Aliphatic saturated alcohols, such as t-butanol, are more drastically transformed by sulphuric acid. The initial ionisation is followed

[64] For a review of this subject, see N. N. Lichtin, *Progr. Phys. Org. Chem.*, **1**, 75 (1963).

[65] A. Bentley and G. Evans, cited in E. Gelles, E. D. Hughes and C. K. Ingold, *J. Chem. Soc.*, 2918 (1954).

[66] P. D. Bartlett and J. D. McCollum, *J. Amer. Chem. Soc.*, **78**, 1441 (1956).

by alkylations, polymerisation and hydride transfer reactions [54], the single identified ion being a cyclopentenyl ion (see p. 156). However, stable enyl and dienylcarbenium ions can be generated from the corresponding alcohols in sulphuric acid solution [55]. A better method involves the treatment of the aliphatic saturated alcohols with superacids containing SbF_5 [67]. The presence of SbF_5 prevents further alkylation and dealkylation reactions of the carbocation primarily formed.

(*c*) *Hydrocarbon protonation*. In strong acid solutions, nucleophilic hydrocarbons generate carbocations reversibly:

$$>C{=}C< \; + \; HA \rightleftharpoons >CH{-}\overset{+}{C}< \qquad (4.29)$$

$$C_6H_6 \; + \; HA \rightleftharpoons C_6H_7^+ \qquad (4.30)$$

In sulphuric acid solutions, the protonation can be accompanied by alkylation and dealkylation [68]. However, the use of superacids prevents the occurrence of undesirable side reactions. Moreover, superacids are advantageous in that they may generate carbonium ions by protonation of C—H and C—C bonds (see p. 307).

(*d*) *Deamination reactions*. The deamination reaction of primary amines with nitrous acid is a convenient route for the generation of carbocations (particularly carbenium ions):

$$R{-}NH_2 + HNO_2 \longrightarrow R{-}N_2^+ \longrightarrow R^+ + N_2 \qquad (4.31)$$

It is usually assumed that in reaction (4.31) carbenium ions R^+ result in a higher vibrational level (*hot ions* [69]), possessing higher reactivity than the carbenium ions generated by other methods. Consequently, the tendency to rearrange (to other carbenium ions or to carbonium ions) will be less pronounced. Deamination reactions are sometimes used to demonstrate by comparison the formation of carbonium ions by simple dissociations of type (4.27) [70].

(*e*) *Hydride transfer reactions*. Equilibrium (4.25), mentioned above, is a convenient route to generate carbonium or carbenium ions

[67] G. A. Olah, M. B. Comisarow, C. A. Cupas and C. U. Pittman, Jr., *J. Amer. Chem. Soc.*, **87**, 2997 (1967).

[68] See, for example, H. Pines and J. M. Mavity, *The Chemistry of Petroleum Hydrocarbons*, vol. 3, Reinhold (1955), p. 9.

[69] R. Huisgen and C. Ruchardt, Liebigs Ann. Chem., **601**, 1 (1956); H. Zollinger, *Diazo and Azo Chemistry*, vol. 1, Interscience, New York (1963), p. 593.

[70] E. J. Corey, J. Casanova, Jr., P. A. Vatakenchery and R. Winter, *J. Amer. Chem. Soc.*, **85**, 169 (1963).

if the differences in pK_{R^+} value of the ions involved are sufficiently large. This method has been applied by Dauben and his co-workers[71] to obtain tropylium ions (see also Table 4.8):

$$Ph_3C^+ + C_7H_8 \longrightarrow Ph_3CH + C_7H_7^+$$

The hydride transfer is a characteristic reaction of carbocations (see also p. 154).

(E) Reactions

Most of the reactions mentioned here will be discussed in more detail in the following chapters. The aim of this section is to give a brief account of the reactivity of carbocations.

(*a*) *Rearrangement reactions.* One of the most characteristic reactions of carbocations is the rearrangement. Carbenium ions **21*a*** may equilibrate either with other carbenium ions **21*b*** by 1,2-shifts:

$$\underset{\mathbf{21}a}{>\overset{R}{\overset{|}{C}}-\overset{+}{C}<} \rightleftharpoons \underset{\mathbf{21}b}{>\overset{+}{C}-\overset{R}{\overset{|}{C}}<} \tag{4.32}$$

or with carbonium ions **22**:

$$>\overset{R}{\overset{|}{C}}-\overset{+}{C}< \rightleftharpoons \underset{\mathbf{22}}{>C\overset{R}{\overset{+}{\cdots}}C<} \tag{4.33}$$

where R = H, alkyl, aryl, halogen, RO, R_2N, etc.

The skeleton modification can be experimentally proved by trapping the intermediate ions. Indeed, in reaction (4.32) only two positions are equilibrated, whereas in reaction (4.33) the charge is distributed between three centres.

Theoretically, a carbonium ion, **22**, may be more stable than the corresponding carbenium ions. Thus, MO calculations

[71] H. J. Dauben, Jr., F. A. Gadecki, K. M. Harman and D. L. Pearson, *J. Amer. Chem. Soc.*, **79**, 4557 (1957).

indicated [72] that in equilibria (4.34) the primary carbenium ion is the less stable species:

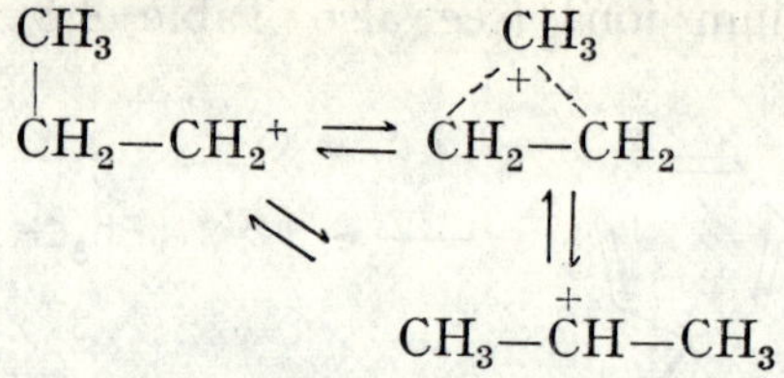

$$\begin{array}{c} CH_3 \\ | \\ CH_2{-}CH_2^+ \end{array} \rightleftharpoons \begin{array}{c} CH_3 \\ \overset{+}{\wedge} \\ CH_2{-}CH_2 \end{array} \rightleftharpoons CH_3{-}\overset{+}{C}H{-}CH_3 \qquad (4.34)$$

The bridged ion even if less stable than the secondary 2-propenium ion is far more stable than the primary ion. The theoretical calculations are borne out by experimental data [73].

The greater stability of carbonium ion **22** with respect to the unbridged carbocations is due to the overlap interactions between the vacant p orbital in **21** with the neighbouring σ bonds. Such interactions are known to occur in boron derivatives [74], such as trimethylboron **23**. This compound actually exists only as a dimer, involving overlap interactions between the empty p orbital of boron and the adjacent B—C bonds.

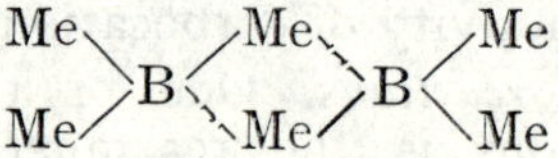

$$\begin{array}{c} Me_2B(\mu\text{-}Me)_2BMe_2 \end{array}$$

23

One may expect a difference between equilibria involving carbenium and carbonium ions, as a consequence of solvent interactions. The solvation should be more significant for carbenium ions, with concentrated charge, than for the delocalised carbonium ions. Brown [75] has thus inferred that in polar solvents equilibrium (4.32) must prevail over (4.33). Nevertheless, in superacid solutions bridged ions seem to be the only observable cationic species (see also p. 197).

1,2-Shifts of hydride ion or other groups are involved in the well-known Wagner-Meerwein rearrangements[76,77]. It should be mentioned that in such reactions **22** may represent either an intermediate or a transition state [78].

(*b*) *Reactions with nucleophiles*. Carbocations are electrophilic species and therefore react easily with nucleophilic reagents. The

[72] L. L. Radon, J. A. Pople, V. Buss and P. von R. Schleyer, *J. Amer. Chem. Soc.*, **94**, 311 (1972).

[73] D. J. McAdoo, F. W. McLafferty and P. F. Bente III, *J. Amer. Chem. Soc.*, **94**, 2027 (1972).

[74] H. Steinberg, *Organoboron Chemistry*, vol. 1, J. Wiley, New York (1964).

[75] H. C. Brown, *Chem. Soc. (London) Special Publ.*, **16**, 140 (1962).

[76] H. Meerwein and K. van Emster, *Ber. dtsch. chem. Ges.*, **55**, 2500 (1922).

[77] F. C. Whitmore, *J. Amer. Chem. Soc.*, **54**, 3274 (1932).

[78] See, for example, H. C. Brown, *Accounts Chem. Res.*, **6**, 377 (1973); D. J. Cram, *J. Amer. Chem. Soc.*, **86**, 3767 (1964); H. C. Brown, K. J. Morgan and F. J. Chloupek, *J. Amer. Chem. Soc.*, **87**, 2137 (1965).

reaction with nucleophiles is currently used for trapping carbocations in view of their identification:

$$R^+ + Nu: \longrightarrow \underset{\mathbf{24}}{R{-}Nu^+} \qquad (4.35)$$

The positive charge in **24** merely indicates that the negative charge of the nucleophile Nu: diminishes by one unit. Depending on the nucleophile, the reaction (4.35) can yield either stable products or other cationic intermediates.

Alkylation reactions of aromatic hydrocarbons (see Friedel-Crafts reactions):

$$R^+ + C_6H_6 \rightleftharpoons [C_6H_6R]^+ \text{ (arenium ion, H and R on the same carbon) or } [C_6H_6 \cdot R^+] \text{ (π complex)} \qquad (4.36)$$

or of alkenes:

$$R^+ + \rangle C{=}C\langle \longrightarrow \rangle \overset{R}{C}{-}\overset{+}{C}\langle \text{ or } \rangle C \overset{\overset{R}{+}}{-} C\langle \qquad (4.37)$$

are examples in which carbenium or carbonium ions are formed by combination of a nucleophile with a carbocation. The formation of a non-classical carbonium ion is favoured if the two reacting moieties (carbenium and the nucleophilic π bond) are present in the same molecule (see π-route reactions).

(*c*) *Equilibration with alkenes.* Carbocations can form stable products by elimination of a neighbouring proton. This reaction occurs in the unimolecular elimination (E1) and will be discussed in detail in a later chapter.

(*d*) *Hydride transfer reactions.* Hydride transfer may occur either intermolecularly [79,80] or intramolecularly [81]. The latter one is involved in the equilibration of carbenium ions (see p. 154). The intermolecular hydride transfer is used in the preparation of carbocations (see p. 160) and constitutes an important step in many petrochemical processes.

4.2.2 *Free radicals*

The first free radical to be discovered was triphenylmethyl, $(Ph)_3\dot{C}$. Gomberg, in 1900 [27], in an attempt to prepare hexaphenylethane from triphenylmethyl chloride, obtained a colourless solid, whose

[79] N. C. Deno, H. J. Peterson and G. S. Saines, *Chem. Rev.*, **61**, 7 (1961).
[80] C. D. Nenitzescu in Reference 31, vol. 2 (1969), p. 463.
[81] G. J. Karabatsos and J. L. Fry, in Reference 31, vol. 2 (1969), p. 521.

solution in benzene was yellow and showed unexpectedly high reactivity. Oxygen was readily absorbed and an insoluble peroxide **25** separated from the solution. He interpreted these results by considering that the hexaphenylethane formed equilibrates in benzene with the triphenylmethyl radical, the latter undergoing further reaction with oxygen:

$$\begin{array}{ccl} Ph_3C{-}CPh_3 & \rightleftharpoons & 2Ph_3C\cdot \\ & & \downarrow O_2 \\ & & Ph_3C{-}O{-}O{-}CPh_3 \\ & & \qquad \mathbf{25} \end{array} \tag{4.38}$$

It was later proved that the dimerisation product of the triphenylmethyl radical was not hexaphenylethane [82]. Many attempts to prepare this compound have ever since failed [83]. Surprisingly, Gomberg's erroneous formulation (4.38) was accepted in most of the textbooks published during about 70 years following his paper. Nevertheless, Gomberg was correct in the identification of the free radical and in the interpretation of many of its properties.

The interest in free radicals has grown rapidly after it was recognised by Staudinger [84], in 1920, that free radicals are involved in polymerisation. Some years later, Panneth [85] found a method to generate methyl and other simple alkyl radicals in gaseous phase which were unknown at that time (see also p. 548). He determined also their lifetime and found reactions by which they are transformed into stable molecules. In 1937, Kharasch *et al.* [86] demonstrated the radical addition of hydrogen bromide, while Flory [87] established the mechanism of radical polymerisation.

The chemistry of free radicals will be treated in Chapter 11. In this section we shall only examine their structure.

In the methyl radical (**26**) the carbon atom has an sp^2 hybridisation. It will be therefore planar and highly symmetric (D_{3h}). Substitution of hydrogen atoms with alkyl groups will disturb its symmetry. However, due to rapid molecular rotations (see Section 4.3), the symmetry plane will be preserved, free radicals being achiral compounds.

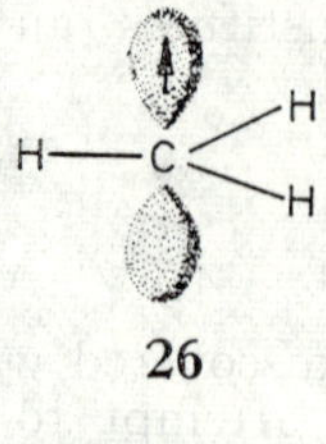

26

[82] H. Lan Kamp, W. T. Nauta and C. McLean, *Tetrahedron Letters*, 249 (1968).
[83] For a review, see W. B. Smith, *J. Chem. Educ.*, **47**, 535 (1974).
[84] H. Staudinger, *Ber. dtsch. chem. Ges.*, **53**, 1073 (1920).
[85] F. A. Panneth and W. Hoffeditz, *Ber. dtsch. chem. Ges.*, **62**, 1335 (1929).
[86] M. S. Kharasch, H. Engelmann and F. R. Mayo, *J. Org. Chem.*, **2**, 288 (1937).
[87] P. J. Flory, *J. Amer. Chem. Soc.*, **59**, 241 (1937).

Stabilisation by hyperconjugation is less important for alkyl radicals than it is for carbenium ions [37]. Thus, the substitution with one methyl group corresponds to an increase of about 5 kcal mol^{-1} in stabilisation energy, the tertiary radical $Me_3C\cdot$ being only 16 kcal mol^{-1} more stable than the methyl radical (compare with the data from Table 4.6). However, p-π conjugation is more effective, the allyl radical, for example, having the same DE as the allyl cation, in agreement with the values determined experimentally [88].

Triphenylmethyl radical has an appreciable stabilisation energy, as has the corresponding triphenylmethyl cation (see p. 154). Thus, it is not surprising that they were the first known electron-deficient species. Like the trityl ion, the triphenylmethyl radical is not planar (**27**), the aromatic rings being twisted with respect to the plane of the sp^2 bonds (propeller structure [89]). The torsional angle τ is high enough ($\tau > 0$) to relieve the steric strain associated with bulky phenyl groups, but is also low enough ($\tau < 90°$) to allow overlap interactions. Considering the geometry **27**, it can be easily seen that two bulky triphenylmethyl radicals cannot approach each other to form hexaphenylethane. They will instead combine in the less sterically demanding way to form the quinonic dimer **28**.

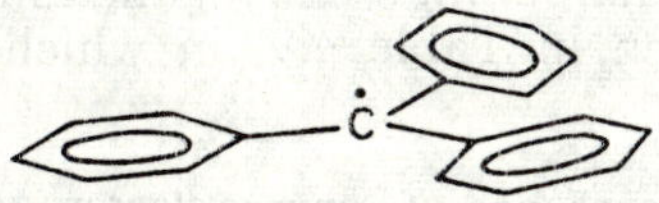

27 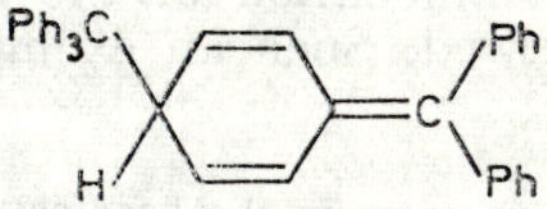 **28**

This dimer structure was proposed by Jacobson [89] shortly after Gomberg's work, but gained no acceptance until recently [82]. It is strange that great personalities, such as Gomberg, sometimes shadow the correct ideas of less renowned workers in the same field.

Free-radical structure is confirmed by modern physical methods such as electron-spin resonance (e.s.r.). This and other methods will be mentioned in the corresponding chapter dealing with free-radical reactions (see Chapter 11).

Free-radicals are formed by homolysis, involving either breaking of a covalent bond or electron-transfer processes. Compared with the corresponding carbocations, they are less regiospecific and stereospecific in their reactions, this feature being used as a decisive argument in the interpretation of organic reactions.

4.2.3 *Carbanions*

Carbanions are ionic species with the negative charge either localised at a carbon atom or delocalised over more atoms, one of which is a carbon atom. Carbanions occur in organometallic compounds and

[88] D. M. Golden, K. W. Egger and S. W. Benson, *J. Amer. Chem. Soc.*, **86**, 5416, 5420 (1964).

[89] For a review, see K. Mislow, D. Gust, P. Finocchiaro and R. J. Boetcher, *Fortschr. chem. Forsch.*, **47**, 1 (1974).

are viewed as ion pairs R^-M^+ even if many such compounds have a significant covalent character of the carbon-metal bond. Despite their frequent occurrence in organic reactions, they received less attention than the corresponding carbocations.

(A) Generation and stability

Carbanions are usually generated in acid-base equilibria (4.39). Both the equilibrium rate constants and the rate constant of proton abstraction (kinetic acidity of hydrocarbon R—H) may indicate the stability of the resulted carbanion [90,91]. As has been already shown (see. p. 137), such a correspondence between the thermodynamic and kinetic acidity indicates a linear relationship between the free energy of reaction ΔG and the free energy of activation, $\Delta G^{\ddagger}$.

$$\text{R—H} + \text{B:}^- \underset{}{\overset{k,\,K_a}{\rightleftharpoons}} \text{R:}^- + \text{B—H} \tag{4.39}$$

Linear relationships between pK_a values and the logarithm of the relative rates of hydrogen abstraction have been currently observed [90–92] for structurally related compounds. However, if the groups directly attached to the anionic carbon have different abilities in the accommodation of the negative charge, high discrepancies may be observed. Such an example is given in Table 4.9, in which the

TABLE 4.9

Relative rates in the base-catalysed H/D exchange of some cyclopropane and cyclopropene derivatives

No.	*Cyclopropene*	*Cyclopropane*	k_S/k_U	*Literature*
1	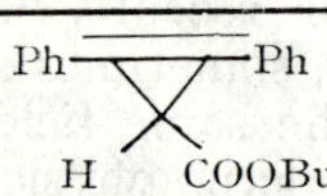 Ph, Ph; H, COOBu^t	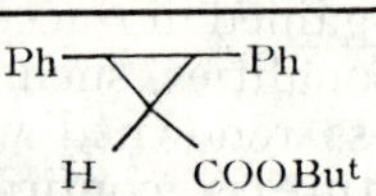 Ph, Ph; H, COOBu^t	300	93,94
2	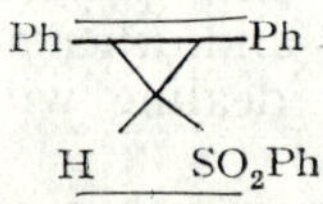 Ph, Ph; H, SO_2Ph	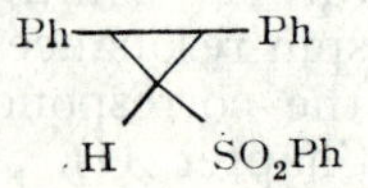 Ph, Ph; H, SO_2Ph	1.000	95
3	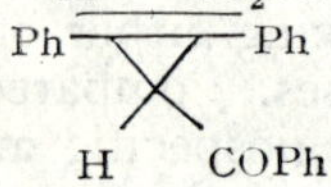 Ph, Ph; H, COPh	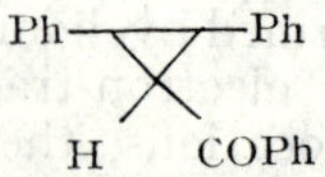 Ph, Ph; H, COPh	6.000	95
4	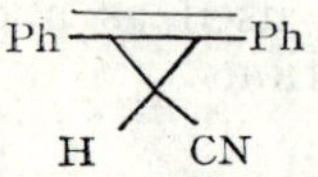 Ph, Ph; H, CN	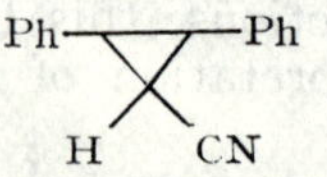 Ph, Ph; H, CN	10.700	96

[90] D. J. Cram, *Fundamentals of Carbanion Chemistry*, Academic Press, New York (1965).

[91] A. Streitwieser, Jr., J. H. Hammons, *Progr. Phys. Org. Chem.*, 3, 41 (1965).

[92] A. Streitwieser, Jr., M. R. Grange, F. Mares and R. A. Wolf, *J. Amer. Chem. Soc.*, **95**, 4257 (1973).

[93] R. Breslow and M. Battiste, *Chem. Ind.*, 1143 (1958).

[94] R. Breslow, *Angew. Chem.*, **80**, 573 (1968).

[95] R. Breslow, J. Brown and J. J. Gajewsky, *J. Amer. Chem. Soc.*, **89**, 4383 (1967).

[96] R. Breslow and M. Douek, *J. Amer. Chem. Soc.*, **90**, 2698 (1968).

kinetic acidity of some cyclopropenes is compared with the kinetic acidity of the corresponding cyclopropanes. It must be remembered (see Table 4.7) that the cyclopropenyl anion does not obey Hückel's rule and should therefore be *antiaromatic*. Large differences in stability between cyclopropenyl and cyclopropyl anions may thus be observed.

The highest relative rate (k_S/k_U) is observed with cyanoderivatives. Even for these compounds only a difference in activation energy of about 5.5 kcal mol^{-1} between the saturated and unsaturated derivative can be calculated. However, this probably does not reflect the whole difference of stability between the two species. Relatively stable cyclopropyl anions have actually been isolated[97], while cyclopropenyl anions could not be demonstrated to be even transient intermediates in organic reactions. Nevertheless, the relative rates in Table 4.9 give a good correlation with the σ^- values of the substituents from position 1 in the corresponding cyclopropanes[98]. This relationship demonstrates that the stability of cyclopropenyl anions is not drastically changed by substituents in the three-membered ring.

pK_a values can be determined as has been described in Chapter 3 (see also Table 3.2). For hydrocarbons with very low acidity, several other methods are available.

In one of them, used by Streitwieser *et al.*[91, 92, 99, 100], the kinetic acidity towards lithium cyclohexylamine (LCHA) is determined, the pK_a values of the corresponding hydrocarbons being appreciated from the corresponding plots of log k *vs.* pK_a. Table 4.10 lists some relative rates of hydrocarbons toward LCHA.

The relative kinetic acidities in Table 4.10 can be used as a criterion for the stabilities of the corresponding anions. Indeed,

TABLE 4.10

Kinetic acidities of hydrocarbons toward lithium cyclohexylamine *

Hydrocarbon	*Relative kinetic acidity*	*Hydrocarbon*	*Relative kinetic acidity*
Higher cycloalkanes	0.5–1.0	Methane	10^4
Cyclohexane	1.00	Cyclopropane	7.0×10^4
Cyclopentane	5.72	Benzene	9.1×10^7
Cyclobutane	28.0		

* Data of Streitwieser *et al.*, see text.

[97] H. M. Walborsky and A. E. Young, *J. Amer. Chem. Soc.*, **86**, 3288 (1964).

[98] F. Badea, personal observation.

[99] A. Streitwieser, Jr., W. B. Holleyhead, G. Sonnichsen, A. H. Pudjaatmaka, C. J. Chank and T. L. Kruger, *J. Amer. Chem. Soc.*, **93**, 5096 (1971).

[100] A. Streitwieser, Jr., see Reference 23, p. 413; *Lecture on Carbanions*, Bucharest, July, 1969.

excepting benzene, the hydrocarbons have similar substituents at the reaction site, so that a linear relationship between ΔG and $\Delta G^{\ddagger}$ can be expected.

Another method was recently proposed by Breslow and his co-workers [101]. The method is based on the fact that the pK_a values for the conversion of a low acidic compound to the corresponding anion can be determined from the equilibrium constant for the conversion of the hydrocarbon to the cation, together with the electrochemical potential for reversible reduction of the cation through radical to anion. From such data, pK_a values of very weak hydrocarbon acids have been calculated. Thus, for trimethyl- and triphenyl-cyclopropene (**18**, R = Me; **18**, R = Ph) the highest pK_a values were found, 62 and 50 ± 4, respectively [102]. These values clearly demonstrate the antiaromatic character of cyclopropenyl anions.

Other methods for the preparation of carbanions will be discussed in Chapter 8.

(B) Classification

According to their structure and stability, carbanions can be classified in the following groups.

(*a*) *Pyramidal carbanions*. This class includes the most simple carbanions, derived from alkanes and cycloalkanes. Their structure is similar to ammonia and amines (Figure 4.4), the most symmetrical compound $CH_3:^-$ having C_{3v} symmetry. If the substituents on the central carbon are different, the molecule is chiral, corresponding to the point group C_1. However, as in amines, simple carbanions may suffer pyramidal inversion (see Figure 4.4), through rehybridi-

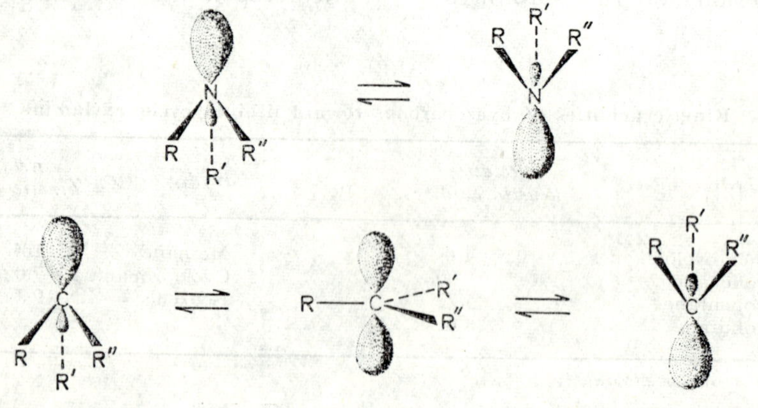

Figure 4.4 Pyramidal inversion of a tertiary amine and of a carbanion.

[101] R. Breslow and K. Balasubramanian, *J. Amer. Chem. Soc.*, **91**, 5183 (1969).

[102] R. Breslow and W. Chu, *J. Amer. Chem. Soc.*, **95**, 411 (1973).

sation of the central atom, which leads to an equilibrium between two enantiomeric forms.

If antiaromatic carbanions are excepted, pyramidal anions can be said to be the least stable anions. Their stability was inferred from kinetic acidities, equilibrium (4.39) being reached very slowly. However, cyclopropanes have hight kinetic acidity (see Table 4.10), as the corresponding anions cannot easily undergo inversion. It is assumed that the acidity of cyclopropane is determined by the partial π-character of the C—C bonds.

(*b*) *Vinyl and alkynyl acetylene carbanions*. The C—H bond involving an sp^2 or an sp carbon is more acidic than the C_{sp^3}—H bond, due to the increasing contribution of the *s* component in the hybridised MO. Thus, acetylene has a pK_a of 25, quite unusual for simple hydrocarbons (see Table 3.2). Vinyl anions are easily obtained and are stable species[103]. Benzene, having the same type of C—H bonds as alkenes, has a high kinetic acidity (see Table 4.10). Even carbenium ions, with sp^2 hybridisation at the central atom, may behave as Brönsted acids (see p. 149).

(*c*) *Carbanions with heteroatom stabilisation*. Substituents at the carbanion centre may readily accommodate the negative charge if the directly bonded atom can use *d* orbitals. Illustrative in that respect are the relative rates of proton abstraction from onium salts[104]:

	relative rate
$Me_4N^+ \longrightarrow Me_3\overset{+}{N}{-}CH_2{:}^-$	1
$Me_4P^+ \longrightarrow Me_3\overset{+}{P}{-}CH_2{:}^- \longleftrightarrow Me_3P{=}CH_2$	2.4×10^6

In the tetramethylammonium salt, the central atom cannot use orbitals, whereas in the corresponding phosphonium salt phosphorus is able to make use of them. The resulted *ylide* will be stabilised by conjugation. Other typical examples are trihalogenomethyl carbanions studied by Hine and his co-workers (see α-Elimination reactions).

(*d*) *Carbanions with p-π conjugation*. The *p* orbital in a carbanion may conjugate with neighbouring π bonds. The most simple examples are allyl and benzyl anions.

$$\underset{\mathbf{29}}{CH_2{=}CH{-}CH_2{:}^- \longleftrightarrow {}^-{:}CH_2{-}CH{=}CH_2}$$

Simple HMO theory predicts the same DE for **29** as for the corresponding cation. It must be however observed that the total π-energy of the cation is $2\alpha + 2.82\beta$, whereas for the anion is $4\alpha + 2.82\,\beta$. The ratio of the coefficients of α and β indicates that the cation is more stable than the anion. This accounts for the low acidities of phenylmethanes (see Table 3.2). Heteroatomic π bonds

[103] E. A. Braude *et al.*, *J. Chem. Soc.*, 2000, 2007, 2012, (1950); 2078, 2085 (1951).

[104] W. von E. Doering and A. K. Hoffman, *J. Amer. Chem. Soc.*, **77**, 521 (1955).

are more efficient in the stabilisation of negative charges; Table 3.2 shows that cyano groups are better than nitro and carbonyl groups in the accommodation of a negative charge.

Carbanions with p-π conjugation must be planar and therefore achiral.

(*e*) *Configuratively stable carbanions.* This class is related to heteroatomic substituted carbanions, but have high barriers of racemisation, retaining thus the configuration of their precursors. Such heteroatomic groups, involving S or P, have been extensively studied by Cram and his co-workers [105] (see also Chapter 8). Groups like RSO_2 confer high acidity to the corresponding *carbon acids*.

(*f*) *Aromatic anions and antiaromatic anions.* Table 4.7 shows that Hückel's rule predicts that cyclic (CH)-annulenes, having a negative charge, such as cyclopentadienyl anion, are aromatic and stable. Conversely, if they do not satisfy this rule ($4n$ electrons), they must be antiaromatic, with a very low stability. This theoretical prediction has been confirmed by many workers [106].

(*g*) *Bicycloaromatic and bicycloantiaromatic anions.* According to Goldstein's theory [62], bicycloanions should be stable if they have $4n$ electrons and antiaromatic if the total π-electrons satisfy the $4n + 2$ rule. Thus, carbanion **30** with 8 π-electrons has been directly observed [107].

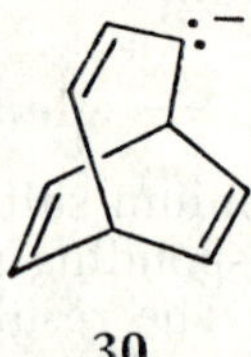

30

(*h*) *Dianions.* Dicarbocations are only rarely observed, due to the high reactivity of the first formed monocation. However, dicarbanions can be generated if the base used is strong enough. Stable dianions, which have acquired interest in organic synthesis [108], are for instance those resulting from β-dicarbonyl compounds. The dimetallation occurs readily with potassium amide in liquid ammonia [109]:

$$PhCOCH_2COMe + 2\ KNH_2 \longrightarrow PhCOCHKCOCH_2K + 2NH_3$$

[105] See, for example, D. J. Cram, R. D. Trepka and P. S. Janiak, *J. Amer. Chem. Soc.*, **88**, 2749 (1966).

[106] See, for example, G. Schröder, *Cyclooctatetraene*, Verlag Chemie, Weinheim (1965); S. W. Staley, G. M. Gramer and A. W. Orvedal, *J. Amer. Chem. Soc.*, **96**, 7433 (1974), and references quoted therein.

[107] J. B. Grutzner and S. Winstein, *J. Amer. Chem. Soc.*, **89**, 6357 (1967).

[108] T. M. Harris and C. M. Harris, *Org. Reactions*, **17**, 155 (1969).

[109] C. R. Hauser and T. M. Harris, *J. Amer. Chem. Soc.*, **80**, 6360 (1958).

A dianion of special interest is that derived from cyclooctatetraene [106]:

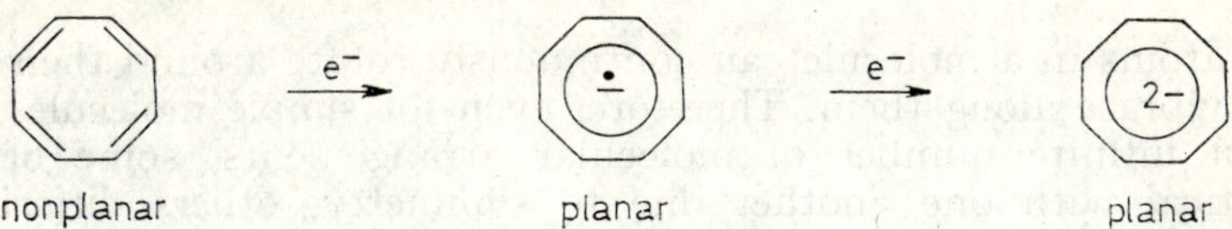

Cyclooctatetraene itself is a highly unsaturated compound with 8 π-electrons. The following energies of the delocalised π orbitals can be calculated by the HMO method:

$$\varepsilon_1 = \alpha + 2\beta; \quad \varepsilon_2 = \varepsilon_3 = \alpha + 1.41\beta; \quad \varepsilon_4 = \varepsilon_5 = \alpha;$$

$$\varepsilon_6 = \varepsilon_7 = \alpha - 1.41\beta; \quad \varepsilon_8 = \alpha - 2\beta$$

The total π-energy is $8\alpha + 9.66\,\beta$, indicating a certain amount of DE for cyclooctatetraene. Nevertheless, the theory predicts an unstable diradical structure. Therefore, this annulene will adopt a non-planar, less strained structure, losing its DE, but gaining in stability. Transfer of two electrons will force the molecule into planarity. The dianion will have the same DE as the parent hydrocarbon, but will differ from it by having a closed shell of $4n + 2$ electrons ($n = 2$).

(*i*) *Radical-anions*. Reactions of electron transfer to π-compounds produce in the first step radical-anions, possessing reactivity characteristic to both reacting centres. Such intermediates will be occasionally encountered and discussed in later chapters.

(C) Reactivity

Carbanions are nucleophilic reagents. Their most characteristic property is the reaction with electrophiles such as carbocations, halogens, carbonyl compounds etc.

Carbanions are also involved in rearrangements (however to a lesser extent than carbocations), like the Favorsky rearrangement (see p. 331), in elimination reactions (see Elimination ElcB), electron transfer reactions, etc. The alkylation of β-dicarbonyl compounds, the condensation reactions of carbonylic compounds and of carboxylic acid derivatives are extensively used in preparative organic chemistry.

4.3 Rotational and rehybridisation modifications

Prior to, or after, the actual bond transformation, organic molecules may suffer modifications of their geometry, with consequences to their reactivity and to the reaction stereochemistry. Such modifications are conformational equilibrations and pyramidal interconversions, which do not involve bond breaking or formation.

4.3.1 *Thermodynamics and kinetics of molecular geometry equilibrations*

The atoms in a molecule can continuously rotate around their bonds and vibrate along them. Therefore, even for simple molecules, there is an infinite number of molecular arrangements, some of them identical with one another due to symmetry, others differing by internal parameters (bond lengths and angles, torsional angles). Actually, only a few molecular arrangements may exist, due to the potential energy barriers separating them. The prediction of such stable arrangements is of utmost importance in appreciating the reactivity of an organic substrate and in understanding the stereochemistry of the reaction.

The modification of an internal molecular parameter is associated with the variation of the potential energy of the molecule. Such variations may be in some instances calculated or determined experimentally, and the change in energy plotted *vs.* the change in the internal parameter. A plot like that represented in Figure 4.5 can

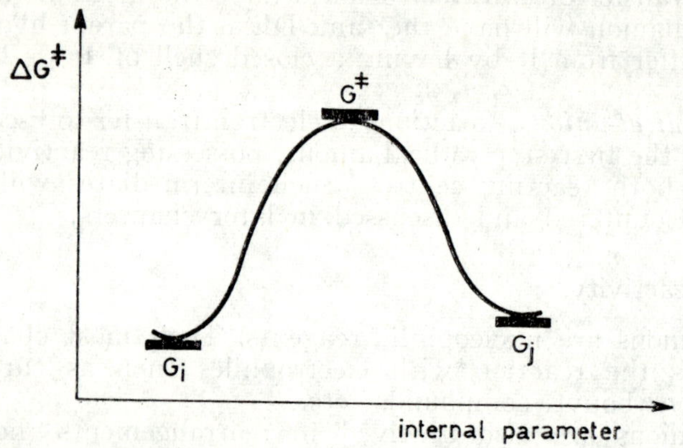

Figure 4.5 Potential energy variation in the modification of a stable molecular geometry (G_i) into another stable geometry (G_j). The coordinate represents the variation of a characteristic internal parameter.

be thus obtained, in which the minima will correspond to the stable molecular arrangements and the maxima to the geometries of highest energy adopted during the change of one stable geometry (G_i) into another stable geometry (G_j). Actually, each point on the plot in diagram 4.5 will correspond to a certain molecular geometry. However, only maxima and minima must be considered in the characterisation of the molecule. Any other point on the diagram is of no interest for the following discussion. Indeed, if we consider the unimolecular equilibration of only two geometries (4.40), the energy difference between the stable geometries G_1 and G_2 will be related to K and thus to the population of the corresponding molecular species:

$$G_1 \underset{k_{-1}}{\overset{k_1, K}{\rightleftharpoons}} G_2 \qquad (4.40)$$

On the other hand, the energy differences between those geometries and the energy of the maximum ($G^{\ddagger}$) will indicate the rates at which G_1 and G_2 are transformed into one another.

Table 4.11 illustrates the relation between the differences in the free energy, the equilibrium constant K and the concentration of the stable isomer calculated for the equilibrium (4.40) at 300 K.

TABLE 4.11

Relationship between standard free-energy difference, equilibrium constant and the concentration of stable isomer in a unimolecular equilibration $G_1 \rightleftarrows G_2$, at 300 K

G^0 kcal mol^{-1}	K^a	% *more stable isomer* [b]
0.0	1.00	50.0
0.1	1.20	54.0
0.2	1.40	58.5
0.4	1.95	66.1
0.8	3.80	79.8
1.0	5.31	84.1
1.5	12.20	93.5
2.0	28.20	96.6
2.3	49.00	98.0
2.7	99.00	99.1
4.0	795.0	99.9
6.0	22400	99.99

a) Calculated from $\Delta G^0 = -2.303\,RT \log K$; b) calculated from $K = G_1/G_2$ and $100 = G_1 + G_2$.

Occasionally, two conformers with identical potential energy do not equilibrate. In those cases, the energy barrier is too high to be surmounted; the molecules are trapped in potential energy wells. An example is encountered in the case of biphenyl derivatives *ortho*-substituted with bulky groups.

Despite the higher DE in planar biphenyls, the bulkiness of the substituents prevents the molecule from adopting the planar conformation (**31***b*); instead, the enantiomeric non-planar arrangements **31***a* and **31***c* will be preferred. The larger the substituents A and B in the aromatic rings, the higher the energy barrier and the slower the equilibration (racemisation) will be.

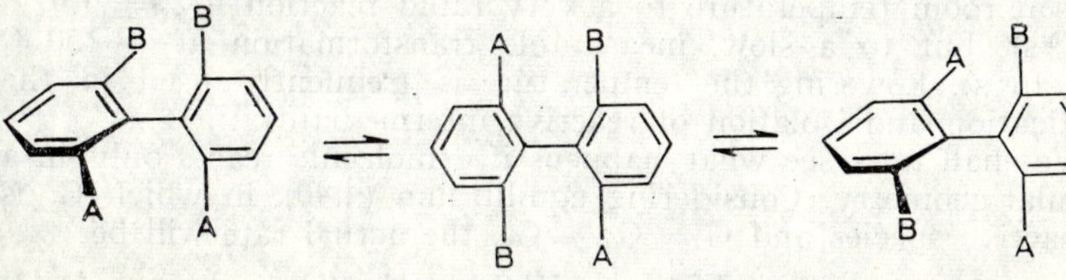

B B A B A B

A A B A B A (4.41)

31*a* **31***b* **31***c*

In order to evaluate the ease with which a particular geometrical isomer can be isolated by chemical methods or identified by physical methods, the correspondence between the free energy of activation ($\Delta G^{\ddagger}$) and the rate constant, together with the corresponding half-time ($t_{1/2}$), are given in Table 4.12.

TABLE 4.12

Relationship between free energy of activation ($\Delta G^{\ddagger}$), rate constant (k) and half-time ($t_{1/2}$) for a unimolecular reaction: $G_1 \longrightarrow G_2$, at room temperature

$\Delta G^{\ddagger}$ kcal mol^{-1}	k s^{-1} [a]	$t_{1/2}$ s [b]
0.3	3.65×10^{12}	1.9×10^{-13}
0.6	2.21×10^{12}	3.1×10^{-13}
1.2	8.15×10^{11}	3.2×10^{-13}
2.0	2.15×10^{11}	3.2×10^{-12}
4.0	7.71×10^{9}	8.9×10^{-11}
10.0	3.51×10^{5}	1.0×10^{-6}
15.0	8.44×10^{1}	8.2×10^{-2}
20.0	2.02×10^{-2}	3.4×10(0.5 min.)
23.0	1.37×10^{-4}	5.0×10^{3}(1.5 h)
24.0	2.58×10^{-5}	2.7×10^{4}(7.5 h)
26.0	7.52×10^{-7}	9.0×10^{5}(10 days)

a) Calculated from: $k = \mathrm{k}(T/h)\, e^{-\Delta G^{\ddagger}/RT}$
b) Calculated from: $t_{1/2} = 0.69/k$

Tables 4.11 and 4.12 permit a rapid estimation of thermodynamic and kinetic characteristics of a unimolecular reaction. Thus, isomers differing from the most stable isomer by more than 3.0 kcal mol^{-1} will only have a transient existence in the equilibrium mixture and can be neglected as constituent elements determining the macroscopic physical properties of the substance. If the activation energy separating two isomers is less than 20—23 kcal mol^{-1}, their interconversion will be too rapid to allow the separation ($t_{1/2} <$ 1 h) or the chemical identification say by titration ($t_{1/2} <$ 0.5 min). Smaller energies of activation ($>$ 10 kcal mol^{-1}) will permit however an identification by n.m.r. spectrometry at room temperature, the relaxation process having a half-time of about 10^{-6} s (see p. 178). For very small energy barriers ($<$ 10 kcal mol^{-1}) even such methods are prohibitive. In these cases the vibrational spectra are useful, the transition to a higher vibrational level being nearly instantaneous.

It must be reminded that equilibrium constants and rate constants are temperature-dependent. Thus, a barrier of 10 kcal mol^{-1} corresponds at room temperature to a very rapid reaction ($t_{1/2} = 1.9 \times$ $\times$ 10^{-6} s), but to a slow, measurable transformation at -150 °C ($t_{1/2} = 10^5$ s). Lowering the temperature is frequently used in the identification and isolation of reactive intermediates.

We shall next see what happens if a molecule reacts only in a particular geometry. Considering equilibrium (4.40), in which G_2 is the reactive species and $G = G_1 + G_2$, the actual rate will be:

$$\text{rate} = k\,[G_2] = Kk/(1 + K)\,[G] \qquad (4.42)$$

The observed rate constant k_{obs} will thus be:

$$k_{obs} = Kk/(1 + K) \tag{4.43}$$

The observed rate constant will not differ significantly from the actual rate constant k of the reactive isomer if K is large (G_2 the stable isomer). However, if the two isomers have about the same stability ($K \approx 1$) or G_1 is more stable ($K < 1$), the actual rate constant will be seriously affected by equilibrium (4.40). Other more complicated situations are discussed elsewhere [110].

Relation (4.43) indicates the importance of the geometry equilibration in the appreciation of relative reactivities of different substrates. Since a particular transition state may resemble the geometry of either the reactant molecules or the reaction products, the geometry of the actual molecular species undergoing the chemical transformation is of utmost importance in studies of reaction mechanism.

4.3.2 *Molecular rotations about single bonds*

The most simple transformations of the molecular geometry are the rotations about molecular axes. For the ethane molecule, the C—C bond is a molecular axis. The torsional angle τ with respect to this internal element can vary between 0 and $2n\pi$. An infinite number of molecular geometries are possible. However, there are only three types differing by their symmetry. If $\tau = 0$, 180° or 240°, the symmetry will be D_{3h}; for $\tau = 60°, 180°$ or 300°, the symmetry is D_{3d}. All other possible geometries have a D_3 symmetry and are chiral. It is now known that the three identical D_{3d} forms are the stable forms (*staggered* ethane), the D_{3h} geometries (*eclipsed* ethane) correspond to the maxima in the energy diagram (see Figure 4.6) and the D_3 structures represent the infinite possible arrangements in passing from one stable conformation to the high-energy conformation. Ethane will therefore be characterised by only two geometries, staggered

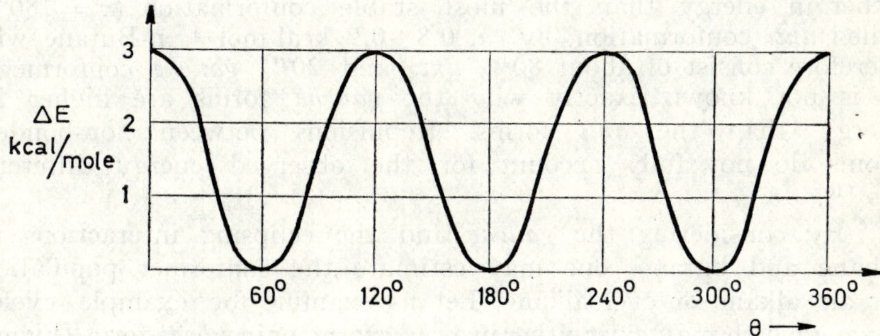

Figure 4.6 Potential energy variation *versus* the torsional angle θ in ethane.

[110] E. L. Eliel, N. L. Allinger, S. J. Angyal and G. A. Morrison, *Conformational Analysis,* Interscience, New York (1965); M. Hanack, *Conformational Theory,* Academic Press, New York (1965).

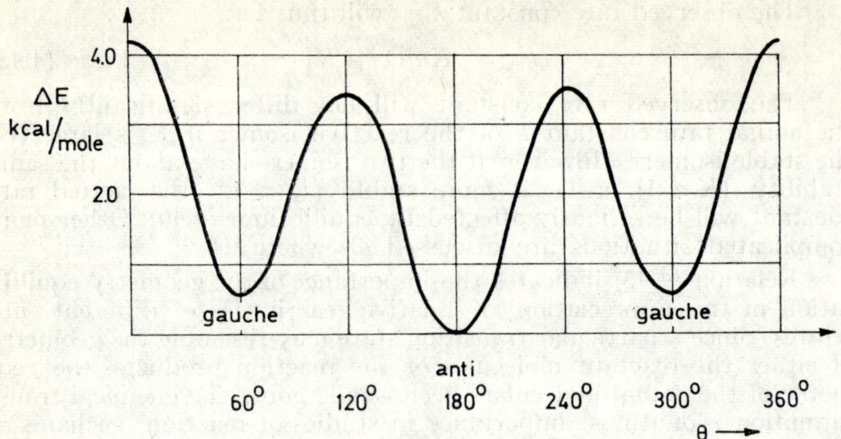

Figure 4.7 Potential energy variation *versus* the torsional angle θ in butane.

ethane and eclipsed ethane. The two molecular species cannot be separated, due to rapid interconversion. Such an interconversion, brought about by a rotation about the molecular axis, has been called *free rotation*. This term is however inadequate, suggesting that no free-energy barrier exists between the possible conformations. Actually, even for ethane this barrier is *ca.* 3 kcal mol^{-1} [110], so that the phenomenon should be more properly referred to as *molecular rotation*. For the butane molecule, if the molecular rotations about C—H bonds are ignored, three stable and three high-energy conformations can be imagined. This has actually been found [110] by i.r. spectroscopy, the potential energy diagram being represented in Figure 4.7. From the three staggered conformations (corresponding to minimum energy in Figure 4.7), two are equivalent in energy and are characterised by a torsional angle about the C(2)—C(3) bond of $\pm 60°$. They are called *gauche* conformations and are richer in energy than the most stable conformation ($\tau = 180°$), called *anti* conformation, by *ca.* 0.8—0.9 kcal mol^{-1}. n-Butane will therefore consist of about 80% *anti* and 20% *gauche* conformers. It is not known exactly why the *gauche* forms are higher in energy than the *anti* forms. Repulsions between non-bonded atoms do not fully account for the observed energy differences [110].

By considering the *gauche* and the eclipsing interactions in ethane and butane, one may estimate the conformer population for an alkane or cycloalkane. Let us examine for example cyclohexane which may exist either in a *chair* form, or in a *boat* form (Figure 4.8). The chair form has six *gauche* interactions between C—C bonds and will be therefore higher in energy by 6×0.8 kcal mol^{-1} than the completely staggered hexane molecule (zig-zag conformation). The boat form has four *gauche* interactions ($4 \times 0.8 = 3.2$ kcal mol^{-1})

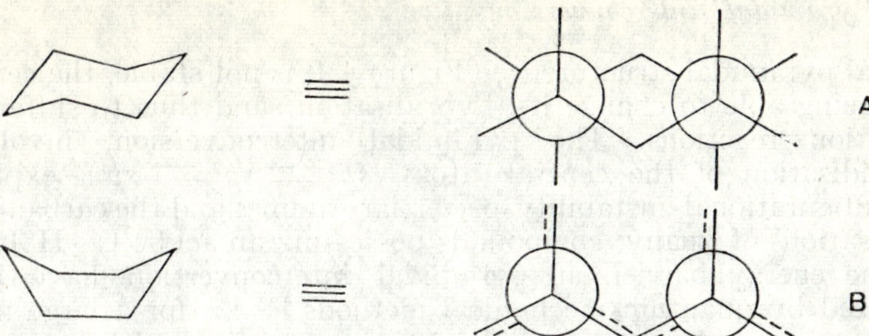

Figure 4.8 Conformations of cyclohexane: A. chair; B. boat.

and two eclipsed interactions ($2 \times 4.2 = 8.4$ kcal mol^{-1}), which means that it will be 6.8 kcal mol^{-1} less stable than the chair conformation. This simple evaluation using the data of Figure 4.7 for butane molecules is in perfect agreement with experimental data [111] and theoretical calculations [112].

Calculation of conformational equilibration was first successfully achieved by Westheimer [113, 114]. Considering that the energy difference between two conformations is given by the modifications of the equilibrium bond length, the valence angle, the torsional angle and the interactions between non-bonded atoms, Westheimer was able to estimate correctly the activation energies in the racemisation of some biphenyl derivatives. Using more sophisticated methods, Hendrickson [111, 115–117] has calculated the stable conformations of some cycloalkanes. Pople and his co-workers have determined the stable geometries of ethane derivatives [118–120]. Such calculations need high-speed computers; in order to obtain reliable data in an economical way (short times in computers), some basic assumptions must be made. These assumptions usually refer to the equilibrium bond lengths and bond angles [112–120], the variable being the torsional angle τ. Such calculations are in perfect agreement with an empirical evaluation considering the n-butane molecule or cyclohexane derivatives [121].

[111] F. A. L. Anet and A. J. Brown, *J. Amer. Chem. Soc.*, **89**, 760 (1967); F.A.L. Anet and his co-workers; *J. Amer. Chem. Soc.*, **97**, 3244 (1975).

[112] J. B. Hendrickson, *J. Amer. Chem. Soc.*, **83**, 4357 (1961).

[113] F. H. Westheimer, *Steric Effects in Organic Chemistry*, (M. S. Newman, editor), J. Wiley, New York (1956), p. 523.

[114] J. Fournier, *Bull. Soc. chim. France*, 1954 (1973).

[115] J. B. Hendrickson, *J. Amer. Chem. Soc.*, **86**, 4854 (1964).

[116] J. B. Hendrickson, *J. Amer. Chem. Soc.*, **89**, 7036, 7043, 7047 (1967).

[117] J. B. Hendrickson, R. B. Boeckman, Jr., J. D. Glidson and E. Grunwald, *J. Amer. Chem. Soc.*, **95**, 494 (1973).

[118] L. Radon and J. A. Pople, *J. Amer. Chem. Soc.*, **92**, 4786 (1970).

[119] L. A Radon, W. A. Lathan, W. J. Hehre and J. A. Pople, *J. Amer. Chem. Soc.*, **95**, 693, 699 (1973).

[120] See also W. G. Dauben und K. S. Pitzer, in *Steric Effects in Organic Chemistry* (M. S. Newman, editor), J. Wiley, New York (1956), p. 53.

[121] E. L. Eliel, *Angew. Chem.*, **77**, 784 (1965).

4.3.3 *Pyramidal interconversions*

A chiral pyramidal structure (see Figure 4.4) is not stable, the central atom being able to change its hybridisation, and thus to suffer racemisation reactions. The pyramidal interconversion, involving rehybridisation of the central atom ($sp^3 \rightleftharpoons sp^2 \rightleftharpoons sp^3$) explains the configurational instability of tertiary amines and the carbanionic racemisation of many compounds possessing an acidic C—H bond.

The energy barriers in pyramidal interconversion have been calculated by quantum mechanical methods [122, 123] for a large number of pyramidal structures. For tertiary amines the activation energy is *ca.* 6—7 kcal mol^{-1}, indicating a very rapid reaction (see Table 4.12), not measurable by usual methods. In order to make it observable at room temperature, Saunders and Yamade [124] have devised a very ingenious method. They recorded the n.m.r. spectrum of an aqueous hydrochloric solution of dibenzylmethylamine **33**. Actually, they viewed the proton directly bonded to the nitrogen atom in the ammonium salt **32**, in the two possible positions, due to the slow equilibration rate of the ammonium salt with the free amine,

(4.44)

PhH_2C, Me, PhH_2C $\overset{+}{N}$—H $\rightleftharpoons$ R_3N: $\rightleftharpoons$ R^1, R^1, R^2 N $\rightleftharpoons$:NR_3 $\rightleftharpoons$ H—$\overset{+}{N}$ CH_2Ph, Me, CH_2Ph

32*a* **33*a*** **34** **33*b*** **32*b***

in acid solution. From their measurements, an activation energy of 11 kcal mol^{-1} for the pyramidal interconversion of the free amine can be calculated.

The new techniques used now in n.m.r. spectrometry allow the direct determination of this energy barrier at low temperatures. Thus, Dewar and Jennings [125] have found at −135 °C an energy barrier of 6.7 kcal mol^{-1}, in perfect agreement with theoretical estimations. It is assumed that the higher value obtained by the earlier workers is determined by the solvation of the nitrogen atom by hydrogen bonds in aqueous solutions. However, the data of Saunders and Yamade are more pertinent to the racemisation of carbon acids R—H by carbanionic mechanism (see p. 323). The energy barrier for carbanion interconversion (see Figure 4.4) has been calculated to be *ca.* 6—12 kcal mol^{-1} [122], much lower than the values found experimentally [126, 127].

[122] G. W. Koeppl, D. S. Sagatys, G.S. Krisnamurthy and S. I. Miller, *J. Amer. Chem. Soc.*, **89**, 3396 (1967).

[123] M. J. S. Dewar and M. Shanshal, *J. Amer. Chem. Soc.*, **91**, 3654 (1969).

[124] M. Saunders and F. Yamade, *J. Amer. Chem. Soc.*, **85**, 1882 (1963).

[125] M. J. S. Dewar and W. B. Jennings, *J. Amer. Chem. Soc.*, **93**, 401 (1971).

[126] G. M. Whitesides, M. Witowsky and J. D. Roberts, *J. Amer. Chem. Soc.*, **87**, 2854 (1965); **88**, 737 (1966).

[127] G. Fraenkel and D. T. Dix, *J. Amer. Chem. Soc.*, **88**, 979 (1966).

Considering that in such measurements organometallic compounds are used (existing as ion pairs $R:^{-}M^{+}$), the agreement can be regarded as satisfactory (compare also with the pyramidal inversion of amine **32** in water solution). Anyway, such data clearly indicate the configurational instability of pyramidal structures in amines and carbanions.

Stable configurations of pyramidal structures are encountered when the rehybridisation of the central atom is prevented by steric and electronic effects. Thus, cyclopropyl anions are configurationally stable [128]. The rehybridisation of the carbanionic atom is opposed by the ring strain produced by the increase of the normal sp^3 angle (109°) to the sp^2 angle (120°). Aziridines **35** have energy barriers of about 11—12 kcal mol^{-1} [123] if the substituent R is H or alkyl, but higher than 23 kcal mol^{-1} if R is halogen, in perfect agreement with experimental data [129]. Steric and electronic effects are responsible for such behaviour.

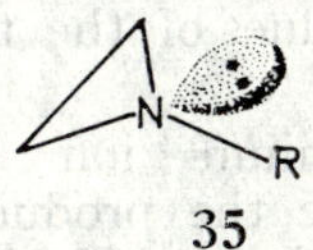

35

Pyramidal interconversions are encountered in the unimolecular electrophilic substitution at the saturated carbon atom (see S_E1 reactions). The stereochemistry of these reactions can be well understood if the phenomena discussed in this section are known.

4.4 Problems

(1) Using the data of Table 4.2 and Table 4.3 show that in DMF the order of reactivity of the halide ions should be $I^- < Br^- < Cl^-$.

(2) Using the same data as above, determine if the acetate ion is more or is less nucleophilic than the chloride ion in DMF.

(3) Find an explanation for the correlation of kinetic data in the Hammett equation:

$$\log k_X/k_H = \rho\sigma_X$$

considering that σ_X values have been determined from thermodynamic measurements. Applying Hammond's postulate, determine in what kind of experiments such equations cannot be observed.

(4) Imagine the elementary steps in the homolytic reaction (4.22) mentioned on p.147.

[128] H. M. Walborsky and A. E. Young, *J. Amer. Chem. Soc.*, **86**, 3288 (1964).
[129] S. J. Brois, *J. Amer. Chem. Soc.*, **90**, 506, 507 (1968).

(5) The allyl cation has the total π-energy $2\alpha + 2.82\beta$, whereas for the benzyl cation this energy is $6\alpha + 8.72\beta$. Show that the allyl cation should be more stable than the benzyl cation.

(6) Hückel's rule predicts that the dication A is aromatic.

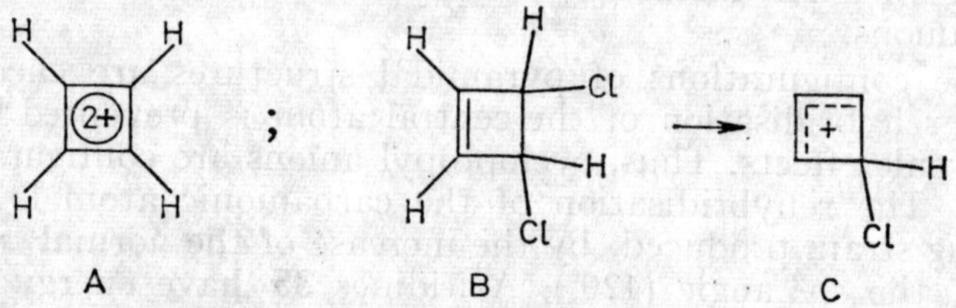

A ; B → C

How do you explain the fact that the dichloride B with Lewis acid catalysts gives as the single observable ion the allylic cation C?

(7) Decide which cationic species derived from the $(CH)_n$-annulene should be aromatic.

(8) Explain the pK_{R^+} values of the tropylium ions listed in Table 4.8.

(9) Both triphenylcarbenium ion and triphenylcarbanion can be obtained. What will be the product of their combination?

(10) Comparing the pK_a values of Table 3.2 with the structures of cyclopropane derivatives from Table 4.9, find a plausible interpretation for the k_S/k_U ratios given in the latter table.

(11) Cram and co-workers have found that in the decarboxylation reaction:

$$R^1R^2R^3C{-}COO^-M^+ \longrightarrow R^1R^2R^3C{:}^-CO_2M^+ \xrightarrow{SH} R^1R^2R^3CH$$

the carbon acid $R^1R^2R^3CH$ is completely racemised if $M = Na$, K, but preserves some optical activity of the initial salt if $M = NH_4$. How do you explain that result?

(12) Give examples of dianions with possible aromatic character.

(13) Considering the stable conformations of ethane and n-butane, show that:

(a) cyclopentane has a ring strain (Pitzer strain), as compared to cyclohexane;

(b) methylcyclohexane is more stable in the equatorial conformation D than in the axial conformation E;

(c) using the n-butane model, estimate the free-energy difference between the axial and equatorial methylcyclohexane.

D E

(14) In the equilibrium reaction:

$$\mathrm{A} \underset{k_\mathrm{B}}{\overset{k_\mathrm{A}, K}{\rightleftarrows}} \mathrm{B}$$

k_A is 1×10^{-6} and $k_\mathrm{B} = 1 \times 10^{-2}\ \mathrm{s}^{-1}$. Represent in an energy diagram the characteristic points in this transformation. If B is further transformed into a reaction product C with $k_\mathrm{C} = 1 \times 10^{-3}\ \mathrm{s}^{-1}$, calculate the observed rate constant if A is introduced as a reagent.

PART 3
Heterolytic reactions

(A) SUBSTITUTION REACTIONS (ALIPHATIC AND AROMATIC)

The simplest heterolytic reaction may be represented as:

$$A{-}B \rightarrow A^{+} + B{:}^{-} \qquad (1)$$

Such a process is involved both in the unimolecular nucleophilic substitution (2) and the in unimolecular electrophilic substitution (3):

$$R{-}X \rightarrow R^{+} + X^{-} \qquad (2)$$

$$R{-}Y \rightarrow R{:}^{-} + Y^{+} \qquad (3)$$

Both mechanisms are encountered very frequently in aliphatic substitutions at a tetrahedral carbon atom and less frequently found in displacement reactions at a trigonal carbon atom (aromatic or olefinic carbon atom). For this reason, aliphatic substitution will be treated first. There are also other reasons for this approach to the subject. Indeed,

(a) one of the most discussed mechanisms of nucleophilic substitution, involving participation of neighbouring groups in ionisation, is related to the bimolecular electrophilic substitution;

(b) the intermediates of some nucleophilic substitutions (nonclassical ions) are closely related to the intermediates (transition states) of the bimolecular electrophilic substitution. The unusual structural and stereochemical features are identical in both cases;

(c) the intermediates of the aromatic substitution, either electrophilic or nucleophilic, have a similar structure, differing only in the number of electrons in the delocalised system. The formation of such intermediates, as well as their further reactions, raise similar problems, reflected in the methods used for their study.

This classification is close to that adopted in Ingold's treatise (*Mechanism and Structure in Organic Chemistry*, 2nd ed., Cornell University Press, New York (1969)) and is not essentially different from that found in other books devoted to organic mechanisms (*e. g.* R. W. Alder, R. Baker, and J. M. Brown, *Mechanism in Organic Chemistry*, Wiley-Interscience, New York (1971)).

5
Aliphatic Nucleophilic Substitution

(A) METHODS FOR THE INVESTIGATION OF THE MECHANISM

5.1 Survey of reactions

In any nucleophilic substitution, a nucleophile displaces a *leaving group*, X, from a substrate R—X, X leaving the molecule with its bonding electrons:

$$\mathrm{Nu{:}^- + R{-}X \longrightarrow Nu{-}R + X{:}^-} \qquad (5.1)$$

The negative signs in equation (5.1) merely show that the charge of the reagent (nucleophile) decreases with a negative unity, whereas the initial charge of the leaving group increases with a negative unity during the reaction. Evidently, Nu:⁻ may be negative or neutral, whereas X in the substrate R—X may be neutral or positive. Table 5.1 presents some examples of the nucleophilic reactions encountered more frequently.

TABLE 5.1
Nucleophilic substitution

Type of substitution	*Examples*	
(1) Negative reagent, neutral leaving group:	$\mathrm{I^- + R{-}Br \rightarrow I{-}R + Br^-}$ Finkelstein reaction	(a)
$\mathrm{Nu{:}^- + R{-}X \rightarrow Nu{-}R + X{:}^-}$ (5.1)	$\mathrm{CN^- + R{-}X \rightarrow CN{-}R + X^-}$	(b)
	$\mathrm{R^2O^- + R^1{-}X \rightarrow R^2O{-}R^1 + X^-}$ ($\mathrm{R^1}$ = alkyl, aryl, H).	(c)
	$\mathrm{R^2S^- + R^1{-}X \rightarrow R^2S{-}R^1 + X^-}$	(d)
	$\mathrm{R^2_2N^- + R^1{-}X \rightarrow R^2_2N{-}R^1 + X^-}$	(e)
	$\mathrm{SO_3^{2-} + R{-}X \rightarrow R{-}SO_3^- + X^-}$	(f)
(2) Neutral reagent, neutral leaving group:	$\mathrm{H_2\ddot{O}{:} + R{-}X \rightarrow R{-}\overset{+}{O}H(H) \rightarrow R{-}OH + H^+}$	(a)
$\mathrm{Nu{:} + R{-}X \rightarrow \overset{+}{N}u{-}R + X{:}^-}$ (5.2)	$\mathrm{\ddot{N}H_3 + R{-}X \rightarrow R{-}\overset{+}{N}H_2(H) \rightarrow R{-}NH_2 + H^+}$	(b)
	$\mathrm{R^2\ddot{O}(R^3){:} + R^1{-}X \rightarrow R^2\overset{+}{O}(R^3)R^1X^- \rightarrow R^2OR^1 + R^3X}$	(c)
	$\mathrm{R^2_3N{:} + R^1{-}X \rightarrow R^2_3\overset{+}{N}{-}R^1\,X^-}$ Menshutkin reaction	(d)

TABLE 5.1 (continued)

Type of substitution	Examples	
(3) Negative reagent, positive leaving group:	$X:^- + R^1{-}\overset{+}{N}R^2_3 \rightarrow X{-}R^1 + :NR^2_3$	(a)
$Nu:^- + R{-}X^+ \rightarrow Nu{-}R + X$ (5.3)	$X:^- + R^1{-}\overset{+}{S}R^2_2 \rightarrow X{-}R^1 + :SR^2_2$	(b)
	$X:^- + R^1{-}\overset{+}{X}R^2 \rightarrow X{-}R^1 + :XR^2$	(c)
	$X:^- + R^1{-}N_2^+ \rightarrow X{-}R^1 + N_2$	(d)
(4) Neutral reagent, positive leaving group:	$R^2OH + R^1{-}N_2^+ \rightarrow R^2\overset{+}{O}(H){-}R^1 + N_2$ Deamination reactions ($R^2 = H$, alkyl, acyl)	(a)
$Nu: + R{-}X^+ \rightarrow Nu^+{-}R + X:$ (5.4)	$R^2OH + R^1{-}OCO^+ (R^1OOCCl + Ag^+) \rightarrow$ $\rightarrow R^2{-}\overset{+}{O}(H){-}R^1 + CO_2$ Production of unstable cations, etc.	(b)

This classification of nucleophilic displacement reactions was first discussed by Ingold [1] in connection with solvent effects in the substitution. It is still useful in the systematisation of the common reactions proceeding by nucleophilic attack at a saturated carbon atom.

Soon after the appearance of the electronic theory, numerous attempts have been made to explain the mechanism of the nucleophilic substitution *. The full merit in the development of a consistent theory of the aliphatic substitution is due to the English School of Hughes and Ingold [1b], who at the end of the 1930s [2] had established the two main mechanisms. Later development of the subject [3], and the passionate discussions and disputes on different topics related to details of the nucleophilic substitution, have clearly originated in this period of pioneering studies of organic mechanism.

[1] C. K. Ingold, *Structure and Mechanism in Organic Chemistry*, Cornell University Press, New York (1953); 2nd ed. (1969). *(a)* p. 346 (458); *(b)* p. 309 (422).

* A historical outlook of the earlier works of Hughes, Ingold *et al.* is found in Reference 1*(b)*. The subsequent main contributions of American workers have been reviewed by Streitwieser [3].

[2] L. C. Bateman, M. G. Church, E. D. Hughes, C. K. Ingold, N. A. Taher, *J. Chem. Soc.*, 979 (1940).

[3] A. Streitwieser, Jr., *Solvolytic Displacement Reactions*, McGraw-Hill, New York (1962); *Chem. Rev.*, **56**, 571 (1956).

In one of the mechanisms of the nucleophilic substitution, bonds are formed and broken synchronously, the energy necessary to break the old bond being partly supplied by the energy released in the formation of the new bond. This mechanism has been designated as *bimolecular mechanism* and labelled S_N2 (substitution, nucleophilic, bimolecular). Clearly, in the transition state **1** of the reaction, two molecules are involved:

$$\mathrm{Nu}\!:^- + \mathrm{R{-}X} \rightarrow [\overset{\delta^-}{\mathrm{Nu}} \cdots \mathrm{R} \cdots \overset{\delta^-}{\mathrm{X}}] \rightarrow \mathrm{Nu{-}R} + \mathrm{X}\!:^- \qquad (5.5)$$

$$\mathbf{1}$$

If the reagent is negative, the charge will be dispersed in the transition state. The partial negative charge on X depends on the degree of bond formation and bond rupture in the transition state, that is on the *early* or *late* character of **1**. Lately, this type of mechanism was also designated as *reagent-assisted substitution* (as opposed to neighbouring group- and solvent-assisted substitutions, see later).

The other mechanism demonstrated by Ingold, Hughes and their school[3] is the *unimolecular mechanism*, labelled S_N1 (substitution, nucleophilic, unimolecular). In the rate-determining step of such a reaction only the substrate molecule is involved. A carbenium ion is formed slowly, which reacts then rapidly either with the nucleophile or with the nucleophilic solvent, HS: :

$$\mathrm{R{-}X} \xrightarrow{\text{slow}} \mathrm{R^+} + \mathrm{X^-} \qquad (5.6)$$

$$\mathrm{R^+} + \mathrm{Nu}\!:^- \text{ (or HS:)} \xrightarrow{\text{rapid}} \mathrm{R{-}Nu} \text{ (or } \mathrm{R{-}\overset{+}{S}H} \rightarrow \mathrm{R{-}S} + \mathrm{H^+}) \qquad (5.7)$$

The more stable the carbenium ion R^+, the higher is the tendency of the substrate RX to follow the unimolecular pathway.

At first sight it would seem that the unimolecular mechanism is highly unfavoured with respect to the bimolecular mechanism, as there is no external assistance for the breaking of the bond (5.6), or no compensation as in the bimolecular mechanism. Indeed, a reaction diagram (Figure 5.1) shows that in the case of the same substrate the activation enthalpy would favour the S_N2 path.

However, what determines the actual course of a particular reaction is not its activation enthalpy (related to the potential energy), but its activation free-energy, $\Delta G^{\neq}$, which takes also into account the activation entropy. In a bimolecular reaction, the activation entropy must be highly unfavourable, since two molecular species, which are initially very distant, are brought close together in the transition state. Obviously, this unfavourable situation does not arise in the unimolecular reaction. However, in the transition state **2** of the heterolysis, charge separation occurs:

$$\mathrm{R{-}X} \rightarrow [\overset{\delta_+}{\mathrm{R}} \cdots\cdot \overset{\delta_-}{\mathrm{X}}] \rightarrow \mathrm{R^+} + \mathrm{X^-}$$

$$\mathbf{2}$$

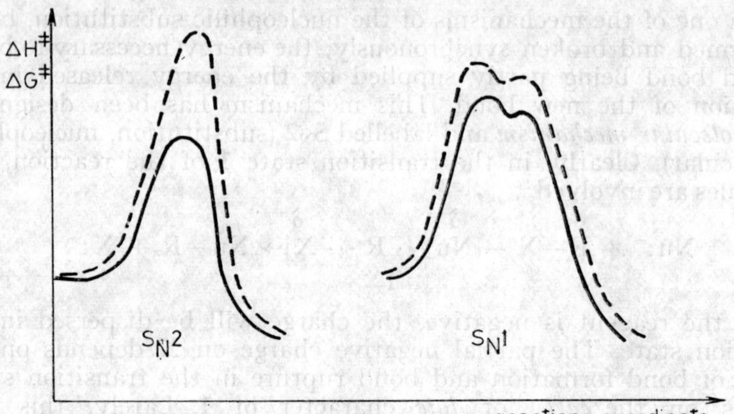

Figure 5.1 Potential energy diagram for nucleophilic substitution: Nu: + R−X→ →Nu−R + X: A. in S_N2 mechanism; B. in S_N1 mechanism.
———— Activation enthalpy
------------ Activation free-energy

Solvent molecules must be involved either in facilitating or in opposing the reaction. Solvent effects in the unimolecular mechanism have been recognised [4] early. However, even if aware of the overwhelming role of the solvent in enabling the S_N1 mechanism, neither the earlier works [3,4], nor later works[5] have clarified the factors (internal and external) which may assist a unimolecular ionisation. The end of a research period is thus marked by new orientations of the investigations:

(a) Bartlett and Nebell [6] observed that in the solvolysis of *p*-methoxybenzhydryl bromide in nitrobenzene with measurable variations in the methanol added, methanol entered the rate equation:

$$\text{rate} = k_1\,[\text{RBr}] + k_2[\text{RBr}]\,[\text{MeOH}] \tag{5.8}$$

It was also discovered that Wagner-Meerwein rearrangements may proceed in acid catalysis [7]. These two observations were the first indications about the role of the reaction medium in ionisation reactions. The subject was later systematically explored by Swain and his co-workers [8], who studied the solvolysis of triphenylmethyl chloride in benzene with methanol. The authors found third-order kinetics:

$$\text{rate} = k_3\,[\text{RCl}]\,[\text{MeOH}]^2 \tag{5.9}$$

If phenol is also added to the reaction medium, it substitutes one methanol molecule in the rate equation (5.9), the rate increasing

[4] E. D. Hughes and C. K. Ingold, *J. Chem. Soc.*, 252 (1935).
[5] K. A. Cooper, M. L. Dhar, E. D. Hughes, C. K. Ingold, B. J. McNulty and L. I. Wolf, *J. Chem. Soc.*, 2043 (1948).
[6] P. D. Bartlett and R. W. Nebel, *J. Amer. Chem. Soc.*, **62**, 1345 (1940).
[7] P. D. Bartlett and I. Pöckel, *J. Amer. Chem. Soc.*, **59**, 820 (1937).
[8] C. G. Swain and R. W. Eddy, *J. Amer. Chem. Soc.*, **70**, 2989 (1948).

subsequently. Increasing concentrations of phenol do not eliminate completely the methanol from the rate equation, which leads to the conclusion that the solvent may have a double effect in ionisation. These effects, called by Swain the *push-pull mechanism,* were later confirmed by other similar studies [9], even if the third-order kinetics in the particular case of triphenylmethyl chloride was criticised by Ingold and his co-workers [10]. By labelling the oxygen in methanol, Swain and Pegues [11] confirmed the third-order kinetics and the double role of the protic solvent in S_N1 reactions. The solvent is thus involved in solvation by hydrogen bonds of the leaving group, in the transition state of the ionisation (3), as well as in a backside solvation of the incipient carbocation:

$$\mathrm{HS{:}} \cdots\cdots \overset{\delta+}{\mathrm{R}} \cdots\cdots \overset{\delta-}{\mathrm{X}} \cdots\cdots \mathrm{HS}$$

3

Thus, *solvent assistance* in ionisation consists in a nucleophilic (rear-attack) and an electrophilic catalysis. The nucleophilic catalysis in the ionisation, comparable to a nucleophilic attack on the substrate, shows that there is no clear-cut difference between the two mechanisms, S_N2 and S_N1, but rather a continuous change. The solvation of carbocations was stressed by Doering and Zeiss [12], but was not neglected either in the earlier works of Ingold, Hughes and their co-workers [13, 14].

About the time when Bartlett produced the first evidences regarding medium effects in ionisation, the first steps in the foundation of the neighbouring-group participation theory were undertaken. The story began with Roberts and Kimball's proposal în 1937 [15] of the first bridged ions, the halogenium ions, which had to account for the stereochemistry of the addition reactions. This was shortly followed by the proposal of Winstein and Lucas [16] regarding the occurrence of halogenium ions in substitution reactions, such as (5.10):

$$CH_3-\underset{|}{\overset{\overset{\displaystyle Br}{|}}{C}}H-\underset{\underset{\displaystyle OH}{|}}{C}H-CH_3 + HBr \xrightarrow[-H_2O]{} CH_3-\overset{\overset{\displaystyle Br}{\diagup + \diagdown}}{CH-CH}-CH_3 \xrightarrow{Br^-} \text{dibromide} \qquad (5.10)$$

Neighbouring-group participation in ionisation was later shown to be a general phenomenon, reflected both in the kinetics and in

[9] E. D. Hughes, C. K. Ingold, S. F. Mok and Y. Pocker, *J. Chem. Soc.*, 1238 (1957).
[10] E. Gelles, E. D. Hughes and C. K. Ingold, *J. Chem. Soc.*, 2918 (1954).
[11] C. G. Swain and E. E. Pegues, *J. Amer. Chem. Soc.*, **80**, 812 (1958).
[12] W. von E. Doering and H. H. Zeiss, *J. Amer. Chem. Soc.*, **75**, 4733 (1953).
[13] E. D. Hughes, C. K. Ingold *et al.*, *J. Chem. Soc.*, 236 (1935); 225 (1936).
[14] M. L. Bird, E. D. Hughes and C. K. Ingold, *J. Chem. Soc.*, 634 (1954).
[15] I. Roberts and G. E. Kimball, *J. Amer. Chem. Soc.*, **59**, 947 (1937).
[16] S. Winstein and H. J. Lucas, *J. Amer. Chem. Soc.*, **61**, 1576, 2845 (1939).

the stereochemistry of the reaction. The effect upon the reaction rates is due to the direct involvement of the neighbouring group at the reaction site.

$$\begin{array}{c}Z \\ | \\ >C-(C)_n-C< \\ | \\ X\end{array} \longrightarrow \begin{array}{c}\overset{+}{Z} \\ >C-(C)_n-C< \\ X^- \\ \mathbf{4}\end{array} \qquad (5.11)$$

Ionisation (5.11) formally corresponds to a unimolecular ionisation. However, it is more related to the bimolecular process, the neighbouring group Z displacing intramolecularly the leaving group X. This internal assistance in ionisation was called by Winstein *anchimeric assistance* [17].

Solvent and neighbouring-group assistance in ionisation were the major contributions to the field of nucleophilic substitution during post-war years. The bridged ions **4** assumed by the theory of neighbouring-group participation have produced the most feverish dispute of our times in organic chemistry.

In the study of a nucleophilic substitution, three types of arguments are most significant: the kinetic, stereochemical and product criteria. The most convincing of the three, when deciding between the possible mechanisms, are the kinetic data. This is the reason why in our treatment we shall start with the kinetic criterion, even if any solution of a chemical puzzle starts with product identification. Another reason for this reversal of the usual order of steps in chemical research is that products are formed in *later stages of the reaction,* while the differentiation in mechanisms begins in the rate-determining step.

5.2 The kinetic criterion

The correct reaction mechanism in a nucleophilic substitution is ascertained mostly on the basis of reaction kinetics. However, a complete kinetic study is not confined to establishing the rate equation, but must be extended to the study of the internal and external factors which may affect the rates.

5.2.1 *Rate equations*

The rate equations, with only a few exceptions, rarely give precise indications about the reaction mechanism in nucleophilic substitution. If the changes in concentration of both reactants are measur-

[17] S. Winstein, C. R. Lindegren, H. Marshall and L. L. Ingraham, *J. Amer. Chem. Soc.*, **75**, 147 (1953).

able, second-order kinetics should be expected for the bimolecular reaction:

$$-\mathrm{d}[RX]/\mathrm{d}t = k_2\,[RX]\,[Nu:] \tag{5.12}$$

and a first-order rate equation for the unimolecular reaction:

$$-\mathrm{d}[RX]/\mathrm{d}t = k_1\,[RX] \tag{5.13}$$

Even in such cases, an equation of type (5.12) is not limited only to S_N2 reactions, while a rate equation (5.13) may very well correspond to a bimolecular process. Indeed, if the reaction mechanism is given by equations (5.14) — (5.15), the reaction intermediates being tight ion-pairs:

$$R{-}X \longrightarrow R^+X^- \tag{5.14}$$

$$R^+X^- + Nu{:}^- \xrightarrow{\text{slow}} NuR + X{:}^- \tag{5.15}$$

the rate equation is best expressed by equation (5.12), even if the mechanism involves an ionisation.

In a solvolysis, the change in the concentration of solvent is not detectable and can be thus considered constant:

$$-\mathrm{d}[RX]/\mathrm{d}t = k_2\,[RX]\,[SH] = k'\,[RX] \tag{5.16}$$

Even if a bimolecular solvolysis were to occur, first-order kinetics is observed.

Acidolyses are other examples in which the simple determination of the rate equation cannot solve the question of the mechanism. Indeed, considering for example the transformation:

$$R{-}OH + HBr \longrightarrow R{-}Br + H_2O$$

the concentration of the reagent, HBr, may enter the rate equation both in the bimolecular reaction:

$$ROH + HBr \longrightarrow R{-}\overset{+}{O}H(H) + Br^-$$

$$Br^- + R{-}\overset{+}{O}H(H){-}[\overset{\delta-}{Br}\cdots R\cdots\overset{\delta+}{O}H_2] \rightarrow Br{-}R + H_2O$$

and in the unimolecular reaction:

$$R{-}\overset{+}{O}H(H) \xrightarrow[-H_2O]{\text{slow}} R^+ \xrightarrow[+Br^-]{\text{rapid}} R{-}Br$$

Both mechanisms are possible, the oxonium salt $R{-}\overset{+}{O}H_2$ being the intermediate and the bromide ion the nucleophile. It is interesting to mention that for acid-catalysed reactions of alcohols and their derivatives, even zeroth-order rate equations may be found [18]. Indeed, if a substrate A may form a reactive intermediate I with a catalyst C (5.17), which then generates the products by reaction with solvent molecules (5.18):

$$A + C \overset{K}{\rightleftharpoons} I \tag{5.17}$$

$$I + S \xrightarrow{k_1} \text{Products} + C \tag{5.18}$$

the reaction rate depends upon the concentration of the intermediate I and thus upon the equilibrium constant K:

$$\text{rate} = k_1\,[I] = k_1 K\,[A]\,[C] \tag{5.19}$$

Clearly, the equilibrium constant is given by the ratio:

$$K = \frac{[I]}{[A]\,[C]}$$

If the equilibrium constant is very large, as [I] must be small, the denominator must vanish, and since [A] is finite, it follows that:

$$[C] = [C]_0 - [I] = 0, \text{ or } [C]_0 = [I] = \text{constant}$$

Evidently, in such a case equation (5.19) becomes a zeroth-order equation:

$$\text{rate} = k_1\,[I] = k_1\,[C]_0 = k'_0$$

In the previous section it has been shown that the methanolysis of triphenylmethyl chloride in benzene may present third-order kinetics [8]. Similar orders have been determined both for typical unimolecular [9, 19] and bimolecular reactions [20]. Thus, any order between zero and three may correspond to both mechanisms of the aliphatic substitution. Evidently, the order of the equation rate can scarcely be of any use in deciding the correct reaction scheme. However, according to Bateman, Hughes and Ingold [3, 21], there is a situation in which the form of the rate equation may be significant. Their argument relies on the reversibility of the ionisation process:

$$R{-}X \underset{k_{-1}}{\overset{k_1}{\rightleftharpoons}} R^+ + X^-$$

$$R^+ + Nu: \xrightarrow{k_2} R{-}\overset{+}{N}u$$

[18] A. Ehret and S. Winstein, *J. Amer. Chem. Soc.*, **88**, 2048 (1966).
[19] C. G. Swain, *J. Amer. Chem. Soc.*, 72, 2794 (1950).
[20] Y. Pocker, *J. Chem. Soc.*, 1279 (1957).
[21] L. C. Bateman, E. D. Hughes and C. K. Ingold, *J. Chem. Soc.*, 960 (1940).

If the intermediate carbenium ion is sufficiently stable, it could discern between different nucleophiles, that is its counter-ion $X:^-$ (also called gegen-ion from German "gegen" = against, counter) and the nucleophile, Nu:. The observed rate will then be:

$$\text{rate} = \frac{k_1\,[\mathrm{RX}]}{1 + \dfrac{k_{-1}\,[\mathrm{X}^-]}{k_2\,[\mathrm{Nu:}]}} \tag{5.20}$$

In the case of a solvolytic reaction, where Nu: is the solvent itself, equation (5.20) may be written:

$$\text{rate} = \frac{k_1\,[\mathrm{RX}]}{1 + \alpha\,[\mathrm{X}^-]} \tag{5.21}$$

where $\alpha = k_{-1}/k_s$ and $k_s = k_2$ [Nu:].

Ingold, Hughes and their co-workers were able to verify this equation in certain cases [3,21-24], by adding to the reaction mixture a salt containing the counter-ion X^-. The depressing of the rate is an indication that (5.21) operates. There are however other, more precise means for detecting a unimolecular mechanism from its kinetics.

5.2.2 *Correlation of rate constants with structure*

The comparison of the rate constants observed (k_{obs}) in the nucleophilic substitution of certain compounds, with other rate constants of related compounds whose mechanism is known, or simply the study of the variation of k_{obs} in series of compounds differing by small structural modifications, constitute the most frequent procedures applied in the mechanism study. They actually rely on the verification of free-energy relationships.

The early identification of the S_N2 and S_N1 mechanisms was based on this method, even if no direct free-energy relationships were established. Thus, Table 5.2 presents after Bateman, Hughes and Ingold [25] the relative specific rates observed in the base hydrolysis of some bromides, constituting the so-called α-series. The members of this series are derived from methyl bromide by substitution of the hydrogen atoms with methyl groups. It is seen that the rate decreases about 12.5 times from methyl bromide to ethyl bromide, but only about six times from ethyl bromide to isopropyl bromide. Contrary to this trend, a steep increase occurs from the secondary to the tertiary bromide.

[22] L. C. Bateman, E. D. Hughes and C. K. Ingold, *J. Chem. Soc.*, 974 (1940).
[23] M. K. Church, E. D. Hughes, C. K. Ingold and N. A. Taher, *J. Chem. Soc.*, 971 (1940).
[24] M. K. Church, E. D. Hughes and C. K. Ingold, *J. Chem. Soc.*, 966 (1940).
[25] L. C. Bateman, K. A. Cooper, E. D. Hughes and C. K. Ingold, *J. Chem. Soc.*, 925 (1940).

TABLE 5.2

Relative specific rate constants in the base hydrolysis of alkyl bromides in 80% aqueous ethanol at 55 °C (after Bates, Hughes and Ingold[25])

Alkyl bromide:	CH_3-Br	CH_3CH_2-Br	$(CH_3)_2CH-Br$	$(CH_3)_3C-Br$
Relative specific rate constant[a]	1	0.0795	0.0135	47.5

a) Observed rate constants divided by the concentration of base.

It is clear that the first two members of this series (primary halides) must react by a completely different mechanism than the tertiary halide. If this observation is correlated with the second-order kinetics determined for methyl and ethyl bromide, and the first-order kinetics for t-butyl halide, the nature of the two reaction mechanisms is clear.

The data of Table 5.2 may be reconsidered using the Taft relation (see p. 124). Thus, taking into account the 12.6-fold decrease observed for ethyl bromide, a 150-fold decrease for isopropyl bromide and a 2000-fold decrease for t-butyl bromide is expected. Actually, isopropyl bromide reacts only 75 times slower than methyl bromide, whereas the tertiary halide is about 50 times more reactive than methyl bromide. Thus, t-butyl bromide reacts $50 \times 2000 = 10^5$ faster than expected if a common mechanism were operating in the series.

The case of isopropyl bromide needs further discussion. From this and other similar results[26, 27] it was concluded that the reaction may change the mechanism when secondary halides are involved. Indeed, a rate equation containing first- and second-order terms (5.22) is usually found in such cases:

$$\text{rate} = k_1 [\text{RX}] + k_2 [\text{RX}] [\text{Nu:}] \tag{5.22}$$

The intervention of either mechanism will depend on the ratio between the two rate constants. For $k_1 > k_2$ [Nu:], the reaction would appear unimolecular; conversely, if k_2 [Nu:] $> k_1$ the bimolecular reaction prevails. Evidently, there are situations in which

[26] J. L. Gleave, E. D. Hughes and C. K. Ingold, *J. Chem. Soc.*, 236 (1935).

[27] E. D. Hughes and C. K. Ingold, *J. Chem. Soc.*, 244 (1935); *Trans. Faraday Soc.*, 37, 657 (1941).

$k_1 \approx k_2$[Nu:]. Such cases correspond to the *border-line* nucleophilic substitution [28–32].

Bird, Hughes and Ingold [28] demonstrated that isopropyl halides cannot react by two different mechanisms. According to them, there is no sharp difference between the unimolecular and bimolecular substitution. One may imagine that when the C—X bond begins to weaken, a molecule of nucleophile, Nu:, is always close enough to interact with the partially vacant orbital. The stronger this interaction, the more predominant is the bimolecular character. Thus, equation (5.22) may illustrate a single type of mechanism, in which the interaction with the nucleophile is neither strong enough to determine a bimolecular process, nor weak enough to be not felt in the rate. However, as will be shown later in this chapter, such a rate equation may indeed correspond to different reaction mechanisms.

The reactions of primary alkyl derivatives, in the absence of nucleophilic groups in the molecule, are considered now with certainty to be bimolecular processes [33–38]. There is strong evidence that these compounds undergo bimolecular solvolysis in usual solvents, such as water, alcohols, carboxylic acids [2, 39], but quite different reaction mechanisms may be involved if strong acids are used [40, 41]. Table 5.3 gives some examples involving members of the β-series. The β-series is derived from ethyl bromide by substitution of β-hydrogen atoms with methyl groups.

In ethanol, the reaction is a typical bimolecular process, showing the expected trend in the decrease of the rate constants (see also Section 5.5). This order is completely reversed in the more acidic solvents, the neopentyl tosylate reacting about 10^6 faster than methyl tosylate, which is not in agreement with the bimolecular mechanism. It must be remembered that the role of the solvent may be decisive, to the point of changing the reaction mechanism even for primary alkyl derivatives. Anyway, the reactions in solvents

[28] M. L. Bird, E. D. Hughes and C. K. Ingold, *J. Chem. Soc.*, 634 (1954).
[29] S. Winstein, E. Grunwald and H. W. Jones, *J. Amer. Chem. Soc.*, 73, 2700 (1951).
[30] V. Gold, *J. Chem. Soc.*, 4633 (1956).
[31] P. Casapieri and E. R. Swart, *J. Chem. Soc.*, 4342 (1961); 1254 (1963).
[32] V. J. Shiner, Jr., M. W. Rapp and H. R. Pinnick, Jr., *J. Amer. Chem. Soc.*, 92, 232 (1970).
[33] P. B. D. de la Mare, *J. Chem. Soc.*, 3169, 3180, 3197 (1955).
[34] E. D. Hughes, C. K. Ingold and J. D. H. Mackie, *J. Chem. Soc.*, 3173, 3177 (1955).
[35] L. Fowden, E. D. Hughes and C. K. Ingold, *J. Chem. Soc.*, 3187, 3193 (1955).
[36] F. Aziz and E. A. Moelwyn-Hughes, *J. Chem. Soc.*, 2635 (1959).
[37] A. Streitwieser, Jr., *J. Amer. Chem. Soc.*, 77, 1117 (1955).
[38] A. Streitwieser, Jr., C. L. Wilkins and E. Kiehlmann, *J. Amer. Chem. Soc.*, 90, 1598 (1968).
[39] W. Pritzkow and K. H. Schöppler, *Chem. Ber.*, 95, 834 (1962).
[40] A. Diaz, I. L. Reich and S. Winstein, *J. Amer. Chem. Soc.*, 91, 5635, 5637 (1969).
[41] P. C. Myhre and K. S. Brown, *J. Amer. Chem. Soc.*, 91, 5639 (1969).

TABLE 5.3

Relative rates in the solvolysis of alkyl tosylates [a] (R—OTs) in solvents with various nucleophilicities and acidities

Solvent		*R in R—OTs:*				
		CH_3	CH_3CH_2	$CH_3CH_2CH_2$	$(CH_3)_2CHCH_2$	$(CH_3)_3CCH_2$
CH_3CH_2OH	(b)	4 040	1 750	1 140	8	1
CH_3COOH	(b)	10	9	7.3	2.8	1
HCOOH	(c)	0.56	1	1.2	1.22	1
CF_3COOH	(c)	1	13	93	3 060	6 000
H_2SO_4	(d)	1	26	530	7 500	50 000
$HFSO_3$	(e)	1	118	33 000	540 000	1 140 000

(a) Tosylate = toluene-*p*-sulphonate; Ts = *p*-$MeC_6H_4SO_2$. (b) After Winstein and Marshall [42]. (c) Determined by Diaz, Reich and Winstein [40]. (d) Values of Myhre and Brown[41]. (e) Determined by Reich, Diaz and Winstein [40].

such as methanol and acetic acid [39] may be considered as reference reactions with definite bimolecular mechanism. Their rate constants may serve as a basis of comparison in determining other nucleophilic mechanisms.

The verification of a Taft relationship is very useful in establishing the reaction mechanisms. Table 5.4 indicates some ρ^* parameters determined in the nucleophilic substitution.

TABLE 5.4

Taft's $\rho^*\sigma^*$ correlations in aliphatic nucleophilic substitution

Reaction	ρ^*
(1) $R-CH_2OTs$ + AcOH (30 °C)[a] (R = H, alkyl)	0
(2) $R-CH_2OTs$ + NaOAc (AcOH, 110 °C)[a] (R=H, alkyl, cycloalkyl)	+1
(3) $R-CH_2OTs$ + EtOH (EtOH, 100 °C)[b]	−0.740
(4) $R-CH_2Br$ + PhS^- (MeOH, 20 °C)[b]	−0.606
(5) $R(R')CH-OTs$ + AcOH (70 °C)[a]	−2.58
+HCOOH (70 °C)[c]	−3.5
+CF_3COOH (25 °C)[c]	−7.05
(6) $R(R')CH-OBs$ + AcOH (70−75 °C)[b, d]	−3.49

[42] S. Winstein and H. Marshall, *J. Amer. Chem. Soc.*, **74**, 1120 (1952).

TABLE 5.4 (continued)

Reaction	ρ*
(7) $R^1R^2C{-}CR^3R^4$ (epoxide, bridged by O) + HOH (0.7 M $HClO_4$)	−1.95
(8) $R_3C{-}X$ + EtOH (80%, 25 °C)[b]	−3.29

(a) Data of Pritzkow and Schöppler [39]. (b) Data reported by Streitwieser [37]. (c) After Petersen *et al.* [43], considering a transformation factor $\rho^* = \rho_I$ 0.45. (d) Bs = $p{-}BrC_6H_4SO_2$. (e) After J. G. Pritchard and F. A. Long[44].

Table 5.4 shows that substrates reacting by an S_N2 mechanism have small absolute values of ρ* parameter (either positive or negative), whereas in unimolecular reactions high negative values of $\rho^*(|\rho^*| \geqslant 2)$ are found. The determination of the slope in a Taft correlation may constitute a useful method for deciding a specific mechanism in a series of related reactions [45].

Clearly, for benzyl derivatives either the Hammett ρσ correlation or the Brown $\rho\sigma^+$ relationship must be checked. The verification of the latter relationship is a convincing argument for the unimolecular reaction. Thus, the logarithms of the relative rates observed in the solvolysis of benzyl tosylates [46] are correlated with the σ^+ values, the slope of the correlation line having the value $\rho = -6.15$. If the definition of the σ* parameter is considered, the former reaction parameter can be converted to $\rho^* = \rho/2.48 = -2.48$, in perfect agreement with the values determined for the unimolecular solvolyses in acetic acid (Table 5.4). However, the presence of a phenyl group does not always involve a good correlation of the reaction rates of a nucleophilic substitution in the side-chain. In many cases, the absence of correlation corresponds to aryl participation in the ionisation step. Other group participations in the ionisation step may also determine significant deviations from the linear correlations of the Hammett and Taft type. Thus:

(a) In the solvolysis of β-arylalkyl derivatives, Cram [47, 48] has proposed an intermediate phenonium ion, **5**:

[43] P. E. Petersen, R. E. Kelley, Jr., R. Belloli and K. A. Sipp, *J. Amer. Chem. Soc.*, 87, 5169 (1965).

[44] J. G. Pritchard and F. A. Long, *J. Amer. Chem. Soc.*, 78, 2667 (1956).

[45] See, for example, J. E. Leffler and E. Grunwald, *Rates and Equilibria of Organic Reactions*, J. Wiley, New York (1963), p. 226.

[46] H. C. Brown, R. Bernheimer, C. J. Kim and S. E. Sheppele, *J. Amer. Chem. Soc.*, 89, 370 (1967).

[47] D. J. Cram, *J. Amer. Chem. Soc.*, 71, 3863, 3875, 3883 (1949); 74, 2129 (1952).

[48] D. J. Cram, *J. Amer. Chem. Soc.*, 86, 3767 (1964).

$$R_2C_\beta(Ph)—C_\alpha R_2—X \longrightarrow R_2C—CR_2 \; X^- \xrightarrow{Nu:^-} R_2C(Ph)—CR_2(Nu) \qquad (5.23)$$

5

Cram's theory of phenonium ions was criticised due to the very small acceleration factors observed. Thus, the acetolysis of 3-phenyl-2-butyl tosylate [47] proceeds about three times slower than the acetolysis of 2-butyl tosylate. By coincidence, this corresponds to the rate depression predicted on inductive grounds by the Taft relation:

$$\log k_H/k_{Ph} = \rho^*\Sigma\sigma^* = -2.6\,(\sigma_{Me} + \sigma_H + \sigma_{Et} - \sigma_H - \sigma_{Me} - \\ -\sigma_{PhCHMe}) = (-2.6)\,(-0.21) = +0.546$$

Thus,

$$k_H/k_{Ph} = 3.5$$

which is in agreement with experience. Moreover, the same depression was observed for 2-phenylethyl and 1-phenyl-2-propyl tosylates. It can be demonstrated, considering the additivity of σ^* constants, that for all systems:

$$\log k_H/k_{Ph} = \rho^*\,\sigma^*_{PhCH_2} = 0.546$$

However, it has been shown [47] that 3-phenyl-2-butyl tosylate ionises three times more rapidly than 2-butyl tosylate, so that a rate enhancement of at least $3 \times 3.5 = 10.5$ seems to be involved. More precise calculations indicate a value of about 24—30 for neighbouring phenyl rate acceleration [49], which is generally accepted for solvolysis. However, even this higher value is too small to serve as definite proof for the scheme (5.23). The decisive kinetic argument for phenyl participation in the case of β-arylalkyl systems was advanced by Norlander and Deadman [50], who have shown that 2-phenylethyl tosylate, even if reacting slower with acetic acid than ethyl tosylate, in trifluoracetic acid is about 3000 times more reactive than the latter compound.

Another demonstration of the process (5.23), due to Schleyer and Lancelot [51], was based on Winstein's procedure [52a] to decompose

[49] C. J. Lancelot, D. J. Cram and P. von R. Schleyer, in *Carbonium Ions*, vol. 3 (G. A. Olah and P. von R. Schleyer, editors), Wiley-Interscience, New York (1972), p. 1347.

[50] J. E. Nordlander and W. G. Deadman, *J. Amer. Chem. Soc.*, **90**, 1590 (1968).

[51] P. von R. Schleyer and C. J. Lancelot, *J. Amer. Chem. Soc.*, **91**, 4291, 4294, 4296, 4297 (1969).

[52a] E. J. Jenny and S. Winstein, *Helv. Chim. Acta*, **41**, 807 (1958).

the observed constants into solvent-assisted (k_s) and anchimerically-assisted (k_Δ) rate constants. Although the observed rates for β-arylalkyl systems do not obey a $\rho\sigma$ or a $\rho\sigma^+$ relationship [52b], the partial rate constants show excellent linear correlations [49, 51].

(b) A different example is constituted by the solvolysis of 2-(3-cyclopentenyl)-ethyl *p*-nitrobenzenesulphonate (nosylate) **6**, whose solvolysis was compared with the acetolysis of the saturated analog [53, 54], **7**:

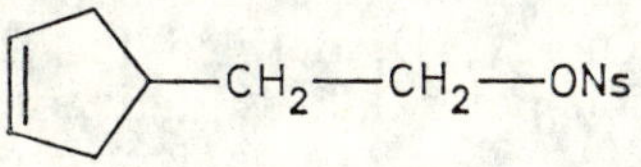

3-cyclopentenyl–CH_2–CH_2–ONs **6** cyclopentyl–CH_2–CH_2–ONs **7**

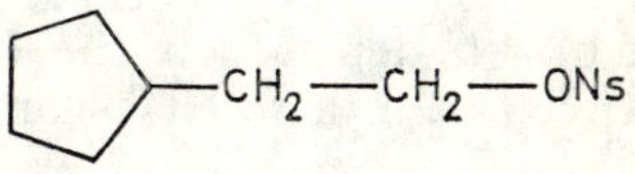

Ns = O_2N–C_6H_4–SO_2

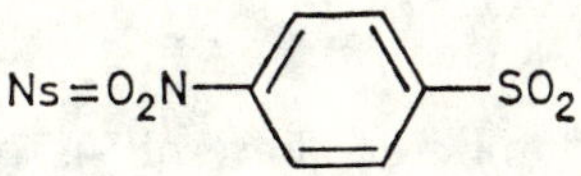

Other data [55] indicate that a remote double bond has only a very weak retarding effect, which is due to its inductive effect. However, in the case of **6** and **7** the ratio k_{unsat}/k_{sat} is 6 in ethanol, 70 in acetic acid and about 640 in formic acid [54]. It is clear that the double bond must be involved in the reaction.

(c) A typical example, demonstrating neighbouring-group participation in chemical reactions, is the comparative solvolysis of 7-substituted bicyclo[2.2.1] derivatives, **8**–**11** [56–58].

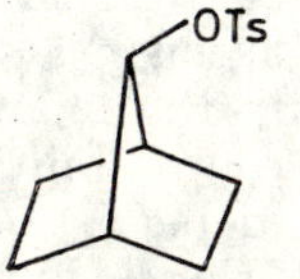

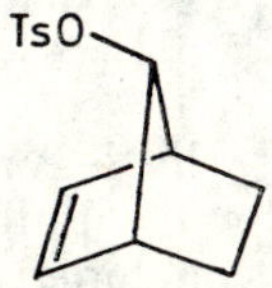

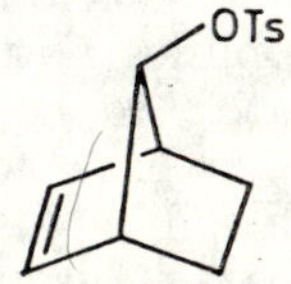

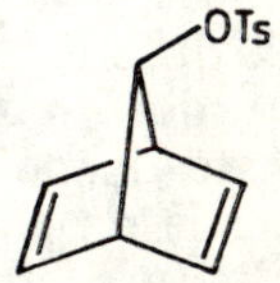

	8	9	10	11
	OTs	TsO	OTs	OTs
k_{rel}	1	10^4	10^{11}	10^{14}

The kinetic evidence in this case leaves no doubt about the assisted mechanism. The nature of the solvolysis products, either of **6** or of **9**–**11**, reinforces the kinetic argument.

[52b] H. C. Brown, K. J. Morgan and F. J. Chloupek, *J. Amer. Chem. Soc.*, 87, 2137 (1965).
[53] R. G. Lawton, *J. Amer. Chem. Soc.*, 83, 2399 (1961).
[54] P. D. Bartlett and S. Bank, *J. Amer. Chem. Soc.*, 83, 2591 (1961); P. D. Bartlett, S. Bank, R. J. Crawford and G. H. Schmid, *J. Amer. Chem. Soc.*, 87, 1288 (1965).
[55] P. D. Bartlett and M. R. Rice, *J. Org. Chem.*, 28, 3351 (1963).
[56] S. Winstein, M. Shatavsky, C. Norton and R. B. Woodward, *J. Amer. Chem. Soc.*, 77, 4183 (1955).
[57] S. Winstein and M. Shatavsky, *J. Amer. Chem. Soc.*, 78, 592 (1956).
[58] S. Winstein and C. Ordreneau, *J. Amer. Chem. Soc.*, 82, 2084 (1960).

The correlation of the reaction rates with the structure of the organic radical is explained by the fact that in the unimolecular mechanism, a positive charge is created at the carbon atom, during the ionisation. High negative values of ρ^* are thus expected. In the S_N2 mechanism, depending on the gap between bond formation and bond breaking, that is on the concertedness of these two processes, small ρ^* values, either positives or negative, are expected.

Another structural correlation refers to the nature of the leaving group. In the S_N1 reaction, the ease with which a C—X bond is broken will depend on the dissociation energy, D, defined for the reaction (5.24), as well as on the ionisation potential, as defined in (5.25).

$$R{-}X \rightarrow R\cdot + X\cdot \qquad \Delta H = D \tag{5.24}$$

$$X\cdot + e^- \rightarrow X{:}^- \qquad \Delta H = I \tag{5.25}$$

Iodides will therefore undergo heterolysis more easily than bromides, and the latter more easily than chlorides. On the other hand, in S_N1 reactions arenesulphonates will be more reactive than the bromides. Conversely, in the S_N2 reaction, the energy necessary to break the C—X bond is partly supplied by the formation of the new bond. In that case, the polarisability of the C—X bond is more important than the dissociation energy in determining the *leaving group ability (nucleofugicity)*. A reverse order of reactivity, with respect to S_N1, may be observed, as is the case with bromides and arenesulphonates. It has been proposed [59, 60] that the ratio k_{ROTs}/k_{RBr} be taken as a criterion for the substitution mechanism. Recently, however, studying the reaction of CH_3X derivatives (X = Cl, Br) with different nucleophiles (Nu: = F^-, Cl^-, CH_3S^-), in gaseous phase (in order to eliminate the complications of solvation), Brauman, Olmstead and Lieder [61] have found that the leaving group ability depends on the nucleophile, just as the nucleophilicity depends on the nature of the leaving group. They concluded that the leaving group ability cannot be treated as an intrinsic property, and should be carefully considered from case to case. Anyway, in the usual solvents, the k_{ROTs}/k_{RBr} criterion seems to be valid, the bromides reacting more slowly than the tosylates [62].

Considering that the carbonyl derivatives may be regarded as carbenium ions stabilised by conjugation, Brown and his co-workers [63–65] have proposed the correlation of solvolysis rates of secondary

[59] H. M. R. Hoffmann, *J. Chem. Soc.*, 6753, 6762 (1965).

[60] R. G. Pearson and J. Songstad, *J. Org. Chem.*, 32, 2899 (1967).

[61] J. I. Brauman, W. N. Olmstead and C. A. Lieder, *J. Amer. Chem. Soc.*, 96, 4030 (1974).

[62] For a more recent example, see M. Banciu, F. Badea, R. Jelescu and E. Ciorănescu, *Rev. Roumaine Chim.*, 20, 121 (1975).

[63] H. C. Brown and K. Ichikowa, *Tetrahedron*, 1, 221 (1957).

[64] H. C. Brown, H. M. Bell, *J. Org. Chem.*, 27, 1928 (1962).

[65] H. C. Brown, R. Bernheimer and K. J. Morgan, *J. Amer. Chem. Soc.*, 87, 1280 (1965).

tosylates R^1R^2CHOTs with the rates of reduction with sodium borohydride of the corresponding ketones R^1R^2CO. It must be however observed that in the rate-determining step of the ionisation of the tosylate, the hybridisation of the central carbon atom changes from sp^3 to sp^2, whereas the reverse process is associated with carbonyl reduction. It is believed [46, 65] that the verification of such a correlation is a clear indication for a classical S_N1 mechanism, excluding participation of neighbouring groups.

Based on the same type of related phenomena, Foote [66] established a linear relationship between the logarithm of the rate constants of a large number of secondary tosylates and the vibration frequency ν_{CO} of the corresponding ketones. An excellent correlation line was obtained, covering a range of about 10^{10} in the rate constants. It was assumed [67] that rate constants higher than the values predicted from the carbonyl frequency are a clear indication of an anchimeric assistance. However, this view is not generally accepted [68].

5.2.3 *Correlation of rate constants with the solvent*

The importance of the solvent in the nucleophilic substitution was early recognised. Menshutkin [69] observed the effect of the polarity of solvent in the reaction of halides with tertiary amines, but he also noticed that the rates cannot be directly correlated with the solvent polarity. The first systematic study was undertaken by Ingold, Hughes and their co-workers [70, 71]. Considering the reaction types in the nucleophilic substitution, they examined the corresponding transition states, predicting the effect of the increasing polarity of the solvent on the reaction rates. Thus, for bimolecular reactions of type I (see Table 5.1), charge is dispersed in the transition state **1**. A small decrease with increasing polarity of the solvent is expected. If the reaction is of type II, the examination of transition state **12** shows that a slight increase should be observed (charges are created), whereas transition state **13** predicts a large decrease (charges are destroyed).

$$\overset{\delta+}{\mathrm{Nu}} \cdots\cdots \mathrm{R} \cdots\cdots \overset{\delta-}{\mathrm{X}} \qquad\qquad \overset{\delta-}{\mathrm{Nu}} \cdots\cdots \mathrm{R} \cdots\cdots \overset{\delta+}{\mathrm{X}}$$

12 **13**

Large rate enhancements are predicted to occur with increasing polarity of the solvent in the S_N1 reaction, if the leaving group

[66] C. S. Foote, *J. Amer. Chem. Soc.*, **86**, 1853 (1964).
[67] P. von R. Schleyer, *J. Amer. Chem. Soc.*, **86**, 1854, 1856 (1964).
[68] H. C. Brown, *Boranes in Organic Chemistry*, Cornell University Press, Ithaca (1972), Part III.
[69] N. Menshutkin, *Z. physik. Chem.*, **5**, 589 (1890).
[70] E. D. Hughes and C. K. Ingold, *J. Chem. Soc.*, 244 (1935).
[71] K. A. Cooper, M. L. Dhar, E. D. Hughes, C. K. Ingold, B. J. McNulty and L. I. Wolf, *J. Chem. Soc.*, 2043 (1948).

has no charge in the substrate (charges are created). However, if X is positively charged, the decrease of the reaction rate is possible.

The dependence of the reaction rates on the dielectric constant of the solvent may be understood if we write the rate constant as a function of the activity coefficients of the reagents and transition state[72]:

$$k = \frac{\mathbf{k}T}{h} K^{\ddagger} = \frac{\mathbf{k}T}{h} \frac{C^{\ddagger}}{C_{RX} C_{Nu:}} \frac{\gamma^{\ddagger}}{\gamma_{RX} \gamma_{Nu:}}$$

or:

$$\ln k = \ln \frac{\mathbf{k}T}{h} - \ln \frac{C_{RX} C_{Nu:}}{C^{\ddagger}} - \ln \frac{\gamma_{RX} \gamma_{Nu:}}{\gamma^{\ddagger}} =$$

$$= \ln k_0 - \ln \frac{\gamma_{RX} \gamma_{Nu:}}{\gamma^{\ddagger}} \tag{5.26}$$

in which h is Planck's constant, **k** Boltzmann constant, k_0 the rate constant in a reference solvent for which the ratio of the activity coefficients is equal to unity. This ratio is however dependent on the dielectric constant, the observed rate being in its turn directly related to the latter.

Thus, solvents affect both the activities of the reagents and the transition state. Evidently, the activity of the nucleophile is important only in the bimolecular mechanism. It has been thus observed that the rates of typical bimolecular reactions, such as Finkelstein reaction (see Table 5.1), are only little affected when a protic solvent is replaced by another, even if they differ largely in their dielectric constants (compare Table 5.5 with Table 5.6). On the other hand, if the reaction is transferred from a protic solvent, such as methanol, to an aprotic dipolar solvent[73, 74], with the dielectric constant of about the same order of magnitude, a very large rate enhancement may be observed (Table 5.5).

TABLE 5.5

Relative rates in the nucleophilic substitution: $MeI + Li^+Cl^- \rightarrow MeCl + Li^+I^-$ **(according to Parker[73])**

Solvent	MeOH	H_2N-CHO	$MeNH-CHO$	Me_2NCHO	MeCOMe	Me_2NCOMe
Relative rate	1	12	14	1.2×10^6	1.6×10^6	7.4×10

This rate enhancement is determined by the higher activity of nucleophiles in the aprotic dipolar solvents, as compared to their activity in protic solvents. The lower activities in the latter solvents is explained by strong hydrogen bonds between the protic molecules

[72] E. S. Amis, *Solvent Effects on Reaction Rates and Mechanism*, Academic Press, New York (1966).

[73] A. J. Parker, *Quart. Rev.*, **16**, 163 (1962).

[74] B. Tchoubar, *Bull. Soc. chim. France*, 2069 (1964).

and the anion, which are, obviously, absent in an aprotic solvent. The effects of solvent on the activity of nucleophiles may be measured by *solvent activity coefficients* [75-77] (Chapter 4). If these are known, from rate constants measured in the reference solvent (methanol) and the aprotic solvent, the effect of the solvent on the transition state can be appreciated [77]. It is thus concluded that with few exceptions, solvent activity coefficients are not significantly modified by the transference from methanol to another protic solvent.

$$\log \frac{k_s}{k_0} = \log {}^{S}\gamma^0_{Nu:} + \log {}^{S}\gamma^0_{RX} - \log {}^{S}\gamma^0_{NuRX\ddagger} \quad (5.27)$$

Similar results were obtained by other researchers [78]. The slight effect of the solvent on the transition state of a bimolecular substitution was also predicted by the original theory of Hughes and Ingold [70,71].

More drastic effects are however anticipated for the transition state of a unimolecular reaction. We mentioned previously that the works of Bartlett and Nebel [6], Swain and Eddy [8], Gelles, Ingold and Hughes [9], have demonstrated the double effect of the solvent in an ionisation reaction, that is its nucleophilic assistance, on the one hand *(push mechanism)*, and its electrophilic assistance *(pull mechanism)*, on the other hand. Evidently, the latter effect will be related to the acidity of the solvent. Table 5.6 lists some pertinent values for a few usual solvents.

TABLE 5.6

Polar and acid properties of some solvents

Solvent	*Dielectric constant* [a]	pK_a^b	*Solvent*	*Dielectric constant* [a]	pK_a^b
N-methylformamide	190.5	—	Ethanol	25.1	18
Hydrogen cyanide	115	9.5	Acetone	21.2[c]	20
Sulphuric acid	110	−3	n-Butanol	17.8	18
Formamide	109	—	Ammonia	17.3	35
Water	80.4	14	Sulphur dioxide	14.1	—
Hydrogen fluoride	84	3	t-Butanol	10.9	19
Formic acid	57.9	3.75			
N,*N*-dimethylformamide	37.6	—	Trifluoroacetic acid	8.32[c]	−3
Nitromethane	37.5	10.2	Acetic acid	6.15	4.75
Acetonitrile	37.5	25	Diethylether	4.34	—
			Benzene	2.28	37
Methanol	33.6	16	n-Hexane	1.89	45

(a) At 20 °C; largely from J. Hine, *Physical Organic Chemistry*, McGraw-Hill, New York (1962), p. 39. (b) Largely from R. P. Bell, *The Proton in Chemistry*, Methuen (1959), and from D. J. Cram, *Fundamentals of Carbanion Chemistry*, Academic Press, New York (1965), Chapter 1. (c) E. L. Mackor, P. J. Smit, J. H. van der Waals, *Trans. Faraday Soc.*. **53**, 1309 (1957).

75 R. Alexander and A. J. Parker, *J. Amer. Chem. Soc.*, **89**, 5549 (1967).
76 A. J. Parker and R. Alexander, *J. Amer. Chem. Soc.*, **90**, 3313 (1968).
77 L. Alexander, E. C. F. Ko, A. J. Parker and T. J. Broxton, *J. Amer. Chem. Soc.*, **90**, 5049 (1968).
78 I. Bacaloglu, Ph. D. Thesis, Polytechnic Institute Timişoara (Romania) (1975).

Another approach to the solvent effect in the unimolecular reactions is due to Winstein, Grunwald and Fainberg. Considering the results of Swain, the dependence of the rate constant on solvent properties may be expressed by the equation (5.28):

$$\mathrm{d} \log k = \left(\frac{\partial \log k}{\partial Y}\right)_N \mathrm{d} Y + \left(\frac{\partial \log k}{\partial N}\right)_Y \mathrm{d} N \tag{5.28}$$

The first member of this equation represents the variation of the rate constant with the ionisation power of the solvent, Y, the last term corresponding to the variation due to the nucleophilicity of the solvent. For most unimolecular reactions, the last term may be neglected. If the term $(\partial \log k/\partial Y)_N$ is denoted by m, the integration of equation (5.28) gives the so-called *Grunwald-Winstein equation*[79]:

$$\log \frac{k}{k_0} = m\, Y \tag{5.29}$$

where k is the observed rate constant, k_0 is the rate constant in a reference solvent, and m the slope of the correlation line, which measures the substrate susceptibility to the ionisation power Y of the solvent.

When defining the Y values, as reference substrate ($m \equiv 1.00$) is taken t-butyl chloride and as reference solvent 80% aqueous ethanol. The measurement of the rate constants in the solvolysis of t-butyl chloride in different solvent mixtures, at 25 °C, has enabled the determination of Y values (Table 5.7). The reactions verifying equation (5.29) have been designated by Winstein as admitting a *limiting mechanism*.

The Grunwald-Winstein relationship is an example of linear free-energy relationship. The plot of the logarithm of the observed rate constants in a mixture of solvents, *versus* the Y values, may yield a straight line, the slope of it representing m. This parameter has a value of about 1.0 for a limiting solvolysis, and low values (0.1—0.5) for bimolecular reactions [2]. Winstein and his co-workers [79–82]

TABLE 5.7

Y **values**[80]

A—B solvent system	*% Vol. of A*	*Y*
Ethanol-water	100	−2.033
	95	−1.285
	90	−0.747
	80	0.000
	60	1.124
	40	2.196
	20	3.051
	0	3.493

[79] E. Grunwald and S. Winstein, *J. Amer. Chem. Soc.*, 70, 846 (1948).

[80] A. H. Fainberg and S. Winstein, *J. Amer. Chem. Soc.*, 78, 2770 (1956).

[81] A. H. Fainberg and S. Winstein, *J. Amer. Chem. Soc.*, 79, 1597, 1602, 1608 (1957).

[82] S. Winstein, A. H. Fainberg and E. Grunwald, *J. Amer. Chem. Soc.*, 79, 4146 (1957).

TABLE 5.7 (continued)

A—B solvent system	*% Vol. of A*	*Y*
Methanol-water	100	−1.090
	80	0.381
	60	1.492
	40	2.391
	20	3.025
Formic acid-water	100	2.054
	50	2.644
	25	3.100
Acetic acid-water	100%	−1.650
	2.00 M H_2O	−0.863
	8.00 M H_2O	+0.193
	16.00 M H_2O	0.984
	50%	1.938
	25	2.843
Acetone-water	95.2	−2.76
	90	−1.856
	80	−0.673
	70	+0.130
	60	0.796
	50	1.394
	40	1.981
	20	2.913

have demonstrated the utility of their relation for ascertaining the correct reaction mechanism. The following more recent examples stress in their turn this utility.

Streitwieser, Wilkins and Kiehlmann [83] have studied the solvolysis of ethyl trifluoromethanesulphonate EtO_3SCF_3 in order to find if simple primary derivatives may undergo unimolecular reactions. The trifluoromethanesulphonate group was chosen as one of the most efficient neutral leaving groups. From the low *m* value ($m = 0.15$), as well as from other kinetic data, it was actually concluded that the reaction is unimolecular.

Cram and his co-workers [84] studied the racemisation of cyclopropane derivative **14** as a function of the polarity of the solvent. It has been later proved [85] that the activation parameters observed in benzene, methanol, acetic acid, dimethylformamide (DMF), and DMF with added LiBr (but not in formic acid) satisfy an isokinetic relationship, indicating that in all solvents, with the exception of formic acid, a common racemisation mechanism is involved:

[83] A. Streitwieser, Jr., C. L. Wilkins and E. Kiehlmann, *J. Amer. Chem. Soc.*, **90**, 1598 (1968).

[84] E. W. Yankee and D. J. Cram, *J. Amer. Chem. Soc.*, **92**, 6328, 6329, 6331 (1970).

[85] E. W. Yankee, F. D. Badea, N. E. Howe and D. J. Cram, *J. Amer. Chem. Soc.*, **95**, 4210 (1973).

14-R ⇌ **15** ⇌ **14-S**

In order to verify this conclusion, log k_{obs} in a mixture of formic and acetic acid were plotted *vs.* the corresponding Y values. A broken line was obtained, with the slope $m = 0.5$ in the case of low formic acid concentration, and with $m = 1.2$ at high formic acid concentrations (similarly to the slope observed in the solvolysis of benzhydryl derivatives[86]). Thus, the mechanisms in the two solvents (acetic acid and formic acid) are not identical, in formic acid the reaction being very similar to the ionisation of benzyl derivatives. Intermediate **15** was thus suggested.

For such a solvolysis, in which the nucleophilicity of the solvent cannot be neglected, a relation with four parameters should be applied. Even if such relationships are known[72], they are not convenient for practical purposes. It is therefore preferable to make rough estimations of substrate susceptibility to the nucleophilicity of the solvent by measuring $k_{aq.alc}/k_{RCOOH}$, where k_{RCOOH} is the rate constant measured either in formic acid (R = H) or in acetic acid (R = Me), as poor nucleophiles, and $k_{aq.alc}$ is the rate constant measured in aqueous ethanol, having the same Y value as the acid (as nucleophilic solvent). A few typical values are presented in Table 5.8.

TABLE 5.8

Values of the ratio $k_{aq.alc}/k_{RCOOH}$ for different alkyl compounds

Compound	*Temperature*/°C	$k_{aq.alc}/k_{RCOOH}$	
		R=H	R=Me
MeBr	50	200	—
MeOTs	25	55	97
EtOTs	50	18	80
EtO_3SCF_3	25	—	79[b]
Pr^{i}Br[c]	50	—	40
Pr^{i}OTs[c]	25	—	7.8
2-Adamantyl bromide[c]	25	—	4.8
2-Adamantyl tosylate[c]	25	—	0.13
2-C_5H_{11}OTs	58.2	0.4	4
Bu^t Br	25	10	—

(a) From Reference 2, unless otherwise stated. (b) Reference 83. (c) Reference 126.

[86] J. E. Leffler and E. Grunwald, *Rates and Equilibria of Organic Reactions*, J. Wiley (1963), p. 300.

It can be seen that for secondary and tertiary derivatives undergoing unimolecular solvolysis, the ratio $k_{aq.alc}/k_{RCOOH}$ is small (< 10), but may attain high values for primary derivatives.

Other relationships [87], correlating rate constants with some defined properties of the solvent, are known. It has been however shown [88] that they are equivalent to relation (5.28). Both the mY equation and the ratio $k_{aq.alc}/k_{RCOOH}$ are very useful in the study of unimolecular reactions, either assisted or not.

5.2.4 *Salt effects. Ion pairs in the nucleophilic substitution*

The addition of a salt to the reaction medium may affect the reaction rates in different ways, depending on the particular mechanism which operates. First, it increases the ionic strength μ, defined by the relation:

$$\mu = \frac{1}{2}\sum C_i Z_i^2 \tag{5.30}$$

where C is the concentration of the ionic species and Z_i is the electric charge of the ion. The increase of the ionic strength modifies the activity coefficients, the following relation being applied in water, at low ionic strength:

$$\log \gamma_c = - aZ_i^2\sqrt{\mu} \tag{5.31}$$

where a is given by the relation:

$$a = \frac{1.825\times 10^6}{(\varepsilon T)^{3/2}} \tag{5.32}$$

ε being the dielectric constant.

Just as the rate constants observed are related to the activity coefficients by equation (5.26), the variation of the ionic strength is also reflected in the rate constants. Thus, the addition of a salt containing a nucleophile may increase the rate of a bimolecular reaction (the concentration of the salt entering directly into the equation rate), as well as the rate of a unimolecular reaction, as a result of the increase of the ionic strength. In order to avoid such rate variations, kinetic measurements are usually performed at constant ionic strength, by adding an electrolyte which is unreactive towards the substrate. On the other hand, ionic strength is related to the medium polarity and will therefore affect the rate in the same way as the polarity of the solvent does. The salt effects resulted from the increase of the ionic strength are called *primary* or *normal* effects.

Considering relations (5.26) and (5.31), a linear dependence between the logarithm of the observed rate constants and the square

[87] E. M. Kosower, *J. Amer. Chem. Soc.*, **80**, 3253, 3267 (1958).
[88] S. G. Smith, A. H. Fainberg and S. Winstein, *J. Amer. Chem. Soc.*, **83**, 618 (1961).

root of the ionic strength μ should be expected. Even if such relations are known[89] to apply, the exceptions are very frequent. Thus, Bateman, Church, Hughes, Ingold and Taher [3] have found that in the unimolecular solvolysis of alkyl halides in aqueous acetone, the following relationship holds:

$$\log k = \log k_0 + \frac{0.912 \times 10^{16}}{(\varepsilon T)^2} \sigma\mu \tag{5.33}$$

where k_0 is the rate constant in the absence of salt and σ is an experimental value having the dimension of a length and representing the dipole moment of the transition state. The relation seems to be limited to aqueous solvents.

In non-polar and acidic solvents (for example acetic acid), Fainberg and Winstein[90, 91] have found a non-logarithmic linear relationship (5.34) between the observed rate constants and the stoichiometric concentration of the salt, Z:

$$k = k_0(1 + b[Z]) \tag{5.34}$$

The parameter b (the slope of the correlation line) measures the salt effects. It depends upon the nature of the salt, of the substrate and even of the cation (Table 5.9). The validity of this relation has been challenged by several authors [68, 92], as no clear theoretical interpretation has been found as yet (see however p. 215). Anyway,

TABLE 5.9

Values observed for the b parameter (according to Winstein *et al.* [90, 91, 94, 95])

Electrolyte added	*b*	*Reaction*
$LiClO_4$	28–30	Acetolysis of cholesteryl arenesulphonates
	37–38	Acetolysis of 2-*exo*-norbornyl brosylate, benzyl tosylate etc.
	12	Acetolysis of 2-(2,4-dimethoxyphenyl)ethyl brosylate
$HClO_4$	28	Acetolysis of cholesteryl tosylate
LiOAc	2–3	Acetolysis of cholesteryl tosylate and 2-(2,4-dimethoxyphenyl)ethyl tosylate, pinacolyl arenesulphonates etc.
	1.2	Acetolysis of benzyl tosylate
HOTs, LiOTs	3.4	Acetolysis of pinacolyl and benzyl derivatives

[89] See, for example, L. P. Hammett, *Physical Organic Chemistry*, McGraw-Hill, 2nd ed., New York (1970), Chapter 9.

[90] A. H. Fainberg and S. Winstein, *J. Amer. Chem. Soc.*, 78, 2763, 2780 (1956).

[91] A. H. Fainberg, G. C. Robinson and S. Winstein, *J. Amer. Chem. Soc.*, 78, 2777 (1956).

[92] E. D. Hughes, C. K. Ingold, S. F. Mok, S. Patai and Y. Pocker, *J. Chem. Soc.*, 1265 (1957).

[93] A. F. Diaz and S. Winstein, *J. Amer. Chem. Soc.*, **91**, 4300 (1969).

[94] A. H. Fainberg and S. Winstein, *J. Amer. Chem. Soc.*, **78**, 2767 (1956).

[95] S. Winstein and E. Clippinger, *J. Amer. Chem. Soc.*, **78**, 2784 (1956).

it must be said that the equation fits well the facts. It is always obeyed in the solvolysis in acetic acid and in trifluoroacetic acid[93], which are solvents with low dielectric constant (Table 5.6), where undissociated salts are expected. Equation (5.34) does not exclude the specific effect of a salt. Indeed, according to equation (5.30), lithium perchlorate and lithium acetate as monovalent salts should display the same salt effect, but this is actually contradicted by experiments.

Another type of salt effect is the mass-law effect[3], discovered by Hughes, Ingold and their co-workers. As already shown (Section 5.2.1), if the carbocation is sufficiently stable to discern between different nucleophiles, as the counter-ion concentration is built up by the direct heterolysis, the reverse reaction gains in importance, so that the ionisation step is being retarded by its reversibility. On the other hand, the progress of the solvolysis produces the counter-ion, and thus the ionic strength will increase. The net effect observed will be either acceleration, or retardation, or no observable change. Some values collected from the classical works of Hughes, Ingold and their co-workers are presented in Table 5.10.

TABLE 5.10

Variation of integrated rate constants (s^{-1}) with conversion in some solvolyses

	Bu^tBr (90% aqueous acetone, 50° C)[a]							
% Conversion:	9.9	18.9	27.3	38.9	49.2	59.0	74.4	91.0
10^5k_1	19.3	19.4	19.7	20.5	20.5	21.2	21.7	23.0
	Ph_2CHCl (90% aqueous acetone, 50 °C)[b]							
% Conversion:	10.1	19.5	31.7	39.5	50.4	61.5	76.0	
10^5k_1	5.94	6.50	6.07	6.05	6.30	6.67	6.83	
	Ph_2CHCl (80% acetone, 25 °C)[b]							
% Conversion:	13.3	21.8	30.0	46.2	53.5	62.0		
10^5k_1	7.24	7.32	7.34	7.32	7.35	7.36		
	Ph_2CHBr (80% aqueous acetone, 25 °C)[b]							
% Conversion:	19.2	28.6	38.0	46.2	65.7	73.8	80.8	
10^5k_1	178	173	174	172	169	172	167	
	$(p\text{-}MeC_6H_4)_2CHCl$ (90% aqueous acetone, 0 °C)							
% Conversion:	8.0	16.2	27.0	35.7	57.0	69.2	79.7	
10^5k_1	8.68	8.30	7.83	7.53	7.05	6.87	6.70	

(a) Data of L. C. Bateman, E. D. Hughes and C. K. Ingold, *J. Chem. Soc.*, 960 (1940). (b) Data of M. J. Church, E. D. Hughes and C. K. Ingold, *J. Chem. Soc.*, 966 (1940). (c) Data of L. C. Bateman, E. D. Hughes and C. K. Ingold, *J. Chem. Soc.*, 974 (1940).

It can be observed that the mass-law effect is completely absent in the case of t-butyl bromide. If benzhydryl ion is generated, the observed effect depends on the leaving group (chloride or bromide) and even on the composition of the aqueous solvent. The lower the water ratio in the aqueous acetone, the more visible is the effect produced by the progressive rise of the ionic strength. In the case of the more stable cation, the bis-*p*-methylphenylmethyl carbocation, even with the poorest leaving group and in the solvent with the lowest ionisation power, the mass-law effect is visible.

In such and other cases, Hughes, Ingold and their co-workers[21–24] were able to demonstrate the typical S_N1 rate equation (5.21). Some significant values of the α parameters are: zero for t-butyl bromide; 10 for benzhydryl chloride; 60 for the corresponding bromide; 74 for *p,p'*-dimethylbenzhydryl chloride and 3×10^3 for triphenylmethyl chloride[96]. The α parameter, measuring the ability of the counter-ion to discern between different nucleophiles, is also a measure of the carbocation stability.

Winstein and their co-workers[97] have shown that Ingold's scheme regarding the unimolecular solvolysis (5.6) and (5.7) is incomplete, at least with solvents of low polarity, such as acetic acid. According to them, the following scheme is more complete:

$$\text{R—X} \underset{k_{-1}}{\overset{k_1}{\rightleftharpoons}} \underset{\substack{\textbf{16}\\ \text{Intimate}\\ \text{ion pair}}}{\text{R}^+\text{X}^-} \underset{k_{-2}}{\overset{k_2}{\rightleftharpoons}} \underset{\substack{\textbf{17}\\ \text{Solvent-}\\ \text{separated}\\ \text{ion pair}}}{\text{R}^+//\text{X}^-} \underset{k_{-3}}{\overset{k_3}{\rightleftharpoons}} \underset{\substack{\text{Dissociated}\\ \text{ions}}}{\text{R}^+ + \text{X}^-}$$

$$\text{SH}\downarrow k_s \qquad \text{SH}\downarrow k_s' \qquad \text{SH}\downarrow k_s''$$

$$\text{Products} \qquad \text{Products} \qquad \text{Products}$$

Scheme 1

The mechanism in Scheme 1 differentiates the *ionisation* from *dissociation*. The first step leads to an intimate ion pair (tight ion-pair, contact ions), which are further transformed into solvent-separated ion pairs. During these steps, both ions, the carbenium ion and its counter-ion, are not free and cannot move independently in solution. The collapse to the original covalent substrate RX is called *return*, namely *internal return* if the return is from the tight ion pairs and *external return* if the intermediates suffering collapse are either the solvent-separated ion pairs or the dissociated ions. The importance of ion pairs as intermediates in organic reactions is now definitely demonstrated, in nucleophilic substitution[98,99] as well as in other types of reactions[100].

[96] C. G. Swain, C. B. Scott and K. H. Lohmann, *J. Amer. Chem. Soc.*, **75**, 136 (1953).

[97] S. Winstein, E. Clippinger, A. H. Fainberg, R. Heck and G. C. Robinson, *J. Amer. Chem. Soc.*, **78**, 328 (1956).

[98] S. R. Hartshorn, *Aliphatic Nucleophilic Substitution*, Cambridge University Press, London (1973).

[99] R. A. Sneen, *Accounts Chem. Res.*, 6, 46 (1973).

[100] M. Szwarc (editor), *Ion Pairs*, J. Wiley, New York (1974).

Scheme 1 was originally demonstrated by the *special salt effect*. Fainberg and Winstein[94] have observed that adding variable amounts of lithium perchlorate in the acetolysis of 2-(*o*- or *p*-)anisylethyl tosylates, large non-linear variations of the rate constant occur at

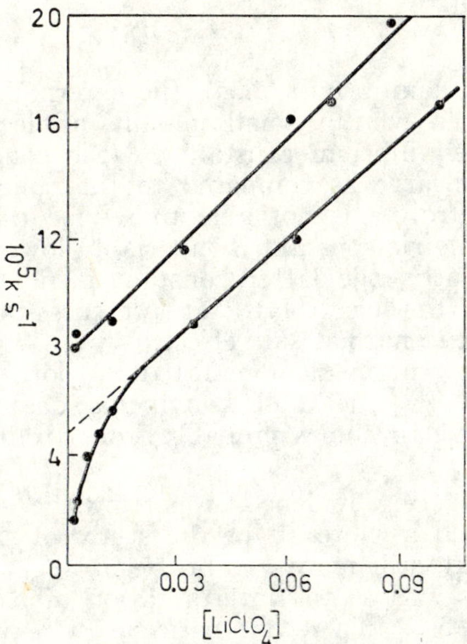

Figure 5.2 Lithium perchlorate salt effect in the acetolysis of *threo*-3-*p*-anisyl-2-butyl *p*-bromobenzenesulphonate (after Winstein and Robinson [101]).

small concentrations of the electrolyte, whereas at higher concentrations the rate increase is moderate and linear. A similar but more interesting example was found in the case of optically active *threo*-3-*p*-anisyl-2-butyl-*p*-bromobenzenesulphonate (Figure 5.2). The reaction could be followed both titrimetrically and polarimetrically. The polarimetric constant was much higher than the titrimetric constant ($k_p > k_t$), which is easily explained knowing that the polarimetric constant measures the ionisation rate, whereas the titrimetric constant corresponds to the rate of product formation (see however Section 5.3.2) and is thus affected both by internal and external return. When variable amounts of lithium perchlorate are added, the polarimetric rate constant is found to increase linearly with the salt concentration, according to the equation (5.34), whereas the titrimetric rate constant behaves like in the example studied by Fainberg and Winstein. The large rate enhancement produced at small concentrations of the salt was named *special salt effect* and proved to be a general phenomenon in neighbouring-group assisted reactions. Its interpretation is that at small amounts of electrolyte, the

[101] S. Winstein and G. C. Robinson, *J. Amer. Chem. Soc.*, 80, 169 (1958).

solvent-separated ion pair **17** suffers equilibration (5.35), with the formation of another ion pair **18** unable to collapse to a covalent species[102]:

$$\underset{}{LiClO_4} + \underset{\mathbf{17}}{R^+//X^-} \overset{K}{\rightleftharpoons} \underset{\mathbf{18}}{R^+//ClO_4^-} + LiX \tag{5.35}$$

As a result of equilibrium (5.35), the external return will be reduced or annihilated even by small amounts of electrolyte, whatever the value of the equilibrium constant K (the concentration of the electrolyte is very large as compared to the concentration of **17**). However, the electrolyte is not able to reduce the internal return and thus the titrimetric constant is invariably lower than the polarimetric one. Evidently, the latter constant is not affected directly by the addition of the electrolyte; it presents the usual increase corresponding to the normal salt effect.

Scheme 1 is now amply demonstrated by both kinetic and non-kinetic methods[103–105]. The most convincing demonstration of the return from an ion pair comes probably from product analysis (Section 5.4).

Despite the existing evidence supporting the general Scheme 1, which was proposed as a result of the study of salt effects in the acetolysis, there are still many doubts regarding the correctness of equation (5.34), as well as the interpretation of plots like those represented in Figure 5.2. In a recent criticism[106, 107], it was found that in the hydrolysis of 2-octyl methanesulphonate in 30% aqueous dioxane, at 43.5 °C, a better fit is obtained if in equation (5.34) the salt activity is introduced, instead of the stoichiometric concentration:

$$k = k_0(1 + b'\alpha_{salt}) \tag{5.36}$$

The authors observed that while b values decrease with increasing concentration of the salt, the b' parameter in (5.36) is fairly constant. With respect to the acetic acid solvent, even if the studied salts exhibited substantially different salt effects, the b' value increased with increasing nucleophilicity of the salt, $LiClO_4$ and $NaClO_4$ having negligible effects.

In the opinion of the present author, equation (5.34) is restricted to solvents with low polarity and cannot be used for aqueous solvents, as in the case discussed before. However, it seems that the phenomena explaining both equations (5.34) and (5.36), as well as the special

[102] The covalent perchlorate, even if formed, is rapidly solvolysed[18], the return being unsignificant for the overall kinetics.

[103] D. J. Raber, J. M. Harris, R. E. Hall and P. von R. Schleyer, *J. Amer. Chem. Soc.*, 93, 4821 (1971).

[104] C. D. Ritchie, *Accounts Chem. Res.*, 5, 348 (1972).

[105] F. G. Bordwell, *Accounts Chem. Res.*, 5, 374 (1972).

[106] G. A. Gregoriou, *Tetrahedron Letters*, 233 (1974).

[107] P. J. Dais, G. A. Gregoriou, *Tetrahedron Letters*, 3827 (1974).

salt effects are related. Indeed, let us reduce Scheme 1 and equilibrium (5.35) to Scheme 2:

$$\mathrm{R{-}X} \underset{k_{-1}}{\overset{k_1}{\rightleftharpoons}} \mathrm{R^+X^-} + \mathrm{MY} \underset{k_{-2}}{\overset{k_2}{\rightleftharpoons}} \mathrm{R^+Y^-} + \mathrm{MX}$$

16 or **17** $\xrightarrow{k_s}$ Reaction products; **18** $\xrightarrow{k_s}$ Reaction products

Scheme 2

Assuming that all intermediates have a constant small concentration and are rapidly consumed either in the reaction with the solvent or in the reaction with the *undissociated* salt, the steady-state treatment leads to the following rate law:

$$k = \frac{k_s k_1 (k_{-2}[\mathrm{MX}] + k_s) + k_s k_1 k_2 [\mathrm{MY}]}{(k_{-1} + k_s)(k_{-2}[\mathrm{MX}] + k_s) + k_2 k_s [\mathrm{MY}]} \tag{5.37}$$

If the last term in the denominator of (5.37) is neglected, relation (5.38) is obtained, which has the same form as relation (5.34):

$$k_{\mathrm{obs}} = \frac{k_s k_1}{k_{-1} + k_s}\left(1 + \frac{k_2}{k_{-2}[\mathrm{MX}] + k_s}[\mathrm{MY}]\right) \tag{5.38}$$

This term can be neglected if $k_{-1} \gg k_2\,[\mathrm{MY}]$, *i.e.* if the internal return dominates the ionisation process. However, the latter process is not important enough to allow the neglect of the other terms. Equation (5.38) would correspond to the reaction of a carbocation sufficiently stable to be transformed into a solvent-separated ion pair and to discern between the solvent and other nucleophiles present in the solution.

It is easily seen that b is given by the relation:

$$b = \frac{k_2}{k_{-2}[\mathrm{MX}] + k_s} \tag{5.39}$$

whereas the rate constant in the absence of any electrolyte is:

$$k_0 = \frac{k_s k_1}{k_{-1} + k_s} \tag{5.40}$$

The inspection of these equations shows that:

(i) b depends on the structure of the substrate and even more on the stability of the corresponding carbocation (through k_2/k_{-2});

(ii) b depends on the salt structure (through k_2), and on the nature of the counter-ion. Evidently, in solvents of low polarity in which dissociated ions are less probable than ion pairs[108], the undissociated salt must be involved. No direct relation is to be expected between the nucleophilicity of the salt and the value of b;

(iii) b must depend on solvent. If the solvent is highly nucleophilic b will vanish, and the solvolysis will not exhibit Winstein's salt effect. If the electrolyte has a nucleophilicity close to that of the solvent, dependence on the salt concentration is also to be expected.

It seems that the experimental data are in fairly good agreement with this interpretation. Moreover, if the carbocation is very stable, and thus possesses a high selectivity, the terms of equation (5.37) cannot be neglected, and no linear representation is to be expected. Anyway, the curve in Figure 5.2 seems to correspond to the plot of a function like (5.37).

A different salt effect is the *secondary salt effect*, observed in bimolecular reactions. The secondary effect is the result of the addition of a salt MZ, having the same cation as the nucleophilic salt, $Nu:^{-}M^{+}$. This will shift the equilibrium of the dissociation towards undissociated ion pairs:

$$Nu:^{-}M^{+} \overset{K}{\rightleftharpoons} Nu:^{-} + M^{+} \quad (5.41)$$

We must note that undissociated ion pairs are less nucleophilic[109–111] than the corresponding dissociated ions.

A specific effect of added electrolytes in a unimolecular reaction is the electrophilic catalysis. The accelerating effect of silver and mercury salts is well known [112] and is currently used in the reactions of less reactive substrates, such as chlorides. Loose ion pairs are generally assumed to result, as in the example of the optically active chloride **19** [113]:

$$\underset{\textbf{19}}{(p\text{-}ClC_6H_4)(C_6H_5)C(H)Cl} \xrightarrow{HgCl_2} p\text{-}ClC_6H_4(C_6H_5)CH^{+} + HgCl_3^{-} \quad (5.42)$$

The exceptionally high rates in the unimolecular solvolysis in trifluoroacetic acid and in other strong acids seem also to be due to electrophilic catalysis. Thus, if the values of the parameter ρ^*

[108] L. D. Petit and S. Bruckenstein, *J. Amer. Chem. Soc.*, 88, 4783, 4790 (1966).

[109] W. M. Weaver and J. D. Hutchison, *J. Amer. Chem. Soc.*, 86, 261 (1964); F. Guibe and G. Bram, *Bull. Soc. chim. France*, 933 (1975).

[110] H. Elias, O. Christ and E. Rosenbaum, *Chem. Ber.*, 98, 2725 (1965).

[111] H. Elias and H. Strecker, *Chem. Ber.*, 99, 1019 (1966).

[112] Y. Pocker and D. N. Kevill, *J. Amer. Chem. Soc.*, 87, 4760, 4771, 4778 (1965).

[113] A. Diaz, I. L. Reich and S. Winstein, *J. Amer. Chem. Soc.*, 92, 7598 (1970).

in Table 5.4 are plotted *versus* the pK_a values of the solvent, a straight line is obtained, demonstrating that the development of the positive centre is related to the acidity of the solvent.

In poor-ionising solvents, any cation seems to offer electrophilic assistance in ionisation. The controversy between Ingold *et al.* [114, 115], and Swain and his co-workers [8,11,116], regarding the nucleophilic substitutions of triphenylmethyl (trityl) chloride in benzene, seems to have arisen in connection with such a phenomenon. The latter research group observed that the reaction of trityl chloride with quaternary ammonium salts in benzene is unimolecular, the rate constant depending however on the nature of the anion. Moreover, the reaction of the azide was found to be bimolecular [116]. This result appeared at first sight to be erroneous, the trityl cation being one of the most stable carbocations, and the corresponding chloride ionising easily in a large variety of solvents. Hughes, Ingold and their co-workers challenged these data, but their own results, even if different from Swain's, contained the same puzzling observation (Table 5.11),

TABLE 5.11

Kinetics of the reaction of triphenylmethyl chloride in benzene with different nucleophiles

Salt		*Temperature* °C	*Rate constant* s^{-1}	*Literature*
Cation	*Anion*			
$Me_2\overset{+}{N}(C_{16}H_{32})_2$	Cl^-	50	7.9×10^{-5}	8, 11, 116
		30	3.3×10^{-5}	
	F^-	50	9.5×10^{-5}	
	N_3^-	50	bimolecular	
Bu_4N^+	Cl^-	30	0.505×10^{-5}	114, 115
		44.6	1.68×10^{-5}	
	N_3^-	30	2.5×10^{-5}	

namely that the first-order constants depended on the nucleophile. However, the reaction with azide appeared in their hands to be first-order. In their work, Hughes and Ingold demonstrated [117] that in benzene higher aggregates may be present, beside ion pairs, even at the low concentrations used in the experiments. They suggested triplets as intermediates which could account for the different rates. However, as two different quaternary ammonium salts were used in the two laboratories, it seems more likely that an electrophilic catalysis was involved. The first step would thus lead to a quadru-

[114] E. D. Hughes, C. K. Ingold, S. E. Mok, S. Patai and Y. Pocker, *J. Chem. Soc.*, 1220 (1957).

[115] E. D. Hughes, C. K. Ingold, S. Patai and Y. Pocker, *J. Chem. Soc.*, 1265 (1957).

[116] C. G. Swain and M. M. Kreevoy, *J. Amer. Chem. Soc.*, 77, 1122 (1955).

[117] E. D. Hughes, C. K. Ingold, S. Patai and Y. Pocker, *J. Chem. Soc.*, 1206 (1957).

plet (**20**) instead of an ion pair; **20** may further react as shown in Scheme 3:

$$R{-}Cl + M^+X^- \underset{k_{-1}}{\overset{k_1}{\rightleftharpoons}} R^+Cl^-M^+X^- + M^+X^- \underset{k_{-2}}{\overset{k_2}{\rightleftharpoons}} M^+X^-R^+Cl^- + M^+X^-$$

20 a $X^- = Nu{:}^-$ **21**

b $X^- = Cl^-$

$$\mathbf{21} \xrightarrow{k_3} R{-}X + M^+Cl^-$$

Scheme 3

Clearly, in the first step either the salt containing the nucleophile ($X^- = Nu{:}^-$) or the chloride progressively formed may act as catalysts. The rate constants of the formation of **20***a* and **20***b* are however equal, the same cation being involved in the catalysis. The second step corresponds to the formation of a new quadruplet **21** by an ion-pair displacement, the leading part in this process being probably played by the anion. The first step and the last step (the collapse of **21** to an unreactive covalent compound), must be independent of the nature of X. By noting $[M^+Nu{:}^-] + [M^+Cl^-] = [M^+X^-] =$ constant, the following rate equation is found:

$$\text{rate} = \frac{k_1k_2k_3[M^+X^-]\,[RCl]\,[M^+Nu{:}^-]}{k_{-1}k_{-2}[M^+X^-]^2 + k_3\{k_{-1}[M^+X^-] + k_2[M^+Nu{:}^-]\}} \tag{5.43}$$

Intermediate **20** may be regarded as an ion pair composed of the carbocation R^+ and the counter-ion $Cl^-M^+X^-$. If R^+ is bulky, as is the case of the triphenylmethyl cation [118], **20** should be more stable than **21**, *i.e.* $k_{-2} \gg k_2$. The last term in equation (5.43) can be then neglected, which gives:

$$\text{rate} = \frac{k_1k_2k_3[RX]\ [Nu{:}^-M^+]}{k_{-1}(k_{-2} + k_3)\ [M^+X^-]} \tag{5.44}$$

Equation (5.44) is a second-order rate equation, having the second-order rate constant (5.45), the value of which depends upon the initial salt concentration:

$$k_{obs} = \frac{k_1k_2k_3}{k_{-1}(k_{-2} + k_3)\,[M^+X^-]} \tag{5.45}$$

The data of Swain and Kreevoy [116] seem to confirm this reasoning. On the other hand, Hughes, Ingold and their co-workers [114,115] have observed that the rates measured decrease with conversion, the tabulated values being obtained by extrapolation, that is for

[118] See, for example, D. Gust and K. Mislow, *J. Amer. Chem. Soc.*, 95, 1535 (1973), and references quoted therein.

$[Nu{:}M] = [M^+X^-]$ = constant. In that case, the salt concentration disappears from relation (5.44), which reduces to a first-order rate, having the observed rate constant (5.46):

$$k'_{obs} = \frac{k_1 k_2 k_3}{k_{-1}(k_{-2} + k_3)} \tag{5.46}$$

Clearly, this rate constant should depend on the nucleophile (through k_2 and k_3) and on the cationic partner of the nucleophile (through k_1). Table 5.11 shows that this is actually the case. This interpretation seems to indicate that the results obtained in both laboratories were correct and not contaminated by some undetectable catalysis.

The concept of ion pairs in the nucleophilic substitution will be discussed in Section 5.4.

5.2.5 *Kinetic isotope effects*

The kinetic isotope effects are currently used in the study of other types of mechanisms and to a lesser extent in nucleophilic substitution. The reason lies partly in the fact that the broken bond C—X involves the heavy atom X (X = halogen, oxygen, etc.), giving rise to negligible isotope effects (see p. 88). However, modern techniques have allowed the detection of heavy-atom kinetic isotope effects. Thus, Groczyk and Taylor [119] have studied chlorine isotope effect, k_{35}/k_{37}, in the reaction of *p*-methoxybenzyl chloride with sodium azide, in 70% aqueous acetone. The benzyl chloride with ^{37}Cl presents an isotope effect between 1.00786 and 1.01049, the effect increasing with increasing concentration of azide. It is believed that this result indicates the intervention of ion pairs in the reaction, although different opinions [120] have also been expressed (see also Section 5.4).

Secondary isotope effects [121], namely α- *and* β-*deuterium isotope effects*, are used more frequently.

The secondary α-deuterium isotope effect [122] arises because the change in the vibrational frequency of the bond, in going from the ground state to the transition state, is not the same for C—H and for C—D (due to the rehybridisation of the central carbon atom).

For most unimolecular solvolyses and radical reactions, fairly constant values of the α-deuterium isotope effect, $k_H/k_D = 1.12$,

[119] D. G. Groczyk and J. W. Taylor, *J. Amer. Chem. Soc.*, **96**, 3255 (1974).

[120] V. F. Raaen, T. Iuhlke, F. J. Brown and C. J. Collins, *J. Amer. Chem. Soc.*, **96**, 5928 (1974).

[121] Reviews: W. H. Saunders, Jr., *Survey of Progress in Chemistry*, vol. 3, Academic Press, New York (1966), p. 109; E. A. Halevi, *Progr. Phys. Org. Chem.*, **1**, 109 (1963).

[122] A. Streitwieser, Jr., R. H. Jagow, R. C. Fahey and S. Suzuki, *J. Amer. Chem. Soc.*, **80**, 2326 (1958).

have been found [123]. However, it has been shown [124, 125] that such values correspond to non-limiting (border-line) solvolyses and depend on the degree of nucleophilic involvement at the reaction centre, on the leaving group and on ion-pair partitioning effects. For "true limiting solvolyses", involving such substrates as 2-adamantyl derivatives [126] **22** which are not susceptible to backside attack, values as high as 1.26 can be obtained.

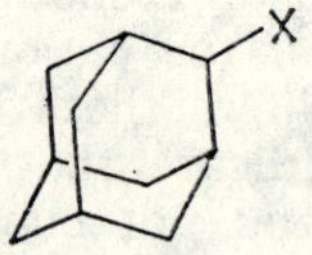

22

The secondary β-isotope effect is believed to be determined by hyperconjugation [127, 128]. Indeed, in the transition state of a unimolecular reaction, a C—H bond (in **23**) would better hyperconjugate with the *p* orbital developing at the reaction centre, than would do a C—D (**24**):

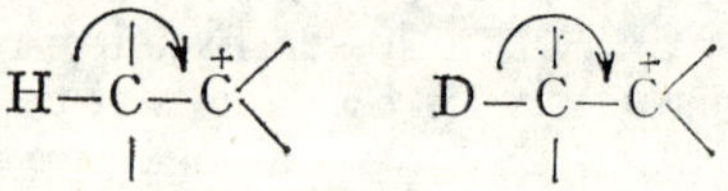

23 **24**

Secondary isotope effects, especially α-deuterium effects, are considered to be very efficient means for detecting the limiting and border-line character, respectively, of a solvolysis [124, 125]. In bimolecular reactions, k_H/k_D is unity [129] (or slightly less), allowing thus a correct diagnosis.

5.3 The stereochemical criterion

The stereochemistry of a reaction is one of the tools which, used adequately, may afford more precise information about the geometry

[123] S. Seltzer, *J. Amer. Chem. Soc.*, **83**, 2625 (1961).
[124] J. M. Harris, R. E. Hall and P. von R. Schleyer, *J. Amer. Chem. Soc.*, **93**, 2551 (1971).
[125] V. J. Shiner, Jr., and R. D. Fischer, *J. Amer. Chem. Soc.*, **93**, 2553 (1971).
[126] J. L. Fry, C. J. Lancelot, L. K. M. Lam, J. M. Harris, R. C. Bingham, D. J. Raber, R. E. Hall and P. von R. Schleyer, *J. Amer. Chem. Soc.*, **92**, 2538 (1970).
[127] E. S. Lewis, *Tetrahedron*, **5**, 143 (1959).
[128] V. J. Shiner, *Tetrahedron*, 5, 243 (1959).
[129] V. J. Shiner, Jr., M. W. Rapp and H. R. Pinnick, *J. Amer. Chem. Soc.*, **92**, 232 (1970).

of the transition state. In studying the stereospecificity of a reaction, the following methods may be used [130]:

(a) The use of optically active reagents. The optical purity of the reagent must be known, even if it is not absolutely necessary to use optically pure substrates.

(b) The use of diastereomers. It has the advantage that it is not necessary to resolve racemic mixtures and to check their optical purity. In many instances it may afford the same information as when the pure enantiomer is used.

(c) If cyclic or alkenic compounds are used, *cis-trans* isomerism may be used for stereochemical purposes.

(d) A relatively new method [131] involves the n.m.r. technique. It has the advantage over the other methods that it has less limitations regarding the reaction centre.

The stereochemistry of the aliphatic nucleophilic substitution was an important type of evidence in the classical works of Hughes and Ingold [132] which demonstrated the S_N2-S_N1 duality.

5.3.1 *The bimolecular mechanism*

The stereochemistry of the concerted bimolecular mechanism depends on the direction of attack of the nucleophile, *i.e.* on the geometry of the transition state. There are two essentially different geometries of the transition states, according to the angle formed by the incoming and outgoing groups with the carbon atom under attack:

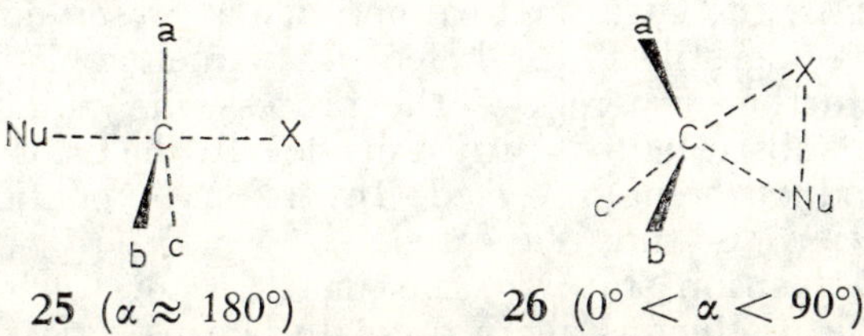

25 ($\alpha \approx 180°$) **26** ($0° < \alpha < 90°$)

In one of the transition states, **25**, the angle is about 180°, and no interaction arises between the incoming nucleophile and the leaving group. In order to build up the transition state from substrate and reagent, the nucleophile must produce a rearside attack. In Figure 5.3, the progress of the attack is followed using a conventional chiral substrate. If the order of priority of the substituents about the chiral centre is X, Nu > a > b > c, configuration R may be attributed to the substrate molecule. As the reaction proceeds to the uphill part of the diagram, the substituents a, b, c move in the plane of the

[130] E. Eliel, *Stereochemistry of Carbon Compounds*, McGraw-Hill, New York (1962), p. 116–119.

[131] C. A. Kingsbury and W. B. Thornton, *J. Amer. Chem. Soc.*, **88**, 3159 (1966).

[132] See Reference 1, p. 372.

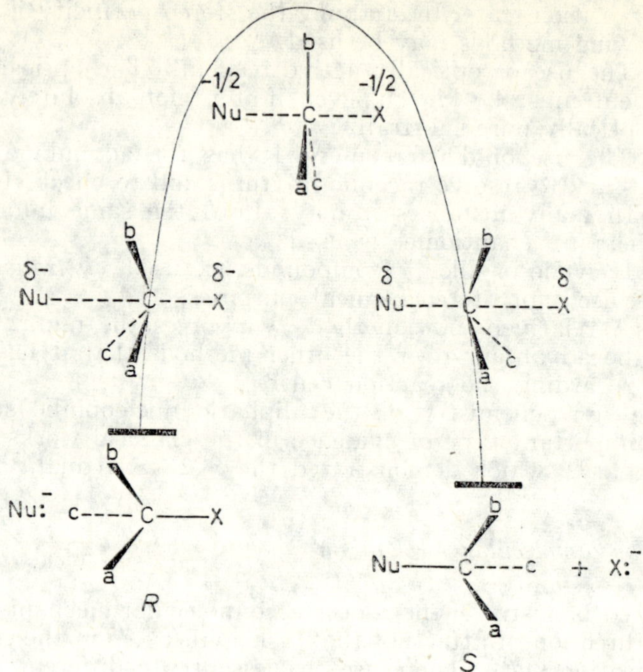

Figure 5.3 Progress of a nucleophilic attack at a chiral centre.

chiral atom. In the limit geometry, with the planar disposition of the three bonds, the entire optical activity is preserved if Nu: $\neq$ X:, the species having still a chiral element, namely a plane of chirality. During the further progress of the process the "umbrella" formed by the three substituents is turned inside out, the priority order in the resultant product being reversed. Inversion of configuration, with no racemisation, is thus expected.

In the transition state **26**, the attack occurs on the side of the old bond. The product is again optically active, but the initial configuration is entirely retained this time. It must be noted that in this transition state the electronic interactions between the incoming and outgoing group may not be neglected.

By using symmetric substitutions, (5.47), Hughes, Ingold and their co-workers [133–136] were able to distinguish between the two transition states:

$$^{*}X^{-} + R{-}X \longrightarrow {}^{*}X{-}R + X^{-}, \; ({}^{*}X^{-} = \text{radioactive halide}) \qquad (5.47)$$

[133] E. D. Hughes, F. Juliusburger, S. Masterman, P. Topley and J. Weiss, *J. Chem. Soc.*, 1525 (1935).

[134] E. D. Hughes, F. Juliusburger, A. D. Scott, B. Topley and J. Weiss, *J. Chem. Soc.*, 1173 (1936).

[135] W. A. Cowdrey, E. D. Hughes, T. P. Nevell, C. L. Wilson, *J. Chem. Soc.*, 209 (1938).

[136] P. Brewster, F. Hiron, E. D. Hughes, C. K. Ingold and P. A. D. Rao, *Nature*, **166**, 178 (1950).

An actual reaction used was for instance:

$$\underset{}{{}^{*}I^- + CH_3(CH_2)_5-\underset{\substack{|\\ I\\ R}}{\overset{\substack{H\\ |}}{C}}-CH_3} \longrightarrow CH_3(CH_2)_5-\underset{\substack{|\\ H\\ S}}{\overset{\substack{*I\\ |}}{C}}-CH_3 + I^-$$

In order to understand better their arguments, let us refer back to Figure 5.3. In the case of a symmetric substitution (Nu: = X:), the transition state, with the three bonds of the central carbon lying in the same plane, has a symmetry plane, which means that chirality is lost. Whatever the further evolution of the transition state, the substrate undergoing nucleophilic attack will lose its activity. Hughes, Ingold and their co-workers found that the rate of racemisation in the above reaction was twice that of radioactive halide exchange. This indicates an inversion of configuration for every molecule of tagged organic halide formed, which is consistent with a bimolecular mechanism of exchange.

Doubts have been expressed with regard to the exclusive preference of transition state **25** in the bimolecular substitution. Electrostatic repulsions occurring in reactions of type I (Table 5.1) should favour **25** over **26**. In that case, reactions of type (5.3), with negative nucleophile and initially positive leaving group, may have a preference for the transition state **26**, in which the configuration is retained. Hughes, Ingold and their co-workers [137, 138] have verified this hypothesis by subjecting the optically active 1-phenyl-ethylchloride to the following transformations:

$$(-)\text{-PhCHClMe} \xrightarrow[\substack{S_N2\\ \text{Inversion of configuration}}]{N_3^-} \underset{\substack{|\\ N_3}}{\text{PhCHMe}} \xrightarrow[\text{No attack at the chirality centre}]{H_2} (+)\text{-}\underset{\substack{|\\ NH_2}}{\text{PhCHMe}}$$

(−)-PhCHClMe $\xrightarrow{SH^-\ (S_N2)}$ (downward; Inversion of configuration)

$$\underset{\substack{|\\ SH}}{\text{PhCHMe}} \xrightarrow[\text{No attack at the chirality centre}]{MeI} (+)\text{-}\underset{\substack{|\\ {}^{+}SMe_2}}{\text{PhCHMe}} \xrightarrow{N_3^-} \underset{\substack{|\\ N_3}}{\text{PhCHMe}}$$

$$\xrightarrow[\text{No attack at the chirality centre}]{H_2} (-)\text{-}\underset{\substack{|\\ NH_2}}{\text{PhCHMe}}$$

[137] F. Hiron and E. D. Hughes, *J. Chem. Soc.*, 795 (1960).

[138] S. H. Harvey, P. A. T. Hoye, E. D. Hughes and C. K. Ingold, *J. Chem. Soc.*, 806 (1960).

From earlier works, an inversion of configuration is known to occur in the reaction of the chloride with azide ion or with hydrogen sulphide ion SH^-. On the other hand, the reduction of the azide to the corresponding amine, as well as the transformation of the thiol into the sulphonium ion, involving no attack at the chirality centre, must preserve the original configuration. If the optical activity of the amine obtained by direct azide attack on the chloride is compared with the activity of the amine obtained from the sulphonium salt, the two amines are found to be enantiomeric. This shows that another inversion of configuration is involved in the latter series of reactions. The inversion can only occur during the azide attack on the sulphonium salt.

These and other data show that the backside attack of the reagent (transition state **25**), and therefore the inversion of configuration, constitute an absolute stereochemical rule for bimolecular substitutions.

This rule could have been anticipated on theoretical grounds. Indeed, either in transition state **25** or in transition state **26** four electrons are distributed between three nuclei. It is easily seen that we deal here with systems resembling the four-electron allyl system (in case of **25**) or the cyclopropenyl system (**26**). It has been shown previously that the cyclopropenyl anion is antiaromatic, whereas the allyl anion is stabilised by conjugation (p. 71). Transition state **26** is thus ruled out, being energetically unfavourable. On the other hand, if only two electrons are to be distributed between the three centres (electrophilic substitution), a preference for transition state **26** is evident, the cyclopropenylium cation having a larger delocalisation energy then the allyl cation. Such qualitative considerations based on MO theory, even if far from the reliability of calculations, are useful in decisions of first approximation[139]. LCAO-MO-SCF calculations for simple displacements [140–142] have confirmed that backside attack is the single possible process for nucleophilic substitution and that frontside attack (transition state **26**) is largely favoured in most of the electrophilic substitutions. Experiments agree entirely with the results of these calculations.

5.3.2 *The unassisted unimolecular mechanism*

Ingold and his co-workers [132] have established that unimolecular reactions occur with partial racemisation, accompanied either by retention or by inversion of configuration. The ratio between the

[139] See, for example, M. J. S. Dewar, *The Molecular Orbital Theory of Organic Chemistry*, McGraw-Hill, New York (1969), p. 290.

[140] H. Fujimoto, S. Yamabe and K. Fukui, *Tetrahedron Letters*, 439, 443 (1971).

[141] A. Dedieu and A. Veilard, *J. Amer. Chem. Soc.*, **94**, 6730 (1972); see D. K. Bohme, G. I. Mackay and J. D. Payzant, *J. Amer. Chem. Soc.*, **96**, 4027 (1974), for a confirmation of the theoretical calculations.

[142] R. F. Bader, A. J. Dake and R. R. Messer, *J. Amer. Chem. Soc.*, **95**, 7715 (1973).

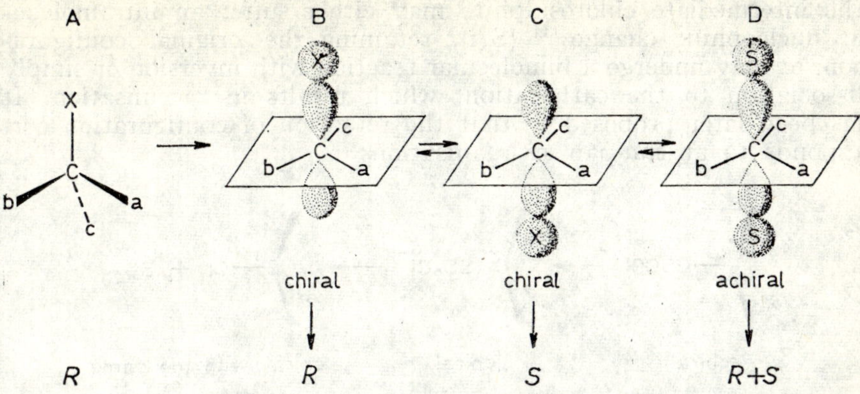

Figure 5.4 Stereochemistry of unassisted unimolecular nucleophilic substitutions; A. Initial state; B., C. ion pairs; D. dissociated ions (symmetrically solvated ions).

racemisation process and the stereospecific reaction (inversion or retention) depends largely on the stability of the carbocation.

However, one finds that this rule is rarely obeyed and may be subject to criticism [3, 100]. The main violations of it refer to participation reactions (anchimerically assisted reactions) and will be treated in the following section. However, exceptions to Ingold's rule are not restricted to anchimeric assistance. It is assumed that other exceptions are accounted for by ion-pair reactions[99]. Let us follow, for instance, the reaction scheme shown in Figure 5.4. The substrate A, with a chiral centre, ionises to the tight ion-pair B, which has a chiral plane. Optical activity is not lost in this step. The ion pair B can racemise by equilibration with the ion pair C. However, if capture of B is more rapid than ion-pair equilibration B $\rightleftharpoons$ C, inversion of configuration will be observed.

Ion-pair reorganisations have frequently been proposed for the retention mechanism. Thus, in the substitution of OH group by chloride, either retention, or inversion, or partial racemisation and partial inversion, are observed if the reagent is thionyl chloride [143]. The following reaction scheme has been proposed in order to account for the observed stereochemistry:

$$\mathrm{RCl} + \mathrm{SO_2} \qquad \text{(a)}$$

$$\uparrow \text{Retention} \qquad (5.48)$$

$$\mathrm{ROH} + \mathrm{SOCl_2} \rightarrow \mathrm{R{-}O{-}S(Cl){=}O} \rightarrow \mathrm{ROSO^+} + \mathrm{Cl^-} \rightarrow \underset{\text{Inversion}}{\mathrm{ClR} + \mathrm{SO_2}} \qquad \text{(b)}$$

$$\downarrow$$

$$\mathrm{R^+} + \mathrm{SO_2} + \mathrm{Cl^-} \rightarrow \underset{\text{Racemisation}}{\mathrm{RCl} + \mathrm{ClR}} \qquad \text{(c)}$$

[143] W. A. Cowdrey, E. D. Hughes, C. K. Ingold, S. Masterman and A. D. Scott, *J. Chem. Soc.*, 1252 (1937), namely p. 1267.

The intermediate chlorosulphite may either suffer an intramolecular nucleophilic change [143] (S_Ni), retaining the original configuration, or may undergo a bimolecular reaction with inversion or simply dissociation to the carbocation, which results in racemisation. It has been later proposed [144] that the retention of configuration corresponds to an ion-pair reorganisation:

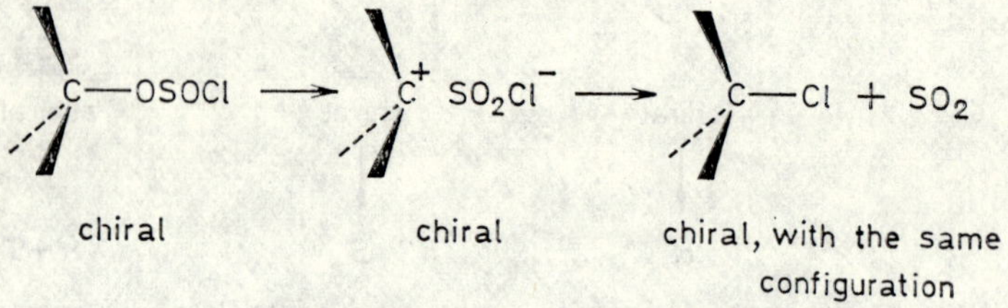

$$\rangle C—OSOCl \longrightarrow \rangle C^+ \; SO_2Cl^- \longrightarrow \rangle C—Cl + SO_2$$

chiral — chiral — chiral, with the same configuration

In the early theory of Hughes, Ingold and their co-workers, the solvation of the intermediate carbocation was not considered in detail with regard to the stereochemistry of the reaction. Later works [3] have indicated that carbocations may either be symmetrically or unsymmetrically solvated, leading thus to non-stereospecific or stereospecific reactions, respectively (Scheme 4):

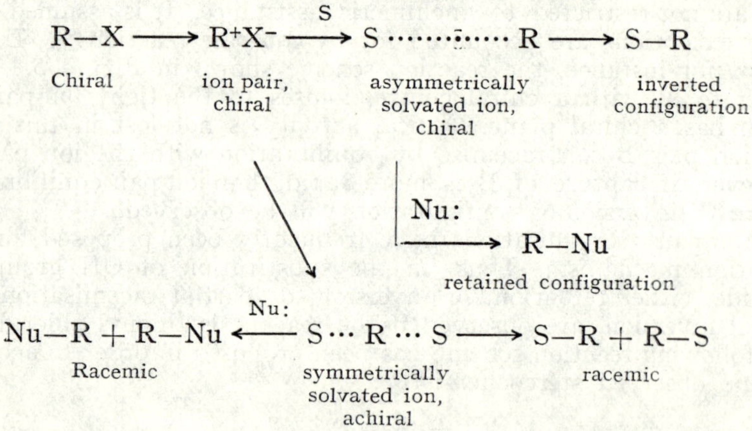

$$R—X \longrightarrow R^+X^- \xrightarrow{S} S \cdots\cdots R \longrightarrow S—R$$

Chiral — ion pair, chiral — asymmetrically solvated ion, chiral — inverted configuration

$$S \cdots\cdots R \xrightarrow{Nu:} R—Nu$$

retained configuration

$$Nu—R + R—Nu \xleftarrow{Nu:} S \cdots R \cdots S \longrightarrow S—R + R—S$$

Racemic — symmetrically solvated ion, achiral — racemic

Scheme 4

Thus, racemisation will occur only if the ion-pair stage has been surpassed by the reacting system. Data supporting this conclusion have been obtained by von Doering and Streitwieser [145], in the methanolysis of the hydrogen phthalate **27**:

[144] D. J. Cram, *J. Amer. Chem. Soc.*, **75**, 332 (1953).

[145] W. von E. Doering and A. Streitwieser, Jr., quoted in Reference 3, p. 656.

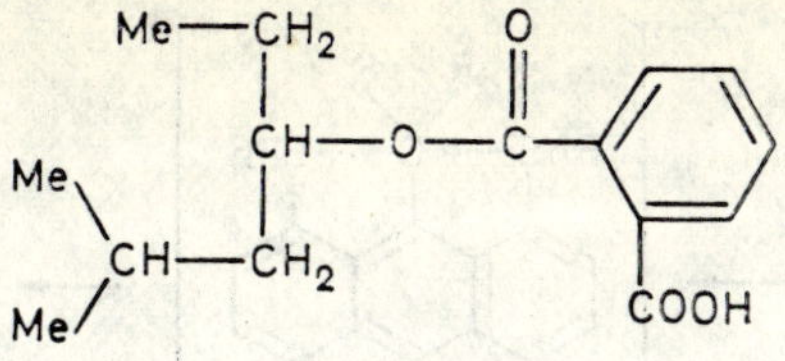

In pure methanol, **27** suffers 60% inversion and 40% racemisation. In methanol-benzene and methanol-carbon tetrachloride mixtures, the stereochemistry is not appreciably affected. However, addition of acetonitrile to the solvent mixture may lead to 100% racemisation. It is believed that only acetonitrile may solvate symmetrically the carbocation; in the other solvent systems the asymmetrically solvated ions predominate.

The question of symmetrically solvated species was first studied by Breslow and his co-workers [146, 147]. Taking into account the fact that such species must be highly disfavoured by the activation entropy, the authors have chosen as a model the cation **28**, where the symmetric solvation may be achieved by an intramolecular process:

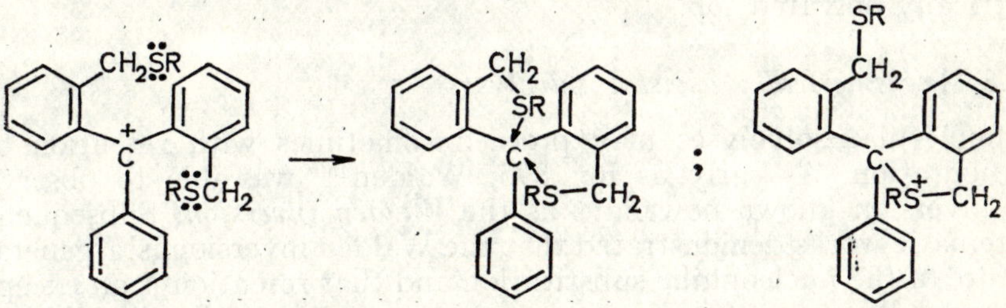

On the other hand, cation **28** must be sufficiently stable to allow the conversion into the symmetrically-solvated species **29**. However, n.m.r. studies have definitely ruled out structures **29** and confirmed the unsymmetrically-solvated ion **30**. It was however thought that the particular geometry of the trityl ion [118], with rotated rings, may prevent the access of mercapto groups to the cationic centre. The anthracene derivative **31** was subsequently studied by Basalay and Martin [148], the peri location of the electron-donor substituents enabling the access to the electron-deficient centre.

[146] R. Breslow, S. Garratt, L. Kaplan and D. La Folette, *J. Amer. Chem. Soc.*, **90**, 4051 (1968).

[147] R. Breslow, L. Kaplan and D. La Folette, *J. Amer. Chem. Soc.*, **90**, 4056 (1968).

[148] R. J. Basalay and J. C. Martin, *J. Amer. Chem. Soc.*, **95**, 2565, 2572 (1973).

PhS C+ SPh (R, R) → [PhS→C←SPh (R, R)]$^+$ → Ph$\overset{+}{S}$—C SPh (R, R)

31 32 33

As in the case of the triphenylmethyl cation, it was found that the symmetrically solvated species **32** cannot be detected. Rapidly equilibrating sulphonium salts **33** may instead account for the observed n.m.r. spectra.

The impossibility of detecting symmetrical species such as **29** or **32** does not alter the S_N1 picture given in Scheme 4. It has been shown that solvent reorganisation about a cation occurs at a constant rate [149] of no less than 10^{10} s^{-1}. Thus, even if symmetrically solvated species are excluded, a rapid equilibrium:

$$S \cdots\cdots \overset{+}{R} \rightleftharpoons \overset{+}{R} \cdots\cdots S \qquad (5.49)$$

must be admitted. Obviously, under the usual conditions of a solvolytic reaction, the equilibrating species are equivalent to a symmetrically solvated ion.

5.3.3 *Anchimerically assisted solvolyses*

Unimolecular solvolyses may proceed sometimes with retention of configuration. As early as in 1895, Walden[150] was able to observe the inversion known nowadays as the *Walden inversion.* Subsequent systematic works demonstrated that the Walden inversion is a general feature of the nucleophilic substitution and that retention is an exception, usually encountered in unimolecular reactions of such substrates as α-halogenoacids. In the case of the base-catalysed hydrolysis of chlorosuccinic acid to malic acid (5.50) carried out originally by Walden, Cowdrey, Hughes and Ingold [151] demonstrated that the retention of configuration is not conditioned by the use of silver ion:

$$\begin{array}{ccc} (+)\text{-chlorosuccinic acid} & \xrightarrow[H_2O]{Ag_2O} & (+)\text{-malic acid} \\ PCl_5 \uparrow \quad \downarrow KOH & & PCl_5 \downarrow \quad \uparrow KOH \\ (-)\text{-malic acid} & \xleftarrow[H_2O]{Ag_2O} & (-)\text{-chlorosuccinic acid} \end{array} \qquad (5.50)$$

[149] J. T. Keating and P. S. Skell, in *Carbonium Ions*, vol. 2 (G. A. Olah and P. von R. Schleyer, editors), Wiley-Interscience, New York (1970), p. 573.

[150] P. Walden, *Ber. dtsch. chem. Ges.*, 28, 1287, 2766 (1895); P. Walden, *Optische Umkehrerscheinungen*, Vieweg, Braunschweig (1919).

[151] W. A. Cowdrey, E. D. Hughes and C. K. Ingold, *J. Chem. Soc.*, 1208 (1937); W. A. Cowdrey, E. D. Hughes, C. K. Ingold, S. Masterman and A. D. Scott, *J. Chem. Soc.*, 1252 (1937).

They proved that the reaction with potassium hydroxide is a typical bimolecular substitution, occurring with complete inversion of configuration, whereas the retention process involves a unimolecular ionisation. The same stereochemistry may be observed either in buffered solution or in presence of silver ion (which catalyses the formation of the carbocation). The abnormal situation is due to the ability of the α-carboxyl group to prevent the rear-side nucleophilic attack on the carbocation. The effect was thought to be weak and mainly of electrostatic character. Ingold [152] concluded that "... it is necessary that the protecting goup should not be so nucleophilic and should not be so favourably positioned, that it intervenes before the heterolysis has passed through its transition state, and thereby converts what might have been a unimolecular, into an internal bimolecular substitution".

This theory regarding the retention mechanism in the solvolysis of α-halogenoacids was considered from another point of view by Grunwald and Winstein [153]. These authors demonstrated that the involvement of the carboxylate group at the reaction centre is more profound, as it participates in the ionisation. The reaction is a true internal S_N2 process, occurring with inversion of configuration. An intermediate unisolable α-lactone is formed. The rapid hydrolysis of this lactone is in its turn a bimolecular reaction, involving a Walden inversion. Thus, the overall stereochemical outcome, the retention of configuration, is explained by two successive Walden inversions at the same carbon atom:

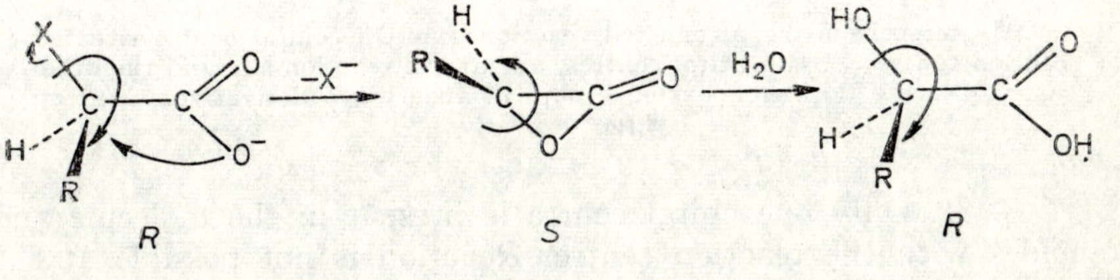

(5.51)

Neighbouring-group participation in nucleophilic displacements is now well demonstrated. Most of the abnormal situations involving retention of configuration are explained by the formation of bridged intermediates. From the stereochemical point of view, three cases deserve special mention (Figure 5.5). For the sake of simplicity, only β-neighbouring groups are considered.

[152] C. K. Ingold, in Reference 1, p. 383.

[153] E. Grunwald and S. Winstein, *J. Amer. Chem. Soc.*, **70**, 841 (1948).

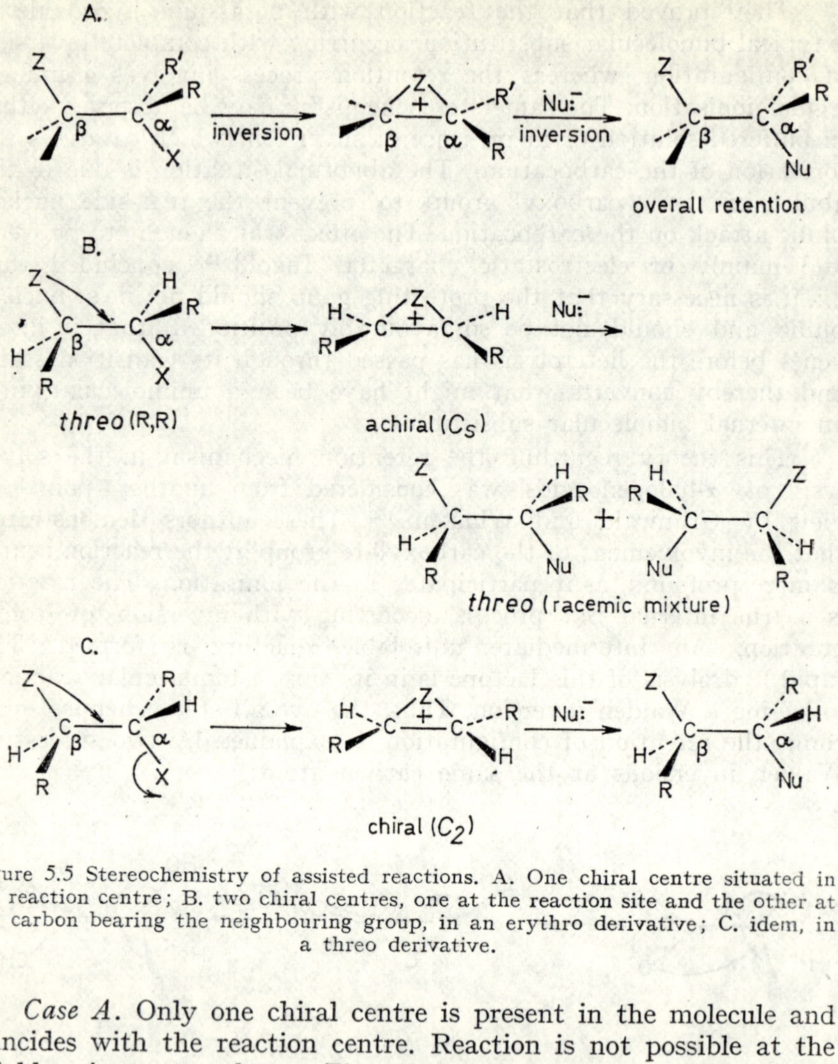

Figure 5.5 Stereochemistry of assisted reactions. A. One chiral centre situated in the reaction centre; B. two chiral centres, one at the reaction site and the other at the carbon bearing the neighbouring group, in an erythro derivative; C. idem, in a threo derivative.

Case A. Only one chiral centre is present in the molecule and coincides with the reaction centre. Reaction is not possible at the neighbouring group locus. The case corresponds to hydrolysis of α-halogenoacids (5.51). *Retention of configuration* is the outcome in this case. The stereospecificity of the reaction (the net retention ratio) will depend on the intervention of other possible mechanisms beside that shown in Figure 5.5 (A) (which is an example of simple assisted ionisation).

In case B and case C reactions are possible at both loci of the bridged carbon atoms. The two cases involve two chirality centres in the molecule, situated at C_α and C_β. The difference between the two examples arises from the fact that the original substrates are diastereomers.

In the case of an erythro derivative, the initial substrate with C_1 symmetry passes over into the bridged intermediate with C_2 symmetry. Chirality will be preserved in this step, the bridged intermediate having chiral properties. The reaction at both centres, α and β, will lead to the same product, having the same absolute configuration as the substrate. *Absolute retention of configuration is predicted for such situations.* However, it must be observed that the retention of configuration at both centres, C_α and C_β, is not specific for the bridging mechanism. Simple ionisation at C_α, with the protecting action (but not chemical interaction) of the neighbouring group, as in Ingold's theory, will also account for this stereochemistry.

In the second case, the *threo*-derivative gives rise to a bridged intermediate having a mirror symmetry plane (group C_s). Chirality is thus lost in this step, and the product of the reaction must be optically inactive. Indeed, the attack at C_α produces the enantiomer of the C_β-attack product, and a racemic mixture is formed. It must be observed however, that the relative configuration is preserved, a *threo*-derivative results from the overall process. The stereochemical rule corresponding to this case is: *loss of the absolute configuration, retention of the relative configuration.* This case is *specific* for a bridging mechanism and affords one of the most convincing arguments supporting the latter, being used by Winstein and Lucas[16] in the first known demonstration of a neighbouring group participation. The same argument was used by Cram[47] for proving the β-phenyl participation in a nucleophilic displacement. In the latter case, the stereochemical result was the single compelling evidence for phenonium ion formation, the kinetic data being of little use.

Other examples related to this problem will be treated in subsequent sections of this chapter.

5.4 Product criterion

Product analysis is the first and foremost convincing piece of evidence in the file of a chemical enigma. However, when used alone, very rarely does it give a definitive answer regarding the reaction mechanism. This is understandable considering that products are formed in the last step of the reaction, which usually is preceded by other steps, either of the same, or of lower, or of higher energy barrier than the activation energy of the product-formation step. The structure of products may not reflect the former steps, and, anyway, even if it does, it gives no hint about the energetics of the steps involved. The combination of the kinetic methods, with *complete* product analysis of the reaction mixture (including the stereochemistry), is largely used in any mechanism study and particularly in the nucleophilic substitution.

5.4.1 *The rearrangement of cations*

The rearrangement of the carbon framework in a nucleophilic substitution is one of the most convincing arguments pointing to a unimolecular mechanism.

Rearrangements in reactions involving carbocations have been known for a long time. Thus, Fittig [154] in 1860 recorded that pinacol treated with sulphuric acid yields pinacolone:

$$\underset{\text{OH}}{\underset{|}{Me_2C}}\text{—}\underset{\text{OH}}{\underset{|}{CMe_2}} \xrightarrow[-H_2O]{H^+} Me\text{—}CO\text{—}CMe_3 \quad (5.52)$$

Soon afterwards it was found that the reaction is general for ditertiary alcohols [155], a related example being the conversion of benzpinacol into benzpinacolone:

$$\underset{\text{OH}}{\underset{|}{Ph_2C}}\text{ —— }\underset{\text{OH}}{\underset{|}{CPh_2}} \xrightarrow[-H_2O]{H^+} PhCO\text{—}CPh_3 \quad (5.53)$$

However, these early examples of *pinacol rearrangement* were far from demonstrating the generality of rearrangement reactions and their involvement in carbocation reactions. More systematic work in this field was begun at the end of the last century by Wagner [156], who studied the conversions of the dicyclic terpenes. The subsequent work of Meerwein [157, 158] on the same subject brought a major contribution to rearrangement reactions. Thus, in the course of a very detailed study of the isomerisation of camphene hydrochloride **34** to isobornyl chloride **35** Meerwein and van Emster [158] established that:

(i) the rearrangement is reversible, camphene hydrochloride and isobornyl chloride being the single equilibrating species at low temperature; at higher temperature, bornyl chloride may also be formed;

(ii) the rearrangement is *unimolecular and intramolecular*;

(iii) the rearrangement takes place in a cation formed from the substrate by ionisation.

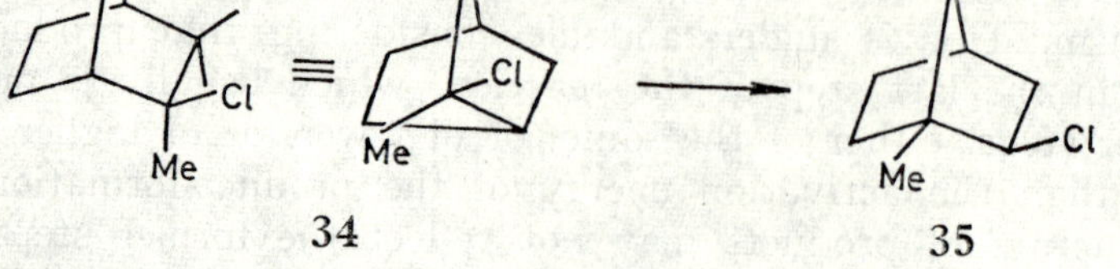

Cl
Me
≡
Me Me
Cl
Me
→
Me Me
Cl
Me

34 **35**

[154] R. Fittig, *Liebigs Ann. Chem.*, **114**, 54 (1860).
[155] W. Thörner and T. Zinke, *Ber. dtsch. chem. Ges.*, **10**, 1473 (1877).
[156] G. Wagner, *J. Russ. Phys. Chem. Soc.*, **31**, 680 (1899).
[157] H. Meerwein and K. van Emster, *Ber. dtsch. chem. Ges.*, **53**, 1815 (1920).
[158] H. Meerwein and K. van Emster, *Ber. dtsch. chem. Ges.*, **55**, 2500, namely p. 2520 (1922).

Subsequent work [159, 160] proved that the theory of Meerwein was essentially correct, the rearrangement of carbocations being a general phenomenon, involved in many important technological processes [161, 162] (see also Chapter 7). Ingold [159], discussing nucleophilic rearrangements (rearrangements occurring in the course of a nucleophilic substitution), showed that the most frequent are the 1,2-shifts of either hydride, alkyl or aryl groups (5.54), some of them having a well-defined stereochemistry at the three reaction centres.

$$\begin{array}{ccccccc} R & & & & & B & \\ | & & & & & | & \\ A & — & B & \longrightarrow & A & — & B + X{:}^- \\ & & | & & | & & \\ Nu{:}^- & & X & & Nu & & \end{array} \qquad (5.54)$$

Thus, 1,2-shifts usually take place with inversion of configuration at the *migrating origin* A and at the *seat of dissociation* B, the *migrating group* R retaining its configuration. It has been later shown that these rules apply to processes involving bridged ions.

The recognition of Wagner and Meerwein's contribution to the understanding of carbenium ion rearrangements is marked by the name of the 1,2-shift (5.54) of an alkyl group, which is called *Wagner-Meerwein rearrangement.*

In order to understand better the processes involved in a 1,2-shift, let us consider Figure 5.6:

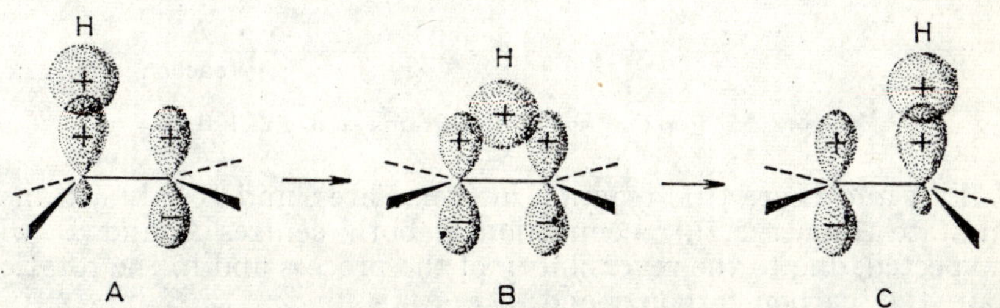

Figure 5.6 1,2-Hydride shift: A. original carbocation; B. transition state or intermediate; C. rearranged carbocation.

One of the C—H bonds in the original cation may have a favourable orientation with respect to the vacant *p* orbital, allowing interaction. If the latter is strong enough, a three-centre, two-electron bond may be formed (B). The system may then suffer collapse either

[159] C. K. Ingold, *Ann. Repts. Progr. Chem.*, **25**, 134 (1928); see also Reference 1, p. 474.

[160] F. C. Whitmore, *J. Amer. Chem. Soc.*, **54**, 3274 (1932).

[161] H. Pines and N. E. Hoffman, in *Friedel-Crafts and Related Reactions*, vol. 2 (G. A. Olah, editor), Wiley-Interscience, New York (1964), p. 1211.

[162] G. J. Karabatsos and J. L. Fry, in *Carbonium Ions*, vol. 2 (G. A. Olah and P. von R. Schleyer, editors), Wiley-Interscience, New York (1970), p. 521.

to the original cation or to the rearranged product C. Now, it is evident that this process corresponds to an intramolecular electrophilic substitution at the A—R bond. If R has a chiral centre directly bonded to A, the process will occur with retention of configuration, as already mentioned.

Another point of interest is the nature of the three-centre, two-electron state B which may correspond either to a transition state, to a high-energy intermediate, or to a low-energy species (Figure 5.7).

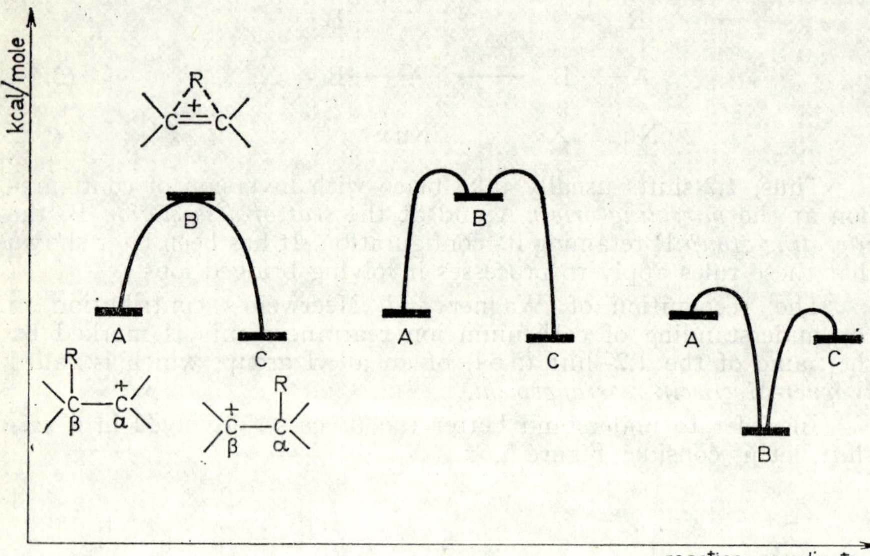

Figure 5.7 Possible reaction diagrams in a 1,2-shift.

If A and C are true reaction intermediates, and B only a transition state (Figure 5.7), racemisation at both centres A and B will be expected, due to the reversibility of the process and to the rotation about the carbon-carbon bond:

retained products ⇌ inverted products (5.55)

The migrating group will obviously preserve its configuration throughout the process.

The same stereochemistry will be anticipated if B also is a reaction intermediate, beside A and C, and has higher energy than the latter intermediates. However, if B is the lower-energy species, the nucleophilic substitution (5.54) may no longer involve A and C, B being the single intermediate:

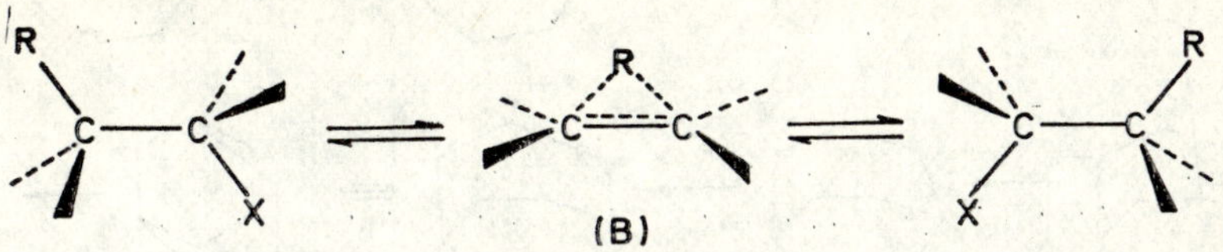

(5.56)

(B)

Evidently, this scheme will correspond to neighbouring-group participation of R in ionisation. Product analysis, coupled with knowledge of the stereochemistry of the 1,2-shift at C_α and C_β, may thus give precious indications with respect to the ionisation mechanism, anchimerically assisted or not. However, caution must be observed in using this experimental evidence. Indeed, if the bridged intermediate in (5.56) is achiral (Figure 5.5, case C), both the equilibration mechanism (5.55) and the bridging mechanism (5.56) are predicted to involve complete racemisation at both termini of the system. Due to this fact, the bridging mechanism was contested , at least for cases in which the kinetic data were not compelling in indicating rate acceleration [68, 163]. It must be however observed that in (5.56) only the absolute configuration may be lost, integral retention of the partial configuration being observed. In (5.55), beside the loss of optical activity, the partial configuration is expected to be at least partly lost. One can see why in the study of a reaction, especially of one involving an ambiguous case, all the tools must be used in order to establish the correct mechanism.

Sometimes, a certain rearrangement may be followed by other similar processes until the most stable species is formed. Probably the most spectacular example of this type is the *adamantane rearrangement* [164]. The reaction was discovered in 1957 by von Schleyer [165], who studied the isomerisation of polycyclic hydrocarbons in presence of aluminium chloride. When *endo*-tetrahydrodicyclopentadiene, **36** was treated with aluminium chloride, a highly-melting isomeric hydrocarbon separated from the reaction mixture. The isomer proved to be a damantane, the reaction being the most drastic rearrangement known at that time. The mechanism of the adamantane rearrangement must involve several Wagner-Meerwein rearrangements [166], leading to the stable adamantyl ion (Scheme 5):

[163] H. C. Brown, *Chem. Scc. (London), Spec. Publ.*, **16**, 140 (1962).

[164] Reviews: P. von R. Schleyer and R. C. Fort, *Chem. Rev.*, **64**, 272 (1964); M. A. McKervey, *Chem. Soc. Rev.*, **3**, 479 (1974).

[165] P. von R. Schleyer, *J. Amer. Chem. Scc.*, **79**, 3292 (1957).

[166] P. von. R. Schleyer and M. M. Donaldson, *J. Amer. Chem. Scc.*, **82**, 4645 (1960).

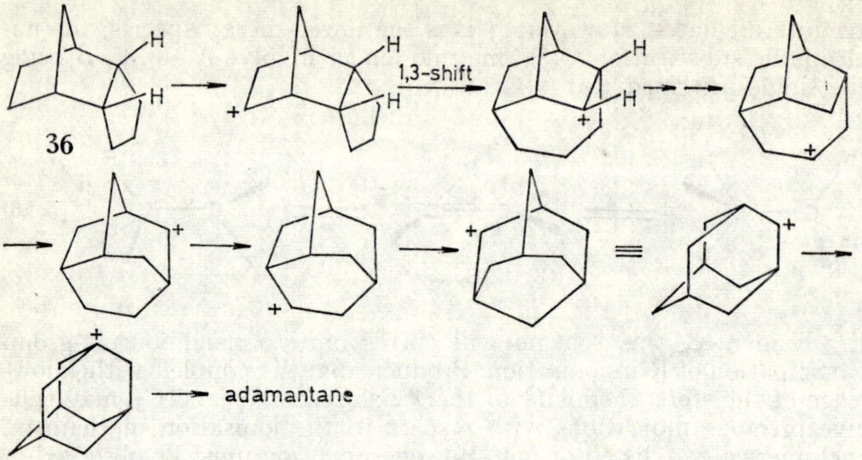

Scheme 5

The high symmetry of adamantoid hydrocarbons, as compared to the other possible isomers, ensures the isolation from the reaction mixture.

Two or more successive rearrangements, occurring in a unimolecular substitution, may sometimes memorise the initial geometry of the substrate, even if no bridged intermediates are involved. This *memory effect* described by Benson[167] is due to the fact that if appropriately rigid systems are involved, the ion formed in the first step has no time to acquire its full symmetry before the Wagner-Meerwein rearrangement starts. The original geometry is thus preserved in a *twisted ion*, favouring the 1,2-shift of the better-situated neighbouring bond. An example is given in Figure 5.8.

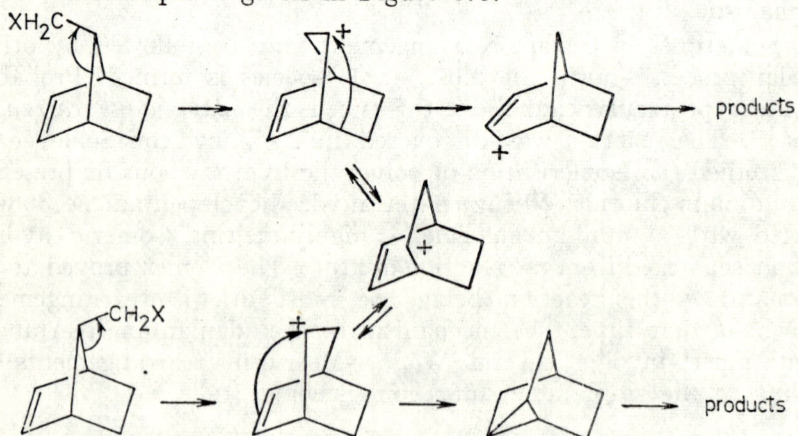

Figure 5.8 Memory effects in the solvolysis of *endo*- and *exo*-7-hydroxymethylbicyclo [2.2.1] hept-2-ene. A, B, twisted ions; C, symmetrical ion.

[167] J. A. Benson *et al., J. Amer. Chem. Soc.*, **91**, 5550, 5578, 5581, 5594, 5601 (1969); ion-pairing is considerered to contribute to memory effects (C.J. Collins, *Chem. Soc. Reviews*, **4**, 251 (1975)).

Examples are known in which the succession of several Wagner-Meerwein rearrangements does not change the structure of the molecules, but makes all its groups undistinguishable. Such processes, called *degenerate rearrangements,* have been observed[168, 169] in the $(CH)_n^+$ ions. An example of a simple degenerate rearrangement is given in Figure 5.9.

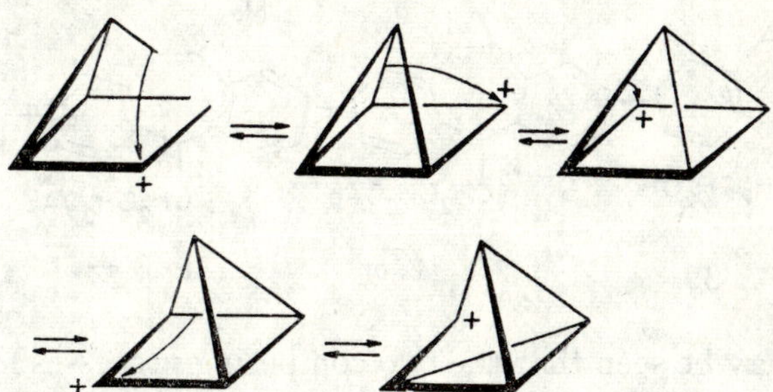

Figure 5.9 Degenerate rearrangement of homotetrahedryl cation (only a few of the 60 possible rearrangements are shown).

The number of the different possible permutations of n objects is n! Considering the polycyclic nature of such ions, $n!/2$ permutations (1,2-shifts) are necessary in order to determine the symmetry on which the n objects are all fully equivalent. Thus, for the homotetrahedryl cation 60 1,2-shifts are required, for the homocubyl ion [169] the number is $9!/2 = 181\,440$. Evidently, the degenerate rearrangement cannot be observed by usual methods (see a discussion in Section 5.4.2).

If two rearrangements at two different reaction sites may compete in product formation, one of them can give an indication regarding the mechanism of the other. An example in that respect is supplied by the acetolyses of compounds **37**, **39** and **40** [170], in which the reference reaction is the acid-catalysed ring opening of cyclopropane, while the solvolysis of tosylate **40** is the reaction under study.

The acid-catalysed ring opening of cyclopropane is known to occur *via* protonated cyclopropane, **38***a* (see Section 10.3):

37 **38***a* **38***b*

[168] P. von R. Schleyer and R. E. Leone, *J. Amer. Chem. Soc.*, **90**, 4164 (1968).

[169] P. von R. Schleyer, J. J. Harper, G. L. Dunn, V. J. DiPasquo and J. R. E. Hoover, *J. Amer. Chem. Soc.*, **89**, 698 (1967).

[170] A. Banciu, Ph. D. Thesis, Bucharest (1975); A. Banciu, M. Elian, A. Bucur, D. Georgescu, E. Ciorănescu and C. D. Nenițescu, *Rev. Roumaine Chim.*, **19**, 213 (1974).

It can be easily seen that the structural differences between **37** and **39** cannot affect the rate of cyclopropane protonation and the rate of formation of the two benzyl ions. Actually, **39** undergoes ring opening at a similar rate as **37**, but this process is paralleled by the solvolysis of the tosylate group:

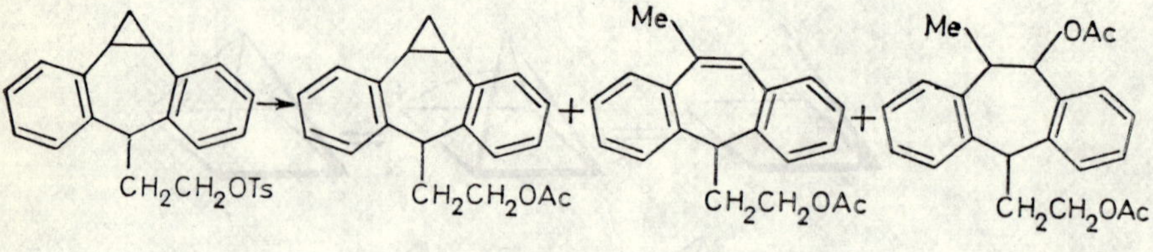

CH_2CH_2OTs → CH_2CH_2OAc + Me, CH_2CH_2OAc + Me, OAc, CH_2CH_2OAc

39 48% 37% 12%

It may be seen that the two competing processes must have about the same rates, as indicated by the ratio of the two types of products. Now, the same rate of ring opening is expected for tosylate **40**. However, when all the initial tosylate **40** has been consumed, under the same conditions as for **39**, no cyclopropane ring opening is found to have occurred:

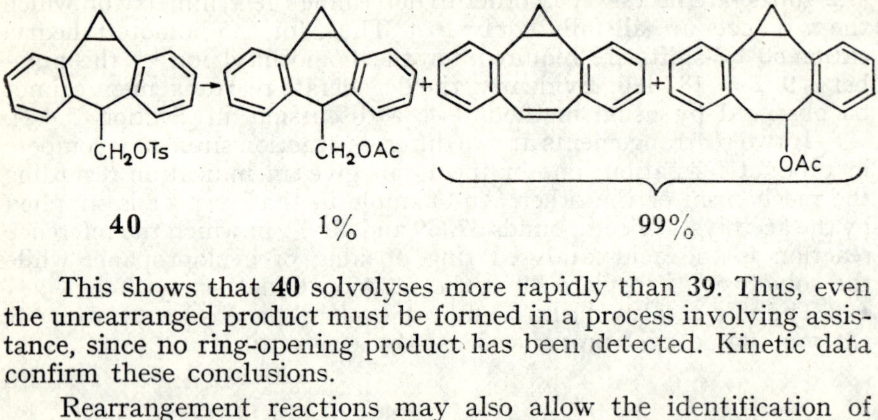

CH_2OTs → CH_2OAc + + OAc

40 1% 99%

This shows that **40** solvolyses more rapidly than **39**. Thus, even the unrearranged product must be formed in a process involving assistance, since no ring-opening product has been detected. Kinetic data confirm these conclusions.

Rearrangement reactions may also allow the identification of the *return* in unimolecular reactions. Actually, what was originally meant by return was the collapse of the ion pairs *to the initial covalent state*. Thus, with the exceptions already discussed, the return should be an invisible reaction. However, if the ion pair involves a bridged species, with two possible sites for nucleophilic attack, the return will afford, beside the initial product (invisible process), also the detectable rearranged product [171]. The two types of return can be distinguished if they proceed by different rates. This is illustrated by the solvolysis of the two bromides **41** and **42** [172]:

[171] S. Winstein and A. H. Fainberg, *J. Amer. Chem. Scc.*, **80**, 459 (1958).
[172] M. Voicu, F. Badea and A. Bucur, *Rev. Roumaine Chim.*, **17**, 1563 (1972).

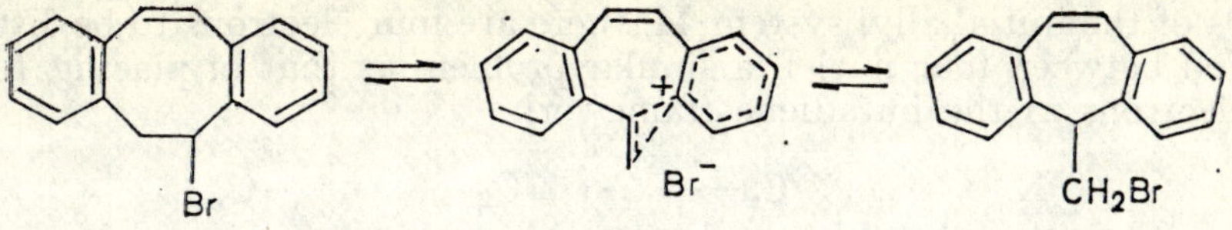

Br CH₂Br

41 43 42

In aqueous acetone (75/25 v/v), bromide **41** is converted to the corresponding alcohol at a constant rate $k = 6 \times 10^{-2}$, at °C. The other bromide, **42**, even at reflux, is recovered unchanged after 52 hours. A rate ratio of about $10^6 - 10^7$ is thus estimated. In acetic acid, after 24 hours, at reflux, only 5% of the bromide **42** has been transformed, whereas the isomer **41** is quantitatively converted into **42** within 3 hours, even at 0 °C. No other product is detected under these conditions. The large rate gap between the two bromides determines the return in the ion pair **43** as the single observable reaction at low temperature, even if the transformation of the more reactive component has achieved completion.

Rearrangements have also been predicted in bimolecular reactions, such as the nucleophilic substitutions of allylic systems [173]:

$$\text{Nu:}^- \; \text{C}=\text{C}-\text{C}-\text{X} \longrightarrow \text{Nu}-\text{C}-\text{C}=\text{C} + \text{X:}^- \qquad (5.58)$$

Thus, while γ-methyl and γ-ethylallyl chlorides behave normally in the reaction with sodium ethyl malonate, the α-isomers yield 10% and 23%, respectively, rearranged products [174]. The reaction conditions rule out a unimolecular mechanism. It has been proved later [175] that the reaction is general for allyl systems and is highly stereospecific, the reagent attacking on the side of the leaving group X:

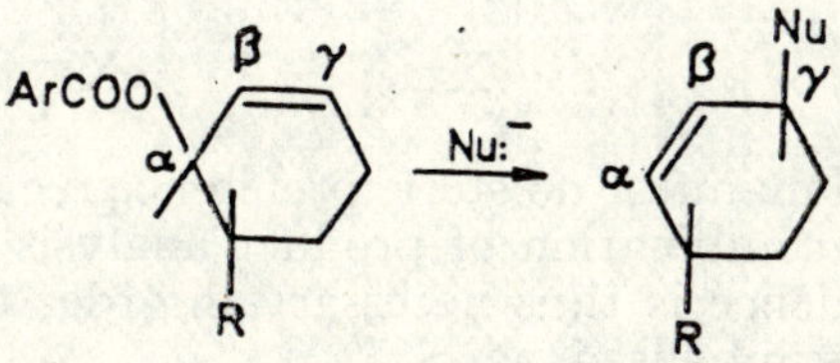

ArCOO β γ α R Nu:⁻ → Nu β γ α R

R = Me, Pr; Nu: = piperidine, sodium malonate.

The stereochemistry of this reaction has been explained using the frontier orbitals theory [176]. Indeed, if one considers only the essential

[173] E. D. Hughes, *Trans. Faraday Soc.*, 34, 185 (1938).
[174] R. E. Kepner, S. Winstein and W. G. Young, *J. Amer. Chem. Soc.*, 71, 115 (1949).
[175] G. Stork and W. N. White, *J. Amer. Chem. Soc.*, 75, 4119 (1953); 78, 4609 (1956).
[176] W. Drenth, *Rec. Trav. Chim.*, 86, 318 (1967).

MOs of the initial allyl system **44**, there are four electrons to be distributed between four nuclei, a similar problem as that of placing four π-electrons in the butadiene framework:

$$\mathrm{X}\cdots\cdots\cdots\cdots\mathrm{C}_\alpha\cdots\cdots\cdots\cdots\mathrm{C}_\beta\cdots\cdots\cdots\cdots\mathrm{C}_\gamma$$

44

Thus, the four corresponding MOs will be constructed in the same way as in butadiene (Figure 5.10).

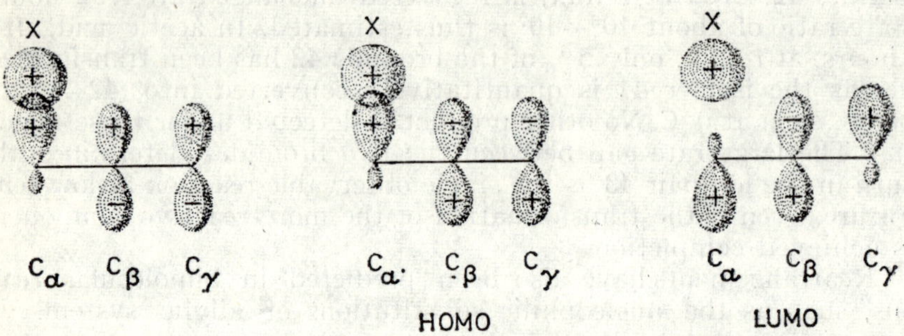

Figure 5.10. The MOs of an allyl derivative involved in a bimolecular substitution.

In building up the transition state of the direct substitution, interaction between the frontier orbitals must be possible, for example, between the highest occupied orbital of the nucleophile and the lowest vacant orbital of the allyl system (Figure 5.10). The nucleophile will thus attack the molecule on the side of the displaced group. Evidently, in a simple allyl rearrangement, involving ionisation:

$$\mathrm{X{-}\overset{|}{\underset{|}{C}}{-}\overset{|}{C}{=}C}\!\langle \longrightarrow \rangle\overset{+}{\overbrace{\mathrm{C{\cdots}\underset{|}{C}{\cdots}C}}}\langle \xrightarrow{\mathrm{Nu:^-}} \rangle\mathrm{C{=}\overset{|}{C}{-}\overset{|}{\underset{|}{C}}{-}Nu} \qquad (5.59)$$

$$\mathrm{X:^-} \qquad\qquad \mathrm{X:^-}$$

the reaction should manifest no steric preference regarding the nucleophilic attack. The combination of product analysis with kinetic and stereochemical evidence is thus necessary in order to prove this mechanism, usually symbolised as S_N2.

5.4.2 *Other transformations of carbocations*

The intermediate formation of carbocations may be also demonstrated by the identification of the specific products of their reactions (see also Chapter 4). As characteristic reactions of carbocations we shall mention:

(*a*) *Carbocation trapping by nucleophiles.* The general process:

$$\mathrm{R^+ + Nu:^- \longrightarrow R{-}Nu}$$

usually occurs in a nucleophilic substitution. However, as reaction partners of nucleophiles, carbocations are less selective than the covalent substrates R—X. Thus, by carrying out a nucleophilic substitution in presence of two or more nucleophiles of different reactivity, one may obtain a precious indication regarding the nature of the partner of the nucleophile (Section 5.4.4). The method of trapping carbocations with reagents like the azide ion [63–65,100] has already been mentioned.

(*b*) *Reversible alkene formation.* A carbocation may eliminate a proton, yielding an alkene (E_1 reaction, see p. 427):

$$R^1\overset{+}{C}H{-}CHR^2_2 \rightleftharpoons R^1CH{=}CR^2_2 + (H^+)$$

The process is reversible and may thus be used for the generation of carbocations.

(*c*) *Intermolecular hydride shifts* [177,178]. Intermolecular hydride shifts may be represented by the general equilibrium (5.60):

$$R^{1+} + R^2{-}H \overset{K}{\rightleftharpoons} R^1{-}H + R^{2+} \tag{5.60}$$

The position of the equilibrium depends on the pK_{R^+} values of the carbocations involved, on the concentration of hydrocarbon R^2—H and on specific medium conditions. Whenever a carbenium ion is generated by heterolysis in presence of precursors of stable carbocations, the equilibrium (5.60) can be observed.

(*d*) *Alkylation of aromatic hydrocarbons and alkenes.* Aromatic hydrocarbons as well as alkenes are nucleophiles able to trap carbocations (see Friedel-Crafts and related reactions, Chapter 7). It should be mentioned that the neighbouring group participation of an aryl group (phenonium route) or of a double bond (π - route) involves an intramolecular alkylation.

All carbocation reactions may be involved in complex processes like the isomerisation of hydrocarbons in presence of aluminium halides. Such reactions may be regarded as nucleophilic rearrangements in which the leaving group is the hydride ion. An example is shown in Scheme 5.

5.4.3 *Isotope labelling*

Isotope labelling of organic substrates is practised either for kinetic reasons (determination of the rate-controlling step in displacements at a C—X bond, Section 5.2.5) or for tagging an undistinguishable group of the molecule. For the latter purpose it has been used in the study of both bimolecular and unimolecular reactions.

177 N. C. Deno, H. J. Peterson and G. S. Saines, *Chem. Rev.*, **61**, 7 (1961).
178 C. D. Nenitzescu, in *Carbonium Ions*, vol. 2 (G. A. Olah and P. von R. Schleyer, editors), Interscience, New York (1971), p. 463.

Thus, Dostrovsky and Hughes [179] have assumed that the reactions of the neopentyl system, involving no rearrangements, are bimolecular. This has been later proved [180] by labelling with deuterium one of the two hydrogen atoms in the CH_2X group. The difference between hydrogen and deuterium, even if small, is sufficient to confer chirality to the molecule. The use of optically active **45** has shown that substitutions without rearrangements in the neopentyl system are typical S_N2 reactions occurring with complete retention of optical activity and inversion at the reaction site.

Me—C(Me)(Me)—C(D)(H)—X $\xrightarrow{Nu^-}$ Me—C(Me)(Me)—C(Nu)(H)—D

45 (*R*) (*S*)

More frequently, isotope labelling is used for demonstrating the nature of the unimolecular mechanism. Thus, β-phenyl derivatives are transformed by solvolysis into products having the same carbon skeleton. The occurrence of a symmetric intermediate **47** cannot be demonstrated either by product or by stereochemical analysis. Moreover, except in a few instances, kinetic measurements are not conclusive (Section 5.2.2). However, by labelling either the hydrogen atoms of the CH_2 groups with deuterium or tritium, or the C_α with radioactive carbon, a distinction between α and β positions can be made. The most interesting example is provided by the trifluoroacetolysis of **46**. 50% of the label introduced at C is found in its original position, while the rest of 50% is found at C_α (100% scrambling).

$C_\beta H_2(Ph)—\overset{*}{C}_\alpha H_2—X \rightleftharpoons H_2C_\beta—\overset{*}{C}_\alpha H_2$ (bridged by $C_6H_5^+$) $\rightleftharpoons X—C_\beta H_2—\overset{*}{C}_\alpha H_2(Ph)$

46 **47**

This result, combined with the kinetic data[50], shows that in trifluoroacetic acid **46** reacts 100% by the assisted route. In less acidic and more nucleophilic solvents, less radioactive carbon is found at C_β [181].

Isotope labelling has played an important role in the study of degenerate rearrangements. Let us consider the solvolysis of the

[179] I. Dostrovsky and E. D. Hughes, *J. Chem. Soc.*, 157, 161, 171 (1946).

[180] P. H. Anderson, B. Stephenson and H. S. Mosher, *J. Amer. Chem. Soc.*, 96, 3171 (1974), and references cited therein.

[181] C. C. Lee, G. P. Slater and J. W. T. Spinks, *Can. J. Chem.*, 35, 1417 (1957).

homocubyl derivative **48**, marked with deuterium in the 9-position. The homocubyl cation is a polymethine cation, $(CH)_9^+$; if it undergoes an infinite number of Wagner-Meerwein rearrangements, its CH groups may become perfectly equivalent. Simple calculations show that in that case only 11.1% of the deuterium would be found in its initial position:

CD—OTs (100%) → $\overset{+}{C}D$ → CH, CD —complete scrambling→ OAc (11,1%)

48 **49***a* **49***b* **49***z*

If only one Wagner-Meerwein equilibration takes place, 50% of the original deuterium will be found at C-9. Actually, the experiment shows that in the buffered acetolysis only 37.5% of the deuterium is located at C-9, whereas in formolysis [169] the initial deuterium at C-9 is only 10 ± 2%. Thus, complete degeneration occurs only in formolysis.

Isotope labelling is gaining importance in product analysis of organic reactions [182].

5.4.4 *Stability-selectivity relationships*

If a substrate A is transformed into B and C by parallel reactions, the overall reaction rate is:

$$k = k_B + k_C \tag{5.61}$$

$$A \xrightarrow{k} \begin{cases} \xrightarrow{k_B} B \\ \xrightarrow{k_C} C \end{cases}$$

It may be easily shown that if both reactions are first-order in A, the product ratio (corrected for initial concentration of B and C, if necessary) will correspond to the rate ratio:

$$[B]/[C] = k_B/k_C \tag{5.62}$$

[182] See, for example, J. G. Burr, Jr., *Tracer Applications for the Study of Organic Reactions*, Wiley-Interscience, New York (1957); W. H. Saunders, Jr., in *Technique of Organic Reactions*, vol. 8 (A. Weissberger, editor), Wiley-Interscience, New York (1966), p. 1519.

From the two equations, k_B and k_C can be calculated. If A represents a common intermediate, like a carbocation R^+, and if k_B and k_C are the rates of its trapping by two different nucleophiles:

$$R{-}X \xrightarrow{k} R^+ \begin{cases} \xrightarrow[k_B]{Nu:} B \\ \xrightarrow[k_C]{Nu':} C \end{cases} \tag{5.63}$$

then the product ratio will be a measure of the selectivity of the carbocation when discerning between the two nucleophiles. It can be demonstrated (see for example Section 7.3) that the higher the stability of an intermediate, the higher is its selectivity. Hughes, Ingold and their co-workers [183] were the first to apply this principle to a systematic study of hydrolysis reactions in aqueous acetone in presence of azide ion. Despite the high nucleophilicity of the azide ion, as compared to water (Table 4.2, p. 137), the ratio of alcohol is usually high, due to the small concentration of azide (see 5.64)), but also to the low selectivity of the reactive species:

$$[B]/[C] = (k_B/k_C)\,([Nu:]/[Nu':]) \tag{5.64}$$

Sneen, Carter and Kay [184] have shown that between the relative rates (k_r) observed in the hydrolysis of chlorides R—Cl in aqueous acetone and the ratio between the resulted azide and alcohol (k_N/k_S), there is a correspondence of the precision of a linear relationship. Significant data quoted in this paper are shown in Table 5.12.

TABLE 5.12

Relative rates (k_r) and selectivity ratios (k_N/k_S) in the hydrolysis of R—Cl in aqueous acetone (0°), in presence of sodium azide[a].

R—Cl	k_r	k_N/k_S[b]
$Me_3C{-}Cl$	1	1.45×10^1
$Ph_2CH{-}Cl$	6.50×10^1	6.10×10^1
$PhCHCl{-}CH{=}CH{-}CH_3$	9.40×10^2	2.10×10^2
$Me_2CCl{-}CH{=}CH_2$	1.17×10^3	3.90×10^2
p-$MeC_6H_4CHCl{-}CH{=}CH{-}CH_3$	9.40×10^3	8.15×10^2
p-$MeC_6H_{42}CH{-}Cl$	3.72×10^4	8.70×10^2
$Ph_3C{-}Cl$	1.95×10^7	1.12×10^4

(a) Data quoted by Sneen *et al.* [184]. (b) Calculated from product ratio, according to equation (5.64).

[183] L. C. Bateman, E. D. Hughes and C. K. Ingold, *J. Chem. Soc.*, 974 (1940); M. C. Church, E. D. Hughes and C. K. Ingold, *J. Chem. Soc.*, 966, 971 (1940); see also Reference 3 for a discussion.

[184] R. A. Sneen, J. V. Carter and P. S. Kay, *J. Amer. Chem. Soc.*, **88**, 2594 (1966).

The authors assume that the substrates listed in the above table are typical examples of compounds involved in limiting mechanisms, the products arising probably from dissociated ions. The derivatives with a higher reactivity (expressed by k_r), as predicted from the correlation line of the reactivity and selectivity (expressed by k_N/k_S), probably react as ion pairs. Considering the following reaction scheme:

$$\mathrm{R{-}X} \underset{k_{-1}}{\overset{k}{\rightleftharpoons}} \underset{\text{ion pair}}{\mathrm{R^+X^-}} \begin{cases} \xrightarrow{\mathrm{SH}(k_S)} \mathrm{R{-}S} \\ \xrightarrow{\mathrm{N_3^-}(k_N)} \mathrm{R{-}N_3} \end{cases} \tag{5.65}$$

Sneen and co-workers [100,185] have derived a relation between the ratio of the observed rates in presence (k) and in absence of added nucleophile (k_{NA}), and the ratio of products:

$$\frac{k}{k_{NA}} = \frac{(x+1)\,(1+m[N_3^-])}{(x+1+m\,[N_3^-])} \tag{5.66}$$

where $x = k_{-1}/k_S$ and $m = k_N/k_S$. The latter parameter is calculated from product data, whereas x is determined from plots of k/k_{NA} *vs.* the concentration of azide. According to Sneen *et al.* [185], the experimental points deviate considerably from a true bimolecular reaction plot, which is in excellent agreement with the ion-pair mechanism and equation (5.66).

Studying the solvolysis of 1-adamantyl bromide and 2-adamantyl tosylate, as well as the solvolysis of 2-propyl tosylate in 80% ethanol and in 75% dioxane, in presence of azide ion, von Schleyer and his co-workers [186] noted the absence of rate product correlation for the adamantyl derivatives. Moreover, the secondary and the tertiary derivative behave similarly in the hydrolysis, the rates and products being very little affected by the sodium azide concentration. According to these authors, nucleophilic participation in these highly hindered adamantyl derivatives is absent. On the other hand, for the less hindered secondary derivatives, 2-propyl and 2-octyl arenesulphonates, the nucleophilic assistance seems to be important. Solvent nucleophilic assistance, according to Schleyer and his co-workers [186,187], explains Sneen's data better.

Sneen's mechanism of ion-pair reactions seems however to be consistent with numerous other data[100]. A more recent confirmation[119] is based on the heavy atom kinetic isotope effect. Thus, if the ion-pair

[185] R. A. Sneen and J. W. Larsen, J. *Amer. Chem. Soc.*, **88**, 2593 (1966).

[186] D. J. Raber, J. M. Harris, R. E. Hall and P. von R. Schleyer, *J. Amer. Chem. Soc.*, **93**, 4821 (1971).

[187] D. J. Raber, J. M. Harris and P. von R. Schleyer, *J. Amer. Chem. Soc.*, **93**, 4829, (1971).

mechanism (5.65) is considered, the increase in the concentration of sodium azide in the solvolysis of R-^{35}Cl and R-^{37}Cl must be associated with an increase of the isotope effect, $^{35}k/^{37}k$, as is actually observed. If the reaction with sodium azide were a bimolecular process, the reverse would be true (kinetic isotope effects are smaller for the bimolecular than for the unimolecular substitution).

A theoretical treatment of stability-selectivity relationships [188] has emphasised the necessity to consider the complete scheme of product formation (Scheme 1, p. 212). Thus, if any selectivity S_i is defined by relation (5.67):

$$S_i = \log(k_B/k_C)_i \tag{5.67}$$

then the over-all observed selectivity S_{obs} can be composed of more terms, depending on the actual intermediate (i) giving rise to products B and C:

$$S_{obs} = \sum_i a_i S_i \tag{5.68}$$

where a_i represents the fraction of products arising from attack on species i. In the unimolecular solvolysis, according to Scheme 1, there are three intermediates (tight ion-pair, solvent-separated ion-pair and dissociated ions). If the nucleophilic attack on the substrate R—X is not excluded, S_{obs} can be made up of four terms. There are indications that S_i varies little with the stability of the substrate [104]. Thus, the ratios of the rate constants of the reactions of azide and cyanide either with tri-(p-anisyl)methyl or with trityl cations are essentially the same, even if the cations differ largely in their stability. Harris and his co-workers [188] conclude that the stability-selectivity relationship *must result because of a shift on the ion-pair spectrum (Scheme 1) with stability and consequently a different blending of selectivities.*

An interesting demonstration of this theory is based on the fact that 2-adamantyl derivatives cannot undergo backside nucleophilic attack. Thus, reaction will occur either from solvent-separated ion pairs or from dissociated ions.

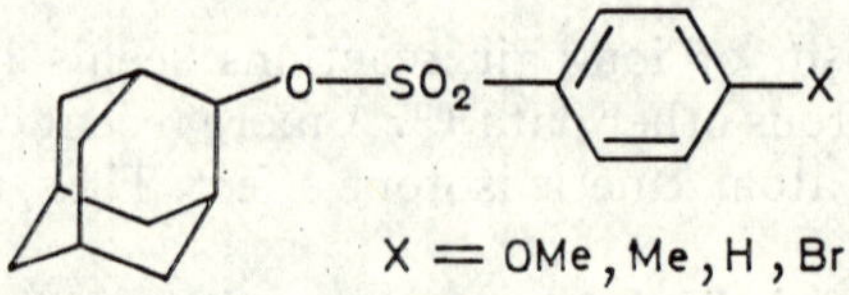

O—SO_2—C_6H_4—X

X = OMe, Me, H, Br, NO_2

[188] J. M. Harris, D. C. Clark, A. Becker and J. F. Fagan, *J. Amer. Chem. Soc.*, **96**, 4478 (1974).

The nearly constant selectivities which are observed [189] are consistent with a collapse from a solvent-separated ion pair **50**:

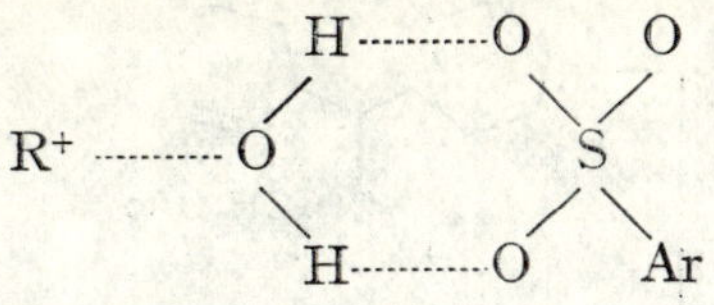

50

A somewhat different demonstration is offered by von Schleyer and his co-workers [187]. In the solvolyses of 1-aryl-2-propyl and 2-aryl-1-propyl derivatives, in the presence of azide ion, the observed rate constants were split into their components, the rate constant of the phenonium route k_{Δ} and the rate constant of the solvent-assisted route k_s:

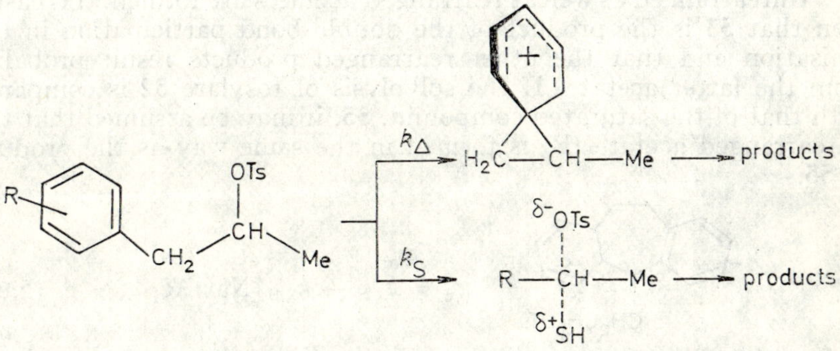

51

The phenonium-route rate constants are little, if at all, affected by azide concentration (for the *p*-methoxy derivative **51**, at 0.00 and 0.06 $_M$ NaN_3, the rate constants are, at 75 °C in 80 % ethanol, $1.01 \times \times 10^{-4}$ and 1.26×10^{-4}, respectively), whereas product formation varies largely (the yields of RN_3 for the same concentrations of azide are 0.0 % and 53 %, respectively). For the solvent-assisted route (R in **51** is *p*-NO_2), a similar increase of the yield is associated with a threefold increase in the rate constants, which is observed at much lower concentrations of sodium azide. The authors conclude that product analysis alone gives little indication on the reaction mechanism, so that it must be combined with a complete kinetic analysis.

A different stability-selectivity relationship was established for the acetolysis of **52**. Ciorănescu, Nenitzescu and their co-workers [190]

[189] J. M. Harris, A. Becker, J. F. Fagan and F. A. Walden, *J. Amer. Chem. Soc.*, **96**, 4484 (1974).

[190] E. Ciorănescu, M. Banciu, R. Jelescu, M. Rentzea, M. Elian and C. D. Nenitzescu, *Rev. Roumaine Chim.*, **14**, 929 (1969).

have identified the reaction products formed in the acetolysis of tosylate **52**:

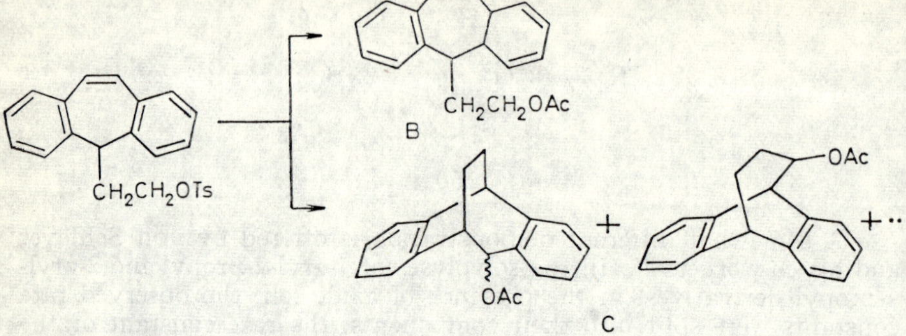

CH_2CH_2OTs → CH_2CH_2OAc B; OAc + OAc C +···

52 **53** **54**

Unrearranged as well as rearranged acetates are formed. It is easily seen that **53** is the product of the double bond participation in the ionisation and that the other rearranged products result probably from the latter acetate. If the solvolysis of tosylate **52** is compared with that of the saturated compound, **55**, it may be assumed that the unrearranged acetate (B) is formed in the same way as the product of **55**.

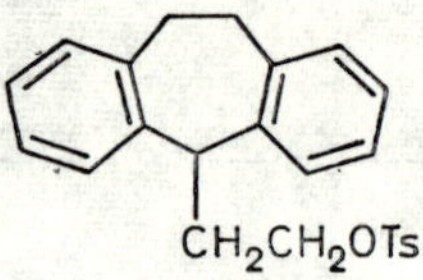

CH_2CH_2OTs

55

$$k_S = k_{1,S} + k_{2,S}\,[\text{NaOAc}] \tag{5.69}$$

The acetolysis rates of both tosylates are affected by the concentration of sodium acetate [190,191]. In the case of the saturated tosylate, which is a primary derivative, the effect of sodium acetate may be attributed to nucleophilic assistance, as shown in (5.69). The observed rates for the unsaturated tosylate **52** should then be given by equation (5.70):

$$k_U = k_{1,U} + k_{2,U}\,[\text{NaOAc}] + k_\Delta \cong k_S + k_\Delta \tag{5.70}$$

Plotting the product ratio (unrearranged/rearranged = B/C) *vs.* sodium acetate concentration, should afford a straight line with the intercept $k_{1,S}/k_\Delta$ and the slope $k_{2,S}/k_\Delta$. Actually, a very good linear correlation is obtained (Figure 5.11).

The intercept of the line agrees well with the kinetic data obtained with both **52** and **55**. However, the calculated slope (0.91) is much larger than the calculated ratio $k_{2,S}/k_\Delta = 0.184$. Therefore, the unrear-

[191] M. Voicu and F. Badea, *Rev. Roumaine Chim.*, **14**, 929 (1969).

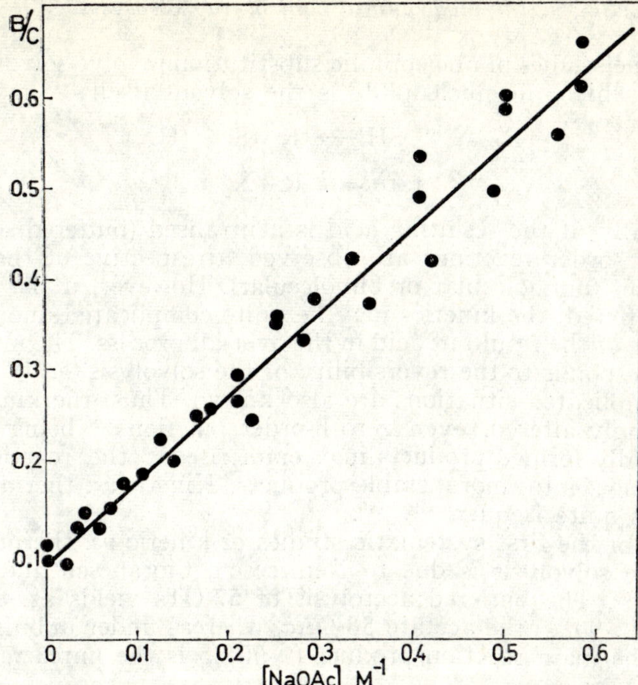

Figure 5.11 Dependence of the ratio unrearranged acetate/rearranged products (B/C) on sodium acetate concentration in the acetolysis of 1,2:5,6-dibenzocycloheptatrien-7-yl-ethyl tosylate (**52**).

ranged acetate must result also by another route, unaccessible to the saturated derivative **55**, and which should depend upon sodium acetate concentration. The most plausible interpretation of that result [191] is that the ion pair formed in the ionisation of **52** by double bond participation (**56**-OTs) forms another ion pair with sodium acetate (**56**-OAc), the return of the latter affording either the unrearranged product (detectable by the plot of Figure 5.11) or the rearranged products. The latter process may not be discerned from the normal solvolysis processes of all the intermediates involved (ion pairs **56**-OTs, **56**-OAc, dissociated ions). This is probably one of the reasons why in the acetolysis the salt effect of the acetate is usually very small (Table 5.9).

Thus, this type of selectivity-stability relationship agrees well with Winstein's theory (Scheme 1) as well as with Sneen's theory of ion pair reactions. Voicu and Badea [192] found that other related tosylates behave in a similar way.

TsO^- + NaOAc ⇌ AcO

56-OTs **56**-OAc

5.4.5 *Kinetic vs. thermodynamic control in solvolysis*

Most of the studies of nucleophilic substitution involve solvolytic reactions, in which the nucleophile is the solvent itself:

$$\text{R—X} + \text{SH} \longrightarrow \text{R—S} + (\text{HX}) \qquad (5.71)$$

$$\text{rate} = k[\text{R—X}]$$

Usually, if the resulting acid is neutralised (buffered solvolysis), clean first-order reactions are observed, irrespective of the reaction mechanism (unimolecular or bimolecular). However, if the solvolysis is not buffered, the kinetics may be quite complicated due to the interference of the resultant acid in the over-all process. The most simple case corresponds to the reversibility of the solvolysis (5.71). However, more complicated situations are also known. Thus, the kinetics may be strikingly altered, even zeroth-order reactions [18] being observed. The initially formed products may epimerise at the reaction centre or rearrange into more stable products. Kinetic *vs.* thermodynamic control is quite frequent [193–196].

One of the first systematic studies of kinetic *vs.* thermodynamic control in solvolysis is due to Nenitzescu, Ciorănescu and their coworkers [193]. The buffered acetolysis of **57**-OTs yields as the major product (>90%) the acetate **58**-OAc, whereas under unbuffered conditions the main reaction product (>90%) is the unrearranged acetate **57**-OAc:

AcOH (TsOH) → **57**-OAc (CH_2OAc) + **59** (OAc) + alkenes

57-OTs (CH_2OTs)

AcOH (AcONa) → **58**-OAc (OAc) + **60** (OAc) + alkenes

[192] M. Voicu, F. Badea and A. Voicu, *Rev. Roumaine Chim.*, **18**, 131 (1973).

[193] E. Ciorănescu, A. Bucur, M. Elian, M. Banciu, M. Voicu and C. D. Nenitzescu, *Tetrahedron Letters*, 3835 (1964); *Rev. Roumaine Chim.*, **10**, 161 (1965).

[194] E. Ciorănescu, A. Bucur, F. Badea, M. Rentzea and C. D. Nenitzescu, *Tetrahedron Letters*, 1867 (1969).

[195] M. Voicu and F. Badea, *Rev. Roumaine Chim.*, **17**, 1551 (1972).

[196] A. Diaz, M. Brookhart and S. Winstein, *J. Amer. Chem. Soc.*, **88**, 3133 (1966); see also S. Winstein and E. M. Kosower, *J. Amer. Chem. Soc.*, **81**, 4399 (1959).

Bridged products are also formed [194]. It has been shown [195] that the kinetics are clean in the buffered acetolysis, the first-order rate constant at 110 °C being $k_1 = 3.2 \times 10^{-5}$ s^{-1}. In the unbuffered acetolysis, the rate constants decrease with conversion from 2.95×10^{-5} (5% conversion) to about 1.0×10^{-5} s^{-1} (93% conversion). It was assumed that in the earlier stages of the unbuffered acetolysis, the reaction is under kinetic control, yielding the rearranged acetate **58**-OAc. Indeed, extrapolation of the observed integrated rate constants at zero time gives the mean value 3.18×10^{-5} s^{-1}, identical to that found in the buffered acetolysis. It was assumed that the product of kinetic control, **58**-OAc, is converted in presence of TsOH to the thermodynamically-controlled product **57**-OAc. It has also been found [195] that under the acetolysis conditions, in presence of TsOH, **58**-OAc is nearly quantitatively converted to **57**-OTs. The following reaction scheme was proposed:

$$\begin{array}{ccccc} & & \textbf{58}\text{-OAc} + \text{TsOH} & & \\ & & k'_2 \downarrow\uparrow k'_{-2} & & \\ \textbf{57}\text{-OTs} & \underset{k_{-1}}{\overset{k_1}{\rightleftharpoons}} & \text{I} & \underset{k'_{-1}}{\overset{k'_1}{\leftrightharpoons}} & \textbf{58}\text{-OTs} \\ & & k_2 \downarrow\uparrow k_{-2} & & \\ & & \textbf{57}\text{-OAc} + \text{TsOH} & & \end{array}$$

Scheme 6

Using the steady-state approximation, the concentration of the intermediate I is found to be:

$$[\text{I}] = \frac{k_1[\textbf{57}\text{-OTs}] + k'_1[\textbf{58}\text{-OTs}] + \{k_2[\textbf{57}\text{-OAc}] + k'_2[\textbf{58}\text{-OAc}]\}\,[\text{TsOH}]}{k_{-1} + k'_{-1} + k_{-2} + k'_{-2}} \tag{5.72}$$

It may be seen from this equation that the rate constants should decrease in time, as the concentrations of both **57**-OTs (starting material) and **58**-OAc (product of kinetic control) are decreasing. Scheme 6 suggests that if perfect equilibration between products is achieved, the disappearance of **57**-OTs must obey a first-order equation. Indeed, if the constant of the equilibrium between the seven- and the eight-membered ring system is independent of the functional groups (acetate or tosylate),

$$K = \frac{[\mathbf{57}\text{-OTs}]}{[\mathbf{58}\text{-OTs}]} \approx \frac{[\mathbf{57}\text{-OAc}]}{[\mathbf{58}\text{-OAc}]}$$

and if the equilibrium constant for acetate $\rightleftharpoons$ tosylate is denoted by K',

$$K' = \frac{[\mathbf{57}\text{-OAc}]\,[\text{TsOH}]}{[\mathbf{57}\text{-OTs}]}$$

equation (5.72) becomes:

$$[\text{I}] = \frac{k_1 + k_1'/K + (k_2 + k_2'/K)\,K'}{k_{-1} + k'_{-1} + k_{-2} + k'_{-2}}[\mathbf{57}\text{-OTs}] \tag{5.73}$$

On the other hand, the formation of toluene-*p*-sulphonic acid obeys the equation:

$$\text{rate} = (k_{-2} + k'_{-2})\;[\text{I}] \tag{5.74}$$

showing that a clean first-order reaction must be observed if perfect equilibration of products and intermediates is achieved.

If all these assumptions are correct, the addition of toluene-*p*-sulphonic acid under the initial conditions will determine a clean first-order reaction. Actually, the solvolysis of **57**-OTs in presence of TsOH is a clear-cut first-order reaction with the rate constant $k_T = 1.56 \times 10^{-5}\ \text{s}^{-1}$, lower than the first-order rate constant of the buffered acetolysis ($k = 3.20 \times 10^{-5}\ \text{s}^{-1}$). An examination of equations (5.73) and (5.74) shows that:

$$k_T = \frac{k_1k'_{-2} + k_1k_{-2} + (k_{-2} + k'_{-2})\,[k_1'/K + (k_2 + k_2'/K)K']}{k_{-1} + k'_{-1} + k_{-2} + k'_{-2}} \tag{5.75}$$

whereas the observed rate constant of the buffered acetolysis must be:

$$k_{\text{extr}} = \frac{k_1k'_{-2}}{k_{-1} + k'_{-2}} \tag{5.76}$$

One can see that the numerator in (5.75) is greater than in (5.76). However, since k_T is smaller than k_{extr}, the denominator in (5.75) must be much greater than the denominator in (5.76). Assuming that k_{-1} and k'_{-2} are the same in both equations and that $k_{-2} \gg k'_{-2}$ (the kinetically-controlled acetate being **58**-OAc), it follows that k'_{-1} must be relatively large, that is that tosylate **58**-OTs is also formed, even if it cannot be obtained. This kinetic analysis demonstrates clearly that solvolysis may be subject to thermodynamic control.

The kinetic *vs.* thermodynamic control in solvolysis reactions was approached by Winstein and his co-workers [196] in a different manner, a few years later. Studying the generation of the ion **61** by two different routes, as well as the ratio between the norbornenyl derivative **62** and the tricyclo compound **63**, the authors have concluded that the product ratio in the two processes (kinetically- and thermodynamically-controlled reactions, respectively) may be determined by two parameters, the *partition factor*, $P = k_1/k_2$, measuring the tendency of the ion **61** to yield either **62** or **63**, and the *reactivity factor*, $R = k_{-2}/k_{-1}$, measuring the reactivity of the two systems in generating the common intermediate.

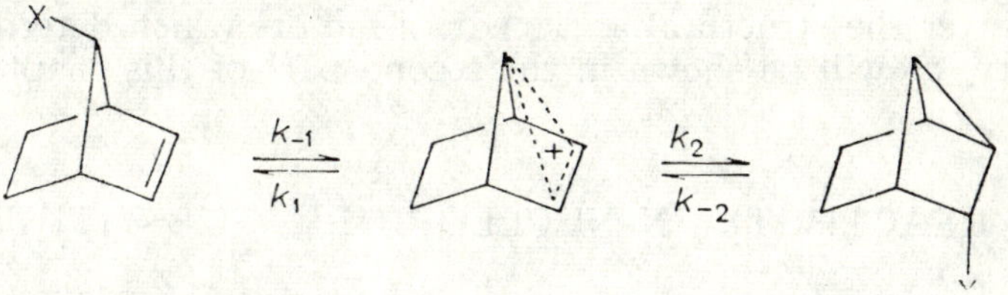

k_{-1}, k_1; k_2, k_{-2}

62-X **61** **63-X**

Both P and R factors are easily determined experimentally, from kinetic data or from product analysis (Section 3.4). It is now well established that the drastic change in product distribution under kinetic control (buffered conditions), as compared to the thermodynamic control (unbuffered conditions), may be determined by the variation of either P or R, or of both.

5.5 Summary of the mechanistic criteria

Using the criteria discussed above, it was possible to identify the mechanism presented in Section 5.1. Moreover, they have permitted the distinction of such details as those shown in Scheme 1, regarding the steps of a unimolecular substitution, and have enabled one to discern between several types of reaction intermediates. The nature of the intermediates revealed further details of the possible mechanisms.

The methods used have confirmed an early theory [1,3] that there is no clear-cut difference between S_N2 and S_N1 mechanisms, the two reactions being characterised only by the degree of the nucleophilic involvement of the reagent or solvent in the transition state. They have also shown that true unimolecular reactions (without nucleophilic assistance from the medium) may only be observed with highly hindered substrates, such as 2-adamantyl derivatives [124,186,197], **49**

[197] J. L. Fry, C. J. Lancelot, L. K. Lam, J. M. Harris, R. C. Bingham, D. J. Raber, R. E. Hall and P. von R. Schleyer, *J. Amer. Chem. Soc.*, **92**, 2538, 2540, 2542 (1970).

where backside attack is impossible for geometrical reasons, and in reactions involving neighbouring group participation Thus, for solvolytic reactions, in which the nucleophile is identical with the solvent, the observed first-order rate constant can be split into four terms:

$$k_{obs} = k + k_{Nu:} + k_S + k_\Delta \quad (5.77)$$

where k is the rate constant of the unassisted ionisation (involving no solvent and no neighbouring group), $k_{Nu:}$ and k_S correspond to the medium-assisted reactions (either by solvent or by nucleophile), and k_Δ represents the neighbouring-group participation. In order to appreciate the weight of each term in equation (5.77), it is necessary to understand better the structural and medium factors which determine the reactivity, as will be shown in the second part of this chapter.

(B) REACTIVITY IN NUCLEOPHILIC SUBSTITUTION

5.6 The S_N2 — S_N1 mechanisms

Since there is no sharp frontier between the S_N1 and S_N2 mechanisms, they will be treated together. It must be remembered that for any organic reaction the reactivity is determined by the structural factors (polar and stereochemical), while for reactions occurring in condensed media external factors (solvent, catalysts, etc.) are also important. The knowledge of the factors affecting the reactivity in the various mechanisms has an enormous practical value, since the nature and the distribution of the products depends on the particular pathway adopted by the reacting system. Considering organic reactions as a whole, the steric factors are by far more important than the polar factors. The former are related to the chiral aspect of our world and played thus a key role in the story of life. However, in the nucleophilic substitution polar factors are more important.

5.6.1 Polar effects

The unimolecular mechanism S_N1 is governed mostly by polar effects. Indeed, in the reaction:

$$R^1R^2R^3C\text{—}X \rightleftharpoons R^1R^2R^3C^+ + X^- \quad (5.78)$$

due to the change of polarity between initial and final state, substituents exert a greater influence upon the transition state than they do in a bimolecular reaction. The effects which stabilise the carbocations may operate in the same manner upon the transition state of reaction (5.78). As was shown in Chapter 4, these effects are inductive, hyperconjugative and electromeric. On the other hand, the unimolecular mechanism will be favoured by low dissociation energies D and high electron-affinity of the leaving group. All these factors are illustrated in Table 5.13 in the formation energies of the carbocations, according to reaction (5.78).

TABLE 5.13

Formation energies of carbocations in the process:
$R{-}X \rightarrow R^+ + X^-$

R^+	ΔH/kcal mol^{-1} [a]		
	X = Cl	X = Br	X = I
CH_3^+	220	215	208
$CH_3CH_2^+$	192	183	176
$CH_3CH_2CH_2^+$	177	166	161
$(CH_3)_2CH^+$	168	150	149
$(CH_3)_3C^+$	149	132	132
$CH_2{=}CH{-}CH_2^+$	158	150	145
$C_6H_5CH_2$	162	141	137

(a) Values quoted in Reference 3

Indeed, the inductive and hyperconjugative effects are reflected in the decreasing energies in the series MeX > EtX > Pr^{i}X > Bu^{t}X. The electromeric stabilisation of the transition state determines the low ΔH values for the formation of allyl and benzyl cations; the effect of the leaving group is obvious from the decrease: RCl > RBr > RI.

It must be noted that the values recorded in Table 5.13 are only rough approximations. Thus, between the isopropyl and the *t*-butyl cation there is an energy gap of about 18 kcal mol^{-1}, corresponding to a ratio in rate constants of 10^{25}, at 27 °C. The largest values recorded in solvolyses are however 10^6 or even less [198]. The discrepancy is still higher with the isopropyl cation and ethyl cation. The calculated ratio of the rate constants is about 10^{24}. The ratios actually observed [3], even in formic acid, are of the order 10^2. Two explanations may be advanced.

First, one may assume that the transition state of the ionisation differs significantly from the carbenium ion. If that were true, the ionisation of the secondary substrates would have an earlier transition state than that of the tertiary derivatives, a situation which, according to Hammond's postulate, cannot be true (the more endothermic reaction has the later transition state). The other alternative is that solvents compensate better the heterolysis energy by solvating the carbocations.

[198] H. C. Brown and M. H. Rey, *J. Amer. Chem. Scc.*, **86**, 5008 (1964).

The latter explanation seems to be correct, and it agrees better with the theory of Doering and Zeiss[12] regarding the mechanism of nucleophilic substitution. Indeed, backside solvent assistance must be hindered to a greater extent in the tertiary than in the secondary compounds. Schleyer and his co-workers [124,186,197] demonstrated that the theory is correct for adamantyl derivatives. On the other hand, steric hindrance (see next section) indicates that the accessibility of the reagent (or solvent) to the reaction site plays a considerable role in determining the type of the reaction.

Electromeric effects are very important in the case of α-substitution with heteroatomic groups having lone pairs of electrons at the atom directly attached to the reaction centre. Thus, the ionisation of dihalogenomethanes proceeds at a higher rate than the ionisation of the corresponding halogenomethanes, due to the stabilisation of the resulted carbocation [199]:

$$R_2CX_2 \longrightarrow \underset{\mathbf{64}}{R_2\overset{+}{C}{-}X \leftrightarrow R_2C{=}\overset{+}{X}} \longrightarrow \text{products} \qquad (5.79)$$

Fluorine is the most efficient halogen in stabilising the positive charge, due to its higher tendency to form double bonds as compared with the heavier atoms. A still greater tendency to ionise will be shown by the α-halogenoalkoxy and α-halogenoamino derivatives [200], the atoms of oxygen and nitrogen, with lower nuclear charge than halogens, being more efficient in stabilising the positive charge:

$$R^1_2\overset{+}{C}{-}\ddot{O}R^2 \leftrightarrow R^1_2C{=}\overset{+}{O}R^2; \quad R^1_2\overset{+}{C}{-}\ddot{N}R^2_2 \leftrightarrow R^1_2C{=}\overset{+}{N}R^2_2$$

The formation energies of carbenium ions with neighbouring heteroatoms [201] illustrate clearly this situation.

Benzyl stabilisation seems to govern the behaviour of $ArCXR_2$, either in unimolecular or in bimolecular reactions [3]. Indeed, both transition states **65** and **66** are stabilised by delocalisation:

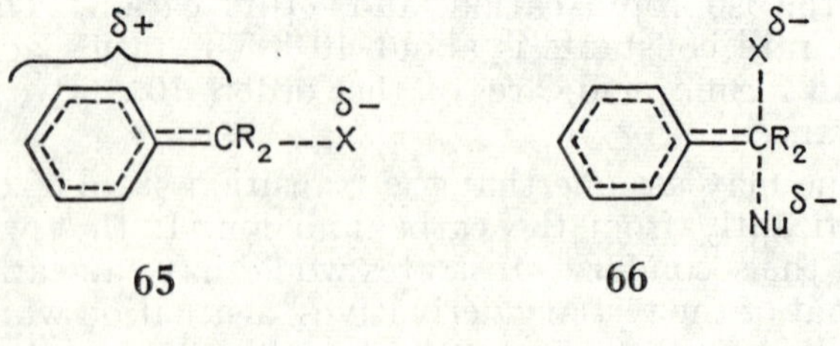

δ+ ; $C_6H_5{=}{=}{=}CR_2{-}{-}{-}X^{\delta-}$ — **65**

$X^{\delta-}$; $C_6H_5{=}{=}{=}CR_2$; $Nu^{\delta-}$ — **66**

[199] J. Hine, C. H. Thomas and S. J. Ehrenson, *J. Amer. Chem. Soc.*, **77**, 3866 (1955).

[200] J. B. Conant, W. R. Kirner and R. E. Hussey, *J. Amer. Chem. Soc.*, **47**, 488 (1925); P. Ballinger, P. B. D. De la Mare, G. Kohnstam and B. M. Prestt, *J. Chem, Soc.*, 3641 (1955).

[201] R. H. Martin, F. W. Lampe and R. W. Taft, Jr., *J. Amer. Chem. Soc.*, **88**, 1353 (1966).

The high stability of the benzyl cation has been recently demonstrated by generating this species in the gaseous phase [202] and submitting it to the attack of different nucleophiles. Even if its selectivity is low, it is still sufficiently high to discern between methanol and ethanol, on the one hand, and bromide and iodide ions, on the other hand.

5.6.2 *Steric effects*

Steric hindrance in organic reactions had been observed [203] before the theory of tetrahedral carbon, of van't Hoff [204] and Le Bel [205], was elaborated. One of the first observations directly related to the nucleophilic substitution was that dimethylanilines *ortho*-disubstituted with bulky groups do not yield quaternary ammonium salts [203]. Even in that early period it was realised that low reactivity must be due to some interference of the groups situated near the reaction centre. Similar observations were made by Kehrmann [206], but Meyer [207] was the first to state that steric hindrance at the reaction centre must account for the low reactivity in highly substituted compounds.

These early conclusions were basically correct. However, it was not realised that this kinetic effect, as steric hindrance certainly is, must be a consequence of a higher compressional energy in the transition state than in the ground state. This point was raised by Hughes, Ingold and their co-workers [208], and later developed by Brown [209].

According to Ingold [208], steric hindrance as a kinetic effect reflects higher non-bonding repulsions in the transition state with respect to the ground state. The concept of *steric retardation*, equivalent to the earlier idea of steric hindrance, seems thus to be more adequate. If the transition states of both S_N2 and S_N1 reactions are examined, it can be seen that in the transition state of a bimolecular substitution the interactions between non-bonding groups are greater compared to the ground state. On the other hand, in the unimolecular reaction there will be less non-bonding energy in the transition state, with the evident result of a *steric acceleration* (Figure 5.12).

202 R. Lorin and L. M. Dorfman, *J. Amer. Chem. Soc.*, **96**, 5715 (1974).
203 A. W. Hofmann, *Ber. dtsch. chem. Ges.*, **8**, 61 (1872).
204 J. H. van't Hoff, *Bull. Soc. chim. France*, **23** 295 (1875).
205 J. A. Le Bel, *Bull. Soc. chim. France*, **22**, 337 (1874).
206 F. Kehrmann *et al.*, *Ber. dtsch. chem. Ges.*, **27**, 510, 3146 (1894).
207 V. Meyer *et al.*, *Ber. dtsch. chem. Ges.*, **27**, 510, 3146 (1894); **28**, 3197 (1895).
208 See Reference 1, p. 401 (2nd edn, p. 547).
209 H. C. Brown, *J. Chem. Soc.*, 1248 (1946); see also Reference 68.

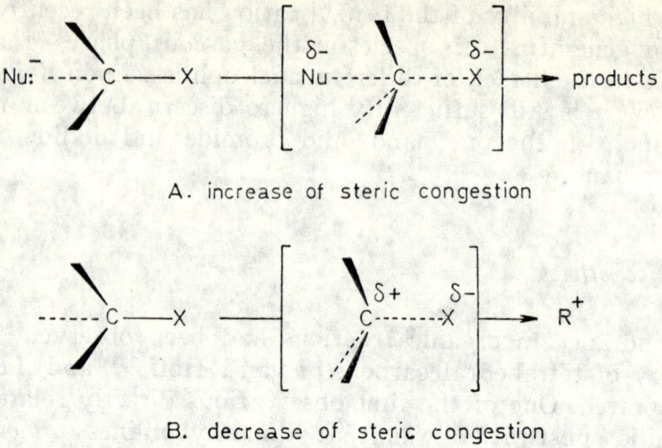

A. increase of steric congestion

B. decrease of steric congestion

Figure 5.12 Rate retardation and acceleration as a consequence of steric congestion in the transition state. A. Bimolecular reactions. B. Unimolecular reactions.

Numerous experimental data confirm these views. Hughes, Ingold and their co-workers[210, 211], in a systematic study of the Finkelstein reaction, observed that methyl groups have a more pronounced effect if situated at β than at α (with respect to the reaction centre). Thus, towards chloride ion, t-butyl bromide is about 2×10^4 less reactive than methyl bromide, whereas neopentyl bromide is about 2.5×10^6 less reactive. Dostrovsky and Hughes have demonstrated[212] the low reactivity of neopentyl derivatives in other typical bimolecular reactions. However, if neopentyl derivatives are solvolysed under conditions favouring the unimolecular mechanism, the order of reactivity may be altered dramatically (Table 5.3). It was supposed[213] and later demonstrated[42,180] that the reactions displaying low reactivity are typical bimolecular reactions, occurring with inversion of configuration, whereas the non-affected and the accelerated reactions are unimolecular substitutions, being accompanied by rearrangement of the neopentyl system:

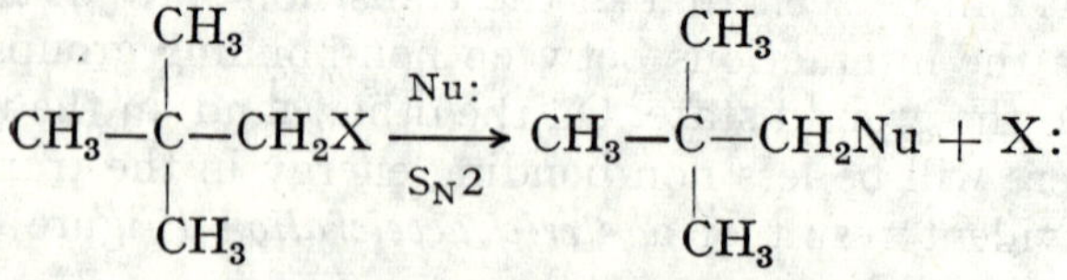

$$CH_3-\underset{CH_3}{\overset{CH_3}{C}}-CH_2X \xrightarrow[S_N2]{Nu:} CH_3-\underset{CH_3}{\overset{CH_3}{C}}-CH_2Nu + X:$$

No rearrangement; steric retardation

[210] For a general account of the works of Hughes, Ingold *et al.*, see C. K. Ingold, *Quart Rev. (Chem. Soc.)*, **11**, 1 (1957).

[211] P. B. D. de la Mare, L. Fowden, E. D. Hughes, C. K. Ingold and J. D. H. Mackie, *J. Chem. Soc.*, 3200 (1955).

[212] I. Dostrovsky and E. D. Hughes, *J. Chem. Soc.*, 157, 161, 171 (1946).

[213] I. Dostrovsky E. D. Hughes and C. K. Ingold, *J. Chem. Soc.*, 173 (1946).

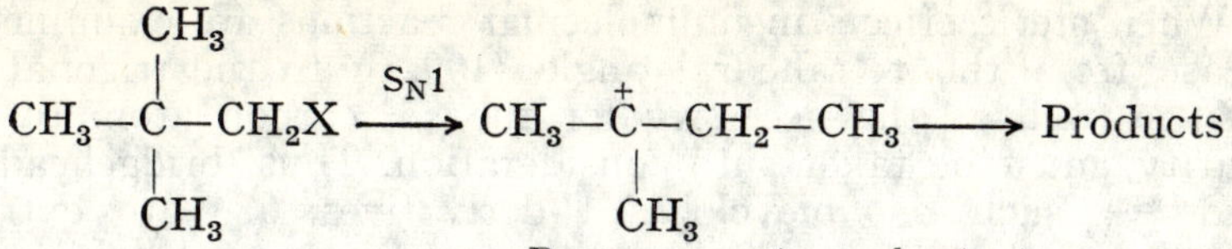

$$CH_3-\underset{CH_3}{\overset{CH_3}{\overset{|}{\underset{|}{C}}}}-CH_2X \xrightarrow{S_N1} CH_3-\underset{CH_3}{\underset{|}{\overset{+}{C}}}-CH_2-CH_3 \longrightarrow \text{Products}$$

Rearrangement; moderate to high steric acceleration

Steric effects in unimolecular reactions have been systematically studied by Brown and his co-workers [209,214]. These authors demonstrated that the relief of steric strain facilitates unimolecular reactions, especially those involving highly branched substrates. On the other hand, it was proved [215] that steric factors are equally important for the substrate and for the reagent. The following example is illustrative:

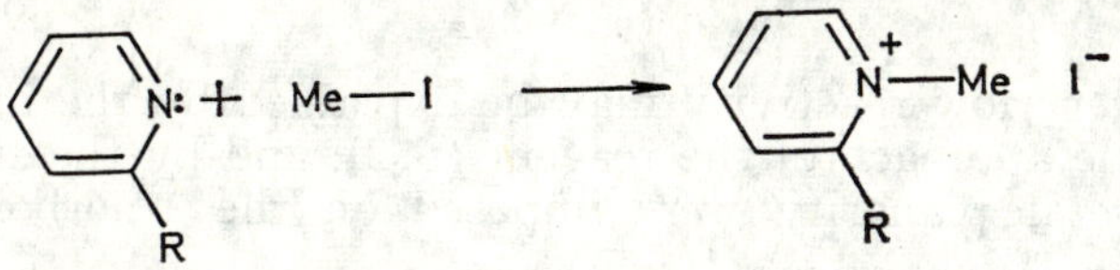

2-R-pyridine (N:) + Me—I ⟶ 2-R-N-methylpyridinium (N^+—Me) I^-

R:	H	Me	Pr^i	Bu^t
relative rate:	1	0.47	0.072	0.0002

It must be noted that in this example the crowding at the reaction site is due to a relatively remote group, as in the case studied by Ingold and Hughes which involved the neopentyl system. This fact is explained by the particular geometry of the S_N2 transition state, where bulky substituents interfere with the reagent in an earlier stage, if situated in β, than if situated in α (Figure 5.13).

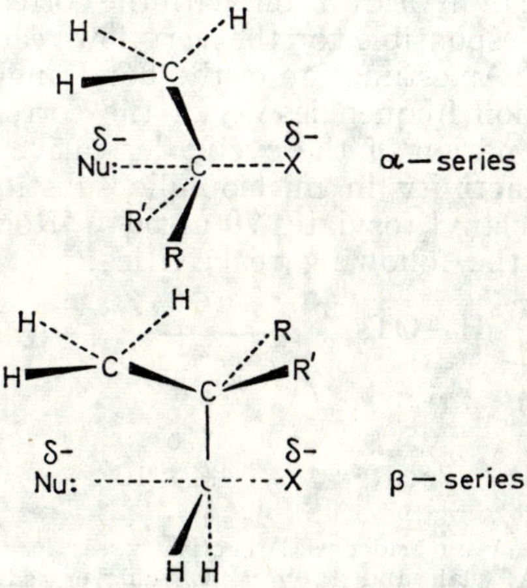

Figure 5.13 Steric retardation in the nucleophilic substitution of alkyl halides (α and β series).

[214] H. C. Brown *et al.*, *J. Amer. Chem. Soc.*, **71**, 1845 (1949); **75**, 10 (1953).

[215] H. C. Brown, D. Gintis and H. Podall, *J. Amer. Chem. Soc.*, **78**, 5375 (1956).

When steric effects in unimolecular reactions are examined, the increase from the tetrahedral angle (109.5°) to the trigonal angle (120°), as well as the necessity for the sp^2 carbon atom to adopt planarity, must be taken into consideration. Thus, bridgehead compounds [216], such as bicyclo[2.2.1]-derivatives **67**, are extremely unreactive both under S_N2 conditions (strong nucleophiles, high concentrations) and under S_N1 conditions (electrophilic catalysis by silver ion, etc.).

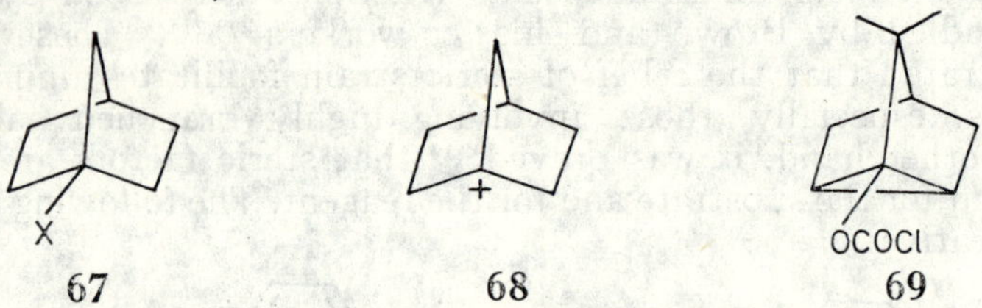

This very low reactivity may be explained by the impossibility of a rearside approach of the reagent (S_N2), and by the strain which would arise if planarity were imposed on the 1-norbornyl cation (S_N1). However, using very efficient leaving groups such as $-\overset{+}{N}_2$ [217] (in deamination reactions), or $-OCO^+$ (generated from the chloroformate group with silver ion [218]), it was possible to obtain bridgehead ions, such as **68**. In reactions with aromatic hydrocarbons they are very reactive, yielding nearly statistical amounts of *ortho*, *meta* and *para* isomers [218]. The more strained 1-apotricyclyl chloroformate **69** does not lead to the expected cation even if treated with silver borofluoride. The corresponding radical is formed instead [219], probably because pyramidal radicals are more stable than pyramidal cations.

The change of hybridisation, with the corresponding change of the bond angle, is responsible for the very low reactivity of 7-norbornyl derivatives[56–58]. An estimation of the bond angle strain may be made using the vibration frequencies ν_{CO} of the corresponding ketones [66,67].

The conformation of the cyclic derivative is in its turn important for the reactivity in nucleophilic substitution. Indeed, in the acetolysis of cyclanyl tosylates **70** at 60 °C, Roberts and Chambers [220] have observed the following reactivities:

$$\underbrace{(CH_2)_{n-1}\ \ CH{-}OTs}_{\mathbf{70}} \xrightarrow[60\ °C]{AcOH} (CH_2)_{n-1}\ \ CH{-}OAc$$

$n =$	3	5	6	7
k_{rel}	1	7×10^5	4.4×10^4	1.3×10^6

[216] For a review on bridgehead reactivity see: R. C. Fort, Jr., in *Carbonium Ions*, vol. 4 (G. A. Olah and P. von R. Schleyer, editors), Wiley-Interscience, New York (1973), p. 1783.

[217] E. H. White, H. P. Tiwari and M. I. Todd, *J. Amer. Chem. Soc.*, **90**, 4734 (1968).

[218] P. Beak, J. T. Adams and J. A. Barron, *J. Amer. Chem. Soc.*, **96**, 2494 (1974).

[219] P. Beak and R. J. Trancik, *J. Amer. Chem. Soc.*, **90**, 2714 (1968).

[220] J. D. Roberts and V. C. Chambers, *J. Amer. Chem. Soc.*, **73**, 5034 (1951).

The low reactivity of the cyclopropyl derivative is easily understood considering the π-character of the cyclopropane bonds. One may also note the decrease in the reactivity of cyclohexyl tosylate as compared to the higher and lower homologues. The difference in this and other similar reactions is determined by the specific problems of conformation arising in each of these cases [221]. Cyclohexyl derivatives **71** may exist either as equatorial (*e*) or as axial (*a*) conformers.

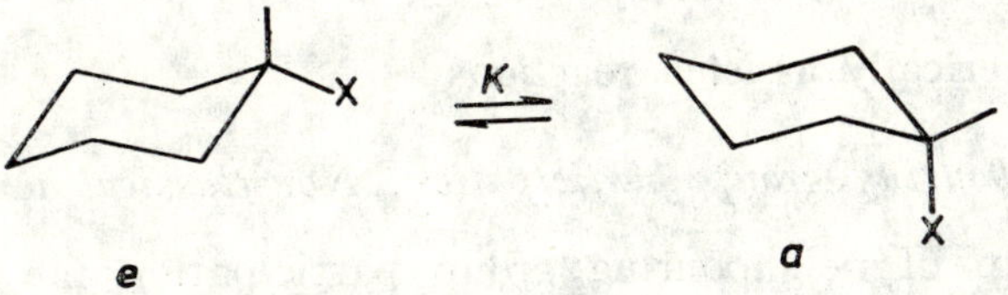

(5.80)

e a

71

The conditions required for the S_N2 reaction are better fulfilled by the axial conformer, despite the fact that the equatorial conformer is more stable. Even if the equilibrium is rapid, the actual concentration of the reactive species will be lower than the overall concentration of the cyclohexyl derivative, which is used to calculate the rate constant (k_{obs}). On the other hand, the axial attack is disfavoured due to 1,3 interactions between the leaving group, the nucleophile and the axial hydrogens. Both equilibrium, (5.80) and 1,3 axial interactions are responsible for the low reactivity of cyclohexane in the bimolecular substitution.

In the unimolecular reaction, the rehybridisation of the central carbon atom (**72**) will cause interactions between the equatorial hydrogens from C-2 and C-6, increasing a little the activation energy. We may thus expect a small rate decrease.

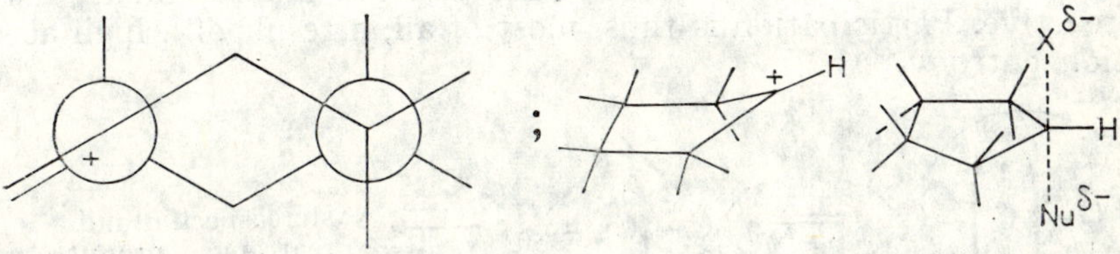

72 ; **73** **74**

In cyclopentane all bonds are eclipsed, which determines the Pitzer strain*. By the formation of cyclopentyl ion the Pitzer strain

221 See, for example, E. L. Eliel N. A. Allinger S. J. Angyal and G. A. Morrison, *Conformational Analysis*, Interscience, New York (1965).

* Pitzer strain or torsional strain (K. S. Pitzer *et al.*, *J. Chem. Phys.*, 4, 749 (1936); *Science*, **101**, 672 (1945)) results from the fact that substituents at two adjacent carbon atoms are not perfectly staggered. In order to avoid the Pitzer strain, cyclopentane derivatives assume non-planar conformations.

is diminished, since the number of eclipsed hydrogen atoms decreases. The same is also true for the S_N2 transition state **74**.

Considering the conformational factors we should thus expect a moderate difference between cyclopentyl and cyclohexyl derivatives in the unimolecular reaction, but much higher differences for the bimolecular mechanism. Indeed, if in acetolysis (S_N1) the rate ratio for the two cyclic derivatives is about 16, in a Finkelstein reaction in acetone (S_N2 conditions) the ratio increases to about 160 [222]

5.7 Anchimerically assisted reactions

5.7.1 *Neighbouring group participation. Non-classical ions*

The problem of neighbouring group participation in nucleophilic substitution presents two different aspects.

The first one is the *kinetic aspect*, unimolecular reactions proceeding by this mechanism being faster than the unassisted ones. The second one is the *thermodynamic aspect;* bridged ions are formed, which may be more stable than the open ions. Evidently, if the substrate undergoes a simple ionisation, the internal assistance being prevented geometrically, a bridged ion may rapidly be formed (Scheme 7) if the latter is more stable. Products will therefore derive from it, even if its formation will not be reflected in the kinetics.

The participation of the neighbouring group in the ionisation at C_α is an intramolecular S_N2 process and should therefore be subject to certain restrictions specific to this reaction. The most important condition is the favourable disposition of the neighbouring group with respect to the reaction centre, in order to realise the rearside attack at C_α. If the group Z is not a strong nucleophile, its involvement at the reaction centre is impossible if it is situated farther away than at C_β (in Scheme 7, $n = 0$). If favourably located, as shown in Figure 5.14, even groups with very low nucleophilicity may be reactive. Participation is thus, most of all, a result of a high activation entropy.

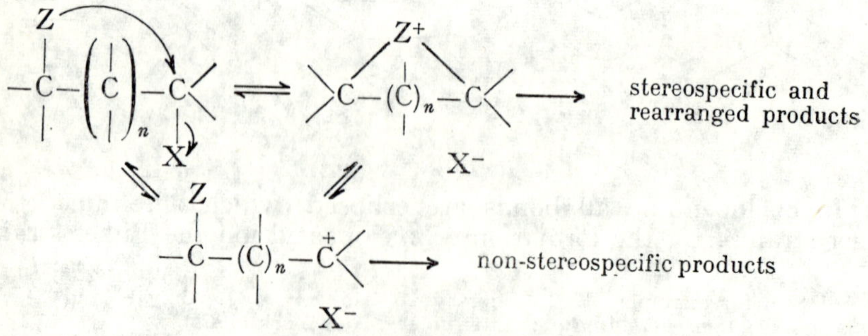

Z–C–(C)$_n$–C–X ⇌ [bridged ion: Z$^+$ bridging C–(C)$_n$–C] X$^-$ ⟶ stereospecific and rearranged products

Z–C–(C)$_n$–C$^+$ X$^-$ ⟶ non-stereospecific products

Scheme 7

[222] P. J. Fierens and P. Vershelden, *Bull. Scc. Chim. Belges*, **61**, 427, 609 (1952).

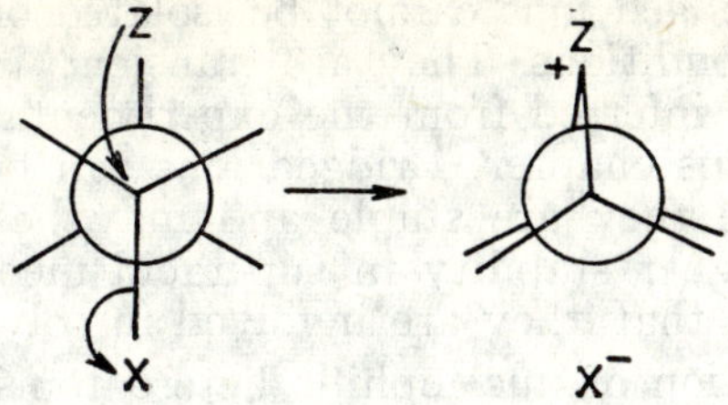

Figure 5.14 Favourable location of neighbouring groups for assisting the reaction centre at C_α.

The importance of the anchimeric assistance is also determined by the nucleophilicity of the neighbouring group. Groups like O^-, R_2N:, etc. which are strong nucleophiles, may be able to assist the reaction even if situated at a greater distance from the reaction site [223].

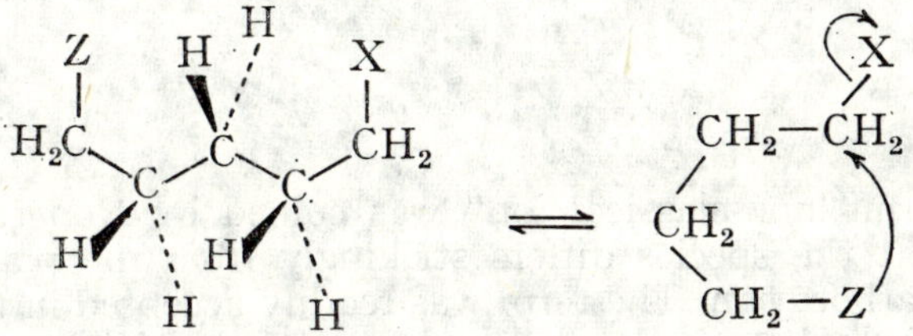

75

76

Let us consider for example the molecule **75**, existing mostly as the completely staggered conformer. The participation of Z is not possible in this conformation, and we must therefore examine the conformation **76**. Using as a first approximation the interactions in the n-butane molecule (Section 4.3), a free-energy difference of 1.3—1.8 kcal mol^{-1} between the two conformers is found*. If this difference is taken into account, it results that the reactivity observed is 10—20-fold lower than that determined only by the nucleophilic ability of Z. If Z=NR_2 and X=Cl [224], rate accelerations of 6×10^3 are observed, the single product formed being the cyclic ammonium salt. On the other hand, if Z is a moderate nucleophile, such as phenyl, even with very efficient leaving groups the reaction seems to be completely unassisted [49,225].

Another consequence of neighbouring group participation is the formation of cyclic intermediates, the bridged ions. With rare

[223] See, for example, B. Capon, *Quart. Rev. (Chem. Soc.)*, **18**, 45 (1964).
* The actual value depends on the nature of Z.
[224] C. A. Grob and F. A. Jenny, *Tetrahedron Letters*, (**23**), 25 (1960).
[225] R. Heck and S. Winstein, *J. Amer. Chem. Soc.*, **79**, 3105, 3114 (1957).

exceptions, the bridged ions cannot be isolated or directly observed under solvolysis conditions. They are transient intermediates, their intervention being inferred from the experimental data discussed in the first part of this chapter. Bridged ions can be studied in superacid media, where they are stable and may be observed directly. However, their higher stability in superacid media does not necessarily demonstrate that they are involved in solvolyses.

The participation of nucleophilic heteroatoms (halogen, oxygen, sulphur, nitrogen, etc.) seems to be generally accepted. Aryl participation, with formation of phenonium ions, even if not beyond controversy [68], is now clearly proved [49]. The disagreement with respect to bridged ions regards the so-called *non-classical ions*, involving σ-bridging in solvolysis.

$$>C(R)-C(X)< \longrightarrow >C\overset{R^{+}}{\overline{\cdots\cdots}}C< \qquad (5.81)$$

77 (R = alkyl)

The term "non-classical ion" was coined by Roberts [226], and it suggests that the species differs strikingly from the planar trigonal (classical) carbocation. The term was readily accepted in the chemical literature, in spite of alternative proposals [227]. What seems to be really contested with regard to non-classical ions is not the existence of bridged ions **77**, but the anchimeric assistance according to (5.81). In the following sections, we shall be mainly concerned with the formation of non-classical ions in solvolyses.

5.7.2 *Halogen participation*

Soon after Roberts and Kimball [15] proposed the halogenium ions as intermediates in addition reactions, Winstein and Lucas [16] proved their occurrence in solvolytic displacements. The efficiency of halogens in the participation was later studied by Winstein and Grunwald [228] in the acetolysis of 2-halogenocyclohexyl brosylates. In order to determine the anchimeric assistance, the rates of solvolysis of both *trans* and *cis* isomers were measured. (Evidently, participation is only possible in diaxial conformation *(trans)* and may not be achieved in the *cis*-derivative.) From the measured values, Win-

[226] J. D. Roberts and C. C. Lee, *J. Amer. Chem. Soc.*, **73**, 5009 (1951).

[227] P. D. Bartlett, *Non-classical Ions, Reprints and Commentary*, Benjamin, New York (1965).

[228] S. Winstein, E. Grunwald and L. L. Ingraham, *J. Amer. Chem. Soc.*, **70**, 84 (1948); E. Grunwald, *J. Amer. Chem. Soc.*, **73**, 5458 (1951).

stein and Grunwald calculated the following factors of anchimeric acceleration by the neighbouring halogen:

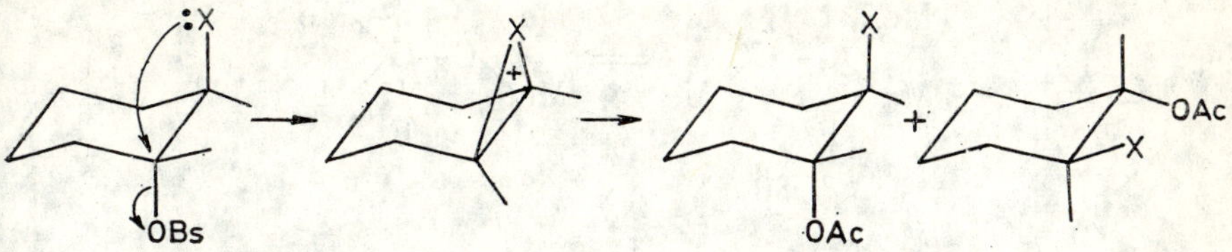

X =	Cl	Br	I
k_{trans}/k_{cis} =	3	810	3×10^6

It may be seen that the participation is more important in the case of the heavier halogens.

Surprisingly enough, it has been found that fluoronium ions may also be involved in solvolysis, in good ionising solvents such as trifluoroacetic acid[229]. An example is the following addition of trifluoroacetic acid to the fluoro alkyne **78**:

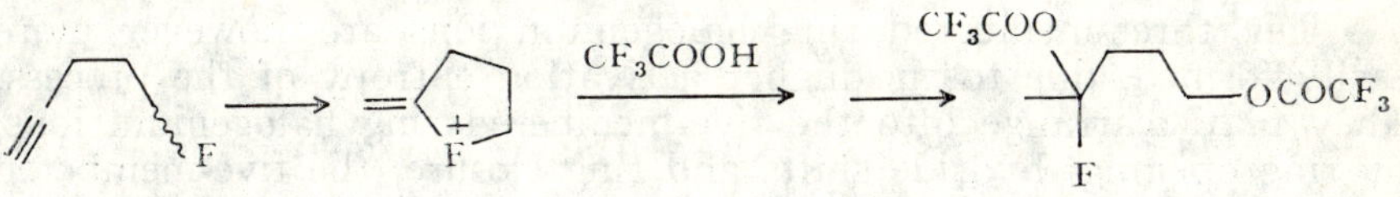

CF_3COOH

78 15%

Halogenium ions have been directly observed in strong acids by Olah and Bollinger[230, 231]. The halogenium ions **79** were thus generated and observed by means of n.m.r., confirming the decreasing order of stability derived by Winstein and Grunwald from their kinetic data:

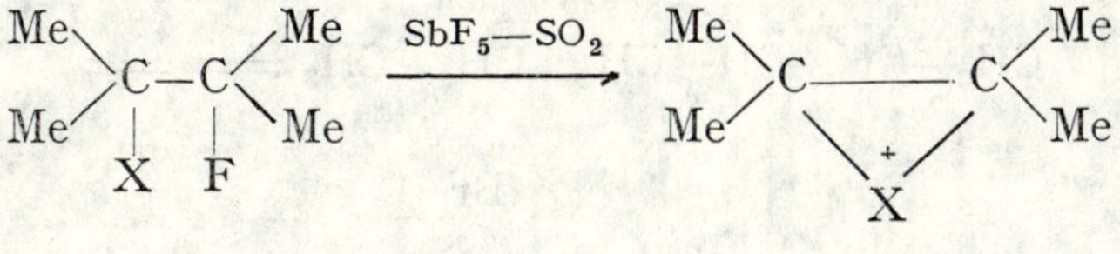

$$Me_2C(X)\text{—}C(F)Me_2 \xrightarrow{SbF_5\text{—}SO_2} Me_2C\text{—}CMe_2 \; (\overset{+}{X} \text{ bridged}) \quad (5.82)$$

79

[229] P. E. Petersen and R. J. Bopp, *J. Amer. Chem. Soc.*, **89**, 1283 (1967).

[230] G. A. Olah and J. M. Bollinger, *J. Amer. Chem. Soc.*, **89**, 4744 (1967); **90**, 947 (1968).

[231] G. A. Olah, J. M. Bollinger and J. Brinich, *J. Amer. Chem. Soc.*, **90**, 2587 (1968).

However, if X = F, no fluoronium ion could be detected. Rapidly equilibrating ions are instead observed:

$$\mathrm{Me_2C(F)-\overset{+}{C}Me_2 \rightleftharpoons Me_2\overset{+}{C}-C(F)Me_2}$$

80

The problem of obtaining halogenium ions with larger rings was considered by Petersen [229, 232], and by Olah and his co-workers [231–233]. They have shown that five-membered ring halogenium ions can be generated easily from molecules with a very efficient leaving group, such as a positive oxygen:

$$\mathrm{CH_3-CH(X{:})-(CH_2)_n-CH_2-\overset{+}{O}R_2 \longrightarrow CH_3-CH-(CH_2)_n-CH_2\ (\text{ring closed through } \overset{+}{X}) + R_2O}$$

The three-membered ring halogenium ions are however more easily formed, due to the higher activation entropy of the process. They may rearrange into the five-membered ring halogenium ions, by ring opening, hydride shifts and ring closure, the five-membered ring being generally more stable than the classical halogenium ion. In this transformation it was not possible to detect a four-membered ring, which is probably disfavoured by both enthalpy and entropy.

$$\mathrm{CH_3-CH-CH-CH_2-CH_2-CH_3\ (bridged\ by\ \overset{+}{Br}) \rightleftharpoons}$$

$$\mathrm{CH_3-CH(Br)-\overset{+}{C}H-CH_2-CH_2-CH_3 \rightleftharpoons \rightleftharpoons}$$

$$\mathrm{\rightleftharpoons \rightleftharpoons CH_3-CH-CH_2-CH_2-CH-CH_3\ (five\text{-}membered\ ring\ closed\ through\ \overset{+}{Br})}$$

[232] G. A. Olah and P. E. Petersen, *J. Amer. Chem. Soc.*, **90**, 4675 (1968).
[233] G. A. Olah, P. W. Westerman, E. G. Melby and Y. K. Mo, *J. Amer. Chem. Soc.*, **96**, 3565 (1974).

With one certain exception [234], four-membered ring halogenium ions are not known (see also Section 10.2).

The generation of halogenium ions in presence of SbF_5 and in FSO_3H at −60 °C has been used to determine by n.m.r. spectroscopy their relative stabilities. Some significant data obtained by Larsen and Metzner [235] are shown in Table 5.13.

TABLE 5.13.

Heats of formation (ΔH_{obs}) of some halogenium ions [a,b]

Precursor	Ion	ΔH°_{obs}/kcal mol^{-1}
$MeCH(Br)-CH_2Br$	$MeCH-CH_2$ bridged by Br^+	-3.29 ± 0.51
$Me_2C(Br)-CH_2Br$	Me_2C-CH_2 bridged by Br^+	-8.39 ± 0.87
$Me_2C(Br)-CHBrMe$	$Me_2C-CHMe$ bridged by Br^+	-13.4 ± 0.90
$Me_2C(Br)-CBrMe_2$	Me_2C-CMe_2 bridged by Br^+	-22.9 ± 4.0
$Me_2C(F)-CHBrMe$	$Me_2C-CHMe$ bridged by Br^+	-15.5 ± 2.15
$MeCH(Cl)-CH_2-CH_2-CH_2Cl$	$MeCH(CH_2-CH_2)CH_2$ ring with Cl^+	-14.1 ± 0.90
$MeCH(Br)-CH_2-CH_2-CH_2Br$	$MeCH(CH_2-CH_2)CH_2$ ring with Br^+	-13.8 ± 1.87
Me_3C-Br	Me_3C^+	-14.5 ± 2.0
Ph_2CHBr	Ph_2CH^+	~ -30

(*a*) Determined by Larsen and Metzner [235]. (*b*) The determinations were performed at −60 °C, in fluorosulphonic acid, containing 11.5% moles SbF_5.

The heats of formation depend on the leaving group. In the case of the t-butyl ion it has been found [236] that the difference in ΔH°_{obs}

[234] J. H. Exner, L. D. Kershner and T. E. Evans, *J. Chem. Soc. Chem. Commun.*, 361 (1973).

[235] J. W. Larsen and A. V. Metzner, *J. Amer. Chem. Soc.*, **94**, 1614 (1972).

[236] E. M. Arnett and J. W. Larsen, unpublished data, quoted by Larsen and Metzner [235].

when starting either from t-butyl chloride or from t-butyl bromide is 7×8 kcal mol^{-1}. It thus results, considering the data of Table 5.13, that tetramethylbromonium ion is by about 7.5 kcal mol^{-1} more stable than the corresponding chloronium ion. Using ion cyclotron techniques in order to determine carbenium and carbonium ion stability, Beauchamp and his co-workers [237] have compared the stabilities of cyclic and acyclic ions, finding that tetramethylenebromonium ions are by far more stable than the dimethylenebromonium ions.

Olah and his co-workers [238, 239] have prepared acyclic halogenium ions. The study has revealed that dimethyl and diethyl halogenium ions are capable of alkylating a variety of n-donor bases. The following processes may take place:

$$R^1{-}\overset{+}{X}{-}R^1 + R^2{-}X \rightleftharpoons R^1{-}X + R^2{-}\overset{+}{X}{-}R^1 \qquad (5.83)$$

$$R^1{-}\overset{+}{X}{-}R^1 + \overset{+}{R^2} \rightleftharpoons \overset{+}{R^1} + R^1{-}\overset{+}{X}{-}R^2 \qquad (5.84)$$

corresponding either to an S_N2 (5.83) or to an S_N1 (5.84) reaction. The question was thus raised [233] whether the equilibration of static alkene halogenium ions with their corresponding halogenoalkyl carbenium ions is not similar to equation (5.84). Anyway, a reaction of type (5.83) between a cyclic halogenium ion and a halide has been positively demonstrated [237].

The interest in halogenium ions concerns not only the solvolytic reactions, but also the addition reactions (Chapter 10).

5.7.3 *Oxygen participation*

The negative oxygen is one of the most efficient neighbouring groups. The formation of epoxides [240] is illustrative:

$$\underset{OH}{CH_2}{-}\overset{Cl}{CH_2} \underset{}{\overset{B:}{\rightleftharpoons}} \underset{O^-}{CH_2}{-}\overset{Cl}{CH_2} \longrightarrow \underset{\diagdown O \diagup}{CH_2{-}CH_2} \;\; Cl^- \qquad (5.85)$$

$$RO^- + R{-}X \longrightarrow R{-}O{-}R + X^- \qquad (5.86)$$

The cyclisation (5.85) is, to a certain extent, similar to the Williamson reaction (5.86) leading to ethers. However, it differs from the latter in that it may be performed in water, even with relatively weak bases such as calcium hydroxide. In contrast, ether formation (5.86) needs

[237] D. W. Wieting, R. H. Staley and J. L. Beauchamp, *J. Amer. Chem. Soc.*, **96**, 7552 (1974).

[238] G. A. Olah, J. R. De Member, Y. K. Mo, J. J. Svoboda, P. Schilling and J. A. Olah, *J. Amer. Chem. Soc.*, **96**, 884 (1974).

[239] G. A. Olah and Y. K. Mo, *J. Amer. Chem. Soc.*, **96**, 3560 (1974).

[240] See, for example, S. Winstein and R. B. Henderson, in *Heterocyclic Compounds*, vol. 1 (R. C. Elderfield, editor), J. Wiley, New York (1950), p. 1.

preformed alkoxide, water being exluded in order to prevent the equilibration between the alcohol and its alkoxide. The formation of a strained ring means that the reaction is not favoured by its activation enthalpy. Rate measurements [241] have actually demonstrated that the reaction is controlled by its activation entropy.

The earliest example of a stereochemical effect of neighbouring groups is the participation of negative oxygen in the hydrolysis of α-halogenoacids (see p. 228). It has been assumed by Winstein[152] that the reaction intermediates are α-lactones, which hydrolyse rapidly under the reaction conditions. However, until recently such intermediates were not known. Chapman[242] and his co-workers were able to prepare the α-lactone **81**, very reactive even at −196 °C.

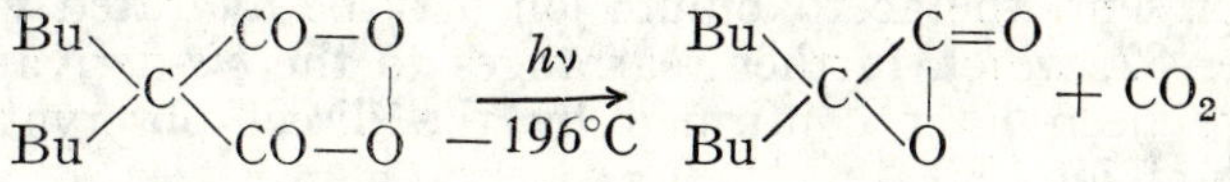

$$Bu_2C(CO{-}O)_2 \xrightarrow[-196^\circ C]{h\nu} Bu_2C(C{=}O)O + CO_2$$

81

The isolation of intermediate **81** confirms the explanation of the retention of configuration in the hydrolysis of α-halogenoacids.

The neighbouring hydroxyl either from alcohols or from acids has a much lower ability of assisting ionisation. Moreover, the observation of the phenomenon is rendered difficult by the inductive effect of OH (OR)[243] and COOH groups[244]. However, significant acceleration factors are observed if the OR group is situated farther from the reaction site. Brosylates **82** and **83**, for instance, are acetolysed 654- and 123-fold, respectively, faster than expected[245].

More efficient than the RO groups are the RS groups[246]. Thus, the hydrolysis of the chloride **84** in aqueous dioxane proceeds 10^4 times faster than the hydrolysis of the corresponding alkoxy chloride[247].

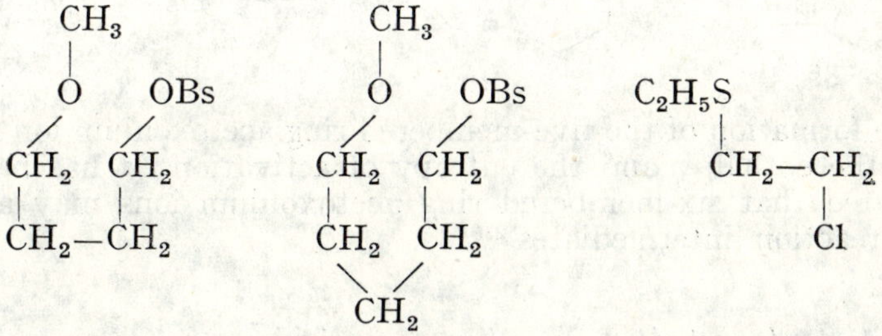

$CH_3O{-}CH_2{-}CH_2{-}CH_2{-}CH_2{-}OBs$ (**82**)

$CH_3O{-}CH_2{-}CH_2{-}CH_2{-}CH_2{-}CH_2{-}OBs$ (**83**)

$C_2H_5S{-}CH_2{-}CH_2{-}Cl$ (**84**)

82 **83** **84**

[241] J. E. Stevens, C. L. McCabe and J. C. Warner, *J. Amer. Chem. Soc.*, **70**, 2449 (1948); W. C. Woodland, R. B. Carlin and J. C. Warner, *ibid.*, **75**, 5840 (1953).

[242] O. L. Chapman, P. W. Wojtkowski, W. Adam, O. Rodriquez and R. Rucktäschel, *J. Amer. Chem. Soc.*, **94**, 1365 (1972).

[243] D. D. Roberts, *J. Org. Chem.*, **33**, 118 (1968).

[244] A. M. Glatz, *Rev. Roumaine Chim.*, **19**, 455 (1974).

[245] S. Winstein, A. L. Albred, R. Heck and C. F. Glick, *Tetrahedron*, **3**, 1 (1958).

[246] K. D. Gundermann, *Angew. Chem.*, **75**, 1194 (1963).

[247] H. Bohme and U. Sell, *Chem. Ber.*, **81**, 123 (1948).

The higher tendency of the sulphur atom, as compared to oxygen, to lose electrons, the longer C—S bonds as compared to C—O bonds, and the lower inductive effect of sulphur, account for these large acceleration factors.

A different participation of neighbouring oxygen is encountered in carbonyl derivatives, first studied by Winstein and Buckles [248]. *trans*-Acetoxy-halogenocyclohexanes **85** in glacial acetic acid containing acetic anhydride, are acetolysed to *trans*-diacetates. Conversely, if small amounts of water are present, *cis*-derivatives are obtained. Winstein assumed the formation of an acetoxonium ion **86**, with inversion of configuration at C_α, followed by a new inversion either at C_β or C_α, with retention of the partial configuration. If water is present, the acetoxonium ion may be hydrated yielding the ortho-ester **87**, which further rearranges to the *cis*-derivative. These ideas have been later confirmed by the direct observation of the acetoxonium ion [249].

The formation of the five-membered ring acetoxonium ion is favoured by the enthalpy and the entropy of activation. It has been demonstrated that six-membered ring acetoxonium ions may also appear as reaction intermediates [250].

5.7.4 *Aryl participation. Phenonium ions*

Studying the acetolysis of 3-phenyl-2-butyl tosylates, Cram [47] observed that the *erythro*-derivative retains the absolute configuration both at C_α and C_β, whereas the *threo*-derivative loses its optical

[248] S. Winstein and R. E. Buckles, *J. Amer. Chem. Soc.*, **64**, 2780 (1942).

[249] R. M. Roberts, J. Corse, R. Boschan, D. Seymour and S. Winstein, *J. Amer. Chem. Soc.*, **80**, 1247 (1958).

[250] M. Banciu, Ph. D. Thesis, Bucharest (1968).

activity but retains the partial configuration. Despite the very small acceleration of the reaction rates (Section 5.2.2), phenyl participation has been invoked, the intermediate being called the phenonium ion. Brown and his co-workers [46, 65, 68, 163] have repeatedly challenged the formation of phenonium ions; the main argument relies on the very small, if any, acceleration. Indeed, in the initial conception, any neighbouring group participation (for example, that of halogen [228]) should be accompanied by significant rate increase. Even if allowance is made for the inductive effect of the phenyl group and for the return, the acceleration factors are lower than about 24—30 [49]. Brown proposed, instead of a symmetric phenonium ion, a pair of rapidly equilibrating carbenium ions (5.87), which would determine nearly the same stereochemical outcome.

$$CH_3-\underset{}{\overset{Ph}{CH}}-\overset{+}{C}H-CH_3 \rightleftharpoons CH_3-\overset{+}{C}H-\overset{Ph}{CH}-CH_3 \qquad (5.87)$$

However, convincing arguments point now to phenonium ions as true reaction intermediates [49]. The main pieces of evidence are the following:

(a) Higher accelerations are observed for β-aryl derivatives substituted in the aromatic ring with strong electron-repelling groups. Thus, 3-*p*-anisyl-2-butyl tosylate is about 80 times as reactive as the phenyl derivative, which corresponds to an accelerating factor of about 1600—2400 as compared with the 2-butyl system. Moreover, the anisyl compound presents special salt effect, whereas the phenyl derivative manifests only normal salt effect [251]. Inductive effects alone cannot account for this large rate difference between the unsubstituted and the substituted derivative. If, however, the formation of an anisonium ion **88** is admitted, the effect of the electron-donating group MeO is easily understood:

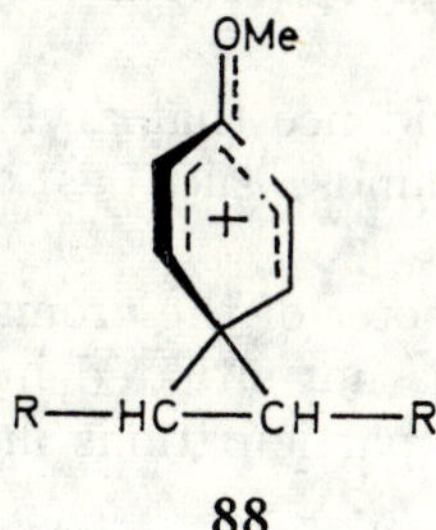

88

[251] S. Winstein, E. Clippinger, A. H. Fainberg, R. Heck and G. C. Robinson, *J. Amer. Chem. Soc.*, **78**, 328 (1956); A. H. Fainberg and S. Winstein, *J. Amer. Chem. Soc.*, **78**, 2780 (1956).

(b) There is a large difference in the reactivity of *threo*- and *erythro*-isomers, the latter being more reactive [252], which is consistent with an intermediate phenonium ion. Indeed, an *erythro*-derivative would yield a *trans*-phenonium ion, which must be more stable than the *cis* (see also Section 5.3.3).

(c) In certain cases the intermediates are stable, isolable products, as for example **89** [253]. Here it can be stressed that phenonium ions are true σ complexes, like those involved in the aromatic electrophilic substitution.

O^- ; CH_2—CH_2—X → O ; H_2C—CH_2

89 (isolated)

(d) High rate accelerations (over 10^3) have been observed even for the simple β-phenylethyl system if the solvent is a good ionising medium [50] (for instance, trifluoroacetic acid). The kinetic evidence for participation is thus not absent.

(e) It has been shown that all the discrepancies regarding the acetolyses of β-aryl derivatives (small acceleration factors, low stereospecificity or no rearrangement) can be explained by assuming that the acetolysis occurs by two different routes, one assisted by the neighbouring aromatic ring, the other by the solvent [51]. As regards the correlation between the stereochemical and the kinetic data, it has been shown [49] that an acceleration factor of only 10 may correspond to 90% neighbouring group participation.

(f) Phenonium ions have been directly observed in solvolyses in strong acid media [49].

(g) There is strong evidence that aryl participation occurs even in highly stressed compounds, such as bicyclic compounds [170] or paracyclophane derivatives [254].

The nucleophilic character of the aromatic ring permits its participation even if the ring is not situated in the close vicinity of the reaction site. Such aryl participations have been designated as

[252] A. H. Fainberg, G. C. Robinson and S. Winstein, *J. Amer. Chem. Soc.*, **78**, 2780 (1956).

[253] R. Baird and S. Winstein, *J. Amer. Chem. Soc.*, **79**, 4238 (1957).

[254] D. J. Cram and L. A. Singer, *J. Amer. Chem. Soc.*, **85**, 1075 (1963); D. J. Cram and F. L. Harris, Jr., *J. Amer. Chem. Soc.*, **89**, 4642 (1967).

Ar_1-5 and Ar_2-6 mechanisms [225], according to the relative position of the aromatic ring and of the reaction centre.

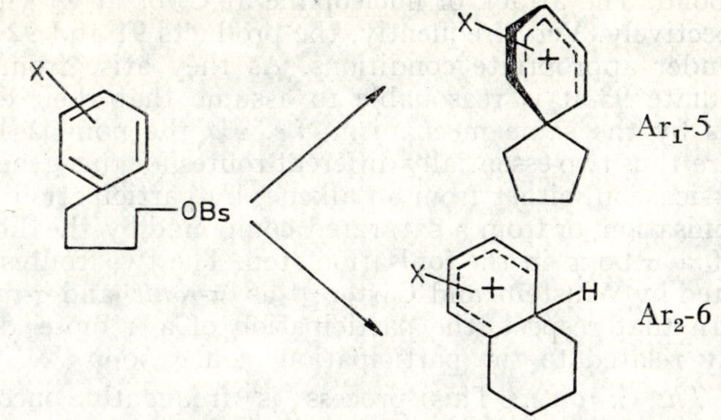

X OBs Ar$_1$-5 X H Ar$_2$-6

Although the rate accelerations are very small, the two mechanisms have been detected and confirmed later [255].

5.7.5 π-*Routes and* σ-*routes in the generation of non-classical ions*

The direct involvement of a double bond at the reaction site during the ionisation step is shown in Scheme 8:

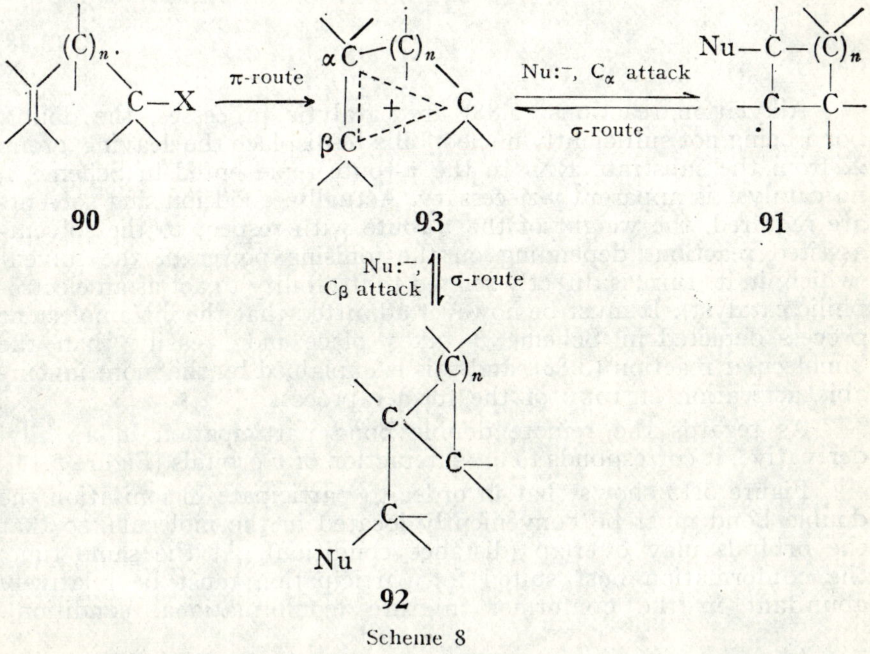

90 π-route **93** Nu:⁻, C_α attack σ-route **91**

Nu:⁻, C_β attack σ-route

92

Scheme 8

[255] L. M. Jackman and V. R. Haddon, *J. Amer. Chem. Soc.*, **96**, 5130 (1974).

In this process, the double bond plays the role of an internal nucleophile. The non-classical ion **93** contains a two-electron, three-centre bond. The attack of nucleophile at C_α or at C_β yields **91** and **92**, respectively. Very frequently, the products **91** and **92** may equilibrate under appropriate conditions. As they arise from a common intermediate **93** it is reasonable to assume that their equilibration proceeds by the same mechanism, *i.e. via* the non-classical ion **93**. There are thus two essentially different routes for the generation of a non-classical ion, either from an alkene, by participation of a π bond in the ionisation, or from a saturated compound by the direct involvement of a σ bond in the ionisation step. The two routes have been designated by Winstein and Carter [256] as *π-route* and *σ-route*, respectively. In that respect, the participation of a remote double bond is closely related to the participation of a σ bond.

(a) *The π-route.* This process is frequently encountered in bimolecular reactions, such as the alkylation (5.88):

$$\mathrm{>C{=}C<} + \mathrm{R{-}X} \rightarrow \mathrm{MX}_n \longrightarrow \left[\mathrm{>C{=}C<} \cdots \overset{+}{\mathrm{R}} \quad \mathrm{MX}_{n+1}^- \right] \longrightarrow \mathrm{>C(R){-}C(X)<} + \mathrm{MX}_n \tag{5.88}$$

One of the possible isomers

Alkylation reactions (5.88) are catalytic processes, the double bond being not sufficiently nucleophilic to displace the leaving group X from the substrate RX. In the π-route represented in Scheme 8, no catalyst is apparently necessary. Actually, good ionising solvents are required, the weight of the π-route with respect to the solvent-assisted reactions depending on the ionising power of the solvent (which, in its turn, is directly related to its ability to act as an electrophilic catalyst). It must be however admitted that the intramolecular process depicted in Scheme 8 takes place more easily than the bimolecular reaction (5.88), and this is explained by the more favourable activation entropy of the former process.

As regards the remote double bond participation in an allyl derivative, it corresponds to a σ interaction of p orbitals (Figure 5.15).

Figure 5.15 shows that in order to participate in ionisation the double bond must be conveniently located in the molecule, so that the orbitals may overlap (distance condition). At the same time, the conformation best suited for participation must be relatively abundant in the conformer mixture (conformational condition).

[256] S. Winstein and P. Carter, *J. Amer. Chem. Soc.*, 83, 4485 (1961).

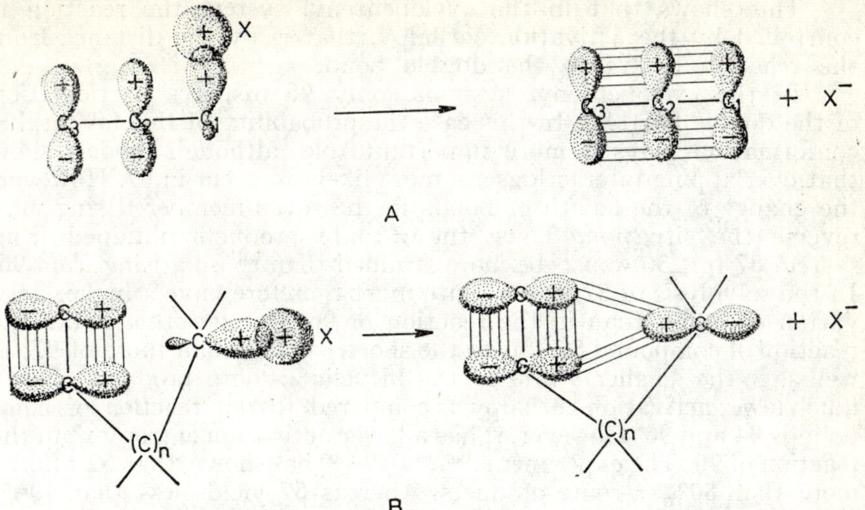

Figure 5.15 The involvement of a double bond in the ionisation process. A. Allyl double bond. B. remote double bond.

These may be regarded as enthalpic and entropic conditions, respectively, for the anchimeric assistance. Significant examples are found in the following series:

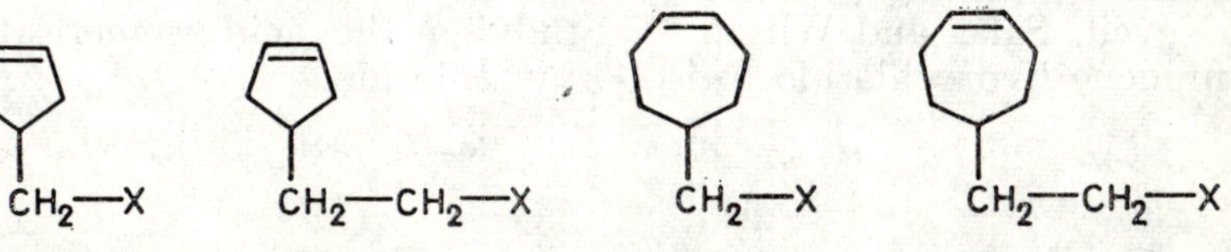

CH_2—X CH_2—CH_2—X CH_2—X CH_2—CH_2—X

94   **95**   **96**   **97**

Of the two 3-cyclopentenyl systems, **94** is more likely to adopt the conformation required for double bond participation. However, Dreiding models show that the distance from the double bond to the ionisation centre is larger in **94** than it is in **95**. The suitable conformation of **95**, on the other hand, should be less stable than other conformations by at least 1,6 kcal mol $^{-1}$. However, only **95** actually reacts by a π-route, leading to the 2-norbornyl system [53, 54]:

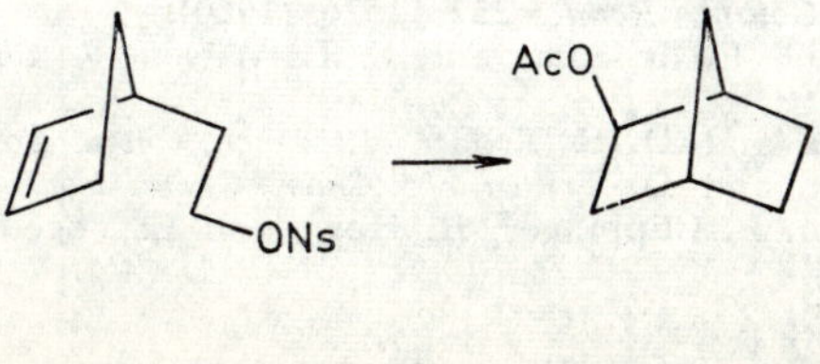

ONs → AcO

This shows that in the cyclopentenyl system the reaction is controlled by the activation enthalpy, that is by the distance from the reaction centre to the double bond.

In the cycloheptenyl systems, only **96** displays participation of the double bond [257]. In this case the probability of the favourable conformation plays a more important role, although models show that overlapping interactions are more likely to occur in **97**. However, the change of the length of bonds in the seven-membered ring may reverse the situation. Thus, the π-route products obtained from system **57** (p. 250) must be more strained than those arising from **96**. In the solvolysis of **57**, the π-route must therefore have a higher activation enthalpy than in the reaction of **96**. On the other hand, the reaction of compound **52**, due to the shorter bonds than those of **97**, as well as to the smaller strain in the bicyclic π-route products, has a much lower activation enthalpy as compared to the reaction of compounds **94** and **96**. However, it has a lower activation entropy than the reaction of **96**. The experiment [190,191,193–195] has shown that **52** affords more than 80% π-route products, whereas **57** yields less than 10%. Both energy (overlap interactions) and probability factors (conformational energies) are responsible for these results.

(b) *The σ-route.* It is rather surprising that the π-route has been so readily accepted, whereas the σ-route was the object of constant and hard criticism. Even more intriguing is the fact that the examples of σ-participation which stirred up more controversy were those related to positively demonstrated cases of π-route. One of them was the formation of the bicyclo[2.2.1]2-heptyl ion. The first indication pointing to a delocalised ion with this structure was due to Nevell, Salas and Wilson [258]. Studying the acid isomerisation of camphene hydrochloride to isobornyl chloride:

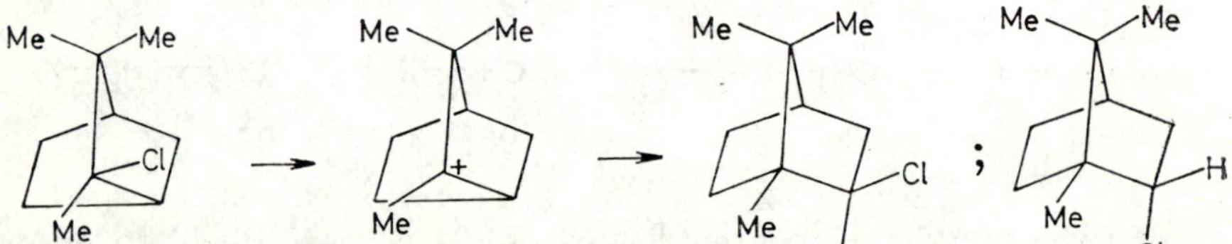

98

they wrote: ..."although the intermediate is represented as having the camphene structure (**98**), it is possible that it is mesomeric between this and the corresponding isobornyl structure." This was the first statement concerning a non-classical carbocation structure. This idea was later formulated in a more precise form by Winstein and Trifan [259] and demonstrated comprehensively [260, 261] for the more

[257] G. LeNy, *Compt. Rend.*, **251**, 1526 (1960).

[258] T. P. Nevell, E. de Salas and C. L. Wilson, *J. Chem. Soc.*, 1188 (1939), namely p. 1192.

[259] S. Winstein and D. S. Trifan, *J. Amer. Chem. Soc.*, **71**, 2953 (1949).

[260] S. Winstein and D. Trifan, *J. Amer. Chem. Soc.*, **74**, 1147, 1154 (1952).

[261] S. Winstein, E. Clippinger, R. Howe and E. Vogelfanger, *J. Amer. Chem. Soc.*, **87**, 377 (1965).

simple system of norbornane. The idea of σ-participation in ionisation, and thus of σ-bond anchimeric assistance, is also found in a paper of Hughes, Ingold and their co-workers [262]. They compared the reactions of bornyl and isobornyl chloride, and found that the latter compound solvolysed about 10^5-fold as fast as the former chloride. They termed this phenomenon *synartetic acceleration* and assumed the intervention of a common *synartetic ion*, an intermediate with the same characteristics as the so-called non-classical ion. The idea of σ-participation in ionisation is therefore not so new, as it may be traced back [258] to the early conceptions of bridged ions [15, 16] or of neighbouring-group participation in displacement reactions [16, 248].

Many examples of σ- and π-routes are well known. One of them involves the non-classical norbornyl ion [263], first identified by Trifan and Winstein [259–261] when generated by a σ-route, and later obtained by Lawton [53] and by Bartlett and his co-workers [54] by a π-route starting from derivatives of type **95**. This ion may be however generated from many other substrates, as has been demonstrated later [264–268] (Scheme 9).

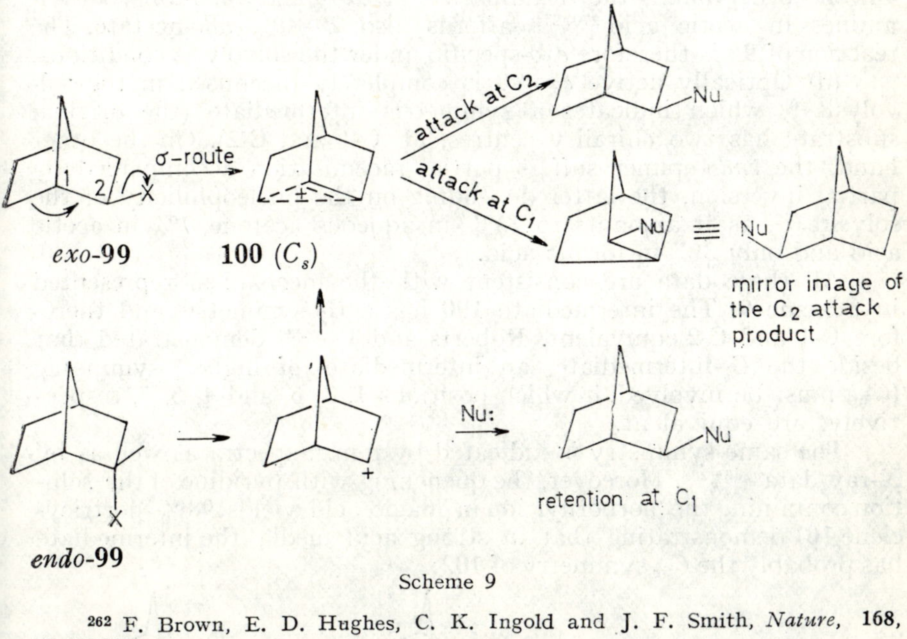

σ-route
attack at C_2
attack at C_1
exo-**99**
100 (C_s)
mirror image of the C_2 attack product
Nu:
retention at C_1
endo-**99**

Scheme 9

[262] F. Brown, E. D. Hughes, C. K. Ingold and J. F. Smith, *Nature*, **168**, 65 (1951).

[263] G. D. Sargent, in *Carbonium Ions*, vol. 3 (G. A. Olah and P. von R. Schleyer, editors), Wiley-Interscience, New York (1972), p. 1099.

[264] P. von R. Schleyer, W. E. Watts, R. C. Fort, Jr., M. B. Comisarow and G. A. Olah, *J. Amer. Chem. Soc.*, **86**, 5679 (1964).

[265] M. Saunders, P. von R. Schleyer and G. A. Olah, *J. Amer. Chem. Soc.*, **86**, 5680 (1964).

[266] G.A. Olah, A. Coomeyras and C. Y. Lui, *J. Amer. Chem. Soc.*, **90**, 3882(1968).

[267] G.A. Olah and A. M. White, *J. Amer. Chem. Soc.*, **91**, 3954, 3956 (1969),

[268] G. A. Olah, G. D. Mateescu and J. L. Riemschneider, *J. Amer. Chem. Soc.*, **94**, 2530 (1972).

The non-classical nature of the norbornyl cation was demonstrated by Winstein and his co-workers [259–261] using kinetic, product and stereochemical data. Their arguments were based on the following observations:

(a) The *exo*-2-norbornyl brosylate *exo*-**99** is by far more reactive than its epimer, *endo*-**99**, the latter having however about the same reactivity as cyclohexyl brosylate. The high ratio $k_{exo}/k_{endo} = 350$ was explained as a σ-assistance in ionisation, the C-6—C-1 bond being favourably situated, with respect to the reaction centre, in the exo isomer, but completely unfavourably in the endo compound. The intermediate was formulated as **100**.

(b) The kinetic data shows that the exo isomer presents special salt effect, the ratio between the polarimetric and titrimetric rate constants being: $k_p/k_t = 4.5$ [261]. On the other hand, the two rates are fairly constant for the *endo*-isomer [260], which shows that it does not undergo ion-pair return, whereas the *exo* does.

(c) Practically, less than 0.05% *endo*-acetate is formed in the reaction of both isomers [261], a result confirmed by other researchers [269]. On the other hand, the deamination of *exo*- and *endo*-2-norbornylamines in acetic acid [270, 271] affords also 2—4% *endo*-acetate. The reaction of **99** is therefore *exo*-specific under the solvolysis conditions.

(d) Optically active *exo*-**99** is completely racemised in the solvolysis [26], which indicates a symmetric intermediate (the original substrate has two chirality centres, at C-1 and C-2). On the other hand, the *endo*-epimer suffers partial racemisation accompanied by partial inversion, the latter depending on the nucleophilicity of the solvent [261], as it amounts to 13% in aqueous acetone, 7% in acetic acid and only 3% in formic acid.

All these data are consistent with the mechanism represented in Scheme 9. The intermediate **100** has a C_s symmetry and therefore C-1 and C-2 equivalent. Roberts and Lee [226] demonstrated that beside the C_s-intermediate, an intermediate of higher symmetry (C_{3v}) must be involved, in which positions 1, 2, 6, and 4, 5, 7, respectively, are equivalent.

The same symmetry is indicated by n.m.r. spectra as well as by X-ray data [266–268]. Moreover, the quenching with pyridine of the solution containing the norbornyl ion in magic acid yields 98% nortricyclene **101** demonstrating that in strong acid media the intermediate has probably the C_{3v} symmetry of **102**:

Precursor —Magic acid→ **102** (C_{3v}) → **101** (C_{3v})

[269] H. Goering and C. B. Schewene, *J. Amer. Chem. Soc.*, **87**, 3516 (1965).
[270] E. J. Corey, J. Casanova, Jr., P. A. Vatakencherry and R. Winter, *J. Amer. Chem. Soc.*, **85**, 169 (1963).
[271] J. A. Berson and A. Remanick, *J. Amer Chem. Soc.*, **86**, 1749 (1962).

On the other hand, nortricyclene is also obtained under solvolysis conditions [261] in about 4% yield, and it may generate the norbornyl ion [266–268]. At the same time, Roberts and Lee [226] observed that the C_{3v} symmetry is not perfect, suggesting an equilibration between **100** and **102**.

Species **100** and **102** may be regarded as protonated cyclopropanes (see Section 10.3). The C_s symmetry corresponds to a corner-protonated species, whereas the C_{3v} symmetry, as represented in **102**, indicates a face-protonated cyclopropane. Theoretical calculations [272, 273] have shown that the corner-protonated species is by far the more stable species. The equivalence of C-1, C-2 and C-6 may thus be explained by rapid 6,2 (6,1) hydride shifts [263], indicated also by the earlier spectral analysis of the ion [264, 265]. Another alternative is the rapid isomerisation of the protonated cyclopropane.

The concept of σ-bridging in the formation of the non-classical norbornyl cation was criticised by Brown [68, 274]. Brown rejected systematically all the arguments advanced in favour of the non-classical ion **100**. Thus,

(a) the high k_{exo}/k_{endo} ratio would not be due to the anchimeric assistance of the C-6—C-1 bond in the *exo*-derivative, since the *exo*-derivative has a rate comparable with that of cyclopentyl derivatives [275]. Anyway, if the norbornyl system is to be compared with a monocyclic one, this should be the cyclopentyl, rather than the cycolhexyl system. The argument is logical, although less attention has been paid to it. Brown concluded that the high k_{exo}/k_{endo} ratio is determined by a particularly low reactivity of the *endo*-isomer. When leaving the molecule, the group from the 2-*endo* position collides with the *endo*-H from the 6-position:

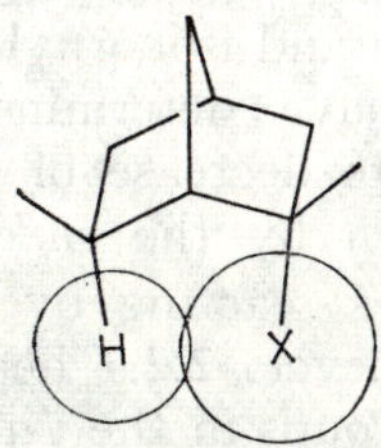

103

In order to check this explanation, Schleyer and his co-workers [276] studied the solvolysis of 6,6-dimethyl-2-norbornyl derivatives, sup-

[272] For a review see: M. Saunders, P. Vogel, E. L. Hagen and J. Rosenfeld, *Accounts Chem. Res.*, **6**, 53 (1973).

[273] G. Klopman, *J. Amer. Chem. Soc.*, **91**, 89 (1969).

[274] H. C. Brown, *Accounts Chem. Res.*, **6**, 377 (1973).

[275] H. C. Brown and M. H. Rei, *J. Amer. Chem. Soc.*, **86**, 5004 (1964).

[276] P. von R. Schleyer, M. M. Donaldson and W. E. Watts, *J. Amer. Chem. Soc.*, **87**, 375 (1965).

posing that bulky substituents in the *endo*-C-position will increase the k_{exo}/k_{endo} ratio. However, no conclusive results were obtained. The ratio k_{exo}/k_{endo} varied only little (220 as compared to 280), whereas a large rate decrease, in both *endo* and *exo* derivatives, was produced by the substitution with methyl groups in the 6-position.

(b) The stereochemical outcome is not specific. Exo attack is characteristic of most of the norbornyl reactions [277], either homolytic or heterolytic. Again, the steric effect of the 6-*endo* substituent is invoked. It has been however shown that a substituent in the 7-position of the bicyclo [2.2.1] heptyl system prevents the access and the leaving of the counter-ion, in the same way as the *endo*-H from the 6-position. In that respect, Winstein and his co-workers [278] studied the solvolysis of apoisocamphyl *(exo)* and apocamphyl *(endo)* derivatives, showing that the ratio k_{exo}/k_{endo} is still higher than for norbornyl derivatives, even if for the latter the value is corrected for the return ($k_{apoiso}/k_{apo} = 4100$, as compared with 350 (corrected 1600) for the norbornyl system).

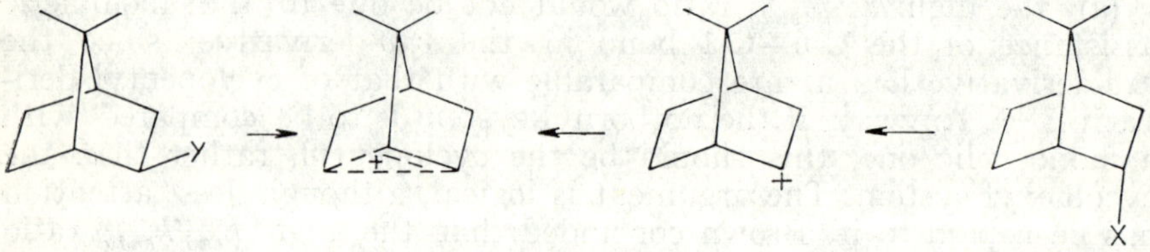

apoisocamphyl

apocamphyl

If the steric effect were the cause of the higher reactivity of the *exo*-isomer, smaller k_{exo}/k_{endo} ratios would be expected. Let us also mention that for bornyl and isobornyl chlorides, Hughes, Ingold and their co-workers [262] have determined a ratio of about 10^5, ruling out completely a rate decrease of the endo compound as a result of a steric retardation by the *endo*-6-H.

However, for substrates reacting by an unassisted path, such as tertiary 2-substituted bicyclo [2.2.1] heptyl derivatives, the introduction of two methyl groups in the 7-position lowers appreciably the k_{exo}/k_{endo} ratio [279,280], dramatic changes being observed if the substituent in the 2-position is a methoxy group [281]. In the latter case, the exceptional ability of this group to stabilise a positive charge rules out completely any assistance by the C-6—C-1 bond. Similar

[277] H. M. Bell and H. C. Brown, *J. Amer. Chem. Soc.*, **86**, 5006, 5007 (1964).

[278] A. Cotter, E. C. Friederich, N. J. Holness and S. Winstein, *J. Amer. Chem. Soc.*, **87**, 378 (1965).

[279] H. C. Brown and S. Ikegami, *J. Amer. Chem. Soc.*, **90**, 7122 (1968).

[280] S. Ikegami, D. L. Vander-Jagt and H. C. Brown, *J. Amer. Chem. Soc.*, **90**, 7124 (1968).

[281] T. G. Traylor and C. L. Perrin, *J. Amer. Chem. Soc.*, **88**, 4934 (1966).

stabilisations of the positive charge by a methyl group have been determined in solvolysis [198].

The fact that the *exo*-attack is not a result of σ-bridging in the norbornyl cation can be demonstrated by using Goering-Schewene diagrams [269]. Thus, the difference between the activation free-energies in the ionisation of the *exo*- and the *endo*-norbornyl derivatives is 4.5 kcal mol $^{-1}$. If allowance is made for the higher ground state of the *endo*-isomer, the energy difference between the transition states of the formation of the two acetates is found to be 5.5 kcal mol $^{-1}$. Evidently, the *exo*-attack will be preferred (Figure 5.16).

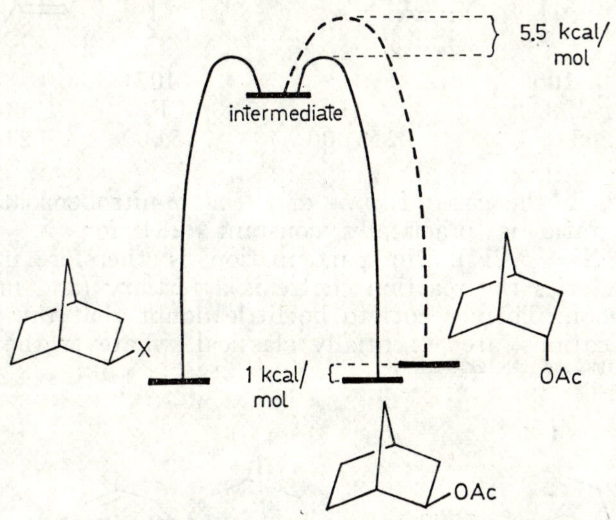

Figure 5.16 Goering-Schewene diagram for the acid-catalysed transformation of *exo*- and *endo*-2-norbornyl derivatives at 75°.

Brown [68, 274] has shown that by similar considerations almost all known cases of highly stereospecific attack in the *exo*-position may be explained.

(c) The effect of increasing electron demands at the reaction centre was systematically studied in three pairs of systems [282]. The first is that of 7-norbornyl and 7-norbornenyl *p*-nitrobenzoates (PNB):

OPNB X **104**   OPNB X **105**

X =	*p*-MeO	H	*p*-CF_3	*m,m'*-$(CF_3)_2$
k_{105}/k_{104}	3.4	41.5	34 500	255 000

[282] E. N. Peters and H. C. Brown, *J. Amer. Chem. Soc.*, **95**, 2397 (1973).

which shows large increases of the ratio k_{105}/k_{104} with increasing electron demands at the reaction centre. In the *p*-methoxy compound, there is no demand from the neighbouring double bond, the ionisation being assisted by benzyl delocalisation. However, the data show that for the electron-poorest group, the double bond is deeply involved in the ionisation (π-route). Similar and even more suggestive results were obtained in the series **106**—**107**:

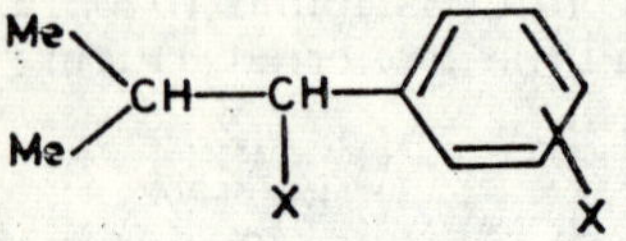

106 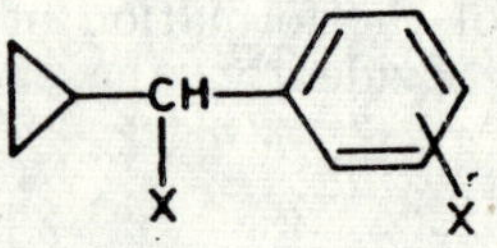 **107**

X:	*p*-MeO	H	*p*-CF_3	*m*,*m*'-$(CF_3)_2$
k_{107}/k_{106}	505	253 000	285 000	1 210 000

However, in the case of *exo*- and *endo*-*p*-nitrobenzoates **108**, the k_{exo}/k_{endo} ratio is practically constant (284 for X = *p*-MeO and 188 for X = *p*-CF_3). No participation is therefore involved, since in this series the reaction centre is a tertiary (and moreover benzylic) carbon. There seems to be little doubt that the tertiary 2-norbornyl cations are essentially classical, whatever the substituent at C-2 [198, 278, 280, 283—285]

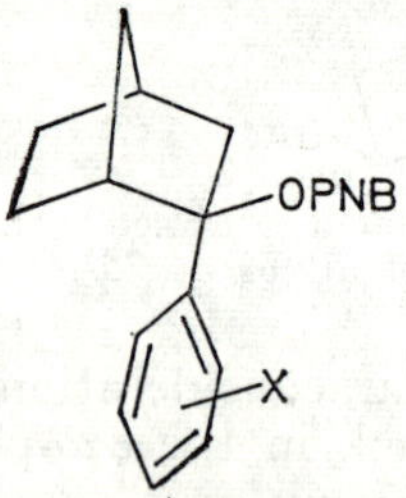

exo-**108** 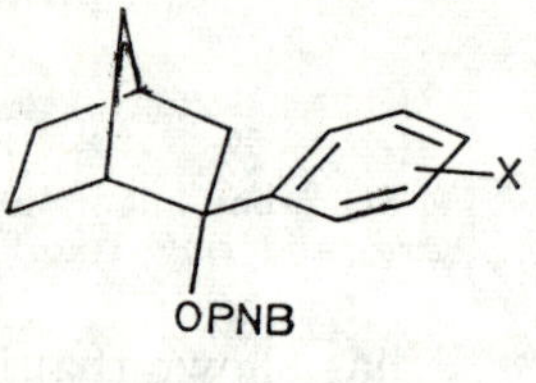 *endo*-**108**

Other arguments, even if not definitely in favour of a delocalised non-classical norbornyl cation, indicate a very stable species occurring as intermediate in solvolysis. Thus, the ratio between the polarimetric and titrimetric rate constants (k_p/k_t) is highly dependent on the nature of the leaving group, being 4.5 for **99**-*exo* with X = *p*-bromobenzenesulphonate [260, 261] (brosylate), 8.5 for X = Br [286]

[283] H. L. Goering and K. Humski, *J. Amer. Chem. Soc.*, **90**, 6213 (1968); **91**, 4594 (1969).

[284] H. L. Goering and J. V. Clevenger, *J. Amer. Chem. Soc.*, **94**, 1010 (1972).

[285] P. von R. Schleyer, D. C. Kleinfelter and H. G. Richey, Jr., *J. Amer. Chem. Soc.*, **85**, 479 (1963).

[286] J. P. Schaeffer, M. J. Dagani and D. S. Weinberg, *J. Amer. Chem. Soc.*, **89**, 6938 (1967).

and 11.4 for X = Cl[287, 288]. The increasing order of the ratio coincides with the increasing order of the nucleophilicity of the ions in an ion pair, which is equivalent to their activity in an aprotic solvent [289]. Thus, the counter-cation must be selective and therefore very stable, which is unusual for a strained secondary carbocation. Moreover, in the acetolysis of 2-*exo*-norbornyl chloride, the ratio k_p/k_t decreases from 11.4 to 9.6 if mercury chloride is used as an electrophilic catalyst and to 1.2 if the catalyst is mercury acetate:

Cl + HgX_2 ⟶ $X_2\bar{Hg}Cl$ (X = Cl, *Cl, OAc)

109

The large difference between the effect of the two salts may be explained by the return within the ion pair. The return in the intermediate **109** with X = Cl will lead to racemised starting material, whereas the return from **109**, X = OAc, will afford the solvolysis product. The large difference between the ratios in the uncatalysed and the catalysed reactions indicates either that the ion pair is loose (allowing rotation of the two counter-ions) or that return occurs from a solvent-separated ion pair, in which the mobility of the ions is greater. In order to decide between these two possibilities, the isotope exchange and the racemisation rate in the reaction of 2-*exo*-norbornyl chloride, in acetic acid with radioactive mercury chloride, were investigated. It was thus found that the rate of racemisation (k_r) is twice as high as the rate of chlorine exchange (k_e); in a loose ion pair (or solvent-separated ion pair), the chlorine atoms in the complex ion are equivalent and the expected k_r/k_e ratio would be 1.5, as it was actually found [113] in cases where internal return is absent or negligible with respect to the external return. The results mentioned here demonstrate internal and external return from **109**, showing at the same time that the small salt effect of acetate salts is not due to the absence of equilibria of type (5.35), but to the return from ion pairs to a product which cannot be distinguished from the solvolysis product [191, 192].

The norbornyl cation is illustrated in Scheme 7, where the two products (**91** and **92**) of the nucleophilic attack on the degenerate ion **93** are enantiomers and thus structurally identical. Other typical

[287] J. P. Hardy, A. Ceccon, A. F. Diaz and S. Winstein, *J. Amer. Chem. Soc.*, **94**, 1356 (1972).

[288] J. P. Hardy, A. F. Diaz and S. Winstein, *J. Amer. Chem. Soc.*, **94**, 2363 (1972).

[289] M. Cocivera and S. Winstein, *J. Amer. Chem. Soc.*, **85**, 1702 (1963).

examples of π- and σ-routes, with (probably) a common intermediate, involve cyclopropylcarbinyl [290] and cyclobutyl [291] cations.

It has been shown that among cyclanyl derivatives **70** [220], the cyclobutyl compound shows an unexpected behaviour in the solvolysis. Even if it has a similar type of bonding as cyclopropane, it reacts about as fast as the cyclopentyl derivative, which in its turn has a higher reactivity than cyclohexyl derivatives. Moreover, if bimolecular reactions are compared, cyclobutyl derivatives show about the same small reactivity as the cyclopropyl compounds [222]. On the other hand, it has been remarked that the hydrolysis in aqueous ethanol of cyclopropylcarbinyl derivatives has a rate constant only 15 times smaller than the rate constant of the hydrolysis of the t-butyl derivative [292]. If a similar mechanism were involved in the solvolysis of the two derivatives, the primary compound would be hydrolysed 10^{10}-fold less slowly than the tertiary derivatives (calculated on the basis of a Taft equation). The equilibration of the cyclopropylcarbinyl, cyclobutyl and allylcarbinyl systems has been clearly demonstrated by Roberts and his co-workers [293–295].

—CH_2X **110-X**; X **111-X** $\xrightarrow{ROH}$ —CH_2OR + OR + CH_2OR

110-OR **111-OR** **112-OR**

(+ **110-X** + **111-X** + **112-X**)

X = halogen, $ArSO_2O$, N_2^+ (deamination of the corresponding amines).

The same products are obtained either from **110** or from **111**. Moreover, the return of the intermediate ion pairs shows that the three systems are related to a common ion. The equilibration of all methylene groups had previously suggested that the probable intermediate is the tricyclobutonium ion, **113** [293]. Later works [294, 295] indicated however that the equilibration of the methylene groups is

[290] H. G. Richey, in *Carbonium Icns*, vol. 3 (G. A. Olah and P. von R. Schleyer, editors), Wiley-Interscience, New York (1972), p. 1201.

[291] K. B. Wiberg, B. A. Hess, Jr., A. J. Ashe, III, in *Carbonium Icns*, vol.3 (G. A. Olah and P. von R. Schleyer, editors), Wiley-Interscience, New York (1972), p. 1295.

[292] H. C. Brown and M. Borkowski, *J. Amer. Chem. Soc.*, **74**, 1894 (1952).

[293] J. D. Roberts and R. H. Mazur, *J. Amer. Chem. Soc.*, **73**, 3542 (1951).

[294] R. H. Mazur, W. N. White, D. A. Semenov, C. C. Lee, M. S. Silver and J. D. Roberts, *J. Amer. Chem. Soc.*, **81**, 4390 (1959).

[295] M. C. Caserio, W. H. Graham and J. D. Roberts, *Tetrahedrcn*, **11**, 171, (1960).

not perfect and, instead of a highly symmetric intermediate **113**, a less symmetric bicyclobutonium ion **114** is more probable.

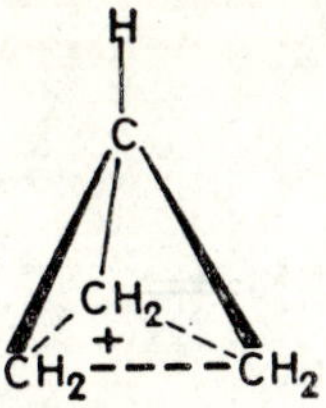

113 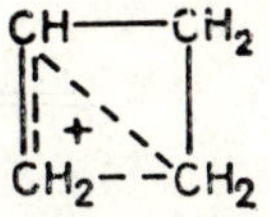 **114**

The ion **114** corresponds to the formation of a delocalised species by a π-route from allylcarbinyl derivatives **112**. Thus, bicyclobutonium ions **114** are related to the covalent species **110—112** in the same way as the norbornyl cation is related to **95** and **99**. The difference between the two systems is that the positions attacked by the nucleophile are equivalent in the norbornyl cation, but differ in the bicyclobutonium ion. Both formulations of the intermediate involved in the solvolysis of cyclopropylcarbinyl and cyclobutyl derivatives, proposed by Roberts and his co-workers, have been subjected to extensive experimental checking [290, 291, 296]. Winstein and Kosower [297] remarked that the difference between a simple homoallyl ion **115** and a bicyclobutonium ion **114** is that beside the 1,3 orbital interactions in the former system, 1,4 orbital interactions are also found in the latter. The 1,4 interactions may be excluded, for geometrical reasons, in cases like the interconversion of cholesteryl (**116**) and i-cholesteryl (**117**) derivatives [298, 299]. However, they are favoured energetically and slightly disfavoured geometrically in the 2-norborn-5-enyl derivatives **118** [300] and favoured in rigid systems [301].

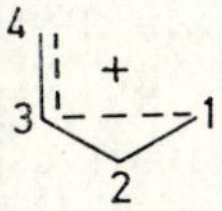

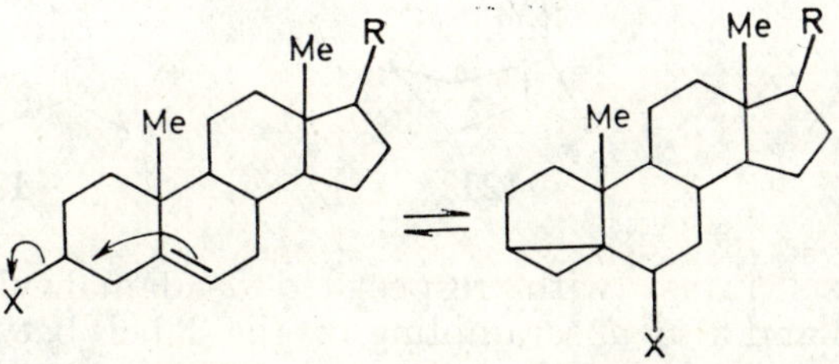

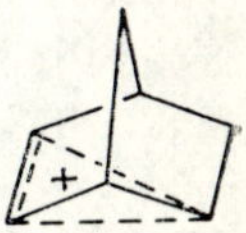

115 **116** **117** **118**

[296] P. von R. Schleyer and G. van Dine, *J. Amer. Chem. Soc.*, **88**, 2321 (1966).
[297] S. Winstein and E. M. Kosower, *J. Amer. Chem. Soc.*, **81**, 4399 (1959).
[298] C. W. Shoppee, *J. Chem. Soc.*, 1147 (1946).
[299] S. Winstein and R. Adams, *J. Amer. Chem. Soc.*, **70**, 838 (1948).
[300] S. Winstein, H. M. Walborsky and K. Schreiber, *J. Amer. Chem. Soc.*, **72**, 5795 (1950); J. D. Roberts, W. Bennett and R. Armstrong, *ibid.*, **72**, 3329 (1950); S. Winstein and M. Shatavski, *ibid.*, **78**, 592 (1956).
[301] M. Simonetta and S. Winstein, *J. Amer. Chem. Soc.*, **76**, 18 (1954).

The equilibration of the methylene groups may be thought to occur *via* the tricyclobutonium ion (5.89) or by a still faster equilibration of homoallyl cations (5.90) :

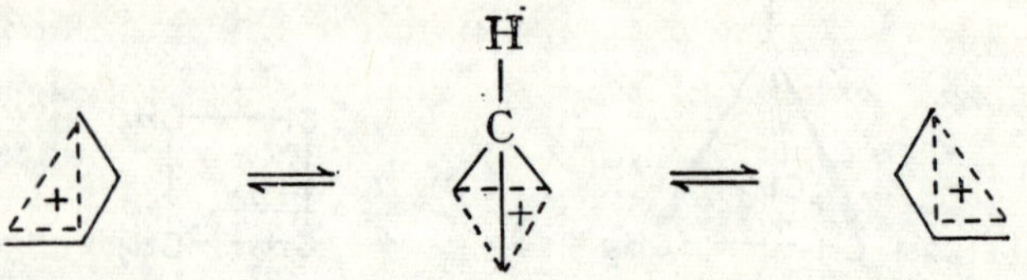

(5.89)

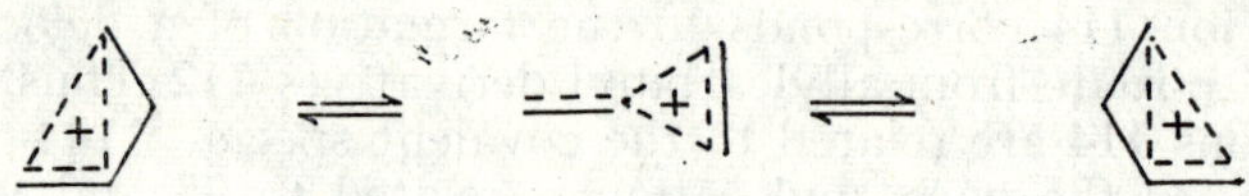

(5.90)

119

In the latter case, the transition state (or intermediate) is a symmetric homoallyl cation **119**. Such intermediates have been also assumed in the generation of stable cyclopropylcarbinyl ions, as for example **120**. The pK_{R^+} value (—2.34 [302]) of the latter indicates a higher stability than that of the triphenylmethyl carbocation (see Table 4.8). In order to prove this structure, Baldwin and Foglesong [303] studied the solvolysis of 8,9-dehydroadamantyl 3,5-dinitrobenzoate **121**:

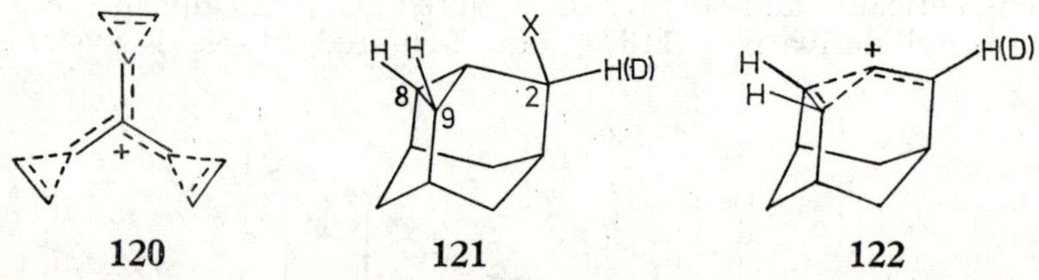

120 **121** **122**

High solvolysis rates (with respect to 2-adamantyl derivatives) were recorded and also a scrambling of the label between C-2, C-8 and C-9. It was concluded that the symmetric homoallyl ion **122** is formed. However, the direct observation of the 8,9-dehydroadamantyl cation by n.m.r. spectral techniques [304] rather indicated classical cyclopropylcarbinyl ions in rapid equilibrium.

[302] N. C. Deno, H. G. Richey, Jr., J. S. Liu, J. D. Hodge, J. J. Houser and M. J. Wistosky, *J. Amer. Chem. Soc.*, **84**, 2016 (1962); H. Hart and P. A. Law, *ibid.*, **84**, 2462 (1964).

[303] J. E. Baldwin and W. D. Foglesong, *J. Amer. Chem. Soc.*, **89**, 6372 (1967).

[304] G. A. Olah, G. Liang, K. A. Babiak and R. K. Murray, Jr., *J. Amer. Chem. Soc.*, **96**, 6794 (1974).

Two different structures can be written for the cyclopropylcarbinyl ion. In one of them, called the *bisected ion*, the plane of the sp^2 carbon atom is perpendicular to and bisects the plane of the cyclopropane ring (**123**). In the other possible structure, the two planes neither bisect nor are perpendicular (**124**). Theoretical calculations agree well with the experimental data as regards the higher stability of the bisected ion [305–308].

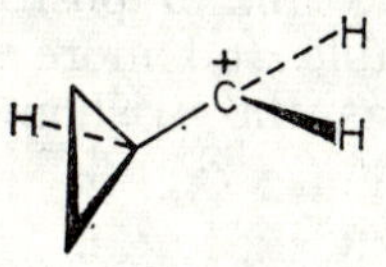

123 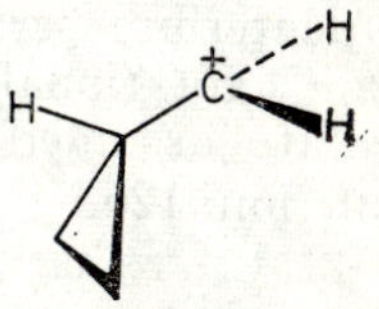 **124**

As in the case of the norbornyl ion, there is still no agreement regarding the delocalised nature of the cyclopropylcarbinyl cation. However, it is generally agreed that cyclopropyl is one of the best groups which stabilise a neighbouring positive charge. Brown and Peters [282], studying the solvolysis of derivatives **106**–**107**, have proved that cyclopropyl is even more efficient than a neighbouring *p*-anisyl group in delocalising a positive charge. Their results were confirmed by ^{13}C n.m.r. spectroscopy [309], as well as by the orientation in the addition to phenylcyclopropyl-alkenes [309]:

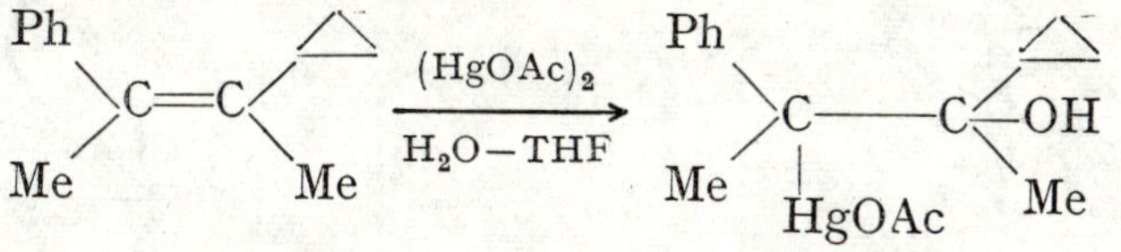

$$\mathrm{Ph(Me)C{=}C(Me)(C_3H_5)} \xrightarrow[\mathrm{H_2O-THF}]{\mathrm{(HgOAc)_2}} \mathrm{Ph(Me)C(HgOAc)-C(OH)(Me)(C_3H_5)} \qquad (5.91)$$

Summarising the facts presented in this section, it must be stressed that remote double bond participation and neighbouring σ-bond involvement in the ionisation step are two connected phenomena (Scheme 8). The anchimeric assistance of a double bond is a well-documented and accepted fact, whereas the σ-bridging is still a matter of controversy. The exceptionally high rates observed sometimes in solvolyses are interpreted by alternative theories, invoking either relief of steric strain [68, 274] in the transition state or synchronous rearrangement [310–312].

[305] C. U. Pittman, Jr., and G. A. Olah, *J. Amer. Chem. Soc.*, **87**, 5123 (1965).
[306] R. Hoffmann, *J. Chem. Phys.*, **40**, 2480 (1964); a difference between the two species of about 9 kcal mol^{-1} has been approximated.
[307] P. von R. Schleyer and V. Buss, *J. Amer. Chem. Soc.*, **91**, 5880 (1969).
[308] J. E. Baldwin and W. D. Foglesong, *J. Amer. Chem. Soc.*, **90**, 4303 (1968).
[309] G. A. Olah, P. W. Westerman and J. Nishimura, *J. Amer. Chem. Soc.*, **96**, 3548 (1974).
[310] K. H. Wiberg and his co-workers, *J. Amer. Chem. Soc.*, **92**, 544, 553, 568, 571 (1970).
[311] Z. Malerski and P. von R. Schleyer, *J. Amer. Chem. Soc.*, **93**, 665 (1971).
[312] K. H. Wiberg and W.-f. Chen, *J. Amer. Chem. Soc.*, **96**, 3900 (1974).

5.7.6 *Homoconjugation, homoaromaticity and related phenomena* [313]

The participation of a remote cyclopropane ring to an ionisation site has been first observed by Winstein and Sonnenberg [314] in the solvolysis of tosylate *cis*-**125**. In contrast with the *trans*-isomer, the *cis* presents higher rates, special salt effects (both demonstrating a more stable intermediate than that arising from the *trans*-compound) and perfect deuterium scrambling between 1, 3 and 5 positions (indicating a C_3 rotational axis). The more stable and more symmetric intermediate assumed in this reaction was the trishomocyclopropenylium ion **126**.

H(D) 5 3 OTs 1 → + → (D)H H(D) OAc H(D)

cis-**125** **126** (C_{3v})

The trishomocyclopropenylium ion, due to the partial *p*-overlap, similar to that in cyclopropenylium ion (Figure 5.17), can be relatively

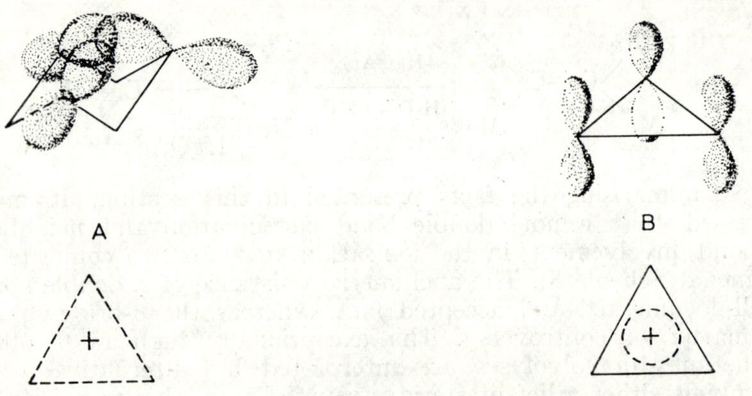

Figure 5.17 Cyclopropenylium and trishomocyclopropenylium ions.

stable in, agreement with HMO calculations [315]. The assisted rate constant, k_Δ, for *cis*-**125**, exceeds the unassisted one only by a factor of nearly 50. Higher acceleration rates in solvolyses have been

[313] S. Winstein, *Quart. Rev. (Chem. Soc.)*, **23**, 141 (1969); S. Winstein, in *Carbcnium Ions*, vol. 3 (G. A. Olah and P. von. R. Schleyer, editors), Wiley-Interscience, New York (1972).

[314] S. Winstein and J. Sonnenberg, *J. Amer. Chem. Soc.*, **83**, 3235, 3244 (1961).

[315] R. J. Piccolini and S. Winstein, *Tetrahedrcn*, **19**, *Suppl.* 2, 423 (1963).

determined for the *endo-anti*-8-tricyclo[3.2.1.0^{2,4}]octyl derivatives[316, 317], **127**, which present enormous rates as compared with the 7-norbornyl analogues or with the *endo-syn* isomer:

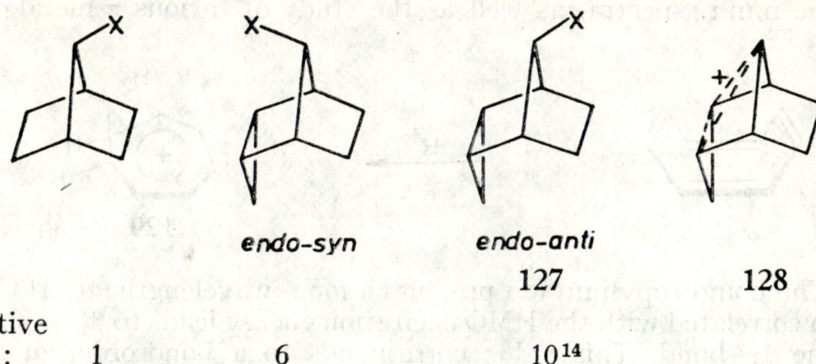

relative rate: 1 6 10^{14}

A trishomocyclopropenylium ion **128** must be involved. The higher accelerations observed with **127** as compared to **125** should be attributed to the more rigid system of the tricyclo-derivative **127** and to the close proximity of cyclopropane ring to the reaction centre.

The remote cyclopropane participation is similar to the remote double bond involvement in ionisation, and is explained by the partial π-character of the Walsh orbitals in cyclopropane. Due to the high ability of the cyclopropane to stabilise the positive charge, if suitably situated, a remote cyclopropane is more efficient than a remote double bond in assisting the ionisation (compare the 7-norbornenyl system[56–58] **10** with **127**). However, remote cyclopropane participation, due to a lower nucleophilicity than that of the double bond, depends to a greater extent on the location and orientation of the three-membered ring with respect to the reaction centre. In **38**-OTs, for instance, despite a favourable geometry, no participation, of the cyclopropane ring has been observed[170], although the unsaturated correspondent **57**-OTs[193–195] shows a moderate participation of the remote double bond. Comparing **10** with **127**, as well as considering the lower efficiency of an aromatic ring than that of a cyclopropyl ring to accommodate a positive charge[282, 309], would have suggested that the reaction of **38**-OTs might benefit from a large anchimeric assistance.

The possibility revealed by Winstein that a cyclopropane may stabilise a positive charge by delocalisation, in a similar way to the double bond, has promoted further research on the generation

[316] H. Tanida, T. Tsuji and T. Irie, *J. Amer. Chem. Soc.*, **89**, 1953 (1967).
[317] M. A. Battiste, C. L. Deyrup, R. E. Pincock and J. Haywood-Farmer, *J. Amer. Chem. Soc.*, **89**, 1954 (1967).

of homoaromatic ions. Thus, homotropylium (homotropenylium ion **129** was generated from cyclooctatetraene by protonation [318, 319] its homoaroamtic character being demonstrated by Winstein and his co-workers [319]. Among the more convincing pieces of evidence are the n.m.r. spectra, as well as the study of various π metal complexes [313].

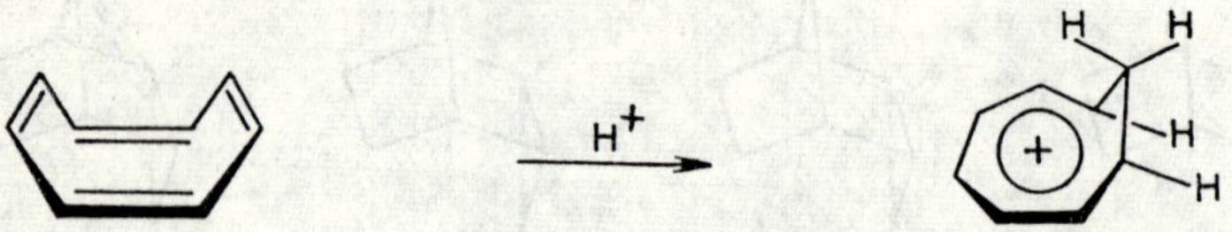

H^+

129

The homotropylium ion presents a long wavelength at 313 nm, which correlated with the HMO excitation energy leads to $\beta_{17} = 0.73\beta_0$ for the 1,7-bond. This value corresponds to a bond order of 0.56, which compares well with the value of 0.64 for the π-bonds in the tropylium ion [313].

Dibenzohomotropylium ions are also known. Winstein and Childs [320] observed that the *cis*-acetate **130** undergoes hydrolysis 2.4×10^2 times faster than the *trans*-isomer, which indicates the formation of the dibenzohomotropylium ion **131**:

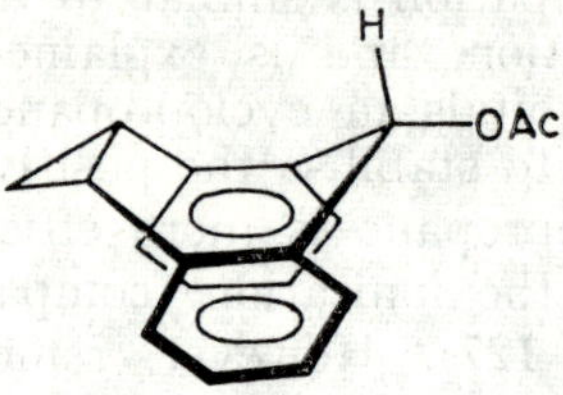

H OAc

130 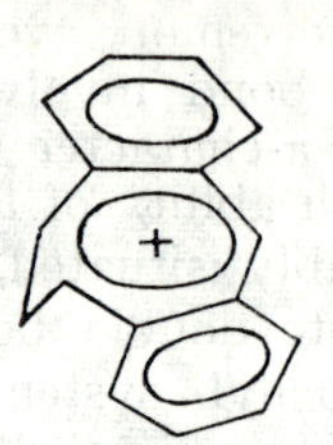 **131**

An isomer of the ion **131** was observed in the acetolysis of **41** and **57** by Nenitzescu, Ciorănescu and their co-workers [171, 194, 195, 321]. The equilibration (5.92) between the dibenzocycloheptatrienylmethyl and dibenzocyclooctatrienyl systems:

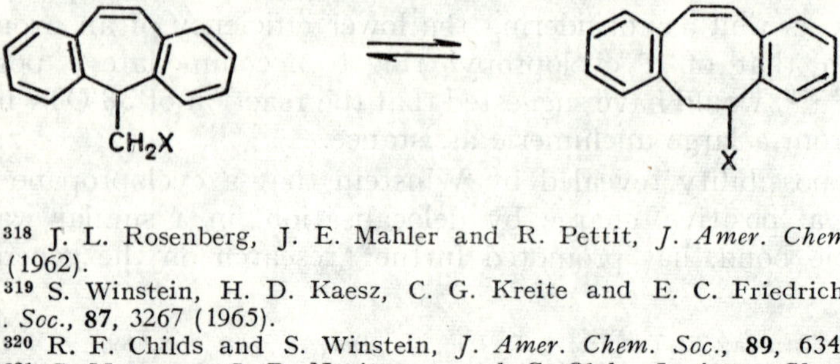

CH_2X ⇌ X (5.92)

[318] J. L. Rosenberg, J. E. Mahler and R. Pettit, *J. Amer. Chem. Soc.*, **84**, 2842 (1962).

[319] S. Winstein, H. D. Kaesz, C. G. Kreite and E. C. Friedrich, *J. Amer. Chem. Soc.*, **87**, 3267 (1965).

[320] R. F. Childs and S. Winstein, *J. Amer. Chem. Soc.*, **89**, 6348 (1967).

[321] G. Mateescu, C. D. Nenitzescu and G. Olah, *J. Amer. Chem. Soc.*, **90**, 6235 (1968).

can be regarded as a transformation of a 2,2-diarylethyl derivative into a 1,2-diaryl compound (5.93). Whilst the former rearrangement is reversible, the latter one is always completely shifted towards the 1,2-diaryl compound [322].

$$\begin{matrix} Ar^1 \\ Ar^2 \end{matrix} \!\!> CH-CH_2-X \longrightarrow \underset{\mathbf{132}}{\overset{Ar^1 \quad X^-}{\underset{Ar^2}{CH}\overset{+}{=\!=\!=}CH_2}} \longrightarrow \underset{\mathbf{133}}{\overset{X^-}{Ar^2\overset{+}{C}H-CH_2-Ar^1}} \longrightarrow$$

$$\longrightarrow Ar^2CHX-CH_2Ar^1 \qquad (5.93)$$

The reactions occur *via* a phenonium ion, which may rearrange into the more stable benzyl ion **133**. The irreversibility of reaction (5.93) may thus be accounted for by the unfavourable process **133**→ **132**. However, tosylate **57** is tranformed into the eight-membered ring acetate, which in acid medium is completely converted into the acetate corresponding to **57**. This unusual behaviour of a 1,2-diarylethyl derivative may be explained by the fact that beside intermediates of type **132** and **133**, a dibenzohomotropylium ion could also arise (Figure 5.18).

Models show that phenonium ion has a geometry similar to that of dibenzohomotropylium ion. Even if the transformation of dibenzohomotropylium ion into phenonium ion is energetically disfavoured, the similar geometries of the two ions makes possible the trapping

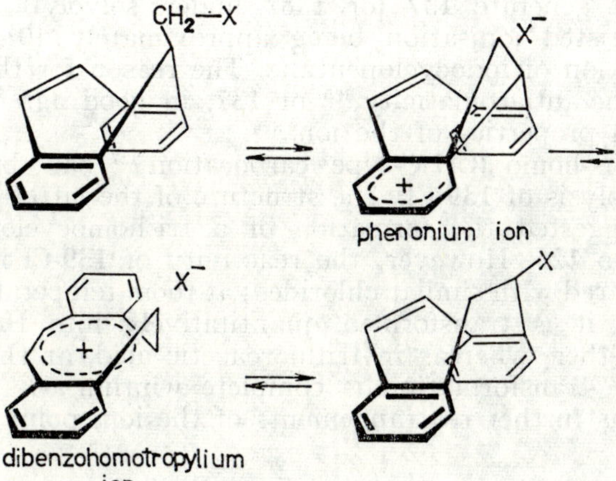

Figure 5.18 Reversible transformation of a 5H-dibenzo [a.d] cyclohepten-5-ylmethyl (dibenzocycloheptatrienylmethyl) derivative into a 5,6-dihydro-dibenzo[a,e]cyclo-octen-5-yl (dibenzocyclooctatrienyl) derivative.

[322] S. Winstein, B. K. Morse, E. Grunwald, K. C. Schreiber and J. Corse, *J. Amer. Chem. Soc.*, **74**, 113 (1952).

of the phenonium ion and transformation into the unreactive 2,2-diaryl compound. If the formation of a dibenzohomotropylium ion is not possible, as for example in the related compound **134**, the back rearrangement (5.93) is not observed.

The formation of dibenzohomotropylium ion was confirmed by Mateescu, Nenitzescu and Olah [321], who generated this ion in superacid media. Its n.m.r. spectra are in complete agreement with the assumed structure.

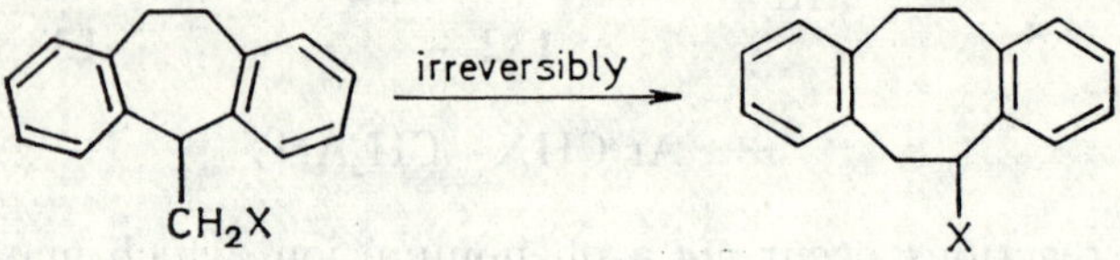

irreversibly

CH_2X X

134 **135**

Theoretical calculations for the $(CH)_5^+$ ions indicate that the square pyramid **136** must be the most stable species [323].

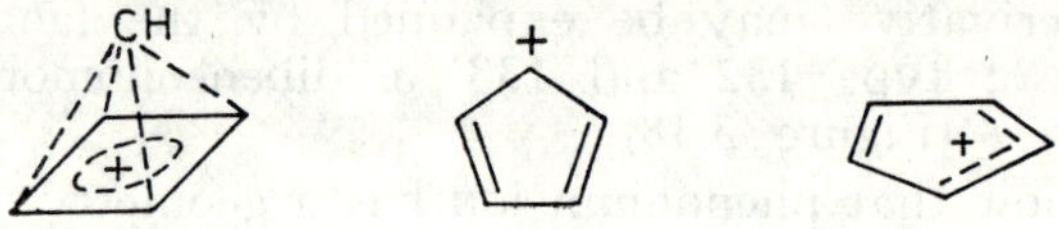

CH

136 **137** **138**

Other possible structures for this ion are **137** and **138**. Breslow and his co-workers [324] demonstrated that monoiodocyclopentadiene does not generate **137** (or **138**) under solvolytic conditions, the silver-assisted ionisation being approximately 10^5-fold as slow as the reaction of iodocyclopentane. The reason for this low reactivity lies in the antiaromaticity [325] of **137**, in good agremeent with other observed properties of the ion [326].

A bis-homo $(CH)_5^+$-type carbocation [327] has been obtained in the solvolysis of **139**-Cl. The structure of the latter compound would have suggested the formation of a trishomocyclopropenylium ion similar to **126**. However, the reactivity of **139**-Cl is unusually high as compared with similar chlorides; at room temperature, in methanol solution, it is transformed quantitatively into the corresponding methyl ether, whereas in trifluoroacetic acid, at the same temperature, the transformation is complete winthin **60** s. N.m.r. spectra, as well as further rearrangements of the ion, point to the symmetric

[323] W.-D. Stohrer and R. Hoffmann, *J. Amer. Chem. Soc.*, **94**, 1661 (1972).

[324] R. Breslow and J. M. Hoffman, Jr., *J. Amer. Chem. Soc.*, **94**, 2110 (1972).

[325] R. Breslow, *Accounts Chem. Res.*, **6**, 393 (1973).

[326] M. Saunders, R. Berger, A. Jaffe, J. M. McBride, J. O'Neill, R. Breslow, J. M. Hoffman, Jr., C. Perchnock, E. Wasserman, R. S. Hutton and V. J. Kuck. *J. Amer. Chem. Soc.*, **95**, 3017 (1973).

[327] H. Hart and M. Kuzuya, *J. Amer. Chem. Soc.*, **96**, 6436 (1974).

structure **140**, which may be designated as a bis-homo derivative of **136**.

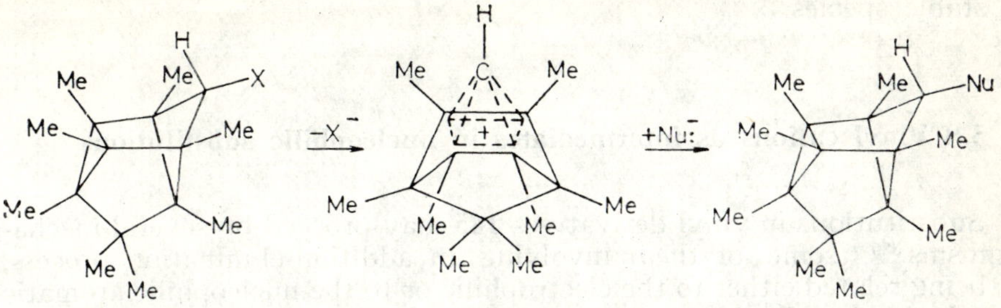

139-X **140** **139**-Nu

Related to ion **136** is the dication **142**, which results by further ionisation of the cation **141** [328]:

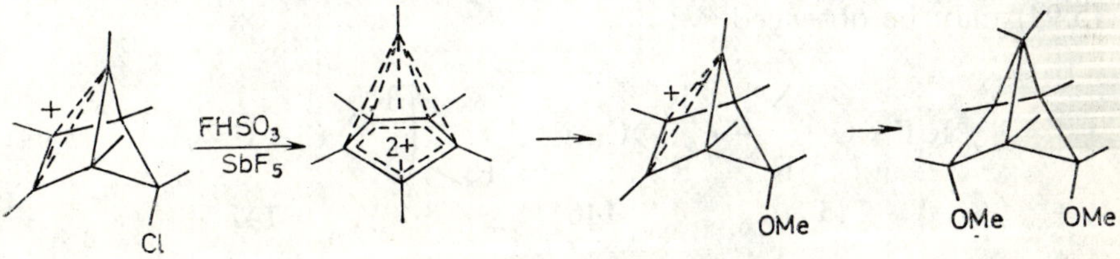

141-Cl **142** **141**-OMe **143**

The solutions of **142** quenched with MeONa—MeOH afford first **141**-OMe, then the stable product **143**. The stereochemistry of the latter compound demonstrates the structure of **142**, which is also confirmed by ^{1}H n.m.r. and ^{13}C n.m.r. spectral data.

It has been noted [327] that the interaction of an antiaromatic $(CH)_n$ polyenic system (neutral, anion or cation) with the hypothetical protonated carbon, :CH^+, would yield stable species. Thus, cation **136** could be regarded as resulting from the antiaromatic cyclobutadiene and :CH^+; tetrahedrane [329] **144** may be thought to result from the antiaromatic $C_3H_3^-$ and a protonated carbon.

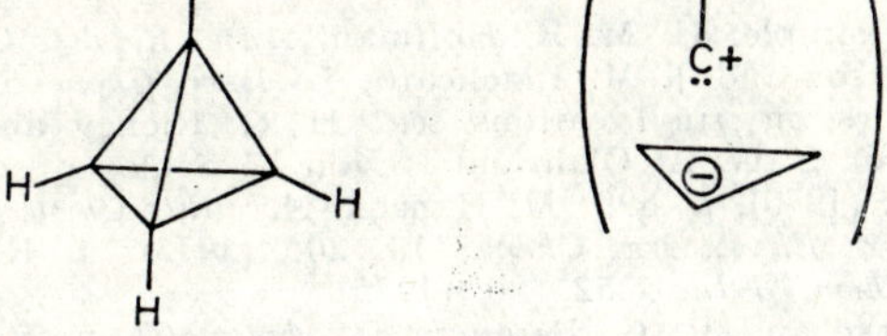

144

[328] H. Hogeveen and P. W. Kwant, *J. Amer. Chem. Soc.*, **97**, 2208 (1974).

[329] See, for example, N. C. Baird and M. J. S. Dewar, *J. Amer. Chem. Soc.*, **89**, 3966 (1967); L. B. Rhodewald and H.-ki Lee, *J. Amer. Chem. Soc.*, **95**, 623 (1973); T. J. Katz and N. Acton, *J. Amer. Chem. Soc.* **95**, 2738 (1973).

The dication **142** would result from the antiaromatic cyclopentadienyl cation and protonated carbon, and must be, as above, a stable species.

5.8 Vinyl cations as intermediates in nucleophilic substitutions

Substitutions in vinyl derivatives **145** may proceed by several mechanisms [330], some of them involving an addition-elimination process, being related either to the electrophilic or to the nucleophilic aromatic substitution. In this section we shall be concerned only with processes related to S_N2 or S_N1 reactions. The direct substitution, S_N2, is disfavoured in the vinyl derivatives as a C_{sp^2}—X bond is stronger than a C_{sp^3}—X bond (higher proportion of *s* orbital). However, vinyl cations **146** can be relatively stable and thus processes like (5.94) may be observed [331]:

$$\underset{\mathbf{145}}{R_2C{=}C{\lower{0.5ex}{\overset{X}{\underset{R}{\langle}}}}} \longrightarrow \underset{\mathbf{146}}{\overset{R}{\underset{R}{>}}C{=}C^{+}\diagdown R} \quad \underset{\mathbf{147}}{\left(\overset{R}{\underset{R}{>}}C=\overset{+}{C}-R\right)} \qquad (5.94)$$

The chemistry of the vinyl cations is very similar to that of other carbocations. Under solvolytic conditions, they can be better generated in presence of an electrophilic catalyst, such as silver or mercury ions. Their structure may be either **146** (sp^2 hybridisation) or **147** (*sp* hybridisation). From kinetic and product data it was concluded that ion pairing plays a significant role in the mechanism of vinyl displacement reactions [322,333]. *m-Y* correlations involve lower *m* values than in usual S_N1 reactions [334].

Participation of neighbouring groups in the ionisation of vinyl compounds is also possible. Thus, Stang and his co-workers [335] demonstrated that in the solvolysis of triflate **148** (Tf=CF_3SO_2) in aqueous ethanol (80%), a phenonium ion is formed. The evidence was obtained

[330] See, for example, H. M. R. Hoffman, *Ann. Reports Chem. Soc.*, **648**, 152 (1967); P. E. Peterson and J. M. Indelicato, *J. Amer. Chem. Soc.*, **90**, 6515 (1968).

[331] For reviews on vinyl cations see: H. G. Richey and J. M. Richey, in *Carbonium Ions*, vol. 2 (G. A. Olah and P. von R. Schleyer, editors), Wiley-Interscience, New York (1970), p. 899; M. Hanack, *Accounts Chem. Res.*, **3**, 209 (1970); P. J. Stang, *Progr. Phys. Org. Chem.*, **10**, 205 (1973); L. R. Subramanian and M. Hanack, *J. Chem. Educ.*, **52**, 80 (1975).

[332] T. C. Clarke and R. G. Bergman, *J. Amer. Chem. Soc.*, **94**, 3627 (1972); T. C. Clarke, D. R. Kelsey and R. G. Bergman, *ibid.*, **94**, 3626 (1972).

[333] R. H. Summerville, C. A. Senkler, P. von R. Schleyer, T. E. Dueber and P. J. Stang, *J. Amer. Chem. Soc.*, **96**, 1100 (1974).

[334] Z. Rappoport, J. Kaspi, *J. Amer. Chem. Soc.*, **96**, 4518 (1974), and references quoted therein.

[335] P. J. Stang and T. E. Dueber, *J. Amer. Chem. Soc.*, **95**, 2683 (1973).

by labelling the methyl groups with deuterium (in order to detect 1,2 shifts), and by carrying out kinetic measurements (to observe the anchimeric assistance).

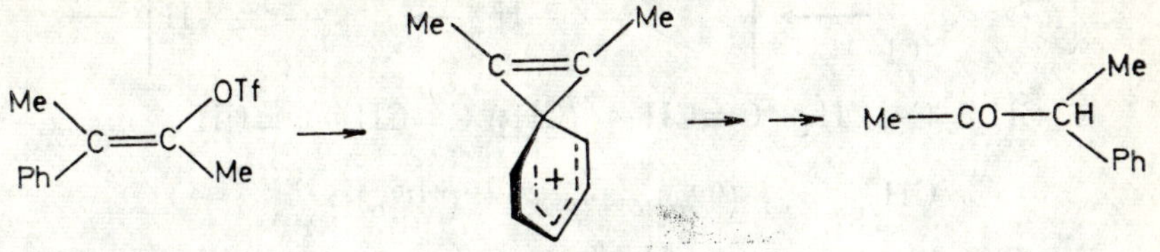

148 **149** **150**

The similarity between the unimolecular ionisations of saturated and of vinyl compounds is also demonstrated by the solvolysis of bromomethylenecyclopropanes **151** [336]. As shown in Scheme 10 the vinyl ion pair **152** may either undergo collapse to the cyclobutenyl derivative **153**—Br, or be trapped by added lithium chloride (a reaction corresponding to the special salt effect), or solvolyse to the cyclobutanone **154**.

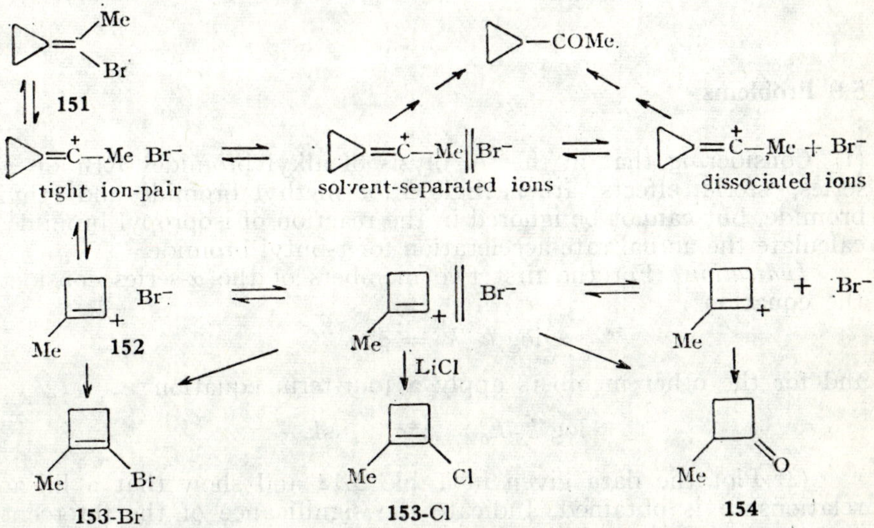

151

tight ion-pair | solvent-separated ions | dissociated ions

152

LiCl

153-Br **153-Cl** **154**

Products of ion-pair return

Scheme 10

The solvolysis of 1-chloromethylene-2,2-dimethylcyclopropane [337] occurs with overwhelming ring opening, demonstrating

[336] M. Hanack, T. Bössler, W. Egmann, W. Heyd and R. Kopp, *J. Amer. Chem. Soc.*, **96**, 6686 (1974).
[337] A. Ghenciulescu and M. Hanack, *Tetrahedron Letters*, 2827 (1970).

that the reverse π-route may also be involved in vinyl solvolyses.

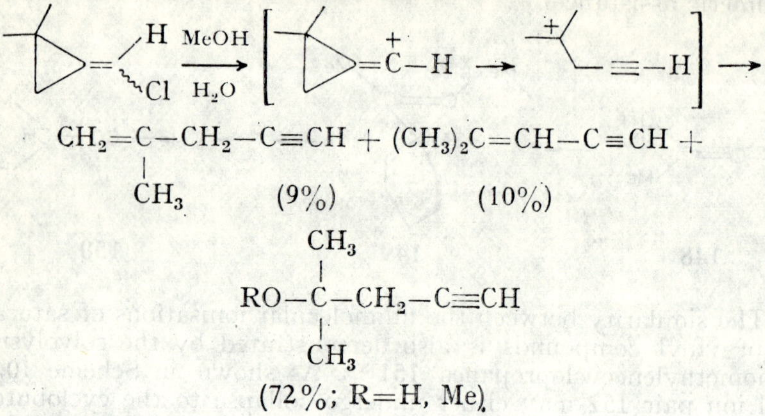

$$\text{(cyclopropylidene)}=C(H)(Cl) \xrightarrow[H_2O]{MeOH} \left[\text{(cyclopropylidene)}=\overset{+}{C}-H \rightarrow \overset{+}{\ }\text{(cation)}-C\equiv C-H \right] \rightarrow$$

$$CH_2=\underset{\displaystyle CH_3}{\underset{|}{C}}-CH_2-C\equiv CH + (CH_3)_2C=CH-C\equiv CH +$$

(9%) (10%)

$$RO-\underset{\displaystyle CH_3}{\underset{|}{\overset{\displaystyle CH_3}{\overset{|}{C}}}}-CH_2-C\equiv CH$$

(72%; R=H, Me)

The unimolecular solvolyses of vinyl compounds are not strikingly different from those of saturated compounds, except for their lower rates.

5.9 Problems

(1) Considering that in the solvolysis of alkyl bromides RBr of α-series, steric effects are negligible for methyl bromide and ethyl bromide, but cannot be ignored in the reaction of isopropyl bromide, calculate the actual rate acceleration for t-butyl bromide.

Indication: For the first two members of the α-series consider the equation:

$$\log k_{\_}/k_0 = \rho^*\sigma^*$$

and for the other members apply a four-term equation:

$$\log k_S/k_0 = \rho^*\sigma^* + sE_s$$

(2) Plot the data given in Table 5.12 and show that a linear relationship is obtained. Indicate the significance of the intercept and slope of the correlation line.

(3) Using the following acetolysis data (at 70 °C):

Substrate:	$k(s^{-1})$
$PhCH_2CH_2OTs$	1.73×10^{-7}
EtOTs	4.55×10^{-7}
Pr^iOBs	1.30×10^{-6}
$k_{R_2CHOBs}/k_{R_2CHOTs}$	~3.5
$\dot{\rho}_{R_2CHOBs}$	−3.5

show that the acetolysis of 2-phenylethyl tosylate must be anchimerically assisted. Comment on the misleading direct comparison between the solvolyses of ethyl tosylate and 2-phenylethyl tosylate.

(4) What explanation can be advanced for the different ρ^* values determined in the acetolysis of toluene-*p*-sulphonates (tosylates) and *p*-bromobenzenesulphonates (brosylates), respectively?

(5) How can you interpret the result of Brown and his co-workers[46], which obtained the same ρ parameter ($\rho \approx -6.0$), of the $\rho\sigma^+$ correlation, in the solvolysis of benzyl derivatives in three different solvents: formic acid, acetic acid and aqueous acetone?

(6) Show the reactivity order of lithium halides in the Finkelstein reaction in the following solvents: dimethylsulphoxide ($\varepsilon_{20^\circ} = 49.5$); diethyl ether ($\varepsilon_{20^\circ} = 4.34$); hydrogen fluoride ($\varepsilon_{20^\circ} = 84$).

(7) Explain the following kinetic isotope effects observed in nucleophilic substitutions:

substrate:	$(CD_3)_2CCl{-}CD_2CH_3$	$(CH_3)_2CCl{-}CD_2CH_3$	$C_6H_5CD_2Cl$
k_H/k_D	2.33	1.41	1.003

$$C_6H_{13}CD(OBs){-}CH_3$$

1.106

Calculate the decrease in activation free-energy for the substitution with one deuterium atom.

(8) Indicate the sign of the specific rotations of the products in the following reactions, explaining your choice:

(a) 3-chloro-5-methylcyclohexene (Cl, CH_3), (−)-*trans* $\xrightarrow{AcOH}$ 3-acetoxy-5-methylcyclohexene (OAc, CH_3), (?)-*trans*

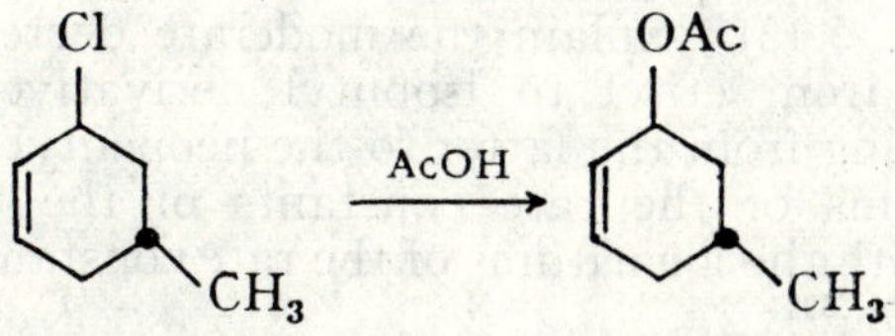

(b) $CH_3CHBrCH(OH)CH_3 \xrightarrow{HBr} CH_3CHBrCHBrCH_3$

(+)-*threo* → (?)-*threo*

$CH_3CHBrCH(OH)CH_3 \xrightarrow{HBr} CH_3CHBrCHBrCH_3$

(+)-*erythro* → (?)-*erythro*

(c) $C_6H_{13}CHClCH_3 \xrightarrow[\text{acetone}]{N_3^-,\ \text{anhydrous}} C_6H_{13}CH(N_3)CH_3 \longrightarrow$

$$\xrightarrow[\substack{\text{(ii) } CH_3I \\ NaOH}]{\text{(i) } H_2} C_6H_{13}-\underset{}{\overset{\overset{+}{N}(CH_3)_3}{\overset{|}{C}H}}-CH_3 \quad (?)$$

(9) Studying the 2-norbornyl cation, Schleyer, Olah and their co-workers [264, 265] have shown that the 2,3-hydride shift is much slower than the 6,2-hydride shift. How can you explain the fact that in this system a 1,3-hydride shift (less frequent) may be faster than the very common 1,2-shifts?

(10) Christ and Elias [338] have determined the following kinetics in the case of the isotope exchange of 1-chlorohexane in diethylene glycol:

$$\text{rate} = k_1[R{-}Cl] + k_2[R{-}Cl][^*Cl^-]$$

The analysis of the reaction product revealed also very small amounts of hexene. Considering that the substrate, as well as the reaction conditions, do not favour a unimolecular reaction, find an explanation for the unimolecular term in the above equation.

(11) When the observed rate constant of the acetolysis of **57**-OTs [195] was split into the rate constants for the phenonium route and for the π-route, the following values were found:

$$k_{Ph} = 2.95 \times 10^{-5}\ s^{-1}$$

$$k_{\pi} = 0.25 \times 10^{-5}\ s^{-1}$$

indicating a more favourable participation of the aromatic ring than of the double bond. Demonstrate that these values, if corrected, indicate however a higher nucleophilicity of the double bond than of the aromatic ring.

(12) Considering the explanation given for the steric effect in the β-series (Figure 5.13), explain the moderate decrease of rate constants in passing from ethyl to isobutyl derivatives, but the steep decrease in passing from the latter to the neopentyl derivative.

(13) The logarithms of the rate constants of the thermolysis (5.95) are correlated with the logarithms of the rate constants observed in the solvolysis (5.96) [339]:

$$R^1R^2R^3C{-}N{=}N{-}CR^1R^2R^3 \xrightarrow[\text{ethylbenzene}]{180\ °C} 2R^1R^2R^3C\cdot + N_2 \quad (5.95)$$

$$R^1R^2R^3C-OOC-C_6H_4-NO_2\text{-}p \xrightarrow[\text{aqueous dioxane}]{100\ °C} R^1R^2R^3C-OH +$$

$$+\ p\text{-}NO_2C_6H_4COOH \quad (5.96)$$

[338] O. Christ and H. Elias, *Chem. Ber.*, **99**, 1 (1966).
[339] W. Duismann and C. Ruchardt, *Tetrahedron Letters*, 4517 (1974).

R^1, R^2 and R^3 having the same significance in both reactions. Some relative rates in the two reactions are given below:

R^1	R^2	R^3	k_{rel} in: *reaction* (5.95)	*reaction* (5.96)
Me	Me	Me	1.00	1.00
Me	Me	Et	1.19	1.28
Me	Me	Bu^t	5.30	3.91
Me	Me	*neo*-C_5H_{11}	247	6.40
Me	Et	Et	1.87	2.66
Me	Et	Pr^i	5.35	4.43
Me	Et	Bu^t	36.5	17.5
Me	Et	*neo*-C_5H_{11}	300	7.89
Et	Et	Et	3.65	3.84
Et	Et	Pr^i	20.2	9.80
Et	Et	Bu^t	107	50
Et	Et	*neo*-C_5H_{11}	365	9.50

Plot for each reaction the logarithms of the relative rates *versus* the σ^* values. Show that in each case two correlation lines, having about the same slope, are obtained. Discuss the significance of the slope and intercept of the two lines. What shows the linear correlation between the logarithms of the rate constants determined in the two reactions?

(14) Applying the heteroatomic model in which: $\beta_X = k_{CX}\beta_0$ (α_X being taken equal to α_0) and considering [313] a value $k_{CX} = 0.5$, calculate the delocalisation energies (Δ_E) for the following systems:

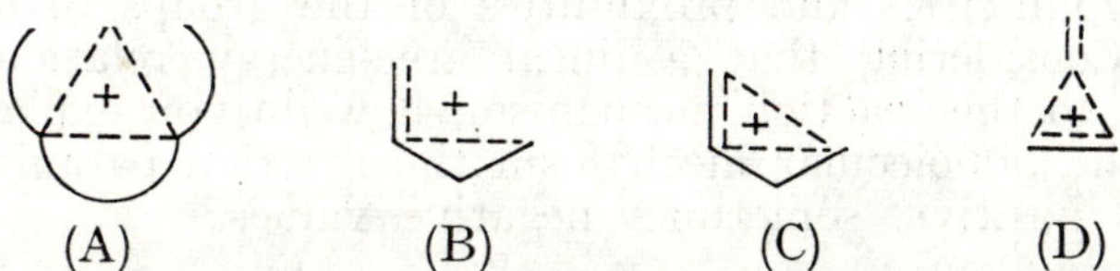

What conclusions can be drawn from the results of such calculations, regarding the participation reactions?

(15) Consider the homoallylic (1,3 interactions), homoaromatic (1,3,5 interactions) and bicyclobutonium (1,3,4 interactions) participations and show:

(a) Which of them should be more sensible to the distance between the reaction centre and the neighbouring bond.

(b) If other parameters are identical, which of them should provide the higher anchimeric assistance (driving force).

Indication: use the calculations made in the preceding Problem.

(16) The n.m.r. spectrum of the 2-norbornyl cation [264, 265] shows at -60 °C three peaks in the ratio 1: 4: 6. The spectrum is not changed by lowering the temperature to -120 °C. At 0 °C a single peak is observed.

(a) Explain the equivalence of the protons at the higher temperature.

(b) Admitting that only 1,2- and 1,3-shifts in classical carbenium ions are responsible for the observed spectra, determine which is the more rapid rearrangement in the system. Calculate an upper limit for the activation free-energies knowing that the rearrangement does not take place at −120 °C. What indicates the calculated value with regard to the controversy around the non-classical ion ?

(c) How can you explain that the same result has been considered by different authors either as a proof for the non-classical structure or as evidence pointing to a rapid equilibration between classical ions?

(17) The monohomocyclopropenyl cation[313, 340] **151** presents an absorption band at 245 nm. Calculate the energy of the electronic transition. Considering the HMO treatment of the system, to what value of the resonance integral $\beta_{1,3}$ will correspond the spectral determination?

(18) The anchimeric assistance observed experimentally may also be interpreted by Scheme 1, in which the rate acceleration is a result of the decrease of the internal return by the transformation of the initially formed classical ion into a bridged ion. Choose a particular example for which pertinent data are available and try to find whether this hypothesis is more plausible than the direct involvement of the neighbouring group in the ionisation.

Indication: for the direct involvement of the neighbouring group in ionisation apply Hammond's postulate, according to which the transition state must be similar in energy and geometry with the intermediate (non-classical ion); for the ion-pair hypothesis consider the relative stabilities of the ions, the kinetic scheme, as well as the nucleophilicities and proximities of the groups involved.

(19) Considering that a linear free-energy relationship must be related to the reaction mechanism, how do you explain the fact that for the bimolecular mechanism the reaction parameter ρ^* has sometimes positive, sometimes negative values?

[340] G. A. Olah, J. S. Staral and G. Liang, *J. Amer. Chem. Soc.*, **96**, 6233 (1974).

6
Aliphatic electrophilic substitution

6.1 Introduction

The aliphatic electrophilic substitution, even if less studied than the aliphatic nucleophilic substitution, is quite common in organic chemistry. Some examples more frequently encountered and better understood are given below:

(1) *Displacement reactions at the C—H bond:*

(a) Hydrogen exchange:

$$D^+ + R{-}H \rightleftharpoons R{-}D + H^+ \tag{6.1}$$

(b) Hydride shifts:

$$R^{1+} + H{-}R^2 \rightleftharpoons R{-}H + R^{2+} \tag{6.2}$$

(c) Other displacements at the C—H bond:

$$E^+ + R{-}H \rightleftharpoons R{-}E + H^+ \tag{6.3}$$

$$(E^+ = X^+\ (X_2),\ NO_2^+,\ \text{etc.})$$

(2) *Displacement reactions at the C—C bond:*

(a) Dealkylation reactions:

$$H^+ + R{-}R \rightleftharpoons R{-}H + R^+ \tag{6.4}$$

(b) Decarboxylation reactions:

$$H^+ + R{-}COO^- \rightleftharpoons R{-}H + CO_2 \tag{6.5}$$

(c) Decondensation reactions:

$$H^+ + R{-}\underset{\underset{Y}{|}}{\overset{\overset{O^-}{|}}{C}}{-}X \rightarrow R{-}H + \underset{\underset{Y}{|}}{\overset{\overset{O}{\|}}{C}}{-}X \tag{6.6}$$

(3) *Displacement reactions at the C—Metal bond:*

$$E^+ + R{-}Met^I \rightarrow R{-}E + Met^+ \tag{6.7}$$

$$(E^+ = H^+(D^+),\ R^+(RX),\ X^+(X_2),\ CO_2,\ \text{etc.})$$

$$(Met^I = Li,\ Na,\ K,\ HgR(HgX),\ MgR(MgX),\ \text{etc.})$$

Both the bimolecular and the unimolecular mechanisms are known. In the transition state of the bimolecular mechanism (S_E2), only two electrons are distributed between the three reaction centres.

This feature of the S_E2 reaction determines some significant differences with respect to the S_N2 reaction.

(a) In contrast with the four-electron, three-centre transition state of the S_N2 reaction, the two-electron, three-centre system [1,2] may correspond to a stable electronic structure. While no intermediates have as yet been positively identified in the nucleophilic bimolecular substitution [3], a great number of carbocations are known which may correspond to intermediates of the S_E2 reaction; they are called *carbonium ions* or *non-classical ions* [4]. Evidently, the carbon atom in such intermediates has higher coordination numbers than four.

(b) Comparing the two-electron, three-centre system with the allyl and cyclopropenyl cations, it may be inferred that the transition state (intermediate) **2** should be preferred to **1**:

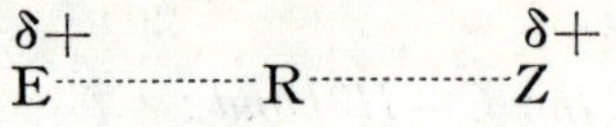

$$\overset{\delta+}{E}\cdots\cdots R\cdots\cdots\overset{\delta+}{Z} \qquad\qquad R\overset{+}{\lhd}\begin{matrix}Z\\ \vdots\\ E\end{matrix}$$

1 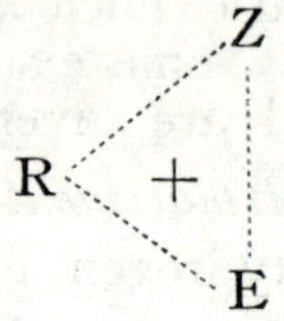 **2**

Thus, the incoming electrophile will enter the molecule on the same side with the displaced group Z. In contrast with the S_N2 reactions, the biomolecular electrophilic substitution must therefore be accompanied by retention of configuration, a conclusion consistent with experimental data.

(c) The close examination of transition state **2** reveals that in an S_E2 reaction the reagent attacks a chemical bond, instead of a nucleus as in the S_N2 reaction [5].

The electrophilic bimolecular mechanism is thus quite different from the direct nucleophilic mechanism.

However, the bimolecular mechanism is related to the nucleophilic substitution S_N1 in which the ionisation step is assisted by a neighbouring σ bond:

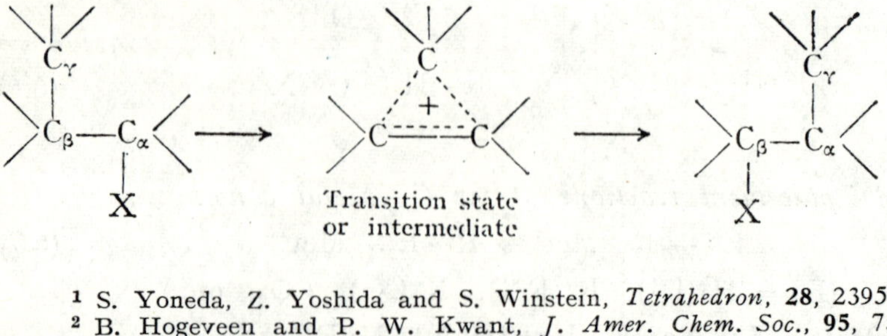

$$\begin{matrix}C_\gamma \\ | \\ >C_\beta-C_\alpha< \\ \quad\ \ | \\ \quad\ \ X\end{matrix} \longrightarrow \begin{matrix}C \\ \overset{+}{\triangle} \\ >C{=}{=}{=}C<\end{matrix} \longrightarrow \begin{matrix}\quad\ \ C_\gamma \\ \quad\ \ | \\ >C_\beta-C_\alpha< \\ | \\ X\end{matrix} \qquad (6.8)$$

Transition state or intermediate

[1] S. Yoneda, Z. Yoshida and S. Winstein, *Tetrahedron*, **28**, 2395 (1972).

[2] B. Hogeveen and P. W. Kwant, *J. Amer. Chem. Soc.*, **95**, 7315 (1973).

[3] Very recently (T. ElGomati, D. Lenoir and I. Ugi, *Nachr. Chem. Techn.*, 22, 499 (1974)), the identification of a penta-coordinated carbanion was claimed.

[4] G. A. Olah, *J. Amer. Chem. Soc.*, **94**, 808 (1972).

[5] G. A. Olah, Y. Halpern, J. Shen and Y. K. Mo, *J. Amer. Chem. Soc.*, **93**, 1251 (1971).

Such reactions will correspond to intramolecular S_E2 reactions at C_γ or to intramolecular S_N2 reactions at C_α. As has been since long observed[6] in Wagner-Meerwein, Beckmann and related rearrangements, the reaction occurs with racemisation and/or inversion at C_α and C_β, but with retention of configuration at C_γ, which is the site of the intramolecular S_E2 reaction.

The unimolecular mechanism S_E1 is more related to the common unimolecular nucleophilic substitution. Both reactions are two-step processes, involving production of reactive intermediates in the slow step:

$$R{-}Z \underset{}{\overset{\text{slow}}{\rightleftharpoons}} R{:}^- + Z^+(\text{Z-Acceptor}) \tag{6.9}$$

$$R{:}^- + E^+ \xrightarrow{\text{rapid}} R{-}E \tag{6.10}$$

As in S_N1 reactions, the heterolysis (6.8) is usually assisted by the reaction medium. Nevertheless, if in S_N1 the electrophilic assistance usually prevails (although rarely being quantitatively expressed), in the S_E1 reaction the nucleophilic assistance is strong enough to determine the appearance of the acceptor (or catalyst) in the rate equation. Better known are carbanion reactions involving the base-catalysed breaking of a C—H bond:

$$R{-}H + B{:} \rightleftharpoons R{:}^- + B{-}H^+ \tag{6.11}$$

The stereochemistry of the S_E1 reaction depends sometimes dramatically on the solvent, as usually happens in S_N1 reactions.

6.2 The bimolecular mechanism (S_E2)

6.2.1 *Development of the basic ideas*

Ingold and Hughes[7], considering the possible pathways of a substitution reaction, recognised in 1935 an associative mechanism, probably the single possible mechanism in electrophilic aromatic substitution, and plausible also for the aliphatic electrophilic substitution. Many of the reactions discussed in this and later papers have proved to occur either by a dissociative mechanism S_E1 or by a completely different mechanism. Thus, in 1936 Ingold, Raisin and Wilson[8] considered the direct introduction of deuterium into aliphatic compounds, with deuterosulphuric acid. They regarded the

[6] See, for example, C. K. Ingold, *Structure and Mechanism in Organic Chemistry*, Cornell University Press, 1st edn (1953), p. 511.

[7] E. D. Hughes and C. K. Ingold, *J. Chem. Soc.*, 244 (1935).

[8] C. K. Ingold, C. G. Raisin and C. L. Wilson, *J. Chem. Soc.*, 915, 1637, 1643 (1936).

reaction as an electrophilic aliphatic substitution identical in pattern with the reaction of the aromatic nucleus:

$$\rangle C{-}H + D{-}OSO_3H \longrightarrow \rangle C\langle{}^{H}_{D}\rangle OSO_3H \longrightarrow \rangle C{-}D + H_2SO_4$$

3

(6.12)

A transition state of type **3** has been assumed [7,8] without any proof, implying retention of configuration. Later development of the theory confirmed that the authors were correct in assuming the electrophilic pathway as an alternative in hydrogen-deuterium exchange with strong acids. They had a remarkable intuition with respect to the geometry of the transition state, but were wrong in considering the simple mechanism (6.12) for the deuteration with sulphuric acid. Indeed, it has been later demonstrated [9,10] that the deuteration of alkanes with sulphuric acid involves the initial formation of a carbenium ion (6.13), followed by intermolecular (and possibly intramolecular) hydride shifts (6.14), (6.16), protonation and deprotonation (6.15). Such a mechanism allows the substitution by deuterium of all hydrogen atoms but one:

$$2R{-}H + 3H_2SO_4 \rightarrow 2R^+ + SO_2 + 2HSO_4^- + 2H_3O^+ \quad (6.13)$$

$$R^+ + (CH_3)_3C{-}H \rightleftharpoons R{-}H + (CH_3)_3C^+ \quad (6.14)$$

$$(CH_3)_2\overset{+}{C}{-}CH_3 \rightleftharpoons (CH_3)_2C{=}CH_2 \underset{-D^+}{\overset{+D^+}{\rightleftharpoons}} (CH_3)_2\overset{+}{C}{-}CH_2D \rightleftharpoons \ldots \rightleftharpoons (CD_3)_3C^+ \quad (6.15)$$

$$(CD_3)_3C^+ + (CH_3)_3C{-}H \rightleftharpoons (CD_3)_3C{-}H + (CH_3)_3C^+ \text{ etc.} \quad (6.16)$$

The distribution of deutero-derivatives, the total absence of the perdeuterated compound, the identification of a hydrogen atom in the tertiary position and the evolution of small amounts of sulphur dioxide support the above mechanism.

Organomercury compounds have played a prominent part in the development of electrophilic substitution. Due to the easy heterolysis of carbon-metal bonds by electrophiles, to the covalent character of the carbon-mercury bond, it was possible, by a systematic study of the kinetics and stereochemistry of the breaking of the carbon-metal bond, to acquire a good understanding of the

[9] O. Beeck, J. W. Otvos, D. P. Stevenson and C. D. Wagner, *J. Chem. Phys.*, **16**, 255 (1948).

[10] J. W. Otvos, D. P. Stevenson, C. D. Wagner and O. Beeck, *J. Amer. Chem. Soc.*, **73**, 5741 (1951).

reaction. Thus, Nesmeyanov, Reutov and their co-workers[11-13] have demonstrated the bimolecular character of substitution reactions at the C—Hg bond, the absence of dissociative processes (6.9) and the retention of configuration at the reaction centre. In the stereochemical works, in order to avoid the rapid racemisation of organometallic compounds, they have used diastereomers, such as **4**, rohich are much more stable. The symmetrisation reaction of (—)-menthyl α-bromomercuryphenyl acetate **4**[11] yielded the optically active mercury derivative **5**:

$$\begin{array}{l} {}^*C_{10}H_{19}OCO \\ \quad H-C-HgBr \\ C_6H_5 \end{array} \underset{HgBr_2}{\overset{NH_3(CHCl_3)}{\rightleftarrows}}$$

4 ($[\alpha]_D^{18} = -86°$)

$$\begin{array}{l} {}^*C_{10}H_{19}OOC \qquad\qquad COOC^*_{10}H_{19} \\ \quad H-C-Hg-C-H \\ C_6H_5 \qquad\qquad\qquad C_6H_5 \end{array}$$

5 ($[\alpha]_D^{18} = -2°$)

The actual demonstration of the reaction stereochemistry consists in the reconversion of the derivative **5** into the original mercury bromide **4**. The retention of configuration in the transformation of diastereomeric organomercury compounds has also been confirmed in other laboratories[14,15].

More convincing arguments were found when optically active organomercury compounds, with a single chiral centre at the carbon atom directly bonded to the metal, were isolated and their reaction studied. The first monochiral mercury derivative, stable enough to be resolved and to preserve its optical activity, was s-butyl-mercury bromide[16, 17] **6**:

$$\text{L}(-)\ \text{Et}-\underset{\text{HgX}}{\overset{\text{H}}{\underset{|}{\overset{|}{\text{C}}}}}-\text{Me} \xrightarrow[\text{Pyridine}\ -65\ °\text{C}]{\text{Br}_2} \text{L}(-)\ \text{Et}-\underset{\text{Br}}{\overset{\text{H}}{\underset{|}{\overset{|}{\text{C}}}}}-\text{Me}$$

6

[11] A. M. Nesmeyanov, O. A. Reutov and S. S. Poddubnaya, *Dokl. Akad. Nauk SSSR*, **88**, 479 (1953).

[12] O. A. Reutov, *Fundamentals of Theoretical Organic Chemistry*, 2nd edn, Appleton-Century-Crofts, New York (1967), Chapter 2.

[13] O. A. Reutov and I. P. Beletskaya, *Reaction Mechanisms of Organometallic Compounds*, North-Holland, Amsterdam (1968).

[14] S. Winstein, T. G. Traylor and C. S. Garner, *J. Amer. Chem. Soc.*, **77**, 3741 (1955).

[15] S. Winstein and T. G. Traylor, *J. Amer. Chem. Soc.*, **77**, 3747 (1955).

[16] H. B. Chapman, E. D. Hughes and C. K. Ingold, *Chem. and Ind.*, 1517 (1958).

[17] F. R. Jensen, *J. Amer. Chem. Soc.*, **82**, 2469 (1960).

Hughes, Ingold and their co-workers[16, 18, 19], on the one hand, and Jensen *et al.*[17, 20, 21], on the other hand, have established the retention of both optical activity and configuration of the carbon atom in the reactions of **6**.

A more detailed discussion of the mechanism of an electrophilic aliphatic substitution was offered by Doering and Prinzbach[22] who have studied the insertion of methylene, $:CH_2$, in a C—H bond. Ruling out a step mechanism, Doering and Prinzbach have demonstrated the concerted character of the reaction. Considering the electrophilic character of the reagent, they have shown that the C—H bond behaves towards methylene in a similar way as the carbon-carbon double bond towards typical electrophiles. Transition state **7** has been proposed, corresponding closely, for example, to halogenium ions:

$$\rangle C{-}H + :CH_2 \longrightarrow \rangle C\begin{smallmatrix} CH_2 \\ \vdots \\ H \end{smallmatrix} \longrightarrow \rangle C{-}CH_3 \qquad (6.17)$$

7

The associative nature of the aliphatic electrophilic substitution, similar to electrophilic additions, is nowadays definitely established[23]. In that respect, the reactions of hydrocarbons in superacids are of utmost interest.

6.2.2 *Electrophilic substitutions in superacids*

Nenitzescu and Cantuniari[24] discovered in the early 1930s that in the isomerisation of alkanes with aluminium chloride a co-catalyst, such as water, is needed in order to start the reaction. It has been later proved in the same laboratory[25] that hydrogen is evolved in the first stages of the reaction, indicating the formation of a strong

[18] H. B. Chapman, E. D. Hughes and C. K. Ingold, *J. Chem. Soc.*, 2523, 2530 (1959).

[19] H. B. Chapman, E. D. Hughes, C. K. Ingold and F. G. Thorpe, *J. Chem. Soc.*, 1121 (1961); 4372 (1964).

[20] F. R. Jensen, L. D. Whipple, D. K. Wedegaertner and J. A. Landgrebe, *J. Amer. Chem. Soc.*, **82**, 2466 (1960).

[21] F. R. Jensen and K. L. Nakamaye, *J. Amer. Chem. Soc.*, **88**, 3437 (1966).

[22] W. von E. Doering and H. Prinzbach, *Tetrahedron*, **6**, 24 (1959).

[23] R. W. Alder and R. Baker, *Mechanism in Organic Chemistry*, Wiley-Interscience, New York (1971), Chapter 5.

[24] C. D. Nenitzescu and I. Cantuniari, *Ber. dtsch. chem. Ges.*, **66**, 1097 (1933).

[25] C. D. Nenitzescu, M. Avram and E. Sliam, *Bull. Soc. chim. France*, 1266 (1955).

acid from aluminium chloride and water, able to promote the reaction by hydride shift:

$$AlCl_3 + H_2O \rightleftharpoons H^+ \, [AlCl_3OH]^-$$

$$R{-}H + H^+ \, [AlCl_3OH]^- \rightarrow R^+ + H{-}H \, [AlCl_3OH]^- \qquad (6.18)$$

The hydride shift (6.18) is an electrophilic substitution at a C—H bond, in which a carbenium ion is displaced from hydrogen by the electrophile, H^+. Later studies in superacids (SbF_5 — HF, $HFSO_3H$, $HFSO_3H$—SbF_5, etc.) performed by the research groups of Hogeveen [26–28] in Europe, and of Olah [29, 30] in the United States, have confirmed the generality of this reaction. It has been also shown in these works that depending on the particular structure of the alkane and the stability of the carbenium ions involved, other reactions such as (6.19b), (6.20), (6.21), may occur:

$$R{-}H + H^+SbF_6^- \underset{b}{\overset{a}{\rightleftharpoons}} R^+ \; H{-}H \;\; SbF_6^- \qquad (6.19)$$

$$R{-}H + D^+SbF_6^- \rightleftharpoons R{-}D + H^+SbF_6^- \qquad (6.20)$$

$$R{-}R + H^+SbF_6^- \rightleftharpoons R^+ + R{-}H + SbF_6^- \qquad (6.21)$$

The reversible reactions (6.19) and (6.21) are of high interest in petrochemistry [31] and are involved in isomerisation, alkylation and dealkylation reactions (Chapter 7). All of them are aliphatic electrophilic substitutions either at a C—H bond or a C—C bond. Olah and his group [4, 29, 30] have unambiguously demonstrated by n.m.r. spectroscopy the intermediate formation of protonated alkanes. Thus, in the deuteration of methane, the species CH_5^+ has been identified. Its properties, as well as theoretical calculations, have indicated a C_s symmetry, **8**, corresponding to a two-electron, three-centre bond between two hydrogen atoms and the central carbon atom.

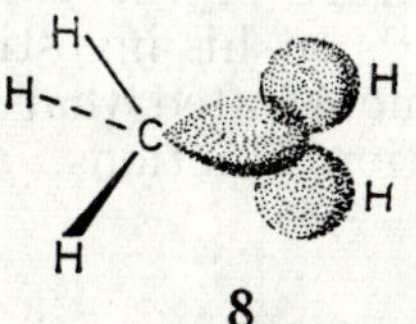

8

The reactivity of the various bonds encountered in alkanes decreases [22] in the order (6.22):

$$C_t{-}H > C{-}C > C_s{-}H \gg C_{pr}{-}H \qquad (6.22)$$

[26] H. Hogeveen, A. F. Bickel, *Chem. Commun.*, 635 (1967).
[27] H. Hogeveen and C. J. Gaasbeek, *Rec. Trav. Chim.*, **87**, 319 (1968).
[28] A. F. Bickel, C. J. Gaasbeek, H. Hogeveen, J. M. Oelderick and J. C. Platteeuw, *Chem. Commun.*, 634 (1967).
[29] G. A. Olah, K. Klopman and R. H. Schlosberg, *J. Amer. Chem. Soc.*, **91**, 3261 (1969).
[30] G. A. Olah *et al.*, *J. Amer. Chem. Soc.*, **95**, 4939, 4952, 4957, 4960 (1973).
[31] G. A. Olah and J. A. Olah, *J. Amer. Chem. Soc.*, **93**, 1256 (1971).

in agreement with dissociation energies of the corresponding bonds and the stability of the carbenium ions displaced. The retention of configuration in a deuteration reaction is illustrated by the hydrogen exchange of adamantane [31]:

H → D

Only the tertiary hydrogen atom is substituted in this example *.

6.2.3 *Other electrophilic reactions at the C—H bond*

Let us again consider the transition state (or intermediate) of type **2**, involved in an electrophilic substitution at the C—H bond (**9**):

$$\gt C \cdots H \cdots E \;(\mathbf{9}) \longrightarrow \gt C{-}E + H^+ \;(\mathbf{10}) \qquad (6.23a)$$

$$\gt C \cdots H \cdots E \;(\mathbf{9}) \longrightarrow \gt C^+ + H{-}E \qquad (6.23b)$$

It can be easily seen that collapse of **9** may occur in two different ways, either to the substituted product **10** or to a carbenium ion and HE. The latter reaction corresponds to an intermolecular hydride shift [32]. The cyclic nature of the transition state in hydride shifts has been accepted [32,33]. In his investigations of hydrocarbons in superacids, Olah has demonstrated that hydride shifts may also be accompanied by alkylation reactions. A common intermediate can be assumed:

$$R^1_3C^+ + HCR^2_3 \rightleftharpoons [R^2_3C \cdots H \cdots CR^1_3] \begin{cases} \nearrow R^2_3C{-}CR^1_3 + H^+ \\ \searrow R^2_3C^+ + R^1_3C{-}H \end{cases}$$

* It may be argued that the stereochemistry is imposed by the geometry of adamantane. In order to avoid an electronically allowed, but sterically hindered route, by an electronically forbidden, but geometrically possible mechanism, more drastic conditions would have been expected.

[32] C. D. Nenitzescu, in *Carbonium Ions*, vol. 2 (G. A. Olah and P. von R. Schleyer, editors), Wiley-Interscience, New York (1970), p. 463.

[33] F. Hawthorne and E. S. Lewis, *J. Amer. Chem. Soc.*, **80**, 4296 (1958).

Hydride shift accompanied by electrophilic substitution at a C—H bond was first demonstrated by Cohen and Lipowitz [34] in the decomposition of the diazonium salt **11**:

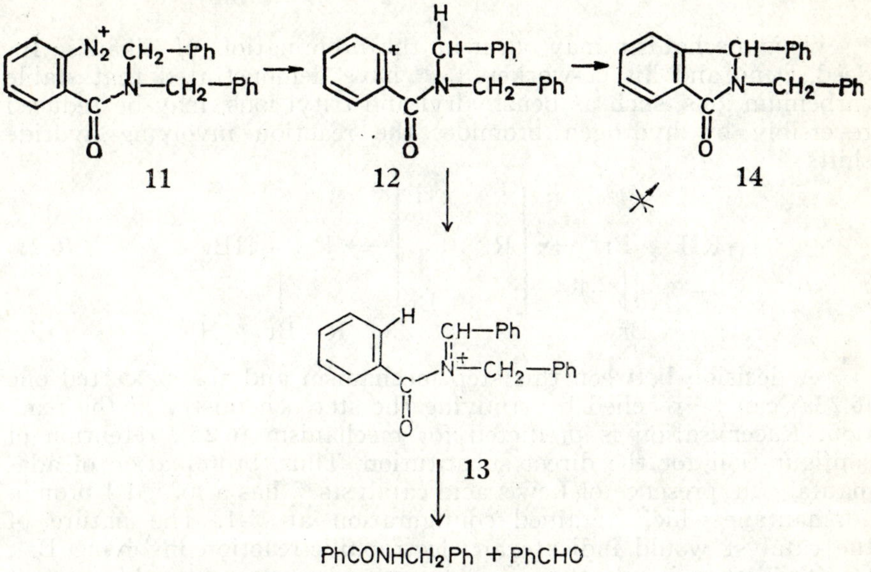

11 **12** **14**

13

$PhCONHCH_2Ph + PhCHO$

An intramolecular hydride shift converts the aryl cation **12** into the delocalised ion **13**. This ion may be the intermediate in the formation of the observed reaction products, the cyclic derivative **14**, benzaldehyde and *N*-benzyl benzamide. The generation of cation **13** by another independent route has shown the easy conversion of it into the splitting products (benzaldehyde and benzyl benzamide) and its reluctance to afford the cyclic product **14**. Moreover, the cyclisation of **13** would have been expected to yield **15** instead of **14**, the aromatic ring being activated in the benzylic and deactivated in the benzoyl moiety. The experimental data thus indicate that both **13** and **14** must be direct products, the most plausible interpretation being that they arise from a common intermediate **16**.

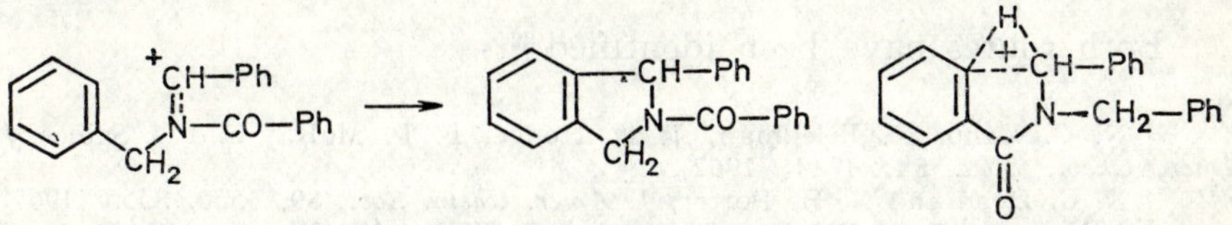

13 **15** **16**

[34] T. Cohen and J. Lipowitz, *J. Amer. Chem. Soc.*, 86, 2514, 2515 (1964).

Scheme (6.23) suggests also another route in the formation of the substitution product **10**. If HE has acidic properties, forming a stable anion E^-, a stepwise mechanism may be involved:

$$\mathbf{9} \longrightarrow\ \rangle C^+ + H{-}E \longrightarrow\ \rangle C{-}E + H^+ \qquad (6.24)$$

This mechanism may occur in the bromination of alkanes. Indeed, Deno and his co-workers [35, 36] have demonstrated that stable carbenium ions, such as benzhydryl and trityl ions, may be reduced reversibly by hydrogen bromide, the reaction involving hydride shifts:

$$\begin{array}{ccccc} RH + Br^+ & \longrightarrow & \left[R \begin{smallmatrix} \cdot\cdot H \\ \vdots \\ \cdot\cdot Br \end{smallmatrix} \right]^+ & \longrightarrow & R^+ + HBr \\ {\scriptstyle -Br^-}\upharpoonleft\downharpoonright{\scriptstyle +Br^-} & & & & \upharpoonleft\downharpoonright \\ Br_2 & & & & R{-}Br + H^+ \end{array} \qquad (6.25)$$

A decision between this step mechanism and the concerted one (6.23a) can be reached by studying the stereochemistry of the reaction. Racemisation is predicted for mechanism (6.25), retention of configuration for the direct substitution. Thus, bromination of adamantane in presence of Lewis acid catalysts [37] has afforded 1-bromoadamantane which retained configuration at C-1. The nature of the catalyst would indicate an electrophilic reaction involving Br^+. On the other hand, the geometry of adamantane would prevent a backside attack of the reagent. As the tertiary cation derived from adamantane is less stable than trityl ions, the direct mechanism is to be accepted.

Cyclic transition states in oxidation reactions with hexavalent chromium have also been proposed [37–39]. The cyclic species **17** may either collapse to an ester **18** or to a carbonium ion:

$$\underset{\mathbf{17}}{\begin{smallmatrix} H\text{--}O \\ R \quad\quad Cr \\ O \end{smallmatrix}} \begin{array}{l} \nearrow \underset{\mathbf{18}}{R{-}\overset{|}{\underset{|}{O}}CrOH} \\ \searrow \underset{\mathbf{19}}{R^+(Cr^{IV})} \end{array}$$

Both routes have been identified [39].

[35] N. C. Deno, N. Friedman, J. D. Hodge, F. P. McKay and G. Saines, *J. Amer. Chem. Soc.*, **84**, 4713 (1962).

[36] N. C. Deno and N. H. Potter, *J. Amer. Chem. Soc.*, **89**, 3550, 3555 (1967).

[37] I. Necșoiu, A. T. Balaban, I. Pascaru, E. Sliam, M. Elian and C. D. Nenitzescu, *Tetrahedron*, **19**, 1133 (1963).

[38] J. Rocek and J. Krupicka, *Chem. Ind.*, 1668 (1957).

[39] For leading references see F. Freeman, *Rev. Reactive Species*, 37 (1973).

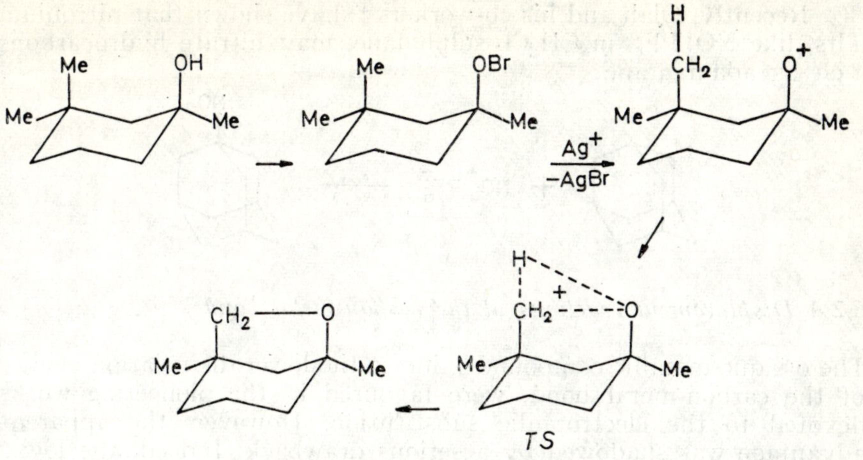

Figure 6.1 The formation of the bicyclic ether **21**. The OCH_3 group is placed axially, having a higher conformational energy than the methyl group (see for example E. L. Eliel, *Angew. Chem.*, **77**, 784 (1965)).

Sneen and Matheny [40] have observed that cyclohexanols having two methyl groups at C-3 (**20**) yield with bromine, in presence of silver salts bicyclic ethers (**21**).

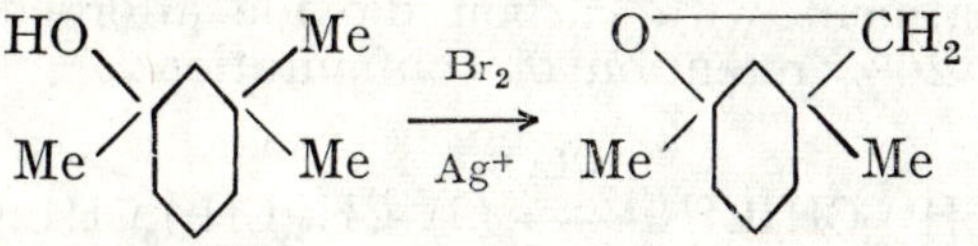

20 $\xrightarrow[Ag^+]{Br_2}$ **21**

It is well known that alcohol treated with bromine may generate alkyl hypohalides, which usually decompose by a radical mechanism (Chapter 11). However, the presence of silver ion favours the heterolysis of the O—Br bond, and thus an electrophilic substitution at the neighbouring methyl (Figure 6.1).

The nitration of alkanes either in solution, with dilute nitric acid, or in gaseous phase, is well known. The reaction has earlier been assumed [7] to involve an electrophilic substitution. The recognition of a homolytic mechanism [41] and the determination of the experimental conditions under which the electrophile NO_2^+ can be produced (Chapter 7), have led to the conviction that alkanes cannot be nitrated by an electrophilic mechanism.

[40] R. A. Sneen and N. P. Matheny, *J. Amer. Chem. Soc.*, **86**, 5503 (1964).

[41] H. B. Hass *et al.*, *Ind. Eng. Chem.*, **31**, 648 (1939); **32**, 427 (1940); **33**, 1138 (1941).

Recently, Olah and his co-workers [42] have shown that nitronium a lts, like $NO_2^+PF_6^-$ in CH_2Cl_2-sulpholane, may nitrate hydrocarbons u ch as adamantane:

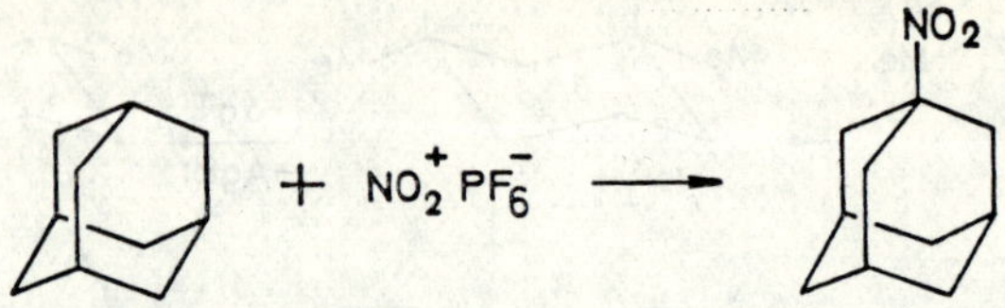

6.2.4 *Displacement reactions at the carbon-metal bond*

The organic metallic compounds, due to the lower dissociation energy of the carbon-metal bond, were favoured in the pioneering works devoted to the electrophilic substitution. However, the apparent advantage was shadowed by a serious drawback. Indeed, the lower the dissociation energy, the higher will be the tendency of the organometallic compound to ionise according to equation (6.26):

$$R - M^I \longrightarrow R:^- + M^{I+} \qquad (6.26)$$

In that respect, lithium and magnesium compounds constitute an illustrative example.

Thus, Letsinger [43] treated optically active 2-iodooctane with s-butyl-lithium at -70 °C, obtaining a lithium compound which on subsequent treatment with carbon dioxide afforded 2-methyloctanoic acid with 20% retention of configuration:

$$(-)\text{-}CH_3(CH_2)_5CHICH_3 \xrightarrow[-70\,°C]{Bu^sLi} (?)\text{-}CH_3(CH_2)_5CHLiCH_3 \xrightarrow{CO_2}$$

$$\longrightarrow CH_3(CH_2)_5\underset{\displaystyle COOH}{\underset{|}{C}}HCH_3$$

Lithium vinylic compounds [44, 45], when generated from vinyl derivatives, retain in their reactions the geometry of the original

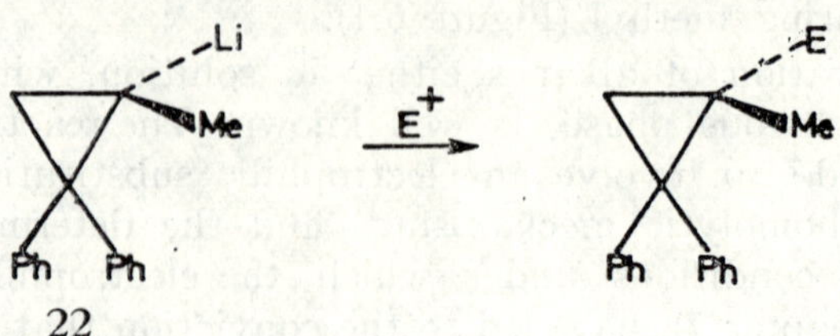

[42] G. A. Olah and H. C. Lin, *J. Amer. Chem. Soc.*, **93**, 1258 (1971).
[43] R. L. Letsinger, *J. Amer. Chem. Soc.*, **72**, 4842 (1950).
[44] E. A. Braude and C. J. Timmons, *J. Chem. Soc.*, 2000, 2007 (1950).
[45] E. A. Braude and J. A. Coles, *J. Chem. Soc.*, 2012 (1950); 2078, 2085 (1951).

substrate. The same is true for the lithium and magnesium derivatives of cyclopropanes [46], such as **22**. These results would suggest a bimolecular mechanism S_E2, in agreement with the covalent character of the C—Li and C—Mg bonds.

However, there are numerous other data which indicate either a reaction between tight ion-pairs or easily ionisable substrates. Thus, Roberts and his co-workers [47] demonstrated that cyclopropylmethylmagnesium bromide is in equilibrium with allylmethylmagnesium bromide:

$$CH_2{=}CH{-}{}^{*}CH_2{-}CH_2MgBr \rightleftharpoons \overset{*}{\triangleright}{-}CH_2MgBr \rightleftharpoons$$

$$CH_2{=}{}^{*}CH{-}CH_2{-}CH_2MgBr$$

The rearrangement is similar with that observed in the cationic series (see p. 284). An anionic intermediate may thus be assumed to occur in the above reaction. An identical observation has been mentioned for a lithium compound [48]:

$$\triangleright{-}CH_2I \xrightarrow{RLi} \triangleright{-}CH_2Li \xrightarrow{PhCHO}$$

$$\triangleright{-}CH_2\underset{OH}{\underset{|}{C}}HPh + CH_2{=}CH(CH_2)_2\underset{OH}{\underset{|}{C}}HPh$$

Lithium and magnesium derivatives, despite their stereoselectivity, may react either as covalent compounds or as ion pairs, the observed stereochemistry being determined by the low racemisation rates of pyramidal carbanions. Theoretical calculations [49, 50] of the energy barriers involved in pyramidal inversion for simple carbanions have indicated low values (6—12 kcal mol^{-1}), in disagreement with the experimental values (15—20 kcal mol^{-1}) determined by Roberts [51] for lithium and magnesium compounds. This discrepancy would indicate an S_E2 reaction at low temperature. For the vinyl and cyclopropyl derivatives, higher energy barriers are predicted for the pyramidal inversion ($>$ 23 kcal mol^{-1}). Consequently, either the direct concerted mechanism or the step mechanism (S_E1) can be involved in the reactions of these organometallic compounds.

[46] H. M. Walborsky and A. E. Young, *J. Amer. Chem. Soc.*, **86**, 3288 (1964).
[47] M. S. Silver, P. R. Shafer, J. E. Nordlander, C. Rüchardt and J. D. Roberts, *J. Amer. Chem. Soc.*, **82**, 2646 (1960).
[48] P. T. Lansbury and V. A. Pattison, *J. Amer. Chem. Soc.*, **85**, 1886 (1963).
[49] G. W. Koeppl, D. S. Sagatys, G. S. Krisnamurthy and S. I. Miller, *J. Amer. Chem. Soc.*, **89**, 3396 (1967).
[50] M. J. S. Dewar and M. Shanshal, *J. Amer. Chem. Soc.*, **91**, 3654, (1969).
[51] M. Witanovski and J. D. Roberts, *J. Amer. Chem. Soc.*, **88**, 737 (1966).

Organomercury compounds, as already stated, react usually according to the bimolecular mechanism. The same is true for organoboron derivatives. Due to the availability of organoboron compounds by the hydroboration reaction [52], the interest in organoboron chemistry has increased in the last few years. With only a few exceptions [53], the stereochemistry of the electrophilic substitutions at the C—B bond are satisfying the general pattern of the S_E2 reaction [54–56], *i.e.* retention of configuration. Exceptions have been observed with rigid systems, like the boronic acids **23**:

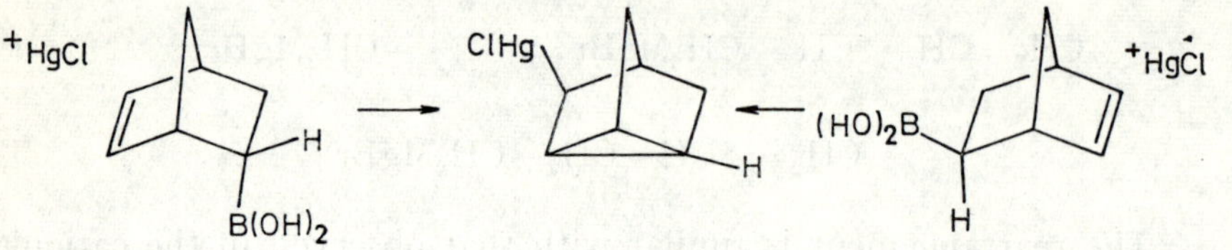

23-*endo* **24** **23**-*exo*

Both stereoisomers yield on treatment with mercuric chloride the same nortricyclene derivative **24**. The reaction occurs with retention of configuration from *endo*-**23**, but with inversion from the *exo*-isomer. The impossibility for the *exo*-isomer to maintain its configuration at C-2 in the cyclisation determines an inversion pathway. It can be thus inferred from this example that the height of the energy barrier of transition state **1** does not differ significantly from the energy corresponding to the transition state **2**.

In the substitution of organomercury derivatives with mercury electrophiles, such as in the symmetrisation reaction, or in the example studied by Winstein and Taylor [14, 15], four-centre transition states **25** have been assumed. Such mechanisms have been designated as S_Ei.

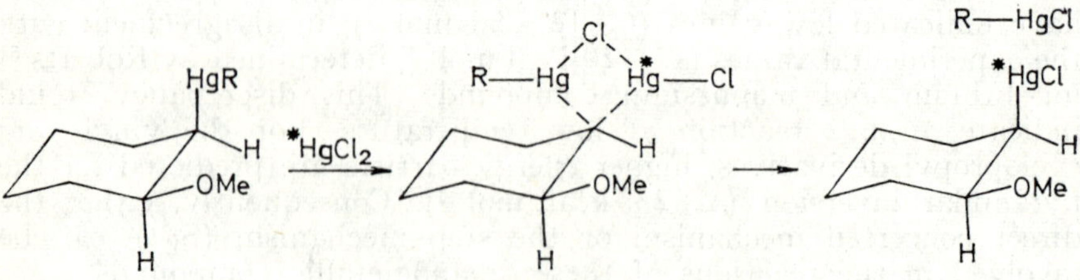

25

[52] H. C. Brown, *Hydroboration*, Benjamin Inc., New York (1962); *Boranes in Organic Chemistry*, Cornell University Press, Ithaca (1972).

[53] D. S. Matteson and J. O. Waldbillig, *J. Amer. Chem. Soc.*, **86**, 3778 (1964).

[54] H. C. Brown and K. J. Murray, *J. Org. Chem.*, **26**, 631 (1961).

[55] R. Dulou, Y. Chrétien-Bessière, *Bull. Soc. chim. France*, 1362 (1959).

[56] G. Zweifel and H. C. Brown, *Org. Reactions*, **13**, 1 (1963).

6.3 The unimolecular mechanism (S_E1)

6.3.1 *Survey of reactions*

The unimolecular mechanism S_E1 (6.9) — (6.10) may occur in three different ways:

(a) *Unimolecular ionisation to carbanions.* The ionisation process leading to carbanions (6.27) is analogous to the ionisation process (6.28) affording carbocations:

$$\mathrm{R{-}Z} \underset{}{\overset{\text{slow}}{\rightleftharpoons}} \underset{\substack{\textbf{26}a\\ \text{tight}\\ \text{ion-pairs}}}{\mathrm{R{:}^{-}Z^{+}}} \rightleftharpoons \underset{\substack{\textbf{26}b\\ \text{solvent-sepa-}\\ \text{rated ion-}\\ \text{pairs}}}{\mathrm{R{:}^{-} \parallel Z^{+}}} \rightleftharpoons \underset{\substack{\textbf{27}\\ \text{dissociated}\\ \text{ions}}}{\mathrm{R{:}^{-} + Z^{+}}} \qquad (6.27)$$

$$\mathrm{R{-}X} \overset{\text{slow}}{\rightleftharpoons} \mathrm{R^{+}X^{-}} \rightleftharpoons \mathrm{R^{+} \parallel X^{-}} \rightleftharpoons \mathrm{R^{+} + X^{-}} \qquad (6.28)$$

The difference between the two ionisation processes lies in the charge localised at the carbon atom. As in the case of the S_N1 reaction, the rate of process (6.27) will depend on the character of the leaving group Z, the structure of the carbanion R^-: and on solvent assistance. The nature of the intermediate actually involved (**26** — ion pairs, or **27** — dissociated, solvated ions), may affect, as in S_N1 reaction, the stereochemical outcome in the rapid step:

$$\textbf{26}\ (\text{or } \textbf{27}) + \mathrm{E^{+}} \xrightarrow{\text{rapid}} \underset{\substack{\text{conservation}\\ \text{of optical}\\ \text{activity (net}\\ \text{retention}\\ \text{or inversion)}}}{\mathrm{R{-}E}} \quad \underset{\text{racemic}}{(\text{or } \mathrm{R{-}E + E{-}R})} \qquad (6.29)$$

The two processes (6.27) and (6.28) may merge in a single reaction in the case of cyclopropane derivatives (**28**)[57–59]:

Ph CN R COOR → Ph CN + − R COOR

28

In order to observe a heterolysis of the C—C bond in cyclopropanes **28**, electron-repelling substituents are needed at the cationic centre and electron-attracting substituents at the anionic centre.

[57] D. J. Cram and A. Ratajczak, *J. Amer. Chem. Soc.*, **90**, 2198 (1968).

[58] E. W. Yankee and D. J. Cram, *J. Amer. Chem. Soc.*, **92**, 6328, 6329, 6331, (1970).

[59] E. W. Yankee, F. D. Badea, N. E. Howe and D. J. Cram, *J. Amer. Chem. Soc.*, **95**, 4220 (1973).

(b) *Unimolecular ionisations to carbanions, preceded by other rapid steps*. Cram and his co-workers [60], in a systematic study of the S_E1 reaction started in 1959, have generated carbanions in different ways, most of them involving rapid reactions (or equilibria) prior to the heterolysis steps:

(i) Carbon as leaving group. This type of reaction is mainly represented by the aldol, Claisen or Michael decondensations [61–65]:

$$\mathrm{R{-}\overset{\displaystyle OH}{\underset{\displaystyle X}{C}}{-}Y} + \mathrm{B{:}} \rightleftharpoons \mathrm{R{-}\overset{\displaystyle O^-}{\underset{\displaystyle Y}{C}}{-}X} \longrightarrow \mathrm{R{:}^-} + \mathrm{XYC{=}O} \qquad (6.30)$$

29

The heterolysis takes place in the ion **29**, formed in an equilibrium involving a base B:. The charge on oxygen and the stability of the leaving group XYCO are the driving forces of the heterolysis.

A similar, but much more simple reaction, is the decarboxylation of carboxylic acids [66–68]:

$$\mathrm{R{-}COOH} + \mathrm{B{:}} \underset{-\mathrm{HB}^+}{\longrightarrow} + \mathrm{RCOO^-} \longrightarrow \mathrm{R{:}^-} + \mathrm{CO_2} \qquad (6.31)$$

(ii) Nitrogen as leaving group [69]. This method of generating carbanions is represented by Kishner-Wolff and McFeyden-Stevens reactions and by the oxidation of hydrazines. In all these processes an unstable compound **30** is formed, which on further reaction with a base yields the carbanion:

$$\left.\begin{array}{l}\mathrm{R{-}NH{-}NH_2} + \mathrm{O_2} \\ \mathrm{R_2C{=}N{-}NH_2{+}B{:}} \\ \mathrm{R{-}NH{-}NHTs} + \mathrm{B{:}}\end{array}\right] \rightarrow \underset{\mathbf{30}}{\mathrm{R{-}N{=}N{-}H}} + \mathrm{B{:}} \longrightarrow \qquad (6.32)$$

$$\longrightarrow \mathrm{R{-}N{=}N{:}^-} \longrightarrow \mathrm{R{:}^-} + \mathrm{N_2}$$

(c) *Bimolecular generation of carbanions*. The ionisation step may be a bimolecular reaction in which the electron-defficient group Z is transferred to an acceptor in a rate-determining step:

$$\mathrm{R{-}Z} + \mathrm{Acc} \longrightarrow \mathrm{R{:}^-} + \mathrm{Z{-}Acc^+} \qquad (6.33)$$

60 D. J. Cram, *Fundamentals of Carbanion Chemistry*, Academic Press, New York (1965).

61 D. J. Cram, A. Langemann and F. Hauck, *J. Amer. Chem. Soc.*, **81**, 5750, 5754 (1959).

62 D. J. Cram, F. Hauck, K. R. Kopecky and W. D. Nielsen, *J. Amer. Chem. Soc.*, **81**, 5767 (1959).

63 D. J. Cram, J. L. Mateos, F. Hauck, A. Langemann, K. R. Kopecky, W. D. Nielsen and J. Allinger, *J. Amer. Chem. Soc.*, **81**, 5774 (1959).

64 D. J. Cram and W. D. Nielsen, *J. Amer. Chem. Soc.*, **83**, 2174 (1961).

65 D. J. Cram and B. Rickborn, *J. Amer. Chem. Soc.*, **83**, 2174 (1961).

66 D. J. Cram and P. Haberfield, *J. Amer. Chem. Soc.*, **83**, 2354 (1961).

67 D. J. Cram and A. S. Wingrove, *J. Amer. Chem. Soc.*, **85**, 1100 (1963).

68 D. J. Cram and P. Haberfield, *J. Amer. Chem. Soc.*, **83**, 2363 (1961).

69 D. J. Cram and J. S. Bradshaw, *J. Amer. Chem. Soc.*, **85**, 1108 (1963).

The simplest example of such reactions is the base-catalysed hydrogen-isotope exchange [70–74]:

$$\text{R—H} + \text{B:} \rightleftharpoons \text{R:}^- + \text{BH}^+ \xrightarrow{\text{SD}} \text{RD} \qquad (6.34)$$

6.3.2 *The stereochemistry of the reaction*

The information presently available concerning the stereochemistry of the unimolecular electrophilic substitution is due to Cram and his co-workers [60]. The most important result of their investigations is that the stereochemical outcome does not depend significantly upon the leaving group, being instead affected by the substituent R^1, R^2 and R^3 at the reaction centre in **31** and by the base-solvent system.

$$R^1\text{—}\underset{R^3}{\overset{R^2}{C}}\text{—Z} \qquad\qquad \text{Ph—}\underset{\text{Et}}{\overset{\text{Me}}{C}}\text{:}^-$$

31 **32**

Groups like CN [68, 75], RCO [75], $CONR_2$ and COOR [71], able to accommodate the negative charge at the reaction centre, favour the complete racemisation of substrates **31** during the reaction. It must be noted however that carbanion **32**, generated in several ways from substrates with various leaving groups, may preserve in a high proportion its original optical activity. This observation is surprising considering that **32** is a benzyl anion. Both chemical[76] and spectral data[77] of the benzyl anion indicate the delocalisation of the charge in the aromatic ring. Benzyl anions should be planar and achiral. However, the solvation of the intermediate carbanion may be asymmetric and thus lead to optically active products.

In the understanding of the stereochemistry of S_E1 reactions a prominent role has been played by hydrogen isotope exchanges at chiral centres. In such reactions, the exchange (rate constant k_e) may be accompanied to a larger or smaller extent by racemisation of the substrate (rate constant k_α). The stereochemical course of the reaction is indicated by the values of the ratio k_e/k_α (Table 6.1).

[70] D. J. Cram, A. C. Kingsbury and B. Rickborn, *J. Amer. Chem. Soc.*, **83**, 3688 (1961).

[71] D. J. Cram, B. Rickborn, C. A. Kingsbury and P. Haberfield, *J. Amer. Chem. Soc.*, **83**, 3678 (1961).

[72] D. J. Cram and L. Gosser, *J. Amer. Chem. Soc.*, **86**, 5445 (1964).

[73] D. J. Cram and A. S. Wingrove, *J. Amer. Chem. Soc.*, **86**, 5490 (1974).

[74] W. D. Kollmeyer and D. J. Cram, *J. Amer. Chem. Soc.*, **90**, 1779, 1784 (1968).

[75] D. J. Cram and L. Gosser, *J. Amer. Chem. Soc.*, **86**, 5457 (1964).

[76] G. A. Russell, *J. Amer. Chem. Soc.*, **81**, 2017 (1959).

[77] B. Bockrath and L. M. Dorfman, *J. Amer. Chem. Soc.*, **96**, 5708 (1974), and literature cited therein.

TABLE 6.1

The ratio of the exchange rate and racemisation rate (k_e/k_α) and the observed stereochemistry of the exchange reaction: RH (or RD) $\xrightarrow{\text{base-solv.}}$ RD (or RH)

k_e/k_α	*Observed stereochemistry*
∞	100% retention of configuration
1	100% racemisation
0.5	100% inversion of configuration
0	100% isoracemisation

Thus, at least four mechanisms should be envisaged [60,74].

(a) *Retention mechanism.* The retention mechanism has been observed in solvents such as Bu^{t}OH, tetrahydrofuran (THF), benzene with added proton donors (PhOH, etc.), all having low dielectric constants and poor ionising properties. According to Cram *et al.*, the high stereospecificity of the reaction ($>$ 90% net retention) can be explained by admitting an asymmetric solvation of the ion pair formed:

33*a* 33*b* retention

In order to maintain the configuration at the reacting centre the potassium cation with its ligands should rotate (**33*a*** $\rightleftharpoons$ **33*b***). This isomerisation of the ion pair and its collapse to a covalent state must be faster than its dissociation. Clearly, the retention mechanism is expected to be dependent on the coordinating ability of the metal cation.

(b) *The inversion mechanism.* In more acidic alcohols, such as methanol and glycol [71], inversions ($>$ 60%) are observed (k_e/k_α between 0.94 and 0.65). As in the retention mechanism, it is believed that the asymmetric solvation (ion pair **34**) is responsible for the observed stereochemistry.

$$\rangle C—D + {}^{-}O—R \xrightarrow{\text{ROH}} R—O—H\text{---}\bar{C}\text{:---}D—O—R \longrightarrow H—C\langle$$

34 Inversion

The values of k_e/k_α, higher than 0.5, but lower than unity, indicate that the collapse of the ion pair **34** to the covalent species occurs at nearly the same rate as the transformation of the asymmetrically solvated ion pair **34** into the symmetrically solvated anion.

(c) *Racemisation mechanism.* The racemisation mechanism ($k_e/k_\alpha = 1$) has been observed in dimethyl sulphoxide [78] (DMSO). This aprotic dipolar solvent seems to be an excellent solvent for carbanions [79]. Thus, a symmetrically solvated species **35***b* should be expected:

$$\rangle C{-}D + \bar{O}{-}R \xrightarrow{DMSO} Me_2\overset{\overset{O}{|}}{S}{}^{+}{-}{-}{-}\overset{|}{\underset{\wedge}{C}}{}^{\bar{:}}{-}{-}{-}D{-}O{-}R \longrightarrow DMSO{-}{-}{-}{-}\overset{|}{\underset{\wedge}{C}}{}^{\bar{:}}{-}{-}{-}DMSO$$

35*a* **35***b*

It has been mentioned in a previous chapter (p. 227) that the symmetric solvation of ionic species has been challenged by different authors. However, with good solvating solvents, such as DMSO, it is likely that the racemisation is produced by a rapid equilibration, such as (6.35):

$$R{-}O{-}H{-}{-}{-}{-}{-}\overset{|}{\underset{\wedge}{C}}{}^{\bar{:}}{-}{-}{-}{-}{-}DMSO \rightleftharpoons DMSO{-}{-}{-}{-}{-}\overset{|}{\underset{\wedge}{C}}{}^{\bar{:}}{-}{-}{-}{-}H{-}O{-}R \quad (6.35)$$

(d) *Isoracemisation mechanism* (racemisation prior to isotopic exchange). This mechanism has been observed in systems formed by a tertiary amine in the absence of any proton pool. An example is offered by the fluorenyl derivative **36** in THF, with Et_3N or Pr_3N as bases.

[Structures: **36** = 9-Me-9-D-fluorene bearing O_2N and $CONR_2$ substituents; **37** = 9-Me-9-D-fluorene bearing $CONR_2$]

36 **37**

Values of k_e/k_α between 0.05 and 0.2 have been observed [75, 79].

[78] D. Martin, A. Weise and H. J. Niclas, *Angew. Chem.*, **79**, 340 (1967).

[79] D. J. Cram and L. Gosser, *J. Amer. Chem. Soc.*, **86**, 2950 (1964); for another racemisation mechanism see Reference 71.

A general mechanism for the isoracemisation pathway is depicted in the following scheme:

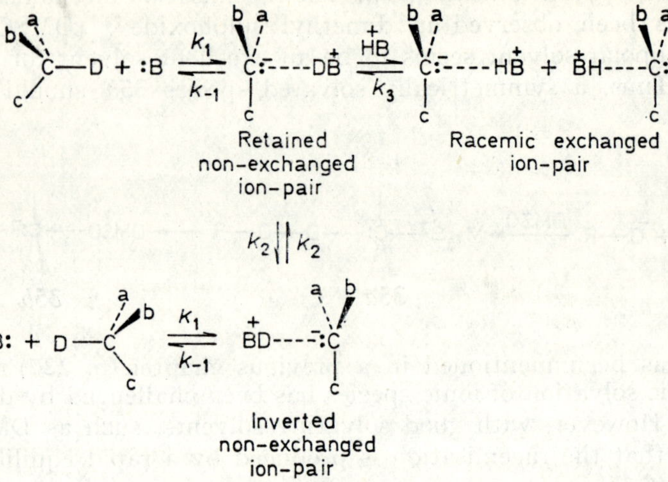

Retained non-exchanged ion-pair

Racemic exchanged ion-pair

(6.36)

Inverted non-exchanged ion-pair

In order to observe the isoracemisation mechanism, both k_{-1} and k_2 must be much greater than k_3. There are several possibilities for the ion-pair reorganisation leading to inverted non-exchanged ion pairs.

In one of them, the resultant ion pair undergoes reorganisation by an intermolecular path. The hydrogen bonds between the carbanion and the protonated base are broken and re-formed, yielding both enantiomeric ion pairs. This mechanism is not different from the racemisation mechanism and requires that re-formation of the hydrogen bond with the original protonated base be faster than the solvation of the carbanion with the deuterated base ($k_2 > k_3$).

Another possible mechanism corresponds to an intramolecular process. The hydrogen bonds are not broken, but merely shifted from one centre to another, and back to the reaction site. During this process, the reaction site becomes planar (and thus achiral), the back solvation being possible both ways (6.37).

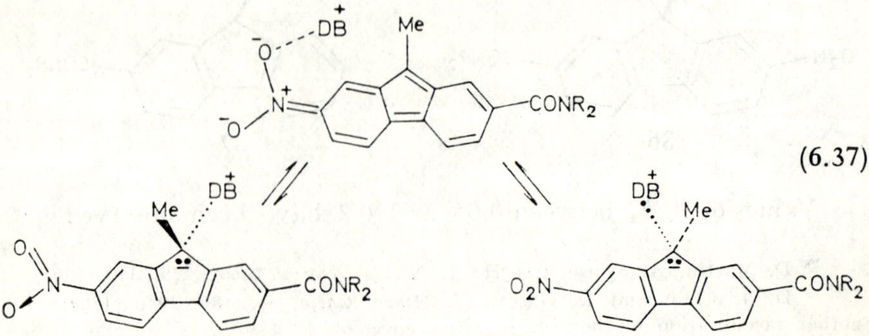

(6.37)

The intramolecularity of the reaction may be proved by adding a proton donor to the reaction medium. This mechanism of isoracemisation has been named the *conducted-tour mechanism* [80].

Clearly, the solvent-base system plays the most important part in the mechanisms implying optical activity conservation. Indeed, as already shown, asymmetrically solvated ion pairs are involved in such processes, and the reaction medium will determine whether these ion pairs reorganise and collapse to covalent products or dissociate to symmetrically solvated carbanions.

However, the structure of the carbanion precursor does in its turn determine the actual mechanism. We shall consider for example the two related fluorenyl derivatives **36** and **37**. The nitro compound **36** treated with $PrNH_2$ in THF yields racemised exchanged product, whereas with Pr_3N in the same solvent (with or without added Bu^tOH) affords only racemised, non-exchanged products. Conversely, compound **37**, both with $PrNH_2$ and with Pr_3N, in THF or BuOH, yields only the retained exchanged product. This shows that substrate **36** has a tendency for the conducted-tour mechanism, leading either to exchanged or non-exchanged racemic products, whereas compound **37** is not likely to undergo isoracemisation. The presence of the strongly electron-attracting group NO_2 in **36** must be responsible for this difference in behaviour. Indeed, hydrogen-bonding at the nitro group must be more important than at the $CONR_2$ group (*aci*-nitro derivatives are more stable compounds than the protonated forms of amides). Moreover, conjugation is more extended in the nitrofluorenyl derivative than it is in the carbanion derived from **37**.

A striking behaviour in S_E1 reactions is shown by the sulphones **38**. The carbanions generated from these carbon acids retain their original configuration on further reactions with electrophiles, both in retaining systems (Bu^tOH-Bu^tOK, THF—base, etc.) and in inverting and racemising solvents (for example MeOH and DMSO, respectively). It seems that functional groups centred around second-row elements stabilise the chiral structure of the carbanion.

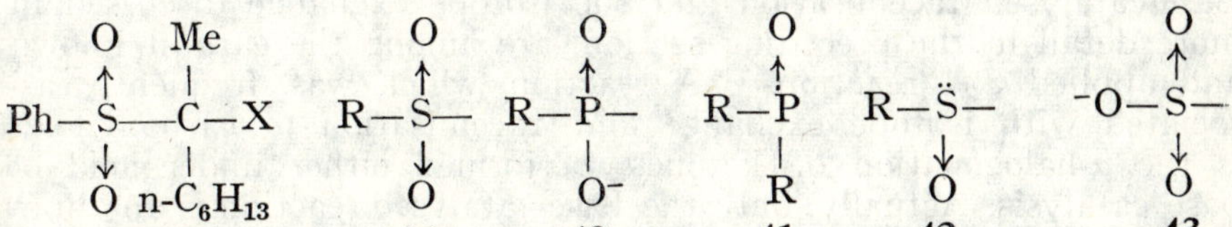

```
O   Me          O         O          O                 O
↑   |           ↑         ↑          ↑                 ↑
Ph—S——C—X    R—S—     R—P—      R—P     R—S̈—      ⁻O—S—
↓   |           ↓         |          |       ↓         ↓
O n-C6H13       O         O⁻         R       O         O
   38           39        40         41      42        43
```

X = D, H, COOH

Indeed, a more attentive survey of carbanion reactions generated from substrates **31**, in which one substituent of the chiral atom is any heteroatomic group with the second-row element bonded to an even number of equivalent oxygen atoms (**39**—**40**), has shown

[80] D. J. Cram, F. Willey, H. P. Fischer and D. A. Scott, *J. Amer. Chem. Soc.*, **86**, 5610 (1964).

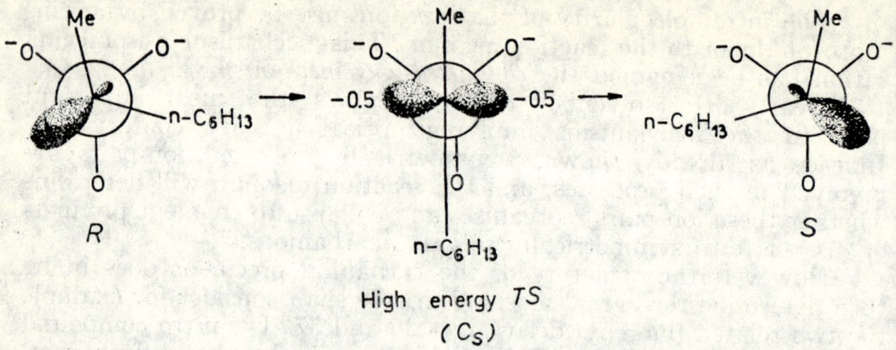

Figure 6.2 Configuration of carbanions with a neighbouring sulphone group.

that the heteroatom (S or P) may induce carbanion asymmetry. Heteroatomic groups, such as **41—43** (odd number of equivalent oxygen atoms), do not seem to be able to induce carbanion asymmetry [60, 64, 81–83]. Two interpretations are offered for the chirality of the carbanion and its high-energy barrier in the racemisation process.

(a) The resultant carbanion is pyramidal (thus chiral). Pyramidal inversion involving rehybridisation of the central atom (Section 4.3.3) will be favoured if the neighbouring second-row element is bonded to an odd number of oxygens, otherwise it is disfavoured (Figure 6.2).

(b) The carbanion centre is planar, but the species has a chirality plane. The racemisation of the planar carbanion would imply a rotation about the C-heteroatom bond, a rotation having higher energy barriers for the first type (heteroatomic group **39—40**) of derivatives (Figure 6.2).

6.3.3 *Simple base-catalysed* S_E1 *reactions*

Base-catalysed racemisation and the isotope exchange discussed in more detail in the preceding section, are among the earliest known and applied S_E1 reactions [7]. A reaction which was frequently associated with isotope exchange and racemisation of carbon acids is the α-halogenation of ketones, performed either under acid or base catalysis. Actually, only the base-catalysed reaction is an aliphatic electrophilic substitution. The two catalytic processes are however related, justifying a common treatment.

Lapworth [84] was the first to observe that the rate of the α-halogenation in acid catalysis is independent of the concentration and

[81] D. J. Cram, R. D. Trepka and P. St. Janak, *J. Amer. Chem. Soc.*, **88**, 2749 (1966).

[82] D. J. Cram, R. D. Trepka and P. St. Janak, *J. Amer. Chem. Soc.*, **86**, 2731 (1964).

[83] H. L. Goering, D. L. Towns and B. Dittmar, *J. Org. Chem.* **27**, 736 (1962).

[84] A. Lapworth, *J. Chem. Soc.*, **85**, 30 (1904).

the nature of the halogen. Later, working also in acid catalysis, Ingold and Wilson[85] demonstrated that α-bromination of 2-*o*-carboxybenzylindan-1-one **44** has the same rate, within the limits of experimental errors, as the acid-catalysed racemisation. Hsü and Wilson[86] noted a similar behaviour under base catalysis. Bartlett and Stauffer[87] observed that phenyl-s-butylketone **45** presents equal racemisation and iodination rates in acid catalysis, whereas Hsü, Ingold and Wilson[88] established for the same ketone identical rates in the base-catalysed halogenation, deuteration and racemisation.

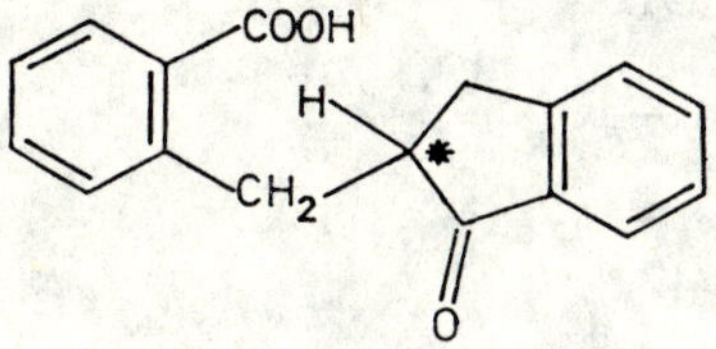

COOH, H, CH_2, *, O

44

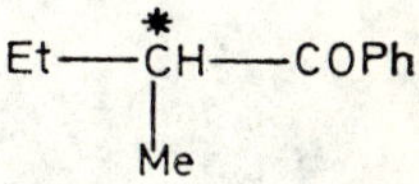

$$\mathrm{Et—\overset{*}{C}H(Me)—COPh}$$

45

A common intermediate in these reactions must therefore be admitted. In the base-catalysed reaction, the intermediate will be the delocalised (and planar) anion **46**, formed in the rate-determining step. The later steps involving the carbanion (protonation leading to racemic products, deuteration or bromination) will not be reflected in the rate equation (Scheme 1).

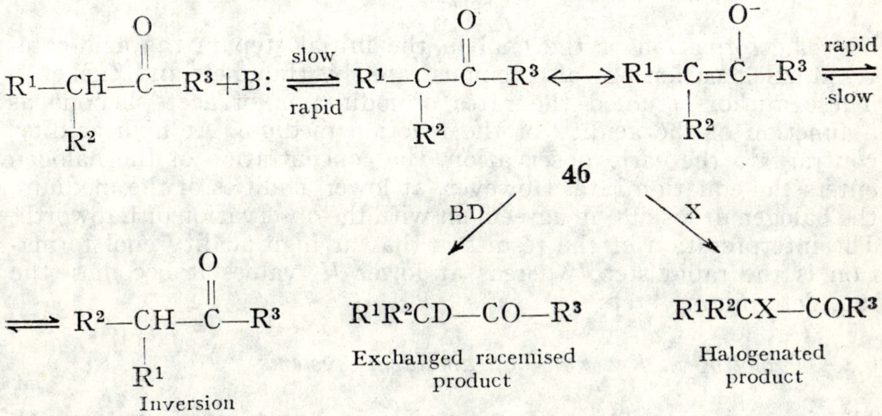

$$R^1—CH(R^2)—C(=O)—R^3 + B: \underset{\text{rapid}}{\overset{\text{slow}}{\rightleftharpoons}} R^1—\bar{C}(R^2)—C(=O)—R^3 \longleftrightarrow R^1—C(R^2)=C(O^-)—R^3 \underset{\text{slow}}{\overset{\text{rapid}}{\rightleftharpoons}}$$

46

BD ↙ ↘ X

$$\rightleftharpoons R^2—CH(R^1)—C(=O)—R^3 \qquad R^1R^2CD—CO—R^3 \qquad R^1R^2CX—COR^3$$

Inversion — Exchanged racemised product — Halogenated product

rate = *k* [Ketone] [Base]

Scheme 1

[85] C. K. Ingold and C. L. Wilson, *J. Chem. Soc.*, 773 (1934).
[86] S. K. Hsü and C. L. Wilson, *J. Chem. Soc.*, 623 (1936).
[87] P. D. Bartlett and C. H. Stauffer, *J. Amer. Chem. Soc.*, **57**, 2580 (1935).
[88] S. K. Hsü, C. K. Ingold and C. L. Wilson, *J. Chem. Soc.*, 78 (1938).

In the halogenation of ketones in acid catalysis, it has been since long assumed the intervention of the corresponding transient enols, arising in a slow step involving protonation and deprotonation of the ketone (Scheme 2).

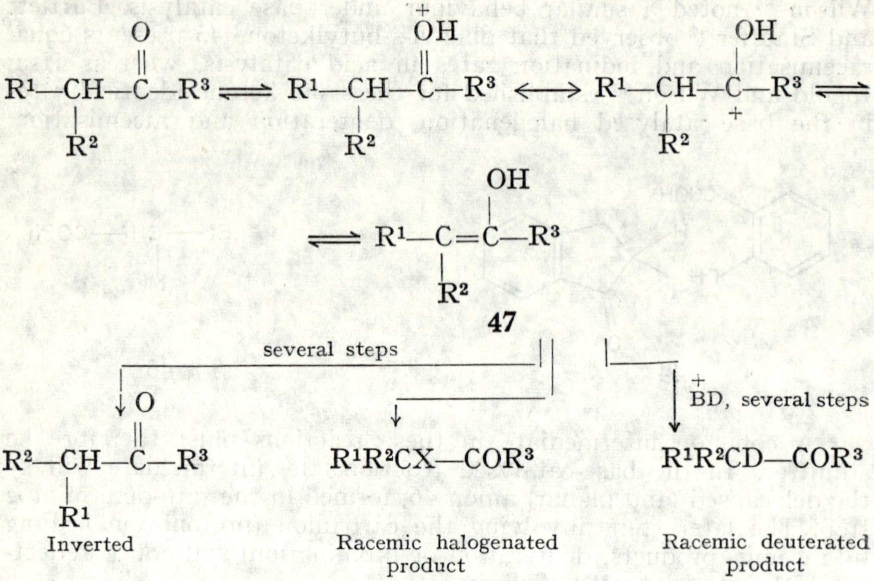

$$R^1\text{—}\underset{\underset{R^2}{|}}{CH}\text{—}\overset{\overset{O}{\|}}{C}\text{—}R^3 \rightleftharpoons R^1\text{—}\underset{\underset{R^2}{|}}{CH}\text{—}\overset{\overset{\overset{+}{O}H}{\|}}{C}\text{—}R^3 \longleftrightarrow R^1\text{—}\underset{\underset{R^2}{|}}{CH}\text{—}\overset{\overset{OH}{|}}{\underset{+}{C}}\text{—}R^3 \rightleftharpoons$$

$$\rightleftharpoons R^1\text{—}\underset{\underset{R^2}{|}}{C}=\overset{\overset{OH}{|}}{C}\text{—}R^3$$

47

several steps; $\overset{+}{B}D$, several steps

$$R^2\text{—}\underset{\underset{R^1}{|}}{CH}\text{—}\overset{\overset{O}{\|}}{C}\text{—}R^3 \qquad R^1R^2CX\text{—}COR^3 \qquad R^1R^2CD\text{—}COR^3$$

Inverted — Racemic halogenated product — Racemic deuterated product

rate = k [Ketone] [H$^+$] or rate = k' [Ketone] [Electrophile]

Scheme 2

The formation of the enol in the initial step in the acid-catalysed halogenation was also demonstrated by Hammett and Zucker[89]. These authors studied the rates of iodination of acetophenone as a function of the acidity of the reaction medium. At high acidity, contrary to the early observation, the concentration of the halogen enters the equation rate. However, at lower acidities of the medium, the halogen drops out, in agreement with the observation of Lapworth. The interpretation of the results is that at high acidity enol formation is the rapid step, whereas at lower H_0 values it becomes the rate-determining step.

6.3.4 *Prototropy. Reactions of ambident systems*

The isomerisation by proton transfer in unsaturated systems **48** may in its turn be base-catalysed. Degenerated allyl or heteroallyl anions must be admitted:

$$\underset{\mathbf{48}a}{A{=}B{-}C{-}H} + B\!: \rightleftharpoons BH^+ + \underset{\mathbf{49}}{\overbrace{A\cdots B\cdots C}^{-}} \rightleftharpoons \underset{\mathbf{48}b}{H{-}A{-}B{=}C} + B\!: \qquad (6.38)$$

[89] L. Zucker and L. P. Hammett, *J. Amer. Chem. Soc.*, **61**, 2785, 2791 (1939).

The phenomenon is called *prototropy*. Evidently, the equilibrating species are tautomers. The delocalised anion **49** in (6.38) can be protonated or can react with other electrophiles, at two different sites, either A or C. Such anions are called *ambident anions* [90]. The most intriguing fact is that depending on the nature of the electrophile, and on the reaction conditions, an ambident anion may be attacked in different manners. A kinetic *vs.* thermodynamic control is assumed [90].

Let us suppose that tautomer **48***b* is more acidic than **48***a*, *i.e.* A is more electronegative than C. In the mesomeric anion **49** the charge will therefore be higher at A than at C. The reaction with a strong electrophile will lead to an early transition state, that is a transition state closer to the allyl anion **49**. Reaction at A will thus have a lower activation energy than the attack at C (Figure 6.3). The kinetic control will therefore favour the isomer in which the electrophile is attached to the more electronegative atom (**48***b*).

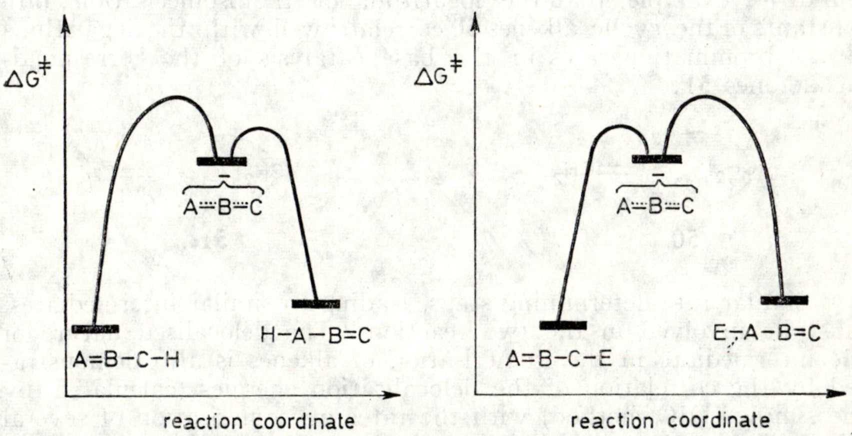

Figure 6.3 The reaction of an ambident anion with electrophiles. A. with proton (strong electrophile, kinetic control). B. with poor electrophiles (thermodynamic control).

Hammond's postulate indicates that very rapid reactions (like the reaction of **49** with strong electrophiles) have an early transition state; since both reactions of **49** are fast, the energy difference between the two transition states involving the allyl anion may be quite low. Now, if there is a high difference in acidity between the two tautomers, it follows that the more acidic component will be the thermodynamically unstable isomer. Kinetic control would therefore produce the thermodynamically unstable isomer. That is the case with simple oxaallyl anions (keto-enol tautomerism).

According to the same postulate, the reaction of **49** with poor electrophiles has a late transition state, resembling the structure and energy of reaction products. Thus, if there is a high free-energy

[90] R. Gompper, *Angew. Chem.*, **76**, 412 (1964).

difference between the two isomers, the transition state leading to the more stable tautomer will be the lower in energy. The reactions with such reagents will therefore be subjected to thermodynamic control. Two reactions involving the same allyl anion, but two different electrophiles, may thus lead to quite different results.

The simplest isomerisations involving allyl anions are the base-catalysed isomerisations of alkenes and acetylenes:

$$\text{H—}\overset{|}{\underset{|}{\text{C}}}\text{—}\overset{|}{\text{C}}\text{=C}\langle + \text{B:} \rightleftharpoons \rangle\overline{\text{C}\cdots\text{C}\cdots\text{C}}\langle + \overset{+}{\text{B}}\text{H} \rightleftharpoons \rangle\text{C=}\overset{|}{\text{C}}\text{—}\overset{|}{\underset{|}{\text{C}}}\text{—H} \quad (6.39)$$

$$\text{H—}\overset{|}{\underset{|}{\text{C}}}\text{—C}\equiv\text{C— + B:} \rightleftharpoons \rangle\overline{\text{C}\cdots\text{C}\equiv\text{C}}\text{— + }\overset{+}{\text{B}}\text{H} \rightleftharpoons \rangle\text{C=C=CH—} \quad (6.40)$$

The intervention of anionic intermediates has been convincingly demonstrated by Schriesheim and his co-workers[91, 92]. They have found, for example, that the logarithm of the isomerisation rate constants of the cyclic alkenes **50** correlate well with the logarithms of the bromination rates in the base catalysis of the corresponding ketones **51**:

$$\underbrace{(CH_2)_{n-1}\;C{=}CH_2}_{\textbf{50}} \qquad\qquad \underbrace{(CH_2)_{n-1}\;C{=}O}_{\textbf{51}}$$

Similar rate-determining steps, leading to similar intermediates, must be involved in the two reactions. The delocalised nature of the intermediate in the isomerisation of alkenes is also demonstrated by the correlation of the delocalisation energies (calculated by the simple HMO method) with the rates of isomerisation of several alkenes and polyenes.

The stereochemistry and the molecularity of the prototropic changes in the isomerisations of alkenes or in the rearrangement of acetylenes to allenes have been extensively studied by Cram and his co-workers[80, 93, 94]. It has been demonstrated by techniques similar to those described in Section 6.3.2 that the base-solvent system is of importance in determining the molecularity and the stereochemistry of the reaction. From rate studies of the isomerisation, deuterium-hydrogen exchange and racemisation reactions, it has been concluded that all mechanisms discussed are possible, depending

[91] A. Schriesheim, R. J. Muller and C. A. Rowe, Jr., *J. Amer. Chem. Soc.*, **84**, 3164 (1962).

[92] A. Schriesheim, C. A. Rowe, Jr. and C. Naslund, *J. Amer. Chem. Soc.*, **85**, 2111 (1963).

[93] D. J. Cram and R. T. Uyeda, *J. Amer. Chem. Soc.*, **84**, 4358 (1962); **86**, 5466 (1964).

[94] J. Almy, D. C. Garwood and D. J. Cram, *J. Amer. Chem. Soc.*, **92**, 4321 (1970).

on the particular reaction conditions. Thus, for the isomerisation of the indene derivative **52** [95] in MeOH—MeOK, the rate of deuterium incorporation is equal, within experimental errors, to the rate of racemisation ($k_e/k_\alpha = 1$). Completely racemised **53** is formed.

H Me B: Me (6.41)

CMe$_3$ H CMe$_3$

52 53

However, if the same reaction is performed in THF with Pr^nND_2, and stopped after 46% isomerisation, for the exchanged non-isomerised **52** one finds $k_e/k_\alpha > 30$, whereas either the non-exchanged or the exchanged (17%) **53** is formed with complete retention of configuration. Ion-pair reorganisation is again suggested. This supposition is confirmed by the effect of added crown ethers (known to solvate specifically metal or ammonium cations). Complete racemisation occurs in that case even in retaining base-solvent systems [94]. The solvation of the cation by the crown ether favours ion-pair dissociation and thus the loss of chirality.

The high stereospecificity in the proton transfer in these systems may be explained by a similar argument as that used by Drenth [95] to explain the stereochemistry of an S_N2' reaction (Figure 6.4). The interaction of the ionised proton attached to the base in the contact ion-pair, or of another proton donor, with the allyl system, may be regarded as the interaction of the bonding orbital of the proton donor with the π-MO of the original four-electron system. Clearly, the new proton will enter on the same side of the allyl system as that on which the old proton has left.

A somewhat related rearrangement is the isomerisation of azomethines, first studied by Ingold and his co-workers [96]. The similarity

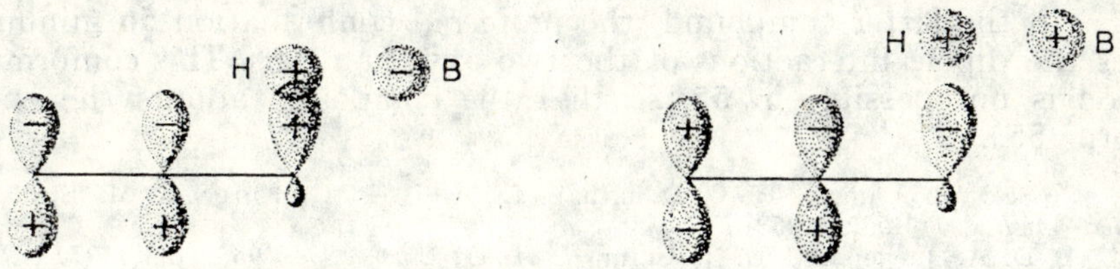

Figure 6.4. Frontier orbitals in the allyl-base system.

[95] W. Drenth, *Rec. Trav. Chim.*, **86**, 318 (1967).

[96] For earlier literature data, see C. K. Ingold, *Structure and Mechanism in Organic Chemistry*, Cornell University Press (1953), p. 572.

between the methylene-azomethine rearrangement and the biochemical transamination [97] was recognised long ago.

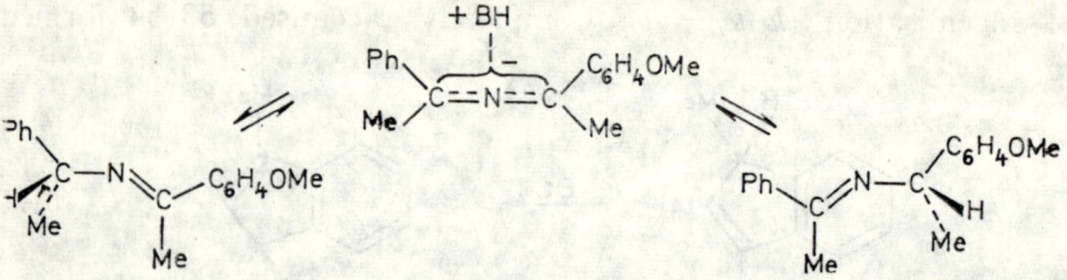

54 (6.41)

Recent studies [98, 99] have confirmed the intervention of a chiral azaallyl ion pair **54**, which may reorganise and suffer collapse to covalent species faster than ion-pair dissociation.

The earliest known and studied rearrangement in a heteroatomic system is the keto-enol equilibration [96, 100]. Simple ketones (such as acetone) contain only very small amounts of enol in the equilibrium mixture (Table 6.2). The enol may be favoured by the following factors:

(a) *The electronic system.* The more extended the π-electron system in the enol, as compared to the carbonyl form, the higher will be the ratio of the enol in the equilibrium mixture. In acetone, the enol represents only 0.00025%, whereas in acetylacetone it amounts to 80%. Phenol (the enol of cyclohexadienol) is practically 100% enol.

(b) *Conformational factors* seem to be responsible for the high ratio of enol in 1,2-cyclopentadione **55** as compared to diacetyl.

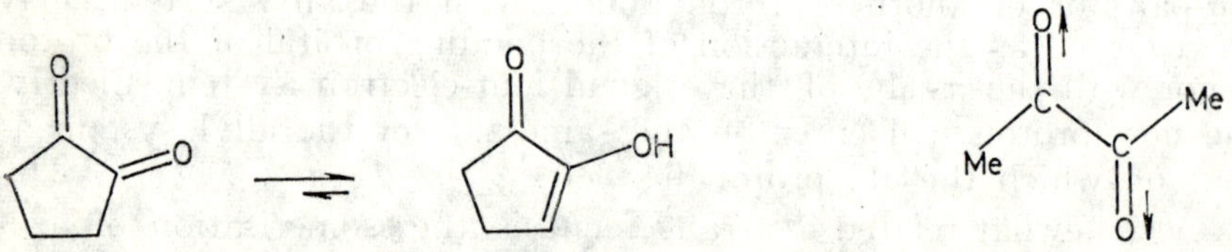

55*a* **55***b* **56**

In the latter compound, the preferred conformation **56** minimises the dipole interactions of the two oxygen atoms. This conformation is not possible in **55**, so that the compound adopts the enol form **55***b*.

[97] See, for example, H. C. Donathan, L. Davis, P. G. Kurg and M. Kaplan, *Biochemistry*, **7**, 4532 (1968).

[98] D. A. Jaeger and D. J. Cram, *J. Amer. Chem. Soc.*, **93**, 5153 (1971).

[99] R. D. Guthrie, D. A. Jaeger, W. Meister and D. J. Cram, *J. Amer. Chem. Soc.*, **93**, 5137 (1971); R. D. Guthrie and J. L. Hedrick, *J. Amer. Chem. Soc.*, **95**, 2971 (1973).

[100] See also J. Hine, *Physical Organic Chemistry*, McGraw-Hill, New York (1962), p. 243, and E. S. Gould, *Mechanism and Structure in Organic Chemistry*, Holt, Rinehart and Winston, New York (1959), p. 380.

(c) *Solvent*. The ratio of enol in ethyl acetoacetate (Table 6.2) indicates that solvent may also shift the equilibrium in favour of the enol. The less polar the solvent, the higher will be the ratio of the less polar enol.

TABLE 6.2

The enol ratio of some ketones *

Compound	*Solvent*	*%Enol*
CH_3COCH_3	Neat	0.00025
$CH_3OOCCH_2COOCH_3$	Neat	0.00
$CH_3COCH_2COOC_2H_5$	Neat	7.5
	Water	0.4
	Acetic acid	5.7
	Benzene	16.2
	Hexane	49
$CH_3COCH(CH_3)COOC_2H_5$	Neat	4
	Hexane	12
$CH_3COCH_2COCH_3$	Neat	80
Cyclopentanone	Neat	0.0048
Cyclohexanone	Neat	0.020
$CH_3COCOCH_3$	Neat	0.0056
1,2-Cyclopentadione	Neat	100

* Values cited in References 96 and 100.

The isomerisation of the enol into the stable ketone may be a stereospecific reaction [101, 102]. Zimmerman [101] observed that in the formation of 1-benzoyl-2-phenylcyclohexane from the corresponding enol **57***a*, only the unstable *cis* isomer **57***b* results.

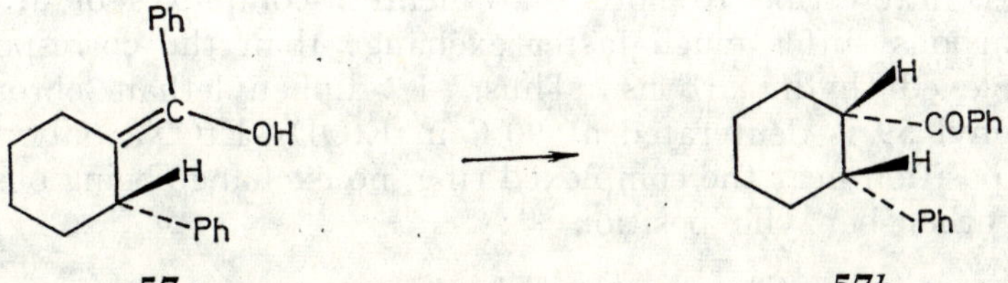

Ph C OH H Ph → H COPh H Ph

57*a* **57***b*

It must be thus assumed that the double bond in **57***a* is protonated from the less hindered part.

A more extended carbanion system is the benzyl anion [76, 77]. Isomerisations by proton transfer in such systems have also been studied [80]. An example is the 1,5-proton transfer in **58** leading to an aromatic compound:

[101] H. E. Zimmerman, *J. Org. Chem.*, **20**, 549 (1955).
[102] H. E. Zimmerman and T. W. Cutshall, *J. Amer. Chem. Soc.*, **81**, 4305 (1959).

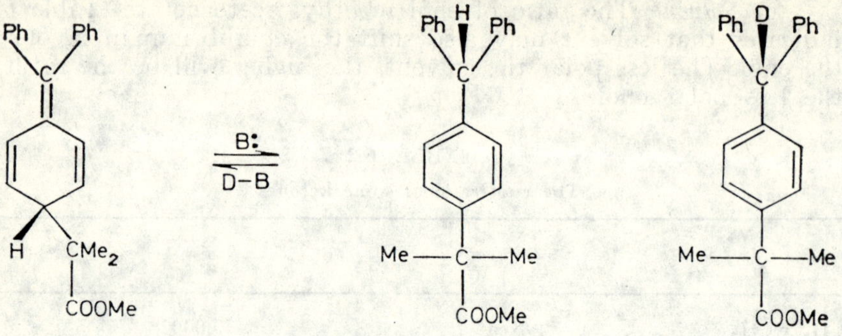

58

The absence of deuterium in the molecule indicates that the reaction possesses to a certain degree an intramolecular character. With highly hindered proton donors (for example Et_3COD) and tertiary amines, the reaction may be, within the limits of experimental errors, 100% intramolecular. Consideration of the symmetry of the highest occupied MO of the benzyl anion leads to the conclusion that the reaction may proceed in a concerted way (see also Section 12.2.1). The larger distance between the termini of the benzylic system may imply a step migration (1,3) of the type already discussed.

Benzylic anions are also involved in simple hydrogen-deuterium exchange in the α-position of arylalkanes. The more extended the conjugated system, the faster is the rate of the isotopic exchange. Thus, triphenylmethane exchanges tritium with hydrogen, in cyclohexyl amine and lithium cyclohexylamide, 10^3 faster than does toluene [103].

It is interesting to note that metal π-complexes of aromatic hydrocarbons suffer much faster exchange than the corresponding uncomplexed hydrocarbons. Thus, 1,2-diphenylethane-chromium-tricarbonyl **59** is deuterated at 90°C in EtOD—EtONa only in the benzyl position near the complexed ring, no exchange being observed in the other benzylic position [104]:

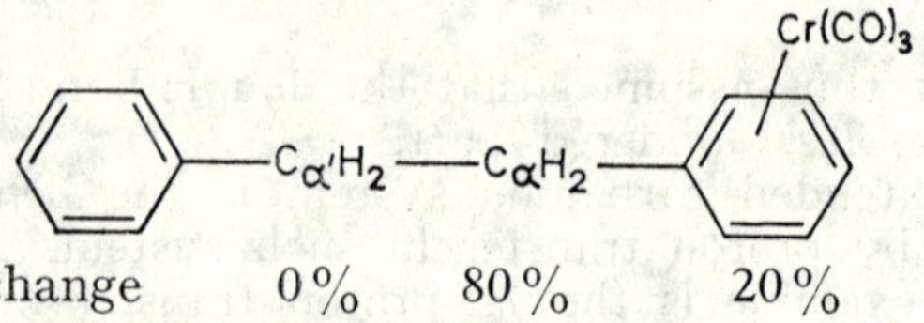

Exchange 0% 80% 20%

59

[103] A. Streitwieser, Jr., M. R. Grange, F. Mares and R. A. Wolf, *J. Amer. Chem. Soc.*, **95**, 4257 (1973), and the literature cited therein.

[104] S. Roșca, Ph. D. Thesis, Bucharest (1970); S. Roșca *et al.*, in the press.

According to Roşca [104], the higher acidity of the complexed hydrocarbons is also determined by a conjugation in which the metal bonds are involved:

$$R{-}H\ddot{C}^{-}{-}C_6H_5 \cdots \overset{-}{\ddot{C}r}{-}C{\equiv}\ddot{O}^{+}$$

60

6.3.5 *Anionic molecular rearrangements*

Molecular rearrangements of carbanions are by far less frequent than those of carbocations.

If in the cationic series a 1,2-shift may be regarded as an intramolecular S_N1 reaction, the anionic 1,2-shift must be considered as an intramolecular S_E1 reaction.

Thus, in the Wittig rearrangement [105, 106] of metallated benzyl ethers (in which α-elimination is not possible), a group R suffers a 1,2-shift from the oxygen atom:

$$PhCH_2{-}O{-}R^1 + R^2Li \rightarrow Ph\overset{Li^+}{\bar{C}}H{-}O(R^1) \rightarrow PhCH{-}O \cdots \bar{R}^1\ Li^+\ (\mathbf{61}) \rightarrow PhCH(R^1){-}O^-Li^+$$

In the transition state **61**, the migrating group R must accommodate partly the negative charge. This probably explains the migration aptitude of radicals R^1, which decreases in the sequence:

allyl > benzyl > methyl > ethyl > *p*-nitrophenyl > phenyl.

A well-known carbanionic rearrangement is the Favorsky rearrangement [107], which consists in the transformation of an α-halogenated ketone **62** into the rearranged acids **63** and **64**, under base catalysis.

$$\underset{\mathbf{62}}{>C_\alpha(X){-}CO{-}C_{\alpha'}H<} \xrightarrow{B:} \underset{\mathbf{63}}{>C_\alpha H{-}C_{\alpha'}(<){-}COOH} + \underset{\mathbf{64}}{>C_\alpha(COOH){-}C_{\alpha'}H<} \qquad (6.42)$$

[105] G. Wittig, *Angew. Chem.*, **66**, 10 (1954).
[106] G. Wittig *et al.*, *Liebigs Ann. Chem.*, **598**, 85 (1956); **605**, 69 (1957).
[107] A. S. Kende, *Organic Reactions*, **11**, 261 (1960).

A semibenzylic mechanism (6.43) was assumed by Tshoubar and Sackur [108] in 1939:

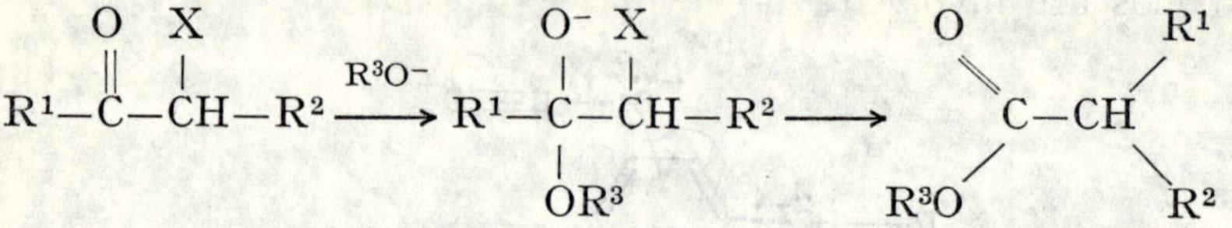

$$R^1-\overset{\overset{O}{\|}}{C}-\overset{\overset{X}{|}}{CH}-R^2 \xrightarrow{R^3O^-} R^1-\underset{\underset{OR^3}{|}}{\overset{\overset{O^-}{|}}{C}}-\overset{\overset{X}{|}}{CH}-R^2 \longrightarrow (O=)(R^3O)C-CH(R^1)(R^2) + X^- \qquad (6.43)$$

Later, by labelling the α-carbon atom in chlorocyclohexanone, Lotfield [109] observed a perfect scrambling of the label in the resulted cyclopentane-carboxylic acid ester. He concluded that a symmetric cyclopentanone must be the reaction intermediate:

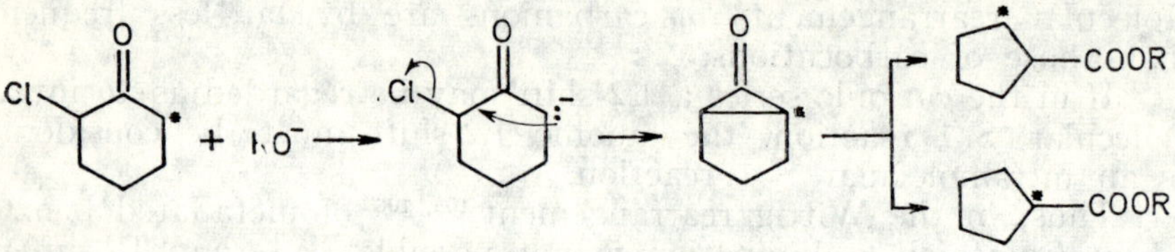

Cl, O, +, RO⁻, Cl, O, O, COOR, COOR

65

This mechanism, however, is possible only for α-halogenated ketones having an α′-hydrogen. Ketones with a trisubstituted α′-carbon atom probably adopt the semibenzylic mechanism. The occurrence of the two mechanisms has been demonstrated by Warnhoff, Wong and Tai [110] which have studied the Favorsky rearrangement of the bicyclic ketones **66**:

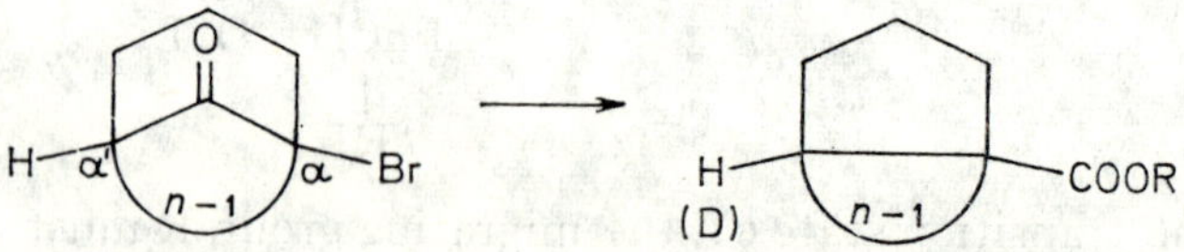

O, H, α′, α, Br, n − 1, H, (D), n − 1, COOR

66 **67**

$n = 6, 7, 8$

These ketones would have been expected not to yield easily carbanions, the conjugated oxaallyl anions violating the Bredt's rule. Indeed, ketone **66** with $n = 6$, treated with $MeOD-MeO^-Na^+$ or with Bu^tOK-Bu^tOD, affords a rearranged product **67** which has no deuterium in the molecule. The reaction mechanism does not seem to involve carbanions, the alternative semibenzylic pathway being probably preferred. If $n = 8$, the rearranged product does contain deuterium. If $n = 7$, exchange is not observed with MeOD—

[108] B. Tchoubar and O. Sackur, *Compt. Rend.*, **208**, 1020 (1939).
[109] R. B. Lotfield, *J. Amer. Chem. Soc.*, **72**, 632 (1950); **76**, 35 (1954).
[110] E. W. Warnhoff, C. M. Wong and W. T. Tai, *J. Amer. Chem. Soc.*, **90**, 514 (1968).

—MeONa, but it does occur with the stronger base system Bu^tOK—Bu^tOD. The latter results indicate a carbanionic mechanism. It has recently been shown [111] that Bredt's rule may be violated in larger ring systems. The stabilisation of the intermediate anions is then possible for the higher members of the **66** ketone series.

Bordwell and his co-workers [112, 113] have questioned the cyclopropanone-mechanism, considering that the stabilisation energy of the intermediate carbanion must be lost in the cyclisation to cyclopropanone. They assumed as more reasonable a concerted mechanism involving a zwitter-ion in which the oxaallyl conjugation is extended. Clearly, the zwitter-ion may collapse to cyclopropanone, or may further react to give the rearranged products (Scheme 3).

$$>CH-\overset{O}{\overset{\|}{C}}-CX< \xrightarrow{\text{base}} >\bar{C}-\overset{O}{\overset{\|}{C}}-CX< \xrightarrow{-X^-} >\bar{C}-\overset{O}{\overset{\|}{C}}-\overset{+}{C}< \longrightarrow \text{products}$$

For example:

$$ArCH_2-CO-CH_2Cl \longrightarrow Ar\bar{C}H-\overset{O}{\overset{\|}{C}}-\overset{+}{C}H_2 \xrightarrow[ROH]{RO^-} ArCH_2CH_2COOR$$

Scheme 3

It should be noted that the zwitter-ion intermediate has an electronic structure **68** similar to that of the singlet trimethylene methane **69**:

$$\underset{\mathbf{68}}{CH_2 \cdots \overset{O}{\overset{\vdots}{C}} \cdots CH_2} \qquad \underset{\mathbf{69}}{CH_2 \cdots \overset{CH_2}{\overset{\vdots}{C}} \cdots CH_2}$$

Bordwell and Scamehorn [112, 113] have checked this mechanism by studying the rearrangement of the aromatic ketones $ArCHCl-CO-CH_3$, substituted in the nucleus. They have found a linear free-energy relationship (6.43):

$$\log k/k_0 = \rho\sigma^+ + \rho'\sigma \qquad (6.43)$$

correlating the logarithms of the observed rate constants with both σ^+ ($\rho = -2\cdot73$) and σ constants. This indicates the development

[111] J. A. Marshall and H. Faubl, *J. Amer. Chem. Soc.*, **89**, 5965 (1967); J. R. Wiseman, *J. Amer. Chem. Soc.*, **89**, 5966 (1967); G. L. Buchanan, *Chem. Soc. Rev.*, **3**, 41 (1974).

[112] F. G. Bordwell and R. G. Scamehorn, *J. Amer. Chem. Soc.*, **90**, 6751 (1968).

[113] F. G. Bordwell, R. G. Scamehorn and W. L. Springer, *J. Amer. Chem. Soc.*, **91**, 2087 (1969).

of a positive charge in the rate-determining step. A similar result has been obtained with $ArCH_2COCH_2Cl$.

Related to the Favorsky rearrangement is the Ramberg-Bäcklund reaction[114, 115]. Instead of a carbonyl group, the substrates involved in this reaction have a sulphonic group. The reaction products are an alkene and sulphur dioxide. A simple ρ-σ correlation is found[116] ($\rho = +0.8$), suggesting the following mechanism:

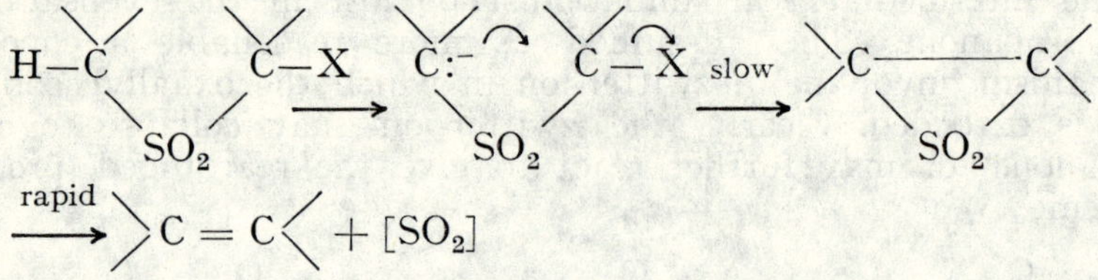

H—C(SO₂)C—X ⟶ C:⁻(SO₂)C—X ⟶(slow) C—C(SO₂) ⟶(rapid) C=C + [SO₂]

The zwitter-ionic intermediates in both Favorsky and Ramberg-Bäcklund rearrangements are supported by the *bishomoconjugative rearrangements*[117] of bromoketone **70** and of chlorosulphone **71**[118]. The rearranged products **72** are best explained[117, 118] by a mechanism in which the intermediate zwitter-ion **73** affords the *bishomobenzene* **74**:

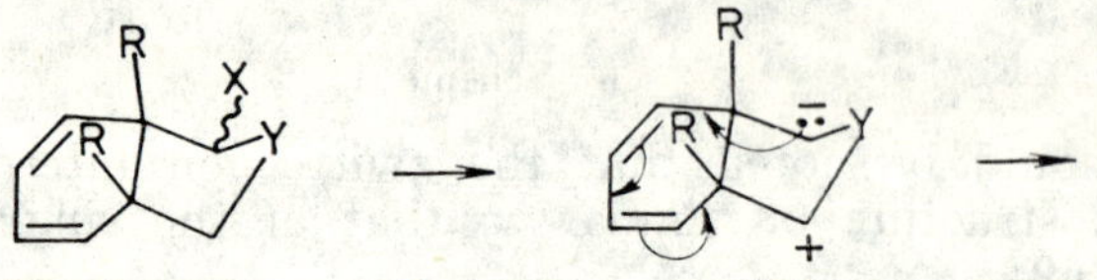

70 X = Br, Y = CO **73**
71 X = Cl, Y = SO_2

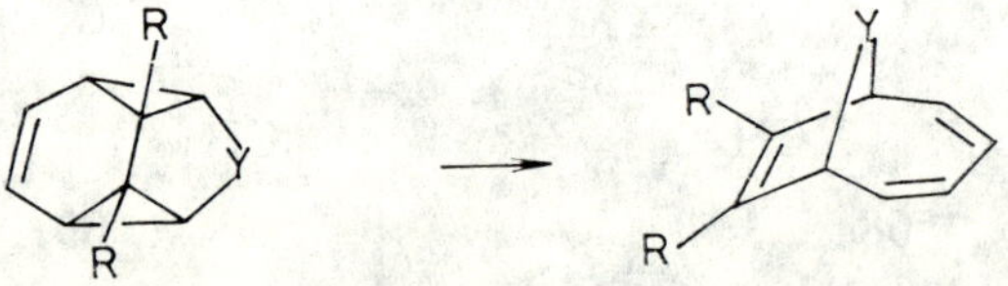

74 **72**

Scheme 4

An aromatic rearrangement, related to the S_E1 mechanism, has been described by Bunnett and his co-workers[119]. These authors have observed that polyhalogenobenzenes, such as 1,2,4-tribromo-

[114] L. Ramberg and B. Bäcklund, *Ark. Kemi, Miner., Geol.*, **13A**, 27 (1940); *Chem. Abstr.*, **34**, 4725 (1940).

[115] L. A. Paquette, *Accounts Chem. Res.*, **1**, 209 (1968).

[116] F. G. Bordwell, *Accounts Chem. Res.*, **3**, 281 (1970).

[117] L. A. Paquette, R. E. Wingard, Jr., and R. H. Meisinger, *J. Amer. Chem. Soc.*, **93**, 1047 (1971); **94**, 2155 (1972).

[118] L. A. Paquette, R. H. Meisinger and R. E. Wingard, Jr., *J. Amer. Chem. Soc.*, **95**, 2230 (1973).

[119] J. F. Bunnett *et al.*, *J. Amer. Chem. Soc.*, **93**, 1183, 1190, 1198, 1201 (1971).

benzene, may rearrange when treated with Bu^{t}OK in DMSO (in the given example to 1,3,5-tribromobenzene). Proton-transfer reactions, as well as bromine transfer, are probably involved in the process:

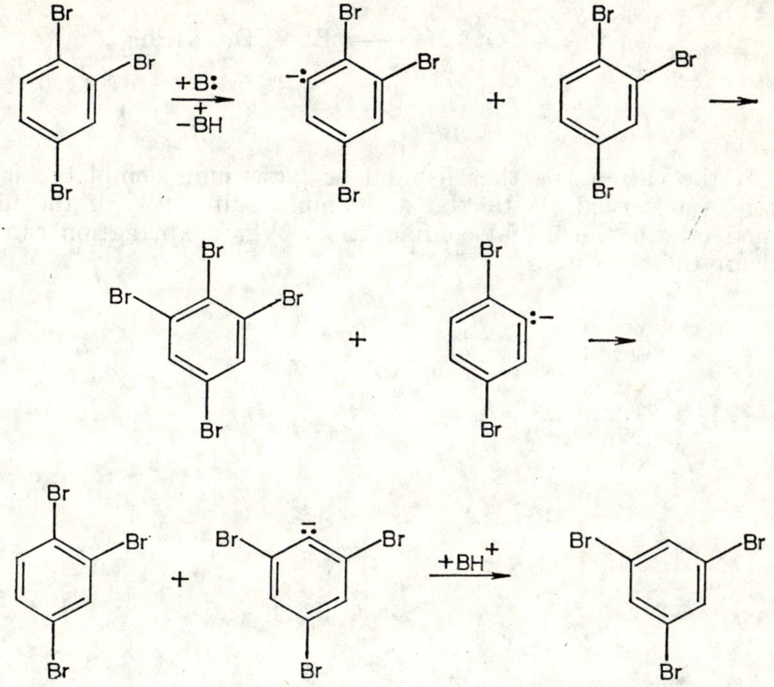

6.4 Problems

(1) Indicate why Winstein and Taylor (see p. 314) have used in their argument regarding the stereochemistry of the S_E2 reaction at the C—Hg bond, the *cis*-mercury derivative. Find an explanation for the fact that the *trans*-isomer was not suitable for the demonstration.

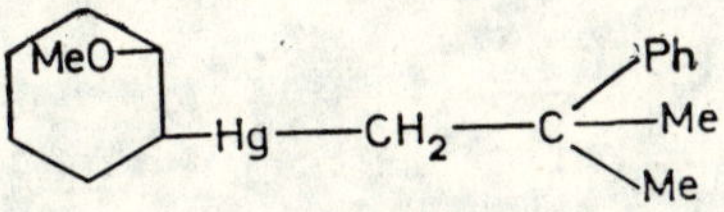

(2) Nortricyclene is dissolved in DF—SbF_5. The solution obtained is quenched with pyridine. Indicate what reactions may take place, the nature of the intermediates involved and the stereochemistry of the processes.

(3) Explain why the treatment of ethane with HF—SbF_5 leads to methane and hydrogen in the ratio 8:1. What conclusion may be drawn regarding an important technological process?

(4) On decarboxylation in a deuterated solvent, the optically active carboxylic salt A yields the deuterated nitrile B (D. J. Cram and P. Haberfield[66]):

$$\underset{\text{(A)}}{\mathrm{Et{-}\overset{\overset{\displaystyle CN}{|}}{\underset{\underset{\displaystyle Ph}{|}}{C}}{-}COO^{-}M^{+}}} \xrightarrow{\text{S–D}} \underset{\text{(B)}}{\mathrm{Et{-}CD(CN)Ph}}$$

If the cation is either lithium or potassium, complete racemisation is observed. With the ammonium salt, 10% of the nitrile formed retains the initial configuration. What explanation can you find for these results ?

7
Electrophilic aromatic substitution

7.1 Introduction

Among the reactions of aromatic compounds, electrophilic substitution was the most thoroughly investigated. The reason lies probably in the fact that electrophilic reactions are highly representative of the aromatic character. This problem, since Kekulé [1] proposed in 1865 the cyclic structure of benzene, has continuously puzzled chemists and challenged their perseverence, application and imagination. However, with rudimentary experimental tools and only valence theory and chemical intuition, even with enthusiasm and perseverence, it was hardly possible at that time to give a complete and correct interpretation of the structure and reactivity of organic compounds. Nevertheless, it must be noted that at the end of the last century, the symmetry of benzene was demonstrated and the orientation in the substitution of its derivatives established [2].

Thiele was the first chemist who derived the aromatic character of benzene from the general properties of unsaturated compounds, obtaining thus a glimpse of what we nowadays regard as positively proved. According to his theory [3], in unsaturated compounds the free affinity of the double bonded atoms is not completely consumed, determining their unsaturation and their addition reactions. In acyclic conjugated polyenes, this affinity can be satisfied within the molecule, between neighbouring atoms, except for the marginal ones. In benzene, conjugation is continuous and uninterrupted, which explains its unusual behaviour. In a sense, Thiele anticipated the different nature of the two bonds, one of them (corresponding to the π bond) being more reactive and being able to extend over more atoms (delocalised bond). Referring to his theory, Thiele wrote: "My theory may seem hazardous. If however one considers that a theory is a point of view which must enable a uniform survey of known facts and the prediction of new facts, a point of view whose meaning and value is allowed to be changed by the natural progress of the science, then I think that my conceptions are

[1] F. A. Kekulé, *Bull. Soc. chim. France*, **3**, 98 (1865); *Liebigs Ann. Chem.*, **137**, 169 (1866).

[2] For a historical outlook see C. K. Ingold, *Structure and Mechanism in Organic Chemistry*, Cornell University Press, Ithaca (1969), Chapters 4 and 6.

[3] J. Thiele, *Liebigs Ann. Chem.*, **306**, 86 (1898).

largely satisfying these claims". Clearly, the same claims are imposed upon the modern theories, as they were earlier.

Electronic theory and quantum mechanical treatments have enabled us to have a more complete and probably more correct interpretation of the reactivity of aromatic compounds based on their structure. These aspects will be dealt with in the following sections of this chapter.

Electrophilic substitutions at the aromatic ring usually imply the displacement of a proton by an electron-deficient species (Table 7.1).

TABLE 7.1

Examples of electrophilic aromatic reactions

Type	*Examples*	*Reagent*	*Carrier*
1	2	3	4
H/D isotope exchange	$ArH + DA \longrightarrow ArD + HA$	D^+	DA
Friedel-Crafts reactions:			
– alkylation	$ArH + RX \xrightarrow{AlCl_3} ArR + HX$	R^+	$[R-X-L]^+$, $R-X$
– acylation	$ArH + RCOX \xrightarrow{AlX_3} ArCOR + HX$	RCO^+	$[RCOXL]^+$, RCOX
Nitration	$ArX + HNO_3 \longrightarrow ArNO_2 + H_2O$	NO_2^+	NO_2X, $NO_2\overset{+}{O}H_2$, NO_2OH.
Halogenation	$ArH + X_2 \longrightarrow ArX + HX$	X^+	$[X-X-L]^+$, X^+OH_2, $X-OY$, $X-X$
Sulphonation	$ArH + H_2SO_4 \longrightarrow ArSO_3H + H_2O$	SO_3	H_2SO_4
Diazo-coupling	$Ar^1H + Ar^2N_2^+ \longrightarrow Ar^1N{=}NAr^2 + H^+$	ArN_2^+	–
Kolbe-Schmitt carbonation	$ArH + CO_2 \longrightarrow ArCO_2^- + H^+$	CO_2	–
Nitrosation	$ArH + HNO_2 \longrightarrow ArNO + H_2O$	NO^+	NOX, $NO\overset{+}{O}H_2$
Mercuration	$ArH + HgX_2 \longrightarrow ArHgX + XH$	HgX^+	HgX_2
Desulphonation	$ArSO_3H + H_2O \longrightarrow ArH + H_2SO_4$	H^+	HA
Oxidative nitration	$ArR + HNO_3 \longrightarrow ArNO_2 + (ROH)$	NO_2^+	NO_2X, etc.
Decarboxylation	$ArCOOH \longrightarrow ArH + CO_2$	H^+	HA

HA = Brönsted acid, L = Lewis acid, X = halogen, Y = H, alkyl, acyl

With only a few exceptions, most of the reactions listed in Table 7.1 are of great technological interest [4,5] in the manufacture of chemicals such as synthetic dyes, pharmaceuticals, monomers, etc. This has also stimulated the fundamental research on electrophilic substitution. Numerous reviews, monographs and books, treating the different theoretical aspects of such reactions, are now available[6–10]. It should also be mentioned that some reactions of alkenes[11,12] are directly related to aromatic electrophilic substitution.

7.2 Particular characteristics of the electrophilic aromatic substitutions

In the study of the electrophilic substitutions, the same type of methods (product isolation and identification, kinetics and thermodynamics, isotope effects, medium effects, etc.) can be used as for the nucleophilic reactions. However, aromatic substitution presents some particular characteristics which must be considered in the planning and interpretation of experiments.

7.2.1 *The electrophile*

In contrast with the nucleophilic substitution, in an electrophilic substitution the actual reagent is pre-formed under the reaction conditions, by rapid or slow equilibrium reactions, as for example:

$$HONO_2 + H^+ \rightleftharpoons H_2O + NO_2^+ \quad (7.1)$$

$$X{-}X + MX_n \rightleftharpoons X^+ + MX_{n+1}^- \quad (7.2)$$

$$R{-}X + MX_n \rightleftharpoons R^+ + MX_{n+1}^- \quad (7.3)$$

Another complication is that the electrophiles are not the only reagents which may be involved in the process. Less reactive

[4] P. H. Groggins, *Unit Processes in Organic Synthesis*, McGraw-Hill, New York (1947).

[5] K. Venkataroman, *The Chemistry of Synthetic Dyes*, Academic Press, New York (1952).

[6] E. S. Gould, *Mechanism and Structure in Organic Chemistry*, Reinhart and Winston, New York (1959), Chapter 11; J. Hine, *Physical Organic Chemistry*, McGraw-Hill, New York (1962), Chapter 16; see also C. K. Ingold [2].

[7] P.B.D. de la Mare and H. H. Ridd, *Aromatic Substitution*, Butterworth, London (1959).

[8] G. A. Olah (editor), *Friedel-Crafts and Related Reactions*, vol. 1, *General Aspects*, Interscience, New York (1963).

[9] G. A. Olah (editor), *Friedel-Crafts and Related Reactions*, vol. 2, *Alkylation*, Interscience, New York (1964).

[10] G. A. Olah (editor), *Friedel-Crafts and Related Reactions*, vol. 3, *Acylation*, Interscience, New York (1964).

[11] B. T. Brooks, C. E. Boord, S. S. Kurtz and L. Schmerling (editors), *The Chemistry of Petroleum Hydrocarbons*, Reinhold, New York (1955).

[12] S. Patai, *The Chemistry of Alkenes*, Interscience, New York (1964).

reagents, preceding the actual electrophile in equilibrium reactions of type (1)—(3), may successfully compete with it for the aromatic substrate. Indeed, reactions (1)—(3) can be transcribed in the general form (4):

$$E{-}X + L^+ \rightleftharpoons E^+ + X{-}L \qquad (7.4)$$

where E—X is a *carrier* of the electrophile E^+, and L^+ is a Lewis or a Brönsted acid. Such reactions usually occur in two or more steps:

$$E{-}X + L^+ \rightleftharpoons \underset{\mathbf{1}}{[E{-}X{-}L]^+} \qquad (7.4a)$$

$$[E{-}X{-}L]^+ \longrightarrow E^+ + X{-}L \qquad (7.4b)$$

where **1** is another carrier formed before E^+. Species E—X and **1** are in their turn electrophilic reagents, able to react with nucleophiles, for example:

$$Nu{:}^- + E{-}X \longrightarrow Nu{-}E + X^- \qquad (7.5)$$

In the series: E—X, $E{-}X{-}L^+$, E^+, the former will be the most stable but the less reactive reagent, whereas E^+ will possess the highest reactivity and the lowest stability. Clearly, **1** may exceed by far E^+ in concentration and, possessing a certain electrophilicity, may replace E^+ in the reactions with aromatic compounds. With highly reactive substrates, even E—X can be a potential reagent in the aromatic substitution.

Another complication in the aromatic substitution is that sometimes the same reagent E—X may react with the aromatic substrate either by a nucleophilic mechanism or by a homolytic mechanism. Indeed, in the absence of any catalyst, the molecule E—X may be broken into free radicals or free atoms which will further substitute the aromatic ring (see Section 11.4.3).

Thus, in any study of an aromatic substitution, first it must be proved that the reaction under investigation is actually an electrophilic substitution and, if so, the substituting reagent (electrophile or carrier) has to be identified. Product identification (orientation) and the determination of the equation rate are the usual means of obtaining these pieces of information.

7.2.2 *Orientation*

Due to its symmetry (D_{6h}), all six positions of the benzene ring are equivalent, each hydrogen atom being substituted with equal probability. In benzene derivatives, however, the potential reaction centres are not equivalent, either on symmetry or on reactivity grounds. It has long been known that the nature of the substituent already present in the molecule may determine the position(s) which a new substituent chooses. There are two broadly different types of substituents, one (first-order substituents) orienting *ortho* and *para*, the other (second-order substituents) orienting *meta*. Hollemann[13] was the first to recognise a relationship between the type of orientation and the reactivity of the aromatic substrate *ortho-para* substitution is usually associated with activation, while *meta* orientation with deactivation of the aromatic compound, as compared with benzene. The interdependence of orientation and reactivity indicates that orientation is closely related to the mechanism involved in the attack of the aromatic ring.

It is reasonable to assume that a substituent may alter the electron distribution in the aromatic ring by way of certain electronic interactions. There are two possible mechanisms of electronic interactions. The *inductive mechanism* (*I*) is the result of a dipole, induced either by an electron-withdrawing substituent (—*I* effect), 2, or an electron-repelling one (+ *I* effect), 3. The effect may be transmitted both through single and double bonds, decreasing, however, with distance:

$$\overset{\delta-}{X} \longleftarrow \overset{\delta+}{C} \longleftarrow \overset{\delta\delta+}{C} \longleftarrow \overset{\delta\delta\delta+}{C} \qquad\qquad \overset{\delta+}{X} \longrightarrow \overset{\delta-}{C} \longrightarrow \overset{\delta\delta-}{C} \longrightarrow \overset{\delta\delta\delta-}{C}$$

2 **3**

The other mechanism, the *conjugative displacement* (*E*), is encountered in unsaturated compounds and involves either electron displacements from the substituent towards the conjugated system (+*E*) **4**, or from the conjugated system towards the substituent (—*E*) **5**.

$$\ddot{X}{-}C{=}C{-}C{=}C \qquad\qquad X{-}C{=}C{-}C{=}C$$

4 **5**

In this case the electronic effect does not depend on the distance which separates the substituent from a position of the conjugated system. The induced charge density (see p. 80) may be estimated by simple HMO calculations, using the heteroatom method. The same method can be applied to the inductive effect of a substituent directly bonded to a conjugated chain.

An electron-seeking reagent will choose the positions having the highest electron density. According to Ingold [2], the observed orientations and reactivities in the aromatic substitution can be explained only by an electrophilic mechanism, which takes into consideration the possible electronic effects of the substituent (Table 7.2). Electron-repellent substituents (+*I*, or +*I* and +*E* effects) are *ortho*- and *para*-orienting and they increase the reactivity of the ring. Electron-withdrawing groups (—*I* or —*E* effects) are *meta*-orienting, with deactivation. A special situation arises when the two electronic effects are opposed (—*I*, + *E*). With weak or moderate inductive effects, the conjugative mechanism is decisive in orienta-

[13] A. F. Hollemann, *Die Direkt Einführung von Substituenten in den Benzolkern*. Veit, Leipzig (1910).

TABLE 7.2

Electronic effects of a substituent in the aromatic electrophilic substitution

Order of the substituent	*Electronic effect*	*Example and electronic mechanism*	*Effect on* orientation	*Effect on* reactivity
First-order substituents (saturated groups, such as alkyl groups, or groups with unshared electrons at the atom directly bonded; π-systems)	$+I$	R; δ−, δ−, δ−	*o, p*	activation
	$+I$; $+E$	$:\ddot{O}:^-$	*o, p*	high activation
	$-I$ (strong) $+E$ (moderate)	:ÖMe	*o, p*	deactivation
	$-I$ (moderate) $+E$ (strong)	:Cl:	*o, p*	activation
Second-order substituents (unsaturated groups with full or formal positive charge at the atom directly bonded)	$-I$	$\overset{+}{N}R_3$; δ+, δ+, δ+	*m*	deactivation
	$-I$; $-E$	O, $\overset{+}{N}$, O^-	*m*	deactivation
	$+I$; $-E$	O, C, O^-	*m* or *o, p*	deactivation or activation, respectively.

tion, even if the inductive effect may be reflected in the reactivity. Halogens are thus *ortho-para* orienting due to a $+E$ effect, opposed to a strong $-I$ effect. The latter, however, determines the decrease of the reactivity.

The type of orientation observed in an aromatic substitution is used even today as a sound argument in the demonstration of the type of substitution. Indeed, in the nucleophilic aromatic substitution (Chapter 8) the orientation is reversed, first-order substituents orienting *meta,* with deactivation, second-order substituents being *ortho-para* orienting, with activation. The homolytic substitution is less regiospecific, *meta* and *para* products being found in about the same amounts (Chapter 11).

In this respect, some examples are illustrative:

(a) The thermal decomposition of 3,5-dihalogenobenzene 1,4-diazo oxides **6** in benzene or other aromatic compounds, ArH, yields 4-hydroxybiphenyls **7** beside polymers of type **8**. A homolytic mechanism has been earlier suggested [14], considering that radicals **9***a* should be stable. However, a more systematic study of the reaction [15] revealed that the orientation in the aromatic compound ArH is *ortho-para* with activation for the first-order substituents in ArH, and *meta* with deactivation for the second-order substituents. This indicates that **6** generates an electrophile **9***b* and that the reaction is an electrophilic substitution:

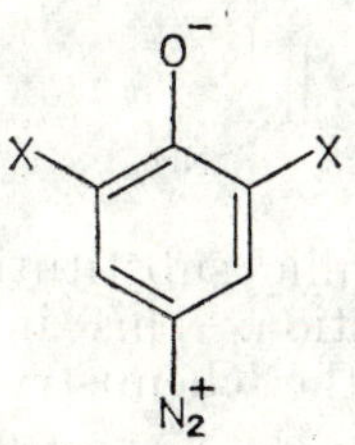

6 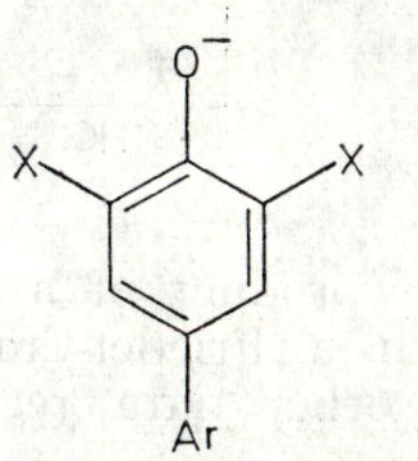 **7** 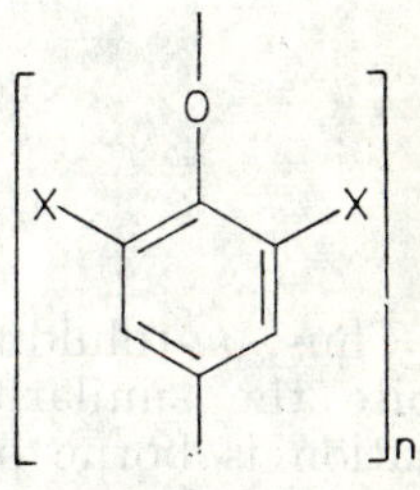 **8**

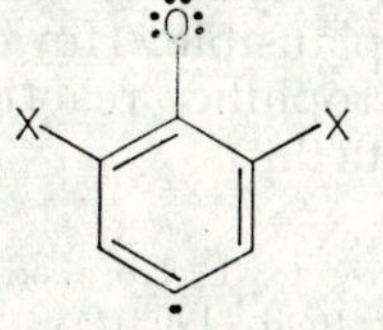

9*a* 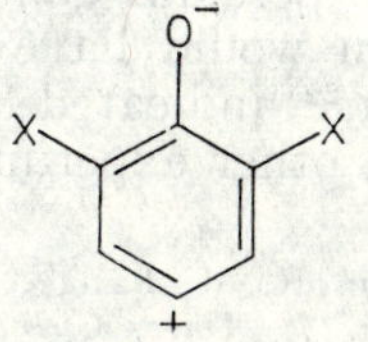 **9***b*

(b) The iodination of aromatic compounds may be performed with iodine in presence of peracetic acid [16]. These conditions would indicate a homolytic reaction, knowing that molecular iodine disso-

[14] M. J. S. Dewar and A. N. James, *J. Chem. Soc.*, 917, 4265 (1958).

[15] M. J. S. Dewar and K. Narayanaswami, *J. Amer. Chem. Soc.*, **86**, 2422 (1964).

[16] Y. Ogata and K. Aoki, *J. Amer. Chem. Soc.*, **90**, 6187 (1968).

ciates easily into free atoms and that peracetic acid is a good initiator for radical reactions. However, the observed orientation points to an electrophilic substitution, in good agreement with the kinetics of the reaction.

(c) Toluene may be aminated with chloramine in presence of aluminium chloride [17]. The catalyst used in the reaction would suggest an electrophilic substitution. However, the amination yields *m*-toluidine, in contrast with the expected orientation in an electrophilic substitution. In view of the fact that *p*-chlorotoluene is also formed in the reaction, the following mechanism has been proposed [18, 19]:

$$NH_2Cl \longrightarrow NH_2^- + Cl^+$$

$$Ph{-}Me + Cl^+ \longrightarrow$$

Me, +, H, Cl

$$\longrightarrow H^+ +$$

Me, Cl

$$+ NH_2^- \quad - HCl \longrightarrow$$

Me, NH_2

Thus, *m*-toluidine is a product of a nucleophilic substitution, despite the similarity with a Friedel-Crafts reaction. This interpretation is borne out by other data regarding the chemistry of chloramines [20].

(d) The oxidation of aromatic compounds with peracids yields phenols[21]. Since peracids generate easily hydroxy radicals, a homolytic mechanism would have seemed plausible. However, the observed orientation[22] indicated an electrophilic reaction, in perfect agreement with other experimental data[21].

[17] P. Kovacic, R. M. Lange, J. L. Foote, C. T. Goralski, J. J. Hiller, Jr., and J. A. Leviski, *J. Amer. Chem. Soc.*, **86**, 1650 (1964).

[18] P. Kovacic, J. A. Leviski and C. T. Goralski, *J. Amer. Chem. Soc.*, **88**, 100 (1966).

[19] P. Kovacic and J. F. Gormish, *J. Amer. Chem. Soc.*, **88**, 3819 (1966).

[20] P. Kovacic *et al.*, *J. Amer. Chem. Soc.*, **95**, 2977, 3646 (1973), and references quoted therein.

[21] (a) H. Hart, *Accounts Chem. Res.*, **4**, 337 (1971); (b) D. Jerina, J. Daly, W. Landis, B. Witkop and S. Udenfriend, *J. Amer. Chem. Soc.*, **89**, 3347 (1969).

[22] (a) For a general review on bridgehead reactivity, see R. C. Fort, Jr., in *Carbonium Ions*, vol. 4 (G. A. Olah and P. von R. Schleyer, editors), Wiley-Interscience, New York (1973), p. 1783; (b) P. Beak and B. R. Harris, *J. Amer. Chem. Soc.*, **96**, 6363 (1974).

(e) Treatment of alkyl chloroformates with silver ion constitutes one of the most efficient methods of generating carbocations, comparable with the deamination reaction:

$$ROCOCl + Ag^+BF_4^- \longrightarrow ROCO^+ \longrightarrow R^+ + CO_2$$

The method is suitable for studying bridgehead carbocations[22a], such as 1-norbornyl carbocation. If the reaction is performed in the presence of an aromatic compound:

OCOCl $\xrightarrow{Ag^+BF_4^-}$ [1-norbornyl cation] $\xrightarrow{PhR}$ C_6H_4R

a mixture of 1-aryl norbornanes is formed. The ratio *o*:*m* is approximately 1:1, regardless of the pre-existing substituent. This indicates an electrophilic mechanism involving a highly reactive electrophile which, in agreement with theory, yields a nearly statistical ratio of isomers.

On the other hand, the generation of bridgehead aponortricyclyl ion **10** by the same method, in presence of aromatic compounds, affords a mixture af aponortricyclenes, with the *o*:*m*:*p* ratio varying from 9:2.5:1 to 2:1:1 [22b]. A net preponderance of the *ortho* isomer with respect to the other isomers (which cannot be explained by the electronic effect of the substituent) has also been observed in the aromatic homolytic substitution. It is thus inferred that a radical reagent must be involved:

OCOCl $\xrightarrow{Ag^+BF_4^-}$ **10***a* $\longrightarrow$ **10***b* $\xrightarrow{PhR}$ C_6H_4R (**10***a* $\longrightarrow$ products)

The formation of a bridgehead radical, more stable than a bridgehead cation, is a consequence of the geometry of the cationic centre located in the three-membered ring [22b].

7.2.3 *Intermediates*

In an aromatic substitution the carbon attacked is an sp^2 atom, able to change its hybridisation into sp^3. As in the nucleophilic aromatic substitution, intermediates should be expected to arise, in contrast with the bimolecular nucleophilic substitution at the saturated atom.

The first reference regarding the structure of the intermediate in the electrophilic substitution was made by Pfeiffer and Wizinger [23]. These authors, resuming earlier works of Gattermann [24], have shown that 1,1-diarylethylenes substituted in the aromatic ring with efficient electron-repelling groups, such as 1,1-bis-(*p*-dimethylamino)ethylene **11** yield, with bromine, salts (**12**), which further break down into products of substitution at the double bond **13**:

$$(\mathrm{Me_2N{-}C_6H_4})_2\mathrm{C{=}CH_2} \longrightarrow \left[\mathrm{Me_2\overset{+}{N}{=}C_6H_4{=}C(Ar){-}CH_2Br}\right]\mathrm{Br_3^-} \longrightarrow$$

11 **12**

$$(\mathrm{Me_2N{-}C_6H_4})_2\mathrm{C{=}CHBr}$$

13

They have concluded that aromatic substitution may proceed by a similar mechanism:

$$\mathrm{C_6H_6} + \mathrm{Br_2} + \mathrm{A} \longrightarrow \left[\mathrm{C_6H_6Br^+}\right]\mathrm{BrA^-} \longrightarrow \mathrm{C_6H_5Br} + \mathrm{HBr} + \mathrm{A}$$

In a more general form, used nowadays, it can be written:

$$\mathrm{ArH} + \mathrm{E^+} \underset{k_{-1}}{\overset{k_1}{\rightleftharpoons}} \left[\mathrm{Ar}\Big\langle\begin{matrix}\mathrm{H}\\ \mathrm{E}\end{matrix}\right]^+ \underset{k_2}{\overset{k_{-2}}{\rightleftharpoons}} \mathrm{ArE} + \mathrm{H^+} \qquad (7.6)$$

14

Treating the reactivity of organic compounds by the MO method, Wheland [25] has considered that species **14** correspond to a transition state. In his calculations, in order to avoid complications due to the nature of the electrophile, he admits that the activation energy should be proportional to the difference between the electronic energy of the monosubstituted derivative **15** undergoing substitution, and that of the models **16** of the transition state **14**. The secular determinants of the transition states were derived from the

[23] P. Pfeiffer and R. Wizinger, *Liebigs Ann. Chem.*, **461**, 132 (1928).
[24] L. Gattermann, *Ber. dtsch. chem. Ges.*, **22**, 1131 (1889).
[25] G. W. Wheland, *J. Amer. Chem. Soc.*, **64**, 900 (1942).

secular determinant of the starting material **15**, by cancelling out the rows and columns corresponding to the substituted positions.

15 **16***a* (*o*-substitution) **16***b* (*m*-substitution) **16***c* (*p*-substitution)

With certain exceptions, a good agreement between calculated and predicted reactivity and orientation has been obtained, by solving out the secular equations.

Brown and Brady[26] considered that species **14** are true intermediates and called them *σ complexes*, stressing on the fact that they arise by the formation of a σ bond between the aromatic compound and the electrophile. Alternative terms for such compounds are Wheland's complexes or benzenium ions (see p. 156).

One of the first experimental proofs for the formation of σ complexes was afforded by the study of the kinetic isotope effect. If the aromatic substitution would proceed in a single step, in which the formation of the new bond and the breaking of the old one are synchronous, deuterium or tritium atoms substituting hydrogen atoms from benzene must determine a kinetic isotope effect (k_H/k_T or $k_H/k_D > 1$). If **14** represents an intermediate instead of a transition state, two different situations may arise: (a) the formation of the intermediate is slower than its collapse into the substitution product; no isotope effect is observed, and (b) the intermediate is formed in a rapid step, but is converted slowly into the reaction products; a significant isotope effect is to be expected. The three possible mechanisms are represented in Figure 7.1. A positive answer regarding the actual

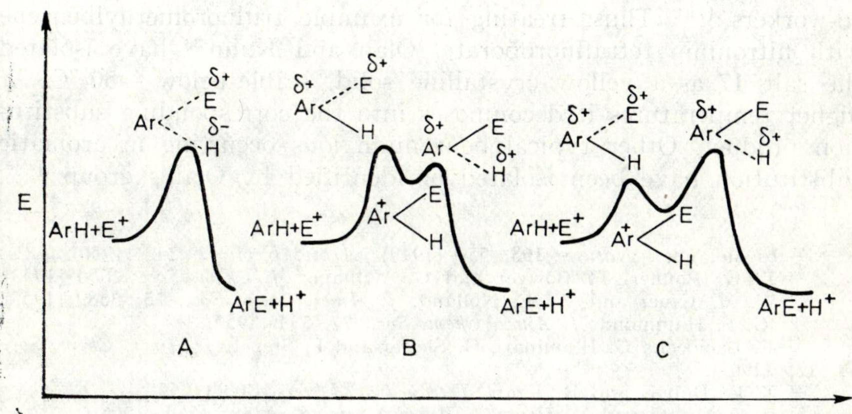

Figure 7.1 Potential-energy diagram of an electrophilic aromatic substitution. A. Direct displacement. B. Formation of a complex in the slow step. C. *Idem*, in a rapid step.

[26] H. C. Brown and J. D. Brady, *J. Amer. Chem. Soc.*, **74**, 3570 (1952).

reaction pathway can be obtained only if no isotope effect is observed.

Melander[27] has shown that the nitration and halogenation of benzene and tritiated benzene presents no isotope effect. That clearly indicates route B in Figure 7.1. Sulphonation, however, manifests kinetic isotope effects which should correspond either to mechanism A or to C depicted in Figure 7.1. It was however considered that sulphonation follows the same reaction pattern as nitration and halogenation, with the difference that the transformation of the intermediate into product is slower than its formation. Similar results have been observed with deuteroarenes[28, 29].

It was argued[30] that the kinetic isotope effects in sulphonation are not consistent with the formation of an intermediate. However, such a reactive intermediate will decompose into the product in a highly exothermic reaction, whose transition state will therefore resemble the intermediate (Hammond's postulate). Thus, the C—H bond is not stretched significantly in the transition state and no isotope effect is observed.

It is now well known that benzenium ions are common intermediates in all electrophilic substitutions. Kinetic isotope effects are determined by the stretching of the C—H bond in the transition state as well as by a more rapid decomposition of the intermediate into the starting materials than into the substitution product ($k_{-1} \gg k_{-2}$). The latter case may be the effect of a steric strain in **14** if a bulky substituent is close to the reaction centre. Steric repulsion will favour the elimination of the large E group instead of H. Adequate substitution of the aromatic substrate evidenced kinetic isotope effects even in halogenation[31, 32] and nitration[33].

Strong and direct evidence for the occurrence of benzenium ions in aromatic substitution has been obtained by Olah and his co-workers[34–37]. Thus, treating for example trifluoromethylbenzene with nitronium tetrafluoroborate. Olah and Kuhn[36] have isolated the salt **17** as a yellow crystalline solid, stable below −50 °C. At higher temperatures it decomposes into the corresponding substitution product. Other typical benzenium ions occurring in aromatic substitution have been isolated or identified by Olah's group[34–37],

[27] L. Melander, *Nature*, **163**, 559 (1949); *Arkiv. Kemi*, **2**, 213 (1950).
[28] T. G. Bonner, F. Bowyer and G. Williams, *J. Chem. Soc.*, 2650 (1953).
[29] W. M. Lauer and W. E. Nolland, *J. Amer. Chem. Soc.*, **75**, 3689 (1953).
[30] G. S. Hammond, *J. Amer. Chem. Soc.*, **77**, 334 (1955).
[31] E. Baciocchi, G. Illuminati, G. Sleiter and F. Stegel, *J. Amer. Chem. Soc.*, **89**, 125 (1967).
[32] J. E. Dubois and R. Uzan, *Tetrahedron Letters*, 309 (1965).
[33] P. C. Myhre and M. Beng, *J. Amer. Chem. Soc.*, **88**, 1569 (1966).
[34] G. A. Olah and W. Meyer, in Reference 8, p. 623.
[35] G. A. Olah, in Reference 8, p. 751.
[36] G. A. Olah and S. J. Kuhn, *J. Amer. Chem. Soc.*, **80**, 6535, 6540, 6541 (1958).
[37] G. A. Olah and S. J. Kuhn, *J. Amer. Chem. Soc.*, **82**, 2380 (1960).

demonstrating the existence of such intermediates and their easy conversion into substitution products.

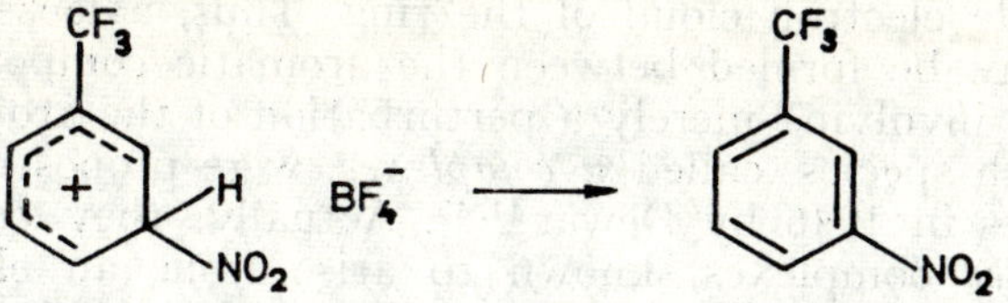

CF_3 ... H, NO_2 ... BF_4^- → CF_3 ... NO_2

17

The most simple benzenium ion **18** the intermediate of the electrophilic exchange of hydrogen, was obtained by protonation [38] in the strongest superacid media below 130 °C. Substituted benzenes [39, 40], anthracenes [40] and naphthalenes [41] are more easily protonated, their properties confirming the benzenium structure of the σ complexes.

A benzenium ion of high stability **19** was isolated by Doering and his co-workers [42] in the exhaustive polymethylation of benzene with methyl chloride and aluminium chloride. The salt is stable even in water containing 4 M hydrochloric acid. At lower acidity it may lose a proton in a reversible reaction, yielding methylenecyclohexadiene **20**.

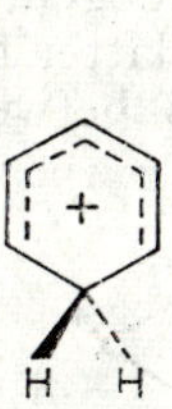

18 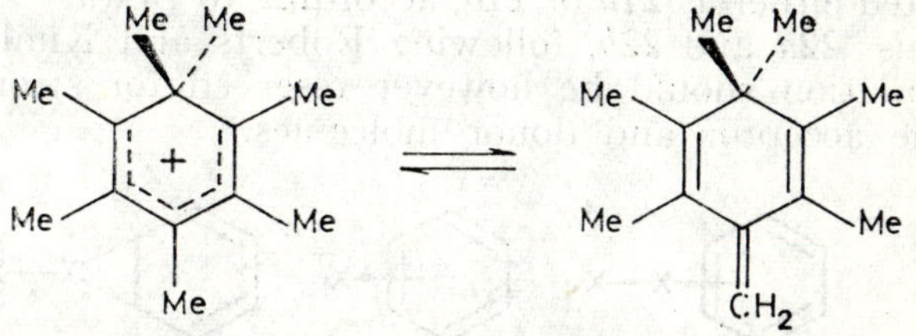 **19** **20**

The formation of σ complexes in the electrophilic aromatic substitution is equivalent to the formation of Meisenheimer's complexes in the nucleophilic mechanism (see p. 413). However, there must be a fundamental difference between the two processes, determined by the nature of the reagent. A nucleophile should be rejected by the π-electron cloud of the aromatic ring. The attack of such a reagent must then be oriented towards the electron-poor positions, the

[38] G. A. Olah, R. H. Schlosberg, D. R. Kelly and G. D. Mateescu, *J. Amer. Chem. Soc.*, **92**, 2546 (1970).

[39] G. A. Olah *et al.*, *J. Amer. Chem. Soc.*, **94**, 2034, 5341 (1972).

[40] D. N. Brouwer, E. L. Mackor and C. McLean, in *Carbonium Ions*, vol. 2 (G. A. Olah and P. von R. Schleyer, editors), J. Wiley, New York (1970), p. 873.

[41] G. A. Olah, G. D. Mateescu and Y. K. Mo, *J. Amer. Chem. Soc.*, **95**, 1865 (1973).

[42] W. von E. Doering, M. Saunders, H. G. Boyton, H. W. Earhart, E. F. Wadly, W. R. Edwards and G. Laber, *Tetrahedron*, **4**, 178 (1958).

geometry of the transition state resembling closely the geometry of the complex. By contrast, the geometry of the transition state of a Wheland complex may not be fixed, an electrophile being attracted by the whole electron cloud of the ring. Thus, a loose un-oriented complex may be formed between the aromatic compound and the electrophile, involving merely a perturbation of the aromatic π electrons[43]. Such species, called π *complexes,* were proposed as reaction intermediates in 1946 by Dewar[44, 45]. Actually, they correspond to *charge-transfer* complexes, known to arise from an electron-donor molecule (D) and an electron-acceptor compound (A)[46]. The charge-transfer bond is relatively weak and does not imply serious modifications of the molecular orbitals either in the acceptor or in the donor molecules:

$$D + A \rightleftharpoons D^+ - A^-$$

However, Brown[47] considers that in the charge-transfer complex the electrophile may be more strongly attracted towards the positions which are more rich in electrons, the π complex being relatively oriented.

The formation of π complexes is reversible and the equilibrium can be sometimes exactly measured[47–51]. Well known examples are the complexes of alkenes and arenes with silver ion and with halogens. The complexes with halogens (see also p. 373) are formulated either as **21***a* or **21***b*, according to Dewar[44, 45], or as halogenium ions, **22***a* and **22***b*, following Roberts and Kimball[52]. The latter formulation should be however reserved for stronger bonds between the acceptor and donor molecules.

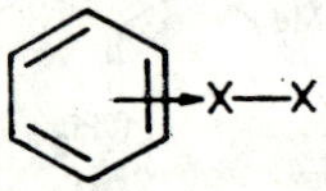

X—X X⁺ X—X X⁺

21*a* 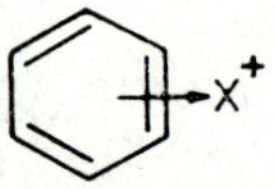 **21***b* 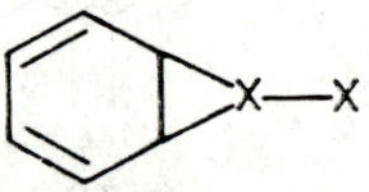 **22***a* 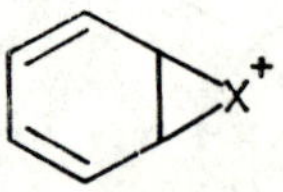 **22***b*

According to Olah[53], the transition state in an aromatic substitution is not fixed, but can vary from an *early* one, resembling the

43 A. Streitwieser, Jr., *Molecular Orbital Theory,* J. Wiley, New York (1961), p. 314.
44 M. J. S. Dewar, *J. Chem. Soc.,* 406 (1946).
45 M. J. S. Dewar, *The Electronic Theory of Organic Chemistry,* Oxford University Press, Oxford (1949).
46 R. S. Mulliken, *J. Amer. Chem. Soc.,* **72**, 600 (1950); **74**, 811 (1952).
47 H. C. Brown and H. Jungk, *J. Amer. Chem. Soc.,* **77**, 5579 (1955).
48 S. Winstein and H. J. Lucas, *J. Amer. Chem. Soc.,* **60**, 836 (1938).
49 N. Ogimachi, L. J. Andrews and R. M. Keefer, *J. Amer. Chem. Soc.,* **78**, 2210 (1956).
50 R. M. Keefer and L. J. Andrews, *J. Amer. Chem. Soc.,* **74**, 640 (1952).
51 G. A. Olah, in Reference 8, p. 853.
52 I. Roberts and G. E. Kimball, *J. Amer. Chem. Soc.,* **59**, 947 (1937).
53 G. A. Olah, *Accounts Chem. Res.,* **4**, 249 (1971).

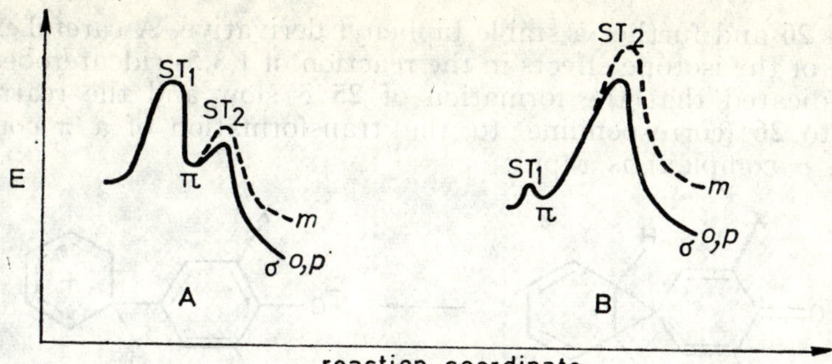

Figure 7.2 Potential-energy diagram of an electrophilic aromatic substitution; A. π complex formed in the slow step, σ complex in the rapid step. B. π complex formed in the rapid step, σ complex in the slow step.

π complex, to a *late* one with the character of a σ complex (Figure 7.2). The two different transition states may be distinguished from one another by the relationship between orientation and reactivity (see p. 366). A close experimental relationship between these types of data is to be expected only for late transition state reactions. This view, even if challenged by other workers (see Section 7.4.2), seems to be supported by numerous experimental data [54].

Quantum mechanical calculations [55] have shown that π complexes may be likely intermediates in the aromatic substitution. However, theoretical and experimental evidence in favour of such complexes is less substantial than for σ complexes [56]. One of the direct proofs supporting π complexes was found by Dewar and his co-workers [15], in the study of the thermal decomposition of 3,5-dihalogeno--1,4-diazo oxides. The electrophilic reagent **23** is a vinylogue of carbene **24**:

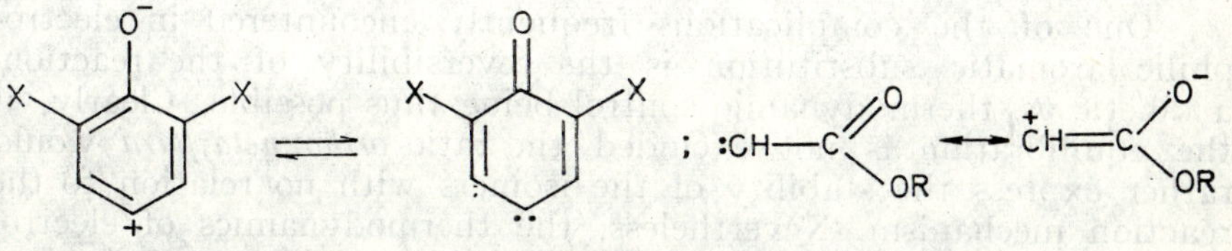

23 ; **24**

It is known [57] that **24** yields with benzene a norcaradiene derivative. A similar reaction may be expected between **23** and ArH, yielding the adduct **25** with the structure of a π complex. Unlike the norcaradiene products of **24**, the adduct **25** may yield the zwit-

[54] G. A. Olah, S. Kobayashi and J. Nishimura, *J. Amer. Chem. Soc.*, **95**, 564 (1973).
[55] R. D. Brown, *J. Chem. Soc.*, 2224, 2232 (1959).
[56] B. C. Chalis, *Annual Reports Chem. Soc.*, **64**, 115 (1967).
[57] M. J. S. Dewar and R. Pettit, *J. Chem. Soc.*, 2026 (1956).

ter-ion **26** and further a stable biphenyl derivative. A careful examination of the isotope effects in the reaction of 1,3,5-trideuterobenzene has indicated that the formation of **25** is slow and the rearrangement to **26** (corresponding to the transformation of a π complex into a σ complex) is rapid:

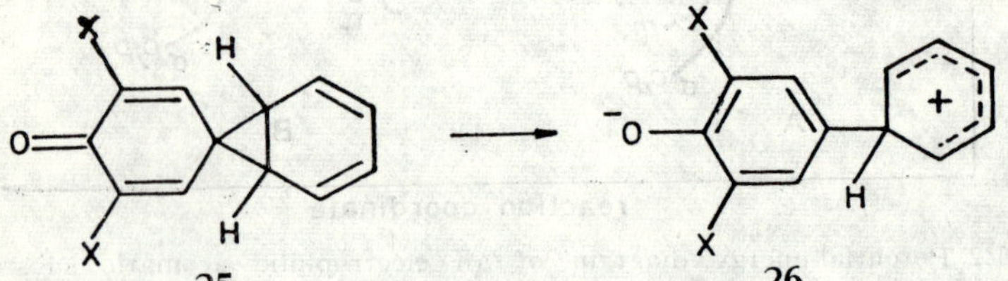

25 → **26**

This example, even if convincing, cannot be generalised to other electrophilic substitutions. Carbenes are known to be efficient reagents in cycloaddition reactions (see Section 12.6.1) and their reactivity cannot be extrapolated to other electrophiles.

7.2.4 *Thermodynamic versus kinetic control*

The determination of the relative reactivities of the different positions of the aromatic substrate *(positional selectivity)*, and of the relative overall reactivities as compared with the reactivity of benzene *(substrate selectivity)*, is of utmost importance in establishing the electrophilic character of the reaction and the type of its transition state (early or late). Thus, the exact determination of the ratios of the isomeric products (*ortho, meta* and *para*) is a necessary step in such studies. However, the ratio *ortho:meta:para* reflects the positional selectivity only if the mixture has resulted in a kinetically controlled reaction.

One of the complications frequently encountered in electrophilic aromatic substitution is the reversibility of the reaction, a kinetic *vs.* thermodynamic control being thus possible. Clearly, if the equilibration is not excluded, the ratio *ortho:meta:para* would rather express the stability of the isomers with no relation to the reaction mechanism. Nevertheless, the thermodynamics of electrophilic substitution is of practical and theoretical interest [58–60], since by choosing suitable reaction conditions either one or other isomer may be produced in high yields. A well-known example is the sulphonation reaction [61], which at short reaction times and moderate temperatures leads to the kinetically controlled product (corresponding to the more reactive position of the nucleus), whereas at higher temperatures and after longer reaction times the more stable product is obtained (thermodynamic control).

58 D. R. Stull, in Reference 8, p. 937.
59 D. A. McCaulay, in Reference 9, p. 1049.
60 H. P. Brandlen and E. T. McBee, in Reference 10, p. 1517.
61 K. LeRoi Nelson, in Reference 10, p. 1355.

One may wonder why certain reactions (such as sulphonation and Friedel-Crafts alkylations) are reversible even under mild conditions, whereas others, like nitration[62], seem to be completely irreversible in all known examples. The careful examination of the general mechanism (7.7) may offer some hints.

$$(L^-)E^+ + C_6H_6 \underset{k_{-E}}{\overset{k_E}{\rightleftharpoons}} \underset{\mathbf{27}}{[C_6H_6E]^+}\, L^- \underset{k_H}{\overset{k_{-H}}{\rightleftharpoons}} C_6H_5E + H^+ \qquad (7.7)$$

(a) Dependence on k_{-E}. The collapse of the intermediate **27** into the aromatic substrate and the electrophile E^+ corresponds to a unimolecular process S_N1. As has been shown before (p. 202), the rate constants of such reactions depend on the dissociation energy D and the ionisation potential I:

$$E\cdot \longrightarrow E^+ + e^- \quad \Delta H = I \qquad (7.8)$$

Highly electronegative groups E in the substitution product will not be removed from the aromatic ring by the electrophilic mechanism (7.7). An example is provided by the dehalogenation of halogenoarenes. It has been since long known that bromoarenes may be disproportionated in presence of Lewis acids such as $AlCl_3$ or $AlBr_3$[63–65]. More recently[66–71], Olah and his co-workers have shown that the halogenobenzenes manifest very different reactivity in the disproportionation:

$$C_6H_5X + H^+[AlBr_3OH]^- \longrightarrow [C_6H_5XH]^+ + C_6H_5X \longrightarrow C_6H_6 + p\text{-}C_6H_4X_2 + H^+$$

X = Cl, Br, I.

Thus, at 240 °C with water-activated $AlCl_3$ or $AlBr_3$, fluorobenzene remains unchanged, chlorobenzene is disproportionated only to an extent of 11%, whereas bromobenzene affords the dispropor-

[62] G. A. Olah and S. J. Kuhn, in Reference 10, p. 1393.
[63] O. von Domreicher, *Ber. dtsch. chem. Ges.*, **15**, 1866 (1882).
[64] C. Friedel and J. M. Crafts, *Ann. Chim. Phys.*, (6), **10**, 411 (1887).
[65] M. Kohn and N. L. Muller, *Monatsh.*, 30, 407 (1909); see, however, M. Ccmisarow and C. N. H. Long, *J. Chem. Soc.*, **119**, 442 (1921).
[66] G. A. Olah, W. S. Tolgyesi and R. E. A. Dear, *J. Org. Chem.*, **27**, 3441 (1962).
[67] G. A. Olah, W. S. Tolgyesi and R. E. A. Dear, *J. Org. Chem.* **27**, 3449 (1962).
[68] G. A. Olah, W. S. Tolgyesi and R. E. A. Dear, *J. Org. Chem.*, **27**, 3455 (1962).
[69] G. A. Olah and M. W. Meyer, *J. Org. Chem.*, **27**, 3464 (1962).
[70] G. A. Olah, J. C. Lepierre and C. G. Carlson, *J. Org. Chem.*, **30**, 541 (1965).
[71] G. A. Olah and T. E. Kiovsky, *J. Amer. Chem. Soc.*, **90**, 2583 (1968).

tionation products in more than 50% yield even at 30 °C[66]. The order of reactivity decreases therefore in the order PhI > PhBr > PhCl ≫ ≫ PhF, corresponding to the decreasing order of the dissociation energy for C—X bonds. The same dependence, but with a smaller gap in reactivity has been observed in the nucleophilic substitution at the saturated carbon. The larger differences in the reactivity of halogenobenzenes must be related to process (7.8). A fluorine atom, for example, has no tendency to lose its electron and form the positive ion F^+. In fact, no electrophilic fluorination has ever been observed and no defluorination reaction (7.7) of fluoro derivatives has been recorded in the literature. In the nucleophilic substitution, the tendency of fluorine to accept an electron compensates in part the energy required to break the C—F bond.

b) Dependence on the k_{-E}/k_{-H} ratio. The examination of the energy profiles in Figure 7.1. shows that once the intermediate **27** is formed, he would choose that pathway which has the lowest activation energy; thus, the higher the ratio k_{-E}/k_{-H}, the easier the reversible reaction will be. This is one of the reasons why sulphonation (with the energy profile C in Figure 7.1) is easily reversible, while halogenation (energy profile B) needs more drastic conditions for equilibration.

Experimentally it is possible to modify the ratio k_{-E}/k_{-H} in the desired direction. Bulky substituents at the reaction centre increase the ratio, as shown by the kinetic isotope effects. The close vicinity of a base to the reaction site will favour the deprotonation of the intermediate. This condition may be realised even in the more acidic solutions by preserving a low polarity of the medium. Indeed, in such systems ion pairs are the most plausible intermediates and elimination reactions from ion pairs are quite common even if the gegen-ion is not a sufficiently strong base. A low temperature will also favour the ion pairs, whereas increasing temperature will enable their dissociation.

Another way by which the ratio k_{-E}/k_{-H} can be increased is by adding to the reaction medium an efficient acceptor for the electrophile E^+. Thus, in desulphonation reactions the water molecules are efficient acceptors for the electrophile sulphur trioxide. In Friedel-Crafts alkylations, both solvent (aromatic hydrocarbon) and reaction products may act as electrophile acceptors, enabling thus disproportionations [72]:

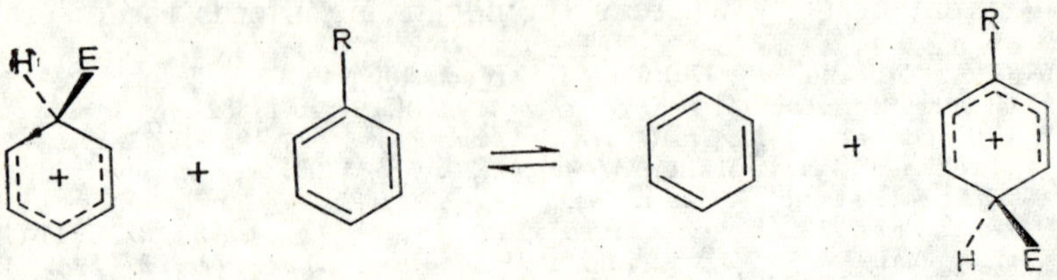

H E + R ⇌ + R H E

(E = alkyl)

[72] A. P. Lien and D. A. McCaulay, *J. Amer. Chem. Soc.*, 75, 2407, 2411 (1953).

(c) The dependence on k_{-H}. Examples are known in which k_{-E}/k_{-H} is significantly high and the reaction is nevertheless irreversible. Thus, a kinetic isotope effect has been observed in the nitration of certain aromatic compounds with bulky substituents near the reaction centre[33], indicating a supraunitary ratio k_{-E}/k_{-H}. However, the reaction is irreversible. Moreover, in the nitration of polyalkylbenzenes with branched chains, the bulky alkyl groups may be removed from the ring[73]. The explanation of this behaviour is found in the protonation step (rate constant k_H) involved in the reversible transformation. We shall first consider the protonation of halobenzenes[39–40]. Their basicity fall in the same range as their ease of disproportionation. The benzenium ion derived from fluorobenzene is quite stable, the fluorine atom having a better orienting power than the methyl group[71]. Only the structure **28** has been identified for this ion, even if equilibrium (7.9) cannot be excluded for heavier halogens:

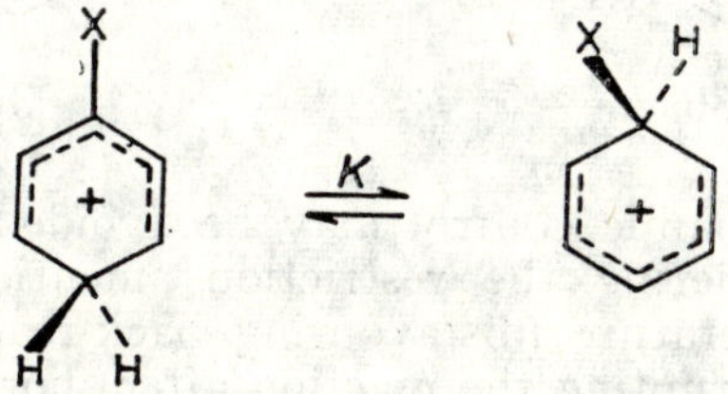

X H H K X H + + 28 29 (7.9)

Clearly, only benzenium ions **29** will be involved in desubstitution reactions (7.7), and thus the equilibrium will also depend on K. For fluoro derivatives, structure **28** should be by far the most stable, due to the strong $+E$ effect of fluorine; for the other halogens, the p-π conjugation decreases with the covalent radius of the heteroatom. Thus, for bromo- and iodo-benzenium ions significant concentrations of **29** should be expected.

An indication in that direction is afforded by the reaction of antimony pentafluoride with benzenoid aromatics. Fluoro- and chlorobenzene react normally with SbF_5, but bromo-benzene undergoes debromination[74], which suggests the following mechanism:

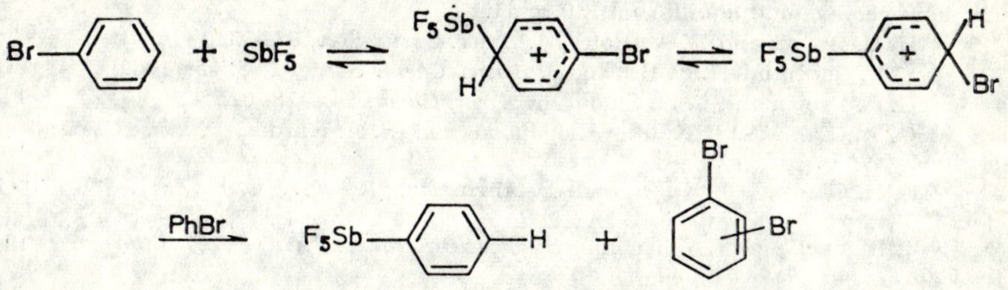

Br + SbF_5 ⇌ F_5Sb H + Br ⇌ F_5Sb + H Br

PhBr → F_5Sb H + Br Br

[73] G. A. Olah and S. J. Kuhn, *J. Amer. Chem. Soc.*, **86**, 1067 (1964).
[74] G. A. Olah, P. Shilling and I. M. Gross, *J. Amer. Chem. Soc.*, **96**, 876 (1974).

Evidently, the rearrangement of the initial intermediate is equivalent to equilibrium (7.9) and demonstrates that a desubstitution reaction of type (7.7) is observed only if the substitution product can be protonated to afford intermediate **27**.

The irreversibility of the nitration can be explained only by this condition. The basicity of the aromatic ring must be very low in nitroarenes. Even for the more basic compounds of this class the protonation is expected to yield aci-nitrobenzenium ions **30** in which the C—N bond would be even stronger than before*.

The thermodynamic control may be avoided more or less successfully by considering the restrictions mentioned above. Thus, lowering the temperature (disfavouring back-reaction by favouring ion pairs) and interrupting the reaction after short times (in order to prevent the equilibrium to be attained) have been currently applied in Friedel-Crafts reactions[75–77]. The utilisation of milder catalysts, such as $FeCl_3$[78, 79], GaX_3[76, 77] and aluminium chloride in nitromethane[78, 80–82] is also recommended.

7.2.5 *Rate equations*

As already shown, an electrophilic substitution can be generally represented by equation (7.7). However, for a kinetic study, the

* We may naturally wonder whether the generally assumed intermediate of the nitration **31**, has not the structure **30**. The actual identification of isotope effects may however exclude equilibrium **30** $\rightleftharpoons$ **31**.

[75] H. C. Brown and B. A Bolto, *J. Amer. Chem. Soc.*, **81**, 3320 (1959).

[76] S. U. Choi and H. C. Brown, *J. Amer. Chem. Soc.*, **81**, 3315 (1959).

[77] H. C. Brown and C. R. Smoot, *J. Amer. Chem. Soc.*, **78**, 6255 (1956).

[78] G. A. Olah, S. J. Kuhn, S.H. Flood and B. A. Hardie, *J. Amer.Chem.Soc.*, **86**, 1044 (1964).

[79] G. A. Olah, S. J. Kuhn and B. A. Hardie, *J. Amer. Chem. Soc.*, **86**, 1055 (1964).

[80] J. Reichel and R. Vîlceanu, *Bul. Ştiinţ. tehnic. Inst. Polit. Timişoara*, **10**, 71 (1965).

[81] G. A. Olah, S. H. Flood, S. J. Kuhn, M. E. Moffatt and N. A. Overchuck, *J. Amer. Chem. Soc.*, **86**, 1146 (1964).

[82] G. A. Olah and C. H. Lin, *J. Amer. Chem. Soc.*, **90**, 6468 (1968).

pre-equilibrium reactions, generating the electrophile, must also be included in the general scheme, as shown in Scheme 1.

$$\text{E—X} \underset{k_{-1}}{\overset{\text{C}, k_1}{\rightleftharpoons}} \text{E—Y} \tag{7.10}$$

$$\text{E—Y(or E—X)} \underset{k_{-2}}{\overset{k_2}{\rightleftharpoons}} \text{E}^+ + \text{Y}^- \text{(or X}^-\text{)} \tag{7.11}$$

$$\begin{matrix}(\text{Y}^-)\ \text{E}^+ + \text{ArH} \\ \text{or E—Y}\end{matrix} \xrightarrow{k_3} \left[\text{Ar}\begin{matrix}\text{H}\\ \text{E}\end{matrix}\right]^+ \text{Y}^- \xrightarrow[\text{—HY}]{\text{rapid}} \text{ArE} \tag{7.12}$$

Scheme 1

where EX, EY are carriers of the electrophile E^+, C a catalyst. It is easily seen that if we consider ArH as the substrate, the aromatic substitution can be related to the S_N1 and S_N2 reactions. The similarity of the two types of processes is even more obvious if EY is an alkyl halide[83, 84]. Applying the steady-state method to Scheme 1, the rate equation (7.13) can be derived:

$$\text{rate} = \frac{k_1 k_3[\text{ArH}]\,[\text{EX}]}{k_{-2}[\text{Y}^-]\,(k_{-1} + k_{-2} - 1) + k_3[\text{ArH}]\,(k_{-1} + k_2)} \tag{7.13}$$

which shows that the reaction rate is a hyperbolic function of the concentration of the aromatic hydrocarbon. In Scheme 1 it was assumed that the first step is a first-order reaction. Actually, the transformation of the reagent EX into the active species EY may have higher orders. However, the general form of the equation, even if more complex, preserves a similar dependence on the concentration of the aromatic substrate[85–88].

Depending on the reactivity of the aromatic substrate, more simple rate equations can be observed[89]. Thus, if practically every molecule of the generated electrophile E^+ combines rapidly only with the aromatic substrate, then $k_3[\text{ArH}] \gg k_{-2}[\text{Y}^-]$ and equation (7.13) becomes:

$$\text{rate} = k'[\text{EX}] \tag{7.14}$$

[83] C. R. Smoot and H. C. Brown, *J. Amer. Chem. Soc.*, **78**, 6245 (1956).

[84] C. R. Smoot and H. C. Brown, *J. Amer. Chem. Soc.*, **78**, 6249 (1956).

[85] C. D. Nenitzescu, S. Tzitzeica and V. Ioan, *Bull. Soc. chim. France*, 1272, 1279 (1955).

[86] V. Ioan, L. Teodorescu, S. Tzitzeica and C. D. Nenitzescu, *St. Cerc. Chim. (Bucharest)*, **7**, 346, 356 (1959).

[87] V. Ioan, D. Săndulescu, S. Tzitzeica and C. D. Nenitzescu, *Tetrahedron*, **19**, 323 (1963).

[88] V. Ioan, D. Săndulescu, S. Tzitzeica and C. D. Nenitzescu, *Tetrahedron*, **19**, 335 (1963).

[89] E. D. Hughes, C. K. Ingold and R. I. Reed, *J. Chem. Soc.*, 2400 (1950).

Conversely, if the aromatic compound has a very low reactivity and its concentration is small, the last term in the denominator of equation (7.13) can be neglected, which leads to equation (7.15):

$$\text{rate} = \frac{k_1 k_3[\text{ArH}][\text{EX}]}{k_{-1}[\text{Y}^-](k_{-1} + k_2 - 1)} \qquad (7.15)$$

Clearly, the observed rate constants in a series of aromatic substrates of very different reactivity cannot be used in the calculation of relative reactivities. Therefore, kinetic measurements are performed especially in order to determine the nature of the electrophile, the mechanism of its formation and only occasionally, when possible, for the determination of relative rates.

7.3 Relative reactivity in aromatic substitution

7.3.1 *Theoretical treatment*

The electronic energy of a π-system consisting of n electrons is:

$$E = n\alpha + \sum n_j m_j \beta = n\alpha + M\beta \qquad (7.16)$$

in which m_j are the coefficients of β in the equation of the MO energy and n_j is the number of electrons which fill orbital j. Wheland [25] has shown that in an electrophilic substitution the reactivity of the aromatic substrate must be proportional to the *localisation energy* L^+, defined by the relation:

$$L^+ = M - M^+, \qquad (7.17)$$

where M and M^+ are the coefficients of β in (7.16) either for the aromatic substrate **15** (M) or for the transition state represented (after Wheland) by structures **16** (M^+).

This dependence has been demonstrated by Streitwieser [90], who has shown that the relative reactivity of a position r in an aromatic hydrocarbon may be expressed by equation (7.18):

$$\log \frac{k_r}{k_{\alpha-\mathrm{C_{10}H_8}}} = \rho^* \sigma_r \qquad (7.18)$$

where σ_r measures the reactivity of the position r with respect to position α in naphthalene. By definition, $\rho^* = 1$ for the protonation of aromatic hydrocarbons in liquid hydrogen fluoride. The σ_r values determined from this reference reaction give excellent linear correlation with the localisation energies L^+ calculated for the correspond-

[90] A. Streitwieser, Jr., *Molecular Orbital Theory for Organic Chemists*, J. Wiley, New York (1961), Chapter 11.

ing hydrocarbons. However, other correlations of L^+ values are less perfect[90, 91].

It has been shown that the correlations of the localisation energies L^+ can be improved by simple thermodynamic considerations regarding the formation of the intermediate[92] (Scheme 2).

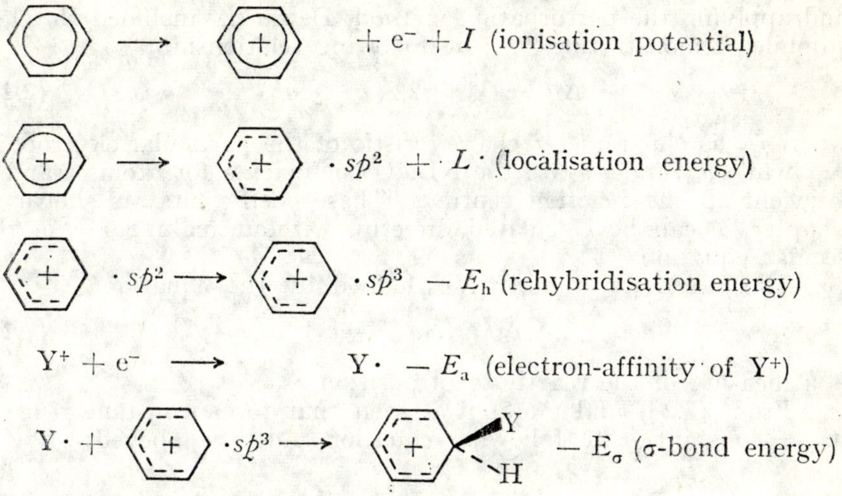

$$\text{C}_6\text{H}_6 \longrightarrow \text{C}_6\text{H}_6^{+} + e^- + I \text{ (ionisation potential)}$$

$$\text{C}_6\text{H}_6^{+} \longrightarrow \text{C}_6\text{H}_5^{+} \cdot sp^2 + L\cdot \text{ (localisation energy)}$$

$$\text{C}_6\text{H}_5^{+} \cdot sp^2 \longrightarrow \text{C}_6\text{H}_5^{+} \cdot sp^3 - E_h \text{ (rehybridisation energy)}$$

$$Y^+ + e^- \longrightarrow Y\cdot - E_a \text{ (electron-affinity of } Y^+)$$

$$Y\cdot + \text{C}_6\text{H}_5^{+} \cdot sp^3 \longrightarrow \text{C}_6\text{H}_5^{+}(Y)(H) - E_\sigma \text{ (}\sigma\text{-bond energy)}$$

Scheme 2

The overall process of Scheme 2 will thus be:

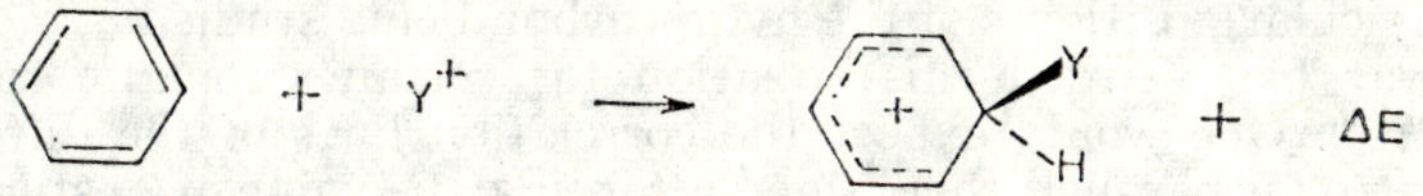

$$\text{C}_6\text{H}_6 + Y^+ \longrightarrow \text{C}_6\text{H}_6Y^+ + \Delta E \quad (7.19)$$

where:

$$\Delta E = I + L\cdot - E_h - E_a - E_\sigma \quad (7.20)$$

By applying relations (16) and (17) to the first steps of the above mechanism we obtain:

$$L\cdot = (M - m) - M^+ = L^+ - m \quad (7.21)$$

where m is the coefficient of β in the energy equation of the highest occupied orbital.

According to Hückel's theory, the ionisation potential I may be expressed as:

$$I = km\beta \quad (7.22)$$

Considering that E_h, E_a and E_σ are constant, relation (7.20) becomes:

$$\Delta E = L^+ + m(k - 1) + \text{const.} \quad (7.23)$$

[91] E. M. Arnett and J. W. Larsen, *J. Amer. Chem. Soc.*, **90**, 792 (1968).
[92] B. G. Ramsey, *J. Amer. Chem. Soc.*, **87**, 2502 (1965).

where $m(k - 1)$ is the correction for the localisation energy L^+. The values of the m parameter are tabulated for a large number of aromatic compounds[93] and constant k can be determined from (7.22) by plotting I *vs.* m.

Wheland's treatment does not take into account the nature of the attacking electrophile. Using the same model of transition state and applying the perturbation method, Dewar has included the electrophile, demonstrating[94, 95] the following relationship:

$$\Delta E = C_X - 2\beta\,(a_{or} + a_{ot}) \tag{7.24}$$

where C_X is a parameter characteristic of the particular electrophile X^+, while a_{or} and a_{ot} are the NBMO coefficients for atoms r and t, adjacent to the reaction centre s. These coefficients, as shown in Chapter 2, can be calculated directly, without making use of the secular equation.

The reactivity number N_s, defined by the equation:

$$N_s = 2(a_{or} + a_{ot}) \tag{7.25}$$

is a measure of the reactivity of position s.

From (7.24) and (7.25) it is seen that for a constant reagent (C_X = constant), the following equation must be obeyed:

$$\log \frac{k}{k_0} = -\frac{\beta}{RT}(N_0 - N_s) \tag{7.26}$$

where k_0 and N_0 are the values for benzene, and k and N_s the values for a particular position s of the hydrocarbon being studied.

Dewar has verified this equation for certain aromatic substitutions[94–96]. One would expect the correlation lines of the different reactions to be parallel, their slope affording the value of β estimated at about -20 kcal mol^{-1} from DE. However, the values calculated from the observed correlations are much lower, varying between 3.2 and 6.0 kcal mol^{-1}. According to Dewar[96], these values in their turn definitely disagree with Wheland's complexes as transition states. In the transition state of an aromatic substitution, only a minor fraction of the β value is lost, namely $\beta'_X = \beta - \beta_X$, where β_X must be directly related to the reagent X^+. Indeed, the activation energy of an aromatic substitution may be considered to result from two different components. An energy component, independent of the electrophile X^+, is determined by the localisation energy. The other energy component, depending on the electrophile X^+, corresponds to the combination of the electrophile with the zwitter-ion. The variation of these two components along a reaction coordinate

[93] E. Heilbronner and J. N. Murrell, *J. Chem. Soc.*, 2611 (1962).
[94] P. M. G. Bavin and M. J. S. Dewar, *J. Chem. Soc.*, 164 (1956).
[95] M. J. S. Dewar, T. Mole and E. W. T. Warford, *J. Chem. Soc.*, 3576, 3581 (1956).
[96] M. J. S. Dewar, *J. Amer.Chem. Soc.*, **74**, 3341 (1952), and following papers in the series.

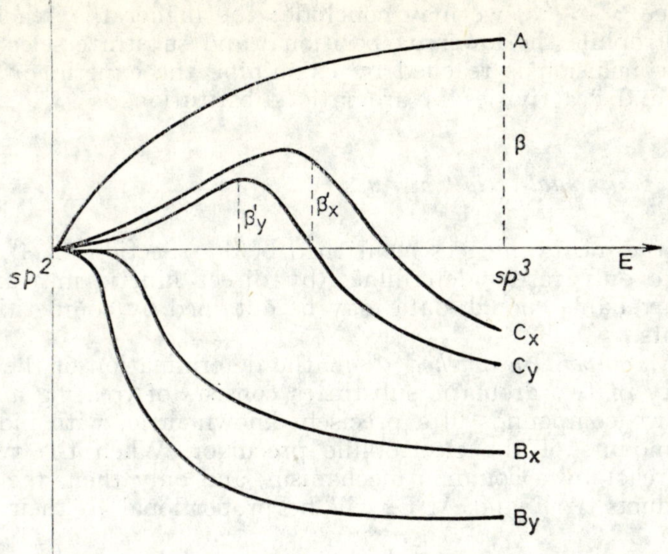

Figure 7.3 Potential-energy plot for two competing aromatic substitutions, involving two electrophiles, X^+ and Y^+, of different reactivities.

(the rehybridisation of the reaction centre) is shown in Figure 7.3. The energy due to the hypothetical formation of the zwitter-ion increases continuously, its maximum value (β) being attained when the benzenium ion is formed (curve A). The second energy component, due to the electrophile X^+ (curve B_X), displays in the earlier stages of the reaction a moderate increase, due to the desolvation of the reagent X^+. As soon as the new bond interactions are established, the energy will be lowered. The energy of the transition state will be given by the maximum of the resultant curve C_X, and will correspond to β'_X. If in the same diagram are plotted the variations of the energy of a system formed by the same aromatic hydrocarbon and another more energetic electrophile Y^+, curve A will remain unchanged, being related to the same process. Curve B_Y and C_Y, however, will be below B_X and C_X and thus the transition state of the latter reaction will be shifted towards the initial state, with $\beta'_Y < \beta'_X$. Otherwise stated, the higher the reactivity of the electrophile, the closer the transition state, in geometry and energy, to the initial system (Hammond's postulate).

If two positions m and n from the same aromatic substrate or from different substrates are compared, either in the reaction with X^+ or with Y^+, one may write:

$$\log\left(\frac{k_m}{k_n}\right)_X = \beta'_X \frac{N_m - N_n}{RT} \tag{7.27a}$$

and

$$\log\left(\frac{k_m}{k_n}\right)_Y = \beta'_Y \frac{N_m - N_n}{RT} \tag{7.27b}$$

Since $\beta'_X > \beta'_Y$, we may conclude: the higher the reactivity of the electrophile the lower its positional and substrate selectivity. A similar conclusion is reached by examining the experimental orientations and reactivities in aromatic substitution.

7.3.2 *Experimental relative rates*

Due to the inconveniences mentioned before (Section 7.2.4), relative rates are only rarely determined by direct kinetic measurements. Usually, reliable enough data may be obtained by competitive measurements.

The *competition method*[2,97] in the determination of the relative reactivity of two aromatic substrates consists of treating a mixture of the two compounds in a precisely known ratio, with indifinitely small amounts of the electrophile precursor. When the two compounds react by a common mechanism, and only then, the ratio of the products, ArE and Ar′E, will be proportional to their relative rates:

$$\frac{[\mathrm{ArE}]}{[\mathrm{Ar'E}]} = \frac{k_{\mathrm{ArH}}}{k_{\mathrm{Ar'H}}} \frac{[\mathrm{ArH}]}{[\mathrm{Ar'H}]} \tag{7.28}$$

The competition method has also been used in the determination of the relative reactivities of the three possible positions in a monosubstituted benzene (*ortho, meta, para*), with respect to one position of benzene. The overall relative rate is then defined by (7.29):

$$\frac{k_{\mathrm{ArH}}}{k_{\mathrm{PhH}}} = 6\mathrm{r} \tag{7.29}$$

considering that benzene has 6 equivalent positions. If the molar fractions of the three isomers in the reaction mixture are denoted by o, m, and p ($o + m + p = 1$), the *partial rate factors* of the three positions are estimated to be:

$$o_f = 3\mathrm{r}o \tag{7.30a}$$

$$m_f = 3\mathrm{r}m \tag{7.30b}$$

$$p_f = 6\mathrm{r}p \tag{7.30c}$$

The partial rate factors (o_f, m_f and p_f) represent the relative rate of the corresponding position in the benzenoid aromatic with respect to one position of the symmetric benzene ring. In relations (7.30) it has been considered that there are two equivalent *ortho* and *meta* positions, but only one *para* position.

The partial rate factors determined by Ingold, Hughes and their co-workers[2] have led to the following conclusions: (a) A first-order

[97] C. K. Ingold and F. R. Show, *J. Chem. Soc.*, 2918 (1927); C. K. Ingold, A. Lapworth, E. Rothstein and D. Ward, *J. Chem. Soc.*, 1959 (1931).

substituent with electron-repelling properties $(+I, +E)$ activates all three possible positions of the aromatic ring, *ortho* and *para* positions being significantly more activated than *meta.* (b) A second-order substituent $(-I, -E)$ deactivates all positions of the ring, but to a smaller extent the *meta* position. (c) Substituents with opposed electronic effects of similar magnitude, such as halogens $(-I, +E)$, deactivate all positions, but the *ortho* and *para* ones to a smaller extent. These conclusions are in perfect agreement with the electrophilic character of the aromatic substitution.

Brown and his co-workers[98-101] have observed that for a large number of substitution reactions of toluene (T) the following linear relationship is obeyed:

$$\log p_f = C \log \frac{p_f}{m_f} \tag{7.31}$$

where p_f and m_f are the partial-rate factors for *para* and *meta* position, respectively. It is considered that p_f is a measure of the substrate selectivity, whereas p_f/m_f refers to positional selectivity. The selectivity factor, S_f, is defined by the relation:

$$S_f = \log \frac{p_f}{m_f} \tag{7.32}$$

and has been used in discussing positional selectivity[102-103]. Considering the experimental data and equation (7.32), relation (7.31) becomes:

$$\log p_f = 1.310\ S_f \tag{7.33}$$

which is usually called the *selectivity relation.* It can be demonstrated that the selectivity relation is a free-energy relationship. Indeed, as partial rate factors are relative rates, we may write:

$$\log p_f = \rho\sigma_p$$

and

$$\log m_f = \rho\sigma_m$$

or

$$\log p_f = \frac{\sigma_p}{\sigma_p - \sigma_m} \log p_f/m_f = \frac{\sigma_p}{\sigma_p - \sigma_m} S_f \tag{7.34}$$

By comparing (7.34) with (7.33) it follows that:

$$\frac{\sigma_p}{\sigma_p - \sigma_m} = 1.310 \tag{7.35}$$

[98] H. C. Brown and C. W. McGarry, Jr., *J. Amer. Chem. Soc.*, **77**, 2300 (1955).
[99] H. C. Brown and K. LeRoi Nelson, *J. Amer. Chem. Soc.*, **75**, 6292 (1954).
[100] H. C. Brown and C. R. Smoot, *J. Amer. Chem. Soc.*, **78**, 2176 (1956).
[101] C. W McGarry, Jr., Y. Okamoto and H. C. Brown, *J. Amer. Chem. Soc.*, **77**, 3037 (1955).
[102] L. M. Stock and H. C. Brown, in *Advances in Physical Organic Chemistry*, vol. 1 (V. Gold, editor), Academic Press, New York (1963), p. 35.
[103] P. Kovacic, R. L. Russel and R. P. Bennett, *J. Amer. Chem. Soc.*, **86**, 1588, (1964).

Identity (7.35) is not verified by the σ_{CH_3} values of Hammett, but a good agreement is observed with $\sigma^+_{CH_3}$ values [104–107]. Thus, aromatic substitution is affected in a similar way by the substituents of the aromatic ring as the S_N1 substitution of benzyl derivatives. This conclusion is supported by the mere inspection of reactions (7.36) and (7.37).

S_N1: R–C$_6$H$_4$–CR$_2$–X → R–C$_6$H$_4$–$^+$CR$_2$ ⟷ R–C$_6$H$_4^+$=CR$_2$ (7.36)

S_EAr: R–C$_6$H$_5$ + E^+ → R–C$_6$H$_5^+$(H)(E) (7.37)

The selectivity relation (7.33) indicates that between the substrate selectivity and positional selectivity there must be a close relation. Such relations have actually been observed in a certain number of aromatic substitutions (Table 7.3). The aromatic substitutions, obeying the selectivity equation, according to Brown and his co-workers [98–100], may fall into three groups, namely:

(a) Substitutions with high substrate selectivity ($k_T/k_B > 100$) and high positional selectivity ($m \leqslant 1\%$). These reactions are observed in the case of poor electrophiles as, for example, halogens in the direct halogenation of aromatics, in absence of catalysts.

(b) Substitutions with moderate substrate and positional selectivity (k_T/k_B between 10 and 100, $m = 1$—5%). The nitration reaction with nitric acid is representative for this group. It involves moderately stable and reactive electrophiles.

(c) Substitutions with very low substrate and positional selectivity ($k_T/k_B < 10$; $m > 5\%$). Very unstable electrophiles and thus very reactive reagents usually fall in this class, as for example carbenium ions in Friedel-Crafts alkylations.

[104] H. C. Brown and Y. Okamoto, *J. Amer. Chem. Soc.*, **79**, 1913 (1957).

[105] H. C. Brown, J. D. Brady, M. G. Grayson and W. H. Bonner, *J. Amer. Chem. Soc.*, **79**, 1897 (1957).

[106] Y. Okamoto and H. C. Brown, *J. Amer. Chem. Soc.*, **79**, 1903, 1909 (1957).

[107] H. C. Brown, Y. Okamoto and G. Ham, *J. Amer. Chem. Soc.*, **79**, 1906 (1957).

TABLE 7.3

Relative reactivity (k_T/k_B) and isomer distributions in some substitution reactions of toluene

No	*Reaction type and reaction conditions*	k_T/k_B	*Isomer distribution* (%)			S_f	*Literature*
			o	*m*	*p*		
	(A) *Halogenations*						
1.	Chlorination, AcOH, 25 °C	344	59.8	0.5	39.7	2.2	a
2.	Chlorination, $MeNO_2$ 25 °C	2445	33.6	—	66.4	—	b
3.	Chlorination, $MeNO_2$—$FeCl_3$, 25 °C	13.5	67.8	2.3	29.9	—	c
4.	Bromination, AcOH, 25 °C	605	32.9	0.3	66.8	—	a
5.	Bromination, CF_3COOH, 25 °C	2580	17.6	—	82.4	3.1	d
6.	Bromination, $MeNO_2$—$FeCl_3$, 25 °C	3.6	68.7	1.8	29.5	—	e
	(B) *Nitrations*						
7.	HNO_3 in $MeNO_2$	26.4 (21)	61.5	3.1	35.4	—	f, g
8.	HNO_3-H_2SO_4 (30%) in tetramethylene sulphone (TMS), 25 °C	28	62.0	3.4	34.6	1.31	f
9.	$NO_2^+BF_4^-$ in TMS	1.6	65.4	2.8	31.8	—	h, i
10.	HNO_3—H_2SO_4 (75%), TMS[3], 25 °C	1.6	56.3	2.6	41.0	—	f
11.	$NO_2^+ClO_4^-$, TMS, 25 °C	1.6	66.2	3.4	30.4	—	h
12.	(pyridinium ring, $\overset{+}{N}$—NO_2, —Me) in MeCN, 25 °C	36.5	63.8	3.2	33.0	—	j
	(C) *Miscellaneous*						
13.	Alkylation, Pr^iBr, $AlCl_3$—$MeNO_2$, 25 °C	1.65	47.7	18.2	35.1	—	k
14.	Alkylation, $PhCH_2Cl$, $AlCl_3$—$MeNO_2$, 25 °C	3.45	44.0	4.5	51.5	—	k
15.	Benzylation, $PhCH_2Br$, $GaBr_3$	4.0	41.0	19.4	39.5	—	l
16.	Alkylation, Bu^tBr, $AlCl_3$—$MeNO_2$, 25 °C	13.7	0	5.9	94.1	—	k
17.	Acetylation, MeCOCl—$AlCl_3$	128	1.17	1.25	97.6	—	m
18.	Benzoylation, PhCOCl — excess	110	9.3	1.4	89.3	—	n
19.	Mercuration, CF_3COOH	9.9	12.2	8.6	19.2	1.26	a, d
20.	Methylation, MeBr—$GaBr_3$, 15 °C	6.3	55.7	9.9	34.4	0.841	p
21.	Ethylation, EtBr—$GaBr_3$, 15 °C	2.55	38.4	21.0	40.6	0.586	p

a) L.M. Stock and H.C. Brown, *J. Amer. Chem. Soc.*, **81**, 5615 (1959); b) L. M. Stock and A. Himoe, *J. Amer Chem. Soc.*, **83**, 4605 (1961); c) G. A. Olah, S. J. Kuhn and B. A. Hardie, *J. Amer. Chem. Soc.*, **86**, 1055 (1964); d) H. C. Brown and R. A. Wirkkala, *J. Amer. Chem. Soc.*, **88**, 1447 (1966); e) G. A. Olah, S. J. Kuhn, S. H. Flood and B. A. Hardie, *J. Amer. Chem. Soc.*, **86**, 1039 (1964); f) G. A. Olah, S. J. Kuhn, S. H. Flood and J. C. Evans, *J. Amer. Chem. Soc.*, **84**, 3687 (1962); g) C. K. Ingold, E. D. Hughes and R. I. Reed, *Nature*, **158**, 448 (1948); h) G. A. Olah and S. J. Kuhn, *J. Amer. Chem. Soc.*, **84**, 3684 (1962); i) G. A. Olah, S. H. Kuhn and S. H. Flood, *J. Amer. Chem. Soc.*, **83**, 4571 (1961); j) C. A. Cupas and R. L. Pearson, *J. Amer. Chem. Soc.*, **90**, 4742 (1968); k) G. A. Olah and N. A. Overchuk, *J. Amer. Chem. Soc.*, **87**, 5786 (1965); l) S. U. Choi and H. C. Brown, *J. Amer. Chem. Soc.*, **81**, 3315 (1959); m) H. C. Brown, G. Marino and L. M. Stock, *J. Amer. Chem. Soc.*, **81**, 3310 (1959); n) H. C. Brown and G. Marino, *J. Amer. Chem. Soc.*, **81**, 3308 (1959); o) H. C. Brown and R. A. Wirkkala, *J. Amer. Chem. Soc.*, **88**, 1453 (1966); p) H. C. Brown and C. R. Smoot, *J. Amer. Chem. Soc.*, **78**, 6255 (1956).

Most reactive and unstable reagents from this class will give nearly statistical ratios ($o, m = 40\%$; $p = 20\%$).

However, Table 7.3 shows that there is a large number of substitution reactions which do not obey the selectivity relation. Thus, applying the selectivity relation (7.33) to the data for the bromination reaction in the system $MeNO_2—FeCl_3$ (Table 7.3, entry 6), a partial-rate factor $p_f = 100$ can be calculated. If this factor is calculated from (7.30 c) considering $r = k_T/k_B = 3.6$, it results that $p_f = 6$. Clearly, this reaction does not obey (7.33). More strikingly, for the nitration of toluene with nitronium salts, the partial-rate factors calculated from these relations would lead to the conclusion that the *meta* position in toluene is deactivated, which is certainly unreasonable[108, 109].

It has been observed by Olah and his co-workers[79, 108–110] that the reactions which obey the selectivity relationship are correlated with the relative stabilities of σ complexes formed in the reaction of hydrogen fluoride with aromatic molecules in presence of boron trifluoride. Conversely, the reactions violating this rule manifest a close relationship with the relative stabilities of π complexes (Table 7.4).

TABLE 7.4

Correlation of relative substitution rates with the relative stabilities of σ and π complexes[a].

Substituent in the benzenoid aromatic	*Relative stabilities*		*Relative reactivities in*			
	of the π complex with I_2	*of the σ complex with* $HF\text{-}BF_3$	*chlorination with* Cl_2	*bromination with* $Br_2FeCl_3—MeNO_2$	*nitration with:* HNO_3	*nitration with:* $NO_2^+BF_4^-$
H	0.48	—	1	0.23	1.0	0.51
Me	0.52	0.01	2 445	0.83	27	0.85
o-Me_2	0.87	2.00	—	0.91	—	0.89
m-Me_2	1.00	20.0	24 700	1.30	—	0.84
p-Me_2	1.00	1.0	14 200	1.00	1 000	1.00
1,3,5-Me_3	2.65	2 800	5 000 000	3.70	—	1.38

a. From the data of Olah *et al.*[79,108–110].

It is thus reasonable to assume that in the reactions of the former type (correlation with σ complexes) the formation of π complexes is rapid and the rate-determining step is their transformation into σ complexes; in reactions of the latter type, π complexes are formed in the slow step (Figure 7.2). More precisely, the former reactions will present a late transition state similar in structure to a Wheland complex, whereas the latter reactions will display an early transition

[108] G. A. Olah, S. J. Kuhn and S. H. Flood, *J. Amer. Chem. Soc.*, **83**, 4571 (1961).

[109] G. A. Olah and S. J. Kuhn, *J. Amer. Chem. Soc.*, **84**, 3684 (1962).

[110] G. A. Olah, S. J. Kuhn, S. H. Flood and B. A. Hardie, *J. Amer. Chem. Soc.*, **86**, 1039 (1964).

state resembling a π complex[53]. These concepts were criticised[111-113], as it was argued that the reactions presenting no substrate selectivity but nevertheless high positional selectivity involve invariably very reactive reagents, the low substrate selectivity being the result of a very fast reaction which precedes the uniform mixing of the reagents. However, recent studies [114] indicate that competitive measurements and the direct determination of the non-competitive reaction rates lead to the same results.

7.3.3 *Ortho versus para orientation*

Table 7.3 reaveals some interesting aspects regarding the ratio between the *ortho* and *para* isomers. There are reactions in which the statistical ratio of 2:1 is observed, indicating no preference between *ortho* and *para* attack; in other reactions the formation of the *ortho* isomer is clearly favoured (for example, entry 6) and in others *para* reactivity exceeds by far the reactivity of the *ortho* position (for example, entry 16). According to Ingold [2], the distribution between *ortho* and *para* positions is determined by steric effects both in reagent and aromatic substrate, by electronic effects in the aromatic substrate and by the reactivity of the reagent. Other factors beside these seem also probable.

(1) *Steric effects.* Steric demands both in the aromatic compound and in the reagent decrease the *ortho*/*para* ratio, due to steric crowding at the reaction site. Thus, the methylation of toluene yields an appreciable amount of *o*-xylene (55.7%), whereas the t--butylation, involving the bulky t-butyl cation, yields no *ortho* isomer[115]. In the substitution reactions of bulky alkylbenzenes, the high steric requirements of such groups may cause either the expulsion of this group from the aromatic ring or the migration of the entering group to other positions[116-118]. Experimentally, such situations are reflected in high kinetic isotope effects even for reactions where they are usually not observed [31-33]. Indeed, if in (7.7) we neglect the protonation of the resulted product ($k_H \approx 0$), the observed rate constant will be given by equation (7.38):

$$k_{obs} = \frac{k_E k_{-H}}{k_{-E} + k_{-H}} = \frac{k_E \cdot \dfrac{k_{-H}}{k_{-E}}}{1 + \dfrac{k_{-H}}{k_{-E}}} \tag{7.38}$$

[111] W. S. Tolgyesi, *Can. J. Chem.*, **43**, 343 (1965).
[112] J. H. Ridd, *Accounts Chem. Res.*, 4, 248 (1971).
[113] P. Rys, P. Skrabal and H. Zollinger, *Angew. Chem. Int. Ed. Engl.*, **11**, 874 (1972).
[114] G. A. Olah and H. C. Lin, *J. Amer. Chem. Soc.*, **96**, 549 (1974).
[115] G. A. Olah and N. A. Overchuck, *J. Amer. Chem. Soc.*, **87**, 5786 (1965).
[116] A. W. Burgstahler, P. L. Chien and M. O. Abdel-Rahman, *J. Amer. Chem. Soc.*, **86**, 5281 (1964); P. C. Myhre and M. Beng, *J. Amer. Chem. Soc.*, **88**, 1568 (1966).
[117] P. C. Myhre, *J. Amer. Chem. Soc.*, **94**, 7921 (1972).
[118] R. C. Hahn and M. B. Groen, *J. Amer. Chem. Soc.*, **95**, 6128 (1973).

If there are bulky groups in the vicinity of the reaction site, then $k_{-E} > k_{-H}$ and the last term in the denominator of equation (7.38) can be neglected. Thus, the observed reaction rate will depend on the rate of proton elimination:

$$k_{obs} = \frac{k_E}{k_{-E}} k_{-H} \quad (7.39)$$

(2) *Electronic effects.* As already mentioned, substituents in an aromatic ring may alter the electron distribution of the parent hydrocarbon by two mechanisms, the inductive effects I and the conjugative effects* E. The action of these mechanisms governs especially the *ortho-para* orientation against *meta* orientation. However, the same mechanisms may affect to a certain extent the *ortho vs. para* attack. The electronic effects of substituents can be treated by MO techniques, either by calculating the charge distribution in the aromatic ring[120,121] or by applying the PMO (Perturbation of Molecular Orbitals) methods[120] and estimating the reactivity numbers for the different possible positions of attack. Simple methods, such as HMO, show no differences between charge distribution in *ortho* and *para* positions (Figure 7.4A). Reactivity numbers calculated in the usual way, in their turn, indicate no preference for *ortho* or *para* attack (Figure 7.4B), even if they indicate the correct relative reactivity with respect to benzene ($N_{benzene} = 2.31$). The insensitivity of simple MO techniques in discerning the differences between *ortho* and *para* reactivity is due to the neglect of electron repulsions in the MOs. However, if electron repulsions are considered, such as in the SCF—MO (Self-Consistent Field MO) methods, a clear-cut discrimination between the two positions is evidenced (Figure 7.4C). It is obvious from the benzyl system used as model for both $+E$ and $-E$ effects (anion and cation, respectively) that the *para* position is more affected than the other positions.

* In the early period of the electronic theory, Lapworth and Robinson[119] termed the latter *electromeric effects*. The signs which they have assigned both to the inductive and electromeric effects differed from those proposed later by Ingold[2] (and used in this book). According to the earlier convention, an electron-repelling group, such as methyl, has a $-I$ effect (instead of $+I$), whereas an electron-withdrawing group such as nitro has a $+E$ effect (instead of $-E$). The earlier authors have stressed thus on the formal charge acquired by the adjacent carbon atom, whereas Ingold stressed on the formal charge which must be attributed to the substituent due to the electron displacement. As mere conventions, except for the widespread use of Ingold's convention, they seem to be equally justified. However, it has been argued[120] that the earlier convention is more logical inasmuch as it reflects the actual charge induced by the electronic mechanism. Moreover, the convention of Lapworth and Robinson agree in sign with the sign convention for σ constants in the free-energy relationships.

[119] A. Lapworth, *J. Chem. Soc.*, **121**, 416 (1922); see also R. Robinson, *Outline of an Electrochemical Theory of the Course of Organic Reactions*, Royal Institute of Chemistry, London (1932).

[120] M. J. S. Dewar, *The Molecular Orbital Theory of Organic Chemistry*, McGraw-Hill, New York (1969), p. 410.

[121] A. Streitwieser, Jr., *Molecular Orbital Theory for Organic Chemists*, J. Wiley, New York (1961), pp. 350, 360.

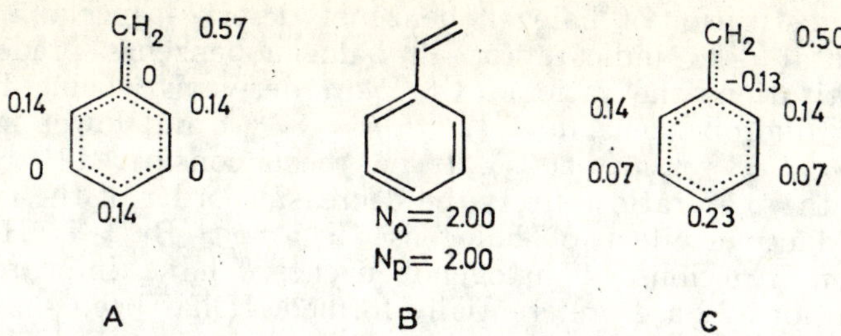

Figure 7.4 Charge distribution and reactivity numbers for certain aromatic systems. A. Charge distribution in the benzyl system calculated by the HMO method. B. Reactivity numbers in styrene. C. Charge distribution in the benzyl system calculated by the SCF method.

(a) *Inductive effects.* Inductive effects decrease in intensity with the distance. Thus, this effect will be more strongly exerted in *ortho* than in *para*. An electron-repelling group ($+I$) will favour *ortho* over *para* substitution, whereas an electron-withdrawing group will decrease *ortho* reactivity to a larger extent than *para* reactivity.

Substitutions of toluenes (Table 7.3), excepting reactions with bulky electrophiles, present high $o:p$ ratios. Even in such cases, the ratio $o:p$ rarely exceeds the statistical ratio 2:1, indicating a certain degree of steric shielding[122]. Substitution reactions of onium salts, such as the nitration of phenyl-trimethyl-ammonium ion[123] (strong $-I$ effect of the NMe_3 group), beside predominant *meta* orientation ($m = 89\%$), indicate that *para* is the next reactive position of the ring (in the example cited $p = 11\%$; $o = 0.05\%$).

In the MO methods, inductive effects are treated by assigning suitable values to the heteroatomic parameter h for the substituent-bearing carbon atom. A $+I$ substituent, such as alkyl, is considered to confer a negative h-value, whereas a $-I$ substituent is associated with a positive h-value[25, 124]. This method, applied to aromatic compounds substituted with groups displaying pure I effects, has led to results consistent with experimental data.

(b) *Conjugative effects.* Considering Figure 7.4C, it may be concluded that $+E$ effects will favour *para* substitution, whereas $-E$ effects will determine a high $o:p$ ratio. Indeed, the halogenation of acetanilide ($+E$ effect of the NHAc group) leads to the *para* isomer[125] in practically quantitative yield. In the nitration of nitrobenzene (strong $-E$ effect of the NO_2 group) carried out in nitromethane with nitronium tetrafluoroborate[114], the ratio $o:m:p$ of the resultant dinitrobenzenes is 10.0: 88.5: 1.5, in perfect agreement with the theoretical predictions.

[122] P. B. D. de la Mare, *J. Chem. Soc.* 2871 (1949); *Progress in Stereochemistry*, 2, 72 (1958).
[123] J. H. Ridd and J. H. P. Utley, *Proc. Chem. Soc.*, 24 (1964).
[124] G. W. Wheland and L. Pauling, *J. Amer. Chem. Soc.*, 57, 2086 (1935).
[125] P. B. D. de la Mare, *J. Chem. Soc.*, 4450 (1954).

The substitution of halogenobenzenes deserves special mention. Experimental data indicate for all halogenobenzenes predominant *para* substitution, the ratio of the *para* derivative being however higher for fluorobenzene ($o = 12\%$; $p = 87\%$) and lower for iodobenzene ($o = 38\%$; $p = 60\%$). It has been considered[126] that the change of the $o:p$ ratio reflects the decreasing order of the intensity of the inductive effect of halogens: $F > Cl > Br > I$. However, protonation reactions of halogenobenzenes[40] have indicated that benzenium ions **32** are preferentially formed. It has been shown that fluorine with its $+E$ effect will better stabilise the positive charge in **32** ($X = F$; $E = H$) than in **33**[66, 71] ($X = F$; $E = H$).

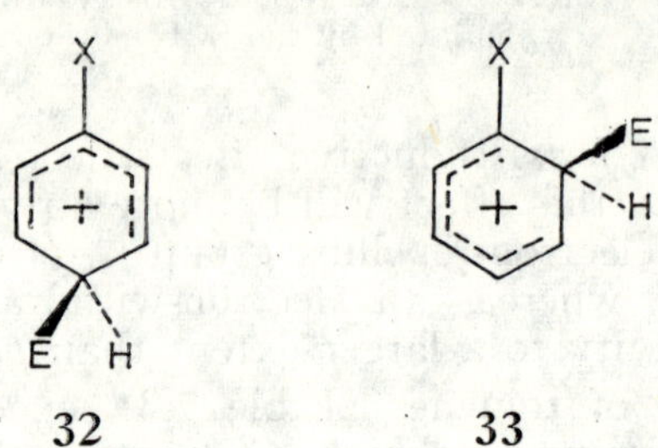

32 **33**

Evidently, the intensity of the $+E$ effect of the halogens will decrease in the same order as the inductive effect and thus the higher *para* ratios for fluorobenzenes than for the other halogenobenzenes may be due to inductive or to conjugative effects.

(3) *The reactivity of the electrophilic reagent.* Positional selectivity is directly related to the reactivity of the electrophilic reagent. Highly reactive reagents are expected to present low selectivity even between *ortho-para* and *meta* positions. A certain degree of selectivity between the *ortho* and *para* positions is to be anticipated only with moderate electrophiles or better with poorly reactive reagents. Thus, in diazo-coupling the reagent is the stable diazonium ion:

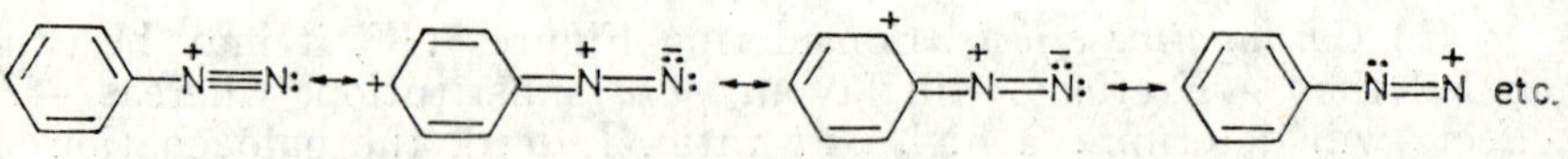

It is well known[5] that this reagent prefers *para* attack. Its high positional selectivity is coupled with a high substrate selectivity (only phenols and amines react with diazonium ions).

(4) *Other factors affecting the* o:p *ratio.* Recent studies have revealed that beside the factors mentioned by Ingold[2] to govern *ortho vs. para* orientation, other factors may be in their turn decisive in increasing the *ortho* ratio in the reaction mixture.

[126] See Reference 2, p. 312.

(a) *Complex formation.* The chelate-complexing of the entering substituent with the pre-existing one may highly favour *ortho* substitution. Thus, the metallation reaction of benzyldimethylamine with lithium tetrachloropalladate practically yields only the *ortho* isomer[127]. Chelate-complexing in **34** may satisfactorily explain the preference for the observed orientation.

CH_2NMe_2 —(Li_2PdCl_4)→ (C_6H_4–Pd(Cl)←; CH_2—NMe_2→Pd)$_2$

34

Chelate-complexing favouring *ortho* substitution is usually involved in reversible reactions. As a rule, thermodynamic control of the reaction determines low *o*:*p* ratios, *ortho* disubstituted products being sterically disfavoured. The formation of chelate complexes may however favour *ortho* substitution. An early, well known example is the Kolbe-Schmitt reaction[128], in which either under kinetic or under thermodynamic control, *para* substitution may be anticipated. The formation of salicylic acid from sodium phenoxide (but not from potassium phenoxide) can be explained reasonably only by chelate formation.

$:\ddot{O}:^- Na^+$ + CO_2 → [$\overset{+}{Na}$ O O^- C=O H] → OH COONa

(b) *Ipso attack. Ipso* substitution (at a carbon already bearing a substituent) in nitration and halogenation is a long-known phenomenon[73, 118, 129, 130]. For example, in the nitration of *p*-cymene, beside the normal products **35** (~ 82%) and **36** (~ 8%), about 10% *p*-nitrotoluene is also formed[73]. Evidence has been advanced for

[127] A. C. Cope and R. C. Friederich, *J. Amer. Chem. Soc.*, **90**, 909 (1968).

[128] D. Cameron, H. Jeskey and O. Baine, *J. Org. Chem.*, **15**, 233 (1950), and the literature cited therein; C. A. Buehler and W. E. Cate, *Org. Synth. Coll.* vol. **2**, 341 (1943); see also G. A. Olah and J. A. Olah, *Friedel-Crafts and Related Reactions* (G. A. Olah, editor), vol. 3, Interscience, New York (1964), p. 1269.

[129] J. Barlow, *Liebigs Ann. Chem.*, **98**, 245 (1856).

[130] E. M. Arnett and G. B. Klingensmith, *J. Amer. Chem. Soc.*, **87**, 1023 (1965).

the formation of intermediate **37**, which is further stabilised by the elimination of a secondary carbocation:

Me, NO_2, CHMe$_2$, **35** (82%); Me, NO_2, CHMe$_2$, **36** (8%); Me, CHMe$_2$ + NO_2^+; Me, +, Me$_2$CH NO_2, **37**; [$\overset{+}{C}HMe_2$] + Me, NO_2 ~10%; COMe$_2$

Low-temperature (0 °C) nitration of *p*-cymene with acetyl nitrate in acetic anhydride[131] has permitted the isolation of the *cis* and *trans*-dienes **39** (41%), beside **35** (41%), **36** (8%) and *p*-nitrotoluene (10%). Treatment of dienes **39** with sulphuric acid led to the quantitative transformation of **39** into **35**:

Me, CHMe$_2$; Me NO_2, +, CHMe$_2$, **38**; Me, H, NO_2, +, CHMe$_2$, **40**; Me, NO_2, CHMe$_2$, **35**; Me NO_2, AcO CHMe$_2$, **39** (*cis* and *trans*)

Thus, the most reactive position in *p*-cymene is C-1, the intermediate **38** being preferentially formed in the nitration. The apparent high

[131] R. C. Hahn and D. L. Strock, *J. Amer. Chem. Soc.*, **96**, 4335 (1974).

reactivity of the C-2, observed in earlier works, is an artifact of the migration of the NO_2 group from C-1 to C-2, demonstrated by the acid-catalysed transformations of **39**. It should be observed that *ipso* attack may alter the actual relative rates of *ortho* and *para* positions (by processes similar to **38**→**40**, which increased apparently *ortho* reactivity), but is not expected to modify *ortho-para vs. meta* substitution.

74 Mechanism and reactivity in some electrophilic substitutions

7.4.1 *Halogenation reactions*

It is well known that aromatic hydrocarbons form reversible complexes with halogens[50, 132–134]. The equilibrium can be easily observed spectrophotometrically. Thus, a benzene solution of iodine, beside the known absorption band at 500 nm, presents another band at 297 nm, which is not observed in hexane solutions [132]. For bromine, this extra band appears at 293 nm. This shows that a *charge-transfer complex* [135] or a π *complex* is formed:

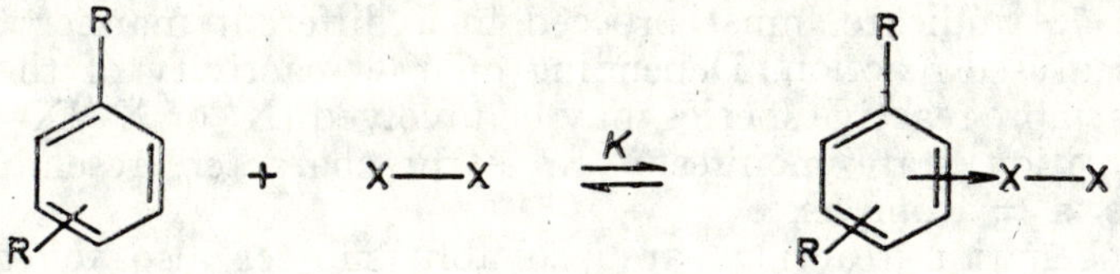

The equilibrium constants K of processes (7.40) vary with the nature of the halogen X_2 (as in the equilibrium of trihalide ion formation[136]), with the nature of substituents of the aromatic compounds (as in the formation of silver-ion complexes[48, 49]) and with the basicity of the aromatic compounds with respect to hydrogen fluoride (in the absence of boron trifluoride). The initial formation of π complexes, at least in the direct halogenation of aromatic compounds with molecular halogen, can thus be regarded as certain. However, the activation energy in process (7.40) must be very low, the actual energy barrier in the halogenation being associated with the transformation of the charge-transfer complex into an oriented σ complex [30].

A kinetic demonstration of the occurrence of π complexes in the uncatalysed halogenation has been advanced by Robertson and de la

[132] H. A. Benesi and J. H. Hildebrand, *J. Amer. Chem. Soc.*, **70**, 2882 (1948); **71**, 2703 (1949).

[133] R. S. Mulliken, *J. Amer. Chem. Soc.*, **72**, 600 (1950); **74**, 811 (1952).

[134] L. J. Andrews and R. M. Keefer, *J. Amer. Chem. Soc.*, **74**, 4500 (1952).

[135] G. Briegleb, *Electronen-Donator-Acceptor-Komplexe*, Springer Verlag, Berlin (1961).

[136] R. L. Scott, *J. Amer. Chem. Soc.*, **75**, 1550 (1953).

Mare[137, 138]. In non-polar solvents, such as acetic acid, at higher bromine concentrations, the equation rate of bromination has higher--order terms (second-order in bromine). Similar results have also been observed in the addition to the olefinic double bond, in the same type of solvents. Higher-order kinetics have also been determined for iodine (but never for chlorine). It has been concluded[138] that the π complex formed needs another halogen molecule in order to pass over into the σ complex:

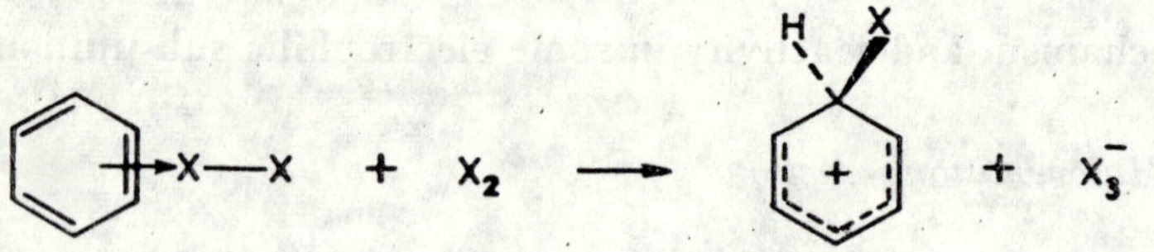

(7.41)

Since the trichloride ion Cl_3^- is unstable[136], the chlorination has no higher-order kinetics. It has been later proved[139] that in good--ionising solvents, like trifluoroacetic acid, the rate equation is second-order even at higher bromine concentrations. In such cases, solvent assistance in the ionisation of the π complex can be sufficient even for unreactive aromatic substrates.

The catalysed halogenation of aromatic compounds, as Tables 7.3. and 7.4 indicate, must proceed in a different manner than in the uncatalysed reaction. Depending on the reactivity of the Lewis catalysts, more reactive species may be involved (X^+ or $X{-}X{\rightarrow}MX_n$). The transition state manifests an early character, resembling in structure a π complex.

Halogenation reactions are probably subject also to catalysis by Brönsted acids (acid is produced in the course of the reaction). It has been recently shown that chlorine and bromine can be easily protonated, yielding positive ions $H{-}\overset{+}{X}{-}X$. Such species are supposed to be involved in the aromatic halogenation[140]:

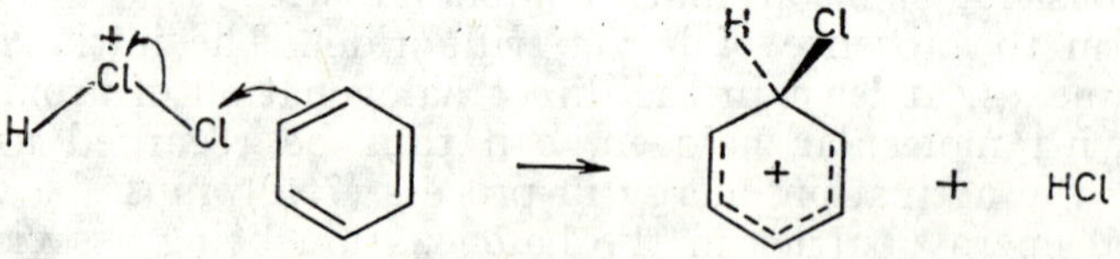

Halogenations with hypohalous acids and their derivatives may involve X^+ or carriers of halogenium ions, $X\overset{+}{O}H_2$, XOY, XOH and even X_2. In aqueous solutions, the halogenium ions are the most

[137] P. W. Robertson, P. B. D. de la Mare and W. T. G. Johnston, *J. Chem. Soc.*, 276 (1943).

[138] For discussions see P. B. D. de la Mare, *Quart. Rev. Chem. Soc.*, **3**, 126 (1949); see also Reference 7, p. 121; see, however, W. M. Schubert and J. L. Dial, *J. Amer. Chem. Soc.*, **97**, 3878 (1975).

[139] H. C. Brown and R. H. Wirkkala, *J. Amer. Chem. Soc.*, **88**, 1447 (1966).

[140] G. A. Olah and J. Shen, *J. Amer. Chem. Soc.*, **95**, 3582 (1973).

probable intermediates. Thus, de la Mare and his co-workers[141-143] have shown that the kinetics of the chlorination of aromatic compounds with hypochlorous acid in acid solution depends upon the reactivity of the aromatic substrate. With moderately reactive compounds, the rate equation includes all the reagents:

$$\text{rate} = k[\text{ArH}][\text{ClOH}][\text{H}^+]$$

With highly reactive derivatives, such as anisole, the rate becomes independent of the concentration of the aromatic compound:

$$\text{rate} = k[\text{ClOH}][\text{H}^+]$$

Moreover, the observed rate constant is identical with the rate constant determined in the additions of hypochlorous acid to reactive alkenes. These data have suggested the following pre-equilibrium reactions:

$$\text{ClOH} + \text{H}_3\text{O}^+ \xrightleftharpoons{\text{rapid}} \text{Cl}\overset{+}{\text{O}}\text{H}_2 + \text{H}_2\text{O} \qquad (7.42)$$

$$\text{Cl}\overset{+}{\text{O}}\text{H}_2 \xrightarrow{\text{slow}} \text{Cl}^+ + \text{H}_2\text{O} \qquad (7.43)$$

In acetic acid, the addition of hypochlorous acid involves chlorine acetate[143]. The halogenation with mixed anhydrides of hypohalous acids has been studied in more detail by Andrews *et al.*[144,145]. They have prepared the halogen acetates by the equilibrium reaction (7.44):

$$(\text{MeCOO})_2\text{Hg} + \text{X}_2 \xrightleftharpoons{K} \text{MeCOOX} + \text{MeCOOHgX} \qquad (7.44)$$

The highest equilibrium constant K is observed for iodine. However, from their own results, as well as from the data of de la Mare[143], it seems that the relative reactivities decrease in the order BrOAc > > ClOAc > IOAc. It is therefore concluded that probably X^+ species are also involved, the ease of the ionisation of the halogen acetates decreasing in the order IOAc > BrOAc > ClOAc, whereas the reactivity of the halogenium ions increases in the reverse order: $I^+ < Br^+ < Cl^+$.

Substrate and positional selectivity in halogenation with hypohalous acids is moderate. Thus, for chlorination and bromination, respectively, of benzene and toluene, the following values have been determined[146]: $k_T/k_B = 60$ and 36.6, respectively; $m = 2.3\,\%$ and

[141] P. B. D. de la Mare, E. D. Hughes and C. A. Vernon, *Research*, 3, 192, 242 (1950).

[142] P. B. D. de la Mare, A. D. Ketley and C. A. Vernon, *J. Chem. Soc.*, 1290 (1954).

[143] P. B. D. de la Mare, I. C. Hilton and C. A. Vernon, *J. Chem. Soc.*, 4039 (1960).

[144] Y. Hatanaka, R. M. Keefer and L. J. Andrews, *J. Amer. Chem. Soc.*, **87**, 4280 (1965).

[145] E. M. Chen, R. M. Keefer and L. J. Andrews, *J. Amer. Chem. Soc.*, **89**, 428 (1967).

[146] a. P. B. D. de la Mare and J. T. Harvey, *J. Chem. Soc.*, **36** (1956); b. P. B. D. de la Mare, J. T. Harvey, M. Hassan and S. Varma, *J. Chem. Soc.*, 2756 (1958).

2.2%, respectively. In both reactions the proportion of *ortho* product is high (74.6% in chlorination and 70.3% in bromination). These data should indicate reactive species, with small steric demands, which is in agreement with the attack of a halogenium ion on the aromatic substrate.

An interesting halogenation can be carried out by treating an aromatic compound with benzoyl peroxide and lithium halide in dimethylformamide (DMF)[147a]. The presence of the benzoyl peroxide would suggest a radical reaction. However, the isomer distribution is similar to that of the halogenation with free halogen. The quantitative formation of benzoic acid rules out a Hunsdiecker reaction, suggesting thus the following mechanism:

$$(PhCOO)_2 + X^- \longrightarrow PhCOOX + C_6H_5COO^-$$
$$PhCOOX + X^- \longrightarrow PhCOO^- + X_2$$

This mechanism is also consistent with the known high reactivity of halide ions in S_N2 reactions occurring in aprotic dipolar solvents, such as DMF.

Two other reaction mechanisms, involving halogens, have been recently discussed.

In one of them[147b], which may be designated as the "electron-transfer mechanism" the intermediate formation of a radical-cation must be admitted. Thus, Illuminati and his co-workers[147c] have found that hexamethylbenzene and chlorine yield chloromethylpentamethylbenzene. Carrying out this reaction in the cavity of an electron spin resonance spectrometer, Kochi[147b] identified the radical-cation **41**, and suggested the following mechanism:

+ Cl_2 ⇌ (+•) + Cl_2^- $\xrightarrow{-H^+}$ •CH_2 $\xrightarrow[\text{or } Cl_2]{Cl_2^-}$ CH_2Cl

41

The electron-transfer mechanism may also be involved in other aromatic halogenations. However, the experimental evidence is still scarce.

Better supported experimentally is the addition-elimination mechanism admitted before for the aromatic substitution by earlier workers[2]. Rejected on energy grounds (unfavourable high-energy route leading to a cyclohexadiene intermediate), it was soon forgotten. However, modern methods have enabled the identification and even the isolation of cyclohexadiene adducts[148]. An interesting example is

[147] (a) J. K. Kochi, B. M. Graybill and M. Kurz, *J. Amer. Chem. Soc.*, **86**, 5257 (1964); (b) J. K. Kochi. *Tetrahedron Letters*, 4305 (1974); (c) E. Baciocchi, A. Ciana, G. Illuminati and C. Pasini, *J. Amer. Chem. Soc.*, **87**, 3953 (1965).

[148] P. B. D. de la Mare. *Accounts Chem. Res.*, **7**, 361 (1974) and references cited therein.

the halogenation of naphthalene. The ratio of the addition products is largely dependent on the solvent, indicating the following mechanism:

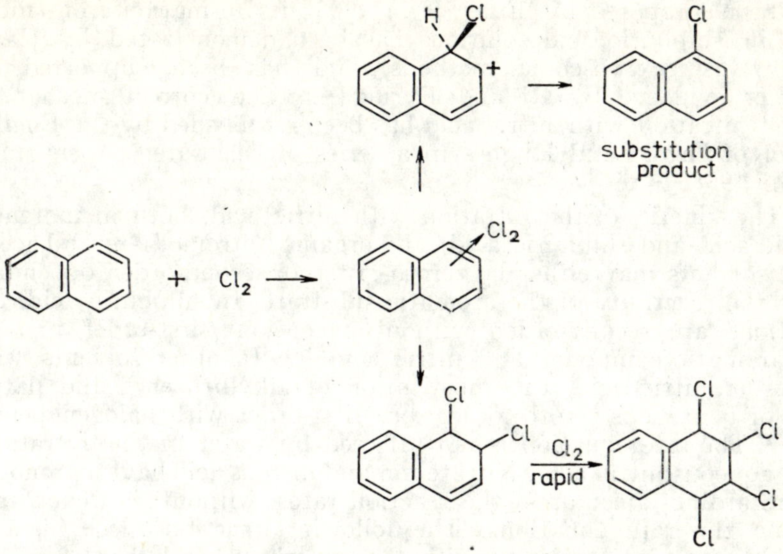

substitution product

$\xrightarrow[\text{rapid}]{Cl_2}$

Solvent:		Addition products (three isomers)
	$MeNO_2$	17%
	AcOH	34%
	$CHCl_3$	43%

Indeed, the increasing ratio of the addition products with decreasing polarity of the solvent suggests that the rearrangement of the π complex to the 1:1 naphthalene-Cl_2 adduct competes with the transformation of the π complex into the σ complex, leading to substitution products. The latter process will be favoured by the polarity of the solvent (charges are created). The stereochemistry of the addition agrees also with this mechanism.

In conclusion, halogenation reactions may involve different electrophiles, the most reactive species being the halogenium ions, X^+, and the less reactive the free halogens X_2 and the hypohalous acids XOH. Depending on the reactivity of the reagent involved, early or late transition states may characterise the particular halogenation.

7.4.2 *Nitration reaction*

The nitration is the most studied reaction of electrophilic substitution of aromatic compounds, partly because it is completely irreversible under the reaction conditions and partly because the introduction of a nitro group deactivates the aromatic ring, thus preventing further substitution. It is generally admitted that all nitration reactions involve the nitronium ion, NO_2^+. This reaction intermediate has been proposed by Euler[149] in 1903; its formation from nitric acid in sulphuric acid solutions has been demonstrated by Hantzsch[150] using cryoscopic methods, and has been supported by indirect evidence by Ri and Eyring[151]. The correct mechanism of the nitration with nitric acid has been established by the English school of Ingold and Hughes in a series of celebrated papers starting in 1938[152–160].

The kinetics of the nitration with nitric acid, both in inorganic (nitric acid and sulphuric acid) and organic (nitromethane and acetic acid) solvents may either be zeroth-, first- or second-order, depending upon the reactivity of the aromatic substrate. In sulphuric acid, the reactions are second-order without exception (first-order in both the aromatic compound and nitric acid)[153]. In other solvents, with excess of nitric acid, the substitution of alkylbenzenes and partly that of benzene is zeroth- order, but first-order with halogenobenzenes[152]. The rate equation is not affected by water in the nitrations with aqueous nitric acid. Nitrate ion and nitrous acid have a pronounced retarding effect upon the reaction rates, without, however, modifying the rate equations. The following reaction mechanism is considered to be consistent with the experimental data:

$$HNO_3 + HNO_3 \underset{k_{-1}}{\overset{k_1}{\rightleftharpoons}} H_2NO_3^+ + NO_3^- \tag{7.45}$$

$$H_2NO_3^+ \underset{k_{-2}}{\overset{k_2}{\rightleftharpoons}} NO_2^+ + H_2O \tag{7.46}$$

$$ArH + NO_2^+ \xrightarrow{k_3} \left[Ar\begin{matrix} H \\ NO_2 \end{matrix}\right]^+ \xrightarrow{\text{rapid}} ArNO_2 \tag{7.47}$$

149 H. Euler, *Liebigs Ann. Chem.*, **330**, 280 (1903).

150 A. Hantzsch, *Ber. dtsch. chem. Ges.*, **58**, 941 (1925).

151 T. Ri and E. Eyring, *J. Chem. Phys.*, **8**, 433 (1940).

152 G. A. Benford and C. K. Ingold, *J. Chem. Soc.*, 929 (1938).

153 E. D. Hughes, C. K. Ingold and R. I. Reed, *Nature*, **158**, 448 (1956); see also Reference 2, p. 320.

154 E. S. Halberstadt, E. D. Hughes and C. K. Ingold, *J. Chem. Soc.*, 2441 (1950).

155 V. Gold, E. D. Hughes, C. K. Ingold and G. H. Williams, *J. Chem. Soc.* 2452, 2467 (1950).

156 R. J. Gillespie, E. D. Hughes and C. K. Ingold, *J. Chem.Soc.*, 2473, 2552. (1950).

157 D. R. Goddard, E. D. Hughes and C. K. Ingold, *J. Chem. Soc.*, 2559 (1950),

158 C. K. Ingold, D. J. Millen and H. G. Poole, *J. Chem. Soc.*, 2576 (1950).

159 C. A. Bunton, E. D. Hughes, C. K. Ingold, D. I. H. Jacobs, M. H. Jones, C. J. Minkoff and R. I. Reed, *J. Chem. Soc.*, 2628 (1950).

160 E. D. Hughes, C. K. Ingold and R. B. Pearson, *J. Chem. Soc.*, 4357 (1958).

Equilibrium (7.45) is demonstrated by the effect produced by the nitrate ion. As this ion only retards the reaction without changing its kinetics, it is considered that the attacking reagent is not $H_2NO_3^+$, but probably its decomposition product, the nitronium ion. Since water does not affect the reaction kinetics, it is assumed that $k_3 > k_{-2}$, *i.e.* the nitronium ion once formed reacts rapidly further with the aromatic substrate. This view has been challenged by Olah[161] who has shown that the relative reactivities of nitronium tetrafluoroborate towards fluorobenzene, water, benzene and toluene are in the ratio 0.55:1.00:1.17:1.77, which indicates that nitronium ion has a very low selectivity. Most probably, the nitronium ion in water is not completely free, its desolvation occurring in the close neighbourhood of the aromatic substrate.

The retarding effect of nitrous acid needs further comment. It is constantly observed in concentrated acid solutions, irrespective of the reactivity of the aromatic compound. The presence of nitrous acid in the nitrations of reactive aromatic compounds, such as phenols and tertiary amines, in dilute nitric acid, has however an accelerating effect upon the reaction rates[159]. It has been demonstrated that in nitric acid, nitrous acid may suffer the following reactions:

$$HNO_2 + HNO_3 \rightleftharpoons N_2O_4 + H_2O \tag{7.48}$$

$$N_2O_4 \longrightarrow NO^+ + NO_3^- \tag{7.49}$$

The nitrosonium ion NO^+ formed in (7.49), is a poor electrophile, due to a conjugation similar to that in carbon monoxide:

$$:\overset{+}{\dot{N}} = \ddot{O}: \longleftrightarrow :N \equiv \overset{+}{O}:$$

Although it may coexist with nitronium ion NO_2^+, it cannot compete with that reactive electrophile, for the aromatic substrate. However, the formation of nitrate ion in (7.49) will shift the equilibrium (7.45), decreasing thus the concentration of the nitronium ion and retarding the nitration. In dilute aqueous solutions, nitronium ion cannot be formed, but nitrosonium ion, as a more stable species, may result even by protonation of nitrous acid:

$$HNO_2 + H^+ \longrightarrow NO^+ + H_2O$$

Unreactive aromatic substrates, with high selectivities, are insensible towards NO^+. Phenols and tertiary amines, however, will be nitrosated and then oxidised to the corresponding nitro derivatives:

$$ArH + NO^+ \longrightarrow ArNO + H^+$$

$$ArNO + HNO_3 \longrightarrow ArNO_2 + HNO_2, \text{ etc.}$$

Nitration reactions with nitric acid are very similar with the halogenations with hypohalous acids. This is demonstrated by the

[161] G. A. Olah, S. J. Kuhn, S. H. Flood and J. C. Evans, *J. Amer. Chem. Soc.*, **84**, 3687 (1962).

comparable values of the ratio k_T/k_B and of the isomer distribution observed in the two types of reactions, as well as by a similar $\rho\sigma^+$ correlation[162]; the values of the parameter ρ are nearly the same, —6.0 and —6.2, respectively, indicating similar transition states. Assuming that halogenium ions are intermediates in the halogenation, the nitronium ion would appear as a plausible reagent in the nitration.

Further insight into the reaction mechanism of the nitration with nitric acid is obtained by inspecting the data of Table 7.3. Nitrations with nitronium ion complexed with pyridine bases[163] certainly do not involve completely free nitronium ions in the transition state. However, the substrate selectivity and the isomer distribution are nearly the same as in the nitration with nitric acid or with nitric acid-sulphuric acid in dilute tetramethylenesulphone solution[161]. On the other hand, nitration with more concentrated sulphuric acid in tetramethylenesulphone presents a striking resemblance with the nitration with nitronium salts. It has been thus concluded[161] that in dilute solutions of either nitric acid or mixed nitric-sulphuric acids, the nitronium ion is not completely free, but solvated, resembling more or less the nitracidium ion, $H_2NO_2^+$.

The effect of acid concentration in mixed acid solutions has been studied in more detail by Combes and his co-workers[164]. They have established (Table 7.5) that in more concentrated acid solutions substrate selectivity is fading out, as a levelling of the alkyl group effects occurs. This levelling in the relative rates is produced by the increasing concentration of NO_2^+ ions. It is considered that the relative rates will be thus determined by the *limiting encounter rates* of the nitronium ion with low selectivity, leading to an *encounter pair* **42**:

$$NO_2^+ + ArH \longrightarrow \underset{\mathbf{42}}{ArH \cdot NO_2^+} \longrightarrow \left[Ar\begin{matrix} H \\ NO_2 \end{matrix} \right]^+ \xrightarrow[-H^+]{} ArNO_2 \qquad (7.50)$$

According to Ridd[112], there is little doubt that such reactions involve the nitronium ion. The encounter rates may also explain the unselectivity of the nitrations with nitronium salts. However, Olah[53] considers that the encounter rates do not satisfactorily account for the observed substrate and positional selectivities. The type of the transition state, early or late, is responsible for the observed substrate selectivity and isomer distribution. In order to support this hypothesis, Olah and his co-workers[114] have determined the rates of nitration of nitrobenzenoid aromatics with nitronium tetrafluoroborate and have compared the relative rates with those obtained in competitive measurements. Perfect agreement has been observed, excluding thus the haphazardous encounter rates. Moreover, a trend has been detected in the rates of these reactions, demonstrating that

[162] P. C. Myhre, *Acta Chem. Scand.*, 14, 219 (1960).
[163] C. A. Cupas and R. L. Pearson, *J. Amer. Chem. Soc.*, 90, 4742 (1968).
[164] R. G. Combes, R. B. Moodie and K. Schofield, *J. Chem. Soc.* (B), 800 (1968); 1 (1969).

transition states in nitrations with nitronium salts are not fixed, but pass smoothly from early to late character.

Steric effects in nitration with nitric acid may be quite important and are reflected in low ratios of the *ortho*-substituted products. Thus, at 0 °C in nitromethane solution, isopropylbenzene and t-butylbenzene yield 23.2% and 10.3%, respectively, *ortho* derivative. Under the same conditions, benzobicyclo [2,2,2] octene forms only 3.3% α-substituted product [165]. This steric shielding of the *ortho* position for the bicyclic derivatives has been called *fused ortho-effect* and is due to the interference of the entering nitro group with the methylene bridge (**43**). It has been recently suggested [166] that fused *ortho*-effects can be better explained considering *ipso* attack at benzobicyclic compounds such as **44**. Such compounds are related to indane, known to give high yields of *ipso* attack products. As has been shown [131], *ipso* adducts may suffer migration of the nitro group to the adjacent *ortho* position, apparently increasing its reactivity. The decrease of the *ortho* reactivity in compounds of type **44** may be thus the result of the decrease of *ipso* attack.

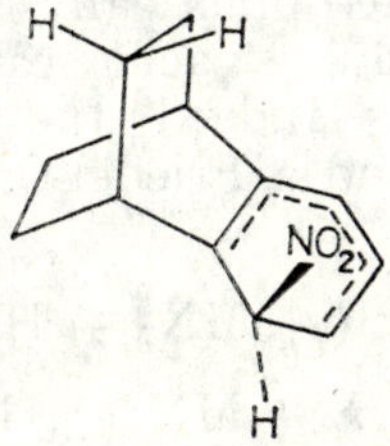

43 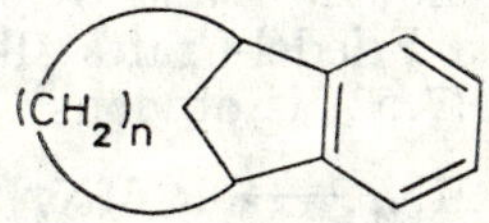 **44**

Quantum mechanical calculations have been performed [167] in order to ascribe the most probable structure to the primary intermediate formed by the aromatic substrate and the nitronium ion. The bisected open form (corresponding to the classical σ complex) is the least stable of all the geometries considered. Cyclic structures **45** and **46** are lower in energy and seem to be equally stable:

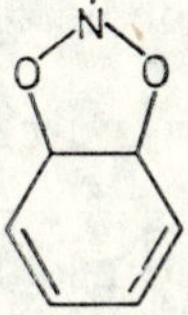

45 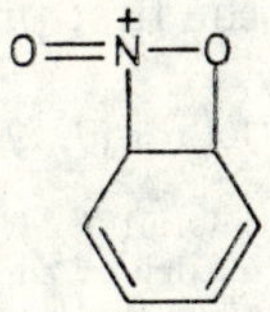 **46**

[165] H. Tanida and R. Muneyuki, *J. Amer. Chem. Soc.*, **87**, 4794 (1965).

[166] R. C. Hahn *et al.*, *J. Amer. Chem. Soc.*, **96**, 4337 (1974), and references cited therein.

[167] F. Bernardi and W. J. Hehre, *J. Amer. Chem. Soc.*, **95**, 3078 (1973).

Ion cyclotron resonance studies of the gaseous reaction between nitronium ion and benzene[168] suggest structure **46**. It should be however stressed that such intermediates have never been observed directly, whereas σ complexes, such as **17**, have been isolated[36]. It is possible that either **45** or **46**, or both, are initially formed (π complexes) and that their conversion into σ complexes is due either to a better solvation of the intermediate or to an equilibrium of type **30**$\rightleftharpoons$**31** (see p. 356). Further studies are expected in this field.

7.4.3 *Friedel-Crafts alkylations*

The alkylation of aromatic hydrocarbons with alkyl halides and aluminium chloride was discovered by Friedel and Crafts in 1877[169,170]. Large applications, both in organic synthesis and in technology, were found shortly after the discovery of the reaction. Even if the name of Friedel-Crafts reactions usually designates alkylations and acylations of aromatic compounds, involving an electrophile precursor (alkyl halide or acyl halide etc.) and a Lewis acid catalyst (7.51), the term may be extended practically to all electrophilic aromatic substitutions[171] involving acid catalysis[172], both by Lewis[173] and by Brönsted[174] acids. Indeed, the similarity between a Friedel-Crafts alkylation (7.51) and a Friedel-Crafts halogenation (7.52) is obvious:

$$\mathrm{R{-}X + MX_n \longrightarrow R^+MX_{n+1}^- + ArH \longrightarrow ArR + MX_n + HX} \quad (7.51)$$

$$\mathrm{X{-}X + MX_n \longrightarrow X^+MX_{n+1}^- + ArH \longrightarrow ArX + MX_n + HX} \quad (7.52)$$

Moreover, Friedel-Crafts reactions of aromatic compounds are related to many other reactions of alkanes and alkenes[175–177]. These reactions, in their turn, are of great theoretical and technological interest. No wonder that Friedel-Crafts and related reactions have acquired such importance in chemical research[8–12], ever since their first applications were found.

The study of Friedel-Crafts reaction is associated with many difficulties due to its reversible character and also to the fact that the newly introduced substituent increases the reactivity of the aromatic ring. Thus, the isolation of the kinetically-controlled reaction is one of the first problems which must be resolved by a researcher

168 G. A. Olah *et al.*, *J. Amer. Chem. Soc.*, **94**, 3667, 6862 (1972), and references quoted therein.

169 C. Friedel and J. M. Crafts, *Compt. rend.*, **84**, 1392 (1877).

170 For a history of the reaction, see G. A. Olah and R. E. A. Dear, Reference 8, p. 1.

171 For definitions and scope, see G. A. Olah, Reference 8, p. 25.

172 G. A. Olah, Reference 8, p. 201.

173 M. Baar and V. Gutmann, Reference 8, p. 367.

174 R. J. Gillespie, Reference 8, p. 169.

175 L. Schmerling, Reference 11, Chapter 54.

176 L. Schmerling, Reference 9, p. 1075.

177 C. D. Nenitzescu and A. T. Balaban, Reference 10, p. 1033.

in this field. Polyalkylation is difficult to avoid due to the higher reactivity of the alkylbenzenes as compared to the parent compounds. The kinetic study may be rendered difficult as the reaction is subject to autocatalysis (acid is produced in the alkylation); the complications due to pre-equilibrium reactions reflected in the rate equations have already been mentioned. However, using mild catalysts, low temperatures, suitable concentrations and molar ratios of the reactants, and short reaction times, these inconveniences can be avoided.

(A) Electrophilic reagents

The similarity between the Friedel-Crafts alkylation and the solvolysis of an aliphatic halide has been early recognised[83–88]. The aromatic compound plays a role similar to that of the nucleophilic solvent, from which it differs, however, by its low polarity and ionising power. It is thus not able to polarise the R—X bond, so that catalysts are needed even for the most efficient leaving groups. Most strikingly, this similarity is observed in the alkylation of benzenoid compounds with alkyl arenesulphonates, catalysed by the corresponding sulphonic acids[178]. Ioan, Nenitzescu and Tzitzeica[95–97] have determined a third-order rate equation:

$$\text{rate} = k\,[PhSO_3R]\,[PhSO_3H]^2 \qquad (7.53)$$

The absence of the concentration of the aromatic substrate in the rate equation is an indication that the formation of the electrophile is rate-determining, as in the common S_N1 reaction. The two molecules of acid catalyst are necessary for the protonation of the alkyl benzenesulphonate and for the solvation of the resulted benzenesulphonate anion in the non-polar reaction medium:

$$PhSO_3{-}R + 2PhSO_3H \longrightarrow (PhSO_3HR)^+ + PhSO_3^- \ldots HSO_3Ph \qquad (7.54)$$

$$(PhSO_3HR)^+ \longrightarrow PhSO_3H + R^+ \qquad (7.55)$$

The acid catalyst assumes the pulling action of an acidic solvent in the unimolecular ionisation. It has also been found that the rate constants of the benzylation of benzene with benzenesulphonates correlate well with σ-values of the substituents from the benzylic moiety, the reaction parameter ρ having a high negative value[95] ($\rho = -4.17$), of comparable magnitude with that of the unimolecular solvolysis of benzylic derivatives[179]. Moreover, the optically active s-butyl benzenesulphonate loses completely its activity during the alkylation process[180]. The reaction thus corresponds to a unimolecular solvolysis in which the intermediate carbenium ion is rapidly captured by the aromatic solvent.

[178] F. A. Drohovzal, in Reference 9, pp. 417 and 641.

[179] C. D. Nenitzescu, V. Ioan and L. Teodorescu, *Chem. Ber.*, **90**, 582 (1957).

[180] H. C. Brown, R. Bernheimer, C. J. Kim and S. E. Scheppele, *J. Amer. Chem. Soc.*, **89**, 370 (1967).

Quite different results have been obtained in the Friedel-Crafts alkylation with Lewis acids, such as aluminium and galium halides. Kinetic studies [76, 77, 83, 84, 181–183] have established a third-order rate equation:

$$\text{rate} = k[\text{RX}]\,[\text{MeX}_3]\,[\text{ArH}] \tag{7.55}$$

the reactivity of the alkyl halides decreasing in the order:

$$\text{Bu}^t > \text{Pr}^i > \text{Et} \gg \text{Me}$$

in agreement with either a unimolecular (7.51) or a bimolecular reaction (7.57), in which the catalyst polarises strongly the R—X bond (7.56), without actually breaking it:

$$\text{R—X} + \text{MX}_n \longrightarrow \underset{\textbf{47}}{\text{R—X}\rightarrow\text{MX}_n} \tag{7.56}$$

$$\text{ArH} + \text{R—X}\rightarrow\text{MX}_n \longrightarrow \text{ArR} + \text{MX}_n + \text{HX} \tag{7.57}$$

An argument in favour of mechanism (7.56)—(7.57) has been afforded by the reaction of n-propyl halides with aromatic hydrocarbons [84, 184, 185], which yields mixtures of n-propylarenes and i-propylarenes. The former are likely to arise by an S_N2 reaction [84].

$$\text{ArH} + \text{CH}_3\text{CH}_2\text{CH}_2\text{X} \xrightarrow{\text{MX}_n} \begin{cases} \xrightarrow{k_2} \text{ArCH}_2\text{CH}_2\text{CH}_3 & (S_N2) \\ \xrightarrow{k_1} (\text{CH}_3\text{CH}_2\text{CH}_2^+ \longrightarrow \text{CH}_3\overset{+}{\text{C}}\text{HCH}_3) \longrightarrow \\ \longrightarrow \text{ArCH(CH}_3)_2 & (S_N1) \end{cases}$$

Brown and his co-workers have calculated the rate constants k_1 and k_2 from the overall reaction rates and the ratio of the observed reaction products. The k_2 rate constants calculated for n-propyl halides have a lower value than the corresponding rate constants for ethyl halides, in agreement with the order of reactivity in an S_N2 reaction. Moreover, the ratio of the unrearranged propyl derivatives increases with the increasing nucleophilicity of the aromatic hydrocarbon.

Olah and his co-workers [186–189] have studied the reaction of organic halides (primary, secondary and tertiary) with antimony

[181] H. C. Brown and M. Grayson, *J. Amer. Chem. Soc.*, **75**, 6285 (1953).
[182] H. C. Brown, L. P. Eddy and R. Wong, *J. Amer. Chem. Soc.*, **75**, 6275 (1953).
[183] H. Jungk, C. R. Smoot and H. C. Brown, *J. Amer. Chem. Soc.*, **78**, 2185 (1956).
[184] V. N. Ipatieff, H. Pines and L. Schmerling, *J. Org. Chem.*, **5**, 253 (1940).
[185] R. M. Roberts and D. Schieng Thong, *J. Amer. Chem. Soc.*, **86**, 2851 (1964).
[186] G. A. Olah, W. S. Tolgyesi, S. J. Kuhn, M. E. Moffatt, I. J. Bastien and E. R. Barker, *J. Amer. Chem. Soc.*, **85**, 1328 (1963).
[187] G. A. Olah, E. R. Barker, J. C. Evans, W. S. Tolgyesi, J. S. McIntire and I. J. Bastien, *J. Amer. Chem. Soc.*, **86**, 1360 (1964).
[188] G. A. Olah, J. M. Bollinger and J. Brinich, *J. Amer. Chem. Soc.*, **90**, 2587 (1968).
[189] G. A. Olah, J. R. DeMember, R. H. Schlosberg and Y. Halpern, *J. Amer. Chem. Soc.*, **94**, 156 (1972).

pentafluoride. In the case of butyl halides, the single observable carbenium ion is t-butyl ion; propyl halides lead only to the i-propyl carbocation, supporting thus the interpretation given to the reactions of n-propyl halides. However, ethyl cation (primary) has also been identified in superacid solutions[189,190]. Initially, at low temperatures, a donor-acceptor complex **47** formed between the halide and the antimony pentafluoride is formed. The complete randomisation of carbon atoms[189] and hydrogen atoms[190], as well as the formation of ethane by a hydride transfer from the solvent[190] are clear indications of the occurrence of the primary ethyl cation. The S_N1 mechanism cannot thus be excluded even for primary derivatives in which no rearrangement has been observed. Another plausible interpretation of the reaction of primary derivatives in presence of Lewis acid catalysts is the formation of dialkylhalogenium ions[191]. Such onium salts have been prepared from alkyl halides and antimony pentafluoride and were identified by n.m.r., laser Raman and i.r. spectroscopy studies. They are highly reactive alkylating reagents in Friedel-Crafts reactions:

$$R{-}X + SbF_5 + R{-}X \longrightarrow R{-}\overset{+}{X}{-}R \quad SbF_5X^- \qquad (7.58)$$

$$ArH + R{-}\overset{+}{X}{-}R \longrightarrow ArR + R{-}X + H^+$$

(B) Transition States. Intermediates

Substrate selectivity in Friedel-Crafts alkylation is low (Table 7.3) and positional selectivity moderate. With efficient leaving groups, the distribution of *ortho*, *meta* and *para* isomers approaches even the statistical distribution[192]. These results would indicate an early transition state similar in structure with a π complex. Evidently, the π complex may arise either from a pre-formed carbenium ion or directly from the donor-acceptor complex **47**:

R-benzene + $R{-}X{\rightarrow}MX_n$ (**47**) ⟶ **48** [or **49**: $R{-}X{\rightarrow}MX_n$ π-complexed to R-benzene]

(7.59)

By analogy with the structure of the π complexes of halogens with the aromatic hydrocarbons, either **48** or **49** may result in the

[190] J. H. Vorachek, G. G. Meisels, R. A. Geanangel and R. H. Emmel, *J. Amer. Chem. Soc.*, **95**, 4078 (1973).

[191] G. A. Olah, J. R. DeMember, Y. K. Mo, J. J. Svoboda, P. Schilling and J. A. Olah, *J. Amer. Chem. Soc.*, **96**, 884 (1974).

[192] G. A. Olah and J. Nishimura, *J. Amer. Chem. Soc.*, **96**, 2214 (1974).

rate-determining step. The distinction between these intermediates should be possible by their further transformation into σ complexes. Indeed, **49** is expected to give lower *o*/*p* ratios than **48**, due to its higher steric requirements. A study of the methylation and ethylation with different alkylating systems[191] suggests that both structures may be involved, and thus mechanism (7.59) cannot be ruled out. A 1:1:1 complex between ethyl fluoride, benzene and boron trifluoride has actually been identified at low temperatures[193] thus supporting scheme (7.59).

The intermediate formation of σ complexes is well evidenced by the isolation and identification of ionic species like **19**[42] and **50**[36].

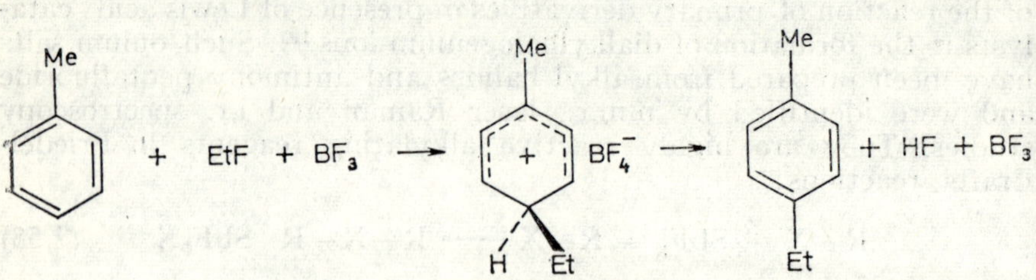

$$\text{PhMe} + \text{EtF} + BF_3 \longrightarrow [\text{MeC}_6H_5(H)(Et)]^+ BF_4^- \longrightarrow p\text{-MeC}_6H_4\text{Et} + HF + BF_3$$

50

Complex **50** is stable at −80 °C and decomposes at −15 °C into the alkylated product.

It has been shown, as for other electrophilic aromatic substitutions, that the transition state is not fixed, displaying either an early or a late character. Thus, in the benzylation of benzene and toluene with substituted benzyl chlorides in presence of Lewis acids[194], the ratio k_T/k_B increases from 2.5 (X = *p*-NO_2) to 97 (X = *p*-OMe_3):

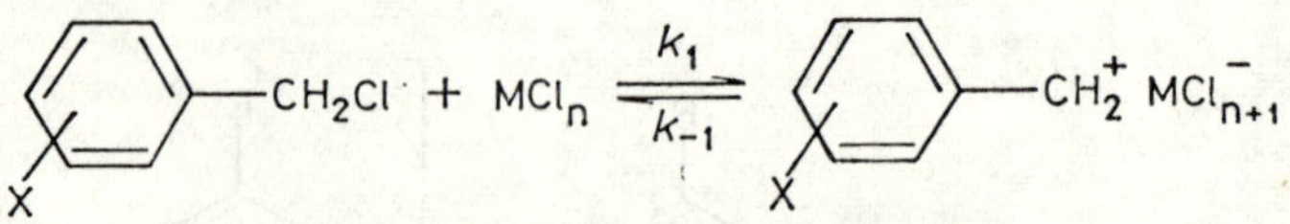

$$X\text{-}C_6H_4\text{-}CH_2Cl + MCl_n \underset{k_{-1}}{\overset{k_1}{\rightleftharpoons}} X\text{-}C_6H_4\text{-}\overset{+}{C}H_2\ MCl_{n+1}^- \quad (7.60)$$

51

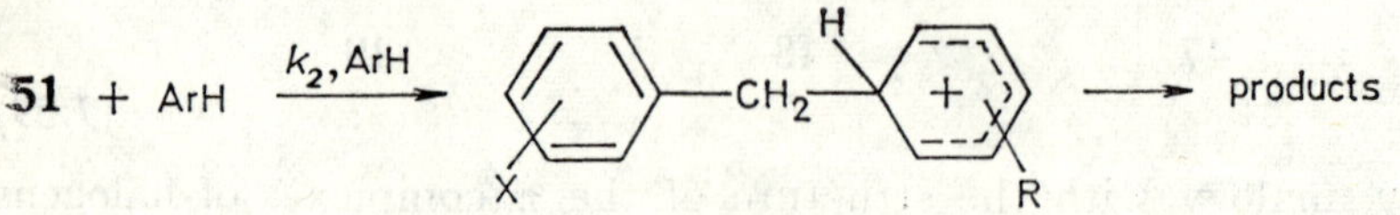

$$\mathbf{51} + \text{ArH} \xrightarrow{k_2,\ \text{ArH}} X\text{-}C_6H_4\text{-}CH_2\text{-}[C_6H_5(H)R]^+ \longrightarrow \text{products} \quad (7.61)$$

52

[193] R. Nakane, T. Oyama and A. Natsubori, *J. Org. Chem.*, **33**, 275 (1968); **34**, 949 (1969).

[194] G. A. Olah, M. Tashiro and S. Kobayashi, *J. Amer. Chem. Soc.*, **92**, 6369 (1970).

It can be demonstrated from the above scheme that k_T/k depends upon the $k_{2,\,ArH}$ values. The following free-energy relationship should then be expected:

$$\log k_T/k_B = \log k_{2,\,T}/k_{2,\,B} = (\rho_{2,\,T} - \rho_{2,\,B})\,\sigma_X = \rho_g\sigma_X$$

where $\rho_{2,\,T}$ and $\rho_{2,\,B}$ are the reaction parameters of the benzylation of toluene and benzene, respectively. If the ionisation step (7.60) is slow as compared to the subsequent steps, and the intermediate **51** is not very stable, $k_{2,\,ArH}$ will have little or no significance for the overall process and ρ_g will have a small negative (if $k_1 > k_{2,\,ArH}$) or positive (if $k_{2,\,ArH} \geqslant k_1$) value. The highest positive value will depend upon the nucleophilicity of the aromatic hydrocarbon (higher for benzene than for toluene). If intermediates **51** are stable and result in a rapid reaction, the ρ parameter will be positive throughout (positive charges are disappearing in the later steps of the reaction), the higher values corresponding to the less nucleophilic hydrocarbon. By plotting the values reported by Olah and his co-workers[194] *vs.* σ^+ values, the curve represented in Figure 7.5 is obtained, which may be considered to consist of two lines of different slopes. The line with the lowest slope ($\rho_g = -0.8$) corresponds to benzyl chlorides with electron-withdrawing substituents (X = *m*-OMe, *m*-F, *m*-Cl, *p*-Cl, *p*-NO_2), whereas the line of higher slope ($\rho_g = -3.9$) includes the benzyl derivatives which afford stable carbenium ions (X = H, *m*-Me, *p*-Me O and *p*-Me). The significant negative slopes of the two branches of the curve in Figure 7.5 demonstrates that benzylic ions are rapidly formed in the system. At the same time, Figure 7.5 demonstrates the oscillating

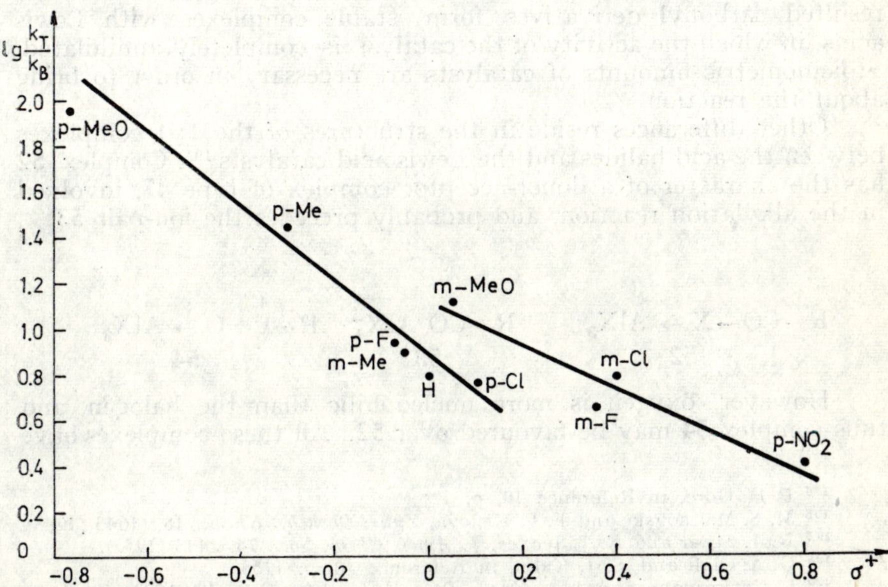

Figure 7.5 The plot of log k_T/k_B *versus* σ_X^+ values in the benzylation of toluene and benzene (see text).

nature of the transition state in the second step of the reaction, very similar to the benzylic cation structure (π complex, high negative ρ_g) or to the σ complex (late transition state, low ρ_g values).

7.4.4 *Friedel-Crafts acylations*

Acid halides and anhydrides[195] and even free carboxylic acids[196,197] in presence of Lewis acids (like AlX_3) or Brönsted acids (like sulphuric acid) yield aromatic ketones, in a reaction (7.62) similar to the alkylation reaction (7.51):

$$RCOX + MX_n \longrightarrow RCO^+MX_{n+1}^- + ArH \longrightarrow (ArCOR + MX_n) + HX \quad (7.62)$$

Aromatic aldehydes cannot be prepared in a similar way. Indeed, formyl chloride is not known and formic acid decomposes in strongly acidic medium. By modifying the general procedure[198], like for example, using a mixture of carbon monoxide and hydrogen chloride in the presence of aluminium chloride, Gattermann and Koch[199] have obtained aromatic aldehydes. More recently, benzaldehydes were obtained from formyl fluoride[37,200] by the general procedure (7.62), due to the higher stability of formyl fluoride[201].

Despite the formal resemblance of (7.62) to the alkylation reaction (7.51), there are notable differences between the two processes. A first difference involves the procedure itself. As is well known, the resulted carbonyl derivatives form stable complexes with Lewis acids, in which the activity of the catalyst is completely annihilated. Stoichiometric amounts of catalysts are necessary in order to bring about the reaction.

Other differences reside in the structures of the 1:1 complexes between the acid halides and the Lewis acid catalysts[195]. Complex **52** has the character of a donor-acceptor complex of type **47**, involved in the alkylation reaction, and probably precedes the ion-pair **53**:

$$\underset{\mathbf{52}}{R{-}CO{-}X \rightarrow AlX_3}; \qquad \underset{\mathbf{53}}{R{-}CO^+AlX_4^-} \qquad \underset{\mathbf{54}}{R{-}\overset{\overset{\displaystyle X}{|}}{C}{=}O \rightarrow AlX_3}$$

However, oxygen is more nucleophilic than the halogen, and thus complex **54** may be favoured over **52**. All these complexes have

195 P. H. Gore, in Reference 10, p. 1.
196 M. S. Malinovski and F. F. Kislova, *Zhur. Obshch. Khim.*, **18**, 1643 (1948).
197 R. J. Speer and K. J. Jeanes, *J. Amer. Chem. Soc.*, **74**, 2443 (1952).
198 G. A. Olah and S. J. Kuhn, in Reference 10, p. 1153.
199 L. Gattermann and J. A. Koch, *Ber. dtsch. chem. Ges.*, **30**, 1622 (1897).
200 G. A. Olah and S. J. Kuhn, *Chem. Ber.*, **89**, 866 (1956).
201 A. N. Nesmeyanov and E. I. Kuhn, *Ber. dtsch. chem. Ges.*, **67**, 203 (1934).

been identified by the usual spectral techniques[186,195,202,203], and each of them may be involved in the substitution. According to Jensen and Goldman[204], acylation reactions cannot be explained by a single mechanism; there must be at least three mechanisms, corresponding to the three intermediates mentioned above (Scheme 2):

Mechanism A:

$$\underset{\mathbf{54}}{\mathrm{R{-}\overset{\overset{\displaystyle X}{|}}{C}{=}O{\rightarrow}AlX_3}} + \mathrm{ArH} \longrightarrow \mathrm{R{-}\overset{\overset{\displaystyle X}{|}}{\underset{\underset{\displaystyle O{-}AlX_3}{|}}{C}}{-}ArH^+} \longrightarrow \text{products}$$

Mechanism B:

$$\mathrm{ArH} + \underset{\mathbf{54}}{\mathrm{RCO{-}X{\rightarrow}AlX_3}} \longrightarrow \mathrm{RCOArH^+AlX_4^-} \longrightarrow \text{products}$$

Mechanism C:

$$\mathrm{ArH} + \underset{\mathbf{53}}{\mathrm{RCO^+AlX_4^-}} \longrightarrow \mathrm{RCOArH^+AlX_4^-} \longrightarrow \text{products}$$

Scheme 2

Mechanism A corresponds to a condensation reaction of carbonyl derivatives with aromatic hydrocarbons (aldol-type addition). The otherwise unreactive carbonyl group from acid derivatives is rendered reactive by complexing with the Lewis acid. This mechanism is expected with the reactive aromatic compounds (phenols, etc.), which are able to trap **54** before its rearrangement into **52** and **53**.

The acylation mechanisms B and C are equivalent to mechanisms (7.51) and (7.59), respectively, of the alkylation reaction. Mechanism C has been demonstrated by Burton and Praill[205] and confirmed by other workers[206,207]. With respect to the latter mechanism, another characteristic of the acylation reaction must be noted. Oxocarbenium ions, occurring in ion pair **53**, are stable ions due to a mesomerism similar to that encountered in carbon monoxide:

$$\mathrm{R{-}C{\equiv}\ddot{O}^+} \longleftrightarrow \mathrm{R{-}\overset{+}{C}{=}\ddot{O}{:}}, \quad \text{as compared with} \quad \mathrm{:C{=}\ddot{O}{:}} \longleftrightarrow \mathrm{{}^-{:}C{\equiv}O{:}^+}$$

[202] G. A. Olah and M. B. Comisarow, *J. Amer. Chem. Soc.*, **88**, 3313, 4442 (1966).

[203] S. E. Rasmussen and N. S. Broch, *Chem. Comm.*, 289 (1965).

[204] F. R. Jensen and G. Goldman, Reference 10, p. 1003.

[205] H. Burton and P. F. G. Praill, *J. Chem. Soc.*, 1203, 2034 (1950); 522, 529, 726 (1951); 827, 837 (1953).

[206] H. C. Brown, G. Marino and L. M. Stock, *J. Amer. Chem. Soc.*, **81**, 3308, 3310 (1959).

[207] G. A. Olah and S. Kobayashi, *J. Amer. Chem. Soc.*, **93**, 6964 (1971).

Moreover, depending on R, conjugation may be extended. Thus, benzoyl cations [208] present large contributions from a "ketene-like" structure, as demonstrated by ^{13}C n.m.r. spectra.

$$C_6H_5-\overset{+}{C}=O \longleftrightarrow {}^{+}C_6H_5=C=O \longleftrightarrow C_6H_5^{+}=C=O \text{ etc.}$$

55

Thus, Friedel-Crafts acylations will have higher selectivities than Friedel-Crafts alkylations (Table 7.3). Indeed, benzoylation reactions of toluene and benzene [207] with substituted benzoyl chlorides and Lewis acids, have selectivity factors (k_T/k_B) ranging between 52 ($X = p\text{-}NO_2$) and 233 ($X = p\text{-}OMe$):

$$XC_6H_4-COCl + MCl_n \longrightarrow XC_6H_4-\overset{+}{CO}\,\overset{-}{MCl}_{n+1} \xrightarrow{ArH} XC_6H_4-COAr + HCl + MCl_n \quad (7.63)$$

The trend observed in the variation of the ratio k_T/k_B is similar to that found in the benzylation reactions (see p. 386) and demonstrates that acylations, as other electrophilic substitutions, do not have a fixed transition state.

In the formylation reactions with HCOF in presence of BF_3, no proof has been found for the intermediate occurrence of the formyl cation [209]. Nevertheless, σ complexes have been identified [36]. Such intermediates may result from protonated aromatic hydrocarbons:

$$\underset{\mathbf{56}}{[MeC_6H_5H_2]^+}\,BF_4^- + CO \longrightarrow \underset{\mathbf{57}}{[MeC_6H_5(H)CHO]^+}\,BF_4^- \longrightarrow MeC_6H_4CHO + BF_3 + HF$$

It has been shown [198] that benzenium ion **56** is transformed into the σ complex **57** on treatment with carbon monoxide. Evidently, the latter complex is converted at higher temperatures into the aromatic aldehyde. It seems thus that aldehyde synthesis with formylfluoride does not differ in its mechanism from the formylation reaction with carbon monoxide and hydrogen chloride in the presence of aluminium chloride.

[208] G. A. Olah and P. W. Westerman, *J. Amer. Chem. Soc.*, **95**, 3706 (1973).
[209] H. Hogeveen, A. F. Bickel, C. W. Hilbers, E. L. Mackor and C. Maclean, *Rec. Trav. Chim.*, **86**, 687 (1967).

7.4.5 *Other aromatic substitutions related to Friedel-Crafts alkylation and acylation*

There is a large number of reactions of aromatic compounds which can be related to Friedel-Crafts reactions. Alkylation of aromatic compounds with alkenes [210] is obviously identical in its later steps with the general reaction, differing from it only in the generation of the carbenium ions:

$$R^1CH{=}CHR^2 + AH \rightleftharpoons R^1\overset{+}{C}H{-}CH_2R^2 \xrightarrow{ArH} ArCH\begin{matrix}CH_2R^2\\R^1\end{matrix} \qquad (7.64)$$

$$\updownarrow$$

$$R^1CH(A){-}CH_2R^2$$

Important side reactions in the course of Friedel-Crafts alkylations are observed if the alkylating agent has two or more potential reaction sites, as for example polyhalogenoalkanes and polyenes [211]. Depending on the reaction conditions, polyarylation of the reactant as well as other reactions of the intermediate carbenium ions (hydride transfer, isomerisations, condensations and dealkylations) can be observed. Such an example is the Anschutz anthracene synthesis [211, 212]:

$$C_6H_6 + CHX_2{-}CHX_2 + C_6H_6 \xrightarrow{AlCl_3} \text{anthracene} \qquad (7.65)$$

In order to explain the formation of anthracene, several elementary steps must be assumed.

The Scholl reaction [212] seems to be related to the Friedel-Crafts alkylation. It has been observed [213] that beside $AlCl_3$, the reaction needs also a co-catalyst, namely hydrogen chloride. According to Nenitzescu and Balaban [212, 214], this is a clear indication of an electrophilic aromatic substitution in which the electrophilic reagent is a benzenium ion:

$$C_6H_6 + HCl + AlCl_3 \longrightarrow C_6H_7^+ \; AlCl_4^- \xrightarrow{C_6H_6} \qquad (7.66)$$

$$H_2C_6H_5{-}C_6H_6^+ \longrightarrow C_6H_5{-}C_6H_5$$

[210] R. Koncos and B. S. Friedman, Reference, 9, p. 289.
[211] S. H. Patinkin and B. S. Friedman, Reference, 9, p. 1.
[212] A. T. Balaban and C. D. Nenitzescu, Reference 9, p. 979.
[213] G. Baddeley, *J. Chem. Soc.*, 994 (1950).
[214] C. D. Nenitzescu and A. T. Balaban, *Chem. Ber.*, **91**, 2109 (1958).

Another possible mechanism[215] involves the formation of a radical-cation, the arylation reaction being thus a homolytic substitution of the aromatic ring:

$$\text{(H, R)-C}_6\text{H}_5^{+} \xrightarrow[-H^+]{-e^-} \text{R-C}_6\text{H}_5^{\bullet +}(\text{H}) \xrightarrow{-H^+} \text{R-C}_6\text{H}_5 \quad (7.67)$$

Further support for the radical-cation mechanism (7.67) was obtained in the study of the reactions of aromatic hydrocarbons with Lewis acids such as $SbCl_5$ and SbF_5[75,216], which are known to have also oxidising ability. Thus, anthracene and naphthalene[216] treated with SbF_5 in SO_2 or SO_2ClF yield dications, due to their very low ionisation potential (see p. 202). On the other hand, benzenoid aromatics with higher ionisation potentials, placed under similar conditions, lead to stibines along with biphenyls and polymers[75] (Scholl-type reactions). The formation of polymers can be completely prevented by adding hydrogen fluoride to the reaction mixture. This indicates a radical mechanism for the dimerisation and polymerisation products and probably also for the Scholl reaction. The preferential formation of stibines from benzene and the rapid transformation of the intermediate biphenyl into polymeric compounds is also in agreement with the higher ionisation potential of benzene and the lower value for biphenyl[217].

Halomethylation reactions:

$$C_6H_6 + CH_2O + HCl \longrightarrow C_6H_5\text{—}CH_2Cl + H_2O$$

involve the stable carbenium ion: $\overset{+}{C}H_2\text{—}OH \leftrightarrow CH_2\text{=}\overset{+}{O}H$. High selectivities are thus expected in this reaction. Indeed, Brown and Nelson[218] have found for the chloromethylation of toluene and benzene with trioxymethylene and HCl—$ZnCl_2$, in acetic acid, that $k_T/k_B =$ $= 113$ and $m < 1\%$.

Fries rearrangement[219] and the Houben-Hoesch reaction[220] are also examples of acylation reactions.

215 J. J. Rooney and R. C. Pink, *Proc. Chem. Soc.*, 142 (1961).
216 D. M. Brouwer and J. A. von Doorn, *Rec. Trav. Chim.*, **91**, 1110 (1972).
217 See Reference 90, p. 191.
218 H. C. Brown and K. L. Nelson, *J. Amer. Chem. Soc.*, **75**, 6292 (1953).
219 A. Gerecs, Reference 10, p. 499.
220 W. Ruske, Reference 10, p. 383.

7.4.6 *Reactions of alkanes and alkenes related to Friedel-Crafts reactions*

In 1930 Nenitzescu [221] tried to prepare acetophenone from a Romanian gasoline fraction, rich in benzene. Beside acetophenone, he observed also the formation of saturated ketones resulted from the cycloalkanes present in the gasoline fraction. Similar but overlooked observations had been made previously [222]; it was however Nenitzescu who realised that alkanes and cycloalkanes in presence of aluminium chloride can act as Friedel-Crafts reagents. Further systematic work in the following years confirmed this view and allowed a better understanding of the activity of Lewis acids towards organic substrates.

Most of the reactions of alkanes and cycloalkanes in the presence of Lewis acids have been treated in Chapter 6. This section will be therefore limited to a brief survey of the reactions of aliphatic compounds related to the typical Friedel-Crafts reactions.

(A) Isomerisation reactions [223, 224]

Alkanes and cycloalkanes are isomerised to branched hydrocarbons by aluminium halides. It has been first demonstrated by Nenitzescu [221] that freshly and carefully purified aluminium chloride is completely inactive in promoting isomerisation. Water molecules may however act as a co-catalyst and it was shown later [225] that hydrogen is a by-product in the reactions catalysed by water-activated aluminium chloride. Heldman [226] has noted that hydrogen halides are also co-catalysts in the isomerisation of n-butane into i-butane. However, Pines and Wackher[227] have shown that hydrogen halides are only efficient in presence of traces of alkenes. In absence of alkenes, hydrogen bromide may act as a co-catalyst if oxygen is also present. Isomerisation reactions are always accompanied by alkylation and dealkylation reactions, leading to products with higher and lower molecular weight, respectively, than the corresponding isomerised hydrocarbon.

The nature of the intermediates in these reactions has been demonstrated by Bartlett, Condon and Schneider [228] in a brilliant paper published in 1944. An apparatus allowing very short contact times between t-butylchloride and aluminium bromide, in isopentane solution, was used. At very short contact times (< 0.001 s), only halogen exchange was observed:

$$Me_3CCl + AlBr_3 \longrightarrow Me_3CBr + AlBr_2Cl$$

[221] C. D. Nenitzescu, C. N. de Bie and I. Cantuniari, *Revista Antigaz*, (7—9), 1 (1930); C. D. Nenitzescu and I. Cantuniari, *Ber. dtsch. chem. Ges.*, **66**, 1097 (1933).

[222] N. D. Zelinsky, *Ber. dtsch. chem. Ges.*, **28**, 1022 (1895); **30**, 387 (1897).

[223] H. Pines and J. M. Muvity, Reference 11, Chapter 39.

[224] H. Pines and N. E. Hoffman, Reference 9, p. 1211.

[225] C. D. Nenitzescu, M. Avram and E. Sliam, *Bull. Soc. chim. France*, 1266 (1955).

[226] J. D. Heldman, *J. Amer. Chem. Soc.*, **66**, 1786, 1789 (1944).

[227] H. Pines and R. C. Wackher, *J. Amer. Chem. Soc.*, **68**, 1642, 2518 (1946).

[228] P. D. Bartlett, F. E. Condon and A. Schneider, *J. Amer. Chem. Soc.*, **66**, 1531 (1944).

At intermediate contact times (0.001—0.22 s) hydrogen-halogen exchange between the organic halide and isopentane occurred:

$$\mathrm{Bu^tBr + HCMe_2Et \xrightarrow{AlBr_3} Bu^tH + BrCMe_2Et}$$

At longer reaction times (contact times > 0.22 s), alkylation reactions leading to products of high molecular weight were observed.

These observations have permitted the understanding of both isomerisation and alkylation reactions[223, 224]. The reaction mechanism (Scheme 3) corresponds to a heterolytic chain reaction.

Initiation step:

(i) water as co-catalyst:

$$\mathrm{AlX_3 + H_2O \longrightarrow [AlX_3OH]^-H^+ + R^0H \longrightarrow R^{0+} + H_2 + [AlX_3OH]^-}$$

(ii) hydrogen halides as co-catalysts:

a. $\mathrm{Alkene + HX \xrightarrow{AlX_3} R^{0+}AlX_4^-}$

b. $\mathrm{2HBr \xrightarrow{[O]} Br_2 + R^0H \xrightarrow{AlBr_3} R^0Br \xrightarrow{AlBr_3} R^{0+}\,AlBr_4^-}$

$\mathrm{R^{0+} + R^1{-}H \longrightarrow R^0H + R^{1+}}$

Propagation step: $\mathrm{R^{1+} \rightleftharpoons R^{2+}}$

$\mathrm{R^{2+} + R^1{-}H \rightleftharpoons R^2{-}H + R^{1+}}$ etc.

Chain termination; $\mathrm{R^+ + X^- \longrightarrow R{-}X}$ etc.

Scheme 3

Isomerisation reactions occur very frequently in the course of Friedel-Crafts alkylations of aromatic compounds. The isomerisation of the aliphatic chain may precede or may follow the aromatic substitution. In the former case, the resulted carbenium ion may undergo a Wagner-Meerwein rearrangement prior to its capture by the aromatic compound (see p. 384 for an example). The subsequent isomerisation of the chain of an alkylated product does not differ from the general mechanism depicted in Scheme 3. However, alkylbenzenes may also suffer an *o—m—p* isomerisation, in which the kinetically controlled product usually is transformed into the thermodynamically stable product. The following mechanisms of *o—m—p* isomerisation may operate:

(a) An intermolecular isomerisation involves an electrophilic aromatic dealkylation (corresponding to the reverse reaction), followed by another reversible alkylation. Such reactions are thermodynamically controlled and the *meta* isomer will be favoured over the *para* isomer.

(b) Bond-to-bond isomerisation [229] was observed in the isomerisation of *ortho*-dialkylbenzenes. Contrary to the theory, the *para* isomer seems to be favoured over the *meta* isomer. According to Olah and his co-workers [229], this unusual reaction may be explained by a rearrangement of the carbonium ions:

$$\text{o-}C_6H_4R_2 + H^+ \longrightarrow \mathbf{58} \tag{7.68}$$

$$\mathbf{58} \longrightarrow \mathbf{59} \xrightarrow[\text{isomerisation}]{\text{bond-to-bond}} \mathbf{60} \tag{7.69}$$

$$\mathbf{60} \xrightarrow{\text{fast}} \mathbf{61} \longrightarrow \text{p-}C_6H_4R_2 \tag{7.70}$$

$$\mathbf{60} \xrightarrow{\text{slow}} \mathbf{62} \longrightarrow \text{m-}C_6H_4R_2 \tag{7.71}$$

[229] G. A. Olah, R. H. Schlosberg, R. D. Porter, Y. K. Mo, D. P. Kelly and G. D. Mateescu, *J. Amer. Chem. Soc.*, 94, 2034 (1972).

The protonated benzenium ion **58** may rearrange into the C-benzenonium ion **59** which will equilibrate with the benzenonium ion **60** (bond-to-bond isomerisation). The collapse of **60** into the benzenium ion **61**, stabilised by inductive and hyperconjugation effects, will be faster than the alternative route to **62**. The formation of the *para*-dialkylated product will be thus favoured.

(B) Alkylation reactions

Alkenes may be alkylated either with alkyl halides (and similar derivatives) or with alkenes[230]. The mechanism of the alkylation with organic halides is obviously similar to the aromatic Friedel-Crafts alkylation:

$$R{-}X + AlX_3 \longrightarrow R^+AlX_4^-$$

$$R^+AlX_4^- + RCH{=}CHR \longrightarrow R_2CH{-}\overset{+}{C}HR\ AlX_4^- \longrightarrow$$

$$R_2CH{-}CHXR + AlX_3$$

$$\downarrow$$

$$R_2C{=}CHR + AlX_3 + HX$$

The alkylation of alkenes with alkanes involves also hydride transfer, for example:

$$CH_2{=}CH_2 + HCl + AlCl_3 \longrightarrow CH_3{-}CH_2^+AlCl_4^- \qquad (7.72)$$

$$CH_3CH_2^+AlCl_4^- + Me_3CH \rightarrow CH_3{-}CH_3 + Me_3C^+\ AlCl_4^- \qquad (7.73)$$

$$Me_3C^+AlCl_4^- + CH_2{=}CH_2 \longrightarrow Me_3CCH_2CH_2^+\ AlCl_4^-,\ \text{etc.} \qquad (7.74)$$

Alkylation reactions are reversible processes and, thus, dealkylation reactions may also accompany them. Usually, a stable carbenium ion is formed by a β-elimination (the reverse of reaction (7.74), for example). Rearrangements of the intermediate carbenium ions are also possible and thus the overall process may be quite complex.

(C) Acylation reactions[173]

The simple acylation reaction (7.75):

$$RCOX + \overset{|}{C}H{=}\overset{|}{C}H \xrightarrow{AlX_3} RCO{-}\overset{|}{C}H{-}\overset{|}{C}H{-}X \qquad (7.75)$$

[230] L. Schmerling, *J. Amer. Chem. Soc.*, **66**, 1422 (1944); L. Schmerling, Reference 9, p. 1075.

is called *Darzens reaction*[231], even though it was first observed by Kondakov[232]. Possible intermediates and reaction products in the acylation of a tertiary alkene are shown in Scheme 4:

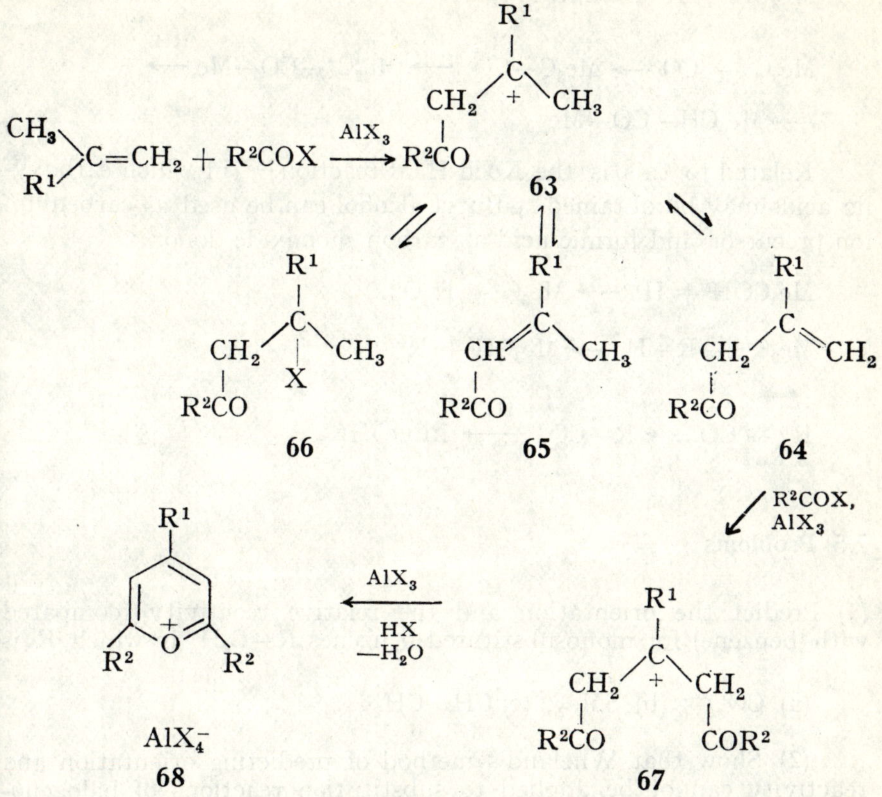

$$\underset{R^1}{\overset{CH_3}{}}\!\!>C{=}CH_2 + R^2COX \xrightarrow{AlX_3} R^2CO{-}CH_2{-}\overset{+}{C}(R^1){-}CH_3 \quad \mathbf{63}$$

$$\mathbf{63} \rightleftharpoons R^2CO{-}CH_2{-}CX(R^1){-}CH_3 \quad \mathbf{66}$$

$$\mathbf{63} \rightleftharpoons R^2CO{-}CH{=}C(R^1){-}CH_3 \quad \mathbf{65}$$

$$\mathbf{63} \rightleftharpoons R^2CO{-}CH_2{-}C(R^1){=}CH_2 \quad \mathbf{64}$$

$$\mathbf{64} \xrightarrow{R^2COX,\ AlX_3} R^2CO{-}CH_2{-}\overset{+}{C}(R^1){-}CH_2{-}COR^2 \quad \mathbf{67}$$

$$\mathbf{67} \xrightarrow[HX,\ -H_2O]{AlX_3} \text{2,6-}R^2_2\text{-4-}R^1\text{-pyrylium } AlX_4^- \quad \mathbf{68}$$

Scheme 4

The initial step of the acylation will produce ion **63**, which may collapse either into the Darzens products **65** and **66** or into the unsaturated ketone **64**, unfavoured if stoichiometric amounts of the reagents are used. However, if an excess of acyl halide is used, **64** will be the more reactive of the two unsaturated ketones, and dication **67** may be formed. The cyclisation of this intermediate leads to pyrylium salts **68**[233], constituting one of the most convenient routes in the synthesis of such derivatives.

The Hopff reaction is the counterpart, in the aliphatic series, of the Gattermann-Koch reaction[234]. The reaction consists in the treatment of an alkane possessing a tertiary carbon atom with carbon

[231] G. Darzens, *Compt. rend.*, **150**, 707 (1910).
[232] I. B. Kondakov, *J. Russ. Phys. Chem. Soc.*, **24**, 309 (1892).
[233] A. T. Balaban and C. D. Nenitzescu, *Liebigs Ann. Chem.*, **625**, 74 (1959).
[234] H. Hopff, *Ber. dtsch. chem. Ges.*, **65**, 482 (1932).

monoxide and an acid, in presence of a precursor of a carbenium ion (alkene or alcohol). Ketones are formed according to the following reactions:

$$Me_3CH + R^+ \longrightarrow Me_3C^+ + RH$$

$$Me_3C^+ + CO \longrightarrow Me_3C{-}CO^+ \longrightarrow Me_2C^+{-}CO{-}Me \xrightarrow{RH}$$

$$\longrightarrow Me_2CH{-}CO{-}Me$$

Related to this is the Koch-Haaf reaction [235], in which carboxylic acids may be obtained. t-Butyl alcohol can be used as carbenium ion precursor and formic acid as carbon monoxide donor :

$$Me_3COH + H^+ \longrightarrow Me_3C^+ + H_2O$$

$$Me_3C^+ + R{-}H \longrightarrow Me_3CH + R^+$$

$$R^+ + CO \longrightarrow R{-}CO^+ \xrightarrow{H_2O} RCOOH$$

7.5 Problems

(1) Predict the orientation and the relative reactivity (compared with benzene) for monosubstituted benzenes $R{-}C_6H_5$ in which R is

(a) COO^-; (b) $\overset{+}{S}Me_2$; (c) $CH{=}CH_2$

(2) Show that Wheland's method of predicting orientation and reactivity cannot be applied to substitution reactions of halogenobenzenes.

(3) How can you explain that nitration is irreversible, whereas sulphonation (involving a group with similar electronic requirements) is reversible?

(4) Calculate the reactivity numbers for the substitution reactions of the following aromatic hydrocarbons:

(a) benzene; (b) naphthalene (α-position); (c) naphthalene (β-position); (d) anthracene (γ-position).

Give an interpretation of the resulting values.

(5) Calculate the reactivity numbers for the different positions of phenanthrene. Show in what ratio the various monosubstitution products of phenanthrene should be expected.

(6) Explain why the acetylation of toluene gives a lower ratio of *ortho* derivative than does the corresponding benzoylation.

[235] W. Haaf and H. Koch, *Liebigs Ann. Chem.*, **638**, 122 (1960).

8
Aromatic nucleophilic substitution

The displacement of a group X with its bonding electrons, from an aromatic carbon atom, by a nucleophile, Nu:, is termed *aromatic nucleophilic substitution* (S_NAr)[1].

$$\text{Nu:} + \text{Ar—X} \rightarrow \text{Nu—Ar} + \text{X:} \qquad (8.1)$$

In the general equation (8.1), the substrate may either be a benzenoid aromatic or a heterocyclic compound. Unlike the aliphatic nucleophilic substitution, the displaced group X may be even a hydride ion.

Many of the aromatic nucleophilic substitutions are of great interest in the organic technology (dyestuffs, pharmaceuticals, etc.). Some of the most frequent reactions are shown below:

(1) *Displacement of hydride ions*

(a) Hydroxylation:

$$\text{Ar—H} + \text{HO}^- \longrightarrow \text{ArOH} + (\text{H:}^-) \qquad (8.2)$$

(b) Amination:

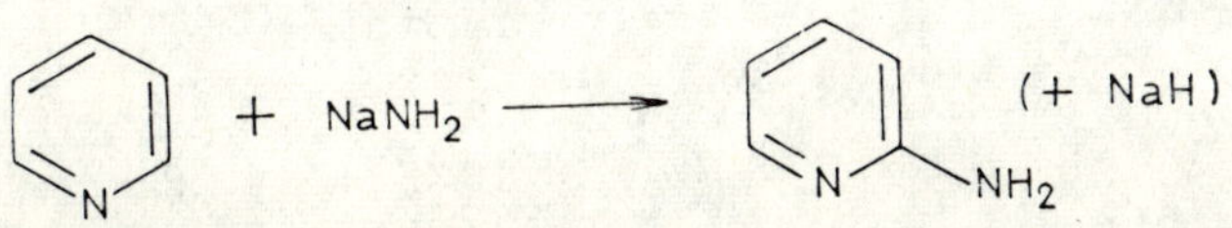

$$\text{pyridine} + NaNH_2 \longrightarrow \text{2-aminopyridine} \; (+ \text{NaH}) \qquad (8.3)$$

(c) Alkylations:

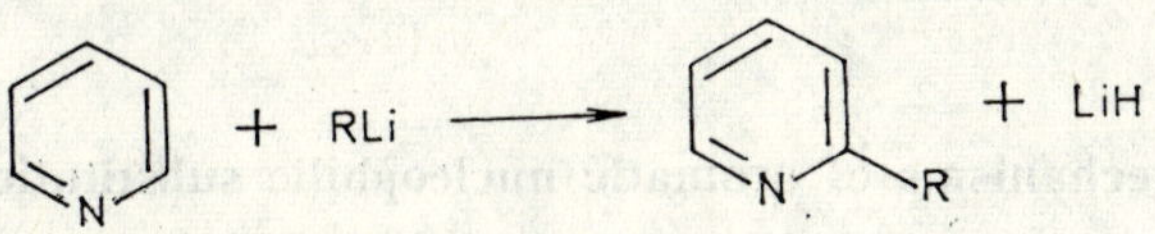

$$\text{pyridine} + \text{RLi} \longrightarrow \text{2-alkylpyridine} + \text{LiH} \quad \text{etc.} \qquad (8.4)$$

(2) *Displacement of halide ions*

(a) Hydroxylation, alkoxylation:

$$\text{Ar—Cl} + \text{RO}^- \longrightarrow \text{Ar—OR} + \text{Cl}^- \;\; (\text{R} = \text{H, alkyl}) \qquad (8.5)$$

[1] J. F. Bunnett, *Quart. Rev. Chem. Soc.*, 12, 1 (1958); J. Miller, *Aromatic Nucleophilic Substitution*, Elsevier, Amsterdam (1968).

(b) Amination:

$$Ar{-}Br + RNH_2 \longrightarrow Ar{-}NHR + (HBr) \qquad \text{etc.} \qquad (8.6)$$

(3) *Displacement of polyatomic groups*

(a) Displacement of a diazonium group:

$$Ar{-}\overset{+}{N}{\equiv}N + Nu{:}^- \longrightarrow Ar{-}Nu + N_2 \qquad (8.7)$$

(b) Displacement of a sulphonic group:

$$Ar{-}SO_3^- + Nu{:}^- \longrightarrow Ar{-}Nu + SO_3^{2-} \qquad \text{etc.} \qquad (8.8)$$

Clearly, the nature of the C—X bond in aromatic compounds must be quite different from that in the corresponding aliphatic derivatives. The differences will be determined by the hybridisation of the carbon atom. Indeed, due to the proportion of *s* orbital in the hybridised orbitals used in the formation of MOs, the carbon atom will attract more strongly the bonding electrons in a C_{sp^2}—X bond than it does in a C_{sp^3}—X bond. As a consequence, the aromatic nucleophilic substitution (8.1) will be disfavoured with respect to the aliphatic substitution. Moreover, if the central atom of the leaving group X has a pair of lone electrons, the strength of the C—X bond will be increased, due to *p*-π conjugation:

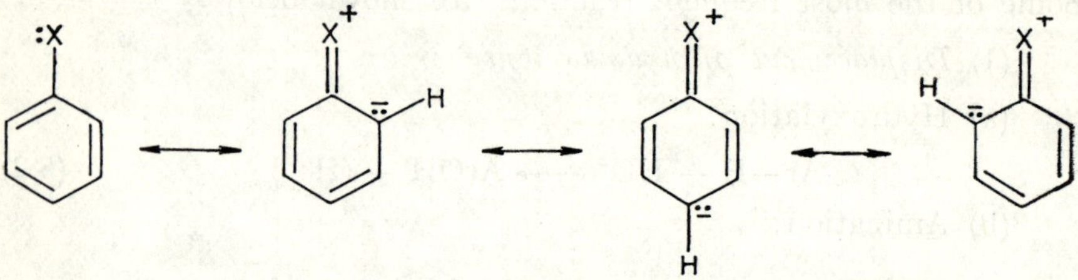

1

Considering these features of the C_{arom}—X bond, it is not surprising that the aromatic nucleophilic substitution has completely different possible mechanistic pathways.

8.1 Reaction mechanisms of aromatic nucleophilic substitutions

The bimolecular (S_NAr2) and unimolecular (S_NAr1) mechanisms are also encountered in the aromatic series. Beside these mechanisms, which may differ however from the classical S_N2 and S_N1 reactions, at least two other kinds of aromatic nucleophilic displacements are known.

(a) *The unimolecular mechanism* (S_NAr1). The unimolecular reactions, which are quite frequent in the aliphatic nucleophilic substitutions, are rather rarely encountered in the aromatic series. Actually, this mechanism is limited to the most efficient leaving-groups, such as the diazo group:

$$Ar-\overset{+}{N}\equiv N \longrightarrow Ar^{+} + N_2 \longrightarrow \text{products}$$

However, even for the diazo derivatives other mechanistic alternatives are possible.

(b) *The bimolecular mechanism* (S_NAr2). As in the aromatic electrophilic substitution, the rehybridisation of the sp^2 aromatic carbon leads to a σ complex, **3**:

(8.9)

2a Transition state **3** Intermediate *2b* Transition state

However, there is a major difference with respect to aromatic electrophilic substitution: the reagent (nucleophile) is rejected by the cloud of π electrons. Activating substituents, acting by electron-withdrawing mechanisms, will be needed in order to enable the substitution. As such substituents modify the distribution of electrons in the aromatic nucleus, the orientation problem must not be neglected. Another feature of the aromatic nucleophilic substitution is that the second step (transition state *2b*) may be assisted by the reaction medium, which is sometimes directly involved in the chemical transformation. That is the case for hydride ion displacements (X=H). Hydride acceptors, such as oxidising reagents and even atmospheric oxygen, may be helpful.

Nucleophilic substitution by the bimolecular pathway is frequently encountered when heteroaromatic compounds are involved, as for example pyridine. The heteroatom replaces in such cases the activating electron-attracting group in the benzenoid series.

(c) *Aryne mechanism*. As in the case of nucleophilic substitutions at the sp^2 carbon atom in alkenes (see p. 294), the nucleophilic aromatic substitution may consist in an elimination, followed by an addition of nucleophile. The generality of this mechanism has been demonstrated by Roberts and his co-workers[2]:

B: elimination; Nu:, BH addition

(8.10)

[2] J. D. Roberts, D. A. Semenov, H. E. Simmons and C. A. Carlsmith, *J. Amer. Chem. Soc.*, **78**, 601 (1956).

Two products are formed, their ratio depending on the other substituents of the aromatic nucleus. Nevertheless, this ratio is independent of the leaving group X.

(d) *The unimolecular, radical-anion mechanism* ($S_{RN}1$). Studying the aryne mechanism in the case of halogenopseudocumenes **4**, Kim and Bunnett[3] have found, as expected, that the ratio of the two resulted pseudocumidines **6*a*** and **6*b*** is fairly constant (~1.5) and independent of the initial position of the halogen and of the nature of it if X = Cl or Br.

Me Me X Me **4*a*** B: Me Me Me **5** B: X Me Me Me **4*b***

NH₂:⁻ BH NH₂:⁻ BH

Me Me H₂N Me **6*a*** Me H₂N Me Me **6*b***

However, if X = I, the ratio **6*b*/6*a*** may differ significantly from the above aryne ratio (1.5), depending on the reaction conditions and on the position of the halogen in **4** (the values vary between 0.63 and 5.86). Clearly, another mechanism must also operate, involving no rearrangement. The addition of radical-trapping agents (for example tetraphenylhydrazine) brings the ratio of the pseudocumidines close to the aryne ratio. A radical-anion mechanism has been proposed:

$$\text{electron-donor} + \text{ArI} \longrightarrow \text{ArI}\cdot^- \quad (8.11)$$

$$\text{ArI}\cdot^- \longrightarrow \text{Ar}\cdot + \text{I}^- \quad (8.12)$$

$$\text{Ar}\cdot + \text{NH}_2^- \longrightarrow \text{ArNH}_2\cdot^- \quad (8.13)$$

$$\text{ArNH}_2\cdot^- + \text{ArI} \longrightarrow \text{ArNH}_2 + \text{ArI}\cdot^- \quad (8.14)$$

The effect of radical-trapping agents demonstrates the radical character of the reaction. The fact that this reaction is observed only with iododerivatives is in agreement with the ease of electron-transfer to iodine. The addition of electron-donors, such as metals, has a favourable effect on the reaction.

[3] J. K. Kim and J. F. Bunnett, *J. Amer. Chem. Soc.*, **92**, 7463, 7464 (1970).

8.2 The unimolecular mechanism. Reactions of the aromatic diazonium salts

The kinetic study of the decomposition of aromatic diazonium salts in aqueous solutions has shown that the rate of formation of the corresponding phenols is independent of the nature and concentration of the added anions [4–7], suggesting a unimolecular mechanism [8]:

$$\mathrm{Ar{-}\overset{+}{N}{\equiv}N \longrightarrow Ar^{+} + N_2 \xrightarrow{H_2O} ArOH + N_2} \qquad (8.15)$$

In agreement with this mechanism, the presence of electron-withdrawing substituents in the aromatic ring decreases the reaction rates [9], whereas electron-donating substituents either with $+E$ or $+I$ effects may accelerate (if situated *meta*) or retard (when *para*), the decomposition of the salt.

The reaction seems to proceed in a manner similar to the corresponding aliphatic substitution, involving ionisation, leading to ion pairs, followed by dissociation. In that respect, the works of Lewis and Insole [10] have brought a substantial contribution. These authors observed that the salts obtained either from aniline or toluidine with labelled nitrous acid, and recovered after an incomplete decomposition of the diazonium salt, have the N-label directly bonded to the aromatic ring. The result may be explained by admitting the formation of an ion pair separated by a nitrogen molecule, suffering return either to the unrearranged salt or to the rearranged labelled compound. A kinetic study of the aqueous decomposition in presence of thiocyanate ion has also indicated the possible formation of the cyclic intermediate **8**, resulted either directly from the original diazonium salt or from the ion pair **7** [11]:

$$\mathrm{Ar{-}\overset{+}{N}{\equiv}{}^{15}NY^{-} \rightleftharpoons \underbrace{(Ar\overset{+}{N}{\equiv}{}^{15}N \rightleftharpoons Ar^{15}\overset{+}{N}{\equiv}N)\,Y^{-}}_{\textbf{7}} \rightleftharpoons Ar{-}{}^{15}\overset{+}{N}{\equiv}NY^{-}}$$

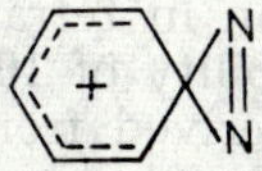

8

Bose and Kugajewsky [12] have challenged the scheme proposed by Lewis and Insole, being unable to detect by the n.m.r. technique

[4] H. Euler, *Liebigs Ann. Chem.*, **325**, 292 (1902).
[5] H. A. H. Pray, *J. Phys. Chem.*, **30**, 1417 (1926).
[6] M. L. Crossley, R. A. Kienle and C. H. Benbrook, *J. Amer. Chem. Soc.*, **62**, 1400 (1940).
[7] D. F. De Tar and A. R. Ballentine, *J. Amer. Chem. Soc.*, **78**, 3916 (1956).
[8] W. A. Waters, *J. Chem. Soc.*, 266 (1942).
[9] E. S. Lewis and W. H. Hinds, *J. Amer. Chem. Soc.*, **74**, 304 (1952).
[10] E. S. Lewis and J. M. Insole, *J. Amer. Chem. Soc.*, **86**, 32 (1964).
[11] E. S. Lewis and J. M. Insole, *J. Amer. Chem. Soc.*, **86**, 34 (1964).
[12] A. K. Bose and I. Kugajewsky, *J. Amer. Chem. Soc.*, **88**, 2325 (1966).

the rearrangement within the diazonium salt. Lewis and Holliday [13] have re-checked their results, ruling out the possibility of experimental errors. Indeed, they were able to establish a correlation between the migrated nitrogen and the decomposed salt, having the precision of a linear relationship, excluding thus any fortuitous fitting of data. The recombination of ion pair **7** to give the original diazo compound has an analogy in the combination of carbon monoxide with carbenium ions (Section 7.4.6):

$$R^+ + CO \longrightarrow RCO^+ \longrightarrow \text{products} \qquad (8.16)$$

Carbon monoxide is isoelectronic with the nitrogen molecule, supporting thus the interpretation of Lewis and his co-workers.

The intervention of carbenium ions in the decomposition of diazonium salts is also confirmed by hydride transfer reactions. Thus, if the decomposition is performed in presence of suitable hydride donors, an aromatic hydrocarbon is formed beside another carbocation:

$$ArN_2^+ \longrightarrow Ar^+ + N_2$$
$$Ar^+ + R{-}H \longrightarrow Ar{-}H + R^+ \qquad (8.17)$$

Meerwein and his co-workers [14] proved the hydride transfer by carrying out the decomposition in aqueous ethers like 1,3-dioxolane:

$$Ar^+ + \underset{\mathbf{9}}{\text{(1,3-dioxolane, C2 bearing H and R)}} \longrightarrow Ar{-}H + \underset{\mathbf{10}}{\text{(2-R-1,3-dioxolan-2-ylium)}}$$

The decomposition of *p*-chlorobenzenediazonium chloride in **9** proceeds 30 times faster than in tetrahydrofuran. This difference is explained by the higher stability of **10** as compared with the stability of the carbenium ion derived from tetrahydrofuran. Caution must however be observed with other reductions of diazonium salts. The reduction with alcohols, for example, has been proved [15, 16] to proceed by a homolytic pathway, in which the intermediate aromatic radical extracts the α-hydrogen atom of the alcohol:

$$Ar{-}\overset{+}{N}{\equiv}N]X^- \longrightarrow Ar\cdot + N_2 + X\cdot$$
$$Ar\cdot + R_2CH{-}OH \longrightarrow ArH + R_2\dot{C}{-}OH \rightarrow R_2CO$$

[13] E. S. Lewis and R. E. Holliday, *J. Amer. Chem. Soc.*, **88**, 5043 (1966).
[14] H. Meerwein, H. Allendörfer, P. Beekmann, F. Kunert, H. Morschel, F. Pawellek and K. Wunderlich, *Angew. Chem.*, **70**, 211 (1958).
[15] L. Melander, *Arkiv. Kemi*, **3**, 525 (1951); *Chem. Abstr.*, **46**, 7534 (1952).
[16] R. Huisgen and A. Tricker, *Liebigs Ann. Chem.*, **573**, 186 (1951).

The diazo group may also be substituted by a bimolecular mechanism (8.18):

$$Ar{-}\overset{+}{N}{\equiv}N + Nu{:}^{-} \rightarrow Ar\begin{matrix} \nearrow N_2^{\delta+} \\ \searrow Nu^{\delta-} \end{matrix} \longrightarrow ArNu + N_2 \qquad (8.18)$$

This mechanism will be favoured [9] by electron-attracting groups and by high concentrations of strong nucleophiles. Usually, the bimolecular process is accompanied by the unimolecular one.

The replacement of the diazo group by a nucleophile may also take place by attack at the nitrogen atom, as it happens in the reactions of the diazonium salts in which the diazo group is preserved (diazo-coupling, reactions with strong nucleophiles like HO^-, CN^-, etc.). An example is the formation of aromatic azides. As azide ion is an efficient carbocation-trapping agent [17], the unimolecular mechanism seemed plausible:

$$Ar{-}\overset{+}{N}{\equiv}N \longrightarrow Ar^+ \xrightarrow[\text{rapid}]{N_3^-} ArN_3 \qquad (8.19)$$

However, Clusius and Hürzeler [18] have ruled out this mechanism, by showing that if in the diazonium salt either the α- or the β-nitrogen atom is labelled isotopically, some of the label is also found in the resultant azide, namely 100% of it if the original label was in α and only 17.5% if it was situated in β. The correct interpretation of this result was given by Huisgen and Ugi [19] who studied the kinetics of azide formation from benzenediazonium chloride and lithium azide in methanol. Surprisingly, at −39.5°C the reaction presents a clean first-order kinetics until 76% of the nitrogen (primary nitrogen) is evolved. The remainder of 24% (secondary nitrogen) is evolved at −0.8°C, also acc aording to clean first-order kinetics. The amount of primary nitrogen depends upon the solvent, being 76% in methanol and 65% in water. A label in the β position is not found in the primary nitrogen evolved. From these results, the authors concluded that phenylpentazole **12** must be an intermediate. The following scheme gives a quantitative account of the results obtained in the decomposition of benzenediazonium salt in water:

$$Ph{-}\overset{+}{N}{\equiv}{}^*N + N_3^- \xrightarrow{\text{rapid}} \underset{\mathbf{11}}{Ph{-}N{=}{}^*N{=}N{=}N{=}N} \xrightarrow[35\%]{k_2}$$

$$\downarrow k_1 \; 65\%$$

$$Ph{-}N{=}{}^*N{=}N + \underset{\text{primary nitrogen}}{N_2}$$

[17] R. A. Sneen, *Accounts Chem. Res.*, **6**, 46 (1973).

[18] K. Clusius and H. Hürzeler, *Helv. Chim. Acta*, **37**, 798 (1954); K. Clusius and H. Craubner, *Helv. Chim. Acta*, **38**, 1060 (1955).

[19] R. Huisgen and I. Ugi, *Chem. Ber.*, **90**, 2914 (1957).

$$\xrightarrow[35\%]{k_2} \text{Ph}-\text{N}\begin{matrix} \diagup {}^*\text{N}=\text{N} \\ \qquad | \\ \diagdown \text{N}=\text{N} \end{matrix} \quad \begin{matrix} \xrightarrow{17.5\%} \text{Ph}-\text{N}={}^*\text{N}=\text{N} + \text{N}\equiv\text{N} \\ \xrightarrow{17.5\%} \text{Ph}-\text{N}=\text{N}=\text{N} + {}^*\text{N}\equiv\text{N} \end{matrix}$$

12 Secondary nitrogen

Scheme 1

The intermediate formation of phenylpentazoles is now well proved [20].

In the reactions of diazo salts, homolytic decompositions are also frequent (Section 11.4.3).

8.3 The bimolecular mechanism

The bimolecular mechanism [1,21–23] is demonstrated by numerous experimental data, such as the reaction kinetics, substituent and medium effects, and especially by the isolation and identification of reaction intermediates.

8.3.1 *Kinetics. Effects of substituents on reactivity*

Since the early 1900s it has been observed that one or more nitro groups in the aromatic ring facilitate the nucleophilic displacement of halogens [24, 25].

$$O_2N-C_6H_4-Cl + H_2N-R \longrightarrow O_2N-C_6H_4-NH-R + HCl \tag{8.20}$$

It has been later established [26–30] that the reaction (8.20) has second-order kinetics (first-order in each reagent). Moreover, it has been observed that the nitro group is not the only group which may promote a nucleophilic substitution at the aromatic carbon. Table 8.1 summarises the relative rate constants measured in the reaction of *para*-substituted *o*-nitrochlorobenzenes with sodium methoxide [31].

[20] R. Huisgen, *Angew. Chem.*, **72**, 359 (1960).
[21] J. Sauer and R. Huisgen, *Angew. Chem.*, **72**, 294 (1960).
[22] R. Foster and C. A. Fyfe, *Rev. Pure Appl. Chem.*, **16**, 61 (1966).
[23] P. Buck, *Angew. Chem.*, **81**, 136 (1969).
[24] A. F. Holleman, W. J. De Mooy and J. Terwell, *Rec. Trav. Chim.*, **35**, 1 (1915).
[25] A. F. Holleman and A. J. De Hollander, *Rec. Trav. Chim.*, **39**, 435 (1920).
[26] J. F. Bunett and R. J. Morath, *J. Amer. Chem. Soc.*, **77**, 5051 (1955).
[27] J. F. Bunnett, R. J. Morath and T. Okamoto, *J. Amer. Chem. Soc.*, **77**, 5055 (1955).
[28] J. D. Reinheimer, W. F. Kiefer, S. W. Frey, J. C. Cochran and E. W. Barr, *J. Amer. Chem. Soc.*, **79**, 1263 (1957).
[29] J. E. Bunnett and J. J. Randall, *J. Amer. Chem. Soc.*, **80**, 6020 (1958).
[30] J. Miller and A. J. Parker, *J. Amer. Chem. Soc.*, **83**, 117 (1961).
[31] J. Miller and A. J. Parker, *Austral. J. Chem.*, **11**, 302 (1958).

TABLE 8.1

Relative rates of *o*-nitrochlorobenzenes, *para*-substituted with electron-withdrawing groups, in the reaction with sodium methoxide [31] (methanol, 0°C).

para-Substituent	Relative rate	*para*-Substituent	Relative rate
$-N_2^+$	3.83×10^8	$-COMe$	8.08×10^3
$-NO$	5.22×10^6	$-N{=}NPh$	1.07×10^3
$-NO_2$	6.73×10^5	$-N_3$	2.41
$-CN$	3.81×10^4	$-H$	1.00

The strongest activating effect is displayed by the positively charged diazonium group. Its activating power has earlier been noted by Hantzsch [32] and Hirsch [33] in the diazotisation of *p*-bromo-amines in concentrated hydrochloric acid:

$$\text{2,4-}Br_2C_6H_3{-}NH_2 \xrightarrow[HCl,\ 0^\circ C]{HONO} \text{2,4-}Br_2C_6H_3{-}\overset{+}{N}{\equiv}N \xrightarrow[0^\circ C]{Cl^-} \text{2-Br-4-Cl-}C_6H_3{-}\overset{+}{N}{\equiv}N \quad (8.21)$$

The activation of the halogen atoms in the nucleophilic substitution is explained by the electron-withdrawing effect of the diazo group. The positive charge of the substituent is delocalised in the ring, enabling thus the nucleophilic attack in the *para* and (to a smaller extent) *ortho* position:

$$Br{-}C_6H_3(Br){-}\overset{+}{N}{\equiv}N: \longleftrightarrow Br{-}\overset{+}{C_6H_3}(Br){=}\overset{+}{N}{=}\ddot{N}: \longleftrightarrow Br{-}C_6H_3(\overset{+}{}Br){=}\overset{+}{N}{=}\ddot{N}:^-$$

Other positively charged substituents have also strong activating effects. Even if no comparable kinetic data are available, it can be assumed that carbenium ions are the reactive reagents in the alkylation of phenols [34] and ambident anions, such as **13** [35]:

$$Ph_3CCl \longrightarrow Ph_3C^+ \longleftrightarrow Ph_2C{=}C_6H_5^+$$

$$Ph_2C{=}C_6H_5^+ + \bar{C}H(COCMe_3)_2 \longrightarrow Ph_2CH{-}C_6H_4{-}CH(COCMe_3)_2$$

13

[32] A. Hantzsch, *Ber. dtsch. chem. Ges.*, **30**, 2334 (1897).
[33] B. Hirsch, *Ber. dtsch. chem. Ges.*, **31**, 1253 (1898).
[34] R. Gompper, *Angew. Chem.*, **76**, 412 (1964).
[35] H. E. Zaugg, R. J. Michaels and E. J. Baker, *J. Amer. Chem. Soc.*, **90**, 3800 (1968).

The nitroso group is more effective than the nitro group in facilitating the nucleophilic substitution. This can be explained by the higher contribution of the mesomeric structure **14*b*** of a nitroso compound, as compared to the mesomeric structure **15*b*** of a nitro derivative:

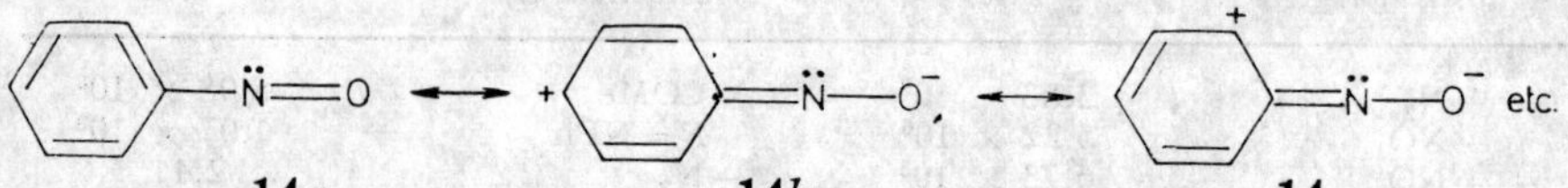

14*a* **14*b*** **14*c***

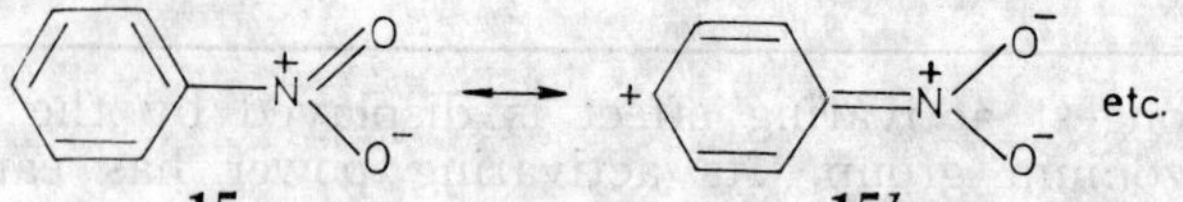

15*a* **15*b***

In the non-benzenoid and heterocyclic aromatic series, the positive systems are the more reactive in the nucleophilic substitution. Thus, pyridinium and pyrylium salts are more reactive than pyridine, which in its turn is more reactive than benzene in the S_NAr2 reaction. Their tendency to undergo attack of nucleophiles may be observed even in the absence of suitable acceptors for the leaving group (as for hydride ion, for example). However, in such instances other products than substitution products are formed. Scheme 2 shows some examples of reactions between pyrylium salts and nucleophiles.

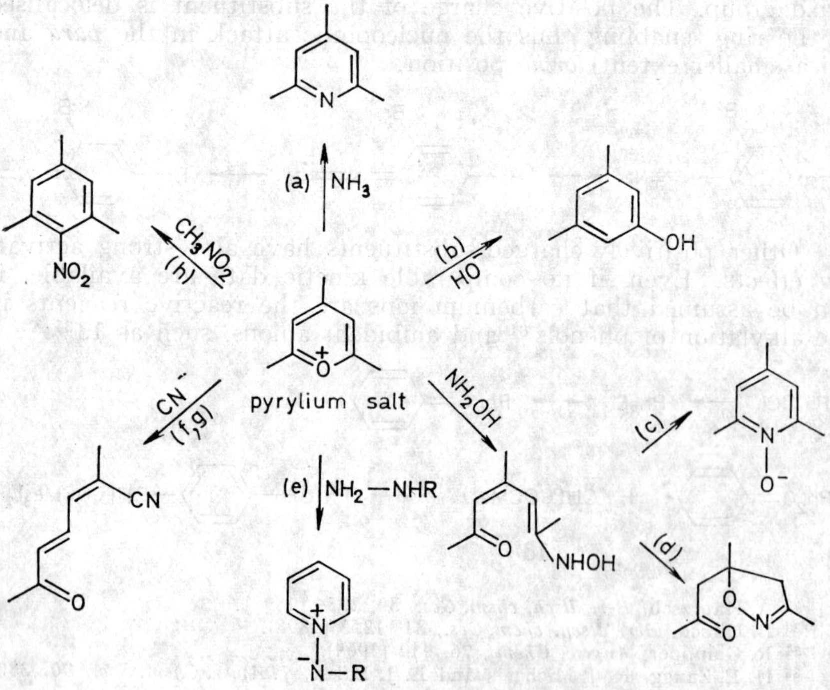

(a) NH_3 (b) HO^- CH_3NO_2 (h) CN^- (f,g) pyrylium salt NH_2OH (c) (e) $NH_2—NHR$ (d)

Scheme 2

Even though some of the reactions of pyrylium salts have been known for a long time [36a], the theoretical and practical interest in these compounds has increased only in recent years [36b]. For all reactions shown in Scheme 2 the following mechanism of ring-opening is admitted:

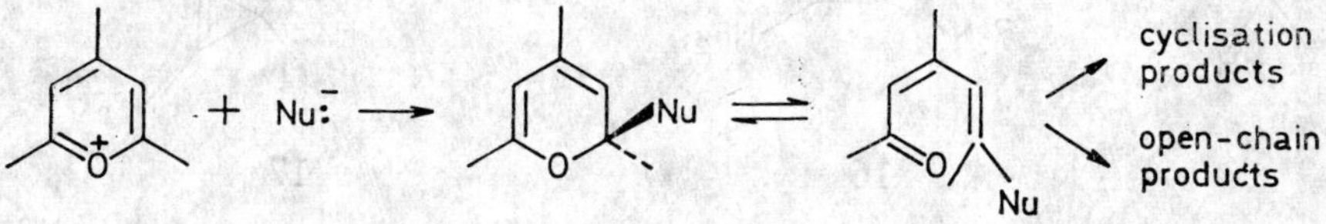

This mechanism is evident for hydroxide and amine attack [36b], can be well fitted to the other reactions of Scheme 2 [36c – 36h] with certain adjustments (for example the generation of the nucleophile in the condensation with nitroalkanes [36h]), and is further supported by other reactions of the open-ring intermediates.

An example involving the aromatic cyclopropenylium ion was studied by Yoshida *et al.* [37]. The treatment of the stable cyclopropenylium ion **16** with cyclopentadienyl anion led to **20**, a derivative of *calicene* *. The formation of the latter compound may be explained by admitting an initial coupling of the reagents to yield the neutral intermediate **17**, which rapidly ionises to give the salt **18**. The further

[36] (a) A. Baeyer and J. Piccard, *Liebigs Ann. Chem.*, **384**, 208 (1911); **407**, 332 (1915); (b) C. Toma and A.T. Balaban, *Tetrahedron*, Suppl., **7**, 9 (1966); (c) A. T. Balaban and C. D. Nenitzescu, *Liebigs Ann. Chem.*, **625**, 74 (1959); E. Schmitz, *Chem. Ber.*, **91**, 1488 (1958); (d) A. T. Balaban, *Tetrahedron*, **24**, 5059 (1968); (e) K. Dimroth, G. Arnoldy, S. von Eicken and G. Schiffer, *Liebigs Ann. Chem.*, **604**, 221 (1957); (f) A. T. Balaban and C. D. Nenitzescu, *J. Chem. Soc.*, 3566 (1961); (g) A. T. Balaban, T. H. Crawford and R. H. Wiley, *J. Org. Chem.*, **30**, 879 (1975); (h) K. Dimroth, *Angew. Chem.*, **72**, 331 (1960).

[37] Z. Yoshida, A. Aroki and H. Ogoshi, *Tetrahedron Letters*, 19 (1975).

* The name of *calicene*, proposed by Prinzbach [38a] for the mesomeric hydrocarbon **21**, suggests its shape which reminds a cup (lat. *calix*). The proposal was very soon followed by numerous syntheses [38b, 39] of substituted representatives.

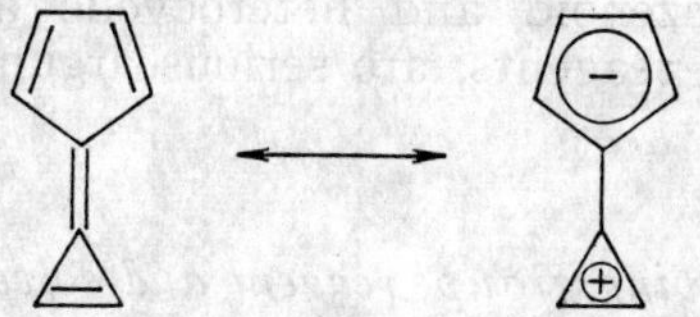

21

[38] (a) H. Prinzbach, *Angew. Chem.*, **76**, 235 (1964); (b) H. Prinzbach *et al.*, *ibid.*, **77**, 258 (1965).

[39] W. M. Jones and R. S. Pyron, *J. Amer. Chem. Soc.*, **87**, 1608 (1965); A. S. Kende and T. Izzo, *J. Amer. Chem. Soc.*, **87**, 1609 (1965); M. Ueno, I. Murata and Y. Kitahora, *Tetrahedron Letters*, 2967 (1965); E. D. Bergmann and I. Agranat, *Chem. Commun.*, 512 (1965).

reaction with the base will afford the calicene **19**, which combines rapidly in a similar way to yield **20**:

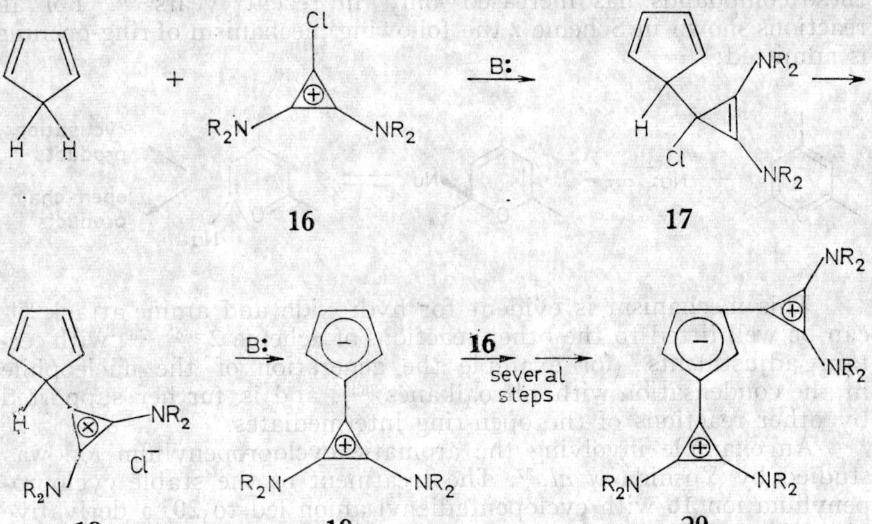

The above example illustrates the fact that the second step in an S_NAr2 reaction corresponds actually to an S_N1 ionisation.

A large amount of experimental data is now available[1] to indicate that both in the benzenoid and in the non-benzenoid or heterocyclic series there is a correlation between substrate reactivity and orientation. For benzenoid compounds the following rules may be formulated. Electron-withdrawing substituents orientate the nucleophilic substitution in *ortho-para* with activation of the aromatic nucleus, whereas electron-repelling substituents are *meta*-orienting with deactivation.

The second-order kinetics, the reactivity of the aromatic substrate closely related to the electronic effect of the substituent, the relationship between reactivity and orientation, as well as the behaviour of non-benzenoid and heterocyclic aromatic compounds towards nucleophilic reagents, are serious arguments supporting the S_NAr2 mechanism.

8.3.2 *Reactivity. Leaving group, reagent and solvent (medium) effects*

Bunnett[40] has established that the leaving group efficiency in the reaction of pyperidine with 2,4-dinitrohalogenobenzenes in methanol decreases in the order F $\gg$ Br, Cl $>$ I. Fluorine is thus one of the most efficient leaving groups, being 3300 times more rapidly

[40] J. F. Bunnett, W. E. Garbisch, Jr. and K. M. Pruitt, *J. Amer. Chem. Soc.*, **79**, 385 (1957).

displaced than iodine and 33 times faster than the toluene-*p*-sulphonic group. If reactivity and orientation are reversed with respect to the aromatic electrophilic substitution, a reversal is also observed when nucleophugicities* in aromatic and in aliphatic nucleophilic substitutions are compared. Clearly, this order of reactivity cannot be explained by a direct displacement reaction, the C—F bond being the strongest of the bonds mentioned. However, considering scheme (8.9), with a rate-determining first step, the sequence order makes sense. The dissociation of the C—X bond will not affect the overall rate. However, a strongly electronegative substituent (such as F) will favour the nucleophilic attack in the first step, and thus enhance the overall rate.

Mechanism (8.9) is also confirmed by solvent effects in the nucleophilic substitutions of fluoro- and iodo-4-nitrobenzenes [41]:

$$X\text{-}C_6H_4\text{-}NO_2 + Nu{:}^- \underset{k_{-1}}{\overset{k_1}{\rightleftharpoons}} [Nu(X)C_6H_4{=}N^+(O^-)_2]^- \underset{k_2}{\overset{k_{-2}}{\rightleftharpoons}} Nu\text{-}C_6H_4\text{-}NO_2 + X{:}^- \qquad (8.22)$$

If Nu: is N_3^-, the ratio of the observed rates k_F/k_I is 200 in methanol and increases to 400 in DMF. If Nu: is PhS^-, the observed k_F/k_I is only 2.5 in methanol and decreases to about 1 in DMF. It is well known [42,43] that in protic solvents anions are solvated by hydrogen bonds. The smaller the ionic radius of the nucleophile, the higher its solvent atmosphere. In aprotic dipolar solvents, anions are less solvated (due to the absence of hydrogen bonds) and therefore may better act as nucleophiles. The increase of rate when the reaction is transferred from a protic into an aprotic solvent will thus depend on the actual size of the ion. The observed trend in (8.22) reflects the higher bulk of thiophenoxide ion as compared with azide, but at the same time the difference in size between I^- and F^-. This difference must be reflected in the second step of the reaction, the expulsion of the leaving group, a protic solvent favouring to a higher extent the elimination of the smaller ion.

The alkylation of the nitrite ion with alkyl halides yields a mixture of nitroalkanes and alkyl nitrites [44]. Nitrite ion is thus an

* The nucleophugicity expresses the ability of a group to leave the molecule with the bonding electrons.

[41] R. Alexander, E. C. F. Ko, A. J. Parker and T. J. Broxton, *J. Amer. Chem. Soc.*, **90**, 5049 (1968).

[42] A. J. Parker, *Quart. Rev. Chem. Soc.*, **16**, 169 (1962).

[43] A. J. Parker, *Chem. Rev.*, **69**, 1 (1969).

[44] N. Kornblum, R. A. Smiley, R. K. Blackwood and D. C. Iffland, *J. Amer. Chem. Soc.*, **77**, 6269 (1955).

ambident ion, the products resulting either by oxygen or by nitrogen attack at the reaction centre. A similar situation was encountered in the aromatic nucleophilic substitution. Dinitrohalogenobenzenes **22** treated with nitrite ion in aqueous acetonitrile at 25 °C and pH 7, afford dinitrophenol, resulted by two independent routes [45]:

$$\mathbf{22}\ (X = F, Cl, Br, I) \xrightarrow{k_{O,X}} \text{ONO-dinitrobenzene} \xrightarrow{\text{rapid}} \text{dinitrophenol}$$

$$\mathbf{22} \xrightarrow{k_{N,X}} \mathbf{23}\ (\text{trinitrobenzene}) \rightarrow \text{ONO-dinitrobenzene}$$

X = F,Cl,Br,I

A careful kinetic study has allowed the estimation of each individual rate constant, for the four halogenobenzenes studied. It has thus been found that dinitrofluorobenzene undergoes only O-attack, whereas the iododerivative yields dinitrophenol exclusively by nitrogen attack ($k_{O,F} = 4 \times 10^{-2}$ mol^{-1} min^{-1}, $k_{O,I} = 10^{-8}$ mol^{-1} min^{-1}; $k_{N,I} = 3.8 \times 10^{-3}$ mol^{-1}min^{-1}). The other dinitrohalogenobenzenes may react by both mechanisms. The behaviour of the four dinitrohalogenobenzenes may be explained either by the high electronegativity of fluorine [45] (rendering the fluoroderivative a better electrophile) or by the reversibility of the oxygen attack as compared to the irreversibility of the nitrogen attack [46]. In the case of the fluoride displacement, the oxygen attack may in its turn be irreversible, due to the good solvation of the leaving ion.

In the nucleophilic displacement of a hydride ion, which occurs for instance in hydroxylation reactions [21,47,48], the presence of hydride acceptors is usually necessary. In their absence, even the substrate or the molecular oxygen may act as hydride acceptors, as for example in the hydroxylation of nitrobenzene:

[45] D. H. Rosenblatt, W. H. Dennis, Jr. and R. D. Goodin, *J. Amer. Chem. Soc.*, **95**, 2133 (1973).

[46] See footnote 14 in Reference 45.

[47] A. Wohl, *Ber. dtsch. chem. Ges.*, **32**, 3486 (1899).

[48] W. Aul, *Ber. dtsch. chem. Ges.*, **34**, 2442 (1901); E. S. Lewis and M. C. R. Symons, *Quart. Rev. Chem. Soc.*, **12**, 230 (1958).

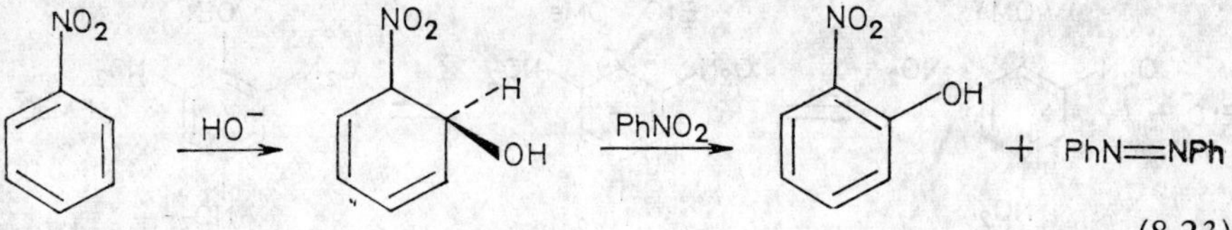

$$\text{PhNO}_2 \xrightarrow{\text{HO}^-} [\text{NO}_2\text{-cyclohexadiene with H and OH}] \xrightarrow{\text{PhNO}_2} o\text{-NO}_2\text{C}_6\text{H}_4\text{OH} + \text{PhN{=}NPh} \qquad (8.23)$$

Beside the substitution product, azobenzene is also formed. The latter results by hydride transfer to the nitro group.

Steric effects are also in agreement with an S_NAr2 mechanism. Indeed, it is known that bulky substituents may affect the reaction rates, either directly, by hindering the access of the reagent to the reaction site *(primary steric effects)* or by modifying certain intramolecular interactions at the reaction site [49] *(secondary steric effects)*. The latter effects may be involved in many nucleophilic substitutions, as for example the substitution of 1,4-dinitro-2,6-dichlorobenzene with methoxide ion [25]:

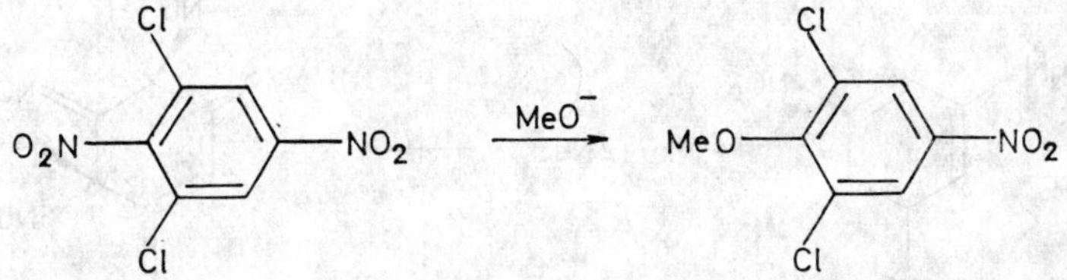

$$\text{1,4-(O}_2\text{N)}_2\text{-2,6-Cl}_2\text{C}_6\text{H}_2 \xrightarrow{\text{MeO}^-} \text{MeO-C}_6\text{H}_2\text{(Cl)}_2\text{-NO}_2 \quad \mathbf{24}$$

The substitution of the nitro group from C-1 may be explained by the fact that it is removed from the plane of the aromatic ring by the neighbouring chlorine atoms, so that it can no longer activate the *ortho* and *para* positions. However, the nitro group at C-4 may still exert its electromeric effect on C-1, favouring thus the formation of the methoxy derivative **24**.

8.3.3. *Isolation of intermediates*

Jackson and Gazzolo [50] in 1900 proposed structures like **25** for the coloured complexes formed from picryl ethers with alkoxides. Soon afterwards Meisenheimer [51] proved that the same complex is formed either from 2,4,6-trinitroanisole with potassium ethoxide, or from 2,4,6-trinitrophenetole with potassium methoxide, confirming thus the structure proposed by Jackson and Gazzolo:

[49] C. K. Ingold, *Structure and Mechanism in Organic Chemistry*, Cornell University Press, Ithaca, New York (1953), p. 60.

[50] C. J. Jackson and F. H. Gazzolo, *Amer. J. Chem.*, **23**, 376 (1900).

[51] J. Meisenheimer, *Liebigs Ann. Chem.*, **323**, 205 (1902).

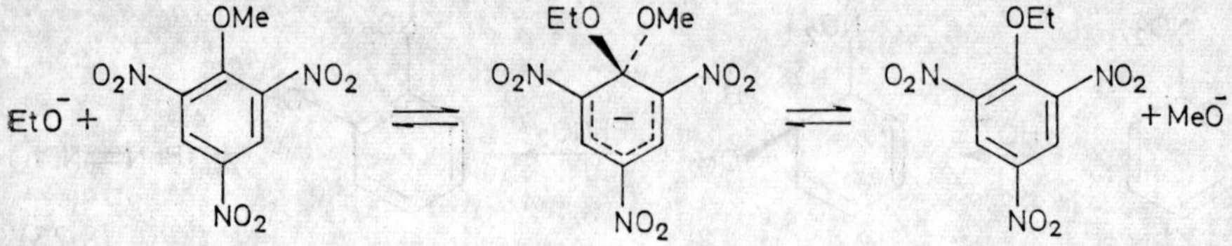

25 (8.24)

σ complexes similar to **25** are therefore called Jackson-Meisenheimer complexes.

A major development in the structural characterisation of Jackson-Meisenheimer complexes occurred in the last 10—15 years as a result of the application of modern physical methods [52]. Theoretical calculations [53] have also contributed to a better understanding of their electronic structure. Despite the negative charge of the complex, the carbon framework is electron-defficient, since the electrons are mostly attracted by the nitro group. This allows the formation of *bis-adducts* (1 : 2-complexes) with nucleophiles [52, 54, 55]:

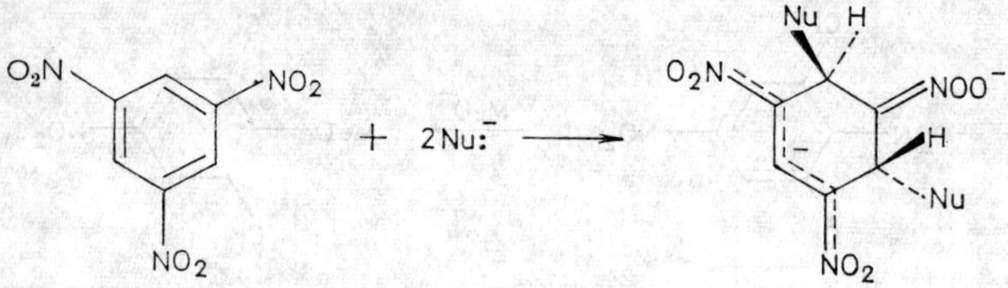

26

With rare exceptions, bis adducts **26** have been identified so far only in solutions and cannot be readily isolated. Surprisingly, *meta*-bridged analogues of complexes **26** may be isolated, and seem to be quite stable [56]. Acetone and trinitrobenzene with triethyl amine, for instance, readily form the ammonium salt **27**, with no tendency to react further.

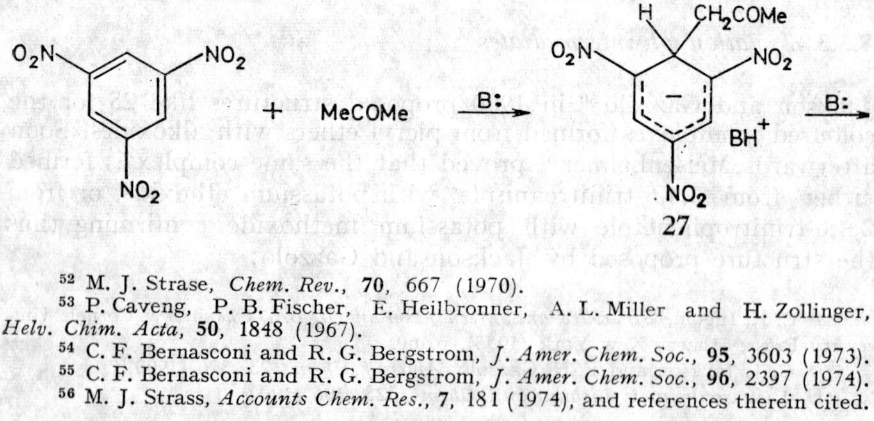

27

[52] M. J. Strase, *Chem. Rev.*, **70**, 667 (1970).

[53] P. Caveng, P. B. Fischer, E. Heilbronner, A. L. Miller and H. Zollinger, *Helv. Chim. Acta*, **50**, 1848 (1967).

[54] C. F. Bernasconi and R. G. Bergstrom, *J. Amer. Chem. Soc.*, **95**, 3603 (1973).

[55] C. F. Bernasconi and R. G. Bergstrom, *J. Amer. Chem. Soc.*, **96**, 2397 (1974).

[56] M. J. Strass, *Accounts Chem. Res.*, **7**, 181 (1974), and references therein cited.

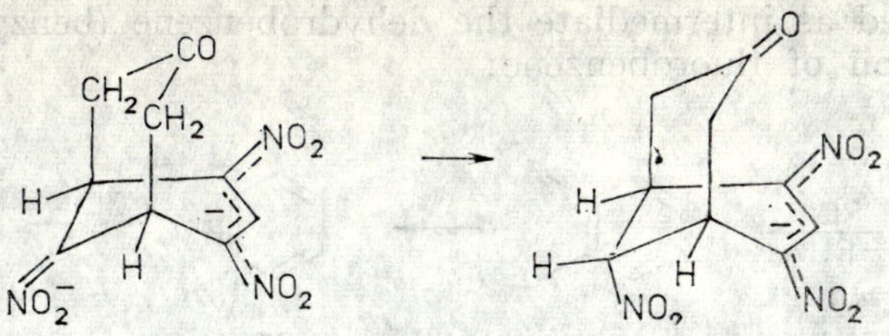

However, stronger bases may convert **27** to the *meta*-bridged dianion **28**, stable enough to be isolated as mono-ion **29**.

The formation of Jackman-Meisenheimer complexes, their sufficient stability to allow the transformation into dianions **26** and **28**, may be regarded as conclusive proofs for the S_NAr2 mechanism.

8.4 The aryne mechanism

8.4.1 *Benzyne as reaction intermediate*

Numerous aromatic substitutions occur with rearrangement, at first sight unexplainable. Thus, even as early as 1875[57] it has been observed that benzene-*p*-disulphonic acid affords by alkaline melting the same product as that resulted from benzene-*m*-disulphonic acid, *i.e.* resorcinol. Similar observations have been made since. The first to have assumed a prior elimination in the formation of the abnormal substitution products were Stoermer and Kahlert[58], who have observed the formation of 2-ethoxy- and 3-ethoxycumarone on treatment of 3-bromocumarone with ethoxide ion:

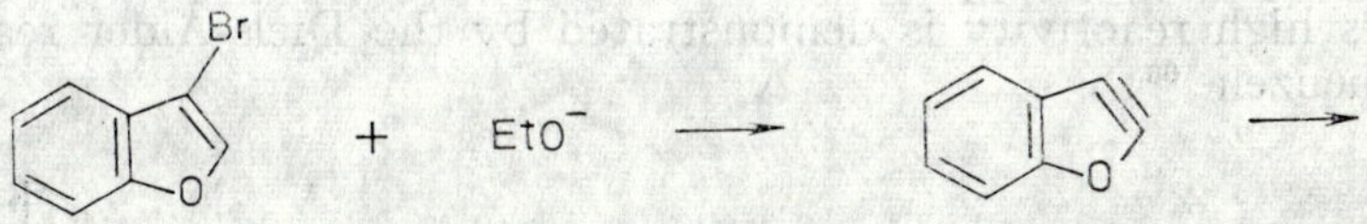

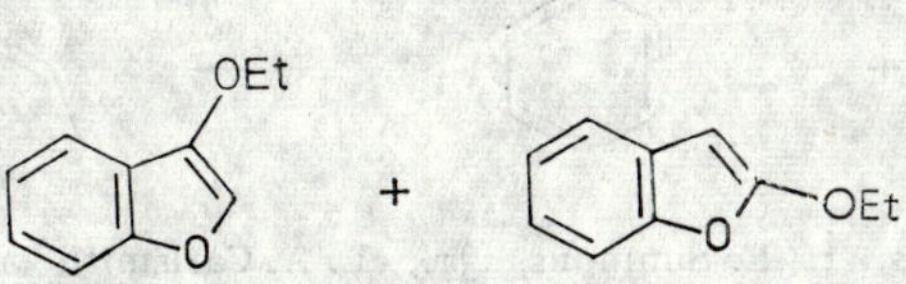

They suggested the formation of the reactive acetylene **30**, which yields further the two reaction products. Much later, Wittig[59]

[57] L. Barth and C. Senhoffer, *Ber. dtsch. chem. Ges.*, **8**, 1477 (1875).
[58] R. Stoermer and B. Kahlert, *Ber. dtsch. chem. Ges.*, **35**, 1633 (1902).
[59] G. Wittig, *Naturwissenschaften*, **30**, 696 (1942).

has proposed as intermediate the dehydrobenzene (benzyne) **31**, in the lithiation of fluorobenzene:

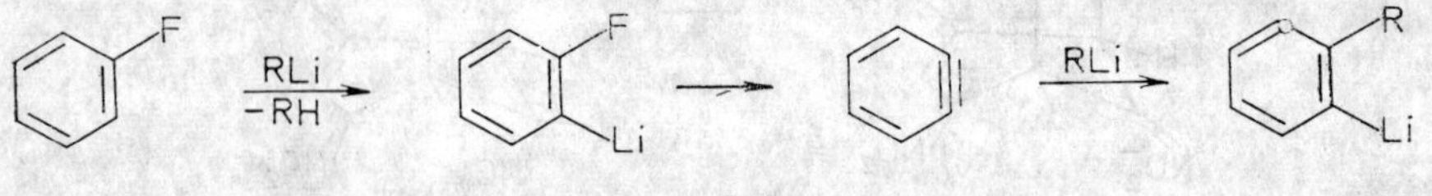

RLi / −RH → RLi →

31

A conclusive proof of the occurrence of benzyne in a substitution reaction has been offered by Roberts and his co-workers[60], who later demonstrated also the generality of the reaction:

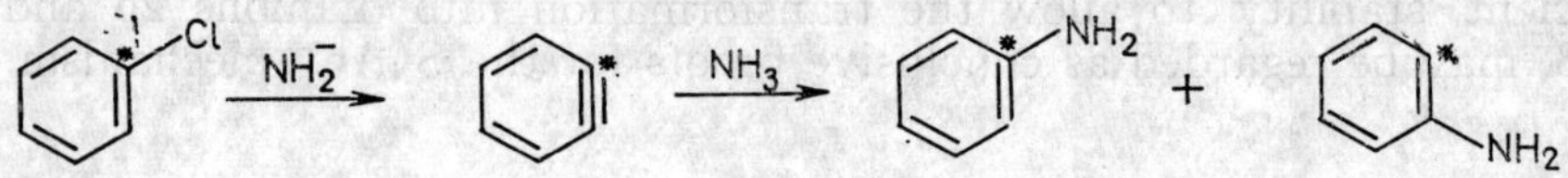

NH_2^- → NH_3 → NH_2 + NH_2

(8.25)

A similar demonstration was almost simultaneously achieved by Huisgen and Rist[61] in the reaction of fluorobenzene with phenyllithium. Soon afterwards, Wittig and Pohmer[62] were able to trap benzyne in a Diels-Alder reaction with furan:

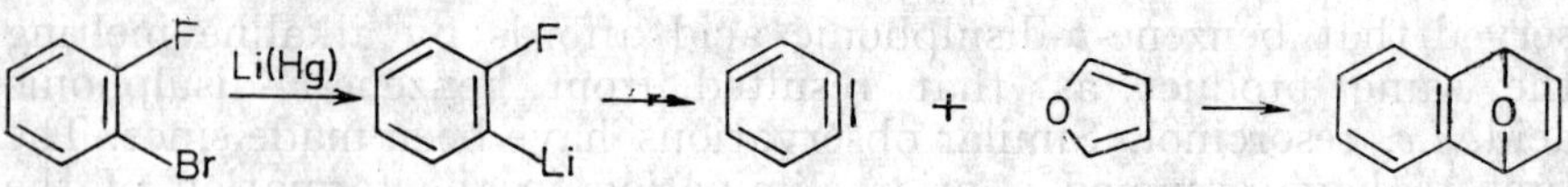

Li(Hg) → + O →

(8.26)

It was later shown in excellent reviews[63, 64] and monographs[65] that benzyne may appear as intermediate in many processes.

Its high reactivity is demonstrated by the Diels-Alder reaction with benzene[66]:

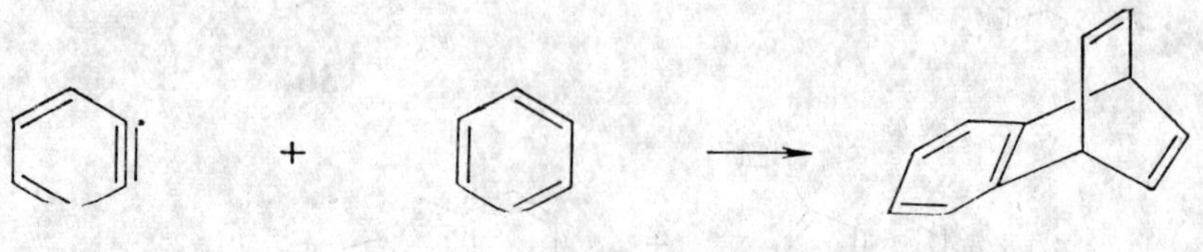

[60] J. D. Roberts, H. E. Simmons, Jr., L. A. Carlsmith and C. W. Vaughan, *J. Amer. Chem. Soc.*, **75**, 3290 (1953).

[61] R. Huisgen and H. Rist, *Naturwissenschaften*, **41**, 358 (1954).

[62] G. Wittig and L. Pohmer, *Chem. Ber.*, **89**, 1334 (1956).

[63] R. Huisgen and J. Sauer, *Angew. Chem.*, **72**, 91 (1960).

[64] R. Haney, *Chem. Rev.*, **62**, 81 (1962).

[65] R. W. Hoffmann, *Dehydrobenzene and Cycloalkynes*, Academic Press, New York (1967).

[66] L. Friedman, *J. Amer. Chem. Soc.*, **89**, 3071 (1967).

Its synthetic applications, comparable to a certain extent with those of carbenes, stimulated the investigation of other generation methods. These methods start from accessible *ortho*-disubstituted derivatives[67–71]. The procedure of Friedman and Logullo[72] seems to be most convenient and susceptible of many variations[73]. The starting material, anthranilic acid, is readily available and the reaction conditions reproducible:

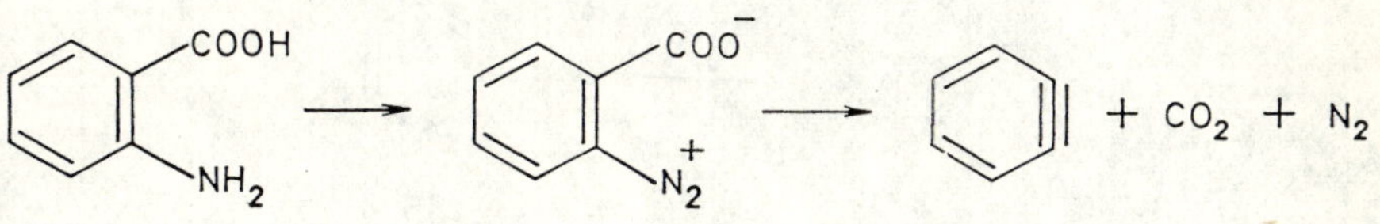

(8.27)

Other diazo compounds, such as **32**[74], have also been used in the generation of arynes. It is interesting to note that the elimination of the neighbouring proton is probably due to the hindered access of the nucleophilic reagent to the cationic centre:

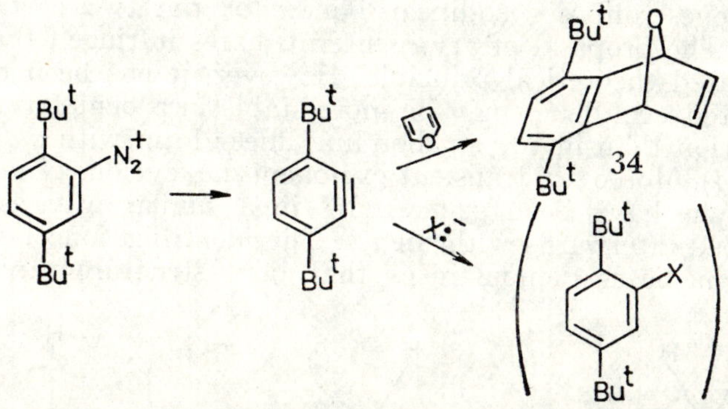

32 33 34

The formation of the aryne **33** was demonstrated by performing the reaction in presence of furan. The adduct **34** is obtained, together with substitution products.

[67] G. Wittig and H. F. Ebel, *Angew. Chem.*, **72**, 564 (1960).

[68] E. Le Gof, *J. Amer. Chem. Soc.*, **84**, 3786 (1962).

[69] E. K. Fields and S. Meyerson, *Chem. Commun.*, 474 (1965).

[70] R. F. C. Brown, D. V. Gardner, J. F. W. McOmie and R. K. Solly, *Chem. Commun.*, 407 (1966).

[71] W. Adam, *Angew. Chem.*, **86**, 683 (1974).

[72] L. Friedman and F. M. Logullo, *J. Amer. Chem. Soc.*, **85**, 1549 (1963).

[73] R. Baker and T. J. Muron, *Chem. Commun.*, 120 (1969).

[74] R. W. Frank and K. Yanagi, *Tetrahedron Letters*, 2905 (1966); *J. Amer. Chem. Soc.*, **90**, 5814 (1968).

Isomers of benzyne have been recently identified in some reactions (see also p. 607). If benzyne may be called 1,2-bis-dehydrobenzene, the others should be designated as 1,3- and 1,4-bis-dehydrobenzenes, respectively. It is interesting to note that the 1,3-intermediate has been generated[75] by a procedure similar to (8.27):

COO^-, $\overset{+}{N}_2$ ⟶ **35**

8.4.2 *Structure of bis-dehydrobenzenes*

The smallest ring in which it was possible to introduce a triple bond was the eight-membered ring[76]. However, even cyclooctyne is not a stable compound, as it isomerises and polymerises rapidly. This behaviour is easily explained considering that the *sp* hybridisation of acetylenes imposes co-linear bonds for the two carbon atoms involved. The proposal of aryne intermediates infringed thus a well-demonstrated theoretical principle. However, it has been recognised that excited acetylenes may be more stable in a bent form with sp^2 hybridisation than in the antibonding linear form with *sp* carbons[77] (Figure 8.1). Moreover, transient cycloheptyne, cyclohexyne and even cyclopentyne have been generated[78] in a similar way as benzyne and trapped afterwards with dienes. The identification of such reactive intermediates demonstrates that bent structures are possible.

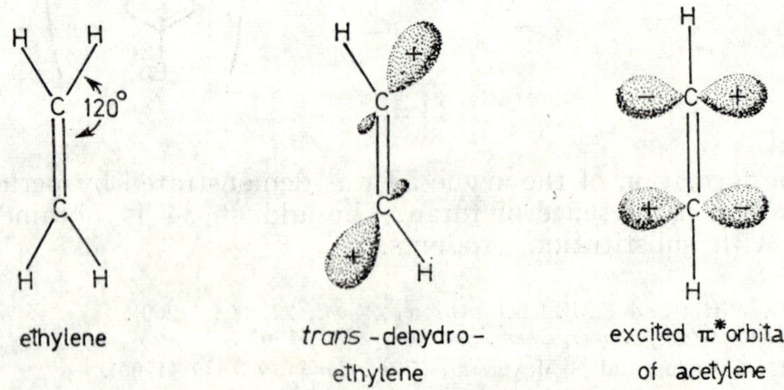

Figure 8.1 Molecular orbitals of excited acetylene.

[75] B. A. Hess, Jr., and L. J. Schaad, *Tetrahedron Letters*, 17 (1971).
[76] A. T. Blomquist and L. H. Liu, *J. Amer. Chem. Soc.*, **75**, 2153 (1953).
[77] C. K. Ingold, *J. Chem. Soc.*, 2991 (1954).
[78] G. Wittig, *Angew. Chem.*, **74**, 479 (1962); L. K. Montgomery and L. E. Applegate, *J. Amer. Chem. Soc.*, **89**, 5305 (1967).

A number of theoretical calculations have been made for the three possible bis-dehydrobenzenes. The *ortho* isomer can be considered to have a common geometry with benzene, the 1,2-interaction being strong enough even if one takes into consideration the length of 1.40 Å of the C-1—C-2 bond[79].

A MINDO/3 study[80], in which the distortion from the benzenoid has also been included, has indicated for **31** a singlet structure, with a short 1,2-bond, large 1-2-3 and 6-1-2 bond angles, the rest of the bonds being similar, which implies the retention of benzenoid character.

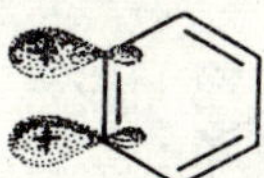

Similar calculations for **35** and **36** have indicated the following order of stability: **31** > **35** > **36**. The most striking prediction is that singlet 1,3-bis-dehydrobenzene should be at least comparable in stability with the *ortho* isomer. Strong binding between the radical centres, in **31** and **35**, is calculated for singlet species.

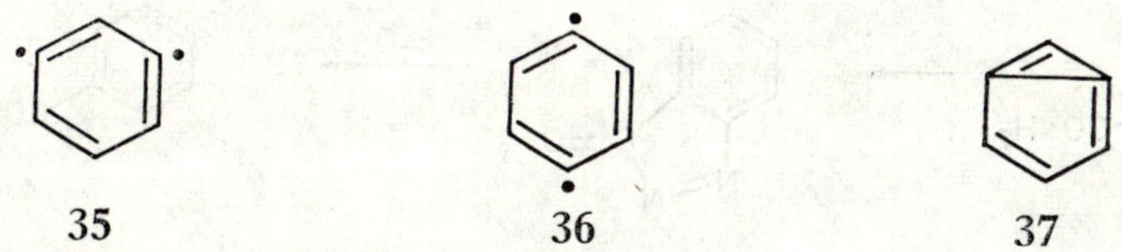

The geometry of the *meta* isomer resembles more **37** than **35**.

8.5 Other reactions related to aromatic nucleophilic substitution

There is a number of other aromatic substitutions, involving nucleophilic attack, but differing essentially in their mechanism from the reactions discussed so far. One of them is von Richter's reaction[81].

$$p\text{-}BrC_6H_4NO_2 \xrightarrow[180^\circ-200^\circ C]{KCN\ in\ EtOH} m\text{-}BrC_6H_4COOH \qquad (8.28)$$

It has been admitted by Bunnett[82, 83] that cyanide attacks the position *ortho* to the nitro group. This interpretation is probably

[79] R. Hoffmann, A. Imamura and W. J. Hehre, *J. Amer. Chem. Soc.*, **90**, 1499 (1968).

[80] M. J. S. Dewar, W. K. Lee, *J. Amer. Chem. Soc.*, **96**, 5569 (1974).

[81] V. von Richter, *Ber. dtsch. chem. Ges.*, 4, 21 (1871).

[82] J. F. Bunnett, J. F. Cormack and F. C. McKay, *J. Org. Chem.*, **15**, 481 (1950).

[83] J. F. Bunnett and M. M. Rauhut, *J. Org. Chem.*, **21**, 934, 939, 944 (1956).

correct, considering that nitroderivatives with blocked *ortho* positions are unreactive.

It has been later observed by Rosenblum [84] that the mechanism proposed by Bunnett and his co-workers [83] does not explain certain experimental data. Thus, in the reaction of *p*-chloronitrobenzene, beside 46.5% *m*-chlorobenzoic acid, 50% nitrogen is also evolved. By isotopic labelling of the nitrogen of the nitro group, of the added ammonia, as well as by oxygen labelling of water, the following mechanism was demonstrated:

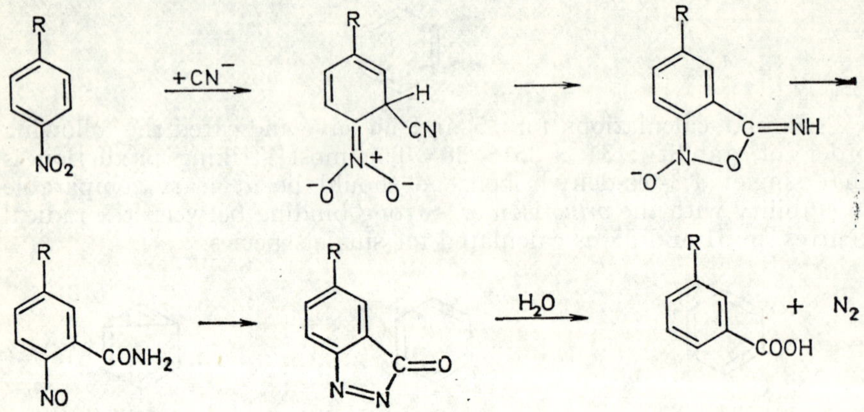

A series of aromatic rearrangements involve nucleophilic substitution by carbanions. An example is the Sommelet rearrangement [85]. The reaction mechanism proposed [86, 87] involves the formation of a nitrogen ylide **38** followed by the nucleophilic attack of the aromatic ring:

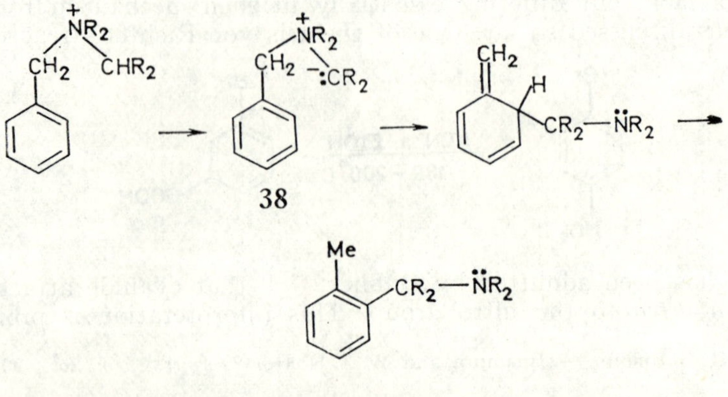

[84] M. M. Rosenblum, *J. Amer. Chem. Soc.*, **82**, 3796 (1960).
[85] M. Sommelet, *Compt. Rend.*, **205**, 56 (1937).
[86] G. Wittig, *Angew. Chem.*, **66**, 10 (1954).
[87] H. J. Shine, *Aromatic Rearrangements*, Elsevier (1967), p. 316.

The neutral intermediate is then stabilised by aromatisation. If the *ortho* positions are occupied, the quinolide structure **39** is preserved [88].

A similar rearrangement was studied by Hauser with the corresponding sulphonium salts [89].

Related to Sommelet and Hauser rearrangements is the transformation suffered by azasulphonium salts **40** [90]:

Azasulphonium salts are readily prepared from aromatic amines **41** by the procedure outlined below:

The rearrangement may be used in the synthesis of indoles [91, 92] and oxindoles [93].

[88] C. R. Hauser and D. N. van Eenam, *J. Amer. Chem. Soc.*, **79**, 5512, 6274 (1957).

[89] C. R. Hauser, S. W. Kantor and W. R. Brasen, *J. Amer. Chem. Soc.*, **75**, 2660 (1953).

[90] P. G. Gassmann and G. D. Gruetzmacher, *J. Amer. Chem. Soc.*, **96**, 5487 (1974).

[91] P. G. Gassmann, T. J. van Bergen, D. P. Gilbert and B. W. Cue, Jr., *J. Amer. Chem. Soc.*, **96**, 5495 (1974).

[92] P. G. Gassmann, G. Gruetzmacher and T. J. van Bergen, *J. Amer. Chem. Soc.*, **96**, 5512 (1974).

[93] P. J. Gassmann and T. J. van Bergen, *J. Amer. Chem. Soc.*, **96**, 5508 (1974).

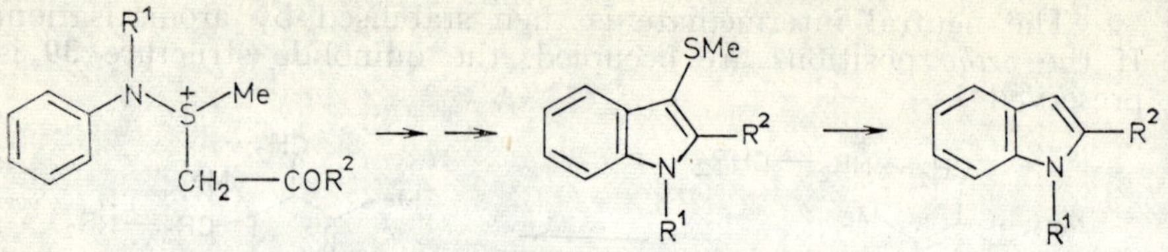

Another rearrangement related to the aromatic nucleophilic substitution is Smiles rearrangement [94]. An example of this reaction [95] is the butyl-lithium-promoted rearrangement of sulphone **42**:

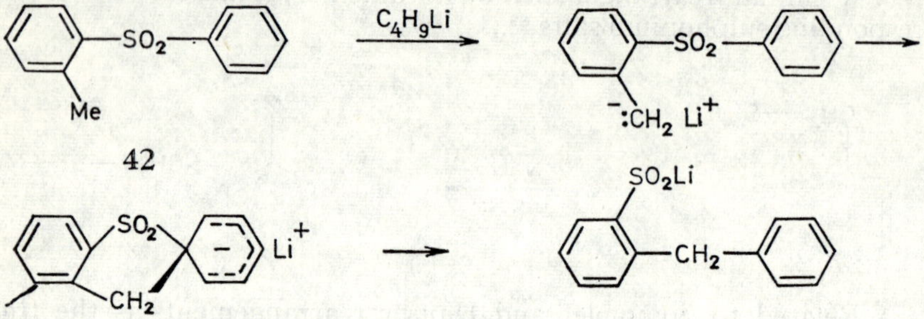

42

43

The reaction intermediate **43**, resulting from nucleophilic attack at the most activated position of the second aromatic ring, collapses to the rearranged product. The latter step corresponds to the reversible step of the aromatic nucleophilic substitution.

8.6 Problems

(1) Suggest reaction mechanisms for the following aromatic substitutions:

(a) $PhN_2^+ + MeOH \longrightarrow PhOMe + N_2 + H^+$

(b) $PhN_2^+ + MeCHTOH \longrightarrow PhT + MeCHO$ (in solution buffered with MeCOONa).

(c)

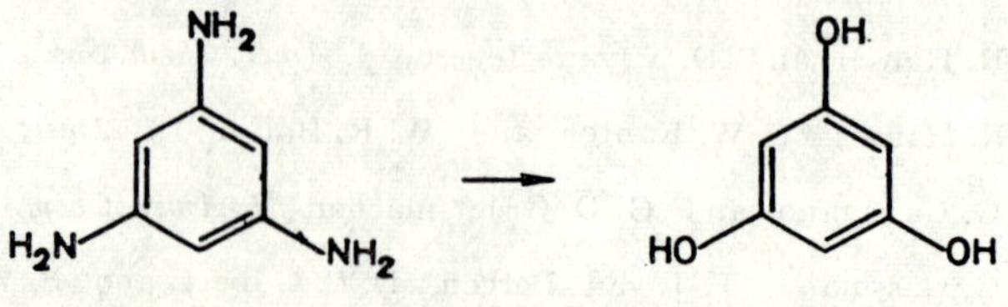

[94] L. A. Warren and S. Smiles, *J. Chem. Soc.*, 914 (1931); W. E. Truce, E. M. Kreider and W. W. Brand, *Org. Reactions*, **18**, 90 (1970).
[95] W. E. Truce and W. J. Ray, *J. Amer. Chem. Soc.*, **81**, 481 (1959).

(2) In the reaction of 2,4-dinitrohalogenobenzenes with hydroxide ion there is a 10^4 ratio between the reaction of the fluoroderivative and of the iodo compound. On the other hand, the ratio $k_{O,F}/k_{O,I}$ in the reaction with nitrite ion exceeds 10^6. How can you explain this difference of relative reactivities?

Indication: see p. 412 for literature and other comments.

(3) Explain the formation of resorcinol in the alkaline melting of benzenesulphonic acid and benzene-p-disulphonic acid.

(4) Explain why the radical mechanism proposed by Kim and Bunnett (see p. 402) is observed only with iodo compounds. Can you propose another reaction scheme and suggest experiments to prove the correct mechanism?

(5) The Tchichibabin reaction may also be explained by an aryne mechanism:

$$\xrightarrow[-NH_3]{+NH_2^-} \quad \xrightarrow{-H:^-} \quad \xrightarrow{+NH_3} \quad NH_2$$

Can you suggest any conclusive experiment which would rule out or confirm this mechanism, as opposed to the direct S_NAr2 mechanism?

(6) Show in a potential energy diagram the conversion of trinitrobenzene and acetone into the σ complexes **27** and **28**. Using the diagram, find reasonable arguments showing that the conversion **27** → **28** needs stronger base than the conversion of the original system to the monoanion **27**.

(7) What products may be formed in the reaction of a pyrylium salt with a primary and with a secondary amine?

(8) Considering the mechanism shown on page 409 and the examples of Scheme 2 (p. 408), suggest the effect of the nature of nucleophile, Nu:⁻ (Nu:) on the reaction of the pyrylium salt (formation of cyclic or open-chain products).

(B) ELIMINATION AND ADDITION REACTIONS

9

Elimination reactions

9.1. Survey of the reactions and mechanisms

The synchronous or stepwise removal of two groups X and Y from a substrate X-A-Y, leading to a molecule A of higher unsaturation than the initial one, is called an *elimination reaction*.

Olefin-forming reactions (9.1), or reactions affording alkynes (9.2), are typical examples of eliminations. The formation of a carbonyl compound by reaction (9.3) or of a carbonyl derivative according to (9.4), fits also the definition of an elimination:

$$X{-}CR_2{-}CR_2{-}Y \longrightarrow CR_2{=}CR_2 + (XY) \qquad (9.1)$$

$$X{-}CR{=}CR{-}Y \longrightarrow CR{\equiv}CR \quad + (XY) \qquad (9.2)$$

$$X{-}\overset{|}{\underset{|}{C}}{-}OH \longrightarrow \rangle C{=}O \qquad + HX \qquad (9.3)$$

$$X{-}\overset{|}{\underset{|}{C}}{-}NHR \longrightarrow \rangle C{=}NR \qquad + HX \qquad (9.4)$$

However, for a number of reactions, either leading to unstable intermediates, (9.5) and (9.6), or to stable products, (9.7), the resemblance to processes (9.1)—(9.4) sometimes is less apparent:

$$R_2CXY \longrightarrow R_2C\colon \ + (XY) \qquad (9.5)$$

$$RNXY \longrightarrow RN\colon \ + (XY) \qquad (9.6)$$

$$(CH_2)_n\begin{matrix} \diagup \rangle C{-}X \\ \diagdown \rangle C{-}Y \end{matrix} \longrightarrow (CH_2)_n\begin{matrix} \diagup C\langle \\ | \\ \diagdown C\langle \end{matrix} \quad + (XY) \qquad (9.7)$$

However, they satisfy also the general formulation:

$$X{-}A{-}Y \longrightarrow A + X + Y$$

It is not difficult to show that the unstable derivatives produced in reactions (9.5) and (9.6) have an unsaturated character, being able to form adducts (Section 9.2). Thus, it remains only to decide if reaction (9.7) is indeed an elimination, whatever the value of n. A correct answer may be obtained if the last part of the definition given above is considered: ... in order to form a molecule A of higher unsaturation... . Processes (9.7) correspond therefore to an elimination only if they afford a strained ring ($n = 1$ or 2). With this limitation for the cyclisations (9.7), elimination reactions may be classified as follows, according to the relative positions of the groups X and Y:

α-Eliminations. The groups being eliminated, X and Y, are bonded to the same atom of the substrate. Reaction (9.5) is an example of an α-elimination at a carbon atom, whereas equation (9.6) shows that α-eliminations may also occur at heteroatoms. The driving force is the result of a kinetic control, as the products are very unstable intermediates.

β-Eliminations. The groups X and Y are attached to adjacent atoms. Olefin- and alkyne-forming reactions are the most common β-eliminations. Due to their practical applications in synthesis, these reactions have been thoroughy studied. β-Eliminations are not restricted to a carbon chain; heteroatomic systems, as shown in (9.3) and (9.4), or as in (9.8), may also be involved:

$$\text{R—NX—NY—R} \longrightarrow \text{R—N=N—R} + (\text{XY}) \qquad (9.8)$$

γ-Eliminations. The reaction may lead to a cyclopropane derivative, if the eliminated groups were formerly bonded to carbon atoms situated in the 1,3 relative positions (equation (9.7) with $n = 1$).

Evidently, if the groups X and Y are farther apart, δ-eliminations etc., may also be encountered. However, higher order eliminations are less related to the usual ones and will not be treated in this chapter. α-, β- and γ-eliminations are related by common mechanisms, justifying thus a common treatment.

Elimination reactions may proceed either by heterolytic mechanisms (treated in this chapter), or by homolytic or by pericyclic mechanisms (dealt with in the last chapters of the book). One may notice that an elimination is the reverse of an addition reaction. Considering the principle of microscopic reversibility, we may assume that addition and elimination reactions have common mechanisms. There are many monographs and reviews [1–14], dedicated to different

[1] C. K. Ingold, *Structure and Mechanism in Organic Chemistry*, Cornell University Press, Ithaca, New York (1953), p. 420; 2nd edn (1969), p. 649.

[2] T. L. Jacobs, *Organic Reactions*, **5**, 1 (1949).

[3] C. K. Ingold, *Proc. Chem. Soc. (London)*, 265 (1962).

[4] G. Wittig, *Angew. Chem.*, **74**, 479 (1962).

[5] J. F. Bunnett, *Angew. Chem.*, **74**, 731 (1962).

[6] D. V. Banthorpe, *Elimination Reactions*, Elsevier, Amsterdam (1963).

[7] W. Kirmse, *Angew. Chem.*, **77**, 1 (1965).

[8] F. Badea, S. Roşca, I. G. Dinulescu, M. Avram and C. D. Nenitzescu, *Rev. Roumaine Chim.*, **10**, 1201 (1965).

[9] W. Hückel and M. Hanack, *Angew. Chem.*, **79**, 555 (1967).

[10] D. J. McLennan, *Quart. Rev. Chem. Soc.*, **21**, 490 (1967).

[11] H. M. R. Hoffmann, *Tetrahedron Letters*, 4393 (1967).

[12] J. F. Bunnett, *Surv. Progr. Chem.*, **5**, 53 (1969).

[13] J. Sicher, *Angew. Chem. Intern. Ed. Engl.*, **11**, 200 (1972).

[14] W. T. Ford, *Accounts Chem. Res.*, **6**, 410 (1973).

aspects of the elimination reactions. All seem to agree on three main types of mechanisms.

(a) *Unconcerted breaking of the C—X and C—Y bonds with the intermediate formation of a carbanion.* This mechanism describes an elimination occurring in two steps:

$$X{-}A{-}Y \xrightarrow{\text{Acceptor}} A{-}Y^- + (\text{Acceptor})^+ - X \qquad (9.9)$$

$$A{-}Y^- \longrightarrow A + Y^- \qquad (9.10)$$

the first step, the formation of the anion, being rate-determining. Most often, X is H and the acceptor is a base. In this case the elimination is symbolised as E1cB (elimination, unimolecular, from conjugate base), the anion A—Y⁻ being the conjugate base of the acid H—A—Y. The E1cB mechanism occurs very frequently in α-eliminations, and plays an important role in the cyclisations of type (9.7). It is less common in the olefin-forming reactions.

It must not be assumed that this type of mechanism is restricted only to E1cB reactions (X = H). Many other examples are known in which the group X is different from hydrogen.

The carbanion mechanism has a variant in which the organometallic compound formed in the first step undergoes further a bimolecular attack, yielding an elimination product [15]:

$$\text{PhCH}{=}\text{CHCl} \xrightarrow[\text{—RH}]{\text{RLi}} \text{PhCH}{=}\text{CClLi} \qquad (9.11)$$

$$\text{PhCH}{=}\text{CClLi} \xrightarrow[\text{—RH, —LiCl}]{\text{RLi}} \text{PhC}{\equiv}\text{CLi} \qquad (9.12)$$

This type of elimination has been symbolised by E2cB.

Carbanions may be trapped, for example, by electrophiles, yielding substitution products. Carbanion eliminations are frequently accompanied by S_E1 reactions (Chapter 6).

(b) *Unconcerted breaking of the C—X and C—Y bonds with the intermediate formation of carbocation* may be represented by equations (9.13)—(9.14):

$$X{-}A{-}Y \longrightarrow X{-}A^+ + Y^- \qquad (9.13)$$

$$X{-}A^+ \longrightarrow A + X^+ \text{ (acceptor)} \qquad (9.14)$$

Depending on Y, the first step may be either a unimolecular or a bimolecular process. The former possibility is more frequent and is usually designated as E1 (elimination, unimolecular). One may note that the first step in the E1 elimination is identical with the first step in an S_N1 reaction. Besides, E1 eliminations are always accompanied by substitution (S_N1) reactions, and may be thus easily recognised.

[15] M. Schlosser and V. Ladenberger, *Chem. Ber.*, **100**, 3877, 3893, 3901 (1967).

The variant with bimolecular first step (9.13) is also encountered often enough, presenting high interest for certain industrial processes, for example, *catalytic cracking*[16];

$$R^{1+} + H{-}C_\beta R^2_2{-}C_\alpha R^2_2{-}R^3 \longrightarrow R^1H + \overset{+}{C}_\beta R^2_2{-}C_\alpha R^2_2{-}R^3 \quad (9.15)$$

carbenium ion,
hydride-ion acceptor

$$\xrightarrow{\beta\text{-scission}} CR_2{=}CR_2{+}R^{3+}$$

In the first step of the reaction a carbenium ion acts as a hydride-ion acceptor, generating a new carbocation. The last step involves the β-scission of a stable carbocation R^{3+}. The reverse of (9.15) corresponds to an alkylation process, and has been treated in Chapter 7.

(c) *Concerted breaking of the two bonds.* According to this mechanism, both C—X and C—Y bonds are broken in the same rate-determining step:

$$X{-}A{-}Y \longrightarrow A + X + Y \quad (9.16)$$

Usually, process (9.16) occurs bimolecularly, requiring a reagent which can accept the electropositive group X. If X is H and the reagent is a base, the reaction is known as E2 elimination (elimination, bimolecular).

$$B\colon + H{-}A{-}Y - [B\cdots H\cdots A\dagger\cdots Y] \longrightarrow BH^+ + A + Y^- \quad (9.17)$$

Olefin-forming reactions occur very frequently according to the E2 mechanism. It must be mentioned however that in the transition state the base may be more bonded either to H or to C, the two corresponding pathways being designated as E2H and E2C[17]. Certain controversies on this point[12, 14] will be discussed in the last section of this chapter. Evidently, since a molecule may be at the same time a base and a nucleophile, E2 eliminations can be accompanied by S_N2 reactions.

Concerted mechanisms are also known in which process (9.16) is a true unimolecular reaction. In that case, X is a stable molecule, as for example CO_2 in the concerted elimination (9.18):

$$ArCHBr{-}CHBrCO_2^- \longrightarrow ArCH{=}CHBr + CO_2 + Br^- \quad (9.18)$$

Considering the two main mechanisms of olefin formation, it is not surprising that eliminations and nucleophilic substitutions occur side by side. Their frequent competition is explained either by a common mechanism (for E1 and S_N1) or by closely related concerted mechanisms[18] (E2 and S_N2).

[16] See, for example, L. Schmerling, in *Friedel-Crafts and Related Reactions* (G. A. Olah, editor), Interscience, New York, vol. 2 (1964), p. 1075.

[17] A. J. Parker, M. Ruane, G. Biale and S. Winstein, *Tetrahedron Letters*, 2113 (1968).

[18] R. W. Alder, R. Baker and J. M. Brown, *Mechanism in Organic Chemistry*, Wiley-Interscience, New York (1971).

9.2. α-Elimination (E_α)

9.2.1. *Products and mechanisms*

In this section we shall be concerned only with α-eliminations at carbon atoms. The elimination products, called *carbenes*, are very unstable intermediates, comparable in their reactions with carbocations or free radicals. The most frequently encountered mechanism for this type of elimination is the E1cB mechanism (Scheme 1):

$$\underset{1}{R_2CXY} \underset{k_{-1}}{\overset{(a)k_1}{\rightleftharpoons}} \underset{2}{R_2\bar{C}{-}X} \underset{k_{-2}}{\overset{(b)k_2}{\rightleftharpoons}} \underset{3}{R_2C:} + X^- \xrightarrow{\text{Acceptor (trapping agent)}} \text{Products}$$

Scheme 1

The reactivity and the reactions of carbenes have been reviewed in many books and monographs [19–26].

Reactions related to those of carbenes, or reactions involving intermediates similar to carbenes, such as nitrenes [27, 28] **4,** silylenes [19, 29] **5,** unsaturated carbenes like vinylidene carbenes **6,** allenidene carbenes [19, 23, 27, 30] **7,** or the carboxocarbene [25] **8,** are encountered in the literature.

$$\underset{\mathbf{4}}{R{-}N:}\ ;\ \underset{\mathbf{5}}{:SiR_2}\ ;\ \underset{\mathbf{6}}{R_2C{=}C:}\ ;\ \underset{\mathbf{7}}{R_2C{=}C{=}C:}\ ;\ \underset{\mathbf{8}}{O{=}C{=}C:}$$

Theoretical treatments regarding the structure of carbenes are outside the scope of this section. We shall however mention that there are two possible structures for carbenes. In one of them, the hybridisation is sp^2, the carbene being called *singlet* (Figure 9.1). The two bonds together with the sp^2 orbital occupied by the lone electrons are situated in the same plane, the vacant p orbital being

[19] W. Kirmse, *Carbene Chemistry*, Academic Press, New York (1964); 2nd edn, Academic Press (1971).
[20] J. Hine, *Divalent Carbon*, Ronald Press Comp., New York (1964).
[21] E. Chinoparos, *Chem. Rev.*, **63**, 235 (1963).
[22] D. Seyferth, *Accounts Chem. Res.*, **5**, 65 (1972).
[23] M. Jones, Jr., and R. A. Moss, *Carbenes*, vol. 1, J. Wiley, New York (1973).
[24] O. M. Nefedov (editor), *News in Carbene Chemistry* (in Russian), Akademia Nauk SSSR, Institut Khim. Fiz., Moscow (1973).
[25] T. Kappe and E. Ziegler, *Angew. Chem.*, **86**, 529 (1974).
[26] M. Jones, Jr., *Accounts Chem. Res.*, **7**, 415 (1974).
[27] T. L. Gilcrest and C. W. Rees, *Carbenes, Nitrenes and Arynes*, Appleton-Century-Crofts, New York (1969).
[28] L. Lee Muller and J. Hammer, *1,2-Cycloaddition reactions, The Formation of Three- and Four-membered Heterocycles*, Interscience, New York (1967).
[29] W. H. Atwell and D. R. Weyenberg, *Angew. Chem. Intern. Ed. Engl.*, **8**, 469 (1969).
[30] G. Köbrich, *Angew. Chem. Intern. Ed. Engl.*, **6**, 41 (1967).

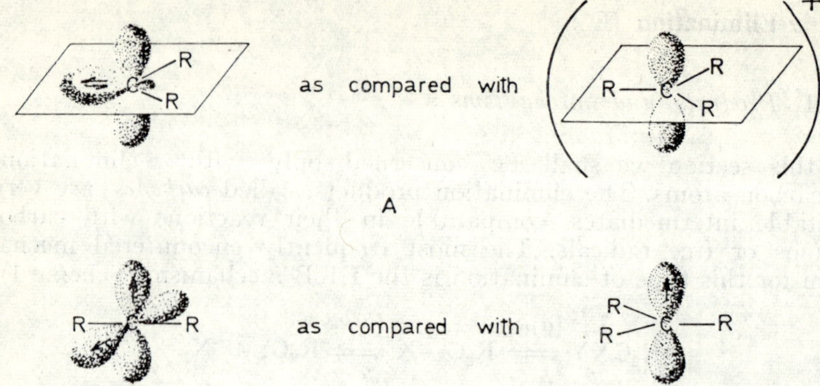

Figure 9.1 Structure of carbenes. A. Singlet carbenes. B. Triplet carbenes.

perpendicular to this plane. Thus, the structure of singlet carbenes is similar to that of trigonal carbenium ions. In the other structure, called *triplet*, the hybridisation of the central atom is *sp*, the three bonded centres in carbene being colinear and the two *p* orbitals, containing each a single electron, perpendicular to each other. Triplet carbenes are therefore similar to diradical species.

Theoretical calculations[31, 32] as well as spectral data indicate that for simple carbenes, such as methylene CH_2, the triplet state is lower in energy than the singlet state. However, all chemical reactions occur with conservation of the overall spin (conservation of the multiplicity) and since the great majority of molecules are singlet in the ground state, α-elimination generates singlet carbenes. The carbenes generated in that way will manifest electrophilic reactivity in a similar way as carbenium ions.

The existence of carbenes was first assumed as early as 1862 by Geuther[33], who suggested the intermediate formation of dichlorocarbene CCl_2, in the base-catalysed hydrolysis of chloroform. Almost ninety years later this view could be confirmed as correct by Hine[34]. However, the explosion of the chemistry of carbenes started in 1954, after Doering and Hoffmann[35] succeeded in trapping dichloro- and dibromo-carbene with alkenes, a reaction which proved to be general[19–26].

$$\gt C{=}C\lt \; + \; :CR_2 \longrightarrow \gt C\underset{CR_2}{\overset{}{—}}C\lt \qquad (9.19)$$

[31] W. B. DeMore and S. W. Benson, *Advances in Photochemistry* (W. A. Noyes, Jr., G. S. Hammond and J. N. Pitts, Jr., editors), Interscience, New York (1964), p. 219.

[32] R. Hoffmann, *J. Amer. Chem. Soc.*, **90**, 1475 (1968); N. Bodor, M. J. S. Dewar and J. S. Warson, *J. Amer. Chem. Soc.*, 94, 9095 (1972).

[33] A. Geuther, *Liebigs Ann. Chem.*, **123**, 121 (1862).

[34] J. Hine, *J. Amer. Chem. Soc.*, **72**, 2438 (1950).

[35] W. von E. Doering and A. K. Hoffmann, *J. Amer. Chem. Soc.*, **76**, 6162 (1954).

The utility of Doering's trapping method in the generation of cyclopropanes, until then not readily accessible, has largely contributed to the rapid acceptance of carbenes as intermediates in organic chemistry. However, it was soon realised that free carbenes are very rare intermediates, probably occurring only in the gaseous phase [26], most of the cyclopropane-forming reactions involving organometallic compounds. The intermediate of the Simmons-Smith reactions [36, 37], in which diiodomethane, for example, is treated with zinc-copper, was first assumed [36] to be methylene, but proved later [37] to be an organozinc compound, ICH_2ZnI. This reagent displays all the properties of carbanions and at the same time has electrophilic properties, which is very unusual for an organometallic compound. Organolithium compounds, generated according to equation (9.20) and trapped with allenes as shown in (9.21), behave in a similar way [38–44]. In order to distinguish such species from common organometallic compounds (which have no electrophilic character) and from free carbenes (with no carbanionic character), the term *carbenoids* was coined, stressing on the fact that the electrophilic character predominates [45]. It is assumed that in such intermediates **9** the cation and the anion of the metal salt interact with the occupied and unoccupied, respectively, orbitals of the carbon atom.

$$CH_2Cl_2 + RLi \longrightarrow \underset{\mathbf{9}}{LiCHCl_2} + RH \qquad (9.20)$$

$$\text{cyclohexene} + \underset{\mathbf{9}}{ClHC{\cdots}(Li)(Cl)} \longrightarrow \text{bicyclo[4.1.0]heptane=CHCl} + LiCl \qquad (9.21)$$

It must be observed that the nature of the intermediate (carbene or carbenoid species) is of no consequence for the α-elimination. There is no doubt that the first step produces a carbanion; the second step may either be an unassisted process (carbene intermediate), or an assisted process in which an acceptor of carbene facilitates the reaction (Scheme 1). Whatever the intermediate, the reaction is an E1cB elimination.

36 H. E. Simmons and R. D. Smith, *J. Amer. Chem. Soc.*, **80**, 5323 (1958).
37 E. P. Blanchard and H. E. Simmons, *J. Amer. Chem. Soc.*, **86**, 1337, 1347 (1964).
38 W. T. Miller and D. M. Whalen, *J. Amer. Chem. Soc.*, **86**, 2089 (1964).
39 G. L. Closs and L. E. Closs, *J. Amer. Chem. Soc.*, **81**, 5991 (1959).
40 G. L. Closs and R. A. Moss, *J. Amer. Chem. Soc.*, **86**, 4042 (1964).
41 G. Köbrich, H. Trapp and I. Hornke, *Tetrahedron Letters*, 1131 (1964).
42 G. Köbrich, K. Flory and W. Drischel, *Angew. Chem.* **76**, 536 (1964).
43 G. Köbrich and H. Trapp, *Chem. Ber.*, **99**, 670, 680 (1966).
44 G. Köbrich, H. Trapp, K. Flory and W. Drischel, *Chem. Ber.*, **99**, 689 (1966).
45 W. Kirmse, *Carbene, Carbenoide und Carbenanaloge*, Verlag Chemie (1969).

9.2.2. *Discussion of the two steps of the α-elimination*

The first step in the formation of a carbene is the generation of a carbanion (step *a* in Scheme 1). The rate of formation of the carbanion will be affected by the system substrate-base-solvent, as shown in Section 3.3.2, Section 4.2.3 and Section 6.3.1. This step was carefully studied by Hine and his co-workers [34, 46–49], in the case of the base-catalysed hydrolysis of haloforms:

$$CHXYZ \xrightarrow[H_2O]{HO^-} CO + HCOO^- \tag{9.22}$$

Hine established that the order of reactivity of halogenomethanes decreases in the following sequence: $CHX_3 \gg CH_2X_2 < CH_3X$, which is not in agreement with an usual S_N2 reaction, for which the reactivity decreases with the increasing number of halogen atoms. It was thus admitted [34, 46, 47] that the first step in the hydrolysis must be a proton extraction:

$$HO^- + CHXYZ \longrightarrow H_2O + {}^-:CXYZ \tag{9.23}$$

The proton abstraction has been demonstrated in several ways, as for example by deuterium exchange:

$$HO^- + CDXYZ \longrightarrow HOD + {}^-:CXYZ \xrightarrow{H_2O} HCXYZ \tag{9.24}$$

By this method it was shown that the reactivity of halogenomethanes decreases in the order [49]:

$$CDI_3 \cong CDBr_3 > CDBr_2Cl > CDBrCl_2 > CDBr_2F > CDCl_3 > CDCl_2F$$

which is approximately the order of stability of the corresponding anions.

Doering and Hoffmann [35] found that dichlorocarbene may be generated with strong bases such as Bu^tOK, in aprotic solvents, under very mild conditions. This result prompted the generation of monohalogeno and even simple carbenes from dihalogeno and monohalogenoalkanes, respectively, by use of very strong bases.

[46] J. Hine, A. M. Dowell, Jr. and J. E. Singley, Jr., *J. Amer. Chem. Soc.*, **78**, 479 (1956).

[47] J. Hine and P. B. Langford, *J. Amer. Chem. Soc.*, **79**, 5487 (1957

[48] J. Hine and J. J. Porter, *J. Amer. Chem. Soc.*, **79**, 5493 (1957).

[49] J. Hine, N. W. Burske, M. Hine and P. B. Langford, *J. Amer. Chem. Soc.*, **79**, 1406 (1957).

The second step in the E1cB α-elimination is similar to the unimolecular ionisation encountered in S_N1 and in E1 reactions:

$$R_3C{-}X \longrightarrow R_3C^+ + X^- \quad (S_N1) \tag{9.25}$$

$$R_2\ddot{C}X^- \longrightarrow R_2C\colon + X^- \tag{9.26}$$

As in the S_N1 process, the ionisation must be assisted either by external factors (solvent, carbene acceptors) or by internal, structural factors.

Despite the high interest in carbene chemistry, less effort was directed towards the elucidation of the mechanism of the α-elimination. Nevertheless, from the recorded results, it may be shown that there is a close parallelism between processes (9.25) and (9.26). Thus, both are reversible, and the demonstration was brought about by Hine and his co-workers [34, 50, 51] who found that in the hydrolysis of chloroform the addition of salts like $NaNO_3$, $NaClO_4$ and NaF has no appreciable effect on the reaction rate. Conversely, the addition of NaCl, NaBr and NaI decreases the rates, the effect depending on the concentration of halide. Moreover, when NaBr and NaI are added, bromodichloromethane and dichloroiodomethane, respectively, may be isolated.

With rare exceptions, the α-elimination is performed in aprotic nonpolar solvents, such as hydrocarbons. Evidently, little assistance, if any, may be expected from the solvent in the heterolysis (9.26). Nevertheless, it must be remembered that one component of the solvent is an alkene, usually in excess with respect to the stoichiometric amount of carbene precursor. Since both carbenes [52, 53] and carbenoids [37] are electrophiles, the solvent may actually assist the heterolysis (9.26) by a nucleophilic catalysis, according to process (9.21), for example. Clearly, no other nucleophilic solvent can be used, as in the S_N1 process, since it may produce parallel reactions (nucleophilic substitution, β-elimination), or it may destroy the resultant carbenes. Thus, carbene reactions in protic media are restricted only to a few examples, like the above-mentioned hydrolysis of chloroform to carbon monoxide and formiate ion.

As in the ionisation (9.25), structural factors play a prominent part in the last step of carbene generation. First, in carbanion **2** the orbital occupied by the two electrons acts as a powerful electron-repelling group, facilitating the heterolysis. Contrary to most S_N1 ionisations, occurring in ionising solvents at temperatures usually over 75 °C, α-eliminations may be carried out at low temperatures (mostly about 0 °C, some even at −100 °C) in non-ionising solvents. The carbanionic structure of the molecule undergoing heterolysis accounts for the lower energy-barrier of this process.

[50] J. Hine and A. M. Dowell, Jr., *J. Amer. Chem. Soc.*, **76**, 2688 (1964).
[51] J. Hine, R. Butterworth and P. B. Langford, *J. Amer. Chem. Soc.*, **80**, 819 (1958).
[52] P. S. Skell and A. Y. Garner, *J. Amer. Chem. Soc.*, **78**, 5430 (1956).
[53] W. von E. Doering and W. A. Henderson, *J. Amer. Chem. Soc.*, **80**, 5274 (1956).

α-Eliminations can also be assisted by participation of neighbouring bonds. Thus, C—H bonds, like in (9.27) or in (9.28), may assist the expulsion of the leaving group X from the molecule:

$$\mathrm{R^1_2C_\beta(H)-\bar{\ddot{C}}(X)R^2 \longrightarrow R^1_2C{=}CHR^2 + X^-} \qquad (9.27)$$

$$\mathrm{{>}C{<}(C_\gamma{-}H)(\bar{\ddot{C}}_\alpha{-}X) \longrightarrow {>}C{<}(C)(CH) + X^-} \qquad (9.28)$$

The stereochemistry of the 1,2-hydrogen shifts occurring in the Bamford-Stevens reaction [54, 55] was studied by Nickon *et al.* [56]. They have shown (Figure 9.2.) that the geometry necessary for the

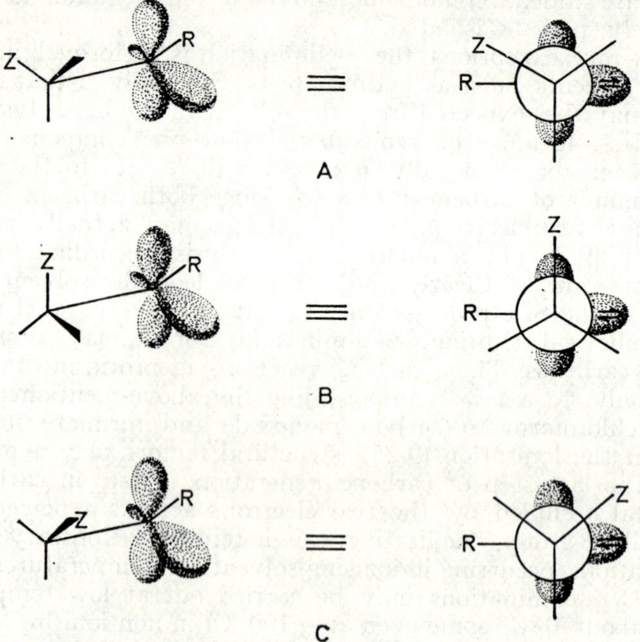

Figure 9.2 Orientations of neighbouring group (Z) in a 1,2-shift in carbene. A. synplanar; B. perpendicular; C. antiplanar.

[54] W. R. Bamford and T. S. Stevens, *J. Chem. Soc.*, 4735 (1952).

[55] For a recent review on this reaction see J. Casanova and B. Waegell, *Bull. Soc. chim. France*, 922 (1975).

[56] A. Nickon, F.-c. Huang, R. Waglein, K. Matsuo and H. Yagi, *J. Amer. Chem. Soc.*, **96**, 5264 (1974).

migration of the neighbouring group is the perpendicular orientation (B). This geometrical condition is similar to the condition imposed on carbocation rearrangements (see Section 5.4.1).

Migration of an aromatic ring in the course of an α-elimination has been observed in the Frietsch-Buttenberg-Wiechel rearrangement[43, 57, 58]. Thus, Köbrich and Trapp[43] noticed that 1-chloro-2,2-diphenylethylene-lithium, **10**, is stable in tetrahydrofuran at −72 °C and gives all the characteristic reactions of carbanions:

$$Ph_2C{=}CClLi \; (\mathbf{10}) \begin{cases} \xrightarrow{CO_2} Ph_2C{=}C(COOLi)Cl \\ \xrightarrow{Br_2} Ph_2C{=}C(Br)Cl \\ \xrightarrow{MeI} Ph_2C{=}C(Me)Cl \end{cases}$$

Over 65 °C, in presence or absence of cyclohexene, it yields diphenylacetylene. The study of substituent effects in the aromatic ring indicates that the latter assists the heterolysis of the C—Cl bond.

$$Ph_2C{=}\ddot{C}Cl \longrightarrow Ph{-}C(\text{bridging Ph}^{\delta+}){=}C^{\delta-}{-}{-}{-}Cl^{\delta-} \; (\mathbf{10a}) \longrightarrow Ph{-}C{\equiv}C{-}Ph + Cl^-$$

The most striking resemblance between S_N1 reactions and α-eliminations lies in the dependence of the reactivity on the leaving group. It is well known that carbocations display a completely different reactivity if generated by the deamination reaction ($X = N_2^+$) instead of other solvolytic routes. The term of *hot ions* has been coined to describe such carbocations[59]. Carbenes may be generated in a similar way from diazoalkanes **11** which may be regarded as stabilised carbanions:

$$R_2C{=}\overset{+}{N}{=}\ddot{N}{:}^- \longleftrightarrow R_2\overset{-}{C}{-}\overset{+}{N}{\equiv}N{:} \quad \text{etc.} \quad (\mathbf{11})$$

$$R_2C{=}N{=}N \; (\mathbf{11}) \longrightarrow R_2C{:} + N_2 \qquad (9.29)$$

[57] G. Köbrich, *Angew. Chem.*, **77**, 75 (1965).

[58] G. Köbrich and H. R. Merkle, *Angew. Chem.*, **79**, 50 (1967).

[59] For a review on deamination reactions and related topics see J. T. Keating and P. S. Skell in *Carbonium Ions*, vol. 2 (G. A. Olah and P. von R. Schleyer, editors), Wiley-Interscience, New York (1970), p. 573.

Diazoalkanes, unlike diazonium salts $R-N_2^+$, are stable species and thus reaction (9.29) may be expected to differ appreciably from the deamination reaction. However, both reactions generate highly reactive species, because the latter are more naked (less solvated) than if produced by other methods. Evidently, the resulted nitrogen molecule will interact to a lesser extent with the carbenium ion or with the carbene than will do an anion or a metal salt, respectively.

As in the case of carbocations, carbenes generated by this method display a specific reactivity, as for example the non-*endo*-stereospecific addition to a double bond [40], and the insertion reaction at a C—H bond (see p. 306). Dichlorocarbene, for example, does not insert at a C—H bond if formed by the conventional methods of α-elimination. When produced by decomposition of diazodichloromethane, CCl_2N_2 [60], it exhibits normal reactivity toward the C—H bond, as does methylene CH_2.

$$\rangle C-H + :CR_2 \longrightarrow \rangle C-CR_2H \qquad (9.30)$$

Reaction (9.29) may be realised either thermally or photochemically. A very practical method is the Bamford-Stevens reaction [54, 55] which consists in treating a tosylhydrazone **12** with a base and heating. The reaction must be regarded as a double α-elimination.

$$\rangle C{=}O \longrightarrow \underset{\mathbf{12}}{\rangle C{=}N{-}NH{-}SO_2Ar} \xrightarrow[\alpha\text{-elimination}]{NaOH} \rangle C{=}N{-}\ddot{\bar{N}}{-}SO_2Ar \longrightarrow$$

$$\rangle C{=}N{=}N \underset{\alpha\text{-elimination}}{\longrightarrow} \rangle C: + N_2 \qquad (9.31)$$

The application of this method to aromatic aldehydes has led to the discovery of *carbene-carbene rearrangement* [61] (9.32), very similar to the 1,2-shifts in carbocations (Section 5.4.1). Actually, the reaction consists in an intramolecular insertion of the carbene at another bond of an unsaturated system:

$$\text{(2-methyl-3-naphthyl)}\ddot{C}{-}H \longrightarrow \underset{\mathbf{93}}{\text{[Me-cyclopropene-fused naphthalene]}} \longrightarrow \qquad (9.32)$$

$$\text{[Me-benzocycloheptatetraene]} \longrightarrow \text{(2-naphthyl)}\ddot{C}{-}Me \longrightarrow \text{2-vinylnaphthalene}$$

[60] G. L. Closs and J. J. Coyle, *J. Amer. Chem. Soc.*, **87**, 4270 (1965).

[61] T. Mitsuhashi and W. M. Jones, Jr., *J. Amer. Chem. Soc.*, **94**, 677 (1972); R. C. Joines, A. B. Furner and W. M. Jones, Jr., *J. Amer. Chem. Soc.*, **91**, 7754 (1969); W. J. Baron, W. M. Jones, Jr., and P. P. Gaspar, *J. Amer. Chem. Soc.*, **92**, 4739 (1970).

The intermediates of this reaction are bicyclo [4.1.0] heptatrienes (13), as demonstrated by trapping with strong nucleophiles[62]:

$$\text{(dichlorocyclopropaphenanthrene)} \xrightarrow{B:} \text{(13)-Cl} \xrightarrow{MeSH} \text{(SMe, Cl adduct)} \quad etc.$$

$$\downarrow$$

$$\text{9-phenanthryl-}\ddot{C}\text{—Cl} \longrightarrow \text{products} \qquad (9.33)$$

Carbene-carbene rearrangement is another example of similarity between carbocations and carbenes.

Carbenes are involved in many other organic reactions[63]. The pyrolysis of organic compounds leading to acetylenes[64] deserves special mention. It is well known that methane heated to over 1200 °C forms acetylene and hydrogen. Fischer[65] assumed that in this reaction radical species, such as $CH_3\cdot$, $\dot{C}H_2\cdot$ and $\cdot\dot{C}H\cdot$, are involved. As the key intermediate in the process Kassel[66] proposed methylene, $:CH_2$, which may react with methane (insertion at a C—H bond) to give ethane. The dehydrogenation of the latter would afford ethylene and in a subsequent step acetylene:

$$CH_4 \xrightarrow{(a)} :CH_2 \xrightarrow{CH_4} CH_3{-}CH_3 \longrightarrow CH_2{=}CH_2 \longrightarrow CH{\equiv}CH$$

A similar mechanism had been proposed by Storch[67].

Even if methylene seems to be the most probable intermediate in methane cracking*, the other intermediates proposed by Fischer

[62] W. E. Billups, I. P. Lin and W. Y. Chow, *J. Amer. Chem. Soc.*, **96**, 4027 (1974).

[63] I. L. Knunyants, N. P. Gambaryan and E. M. Rokhlin, *Usp. Khim.*, **27**, 1361 (1958).

[64] J. A. Nieuwland and R. R. Vogt, *The Chemistry of Acetylene*, Reinhold, New York (1945), p. 20.

[65] F. Fischer, *Brennstoff Chem.*, **9**, 309 (1928).

[66] L. S. Kassel, *J. Amer. Chem. Soc.*, **54**, 3949 (1932).

[67] H. H. Storch, *J. Amer. Chem. Soc.*, **54**, 4188 (1932).

* The term *cracking* is usually applied to the transformation of methane into acetylene. Actually, this term should refer only to step (*a*) in the transformation, and not to the overall process which affords finally a molecule larger than the initial one.

cannot be completely ruled out. Thus, the identification of *carbynes* CR, in extraterrestrial matter[68] may suggest that methyne CH can be involved in the formation of acetylene.

9.2.3 α-*Eliminations from methane derivatives*

α-Eliminations are not easily identified when accompanied by β- and γ-eliminations, a situation which actually arises frequently. However, in the case of halogenomethanes, which have a single carbon atom in the molecule, β- and γ-eliminations are excluded.

Among halogenomethanes, haloforms have been preferred as precursors of carbenes, partly due to their higher reactivity. Hine and his co-workers[34, 46–51] have demonstrated that the hydrolysis of a haloform proceeds according to the mechanism shown in Scheme 2.

$$B: + HCXYZ \rightleftharpoons BH^+ + {}^-{:}CXYZ$$

$${}^-{:}CXYZ \xrightarrow{\text{slow}} {:}CXY + Z^-$$

$$HO^- + {:}CXY \xrightarrow{\text{fast}} HOCXY^- \xrightarrow[H_2O]{\text{fast}} CO + HCOO^-$$

Scheme 2

Thus, according to Scheme 2, carbene formation should depend on the acidity of the haloform. It has been demonstrated however[49, 69] that despite this dependence, the kinetic isotope effects in these reactions are small, due to the specific character of proton transfer reactions[69]. In the transition state of the latter, the proton is entirely transferred to the strong base (late transition state), while the geometry of the haloform is essentially the same as that of the resulted carbanion[70]. As a consequence, primary isotope effects are expected to be small.

Studying the rates of carbanion formation by reaction (9.24), Hine and his co-workers[49, 71] found a correlation between the rate constants of the exchange and the rate constants of the hydrolysis. A high rate of exchange (measuring carbanion formation) is associated with very low rates of hydrolysis. However, fluoro derivatives do not obey this correlation, the rate constants of their hydrolysis increasing faster than anticipated from the carbanion formation. The exceptionally high rates of the difluorohalogenomethanes were interpreted to indicate a concerted α-elimination (E2):

$$HO^- + HCF_2I \text{---} [\overset{\delta-}{H}O \cdots H \cdots CF_2 \cdots \overset{\delta-}{I}] \longrightarrow H_2O + {:}CF_2 + I^- \quad (9.34)$$

[68] P. C. Keenan and W. W. Morgan, *Astrophys. J.*, **94**, 501 (1941); O. P. Strausz, T. DoMinch and J. Font, *J. Amer. Chem. Soc.*, **90**, 1930 (1968), and the literature quoted therein.

[69] Z. Margolin and F. A. Long, *J. Amer. Chem. Soc.*, **95**, 2757 (1973).

[70] R. Hoffmann, quoted in Reference 69.

[71] J. Hine, *Physical Organic Chemistry*, 2nd edn., McGraw-Hill, New York (1962), p. 484.

Mechanism (9.34) is determined by the very low stability of carbanions of type $^{-}{:}CF_2X$ (since fluorine has no *d* orbitals, it cannot accommodate the negative charge of the carbon atom), by the fact that iodine is a very efficient leaving group, and still more by the ability of fluorine to stabilise the dihalogenocarbene, **14**, by the overlap of one of its *p* orbitals with the vacant *p* orbital of the carbon atom:

$$:C(\ddot{X}:)_2 \longleftrightarrow {}^{-}:C(=X^+)(\ddot{X}:) \longleftrightarrow {}^{-}:C(\ddot{X}:)(=\ddot{X}^+)$$

14*a* **14*b*** **14*c***

The last step in Scheme *2* indicates that carbenes **may** react with nucleophiles, being thus electrophilic reagents. The electrophilic character of dicholorocarbene is now demonstrated by all its reactions. Thus, very similar to the reaction with the hydroxide ion is the reaction with alkoxides [72]:

$$CH_3-C(CH_3)_2-CH_2-O^- + :CCl_2 \longrightarrow (CH_3)_3C-CH_2OCCl_2^- \longrightarrow$$

$$\underset{\mathbf{15}}{(CH_3)_3C-CH_2-O-\ddot{C}-Cl} \longrightarrow (CH_3)_3C-CH_2^+ + CO + Cl^- \quad (9.35)$$

$$\xrightarrow{\text{1,2-shift}} \underset{\mathbf{16}}{(CH_3)_2\overset{+}{C}-CH_2CH_3} \longrightarrow \text{products}$$

The above mechanism is supported by the identification of carbon monoxide and of reaction products resulted from cation **16**. This cation must derive from the neopentyl cation, indicating thus decomposition (9.35) of the alkoxychlorocarbene **15**.

A similar reaction of an acyloxychlorocarbene was observed by Ioan *et al.*[73] in aprotic media. The decomposition of silver trichloroacetate in hydrocarbons yields as major product trichloroacetic anhydride. The latter compound is assumed to result according to the following reactions:

$$CCl_3COOAg \longrightarrow :CCl_2 + CO_2 + AgCl \quad (9.36)$$

$$:CCl_2 + CCl_3COOAg \longrightarrow CCl_3COO\bar{\ddot{C}}Cl_2Ag^+ \longrightarrow \underset{\mathbf{17}}{CCl_3COO\ddot{C}Cl} + AgCl \quad (9.37)$$

$$\mathbf{17} \longrightarrow CCl_3COCl + CO \quad (9.38)$$

$$\xrightarrow{CCl_3COOAg} (CCl_3CO)_2O + AgCl$$

[72] P. S. Skell and I. Starrer, *J. Amer. Chem. Soc.*, 81, 4117 (1959).

[73] V. Ioan, F. Badea, E. Ciorănescu and C. D. Nenitzescu, *Angew. Chem.*, **72**, 416 (1960).

The mechanism is consistent with the 1:1 ratio found between carbon dioxide and carbon monoxide, and with the fact that the formation of the gaseous products requires about three moles of silver salt.

Similar decompositions have been later observed by Wagner *et al.*[74]. It must be noted that the decompositions (9.35) and (9.38) are β-eliminations from an alkyl- or acyloxycarbene.

In its reactions with nucleophiles, dichlorocarbene manifests high selectivity. A well-known example is the Reimer-Tiemann reaction [75, 76], which consists in treating phenol with chloroform and an alkaline hydroxide:

$$HCCl_3 \xrightarrow{HO^-} :CCl_2$$

$$\text{C}_6\text{H}_5\text{O}^- + :CCl_2 \longrightarrow \text{(cyclohexadienone, C(H)}\ddot{\text{C}}^-\text{Cl}_2\text{)} \xrightarrow{\text{several steps}} o\text{-}(\text{CHO})\text{C}_6\text{H}_4\text{O}^-$$

Even if in excess, dichlorocarbene prefers to react with the phenoxide ion instead of with hydroxide ion (Scheme 2). Moreover, when dichlorocarbene is generated by Doering's method [35], it prefers to react with the alkene instead of with the alkoxide ion. In the latter case the difference of nucleophilicity between the two reagents is much higher than in the former reaction. However, one would not expect a very reactive species, such as carbene, to discern between different reagents, and even less to choose the less reactive one. These experimental observations contributed largely to the view that carbenoid species, instead of carbenes, occur as intermediates. However, equation (9.21) suggests a nucleophilic assistance in elimination, and thus even carbenoids cannot account for this high selectivity. The quantitative measurements [36, 37, 52, 53, 77, 78] indicate only an electrophilic intermediate, which is not consistent with the observed selectivity.

It seems that the most plausible explanation is that the decomposition (9.39) occurs in a solvent shell:

$$\underset{\mathbf{18}}{CR_2X^-M^+} \longrightarrow :CR_2 + MX \qquad (9.39)$$

The low polarity of the solvent (hydrocarbon) prevents the migration of the species from the solvent shell. In most of the procedures using excess alkene, the shell is made of alkene molecules. The carbene, which equilibrates with the ion pair **18**, can react only

[74] W. M. Wagner, H. Kloosterziel and A. F. Bickel, *Rec. Trav. Chim.*, **81**, 925, 933 (1962).
[75] H. Wynberg, *Chem. Rev.*, **60**, 169 (1960).
[76] J. Hine and J. M. van der Veen, *J. Org. Chem.*, **26**, 1406 (1961).
[77] D. Seyferth and J. Y. P. Mui, *J. Amer. Chem. Soc.*, **88**, 4672 (1966).
[78] D. Seyferth and J. M. Burlitch, *J. Amer. Chem. Soc.*, **86**, 2731 (1964).

with the alkene. It has been however demonstrated[79] that the excess of alkene is not necessary. The same results may be obtained if instead of a high excess of alkene, equimolecular quantities are used, the solvent being an inert saturated hydrocarbon, also of low polarity. The medium being almost the same, the species resulting in (9.39) do not diffuse from the shell. *Solvent sorting*[80, 81] or *specific solvation* of the species resulting in (9.39) may favour the presence of alkene in the neighbourhood of the carbene. In more polar solvents, such as diglyme[79], an excess of trapping agent (alkene) is an absolute necessity.

Dihalogenocarbenes have been generated by many other methods[60, 73, 74, 82, 98], some of them represented in Scheme 3. It should be noted that some methods are not essentially different from Doering's method[35], the carbenes being obtained under base catalysis. A quite different procedure is the decomposition of trichloromethylphenyl-mercury, **19**[88, 89], a reaction which takes place in neutral solution (Seyferth's method).

(A) *Base medium*	Reference
$Bu^tO^-K^+ + HCX_3 \longrightarrow Bu^tOH + :CX_2 + KX$	35
$CCl_3CO{-}R^1 + R^2O^- \longrightarrow R^1{-}COOR^2 + :CCl_2 + Cl^-$ $R^1{=}alkyl,\ R^2O$	82–85
$RLi + CCl_4 \longrightarrow RCl + LiCl + :CCl_2$ (or carbenoid)	89
(B) *Thermal (neutral) processes*	
$PhHgCCl_3 \longrightarrow PhHgCl + :CCl_2$	87, 88
$PhHgCBr_3 \longrightarrow PhHgBr + :CBr_2$	88
$CCl_3COONa \longrightarrow CO_2 + NaCl + :CCl_2$	86
$CCl_3COOAg \longrightarrow CO_2 + AgCl + :CCl_2$	73, 74
$Cl_2C{=}N{=}N \xrightarrow[\text{or thermally}]{h\nu} N_2 + :CCl_2$	60
(C) *Acid medium*	
$AgNO_3 + CHBr_3 \longrightarrow AgBr + :CBr_2 + HNO_3$	92

Scheme 3

[79] W. M. Wagner, K. Kloosterziel and S. van der Ven, *Rec. Trav. Chim.*, **80**, 740 (1961).

[80] P. Debye, *Z. Phys. Chem.*, **130**, 56 (1927).

[81] V. S. Sastri, R. W. Henwood, S. Behrendt and C. H. Langford, *J. Amer. Chem. Soc.*, **94**, 753 (1972).

[82] P. K. Kadaba and J. O. Edwards, *J. Org. Chem.*, **25**, 1431 (1960).

[83] F. W. Grant and W. B. Cassie, *J. Org. Chem.*, **25**, 1433 (1960).

[84] W. E. Parham and F. C. Loew, *J. Org. Chem.*, **23**, 1705 (1958).

[85] W. E. Parham and E. E. Schweizer, *J. Org. Chem.*, **24**, 1133 (1959).

[86] W. M. Wagner, *Proc. Chem. Soc.*, 229 (1959).

[87] O. A. Reutov and A. N. Litsova, *Izv. Akad. Nauk SSSR, Otdel Khim. Nauk*, 1716 (1960).

[88] D. Seyferth, J. M. Burlitch, R. H. Minasz, J. Y. P. Mui, H. D. Simmons, Jr., A. J. Treiber and S. R. Dowd, *J. Amer. Chem. Soc.*, **87**, 4259 (1965).

[89] W. T. Miller and C. S. Y. Kim, *J. Amer. Chem. Soc.*, **81**, 5008 (1959).

Seyferth's method has large applications in the synthesis of dichloro- and dibromocyclopropanes. A disadvantage of the method is the difficult preparation of the organomercury compound. It has been shown recently[90] that tribromomethylphenyl-mercury, $PhHgCBr_3$, may be obtained more easily by a method of Nenitzescu *et al.*[91].

The methods listed in Scheme 3 show a great versatility. The fact that carbenes may be identified by their reactions both in base and acid media points to relatively stable intermediates. Since dichlorocarbene is generated in presence of nitric acid and products of reactions with nucleophiles are obtained, the species may be regarded as stable towards electrophiles.

In passing from haloforms to dihalogenomethanes both the kinetic and the thermodynamic basicities decrease abruptly. Stronger bases are thus necessary to generate monohalogenocarbenes. Closs and Closs [39, 93] obtained monochlorocyclopropanes by treating methylene chloride with butyl-lithium in presence of alkenes, while Köbrich *et al.*[42, 44] isolated dichloromethyl-lithium, **19**, which is stable in tetrahydrofuran between −110° and −65 °C.

$$RLi + CCl_2H_2 \longrightarrow RH + \underset{\mathbf{19}}{Cl_2CH^-Li^+} \qquad (9.40)$$

The lithium compound **19** exhibits all the characteristic properties of carbanions, being at the same time susceptible to electrophilic attack. Carbenoid species must also be considered in that case.

9.2.4 *Other examples of α-eliminations*

The treatment of neopentyl chloride with strong bases yields cyclopropane[94], suggesting an α-elimination. It has been later proved by Doering and Kirmse[95] that in this and other similar cases α- and β-elimination compete in the formation of the reaction pro-

[90] A. Banciu, Ph. D. Thesis, Bucharest (1975).

[91] D. C. Nenitzescu, D. Isăcescu and C. Gruescu, *Bul. Soc. Chim. Roumaine*, **20**, 135 (1938).

[92] F. Badea and C. D. Nenitzescu, *Angew. Chem.*, **92**, 415 (1960).

[93] G. L. Closs and L. E. Closs, *J. Amer. Chem. Soc.*, **82**, 5723 (1960).

[94] F. G. Whitmore and H. D. Zook, *J. Amer. Chem. Soc.*, **64**, 1783 (1942).

[95] W. Kirmse and W. von E. Doering, *Tetrahedron*, **11**, 266 (1960).

ducts. This was demonstrated by using α-deuterated compounds (Scheme 4):

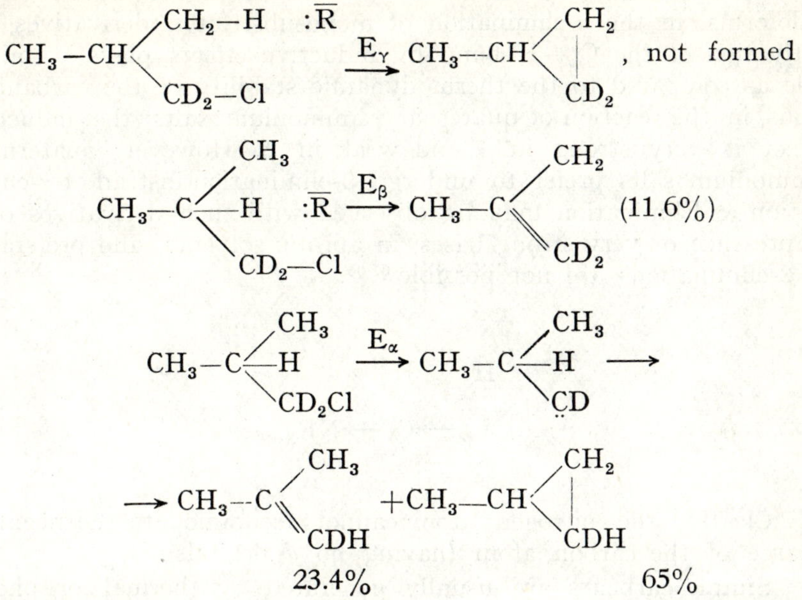

$$CH_3-CH\begin{cases}CH_2-H & :\bar{R}\\ CD_2-Cl\end{cases}\xrightarrow{E_\gamma} CH_3-CH\begin{cases}CH_2\\ |\\ CD_2\end{cases},\ \text{not formed}$$

$$CH_3-C\begin{cases}CH_3\\ H & :\bar{R}\\ CD_2-Cl\end{cases}\xrightarrow{E_\beta} CH_3-C\begin{cases}CH_2\\ \|\\ CD_2\end{cases}\quad(11.6\%)$$

$$CH_3-C\begin{cases}CH_3\\ H\\ CD_2Cl\end{cases}\xrightarrow{E_\alpha} CH_3-C\begin{cases}CH_3\\ H\\ \ddot{C}D\end{cases}\longrightarrow$$

$$\longrightarrow CH_3-C\begin{cases}CH_3\\ \|\\ CDH\end{cases}\ (23.4\%) + CH_3-CH\begin{cases}CH_2\\ |\\ CDH\end{cases}\ (65\%)$$

Scheme 4

The results show that cyclopropane is formed only by α-elimination (no γ-elimination) and that alkene results both by α- and by β-mechanisms. The ratio of cyclopropane (and thus of α-elimination) increases with neopentyl chloride (95%) and decreases with propyl chloride (4%). Thus, α-elimination is favoured by the absence of a neighbouring $C_\beta-H$ bond.

The formation of the cyclopropane derivative in the above examples corresponds to an intramolecular insertion of a carbene at a C—H bond (see Chapter 6). It was possible to isolate the intermediate lithium compounds formed in the cyclisation of neopentyl derivatives [96]. The ratio of the two cyclopropanes **21***a-b*, resulted in the cyclisation of the deutero derivative **20**, depends largely on the nature of the halide, X:

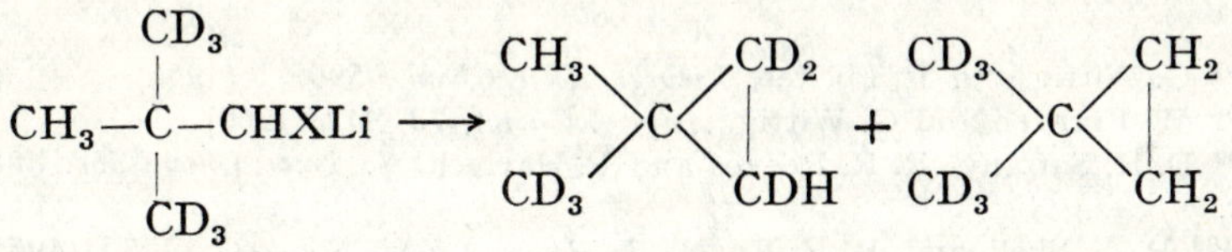

$$\underset{\mathbf{20}}{CH_3-\overset{CD_3}{\underset{CD_3}{C}}-CHXLi} \longrightarrow \underset{\mathbf{21}a}{\begin{matrix}CH_3\\ CD_3\end{matrix}\!>\!C\!<\!\begin{matrix}CD_2\\ |\\ CDH\end{matrix}} + \underset{\mathbf{21}b}{\begin{matrix}CD_3\\ CD_3\end{matrix}\!>\!C\!<\!\begin{matrix}CH_2\\ |\\ CH_2\end{matrix}}$$

[96] M. J. Goldstein and W. R. Dolbier, Jr., *J. Amer. Chem. Soc.*, **87**, 2293 (1965)

This is a clear demonstration of the assistance of the neighbouring C_γ—H bond in the second step of the elimination.

As in the case of the generation of dihalogenocarbenes from haloforms, in the α-elimination of monosubstituted derivatives the activation of the C_α—H bond by inductive effects plays a minor role as compared to the thermodynamic stability of the carbanion. Thus, in the reaction of quaternary ammonium salts the inductive effect is very strong in α and weak in β. However, quaternary ammonium salts prefer to undergo β-elimination instead of α-elimination. α-Elimination may be observed with such derivatives only in presence of very strong bases, in aprotic solvents, and preferably if β-eliminations are not possible[97,98].

$$>C_\beta(H)\longrightarrow C_\alpha(H)\longrightarrow \overset{+}{N}R_3$$

Clearly, the nitrogen atom cannot accommodate the negative charge of the carbon atom (having no *d* orbitals).

Simple carbenes are usually generated by thermal or photochemical decomposition of diazoalkanes. As already mentioned, this reaction may be regarded as an α-elimination from a hypothetical diazonium salt in which the intermediate (diazoalkene) is a stable, isolable compound. The reaction was also used in the generation of monovalent carbon compounds called *carbynes*[68,99,100]. Thus, carbethoxymethyne **22** was generated from diethylmercury-bisdiazoacetate, by short-wavelength photolysis[101].

$$(N_2CCOOEt)_2Hg \xrightarrow{h\nu} 2\ \underset{\mathbf{22}}{EtOOC{-}\dot{C}{:}} + 2\ N_2 + Hg \qquad (9.41)$$

If the photolysis is performed in presence of cyclohexene, the product analysis may show that carbynes have both carbene and radical character:

[97] G. Wittig and R. Polster, *Liebigs Ann. Chem.*, **599**, 1 (1956).

[98] W. Franzen and G. Wittig, *Angew. Chem.*, **72**, 417 (1960).

[99] D. R. Safrany, R. R. Reeves and P. Harteck, *J. Amer. Chem. Soc.*, **86**, 3160 (1964).

[100] P. S. Skell and R. R. Engel, *J. Amer. Chem. Soc.*, **87**, 1135 (1965); **93**, 2699 (1971).

[101] O. P. Strausz, G. J. A. Kennepohl, F.X. Garneau, T. DoMinh, B. Kim, S. Valenty and P. S. Skell, *J. Amer. Chem. Soc.*, **96**, 5723 (1974).

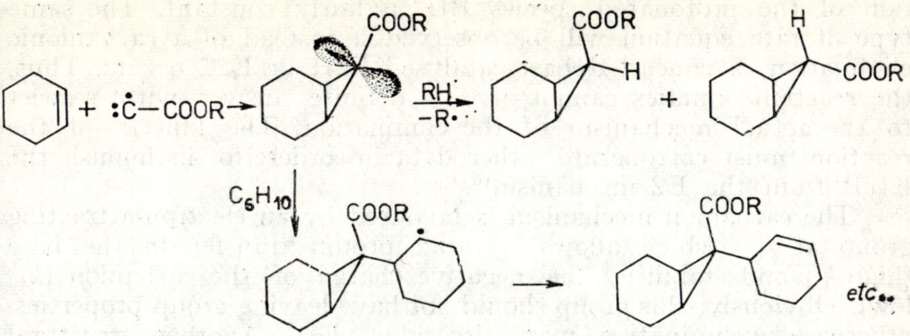

Even if the field of carbene chemistry is not as fertile as it appeared about two decades ago [63], there are still many promising lines of investigation.

9.3 β-Eliminations (E_β)

β-Eliminations lead either to alkenes or alkynes, which are stable products, easily identified and isolated. The reactions may be carried out under a large variety of conditions, characterised by the solvent-base system, and the leaving group. Four different mechanisms are encountered in β-elimination: E1cB, E1, E2H and E2C.

The most frequent, with the larger applicability, is however the E2H mechanism.

9.3.1 *The E1cB mechanism*

A β-elimination occurring by an E1cB mechanism may be represented by equations (9.42)—(9.43):

$$\text{B:} + \text{H—C}_\beta\text{R}_2\text{—C}_\alpha\text{R}_2\text{—X} \underset{k_{-1}}{\overset{k_1}{\rightleftharpoons}} {}^-\text{:C}_\beta\text{R}_2\text{—C}_\alpha\text{R}_2\text{—X} + \text{BH}^+ \quad (9.42)$$

23

$$\mathbf{23} \xrightarrow{k_2} \text{CR}_2\text{=CR}_2 + \text{X}^- \quad (9.43)$$

Using the steady-state treatment, the rate of this process will be given by equation (9.44):

$$\text{rate} = \frac{k_1 k_2\,[\text{Substrate}]\,[\text{B:}]}{k_{-1}\,[\text{BH}^+] + k_2} \quad (9.44)$$

If the rate of carbanion protonation is small as compared to proton elimination (k_2), equation (9.44) is reduced to the simple form (9.45):

$$\text{rate} = k_1\,[\text{Substrate}]\,[\text{B:}] \quad (9.45)$$

A second-order equation is also observed if the rate of removal of the leaving group X from **23** is negligible as compared to that of the protonation reaction ($k_{-1}[\text{BH}^+] \gg k_2$), and the concentration of the protonated species BH^+ is fairly constant. The same type of rate equation will be observed if instead of a carbanionic elimination, a concerted base-catalysed E2H or E2C occurs. Thus, the reaction kinetics cannot give a definite answer with respect to the actual mechanism of the elimination. The kinetics of the reaction must corroborate other data in order to distinguish the E1cB from the E2 mechanism[10].

The carbanion mechanism is favoured by an electron-attracting group at C_β, which allows a rapid proton transfer to the base (high k_1) and stabilises the negative charge of the carbanion (k_{-1} low). Obviously, this group should not have leaving group properties, otherwise α-elimination may also take place. Another structural condition involves the C—X bond, which should not easily ionise, so that it does not break before the transfer of the proton to the base occurs. Such conditions are only rarely met, and thus in the β-elimination the E1cB mechanism is less frequent than the concerted one. Strong bases, such as potassium t-butoxide[102], may also favour the carbanion mechanism.

One of the first examples of an E1cB mechanism, largely discussed in the literature, is the elimination from hexachlorocyclohexanes. In the β-isomer **24** all chlorine atoms are situated equatorially, the geometry being thus unfavourable to the concerted mechanism (see Section 9.4.3.). It has been admitted[103] that the reaction involves the intermediate formation of a carbanion **25** (Scheme 5). However, this carbanion may give rise to the δ-isomer, **26**, which may undergo rapid E2 elimination[104].

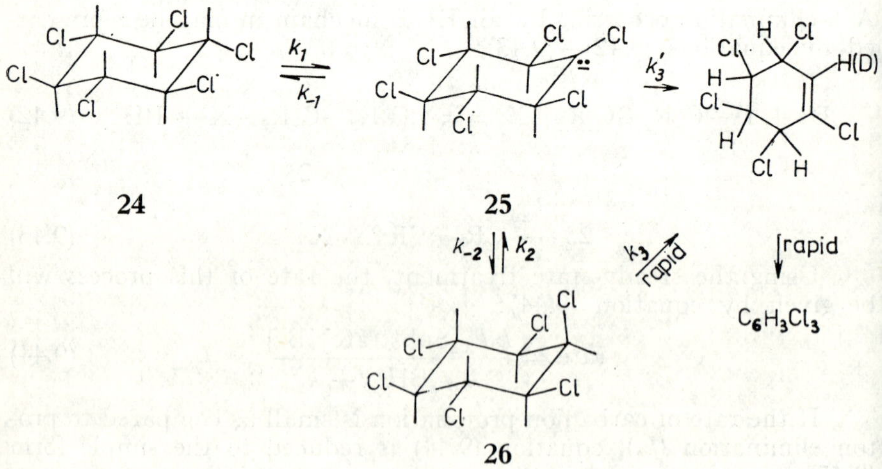

Scheme 5

[102] D. E. Pearson and C. A. Buehler, *Chem. Rev.*, **74**, 45 (1974).
[103] S. J. Cristol, *J. Amer. Chem. Soc.*, **69**, 338 (1947).
[104] S. J. Cristol and D. D. Fix, *J. Amer. Chem. Soc.*, **75**, 2647 (1953).

Evidently, the elimination may occur only from **26** ($k_{-2} \gg k_3'$), and thus, even if a carbanion is formed, the mechanism is not E1cB. Using sodium ethoxide in deuterated alcohol[105], it is found that the cyclohexane recovered after 50% transformation has incorporated only small amounts of deuterium. Moreover, the resulted trichlorobenzenes contain only traces of deuterium. This points to a concerted elimination from the δ isomer **26**, but does not rule out a slow concerted elimination from **24**.

An ambiguous situation is also encountered in the β-elimination from the sulphones **27**. The sulphone group is known to both activate and stabilise the carbanions[106], being at the same time a poor leaving group. A carbanion mechanism[107–109] (9.46) may be thus expected:

SO_2Ar / $(CH_2)_n$ / OSO_2Ar $\underset{k_{-1}}{\overset{k_1}{\rightleftharpoons}}$ $\bar{:}SO_2Ar$ / $(CH_2)_n$ / OSO_2Ar $\xrightarrow{k_2}$ SO_2Ar / $(CH_2)_n$

27 **28** (9.46)

If in the elimination (9.46) an amine R_3N: is used, the addition of a salt $R_3NH^+X^-$ does not affect the reaction rates[108], as expected in the case of a reversible formation of carbanion. However, this result does not rule out completely an E1cB mechanism in which $k_2 \gg k_{-1}[BH^+]$. On the other hand, it may also be consistent with a concerted E2 mechanism. It has been calculated[110] that the rate constant of the isotope exchange must be greater than the rate of elimination from the *trans*-**27** derivatives. The carbanion mechanism is thus possible in this case. However, considering that the arenesulphonate group is an efficient leaving group (Chapter 5), and that the geometrical condition for a concerted mechanism is fulfilled in the *cis* isomer, the E1cB mechanism is certainly ruled out for *cis*-**27**.

In the examples discussed above, the structural conditions imposed by the E1cB mechanism were not fully satisfied by the organic substrate. However, when these conditions are completely fulfilled, and strong bases are used, the carbanion mechanism is

[105] J. Hine, R. D. Weimer, Jr., P. B. Langford and O. B. Ramsay, *J. Amer. Chem. Soc.*, **85**, 3894 (1963).

[106] D. J. Cram, D. A. Scott and W. D. Nielsen, *J. Amer. Chem. Soc.*, **83**, 3696 (1961).

[107] F. G. Bordwell and R. J. Kern, *J. Amer. Chem. Soc.*, **77**, 1141 (1955).

[108] J. Weinstock, R. G. Pearson and F. G. Bordwell, *J. Amer. Chem. Soc.*, **78**, 3468, 3473 (1956).

[109] J. Weinstock, J. L. Barnardi and R. G. Pearson, *J. Amer. Chem. Soc.*, **80**, 4961 (1958).

[110] J. Hine and O. B. Ramsay, *J. Amer. Chem. Soc.*, **84**, 973 (1962).

operative. Such an example is the elimination from the diphosphonium salts **29** [111–113]:

$$\mathrm{Ph_3\overset{+}{P}{-}CH_2{-}CH_2{-}\overset{+}{P}Ph_3} \xrightarrow{\mathrm{B:}} \mathrm{Ph_3\overset{+}{P}{-}\overset{-}{C}H{-}CH_2{-}\overset{+}{P}Ph_3} \longrightarrow$$
$$\mathbf{29} \qquad\qquad \mathbf{30}$$

$$\longrightarrow \mathrm{Ph_3\overset{+}{P}{-}CH{=}CH_2}; \quad \mathrm{Ph_3\overset{+}{P}{-}\overset{-}{C}H{-}\overset{-}{C}H{-}\overset{+}{P}Ph_3}$$
$$\mathbf{31} \qquad\qquad \mathbf{32}$$

The reaction was originally carried out with a view to obtaining a bis-ylide **32**. However, this compound did not form and a β-elimination was assumed to be involved in order to account for the observed substitution products. More recent studies[113] have confirmed the mechanism proposed earlier, showing that the substitution product is formed by the following sequence of reactions:

$$\mathbf{31} \xrightarrow{\mathrm{H_2O}} \mathrm{R_3\overset{+}{P}{-}CH_2CH_2OH\,X^-} \xrightarrow{\mathrm{HO^-}} \mathrm{R_2P(O)CH_2CH_2OH}$$

The strong electron-attracting effect of the substituent of the double bond in **31** prevents the isolation of the elimination product.

A similar behaviour was observed in the case of the fluorenyl derivative **33**. Streitwieser and his co-workers [114] demonstrated that the reaction product, 9-trimethoxymethylfluorene, is obtained *via* an E1cB-elimination product **35**. The intermediate formation of the carbanion **34** was proved by tritium-hydrogen exchange.

H CF₃ — **33-H** ⇌ CF₃ (carbanion) — **34** → CF₂ — **35** → H CF₂OMe → H C(OMe)₃

34 ⇌ T CF₃ — **33-T**

[111] E. Stilz, Dissertation, Tübingen (1955), quoted by Schöllkopf [112]; G. Wittig, H. Eggers and P. Duffner, *Liebigs Ann. Chem.*, **619**, 10 (1958).
[112] U. Schöllkopf, *Angew. Chem.*, **71**, 260 (1959).
[113] H. Christol, H. J. Cristau and M. Soleiman, *Tetrahedron Letters*, 1385 (1975).
[114] A. Streitwieser, Jr., A. P. Marchand and A. H. Pudjaatmaka, *J. Amer. Chem. Soc.*, **89**, 693 (1967).

In that case, the inductive effect of the CF_3 group both enhances the rate of hydrogen abstraction from **33** and stabilises the resulted carbanion[115]. On the other hand, fluorine is a poor leaving group and allows the identification of the carbanion in the elimination process.

An interesting example of a very poor leaving group in the α-position of a stable carbanion was reported by Zimmerman and his co-workers[116]. The treatment of ketone **36** with Bu^{t}OK in Bu^{t}OH affords, after 7 minutes at 40 °C, the unsaturated bicyclic ketone **38**, in 74% yield. It is known from other works [117] that even if carbon is a very poor leaving group, in cyclopropane derivatives the heterolysis of a C—C bond may occur relatively easily. Ionisation of the C—Br bond may also facilitate the migration of the cyclopropane ring.

36 → **37** → **38**

This reaction can also be regarded as the trapping of the dipolar intermediate of the Favorsky rearrangement (Section 6.3.5).

β-Eliminations leading to acetylenes are assumed to involve the E1cB mechanism. In some instances, the intermediate vinyl carbanions have been trapped[118], as in the following example:

$$\text{D(Cl)C=CCl}_2 \longrightarrow \text{Cl}\bar{\text{C}}\text{=CCl}_2 \xrightarrow{\text{SH}} \text{H(Cl)C=CCl}_2$$

$$\text{Cl}\bar{\text{C}}\text{=CCl}_2 \xrightarrow{-\text{Cl}^-} \text{ClC}\equiv\text{CCl}$$

The formation of arynes, as intermediates in the aromatic nucleophilic substitution (Section 8.4), must also be regarded as a typical example of E1cB mechanism.

Recently [119], the use of heavy atom isotope effects has been recommended as a tool for the identification of the E1cB mechanism.

Thus, polycyanoethanes **39** and **40** transfer easily the β-hydrogen, due to the activating and stabilising effect of the two β-cyano

[115] A. Streitwieser, Jr., and D. Holtz, *J. Amer. Chem. Soc.*, **89**, 692 (1967).

[116] H. E. Zimmerman, D. Döpp and P. S. Huyffer, *J. Amer Chem. Soc.*, **88**, 5352 (1956).

[117] E. W. Yankee and D. J. Cram, *J. Amer. Chem. Soc.*, **92**, 6328, 6329, 6331 (1970).

[118] T. J. Houser, R. B. Bernstein, R. G. Meikka and J. C. Angus, *J. Amer. Chem. Soc.*, **77**, 6201 (1955).

[119] F. M. Fouad, P. G. Farrell and A. G. Abdel-Rehiem, *Tetrahedron Letters*, 3355 (1974).

groups. On the other hand, the poor leaving group CN attached to C_α enables in its turn the formation of a carbanion:

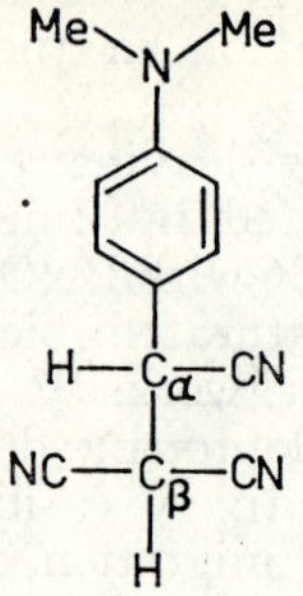

39 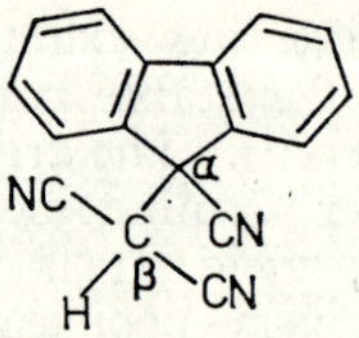 **40**

The formation of carbanions is demonstrated by the rapid exchange of the acidic hydrogen when the cyano derivatives are treated with CD_3OD (S_E1 reaction). If the carbon atom of the leaving group is labelled (^{14}CN), the heavy atom isotope effect (k^{12}/k^{14}) is found to be 1.04 ± 0.01 for **39**, and 1.16 ± 0.03 for **40**. It is considered [119] that these values give an indication concerning the character of the transition state of the second step of the E1cB mechanism*. A late transition state must produce a higher isotope effect than an early transition state. The conjugation of the dimethylamino group with the developing double bond will facilitate the breaking of the C—CN bond (early transition state), whereas the nature of the aromatic system in **40** will oppose the ionisation (late transition state).

The data regarding the E1cB mechanism confirm thus the required structural conditions indicated above. The same data show that the carbanion mechanism is not common in the β-elimination, both structural and external conditions (medium conditions) being rarely encountered in elimination reactions.

9.3.2 *E1 mechanism*

An E1 elimination has the same rate-determining step as an S_N1 substitution:

$$H{-}C_\beta R_2{-}C_\alpha R_2{-}X \xrightarrow{k_1} H{-}C_\beta R_2{-}\overset{+}{C}_\alpha R_2 \; (\mathbf{41})$$

$$\xrightarrow{E_1} CR_2{=}CR_2 + H^+ \quad (9.47a)$$

$$\xrightarrow{S_N1} CHR_2{-}CR_2S + H^+ \quad (9.47b)$$

* An alternative interpretation would be that the heavy atom isotope effects indicate an E2 character for the reaction of compound **39**, considering that in the decarboxylation of malonic acids (with labelled carboxyl) the k^{12}/k^{14} value is in the

Since the formation of the carbenium ion **41** is rate-determining, the overall reaction rate ($k_{E1} + k_{S_N1}$) must not depend on the concentration of the base.

$$\text{rate} = k_1\,[\text{Substrate}] \tag{9.48}$$
$$k_1 = k_{E1} + k_{S_N1} \tag{9.49}$$

This prediction has been confirmed experimentally[121, 122]. Moreover, the racemisation rate of optically active α-chlorophenylethane in liquid sulphur dioxide or in formic acid is practically identical with the elimination rate, which demonstrates that both reactions have a common intermediate [123].

The intermediate formation of carbocations in the rate-determining step would suggest that the E1 elimination must be affected by the same factors discussed for the S_N1 reaction (Chapter 5B). E1 will thus be determined by structural characteristics (the structure of the radical R in the substrate R—X, the nature of the leaving group X), by the ionising power of the solvent, by the presence of electrophilic catalysts, etc.

The E1 mechanism may involve either classical carbenium ions or non-classical carbonium ions. It has been assumed[124] that the elimination may only occur from classical ions, the products indicating that the intermediates are classical. However, elimination products were observed even in such reactions where the occurrence of non-classical ions was clearly demonstrated, as for example in the case of the non-classical norbornyl cation [125].

In many cases, the intermediate carbocation may form two or more alkenes, as for example:

$$CH_3{-}CH_2{-}C(CH_3)_2{-}Br \xrightarrow{-Br^-} CH_3CH_2{-}C^+(CH_3)_2 \xrightarrow{-H^+}$$
$$\underset{72\%}{CH_3CH{=}C(CH_3)_2} + \underset{28\%}{CH_3CH_2{-}C({=}CH_2){-}CH_3} \tag{9.50}$$

range 1.02 — 1.08 [120]. However, the same data would point to an E1 mechanism for **40**, which certainly cannot be true considering the poor leaving group ability of the cyano group. The interpretation of Fouad *et al.* [119] seems thus plausible.

120 A. Fry, in *Isotope Effects in Chemical Reactions* (C. J. Collins and N. S. Bowman, editors), Van Nostrand Reinhold (1970); A. Fry, *Chem. Soc. Rev.*, **1**, 163 (1972).

121 E. D. Hughes, *J. Chem. Soc.*, 255 (1935).

122 E. A. Cooper and E. D. Hughes, *J. Chem. Soc.*, 1183 (1937).

123 E. D. Hughes, C. K. Ingold and A. D. Scott, *J. Chem. Soc.*, 1271 (1937).

124 H. C. Brown and Y. Okamoto, *J. Amer. Chem. Soc.*, **77**, 3619 (1955).

125 P. D. Bartlett, W. S. Trahanowsky, D. A. Bolon and G. H. Schmid, *J. Amer. Chem. Soc.*, **87**, 1314 (1965).

It has been observed [1,2,3] that the more substituted (more stable) alkene is preferentially formed. This indicates that the intermediate carbenium ion is in equilibrium with the corresponding alkene, and that a thermodynamic control is responsible for the observed orientation (Saytzeff rule, see also p. 460).

The exceptions to this orientation are determined by steric effects. Thus, in the solvolysis of **42** alkenes **43** and **44** may be formed. In the more substituted alkene **43** (corresponding to Saytzeff rule), there is a strong repulsion between a methyl and a t-butyl group, whereas the alkene **44** involves only repulsions between hydrogen atoms and bulky groups. Steric effects prevail in the elimination from **42** [126], the less strained alkene resulting predominantly.

$$CH_3-C(CH_3)_2-CH_2-C(CH_3)_2-X \xrightarrow{-HX} (CH_3)_3C-CH{=}C(CH_3)_2 + (CH_3)_3C-CH_2C(CH_3){=}CH_2$$

42 **43** **44**

Other exceptions to the Saytzeff rule may be determined by hydride shifts, occurring in the intermediate carbocation, as for example in the solvolysis of menthyl derivatives[9]:

OTs — MeOH → + + OMe

85.5% 2.5% 13%

As Ingold and his co-workers have observed[1], the unimolecular elimination E1 implies that the ratio between the elimination and the substitution products in (9.47) should not depend on the leaving group X. Otherwise stated, the ratio k_{E1}/k_1 (see equations (9.48) and (9.49) for symbolism) should be constant. Hughes, Ingold and their co-workers [127–129] have confirmed this prediction, in the solvolysis of t-butyl derivatives (Table 9.1).

However, subsequent investigations [130–132] showed that the ratio between the alkene and the substitution product depends largely on the solvent[130] and on the leaving group [131, 132].

[126] H. C. Brown and I. Moritani, *J. Amer. Chem. Soc.*, **77**, 3623 (1955).
[127] E. D. Hughes, C. K. Ingold and U. G. Shapiro, *J. Chem. Soc.*, 1277 (1937).
[128] K. A. Cooper, E. D. Hughes and C. K. Ingold, *J. Chem. Soc.*, 1280 (1937).
[129] K. A. Cooper, E. D. Hughes, C. K. Ingold and B. J. McNulty, *J. Chem. Soc.*, 20, 38 (1948).
[130] M. Cocivera and S. Winstein, *J. Amer. Chem. Soc.*, **85**, 1702 (1963).
[131] D. J. Cram and M. R. V. Sahyun, *J. Amer. Chem. Soc.*, **85**, 1257 (1963).
[132] P. S. Skell and W. H. Hall, *J. Amer. Chem. Soc.*, **85**, 2851 (1963).

TABLE 9.1

The solvolysis rates of some t-butyl derivatives in aqueous ethanol (80%)

X *in* Bu^{t}X	*Temp.* °C	10^5k_1	k_{E1}/k_1	*Ref.*
Cl	25	0.854	0.168	128
Br	25	37.2	0.126	128
I	25	90.1	0.129	122
Cl	65.3	89.7	0.363	129
$\overset{+}{S}Me_2$	65.3	11.8	0.357	129

Thus, in the solvolysis of t-butyl derivatives[130], the larger amounts of isobutene are formed in the less basic solvent (AcOH, see Table 9.2), in the same solvent the more basic counter-ion leading to the higher ratio of alkene.

TABLE 9.2

The ratio of isobutene resulted in the solvolysis of some t-butyl derivatives in various solvents [a]

X *in* Bu^{t}X	*Isobutene* (%) *formed in*:			
	H_2O (25 °C)	H_2O (75 °C)	abs. EtOH (75 °C)	AcOH (75 °C)
Cl	5.0 ± 1.0	7.6 ± 1.0	44.2 ± 1.0	73 ± 2
Br	5.0 ± 1.0	6.0 ± 1.0	36.0 ± 1.0	69.5
I	4.0 ± 1.0	6.0 ± 1.0	32.3 ± 1.0	—
$\overset{+}{S}Me_2$	—	6.5 ± 1.0	17.8 ± 1.8	11.7 ± 1
$\overset{+}{O}H_2$	3	4.7	—	—

[a] According to Cocivera and Winstein [130].

It has been assumed and experimentally verified [133–135] that in non-ionising solvents the elimination occurs within an ion pair, the more basic counter-ion giving rise to higher ratios of elimination products. Thus, in the solvolysis of tosylate **45** in carboxylic acids, the higher ratios of alkenes are observed in pivalic acid [133], which has the lower polarity among the solvents used.

CH_2OTs —RCOOH→ + CH_2

+ solvolysis products

45

[133] E. Ciorănescu, A. Bucur, F. Badea, M. Rentzea and C. D. Nenitzescu, *Tetrahedron Letters*, 1867 (1969).

[134] F. G. Bordwell and G. A. Pagany, *J. Amer. Chem. Soc.*, **97**, 118 (1975).

[135] F. G. Bordwell and T. G. Meca, *J. Amer. Chem. Soc.*, **97**, 123, 127, 132 (1975).

Solvent effects and leaving group ability may affect both the orientation and the stereochemistry of the reaction. Skell and Hall[132] have observed that the elimination from 2-butyl tosylate is *syn* in an aprotic solvent (nitrobenzene), and *anti* in a protic solvent. Ion-pair elimination is assumed to account for the stereochemistry in the aprotic solvent (**46**), whereas a quasi-concerted elimination may be involved in the more basic solvent (**47**):

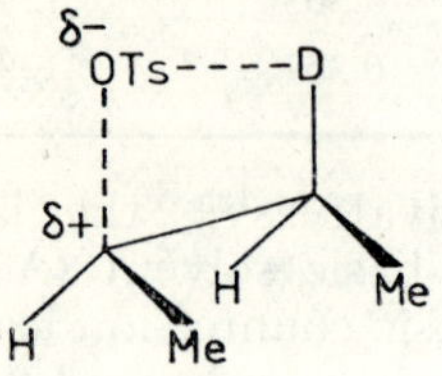

δ− OTs---D δ+ H H Me Me

46

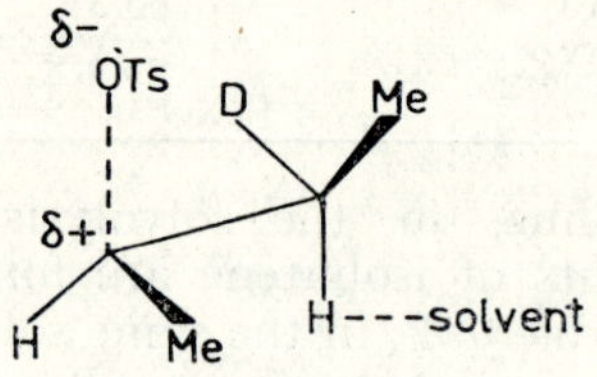

δ− OTs D Me δ+ H Me H---solvent

47

Among E1 eliminations, the dehydration of alcohols deserves a separate discussion. Different mechanisms (Scheme 6) are possible for this reaction.

$$\mathrm{H{-}CR_2{-}CHR{-}OH} \overset{H^+}{\rightleftharpoons} \mathrm{H{-}CR_2{-}CHR{-}\overset{+}{O}H_2} \quad (9.51)$$

48

$$\text{E2: } \mathrm{B\!:} + \mathrm{H{-}CR_2{-}CHR{-}\overset{+}{O}H_2} \longrightarrow \mathrm{BH^+ + CR_2{=}CHR + H_2O} \quad (9.52)$$

$$\text{E1: } \mathrm{HCR_2{-}CHR{-}OH_2^+} \rightleftharpoons \mathrm{H{-}CR_2{-}\overset{+}{C}HR + H_2O} \longrightarrow \mathrm{CR_2{=}CHR} \quad (9.53)$$

49

$$\text{E}\alpha\text{: } \mathrm{H{-}CR_2{-}CHR{-}\overset{+}{O}H_2} \longrightarrow \mathrm{HCR_2{-}\ddot{C}R + H_2O + (H^+)} \quad (9.54)$$

50

$$\mathrm{HCR_2{-}\ddot{C}R} \longrightarrow \mathrm{CR_2{=}CHR} \quad (9.55)$$

Scheme 6

Whatever the mechanism, the first step is the formation of an oxonium salt **48.** Three different routes to the alkene are then possible. In one of them (9.52) the elimination of water from **48** is synchronous with the transfer of the proton to a base. This concerted mechanism (E2) occurs in the case of substrates unable to generate stable carbocations. If stable carbocations **49** may be formed, the E1 mechanism probably operates.

The concerted and the cationic mechanism can be distinguished experimentally by methods involving the identification of carbocations. By means of n.m.r. [136,137], both cationic species, **48** and **49,**

[136] G. A. Olah, M. B. Comisarow, C. A. Cupas and C. U. Pittmann, Jr., *J. Amer. Chem. Soc.*, **87**, 2997 (1967).

[137] G. A. Olah and J. Sommer, *J. Amer. Chem. Soc.*, **90**, 927 (1968).

have been identified in acid media. Kinetic data also support the mechanisms represented in Scheme 6. Thus, in the dehydration of benzyl alcohols[138,139] the logarithms of the rate constants correlate linearly with the acidity function H_0 of the medium, indicating the formation of carbocations in the rate-determining step (Section 3.3.4). On the other hand, in the formation of stilbenes by reaction (9.56), the observed rate constants (k_{XY}) may be correlated with both σ^+ and σ parameters[140]:

$$X\text{-}C_6H_4\text{-}CH(OH)\text{-}CH_2\text{-}C_6H_4\text{-}Y \xrightarrow{k_{xy}} X\text{-}C_6H_4\text{-}CH{=}CH\text{-}C_6H_4\text{-}Y \qquad (9.56)$$

$$\log k_{XY} = -3.78\,(\sigma_X^+ + 0.233\,\sigma_Y) - 3.19 \qquad (9.57)$$

Relation (9.57) points in its turn to an E1 mechanism in the dehydration of benzyl alcohols.

The formation of carbocations is also demonstrated by side reactions. Thus, Deno and his co-workers[141] have observed that in the reaction of t-butanol with 75% sulphuric acid, beside isobutene (dehydration product), dimers and trimers of isobutene are also formed (resulting by alkylation of isobutene with t-butyl cation, see p. 396). With more concentrated acid, saturated hydrocarbons and alkenes with several double bonds are isolated, indicating that intermolecular hydride shifts take place The high number of isomeric hydrocarbons indicate also rearrangement reactions in a carbocation. As alkylation, rearrangement and hydride transfer reactions are characteristic of carbocations, the formation of the latter is well proved.

Rearrangements may also occur in the dehydration of primary alcohols. Formation of non-classical ions of rearrangement synchronous with ionisation were suggested.

A well-known example is the dehydration of neopentyl alcohol, leading to 2-methyl-but-2-ene [142] as the major product:

$$CH_3\text{—}C(CH_3)_2\text{—}CH_2\text{—}OH \xrightarrow[-H_2O]{H^+} CH_3\text{—}\overset{+}{C}(CH_3)\text{—}CH_2\text{—}CH_3 \longrightarrow (CH_3)_2C{=}CH\text{—}CH_3 \qquad (9.58)$$

[138] D. S. Noyce, C. A. Lane, P. A. King and W. L. Reed, *J. Amer. Chem. Soc.*, **84**, 1635, 1638 (1962).

[139] D. S. Noyce, D. R. Hartter and R. M. Pollack, *J. Amer. Chem. Soc.*, **90**, 3791 (1968).

[140] D. S. Noyce, R. D. Hartter and F. B. Miles, *J. Amer. Chem. Soc.*, **90**, 3794 (1968).

[141] N. C. Deno, D. B. Boyd, J. D. Hodge, C. U. Pittman, Jr., and J. O. Turner, *J. Amer. Chem. Soc.*, **86**, 1745 (1964).

[142] cf. P. von R. Schleyer *et. al.*, *J. Amer. Chem. Soc.*, **88**, 4475 (1966)

According to the principle of microscopic reversibility, the E1 mechanism should be identical with the mechanism of electrophilic additions to a double bond. There are experimental indications supporting such a correspondence. Thus, Badea[143] observed that the treatment of *meso*-dibromosuccinic ester with silver acetate in acetic acid and cyclohexene yields, among other products, dimethyl fumarate and bromoacetoxycyclohexane. The observed transbromination demonstrates that the addition and the elimination proceed by a common mechanism:

$$\text{ROOC—CHBr—CHBr—COOR} + \text{Ag}^+ \longrightarrow$$

$$\text{ROOC—CHBr—}\overset{+}{\text{C}}\text{H—COOR} + \text{AgBr}$$

$$\text{ROOC—CHBr—}\overset{+}{\text{C}}\text{H—COOR} + \text{(cyclohexene)} \longrightarrow \text{ROOC—CH=CH—COOR} +$$

$$\text{(2-bromocyclohexyl cation: H, Br, +)} \xrightarrow{\text{AcOH}} \text{(cyclohexane with H, Br, OAc)} + \text{other acetates}$$

Another example is due to Olah and his co-workers[144], who were able to show that the intermediate(s) of halogen addition are in equilibrium either with the alkene or with the addition product, depending on the structure of the original alkene and on the reaction medium.

Thus, the E1 mechanism is closely related to the AdE mechanism. We shall later see that the E2 elimination is in its turn related to an addition mechanism.

9.4 The concerted β-elimination (E2)

9.4.1 *Kinetics. Transition states*

The bimolecular elimination E2 implies the synchronous breaking of the two bonds C_α—X and C_β—Y. Most frequently, with Y = H, the E2 elimination is a base-catalysed reaction:

$$\text{B:} + \text{H—C}_\beta\text{R}_2\text{—C}_\alpha\text{R}_2\text{—X} \longrightarrow \text{BH}^+ + \text{CR}_2\text{=CR}_2 + \text{X}^- \quad (9.59)$$

X must be a poor leaving group, in order to avoid the E1 mechanism, and the β-hydrogen should not be too acid, in order to prevent the E1cB mechanism.

[143] F. Badea, unpublished results.

[144] G. A. Olah, P. Schilling, P. W. Westerman and H. C. Lin, *J. Amer. Chem. Soc.*, **96**, 3581 (1974).

The kinetics of a base-catalysed E2H elimination (9.59) is expressed by the general equation (9.60):

$$\text{rate} = k[\text{Substrate}][\text{B:}] \qquad (9.60)$$

However, this equation is not specific for this mechanism as it may correspond also to an E1cB mechanism (E_α and E_β), or to the concerted mechanisms (E2H and E2C).

If the substrate may in principle undergo both α- and β-eliminations, the former is usually not observed with weak bases in protic solvents. In media of high basicity and in aprotic solvents, however, α-elimination can well compete with β-elimination [145]. Deuterium labelling is conclusive in that respect:

$$\text{R—CD}_2\text{—CH}_2\text{X} \xrightarrow{E_\alpha} \text{R—CD}_2\text{—}\ddot{\text{C}}\text{H} \xrightarrow{\text{D:}^-} \text{RCD=CDH} \qquad (9.61a)$$

$$\xrightarrow{E_\beta} \text{R—CD=CH}_2 \qquad (9.61b)$$

From the ratio between the two deuterated alkenes, considering the isotope effects, the weight of β-elimination in the overall process can be appreciated.

The incursion of the E1cB mechanism in the β-elimination has been established by using a deuterated solvent and measuring the hydrogen-deuterium exchange (side reaction, S_E1). Thus, Skell and Hauser [146] found that the elimination from β-phenylethyl bromide with sodium ethoxide in deuterated ethanol yields styrene free of deuterium Moreover, if the reaction is interrupted at an early stage, the recovered bromide has no deuterium. According to Skell and Hauser, this result rules out a carbanion mechanism.

$$\text{PhCH}_2\text{—CH}_2\text{Br} \underset{\text{EtOH}}{\overset{\text{EtO}^-}{\rightleftharpoons}} \text{Ph—}\ddot{\bar{\text{C}}}\text{H—CH}_2\text{Br} \longrightarrow \underset{\mathbf{51}}{\text{PhCH=CH}_2}$$

$$\upharpoonleft\downharpoonright \text{EtOD}$$

$$\text{PhCHD—CH}_2\text{Br} \rightarrow\rightarrow \text{PhCD=CH}_2$$

In excess as compared to **51**, due to isotope effects

However, it was later shown [10] that the E1cB elimination may proceed either with hydrogen exchange if $k_{-1}[\text{BH}^+] > k_2$, or with no exchange if $k_2 > k_{-1}[\text{BH}^+]$ (see equations (9.42) and (9.43) for symbolism). Thus, if no exchange is observed in a base-catalysed elimination carried out in a deuterium pool, the result does not allow the distinction between the E2 and the E1cB mechanism. However, complete kinetic data may provide essential indications concerning the reaction mechanism. In order to understand the other kinetic arguments, the concerted mechanism must be presented in more detail.

[145] S. M. Luck, D. G. Hill, A. T. Stewart, Jr., and C. R. Hauser, *J. Amer. Chem. Soc.*, **81**, 2784 (1959).

[146] P. S. Skell and C. R. Hauser, *J. Amer. Chem. Soc.*, **67**, 1661 (1945).

The ideal situation in which both bonds, C_α—X and C_β—H, are being broken in a perfectly synchronous way, is probably never achieved. The process starts with the stretching of one bond (either C_α—X or C_β—H), which determines a transition state resembling either an E1 elimination (**52**) or an E1cB elimination (**54**). If the other bond (either at C_α or at C_β) does not differ significantly in its dissociation energy from the polarised bond, the polarisation of the weaker bond will induce the polarisation of the neighbouring bond. Before the C_α—X (or C_β—H) bond is broken, the other bond (C_β—H or C_α—X, respectively) begins to weaken, a double bond being developed between the carbon centres. Evidently, a minimum of energy will correspond to a certain geometry of the transformation and this geometry may be similar either to the transition state of an E1 elimination (**52**, E1-like) or to that of an E1cB elimination (**54**, E1cB-like). Between these limits there is a infinite number of other possibilities determining the somehow intriguing features of the β-elimination.

B: H — R_2C_β—$C_\alpha^{\delta+}R_2$ ⋯ $X^{\delta-}$ (**52**)

$B^{\delta+}$⋯H ⋯ R_2C_β⋯$C_\alpha R_2$ ⋯ $X^{\delta-}$ (**53**)

$B^{\delta+}$⋯H ⋯ $R_2C_\beta^{\delta-}$—$C_\alpha R_2$ — X (**54**)

The structure of the substrate and the experimental conditions (solvent-base system) may determine the actual character of the transition state of an E2 elimination. Kinetic data illustrate significantly the various situations.

Thus, in the ethoxide-promoted elimination from arylethyl derivatives **55**, in ethanol at 30 °C, a Hammett correlation between σ_Z and the logarithms of k_Z is observed[147–149].

$$Z\text{-}C_6H_4\text{-}CH_2\text{-}CH_2\text{-}X + EtO^- \longrightarrow Z\text{-}C_6H_4\text{-}CH{=}CH_2 + EtOH + X^-$$

55 (X = Br, $\overset{+}{S}Me_2$, $\overset{+}{N}Me_3$)

The reaction parameter ρ is +2.1 for bromides[148], +2.7 for sulphonium salts[147] and increases to +3.77 for ammonium salts[149]. The positive sign of the reaction parameter indicates that in the transition state the reaction centre (Ar—$\overset{|}{C}_\beta$—H) acquires a more

[147] W. H. Saunders, Jr., and R. A. Williams, *J. Amer. Chem. Soc.*, **79**, 3712 (1957).

[148] W. H. Saunders, Jr., G. B. Gibbons and R. A. Williams, *J. Amer. Chem. Soc.*, **80**, 4099 (1958).

[149] W. H. Saunders, Jr., D. G. Bushman and A. F. Cockerill, *J. Amer. Chem. Soc.*, **90**, 1775 (1968).

negative charge than in the initial state. This agrees either with an E1cB mechanism, or with an E2 mechanism in which the transition state is variable and has an E1cB-like geometry. The nature of the leaving group in **55** is in perfect agreement with a quasi synchronous mechanism.

The concerted mechanism is not limited to bimolecular reactions or to base-catalysed eliminations. A well-known example of unimolecular concerted elimination leading to alkenes is the bromodecarboxylation (9.18), for which both the carbocationic and the concerted mechanism have been identified[150].

A different reaction is the iodide ion-promoted elimination from dibromo derivatives:

$$3I^- + R_2CBr—CBrR_2 \longrightarrow R_2C{=}CR_2 + I_3^- + 2Br^- \qquad (9.62)$$

The bimolecular character of this elimination has been demonstrated[151]. The kinetic data (Table 9.3) show that if in dibromoethane the hydrogen atoms are substituted with electron-attracting groups, the relative rates increase, whereas substituents with opposite effect produce a decrease of rate. Mulders and Nasielski[155] found however that even if the substitution of one or two hydrogen atoms in dibromoethane with methyl groups decreases the elimination rate, further substitution has the opposite effect. Methyl

TABLE 9.3

Relative rates in the debromination of 1,2-dibromo derivatives with iodide ion

No.	*Dibromo derivative*	*Solvent*	*Relative rate* (60 °C)
1.	$BrCH_2—CH_2Br$	99% MeOH [152]	1
		abs. ETOH [153]	1.63
2.	$CH_3CHBr—CH_2Br$	99% MeOH [154]	0.0375
		abs. EtOH [153]	0.395
		Me_2CO[155]	7.45
3.	*meso*-MeCHBr—CHBrMe	99% MeOH [154]	0.039
4.	±MeCHBr—CHBrMe	99% MeOH[154]	0.0152
5.	$Me_2CBr—CBrMe_2$	Me_2CO [155]	264
6.	PhCHBr—CHBrPh	Me_2CO[155]	382
7.	*meso*-HOOC—CHBr—CHBr— —COOH	EtOH[153] 99% MeOH[154]	12 5.25

150 E. Grovenstein, Jr., and D. E. Lee, *J. Amer. Chem. Soc.*, **75**, 2639 (1953); S. J. Cristol and W. P. Norris, *ibid.*, **75**, 2645 (1953).

151 W. Winstein, D. Pressman and W. G. Young, *J. Amer. Chem. Soc.*, **61**, 1645 (1939).

152 R. T. Dillon, *J. Amer. Chem. Soc.*, **54**, 952 (1932).

153 T. L. Davis and R. Heggie, *J. Org. Chem.*, **2**, 470 (1938).

154 W. G. Young, D. Pressman and C. D. Coryel, *J. Amer. Chem. Soc.*, **61**, 1640 (1939).

155 J. Mulders and J. Nasielski, *Bull. Soc. Chim. Belges*, **72**, 322 (1963).

substitution at the reaction centres in the elimination (9.62) must favour the unimolecular mechanism. Thus, a shift toward an E1-like transition state must occur.

The kinetic data agree well with the oscillation of the E2 transition state between the two limits **52** and **54**. This oscillation is also demonstrated by the orientation and stereochemistry of the reaction.

9.4.2 *Orientation effects in the concerted elimination*

(a) *Saytzeff and Hofmann rules.* More than a hundred years ago it was observed that whenever in an elimination reaction two or more alkenes may be formed, one of them is usually preferred. Thus, the elimination from 2-butyl derivatives may lead either to a highly substituted alkene, but-2-ene (*cis* and *trans*), or to a less substituted one, but-1-ene:

$$\underset{\substack{|\\ \mathrm{X}}}{\mathrm{CH_3CH_2CH{-}CH_3}} \xrightarrow{-\mathrm{HX}} \mathrm{CH_3CH{=}CHCH_3 + CH_3CH_2CH{=}CH_2} \qquad (9.63)$$

Either of the two alkenes may predominate in the reaction mixture, their ratio depending on the nature of the leaving group X. Two rules have been formulated for the eliminations of type (9.63).

One of them refers to eliminations from halides and, as later observed, also to other neutral substrates in which X has a moderate electron-attracting inductive effect. The rule, known as the *Saytzeff rule*[156], may be formulated as follows: whenever two or more alkenes may result in an elimination, the more substituted (thermodynamically more stable) alkene is formed. Thus, the elimination from 2-butyl bromide yields only 19% marginal alkene[157]:

$$\underset{\substack{|\\ \mathrm{Br}}}{\mathrm{CH_3CH_2CH{-}CH_3}} \xrightarrow{-\mathrm{HBr}} \underset{81\%}{\mathrm{CH_3CH{=}CHCH_3}} + \underset{19\%}{\mathrm{CH_3CH_2CH{=}CH_2}}$$

On the other hand, Hofmann[158] established another empirical rule, known as the *Hofmann rule,* concerning the elimination which occurs in the decomposition of quaternary ammonium hydroxides. The olefin formed in higher amount is the one having the smaller number of alkyl substituents. Referring again to reaction (9.63),

[156] A. Saytzeff, *Liebigs Ann. Chem.*, **179**, 296 (1875).
[157] M. L. Dhar, E. D. Hughes and C. K. Ingold, *J. Chem. Soc.*, 2058 (1948).
[158] A. W. Hofmann, *Liebigs Ann. Chem.*, **78**, 253 (1951).

but-1-ene should predominate, according to this rule, as actually is observed[159]:

$$CH_3{-}CH_2{-}\underset{{}^{+}N(CH_3)_3}{\underset{|}{CH}}{-}CH_3 \xrightarrow{-H^+} :N(CH_3)_3 + \underset{95\%}{CH_3CH_2CH{=}CH_2} + \underset{5\%}{CH_3CH{=}CHCH_3}$$

The Hofmann rule is not restricted to quaternary ammonium hydroxides. Other 'onium ions, like dimethyl-2-butylsulphonium ion **56,** undergo elimination according to this rule[160]:

$$CH_3{-}CH_2{-}\underset{{}^{+}S(CH_3)_2}{\underset{|}{CH}}{-}CH_3 \xrightarrow{-H^+} :S(CH_3)_2 + \underset{86\%}{CH_3CH_2CH{=}CH_2} + \underset{14\%}{CH_3CH{=}CHCH_3}$$

It should also be mentioned that the presence of alkyl groups in β has an accelerating effect upon the Saytzeff elimination, but a retarding effect in the Hofmann elimination.

(b) *Dependence of orientation on inductive and electronic effects.* According to Hughes, Ingold and their co-workers[161], the Hofmann orientation is determined by the strong inductive effects of the 'onium group:

$$\overset{}{CH_3}{-}\overset{\delta\delta\delta+}{C_{\beta'}H_2}{-}\overset{\delta+}{C_{\alpha'}H_2}{-}\overset{+}{\underset{CH_3\quad CH_3}{N}}{-}\overset{\delta+}{C_{\alpha}H_2}{-}\overset{\delta\delta+}{C_{\beta}H_2}{-}H$$

57

Thus, in the dimethylethylpropyl ammonium ion **57,** the positive charge of the nitrogen induces a partial positive charge in the neighbouring positions (α and α'). The polarised carbon atoms in α and α', in their turn, induce a partial positive charge in the β and β' positions. However, this charge will be higher in β and lower in β' due to the inductive electron-repelling effect of the methyl group. The base attacks therefore the C_{β}—H bond, leading to ethylene instead of propylene. The same inductive model explains the deactivation effect of the β-alkyl groups in the Hofmann elimination.

Considering inductive effects alone, one may reasonably assume that the base attacks the more acidic C_{α}—H bond, yielding a nitro-

[159] D. V. Banthorpe, E. D. Hughes and C. K. Ingold, *J. Chem. Soc.*, 4054 (1960).

[160] E. D. Hughes, C. K. Ingold and L. I. Wolf, *J. Chem. Soc.*, 2084 (1948).

[161] M. L. Dhar, E. D. Hughes, C. K. Ingold, A. M. M. Mandour, G. A. Maw and L. I. Wolf, *J. Chem. Soc.*, 2093 (1948).

gen ylide. Unlike phosphorous ylides, which are stable intermediates[112], nitrogen ylides **58** are highly unstable, since nitrogen is unable to use *d* orbitals.

$$R^1_3\overset{+}{N}-CH_2-R^2 + :B \underset{k_{-1}}{\overset{k_1}{\rightleftharpoons}} \underset{\mathbf{58}}{R^1_3\overset{+}{N}-\overset{-}{\ddot{C}}H-R^2} + BH^+ \quad (9.64)$$

$$\downarrow k_2$$

$$R^1_3N: + :CH-R^2$$

On the other hand, further elimination would lead to a carbene, another unstable intermediate. Thus, in reaction (9.64) k_{-1} must by far exceed the other rate constants (k_1 and k_2), and hydrogen abstraction from the α position (k_1) may be regarded as an *invisible reaction* leading to nowhere.

As regards the Saytzeff orientation, the inductive effects of the halogen atoms (except fluorine) are too weak to determine the formation of the less substituted alkene. If however it is assumed that the elimination from halogenated compounds is E1-like, one may expect the double bond to be more developed in this case than in the transition state of the Hofmann elimination. Thus, the same factors stabilising the alkene are also stabilising the transition state of the reaction. The rule governing E1 eliminations is also governing E1-like eliminations, as Hughes, Ingold and their co-workers[161] have established. Stabilisation by hyperconjugation of the transition state **59**, which does not occur in the transition state **60**, seems to be responsible for the Saytzeff elimination.

$$\overset{\delta+}{B}\cdots H \quad H \qquad\qquad H$$

$$H-C_\gamma-C_\beta\cdots C_\alpha-C_{\beta'}-H \qquad H-C_\gamma-C_\beta-C_\alpha\cdots C_{\beta'}\cdots H\cdots B^{\delta+}$$

$$X^{\delta-} \qquad\qquad X^{\delta-}$$

59 **60**

(c) *Steric factors in eliminations*. A different interpretation of the orientation rules was proposed by Brown[162]. According to this author, the shift from Saytzeff to Hofmann orientation is determined by steric factors, as for example in reaction (9.65):

$$R-C_\beta H_2-\underset{\displaystyle X}{\underset{|}{C}H}-CH_3 + :B \xrightarrow{-HX} \underset{\mathbf{61}}{R-CH=CH-CH_3} +$$

$$\underset{\mathbf{62}}{R-CH_2-CH=CH_2} \quad (9.65)$$

[162] H. C. Brown, *Boranes in Organic Chemistry*, Cornell University Press, Ithaca, New York (1972), Chapter 7, Section 7.

The base attacks always the less hindered proton, yielding thus the less substituted (usually marginal) alkene. The Saytzeff rule predicts the normal reaction product, whereas the Hofmann rule points to the sterically controlled product. Brown and his co-workers [163–166] have shown that in reaction (9.65) the marginal alkene **62** (product of Hofmann orientation) prevails over the more stable alkene **61** (product of Saytzeff orientation), if R or X are bulky groups, or if the base has large steric demands. Thus, the bulky trimethylammonium group favours the base attack at the marginal position. In an alkoxide-promoted elimination, the amount of marginal alkene increases with the steric demands of the base, in the order: $MeO^- > EtO^- > Me_3CO^- > Et_3CO^-$.

However, Saunders *et al.* [167] found that in the elimination from 2-pentyl derivatives, 2-fluoropentane affords the higher yield of pent-1-ene, a result which does not agree with Brown's theory:

$$\underset{\displaystyle\overset{|}{X}}{CH_3CH_2CH_2CHCH_3} \xrightarrow[EtOH]{EtONa} \underset{(cis \text{ and } trans)}{CH_3CH_2CH{=}CHCH_3} + CH_3CH_2CH_2CH{=}CH_2$$

X = F	18.3%	81.7%
X = Cl	63.8%	36.2%
X = Br	75.4%	24.6%
X = I	80.2%	19.8%

If the steric factor operates, the iodo-derivative, with the bulkier leaving group, would afford larger amounts of marginal alkene than does 2-fluoropentane, which has the smallest leaving group in the series. Similar observations have been recorded for 2-butyl [168] and 2-hexyl [169] halides. It has been argued however that fluorine, even if it is the smallest leaving group, must be the best solvated in the transition state, having thus an actual volume larger than that of the other halogens. Confirming this view, Bartsch and Bunnett [169] have however shown that the solvation of the developing halide ion cannot fully account for the ratio between the two alkenes, which is better rationalised by the inductive effects, larger for fluorine than for iodine.

The fluctuation between the E1-like and the E1cB-like transition states explains satisfactorily the dependence of the orientation on the nature of the leaving group.

[163] H. C. Brown, I. Moritani and M. Nakagawa, *J. Amer. Chem. Soc.*, **78**, 2190 (1956).
[164] H. C. Brown and O. H. Wheeler, *J. Amer. Chem. Soc.*, **78**, 2199 (1956).
[165] H. C. Brown, I. Moritani and Y. Okomoto, *J. Amer. Chem. Soc.*, **78**, 2193 (1956).
[166] H. C. Brown and M. Nakagawa, *J. Amer. Chem. Soc.*, **78**, 2197 (1956).
[167] W. H. Saunders, Jr., S. R. Fahrenholtz, E. A. Caress, J. P. Lowe and M. Schreiber, *J. Amer. Chem. Soc.*, **87**, 3401 (1956).
[168] H. C. Brown and R. L. Klimisch, *J. Amer. Chem. Soc.*, **88**, 1425 (1966); **87**, 5517 (1965).
[169] R. A. Bartsch and J. F. Bunnett, *J. Amer. Chem. Soc.*, **90**, 408 (1968).

9.4.3 *The stereochemistry of the E2 elimination*

In any β-elimination, beside the Hofmann or Saytzeff orientations, one may find other stereochemical features. The structure of the olefins arising under stereochemical control are determined by the relative positions of the groups X and Y being eliminated (in most cases Y = H), in the stable conformations of the substrate. Experimental observations [170–176] led to the conclusion that the β-elimination is an *antiperiplanar* elimination [171], in which the four centres involved, H—C_β—C_α—X, lie in the same plane, and the C_β—H and C_α—X bonds determine a dihedral angle of 180°.

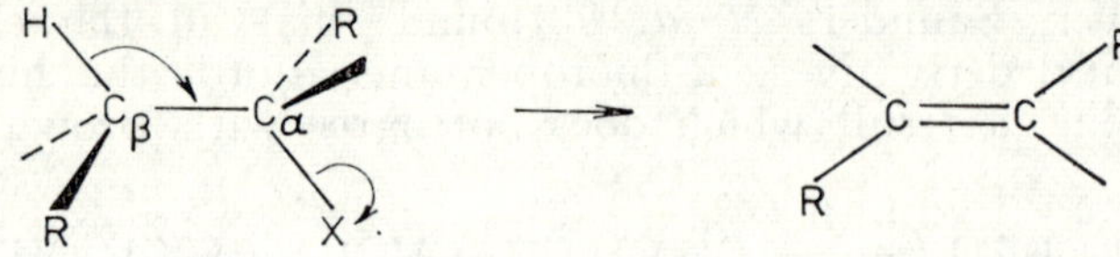

63

The antiperiplanar β-elimination was considered as the general stereochemical rule of the olefin-forming elimination [1, 6]. Sicher *et al.* [13,177–181] demonstrated that *synperiplanar* elimination is another major pathway in the β-elimination. As Figure 9.3 shows, synperiplanar elimination is associated with conservation (or inversion) of configuration at both reaction centres, whereas the antiperiplanar elimination involves one inversion at one reaction centre.

DePuy and his co-workers[182–183] concluded that the elimination has a maximum rate when the dihedral angle is either 180° (antiperiplanar elimination) or 0° (synperiplanar elimination). The minimum rate, if any elimination occurs at all, corresponds to a dihedral angle of 90°. The simple HMO theory can rationalise this rule. Indeed, as Figure 9.4 shows, the overlap between the developing vacant *p* orbital at C_α and the occupied *p* orbital at C_β is maximum when the two developing orbitals are parallel, which corresponds to a dihedral angle between the original bonds of either 180° or 0°.

[170] P. Pfeiffer, *Ber. dtsch. chem. Ges.*, **43**, 3039 (1910); **45**, 1810 (1912).
[171] W. Hückel, W. Toppe and G. Legutke, *Liebigs Ann. Chem.*, **543**, 191 (1940).
[172] D. H. R. Barton and W. J. Rosenfelder, *J. Chem Soc.*, 1048 (1951).
[173] D. H. R. Barton, *J. Chem. Soc.*, 1027 (1953).
[174] D. H. R. Barton and R. C. Cookson, *Quart. Rev.*, **10**, 44 (1956).
[175] F. E. King, J. A. Barltrop and R. J. Walley, *J. Chem. Soc.*, 277 (1945).
[176] F. E. King and H. Booth, *J. Chem. Soc.*, 3798 (1954).
[177] J. Závada, M. Svoboda and J. Sicher, *Tetrahedron Letters*, 1627 (1966).
[178] J. Závada, J. Krupicka and J. Sicher, *Chem. Commun.*, 66, 394, 1626 (1967).
[179] J. Sicher, J. Závada and J. Grupicka, *Tetrahedron Letters*, 1619 (1966).
[180] M. Pancova, J. Závada and J. Sicher, *Chem. Commun*,, 1142 (1968).
[181] J. Sicher, J. Závada and M. Pankova, *Collect. Czech. Chem. Commun.*, **37**, 3456 (1972).
[182] C. H. DePuy, R. D. Thurn and G. F. Morris, *J. Amer. Chem. Soc.*, **84**, 1314 (1962).
[183] C. H. DePuy, C. F. Morris, J. S. Smith and R. J. Smat, *J. Amer. Chem. Soc.*, **87**, 2421 (1965).

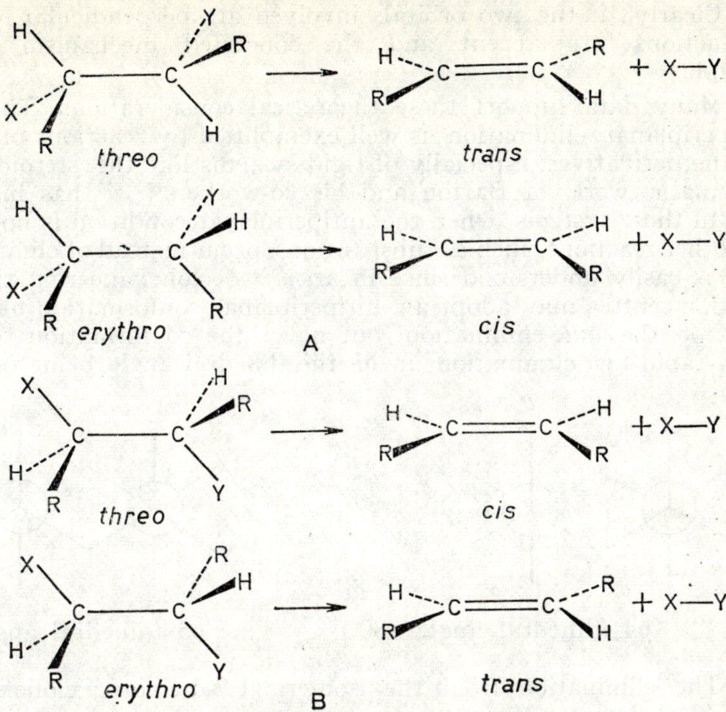

Figure 9.3 Stereochemistry of β-eliminations. A. Synperiplanar eliminations. B. Antiperiplanar eliminations.

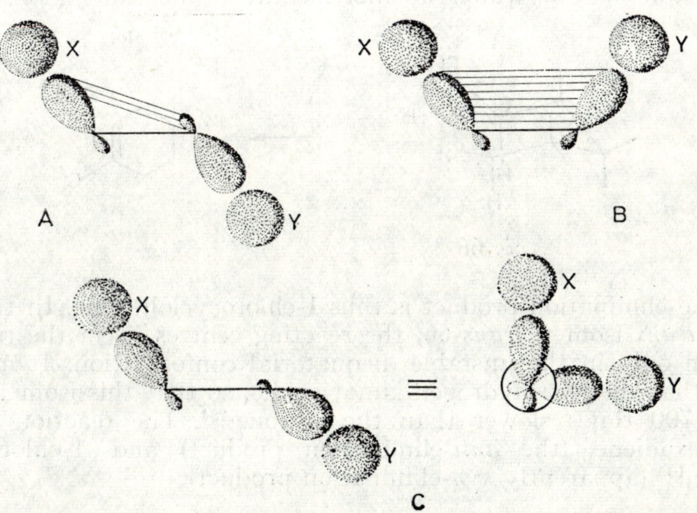

Figure 9.4 The stereochemical rules of β-eliminations in MO representation. A. Dihedral angle 180°. B. Dihedral angle 0°. C. Dihedral angle 90°.

Clearly, if the two orbitals involved are perpendicular, overlap interactions are absent and the concerted mechanism is not possible.

Many data support these theoretical considerations. Thus, the antiperiplanar elimination is well exemplified by reactions of cyclohexane derivatives, especially of rigid systems like the steroids. The systematic work of Barton and his co-workers[172–174] has indicated that in these systems, when the antiperiplanar condition is not fulfilled, other reactions, such as substitutions, occur instead of elimination. This is easily understood since in a *chair* cyclohexane ring the four reaction centres may adopt an antiperiplanar conformation **64** necessary for the *anti*-elimination, but never the conformation required for a rapid *syn*-elimination, in **65** the dihedral angle being 60°:

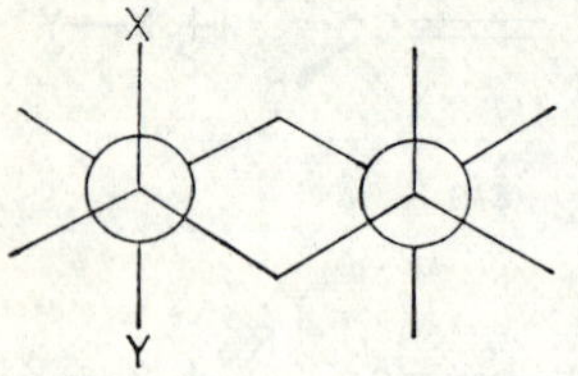

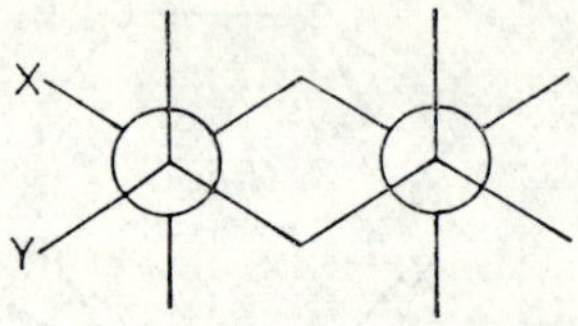

64 (dihedral angle 180°) **65** (dihedral angle 60°)

The elimination from the isomeric 1,2-dichlorocyclohexanes is of special interest. In the *cis* isomer, *cis*-**66**, both *chair* conformations have one chlorine atom lying *anti* with respect to the hydrogen bonded to the carbon which bears the second chlorine atom. Even if another hydrogen atom, at C-6, is available, the base attacks the proton at C-2 which is more acidic.

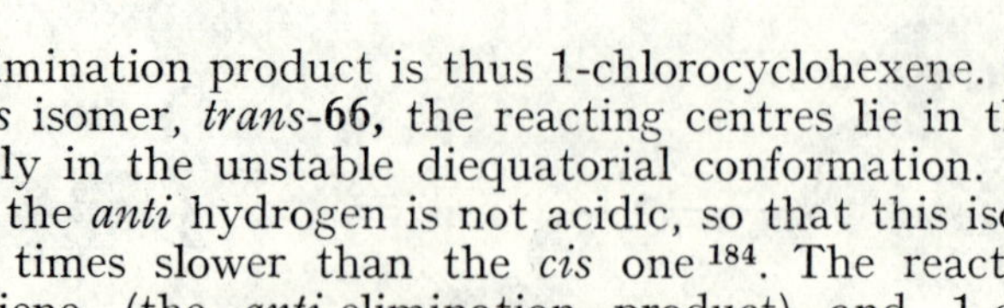

cis-**66**

The elimination product is thus 1-chlorocyclohexene. In the case of the *trans* isomer, *trans*-**66**, the reacting centres lie in the required position only in the unstable diequatorial conformation. It must be noted that the *anti* hydrogen is not acidic, so that this isomer reacts about 700 times slower than the *cis* one[184]. The reaction affords cyclohexadiene (the *anti*-elimination product) and 1-chlorocyclohexene[185] (apparently *syn*-elimination product).

[184] H. L. Goering and H. H. Espy, *J. Amer. Chem. Soc.*, **78**, 1454 (1956).
[185] H. C. Stevens and O. Grummitt, *J. Amer. Chem. Soc.*, **74**, 4876 (1952).

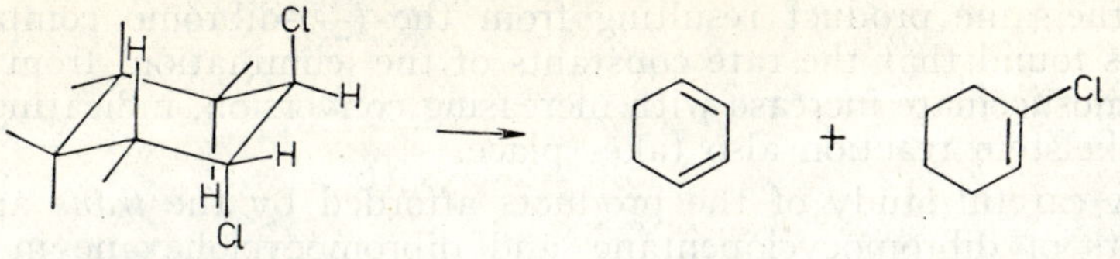

trans-**66** 47% 40%

The reaction of the corresponding dibromo derivatives is still more complex. The *cis* isomer gives as single isolated product (68%) 1-bromocyclohexene, behaving like the *cis*-dichloro compound. The *trans*-dibromo isomer reacts about 340—450 times slower than the *cis*[186]. Baeyer[187] and Zelinsky[188] have found that beside cyclohexadiene, cyclohexene and benzene are also formed, a result which was later confirmed by Cherkasova *et al.* [189].

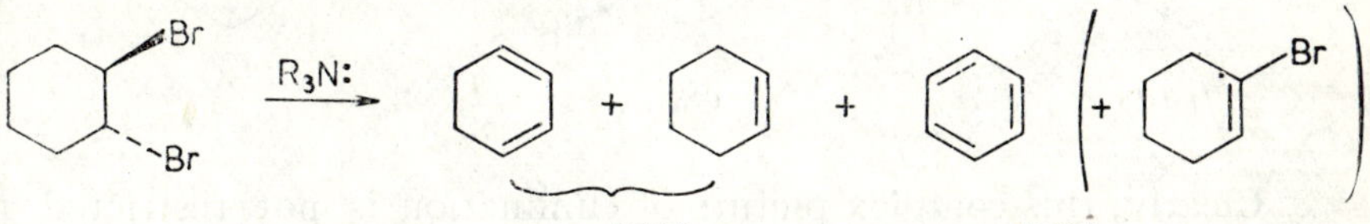

R_3N: 45% 53%

It has been found [8] that cyclohexadiene is formed in the earlier stages of the reaction, whereas cyclohexene results at higher conversions of the dibromo derivative. It was thus concluded that the debromination must be promoted by the bromide ion in a similar way as by the iodide ion (9.62). Later works [190] confirmed this view, demonstrating that in the case of dimethyl α, α′-dibromosuccinate the reaction rate is first-order both in ester and in bromide ion. Moreover, the reaction is perfectly reversible:

$$CH_3OOC{-}CH(Br){-}CH(Br){-}COOCH_3 \overset{DMF}{\rightleftharpoons} CH_3OOC{-}CH{=}CH{-}COOCH_3 + Br^- + Br_2 \quad (9.66)$$

$$Br^- + Br_2 \longrightarrow Br_3^-$$

Equilibrium (9.66) is best observed in aprotic solvents such as dimethyl-formamide (DMF). The *meso* isomer yields dimethyl fuma-

[186] R. Cornubert and A. Rio, *Compt. rend.*, **238**, 867 (1954); *Bull. Soc. chim. France*, 60 (1955).

[187] A. Baeyer, *Liebigs Ann. Chem.* **278**, 108 (1893).

[188] N. Zelinsky and A. Gorsky, *Ber. dtsch. chem. Ges.*, **44**, 2312 (1911).

[189] V. A. Cherkasova, *Zhur. Obsch. Khim.*, **32**, 2792 (1962), and previous works quoted therein.

[190] F. Badea, T. Constantinescu, A. Juvara and C. D. Nenitzescu, *Liebigs Ann. Chem.*, **706**, 20 (1967).

rate, the same product resulting from the (±)-dibromo compound. It was found that the rate constants of the elimination from (±)-dibromosuccinate increase with increasing conversion, indicating that a Finkelstein reaction also takes place.

A careful study of the products afforded by the *trans* and *cis* isomers of dibromocyclopentane and dibromocyclohexane in DMF indicated that dibromoderivatives and halide ions (X = Cl, Br, I) may undergo three types of reactions: debromination, dehydrobromination and substitution. Thus, the apparent *syn* elimination observed by earlier workers is the result of an S_N2 reaction followed by elimination from a more reactive compound[191].

Br Br $\xrightarrow[S_N2]{Br^-}$ Br Br $\xrightarrow[E2]{-HBr}$ Br

trans *cis*

Clearly, this complex picture of elimination is not restricted to dibromo derivatives alone. Any compound with the general formula **67** may undergo two different eliminations, either of X—X (E2X) or of HX (E2H):

$$\text{E2X:}\quad \underset{\mathbf{67}}{\text{X—CHR}^1\text{—CHR}^2\text{—X}} \longrightarrow \text{X}_2 + \text{R}^1\text{CH}{=}\text{CHR}^2 \qquad (9.67)$$

$$\text{E2H:}\quad \underset{\mathbf{67}}{\text{H—CXR}^1\text{—CHR}^2\text{—X}} \longrightarrow \text{HX} + \text{R}^1\text{CX}{=}\text{CHR}^2 \qquad (9.68)$$

Both of them may proceed either *anti* or *syn* (apart from the apparent *syn* elimination mentioned above). The ratios between the possible processes is determined by several factors, which affect the stereochemistry and the reactivity.

9.4.4 *Eclipsing effects*

There are two stereoisomeric ethanes **67**, *erythro* and *threo*, each of them being a mixture of stable conformers (Figure 9.5). One of these conformers may be more reactive in the concerted E2X elimination (9.67), and another one in the E2H reaction (9.68).

[191] M. Popovici, E. Moşanu, T. Constantinescu and F. Badea, *Rev. Roumain. Chim.*, in press.

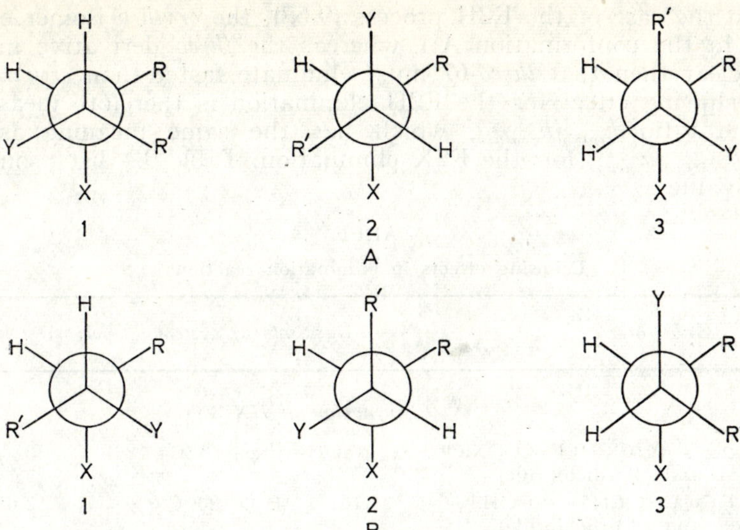

Figure 9.5 Stable conformations for the erythro and threo tetrasubstituted ethanes, RCHX — CHYR′; A. erythro; B. threo.

Let us first consider the E2X reaction. The *anti* elimination from the *erythro* isomer involves only the conformation A.2 (Figure 9.5), in which the dihedral angle between the C—X bonds is 180°. As the reaction proceeds, the dihedral angles between the other bonds decrease from 60° to 0°. In the transition state of the elimination, **68,** the C—R and C—H bonds are nearly eclipsed.

68 **69**

The elimination from the *threo* isomer occurs only in the B.3 conformation (Figure 9.5). In the transition state, **69,** two bulky substituents, R^1 and R^2, are approaching each other, the eclipsing effect being larger than in **68.** Thus, the *threo* derivative eliminates X—X slower than the *erythro* one.

The ratio $k_{erythro}/k_{threo}$ may be used as a measure of the concertedness of the elimination process. Indeed, if the elimination deviates from the ideal E2 situation, being either E1-like or E1cB-like, antiperiplanar conformations are less required, the elimination being possible in all the conformations depicted in Figure 9.5. In the limit, *erythro* and *threo* isomers may display the same reactivity.

In the case of the E2H process (9.68), the *erythro* isomer eliminates in the conformation A.1, whereas the *threo* derivative in B1. It is clear then that *threo*-**67** must eliminate faster than *erythro*-**67**. The eclipsing effect for the E2H elimination is therefore measured by the ratio $k_{threo}/k_{erythro}$, which has the same meaning as the ratio $k_{erythro}/k_{threo}$ for the E2X elimination. Table 9.4 lists some of these values.

TABLE 9.4

Eclipsing effects in elimination reactions

	Substrate	Base-solvent system	Eclipsing effect [a]
	A. Elimination of HX		
1.	NaOOC—CHClCHCl—COONa [b] (*meso* and (±)-dichloride)	NaOH-water, 25 °C	10
2.	HOOC—CHClCHCl—COOH [b] (*meso* and (±)-dichloride)	$HClO_4$—H_2O, 80 °C	2.68
3.	PhCHMe—CHClPh [c] (*erythro* and *threo*)	EtO^-—EtOH, 75 °C	1.0
		Bu^tO^-—Bu^tOH, 75 °C	10.6
4.	PhCHMe—CH($\overset{+}{N}Me_3$)Ph [c]	EtO^-—EtOH, 75 °C	57
		Bu^tO^-—Bu^tOH, 30 °C	1.1
	B. Elimination of X_2		
5.	HOOC—CHBr—CHBr—COOH [d]	I^-, MeOH, 60 °C	5.25
6.	MeOOC—CHBr—CHBr—COOMe [e]	I^-, DMF, 25 °C	10^3

[a] The eclipsing effect is defined as $k_{threo}/k_{erythro}(k_{(\pm)}/k_{meso})$ for the elimination of HX, and as $k_{erythro}/k_{threo}(k_{meso}/k_{(\pm)})$ for the elimination of X_2 (see text). [b] E.D. Hughes and J.C. Maynard, *J. Chem. Soc.*, 4087 (1960). [c] Reference 192,. [d] W.G. Young, D. Pressman and C.D. Coryell, *J. Amer. Chem. Soc.*, **61**, 1640 (1939). [e] F. Badea, T. Constantinescu and C.D. Nenitzescu, *Rev. Roumaine Chim.*, in press.

Table 9.4 shows that the eclipsing effect depends on: (a) the structure of the substrate undergoing elimination; (b) the system base (reagent)-solvent; (c) the leaving group. Obviously, these factors determine the type of the transition state, E1-like **52** or E1cB-like **54.** In the case of the derivatives listed as 3 and 4 in Table 9.4, the change of the system base-solvent has opposite effects. In the better ionising solvent, ethanol, the chloro compound develops a transition state similar to E1, so that eclipsing effects are small. In Bu^tO^- — Bu^tOH, the transition state is shifted toward the ideal situation **53,** and thus the eclipsing effect increases. The quaternary ammonium salt, with the poor leaving group —$\overset{+}{N}Me_3$, has no tendency to ionise even in the better ionising solvent EtOH. At the same time, the base is not strong enough, so that the elimination probably approaches the ideal case **53.** In the stronger-base system

Bu^tOH-Bu^tOK, the geometry of the transition state is closer to that of the E1cB elimination **54,** and the eclipsing effect consequently decreases[192].

The elimination from α,α'-dichlorosuccinic acids[193] shows a preference for E1-like transition states. In presence of an electrophilic catalyst ($HClO_4$), even a poor leaving group like Cl tends to ionise; accordingly[4], the eclipsing effect is small.

Iodide-promoted eliminations deserve a special mention. Badea, Constantinescu and Nenitzescu[194], studying the debromination of dibromosuccinic acids and esters, found that the (±) isomers (*threo*), in DMF, yield fumaric acid derivatives, apparently by *syn* elimination. An S_N2 reaction was demonstrated to precede the elimination, the kinetic study indicating a large eclipsing effect, of at least 10^3.

As data in Table 9.4 shows, this effect depends, even for such reactions, on the solvent and on the substituents at the reaction centres. A carboxyl group, even if electron-attracting, would hardly interfere with the other groups of the molecule, due to its internal stabilisation by conjugation. The COOMe group, however, which is strongly electron-withdrawing, may prevent the ionisation of the C—Cl bond. Thus, for the free acids, an E1-like transition state is expected (small eclipsing effect), whereas a shift towards the ideal E2 transition state is displayed by the esters (large eclipsing effect).

9.4.5 *The base-solvent system*

The base-solvent system is of utmost importance in deciding the type of elimination pathway followed by a given substrate. This has been repeatedly mentioned in the literature[6, 13, 192–195], attention being particularly paid to the *syn-anti* dichotomy. Strong bases may promote carbanion mechanisms (in Eα or Eβ), or the E1cB-like mechanism. In substrates like **67,** the base-solvent system may also determine the type of groups being eliminated.

The strongest bases used in current practice are the alkoxides, RO^-M^+. The system includes thus undissociated species RO^-M^+ (tight ion-pairs), as well as dissociated ions RO^-, the ratio between these species depending on the nature of the solvent. The reactivity may be quite different for the dissociated and the undissociated alkoxide[196]. Moreover, the actual transition states may differ in structure and geometry[13, 195, 196]. Thus, the *syn*-elimination from neutral molecules is favoured with ion-paired base, whereas in the case of

192 S. J. Cristol and F. R. Stermitz, *J. Amer. Chem. Soc.*, 82, 4692 (1960).
193 D. V. Banthorpe, E. D. Hughes and C. K. Ingold, *J. Chem. Soc.*, 4054 (1960).
194 F. Badea, T. Constantinescu and C. D. Nenitzescu, *Rev. Roumaine Chim.*, in press.
195 K. C. Brown and W. H. Saunders, Jr., *J. Amer. Chem. Soc.*, **92**, 4292 (1970.)
196 J. K. Borchardt and W. H. Saunders, Jr., *J. Amer. Chem. Soc.*, **96**, 3910 (1974).

electrically charged molecules, such as ammonium ions, the dissociated bases are more efficient in promoting *syn*-eliminations[13, 196, 197]. The cyclic transition states **70** and **71** may be responsible for these effects:

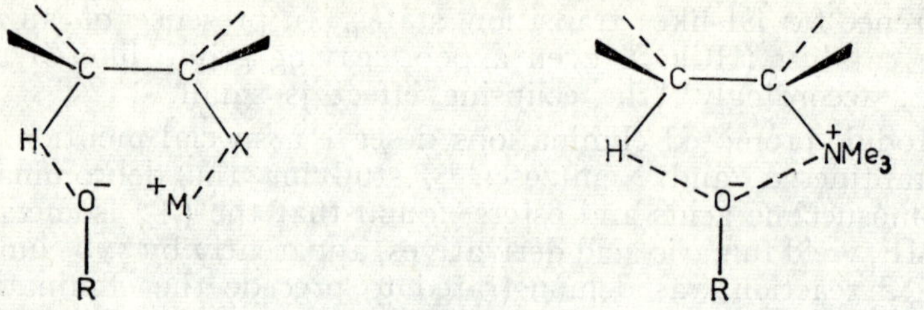

70 **71**

The influence of base-association in elimination is now well established, both by the kinetics of the reaction[190] and by the stereochemistry[13, 196, 197].

An interesting effect on both kinetics and stereospecificity of the elimination is observed when *crown-ethers* are added to the solvent. The study of such compounds is mostly due to Pedersen[198]. Crown-ethers are macromolecular cyclic compounds possessing a cavity in their molecules (see for example **72** and **73**). Due to the presence of the heteroatoms in the cyclic chain, a cation M^+ which enters the cavity is specifically solvated. In an ordinary solvent, a salt X^-M^+ is dissolved only as higher agregates (ion pairs, triplets, etc.). If a small amount of crown-ether is added, the naked anion X^- appears, the cation being trapped in the cavity of the ether. Obviously, this happens only if the cation and the cavity of the crown-ether are compatible in size.

Crown-ethers have found many applications in reactions where the full reactivity of an anion is required[199, 200], that is mostly in nucleophilic substitutions and in elimination reactions. In the bimolecular elimination from tosylate **74**, two products may result, the conjugate alkene **75** (*syn*-elimination) and the unconjugated alkene **76** (*anti*-elimination).

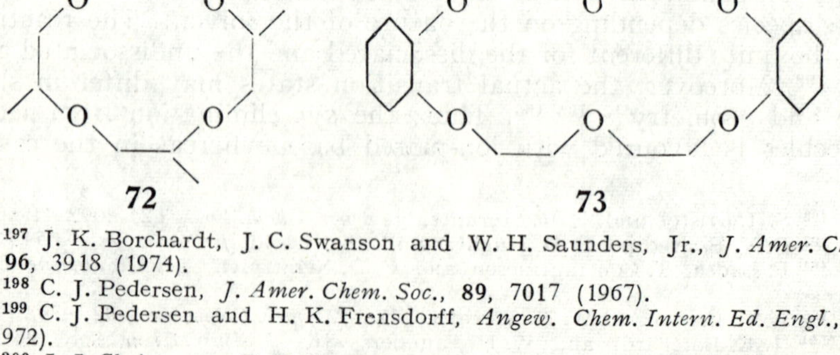

72 **73**

[197] J. K. Borchardt, J. C. Swanson and W. H. Saunders, Jr., *J. Amer. Chem. Soc.*, **96**, 3918 (1974).

[198] C. J. Pedersen, *J. Amer. Chem. Soc.*, **89**, 7017 (1967).

[199] C. J. Pedersen and H. K. Frensdorff, *Angew. Chem. Intern. Ed. Engl.*, **11**, 16 (1972).

[200] J. J. Christensen, D. T. Eatough and R. M. Izatt, *Chem. Rev.*, **74**, 351 (1974.)

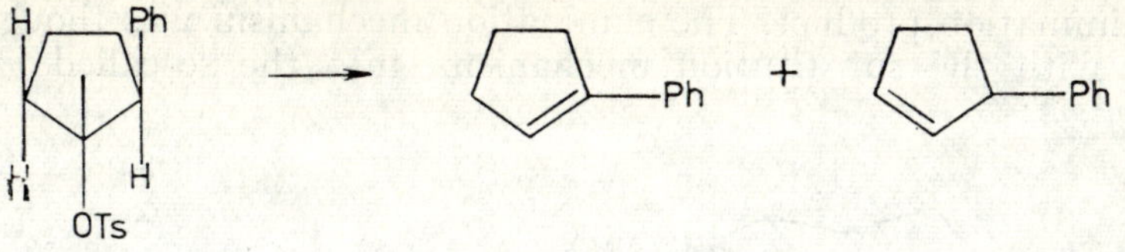

74 75 76

In absence of any crown-ether, alkene **75** predominates in the reaction mixture (89%). The addition of crown-ether **72** has no effect on the stereochemistry and on the kinetics, whereas the crown--ether **73** reduces *syn* elimination from about 90% to about 30%[201] (the latter value is a limit, which remains unchanged between 0.10 and 0.22 M crown-ether added).

Similar results have also been obtained in other eliminations. Thus, tosylate **77** undergoes β-elimination to afford norbornene, and γ-elimination to yield nortricyclene. Deuterium labelling showed that the β-elimination is an exclusive *exo-syn* elimination, in presence of strong bases and absence of crown-ethers[202].

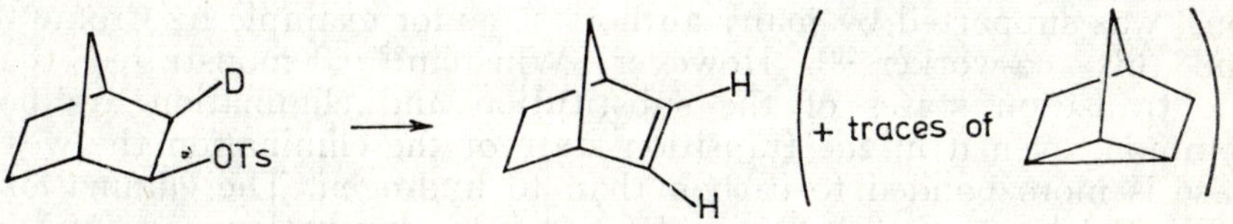

77

The addition of small amounts of a crown-ether reduces the *syn*-elimination from about 99% to about 70%. The explanation is that the cation (potassium or sodium) is complexed by the crown--ether, so that the transition state **70** is disfavoured.

9.4.6 *Elimination reactions catalysed by weak bases*[14]

Studying the reaction of *trans*-4-t-butylcyclohexyl tosylate **78** with halide ions (Cl^-, Br^-, I^-) in acetone, Winstein, Darwish and Holness[203] observed that the main product is 4-t-butylcyclohexene, and that its ratio in the reaction mixture depends on the basicity of the halide ion in acetone ($Cl^- > Br^- > I^-$). Considering that the bulky t-butyl group is situated only in the equatorial position, the authors admitted an intermediate **79**, responsible for both substitution product

[201] R. A. Bartsch, E. A. Mintz and R. M. Parlman, *J. Amer. Chem. Soc.*, **96**, 4249 (1974).

[202] H. C. Brown and K.-J. Liu, *J. Amer. Chem. Soc.*, **92**, 200 (1970); R. A. Bartsch and R. H. Kayser, *ibid.*, **96**, 4346 (1974).

[203] S. Winstein, D. Darwish and N. J. Holness, *J. Amer. Chem. Soc.*, **78**, 2915 (1956).

and elimination product. The elimination mechanism was thought to merge with the substitution mechanism, into the so-called *merged mechanism*:

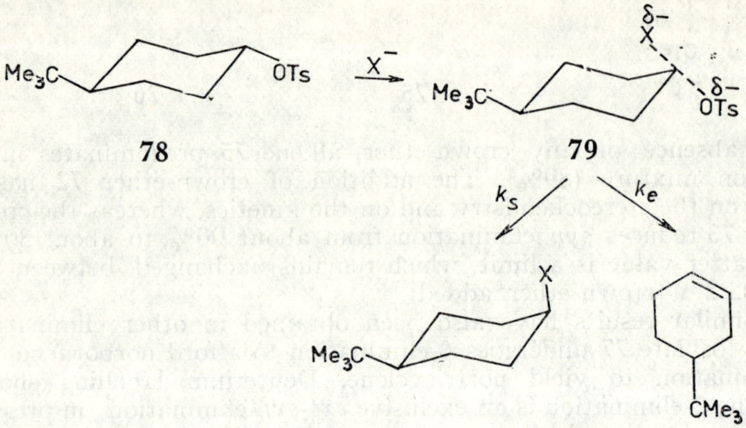

Me$_3$C–(cyclohexyl)–OTs **78** $\xrightarrow{X^-}$ **79** (Me$_3$C–cyclohexyl with $X^{\delta-}$ and $OTs^{\delta-}$); **79** $\xrightarrow{k_s}$ Me$_3$C–cyclohexyl–X; **79** $\xrightarrow{k_e}$ 4-CMe$_3$-cyclohexene

The merged mechanism promoted by weak bases, such as halide ions, was supported by many authors[14], as for example by Cromwell and his co-workers[204]. However, Winstein[205] demonstrated that the transition states of the substitution and elimination are not common, even if in the transition state of the elimination the weak base is more bonded to carbon than to hydrogen. The eliminations promoted by weak bases were termed E2C eliminations, as opposed to the usual eliminations involving strong bases, E2H[17]. Structure **80** represents better the general transition state in a merged mechanism[14], instead of the initially proposed **81**[203].

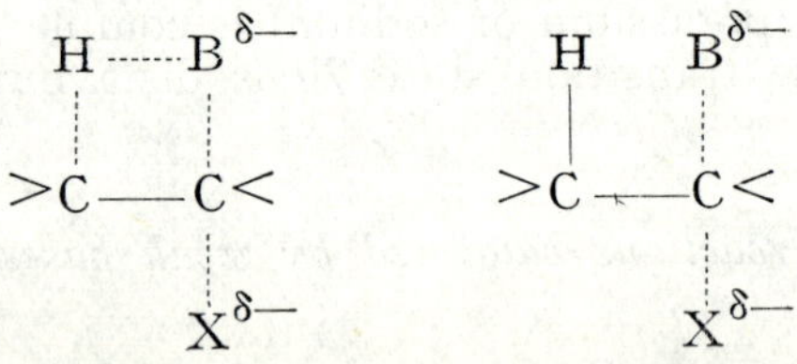

```
H ······ B^δ−        H     B^δ−
:        :           |     :
>C ——— C<          >C ——— C<
         :                 :
         X^δ−              X^δ−

   80                  81
```

One of the strongest arguments in favour of the E2C mechanism is based on the verification of a Brönsted-type equation[206]. Indeed, base-catalysed reactions usually obey the equation (9.69):

$$\log k = \beta\, pK_b + \text{constant} \tag{9.69}$$

[204] N. H. Cromwell *et al.*, *J. Amer. Chem. Soc.*, **79**, 230 (1957); **88**, 4489 (1966) and previous papers in the series cited therein.

[205] S. Winstein, P. Beltrame, G. Biale, R. Cetina, D. Darwish, S. Smith, I.D.R. Stevens and J. Takahashi, 144th National Meeting of The American Chemical Society, Los Angeles, Calif. (1963), cf. Reference 14.

[206] See, for example, L. P. Hammett, *Physical Organic Chemistry*, 2nd edn, McGraw-Hill, New York (1970), p. 317.

where k is the observed rate constant of the base-catalysed reaction, K_b the basicity constant of the reagent and β the slope of the correlation line.

If in the elimination the reagent attacks the carbon atom, a similar equation (9.70) should be obeyed:

$$\log k_e = S \log k_S + \text{constant} \qquad (9.70)$$

where k_e is the observed rate constant of the elimination, k_S the rate constant of the substitution under conditions excluding elimination, and S the slope of the correlation line.

Parker, Winstein and their co-workers [17] found that the reaction of cyclohexyl tosylate in acetone with a series of weak bases (Cl^-, Br^-, I^-, ArS^-, N_3^-, etc.) does not obey equation (9.70). Moreover, the rate constant of the elimination, k_e, varies in the same way as the rate constant of the substitution, k_S, on transferring the reaction from a protic solvent to an aprotic dipolar solvent. The values of the *solvent activity coefficients* (see p. 141) for the transition state, $(^{DMF}\gamma_{\neq}^{MeOH})$, point to a similar distribution of the charges in the transition state of the elimination as in the transition state of an S_N2 reaction [207]. This agrees well with the structure **80.** Other studies confirmed this type of mechanism [14,198–211].

As Ford [14] has shown, the E2 reactions proceed *via* such a wide range of transition states that a single mechanism is not likely to agree with all reliable data. The E2C mechanism may well enter the general spectrum of transition states ranging between **52** and **54.** The elimination starts when the C_α—X bond stretches to a certain extent, and this initial change is accelerated by nucleophilic assistance.

9.4.7 *The ratio between substitution and β-elimination products*

Nucleophilic substitution accompanies both the unimolecular elimination and the concerted elimination. The ratio E/S is usually lower under unimolecular conditions ($E1/S_N1$), a carbocation being faster trapped by a nucleophile than stabilized by proton elimination *. If the alkene is the product sought, conditions favouring the bimolecular mechanism must be used, namely solvents of low polarity and ionising power (for example absolute ethanol is better than aqueous ethanol, acetone better than acetic acid, etc.), strong bases, high concentration of the reagent. For example, the elimination of hydrogen bromide from t-butyl bromide[212] in aqueous ethanol, at 25 °C,

[207] E. C. F. Ko and A. J. Parker, *J. Amer. Chem. Soc.*, **90**, 6447 (1968).
[208] G. Biale, A. J. Parker, S. G. Smith, I. D. R. Stevens and S. Winstein, *J. Amer. Chem. Soc.*, **92**, 115 (1970).
[209] G. Biale, A. J. Parker, I. D. R. Stevens, J. Takahashi and S. Winstein, *J. Amer. Chem. Soc.*, **94**, 2235 (1972).
[210] P. Beltrame, G. Biale, D. J. Lloyd, A. J. Parker and M. Rouane, *J. Amer. Chem. Soc.*, **94**, 2240 (1972).
[211] A. J. Parker *et al.*, *J. Amer. Chem. Soc.*, **90**, 5049 (1968).
* Steric factors operating at the cationic centre usually favour a high E/S ratio.
[212] M. L. Dhar, E. D. Hughes and C. K. Ingold, *J. Chem. Soc.*, 2065 (1948).

in absence of added bases (conditions favouring unimolecular reactions), yields only 19% isobutene. With moderate concentrations of sodium ethoxide (0.05 M), the alkene ratio increases to 46%, whereas at high base concentrations (~ 1 M) isobutene represents 91.3% of the transformed bromide. The mechanism is thus shifted from E1 to E2.

However, even in bimolecular eliminations substitution products may arise beside elimination products. Let us consider the two transition states involved in these parallel reactions:

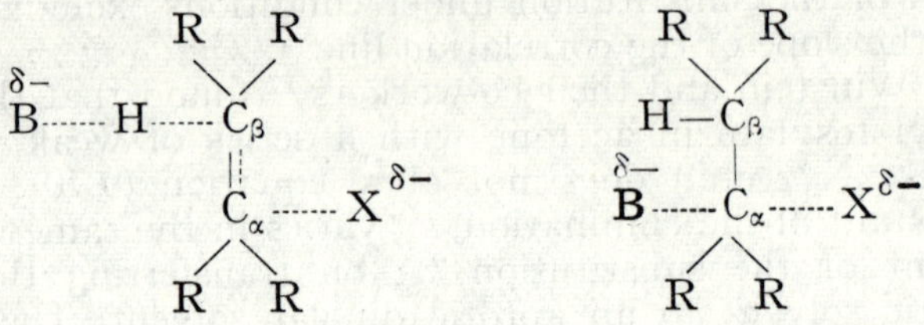

$$\overset{\delta-}{B}\cdots H\cdots C_\beta R_2 \;\; \vdots \;\; C_\alpha R_2 \cdots \overset{\delta-}{X} \qquad \mathbf{82}$$

$$H-C_\beta R_2 - \;\; \overset{\delta-}{B}\cdots C_\alpha R_2 \cdots X^{\delta-} \qquad \mathbf{83}$$

in order to understand the influence of structural and external (medium, solvent-base) factors on the E/S ratio.

α-Substitution. α-Substitution with alkyl groups disfavours the S_N2 mechanism, but favours the E2 elimination. Thus, the elimination from Et-$\overset{+}{S}Me_2$ in abs. EtOH at 45 °C, with sodium ethoxide, yields only 12% ethylene. Under similar conditions, the reaction of Pr^i-$\overset{+}{S}Me_2$ affords 60.6% propylene [213].

β-Substitution. β-Substitution with alkyl groups is known to disfavour strongly the S_N2 mechanism (see p. 259). The same type of substitution disfavours (to a lesser extent) the Hofmann elimination, but has no effect in a Saytzeff-type elimination:

$$R_2CH-CH_2-Br + EtO^-(EtOH) \begin{cases} \xrightarrow{S_N2} R_2CH-CH_2-OEt \\ \xrightarrow{E2} R_2C{=}CH_2 \end{cases} \qquad (9.71)$$

Thus, for the process (9.71) the ratio k_{S_N}/k_E is 107 when R=H, but 0.68 when R=Me[214]. However, in the case of the sulphonium salt:

$$R-CH_2-CH_2-\overset{+}{S}Me_2 + EtO^-(EtOH) \begin{cases} \xrightarrow{S_N2} R-CH_2CH_2SMe + EtOMe \\ \xrightarrow{E2} R-CH{=}CH_2 \end{cases}$$

k_{S_N}/k_E is 4 if R = H, but increases to 800 when R = Bu^t [193]. The high increase of the substitution rate with respect to the elimination

[213] K. A. Cooper, E. D. Hughes, C. K. Ingold, G. A. Maw and J. McNulty, *J. Chem. Soc.*, 2049 (1948).

[214] M. L. Dhar, E. D. Hughes, C. K. Ingold and S. Masterman, *J. Chem. Soc.*, 2055 (1948).

rate is due to the possibility of the reagent of attacking a primary carbon atom.

Solvent polarity. With electrically charged bases, the charge is more dispersed in the transition state of the elimination (**82**) than in the transition state of the S_N2 reaction. The decrease of solvent polarity favours therefore to a larger extent the elimination reaction as compared to the substitution reaction.

Reagent. In the eliminations of halogen promoted by halide ion, the nature of the halide ion has a decisive influence on the type of elimination (dehalogenation over hydrogen halide elimination), as well as on the E/S ratio (the substitution is a Finkelstein reaction). The efficiency of promoting dehalogenation decreases in the order $I^- > Br^- > Cl^-$, whereas the reverse order is found for hydrogen halide elimination and for substitution reactions in aprotic solvents [191].

Temperature. Substitution reactions have lower activation parameters than elimination reactions. This means that the E/S ratio increases with increasing temperature.

9.5 γ-Eliminations (Eγ)

γ-Eliminations are less studied than β-eliminations, even if they have large applications in the synthesis of cyclopropane derivatives [215]. Recent studies have evidenced the occurrence of all types of current elimination mechanisms: E1cB, E1 and E2.

The carbanion mechanism is favoured when in the γ position (with respect to the leaving group location) there is a strong electron-attracting group, as for example in the Perkin, Jr. synthesis of cyclopropanes:

$$Br-CH_2-CH_2-\bar{C}(COOR)_2 \longrightarrow \underset{\text{(cyclic: } CH_2-CH_2-C)}{CH_2-C(COOR)_2}$$

The negative charge at C_γ is obviously accommodated by the two COOR groups. The intramolecular cyclisation leads to the γ-elimination product.

If the substituents in γ position are not very efficient in stabilising the negative charge, the reaction will require a strong base and a poor leaving group [216, 217].

$$\underset{\mathbf{84}}{PhCH_2-CH_2-CH_2-X} \begin{cases} \xrightarrow{k_\beta} PhCH_2CH{=}CH_2 \\ \xrightarrow{k_\gamma} PhCH\text{—}CH_2 \text{ (cyclopropane, ring closed by } CH_2) \end{cases}$$

[215] For leading references in γ-elimination see A. Nickon and N. H. Werstiuk, *J. Amer. Chem. Soc.*, **89**, 3914 (1967).

[216] C. L. Bumgardner, *J. Org. Chem.*, **29**, 767 (1964).

[217] C. L. Bumgardner, *J. Chem. Soc.*, 374 (1965).

Thus, in **84**, if X = F (poor leaving group), practically only cyclopropane is formed (γ-elimination), whereas with X = Br (good leaving group) the major product is benzylethylene (β-elimination).

The unimolecular mechanism is also known to operate in reactions involving σ-bond participation (formation of non-classical ions). Thus, Skell and Starrer [218] found that the deamination of n-propyl amine yields also about 10% cyclopropane. This observation led to the discovery of a new reaction intermediate, the protonated cyclopropane. Soon afterwards, in a careful study of the acetolysis of *exo*-2-norbornyl tosylate, Winstein and his co-workers[219] found approximately 4% nortricyclene, a product of γ-elimination. Again, this result indicated the formation of a protonated cyclopropane **86**:

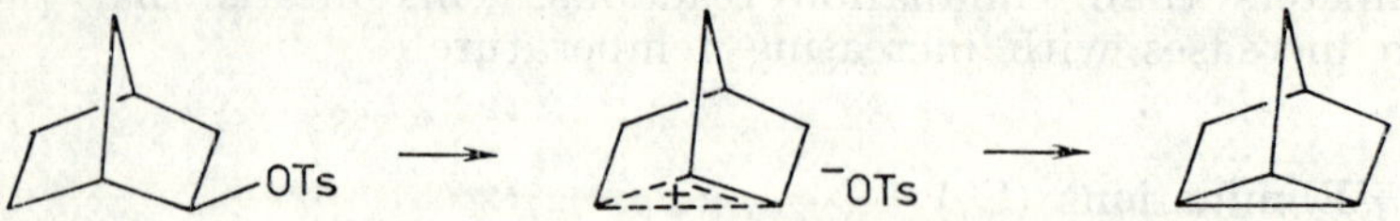

OTs → $^{-}$OTs →

85 **86**

In these examples, the γ-elimination by E1 mechanism involves a σ-route in the ionisation to a non-classical ion. Examples of γ-elimination are also known in which the non-classical ion results by a π-route. Thus, Ciorănescu, Nenitzescu and their co-workers [220], studying the acetolysis of tosylate **87**, found among the reaction products the hydrocarbon **89**, resulted by a γ-elimination. The formation of this hydrocarbon, as well as of bridged acetates, points to an intermediate non-classical ion **88**:

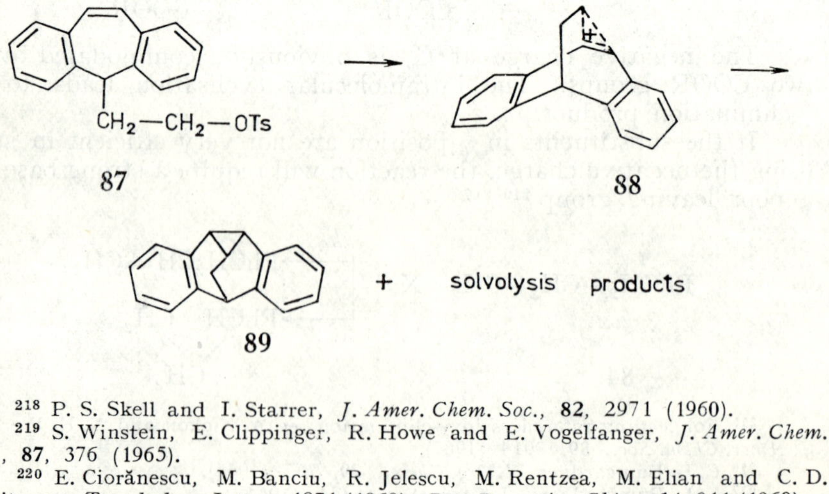

$CH_2—CH_2—OTs$ → →

87 **88**

+ solvolysis products

89

[218] P. S. Skell and I. Starrer, *J. Amer. Chem. Soc.*, **82**, 2971 (1960).

[219] S. Winstein, E. Clippinger, R. Howe and E. Vogelfanger, *J. Amer. Chem. Soc.*, **87**, 376 (1965).

[220] E. Ciorănescu, M. Banciu, R. Jelescu, M. Rentzea, M. Elian and C. D. Nenitzescu, *Tetrahedron Letters*, 1871 (1969); *Rev. Roumaine Chim.*, **14**, 911 (1969).

The amount of hydrocarbon **89** increases when the reaction time exceeds ten half-lives, under unbuffered conditions, and when the polarity of the solvent is lowered (using for example pivalic acid instead of acetic acid). This indicates that the elimination occurs within the ion pair and that the other reaction products (acetates) are in equilibrium with the non-classical ion **88** under unbuffered conditions.

The bimolecular mechanism was observed in the formation of nortricyclene from tosylate **85,** on treatment of the latter with t-butoxide ion in t-butanol. Deuterium labelling in the 6-position of both *exo*- and *endo*-2-norbornyl tosylates (**90**—**93**) allowed the identification of four different stereospecific eliminations, designated as U, W, *exo*-S (S = sickle) and *endo*-S, according to the shape of the line formed by the atoms involved in elimination[221]. In the four compounds shown below **90**—**93** the elimination of TsOD illustrates the four stereochemical possibilities, as indicated in parentheses under each structure.

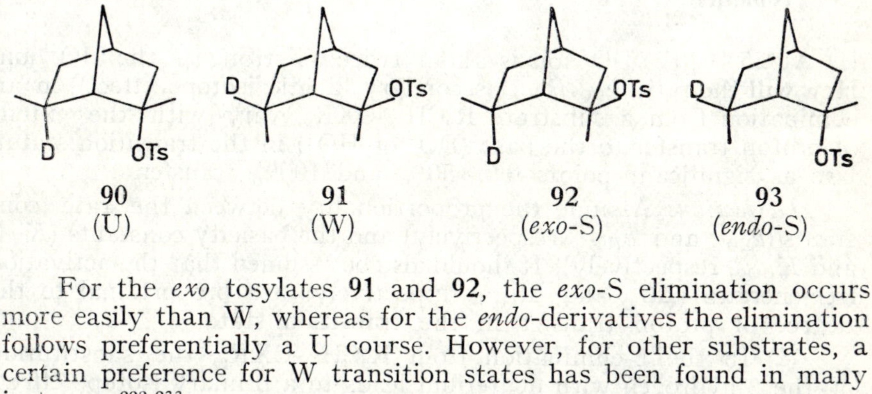

90	**91**	**92**	**93**
(U)	(W)	(*exo*-S)	(*endo*-S)

For the *exo* tosylates **91** and **92**, the *exo*-S elimination occurs more easily than W, whereas for the *endo*-derivatives the elimination follows preferentially a U course. However, for other substrates, a certain preference for W transition states has been found in many instances [222,233].

For the moment, the theoretical interpretations of this stereochemistry of reaction are not clear enough. One of them is based on the *principle of least motion* (PLM). According to this principle[224], the elementary reactions involving the least change in atomic position and electronic configuration will be favoured. The principle was first applied to elimination reactions by Hine [225], in order to explain the *anti*-elimination in ethyl chloride. He assumed that the energy required to stretch or bend a bond is proportional to the extent to which Hooke's law is applicable, that is to the square of the bond

[221] A. Nickon and N. H. Werstiuk, *J. Amer. Chem. Soc.*, **89**, 3915, 3917 (1967).

[222] F. G. Bordwell and B. B. Jarvis, *J. Amer. Chem. Soc.*, **95**, 3585 (1973).

[223] F. G. Bordwell, *Accounts Chem. Res.*, **5**, 374 (1972); in this and the preceding reference it is suggested that most of the 1,3 eliminations regarded as concerted γ-eliminations are actually carbanionic.

[224] F. O. Rice and E. Teller, *J. Chem. Phys.*, **6**, 489 (1938); **7**, 199 (1939).

[225] J. Hine, *J. Amer. Chem. Soc.*, **88**, 5525 (1966).

distance which is modified. The reaction should follow the pathway which minimises the sums of the squares of the atomic displacements. The sum will be a minimum if:

$$\frac{\mathrm{d}\sum D_i^2}{\mathrm{d}\theta} = 0 \tag{9.72}$$

Varying the dihedral angle between the displaced groups, Hine found that the *anti*-elimination (dihedral angle 180°) is favoured over the *syn*-elimination. Applying a similar procedure for the γ-elimination, Yates and his co-workers [226] confirmed the preference for the W-transition state in γ-eliminations and were able to explain the preference for the U-transition state in the case of the norbornyl derivative.

The principle of least motion accounts well for the observed stereochemistry in most elimination reactions.

9.6 Problems

(1) At 25 °C, the DO^- ion is about twice as strong as the HO^- ion. How will the ratio k_{DO^-}/k_{HO^-} (secondary kinetic isotope effect), in an elimination from a substrate $R_2CH{-}CXR_2$, vary with the extent of proton transfer to the base (DO^- or HO^-) in the transition state? Use as significant points 0%, 50% and 100% transfer.

Indication: Assume the proportionality between the rate constants (k_{DO^-} and k_{HO^-}, respectively) and the basicity constants (K_{DO^-} and K_{HO^-}, respectively). It should also be assumed that the activation free-energies ($\Delta G_D^{\ddagger}$ and $\Delta G_H^{\ddagger}$, respectively) are proportional to the degree of proton transfer in the transition state.

(2) In the β-elimination from $R_2CH{-}CXR_2$, the substitution of the β-hydrogen with deuterium leads to a primary isotope effect, k_H/k_D. Assuming that primary isotope effects are determined only by the difference in zero-point vibration energies (see p. 87), how should primary isotope effects vary with the degree of proton transfer to the base? As significant points in your argument consider, as above, 0%, 50% and 100% transfer.

Indication: One must assume, as in the preceding problem, a proportionality between the activation free-energies (ΔG_D and $\Delta G_H^{\ddagger}$, respectively) and the dissociation energies:

$$D_{CD} = \varepsilon_n - \varepsilon_{0,\,D} \text{ and } D_{CH} = \varepsilon_n - \varepsilon_{0,\,H}, \text{ respectively.}$$

(3) The treatment of chloroform in aqueous alcoholic solution with sodium hydroxide and aniline yields phenyl isonitrile.

(a) Write the elementary reactions of this process (reaction mechanism).

[226] O. S. Tee, J. A. Altmann and K. Yates, *J. Amer. Chem. Soc.*, **96**, 3141 (1974).

(b) Explain why the intermediate prefers to react with aniline instead of with the more basic hydroxide ion.

(4) Meerwein obtained tricyclene by the following reaction:

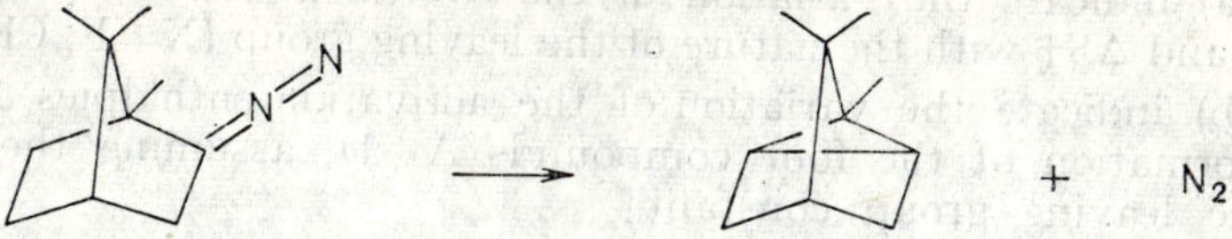

(a) Propose a mechanism for this reaction.

(b) Use the empirical HMO theory to interpret the reaction.

(c) Explain, in terms of HMO theory, why the bornane system does not rearrange, in this reaction, into the camphane system.

(5) The Ramberg-Bäcklund rearrangement (see p. 334) may be regarded as a γ-elimination. A variant of this reaction (see, for example, F. G. Bordwell and B. B. Jarvis, *J. Amer. Chem. Soc.*, **95**, 3585 (1973)) is:

$$\underset{meso \text{ and } (\pm)}{PhCHBr-SO_2-CHBr-Ph} \xrightarrow{Ph_3P} Ph-CH=CH-Ph$$

(a) Write a mechanism for this elimination.

(b) Propose other reagents which may produce the same transformation.

(c) Considering that no racemisation of the chiral centres in the sulphone occurs in the process, indicate the stereochemistry of the resulting alkene, in accordance with the mechanism proposed.

(6) When ketone A is subjected to the Bamford-Stevens reaction, a carbene is obtained (see, for example, R.A. Le Bar and W. M. Jones, *J. Amer. Chem. Soc.*, **95**, 2359 (1973)).

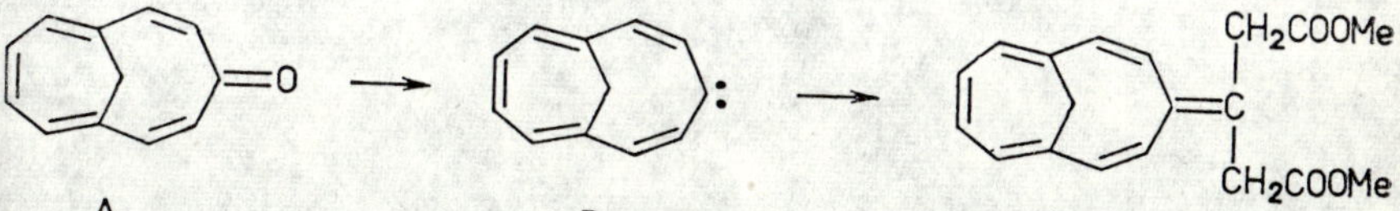

A B C

(a) Examine carbene B and show that it should be a stable species.

(b) Indicate the missing reagent and the mechanism of the transformation of A into C.

(7) In the β-elimination from 2-hexyl halides, with sodium methoxide, four products, A—D, may result:

$$\underset{X}{CH_3-(CH_2)_3\overset{}{CH}-CH_3} \longrightarrow \underset{A}{C_4H_9CH=CH_2} + \underset{B = trans,\ C = cis}{CH_3CH=CHC_3H_7} + \underset{D}{2\text{-}CH_3O-C_6H_{13}}$$

Denoting by $\Delta H_i^{\ddagger}$ and $\Delta S_i^{\ddagger}$ the activation parameters of the reaction leading to one compound (i=A, B, C or D) and with $\Delta H_X^{\ddagger}$ and $\Delta S_X^{\ddagger}$ the overall activation parameters,

(a) indicate the variation of the overall activation parameters $\Delta H_X^{\ddagger}$ and $\Delta S_X^{\ddagger}$ with the nature of the leaving group (X=F, Cl, Br, I);

(b) indicate the variation of the activation enthalpies $\Delta H_i^{\ddagger}$ for the formation of the four compounds A—D, assuming the nature of the leaving group constant;

(c) explain why $\Delta H_B^{\ddagger} \approx \Delta H_C^{\ddagger}$ and $\Delta H_D^{\ddagger} \approx \Delta H_B^{\ddagger}$, when X = F.

(8) Indicate the type of mechanism involved in the following reaction:

$$R_2CX{-}C{\equiv}CH \xrightarrow{HO^-} H_2C{=}C{=}C\colon\ (+\ H_2O + X^-).$$

10
Addition reactions

10.1 Introduction

Addition reactions are associative processes, leading to stable adducts as reaction products. There are many simple reactions in organic chemistry which satisfy this definition, without being usually considered as addition processes. An example is afforded by the acidolysis of cyclic ethers. The halogenohydrines **1** may be regarded as stable adducts formed from the acid reagent HX and the ether:

$$\underset{\text{(ethylene oxide)}}{\overset{CH_2-CH_2}{\diagdown O \diagup}} + HX \longrightarrow \underset{\mathbf{1}}{\underset{OH}{CH_2}}-CH_2-X \qquad (10.1)$$

In a restricted sense, an addition reaction is an associative process involving a rehybridisation of two atoms of an unsaturated substrate molecule (either $sp^2 \rightarrow sp^3$ or $sp \rightarrow sp^2$). Thus, reaction (10.2) of a carbene, even if corresponding to an associative process involving a rehybridisation, will not be considered in this chapter as an addition reaction, the rehybridisation being limited to only one atom of the substrate. By contrast, reaction (10.3) may be considered as an addition in which, however, carbene will be regarded as reagent instead of substrate:

$$:CR_2 + XY \longrightarrow XYCR_2 \qquad (10.2)$$

$$>C=C< + :CR_2 \longrightarrow >C-C< \text{ (bridged by } \underset{R_2}{C}\text{)} \qquad (10.3)$$

Reactions (10.4)–(10.6) are typical examples of additions:

$$>C=C< + X-Y \longrightarrow >\underset{Y}{C}-\underset{X}{C}< \qquad (10.4)$$

$$-C\equiv C- + X-Y \longrightarrow Y-\overset{|}{C}=\overset{|}{C}-X \qquad (10.5)$$

$$\rangle C{=}O + X{-}Y \longrightarrow X{-}\overset{|}{\underset{|}{C}}{-}O^- \ Y^+ \qquad (10.6)$$

The formation of ipso-adducts[1,2] in an aromatic electrophilic substitution, or of Jackson-Meisenheimer complexes[3] in the nucleophilic aromatic substitution, are also clearly related to the examples (10.4)—(10.6):

$$\text{4-}R^1\text{-}C_6H_4\text{-}R^2 + NO_2^+ \xrightarrow{AcOH} \mathbf{2}\ (\text{1-}R^1\text{-1-}NO_2\text{-4-}R^2\text{-4-OAc-cyclohexa-2,5-diene}) \qquad (10.7)$$

$$C_6H_5X + Nu{:}^- \longrightarrow \mathbf{3}\ (\text{1-X-1-Nu-cyclohexadienyl anion}) \quad \textit{etc.} \qquad (10.8)$$

However, such processes will only incidentally be mentioned in the following sections, since they have already been dealt with in previous chapters of this book.

Ring-opening reactions of cyclopropane (10.9) even if similar to the ring-opening acidolysis (10.1) of cyclic ethers, are more related to the addition to an alkene (10.4).

$$\triangle + XY \longrightarrow XCH_2{-}CH_2{-}CH_2Y \quad \textit{etc.} \qquad (10.9)$$

Considering the Walsh orbitals[4] of cyclopropane, this is not surprising, the ring bonds having both π and σ character. Therefore, we shall also deal here with these reactions.

Addition reactions may be either heterolytic, homolytic or concerted (with no definite polar character of the transition state). The heterolytic additions are stepwise reactions, involving either cationic (10.10) or anionic (10.11) intermediates:

$$AB + E^+ \longrightarrow EAB^+ \xrightarrow{Nu:^-} EABNu \qquad (10.10)$$

$$AB + Nu{:}^- \longrightarrow ABNu^- \xrightarrow{E^+} EABNu \qquad (10.11)$$

1 P. B. D. de la Mare, *Accounts Chem. Res.*, **7**, 361 (1974).
2 R. C. Hahn and D. L. Storck, *J. Amer. Chem. Soc.*, **96**, 4335 (1974).
3 M. J. Strauss, *Chem. Rev.*, **70**, 667 (1970); *Accounts Chem. Res.*, **7**, 181 (1974).
4 W. A. Bernett, *J. Chem. Educ.*, **44**, 17 (1967).

According to the character of the initial attack at the multiple bond, the former reaction will be called an electrophilic addition (Ad_E), whereas the latter will correspond to a nucleophilic addition (Ad_N). The carbon-carbon double bond usually reacts with electrophiles, the nucleophilic reagents being repelled by the cloud of π-electrons. Alkenes are thus nucleophilic reagents, their electron-donor character being well recognised in many other reactions. On the other hand, nucleophilic additions are characteristic of heteroatomic double and triple bonds (or other heteroatomic unsaturated π-systems) involving a strong electron-attracting atom X: :

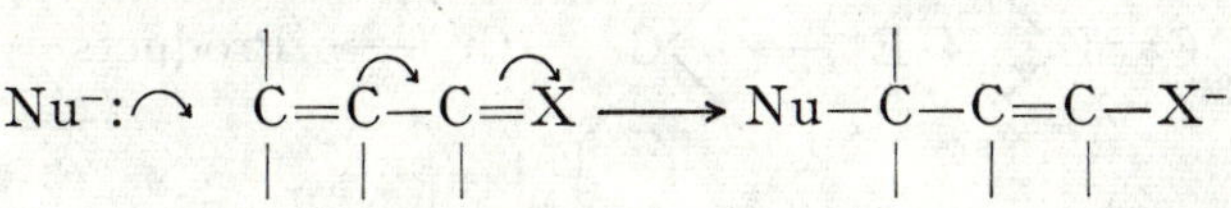

$$Nu^{-}{:}\curvearrowright \; \overset{|}{\underset{|}{C}}=\overset{\frown}{\underset{|}{C}}-\overset{\frown}{\underset{|}{C}}=X \longrightarrow Nu-\overset{|}{\underset{|}{C}}-\underset{|}{C}=\underset{|}{C}-X^{-} \qquad (10.12)$$

Both in (10.6) and in (10.12) the carbon atom at the other terminus of the π-system has a partial positive charge, being thus susceptible to nucleophilic attack. The two possible reaction mechanisms, involving the electrophilic and the nucleophilic attack, will be dealt with separately in this chapter.

10.2 Electrophilic additions to carbon-carbon double bonds

Depending on the nature of the electrophile, the electrophilic additions may be classified into additions of neutral XY molecules (halogens as typical example) and additions of acid reagents [5] (hydrogen halide additions, hydrations, *etc.*). In the latter reactions, the transfer of a proton from an acid to the alkene is characteristic. Irrespective of the type of the electrophilic addition, carbenium ions are intermediates of the reaction.

Considering the carbon atom attacked, the electrophilic addition is related to the aromatic electrophilic substitution. In the later stages of the reaction, the electrophilic addition resembles the unimolecular nucleophilic substitution. The following reaction schemes may therefore be imagined:

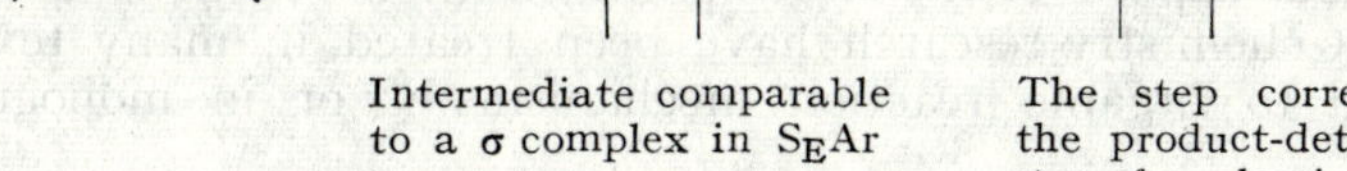

$$>C=C< \; + E^{+} \longrightarrow {}^{+}\overset{|}{\underset{|}{C}}-\overset{|}{\underset{|}{C}}-E \xrightarrow{Nu:} Nu-\overset{|}{\underset{|}{C}}-\overset{|}{\underset{|}{C}}-E$$

Intermediate comparable to a σ complex in S_EAr — The step corresponds to the product-determining step of a classical S_N1 reaction

Scheme 1. Carbenium ion mechanism

[5] C. K. Ingold, *Structure and Mechanism in Organic Chemistry*, Cornell University Press, Ithaca, New York (1953), p. 646; 2nd edn. (1969), p. 946.

$$\rangle C{=}C\langle + E^+ \rightarrow \rangle C{\overset{\downarrow}{=}}C\langle \rightarrow \rangle C{-}C\langle \rightarrow \rangle C(Nu){-}C(E)\langle + \rangle C(E){-}C(Nu)\langle$$

Intermediate comparable to a π complex in S_EAr. — Bridged intermediate. — The step corresponds to an anchimerically assisted nucleophilic substitution.

Scheme 2. Bridged-ion mechanism

$$\rangle C{=}C\langle + E^+ \longrightarrow \rangle C{-}C\langle\,(\overset{+}{E}\ \text{bridged}) \longrightarrow \text{Products}$$

$$\Updownarrow$$

$$\rangle \overset{+}{C}{-}C{-}E \longrightarrow \text{Products}$$

Scheme 3. Mechanism involving two or more intermediates

The following conclusions can be drawn from these schemes:

(a) As in the electrophilic aromatic substitution, the reaction kinetics will reflect to a greater extent the formation of the electrophilic reagent than the reactivity of it. Electrophiles and carriers of electrophiles may be involved. Any mechanism study must establish the nature of the reagent.

(b) Product analysis will be useful in deciding the electrophilic nature of the reagent. The same was true both for the aromatic substitution (in deciding between the electrophilic, nucleophilic or homolytic mechanisms) and for the nucleophilic substitution (in establishing the cationic nature of the intermediates).

(c) The stereochemistry of the addition will offer information concerning the intermediates involved in product formation (in a similar way as in the nucleophilic substitution).

(d) Direct observation of the cationic species in superacid solutions, of interest both in aromatic electrophilic substitution and in aliphatic nucleophilic displacements, may also be valuable in the study of additions.

Various aspects of the electrophilic addition reactions as related to organic chemistry research have been treated in many textbooks dedicated to organic reaction mechanisms [5,6] or in monographs [7].

[6] E.S. Gould, *Mechanism and Structure in Organic Chemistry*, Rinehart and Winston, New York (1959); J. Hine, *Physical Organic Chemistry*, McGraw-Hill, New York (1962); R. Breslow, *Organic Reaction Mechanisms*, 2nd edn., W. A. Benjamin, New York (1969); R. W. Alder, R. Baker and J. M. Brown, *Mechanism in Organic Chemistry* Interscience, New York (1971).

[7] P. B. D. de la Mare and R. Bolton, *Electrophilic Additions to Unsaturated Systems*, Elsevier, Amsterdam (1966).

10.2.1 *Halogen additions*

10.2.1.1 Reaction products

The first to have recognised that a heterolytic mechanism is involved in the addition of halogens was Francis[8]. He proposed the mechanism shown in Scheme 1, supported by the trapping of the intermediate carbocation by other nucleophiles, which was observed either by himself or by other researchers[9]. This step-mechanism was later confirmed by many other data[10–15].

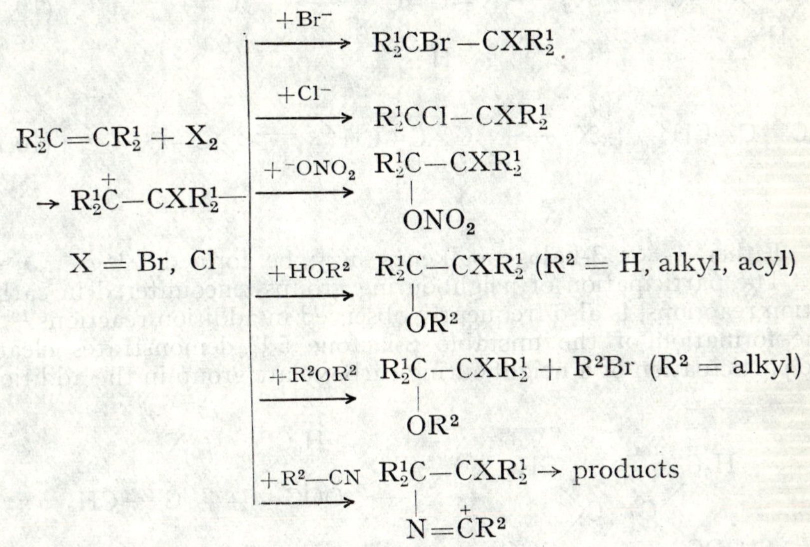

Many of these reactions may be regarded as typical solvolytic processes, as encountered in the nucleophilic substitution.

As in other reactions involving carbenium ion intermediates, rearrangements may accompany the addition. Thus, the addition of bromine to α-pinene affords the bornane bromide **4**[16]:

[8] A. W. Francis, *J. Amer. Chem. Soc.*, **47**, 2340 (1925).
[9] E. M. Terry and L. Eichelberger, *J. Amer. Chem. Soc.*, **47**, 1067 (1925).
[10] E. L. Jackson, *J. Amer. Chem. Soc.*, **48**, 2166 (1926).
[11] P. D. Bartlett and D. S. Tarbell, *J. Amer. Chem. Soc.*, **58**, 466 (1936).
[12] F. C. Weber, G. F. Hennion and R. R. Vogt, *J. Amer. Chem. Soc.*, **61**, 1457 (1939).
[13] D. C. Abbot and C. L. Arcus, *J. Chem. Soc.*, 2934 (1952).
[14] I. G. Dinulescu, M. Avram, C. T. Jijovici, M. Fărcașiu and C. D. Nenitzescu, *Chem. and Ind.*, 840 (1964).
[15] T. L. Cairns, T. J. Graham, T. L. Barrick and R. S. Schreiber, *J. Org. Chem.*, **17**, 751 (1952).
[16] O. Aschan, *Ber. dtsch. chem. Ges.*, **61**, 42 (1928).

$$+ \; Br_2 \longrightarrow \mathbf{4}$$

Eliminations are frequent side reactions of the unimolecular nucleophilic substitution (see E1 eliminations). Proton elimination from intermediate halogenocarbenium ions has also been observed:

$$\mathrm{>C{=}C(H)< + X_2 \longrightarrow X{-}\overset{|}{\underset{H}{C}}{-}\overset{+}{C}< \longrightarrow >C{=}C(X)<} \qquad (10.13)$$

$$\mathrm{>C{=}\overset{|}{C}{-}CH< + X_2 \longrightarrow >\underset{X}{C}{-}\overset{|}{\underset{+}{C}}{-}CH< \longrightarrow >\underset{X}{C}{-}\overset{|}{C}{=}C<} \qquad (10.14)$$

Either 1- or 3-halogenoalkenes may be formed [17,18].

The participation of neighbouring groups, encountered in carbocation reactions, is also frequently observed in addition reactions [19–22]. The formation of the unstable β-lactone **5** [19] demonstrates clearly the participation of a neighbouring carboxylate group in the addition:

$$\mathrm{(H_3C)(^-OOC)C{=}C(CH_3)(COO^-) + X_2 \longrightarrow {}^-OOC{-}\underset{\substack{|\\O}}{\overset{H_3C}{C}}{-}\underset{\substack{|\\CO}}{\overset{X}{C}}{-}CH_3}$$

5

However, classical carbenium ions cannot account for all known data. Thus, the orientation observed in hypohalogenous additions (or solvolytic halogen additions) does not always follow the rules of the electronic theory, which state that the halogen will be bound to the less substituted carbon atom. Actually, beside this product, abnormal products are also formed [23]. The ratio of the latter com-

17 P. Pfeiffer and R. Wizzinger, *Liebigs Ann. Chem.*, **461**, 132 (1928).
18 R. W. Taft, Jr., *J. Amer. Chem. Soc.*, **70**, 3364 (1948).
19 P. D. Bartlett and D. S. Tarbell, *J. Amer. Chem. Soc.*, **59**, 407 (1937).
20 S. Winstein, L. Goodman and R. Boschan, *J. Amer. Chem. Soc.*, **72**, 2311 (1950).
21 R. T. Arnold, M. De Moura Campos and K. L. Lindsay, *J. Amer. Chem. Soc.*, **75**, 1044 (1953).
22 E. E. van Tamelen and M. Schamma, *J. Amer. Chem. Soc.*, **76**, 2315 (1954).
23 P. B. D de la Mare and J. G. Pritchard, *J. Chem. Soc.*, 3910, 3990 (1954).

pounds depends on the substituent of the double bond, as well as on the incoming halogen[23–26]:

$$CH_2{=}CH{-}CH_2X + ClOH \longrightarrow$$

$$\longrightarrow \underset{\text{abnormal orientation}}{\underset{OH}{CH_2}{-}\underset{Cl}{CH}{-}CH_2X} + \underset{\text{normal orientation}}{\underset{Cl}{CH_2}{-}\underset{OH}{CH}{-}CH_2X} + \underset{\text{rearrangement}}{ClCH_2{-}\underset{X}{CH}{-}\underset{OH}{CH_2}} \qquad (10.15)$$

	abnormal orientation	normal orientation	rearrangement
X = H	9%	91%	—
X = OH	30%	70%	—
X = Cl	63%	30%	7%
X = Br	72%		28%

If another halogen is present in the molecule, adjacent to the reaction centre, rearrangements are also possible. Bromine migrates easier than chlorine[24, 25].

In the solvolytic addition of bromine to t-butylethylene in methanol[27], no rearrangement of the neopentyl cation is observed. Moreover, 2-bromo-1-metoxy-3,3-dimethylbutane is abnormally formed:

$$\underset{\mathbf{6}}{CH_3{-}\overset{CH_3}{\underset{CH_3}{C}}{-}CH{=}CH_2} + Br_2 \xrightarrow{CH_3OH} CH_3{-}\overset{CH_3}{\underset{CH_3}{C}}{-}\underset{Br}{CH}{-}\underset{OCH_3}{CH_2} +$$

$$+ CH_3{-}\overset{CH_3}{\underset{CH_3}{C}}{-}CHBrCH_2Br$$

$$CH_3{-}\overset{CH_3}{\underset{CH_3}{C}}{-}\overset{+}{C}H{-}CH_2Br \longrightarrow \text{rearranged products} \qquad (10.16)$$

[24] P. B. D. de la Mare, P. G. Naylor and D. L. H. Williams, *Chem. and Ind.*, 1020 (1959).

[25] P. Ballinger, P. B. D. de la Mare and D. L. H. Williams, *J. Chem. Soc.*, 2467 (1960).

[26] P. B. D. de la Mare, P. G. Naylor and D. L. H. Williams, *J. Chem. Soc.*, 442 (1962).

[27] W. H. Puterbaugh and M. S. Newman, *J. Amer. Chem. Soc.*, **79**, 3469 (1957).

Clearly, Scheme 1 cannot explain the reaction products formed in (10.15) and (10.16). Another scheme is suggested by the reaction stereochemistry.

10.2.1.2 Stereochemistry of halogen additions

In the addition of an XY reagent to an alkene $R^1CH{=}CHR^2$, the original distance relation between the two substituents of the double bond may either be retained or modified (Figure 10.1). In the former case the reaction is called *cis-addition*, while the inversion of this relation corresponds to a *trans-addition*. It has been shown[28] that this designation of the stereochemistry of the reaction may be ambiguous, the same stereochemical prefixes being used either to describe the geometry of the reacting alkene, or the stereochemistry of the addition. In order to avoid confusions, additions with retention of the original stereochemical relationship will be called *syn--additions*, the other possible stereochemistry corresponding to *anti--additions* [28].

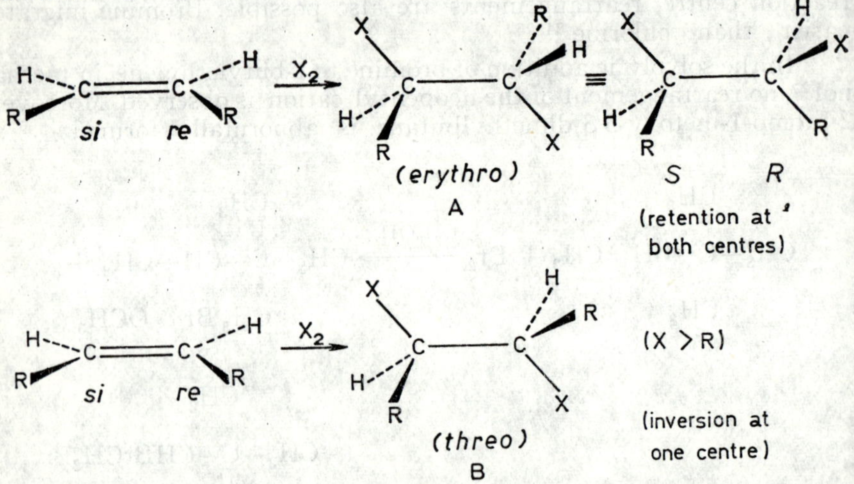

Figure 10.1 Stereochemistry of X_2 additions. A. *syn*-Addition. B. *anti*-Addition.

In any addition to a 1,2-disubstituted alkene, two chiral centres may be created. The stereochemical outcome of the reaction may be thus designated by the Cahn-Ingold-Prelog convention[29]. However caution must be observed in such cases, as the order of priority may differ from reagent to reagent or from one unsaturated compound to another. Figure 10.2 illustrates two different stereochemical designations of two addition reactions proceeding by a common mechanism (and thus having an identical stereochemistry at the reaction centres)

[28] R. C. Fahey, *Topics in Stereochemistry*, **3**, 237 (1968); E. L. Eliel, *Stereochemistry of Carbon Compounds*, McGraw-Hill, New York (1962), p. 355.

[29] R. S. Cahn, C. K. Ingold and V. Prelog, *Angew. Chem.*, **78**, 413 (1966).

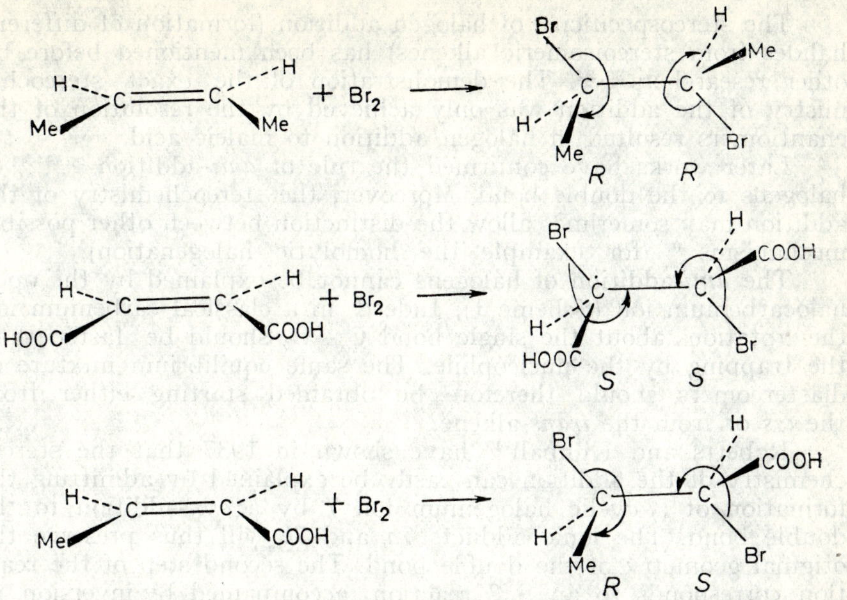

Figure 10.2 Application of the Cahn-Ingold-Prelog rule in the designation of bromine addition at three different *cis*-alkenes.

It has been long observed [30, 31] that halogen additions are stereospecific. Thus, fumaric acid and chlorine yield a dichlorosuccinic acid which cannot be resolved into enantiomers *(meso)*, whereas maleic acid affords a racemic mixture, which may be resolved by the usual methods [32].

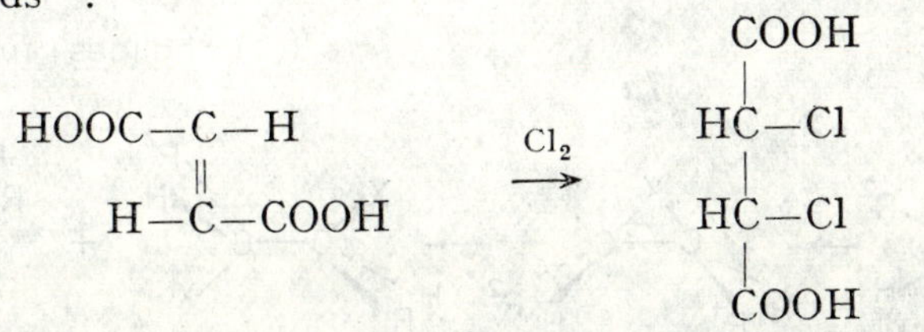

$$\begin{array}{c} HOOC-C-H \\ \| \\ H-C-COOH \end{array} \xrightarrow{Cl_2} \begin{array}{c} COOH \\ | \\ HC-Cl \\ | \\ HC-Cl \\ | \\ COOH \end{array}$$

meso

$$\begin{array}{c} H-C-COOH \\ \| \\ H-C-COOH \end{array} \xrightarrow{Cl_2} \begin{array}{c} COOH \\ | \\ HC-Cl \\ | \\ Cl-CH \\ | \\ COOH \end{array} + \begin{array}{c} COOH \\ | \\ Cl-CH \\ | \\ HC-Cl \\ | \\ COOH \end{array}$$

$(\pm)$

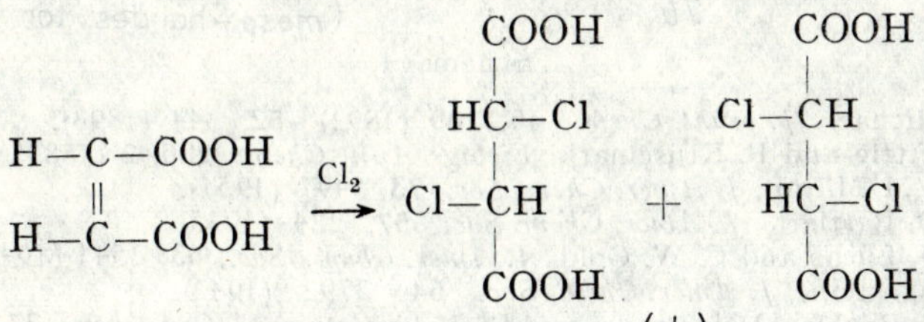

30 B. Holmberg, *Svensk Kemisk Tidskrift*, 5 (1912); *Chem. Zentr.*, II, 1932 (1912).
31 A. McKenzie, *J. Chem. Soc.*, **101**, 1196 (1912); *Chem. Zentr.*, II, 1101 (1912).
32 A. P. Terentiev and V. Potapov, *Usp. Khim.*, **26**, 1152 (1957); S. H. Wilen, *Topics in Stereochemistry*, **6**, 107 (1971).

The stereospecificity of halogen addition (formation of different halides from stereoisomeric alkenes) has been mentioned before by other researchers[33, 34]. The demonstration of the exact stereochemistry of the addition was only achieved by the resolution of the enantiomers resulted in halogen addition to maleic acid.

Later works have confirmed the rule of *anti*-addition[28,35–41] of halogens to the double bond. Moreover, the stereochemistry of the addition may sometimes allow the distinction between other possible mechanisms[40] (for example the homolytic halogenation).

The *anti*-addition of halogens cannot be explained by the open halocarbenium ion (Scheme 1). Indeed, in a classical carbenium ion the rotation about the single bond $C_\alpha—C_\beta$ should be faster than the trapping by the nucleophile. The same equilibrium mixture of diastereomers should therefore be obtained starting either from the *cis* or from the *trans*-alkene.

Roberts and Kimball[42] have shown in 1937 that the stereochemistry of the addition can easily be explained by admitting the formation of a cyclic halogenium ion **7** by a *syn*-addition to the double bond. The ionic adducts **7*a*** and **7*b*** will thus preserve the original geometry of the double bond. The second step of the reaction corresponds to an S_N2 reaction, accompanied by inversion at the atom attacked, the overall process being an *anti*-addition Scheme 4).

cis-alkene — **7*a*** — mixture of *threo*-halides ((±)-halides for $R^1=R^2$)

trans-alkene — **7*b*** — mixture of *erythro*-halides (*meso*-halides for $R^1=R^2$)

Scheme 4

[33] A. Michael, *J. prakt. Chem.*, **46**, 209 (1892); **52**, 344 (1893).

[34] R. Fittig and P. Krusemark, *Liebigs Ann. Chem.*, **206**, 2 (1881); W. R. Vaughan and K. M. Milton, *J. Amer. Chem. Soc.*, **73**, 5497 (1951).

[35] P. D. Bartlett, *J. Amer. Chem. Soc.*, **57**, 224 (1935).

[36] H. J. Lucas and C. W. Gold, *J. Amer. Chem. Soc.*, **63**, 2541 (1941).

[37] S. Winstein, *J. Amer. Chem. Soc.*, **64**, 2792 (1942).

[38] R. E. Buckles, J.M. Brader and R. J. Thurmaier, *J. Org. Chem.*, **27**, 4523 (1962).

[39] R. C. Fahey and C. Schubert, *J. Amer. Chem. Soc.*, **87**, 5172 (1965); R. C. Fahey, *ibid.*, **88**, 4681 (1966).

[40] M. L. Poutsma, *J. Amer. Chem. Soc.*, **87**, 2161, 2172, 4285 (1965).

[41] J. H. Rolston and K. Yates, *J. Amer. Chem. Soc.*, **91**, 1477 (1969).

[42] I. Roberts and G. E. Kimball, *J. Amer. Chem. Soc.*, **59**, 947 (1937).

Roberts and Kimball's theory involves the first example of a bridged ion invoked in order to explain the stereochemistry of an organic reaction [43]. It was very shortly followed [44] by the admission of bromonium ions in the solvolysis of dibromo and bromohydrine derivatives (Chapter 5).

$$\mathrm{Me{-}\underset{H}{\overset{Br}{C}}{-}\underset{OH}{\overset{H}{C}}{-}Me} \xrightarrow{H^+} \mathrm{Me{-}\underset{H}{C}\overset{\overset{Br}{/+\backslash}}{\text{———}}\underset{H}{C}{-}Me} \xrightarrow{Nu:^-} \mathrm{Me{-}\underset{H}{\overset{Br}{C}}{-}\underset{Nu}{\overset{H}{C}}{-}Me} \quad (10.17)$$

The existence of halogenium ions, even if challenged earlier [45] and recently[46], may now be regarded as certain. It is well supported by stereochemical and kinetic data [28,45], thermodynamic data [47], direct observation [48], as well as by theoretical calculations [49]. One may only wonder if Scheme 2 and Scheme 4 represent a universal mechanism for the addition of halogens to alkenes.

A more thorough study of the stereochemistry of halogen additions has revealed that the rule of the *anti*-addition is not strictly obeyed. Even if other types of mechanisms (homolytic, etc.) are ignored, the low stereochemistry may still be explained by the mechanism depicted in Scheme 3 (involving carbenium and halogenium ions). The equilibration of bridged and open-chain ions may be governed by the following factors:

(a) *The nature of the halogen.* It is well known (see p. 267) that the stability of the halogenium ions decreases with the nature of the halogen, in the sequence: I > Br > Cl. The lowest stereospecificity (equilibrium shifted in the direction of open-chain ion) will be thus observed in chlorine additions, whereas for bromine and iodine the rule of *anti*-addition will be better obeyed. An illustration is afforded by the halogen additions to acenaphthylene [50]. Thus, in the addition of chlorine, the only adduct isolated was the *syn*-product (~ 30%), whereas in the bromine addition the isolated *anti*-adduct represented 50%.

[43] See Bartlett's commentary in *Non-classical Ions, Reprints and Commentary* (P. D. Bartlett, editor), W. A. Benjamin, New York (1965).

[44] S. Winstein and H. J. Lucas, *J. Amer. Chem. Soc.*, **61**, 1576, 2845 (1939).

[45] J. G. Traynham, *J. Chem. Educ.*, **40**, 392 (1963), and the literature cited therein.

[46] S. P. McManus and V. W. Ware, *Tetrahedron Letters*, 4271 (1974).

[47] J. W. Larsen and A. V. Metzner, *J. Amer. Chem. Soc.*, **94**, 1614 (1972); see also p. 267.

[48] G. A. Olah, *Science*, **168**, 1298 (1970), and the literature quoted therein.

[49] W. J. Hehre and P. C. Hiberty, *J. Amer. Chem. Soc.*, **96**, 2665 (1974).

[50] S. J. Cristol, F. R. Stermitz and P. S. Ramey, *J. Amer. Chem. Soc.*, **78**, 4939 (1956).

cis trans

(b) *The alkene.* The halogenium ion derived from a *cis*-alkene should be less stable than the corresponding ion resulted from the *trans* isomer (compare **7***a* with **7***b*), due to steric interactions between neighbouring bulky groups. Additions to *trans*-alkenes will be thus more stereospecific than additions to *cis* isomers. It must be noted that the product of *anti*-addition to a *trans*-alkene (*erythro* or *meso*, see Scheme 4) is the product of thermodynamic control. Therefore, the stereospecificity of the reaction must be judged from the addition to the more unstable *cis* isomer. It has thus been observed [39] that the additions to simple alkenes, with no other functional groups in the molecule, are stereospecific. The addition of chlorine (unfavourable choice of halogen) to *cis*-di-t-butylethylene (highly strained alkene) proceeds practically with *anti* stereochemistry [39].

(c) *Relative stabilities of the ions involved.* A carbenium ion stabilised by *p*-π conjugation may be more stable than the corresponding bridged halogenium ion. The equilibrium between the cyclic and the open β-halogeno-carbenium ion may be thus shifted towards the latter species. This seems to be the case in the additions to styrenes and stilbenes [38, 39], which involve benzylic ions **9**. The initial halogenium ions, *cis*-**8** and *trans*-**8**, will equilibrate rapidly with the benzylic ion **9**, due to the rotation about the carbon-carbon bond:

cis – alkene trans – alkene

cis -8 9 trans -8

threo – product erythro – product

Identical reaction products will be expected from both *cis*- and *trans*-alkenes. Yates and his co-workers [41] have shown that the stereochemical outcome in the addition of halogens to styrenes depends largely on the reaction medium. A low polarity of the solvent, coupled with a high concentration of the halide ion, would favour the bridged halogenium ions, which may react before the equilibration with the open-chain carbocations **9** occurs. Conversely, a high polarity of the medium and a low concentration of halide lead to trapping products of the carbenium ion **9.** Another interpretation of these results is that the halogenium ion is not symmetric (**10**), the positive charge being more localised at the benzylic carbon atom:

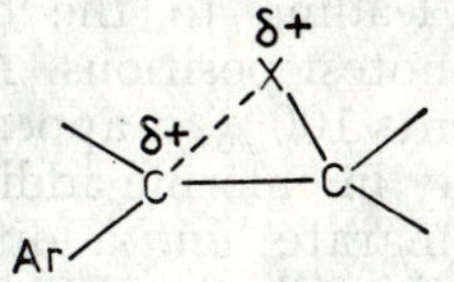

δ+ X δ+ C C Ar

10

Unsymmetrical ions, **10**, have been proposed by de la Mare and his co-workers [23, 26].

(d) *Participation of neighbouring groups.* The participation of neighbouring groups in the opening of bridged bromonium ions is demonstrated by many data. The halogen addition to bicyclic alkenes[51] constitutes an example. Thus, norbornene[52] yields the *anti*-adduct **12** (resulted from halide-ion attack on the halogenium ion **11**), beside the non-classical norbornyl ion products, **14** and **15**, formed by *exo*-attack on the delocalised ion **13**:

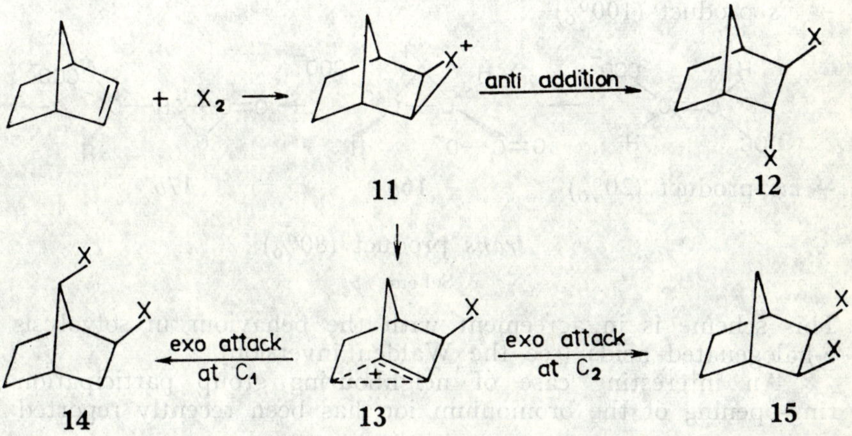

+ X_2 → **11** (X+) —anti addition→ **12** (X, X)

11 ↓ **13**

14 ←exo attack at C_1— **13** —exo attack at C_2→ **15**

[51] T. G. Traylor, *Accounts Chem. Res.*, **2,** 153 (1969).

[52] J. D. Roberts, F. O. Johnson and R. A. Carboni, *J. Amer. Chem. Soc.*, **76,** 5692 (1954).

The C-1—C-2 bond in the norbornyl system is responsible for the anchimeric assistance in the ring-opening of the halogenium ion **11**.

Another example of participation of neighbouring groups is afforded by the carboxyl group. Kuhn and Ebel [53] observed that the addition of chlorine to maleic acid is *anti*, whereas the addition to the disodium maleate is *syn* (identical adducts are obtained from sodium maleate and fumarate). Roberts and Kimball [42] have explained this result by admitting an equilibrium between the bridged and open-chain ions.

Electrostatic repulsions in the chloronium ion derived from sodium maleate will favour the ring-opening and the rotation about the carbon-carbon bond, leading to the carbenium ion with the carboxyl groups in the remotest positions. As the abnormal addition to sodium maleate represents 100%, it appears that the *anti*-addition would be the single pathway in chlorine addition to sodium fumarate. However, for sodium fumarate *anti*-addition actually represents only 80% [53], a figure confirmed also by later works [54]. Moreover, it has been shown [19, 54] that with substituted maleic acids (methyl- and dimethyl-maleic acids) there is no *syn*-addition in base solutions. It seems thus plausible [54] that the initially formed halogenium ions **16***a* and **16***b* are opened with the assistance of the neighbouring carboxylate [55], leading to an unstable α-lactone which is hydrolysed (Scheme 5):

16*a* **17***a*

→ *cis* product (100%)

→ *cis* product (20%) **16***b* **17***b*

↓

trans product (80%)

Scheme 5

This scheme is in agreement with the behaviour in solvolysis of α-halogenated acids (see the Walden inversion).

An interesting case of neighbouring group participation in ring-opening of the bromonium ion has been recently reported by

[53] R. Kuhn and E. Ebel, *Ber. dtsch. chem. Ges.*, **58**, 919 (1925).

[54] F. Badea, Ph. D. Thesis, Bucharest (1964).

[55] W. A. Cowdrey, E. D. Hughes and C. K. Ingold, *J. Chem. Soc.*, 1208 (1937); E. Grunwald and S. Winstein, *J. Amer. Chem. Soc.*, **70**, 841 (1948).

McManus and Ware[46]. The allyl benzamides **18** yield on bromination the open-chain dibromides **19**, beside cyclic derivatives **20** and **21**. One may assume a common intermediate, the bromonium ion **22**:

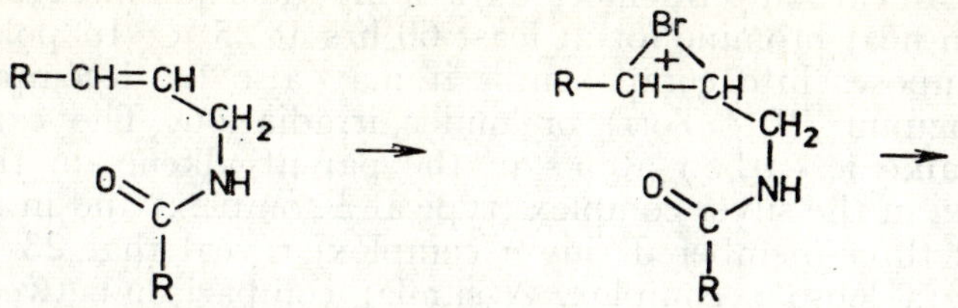

In this case, the oxygen of the carbonyl group assists the ring-opening of the halogenium ion. As in other cases of neighbouring group participation, the formation of the five-membered ring bromo-oxazoline would have been anticipated[56]. However, only the unsubstituted **18** ($R = H$) leads to the oxazoline **20** as the single cyclic product. If $R = C_6H_5$ (cynamylbenzamide), the six-membered ring heterocyclic compound is formed exclusively[57]. With a moderately electron-repelling substituent at the double bond[46] ($R = Me$), both cyclic products are formed, the ratio **20** : **21** varying from 21 : 79 in AcOH to 48 : 52 in CCl_4. This is a clear indication that both the bridged bromonium ion **22** and the acyclic carbenium ions must be involved in the scheme. The equilibrium between these intermediates will largely depend on the substituents at the double bond and on the reaction medium.

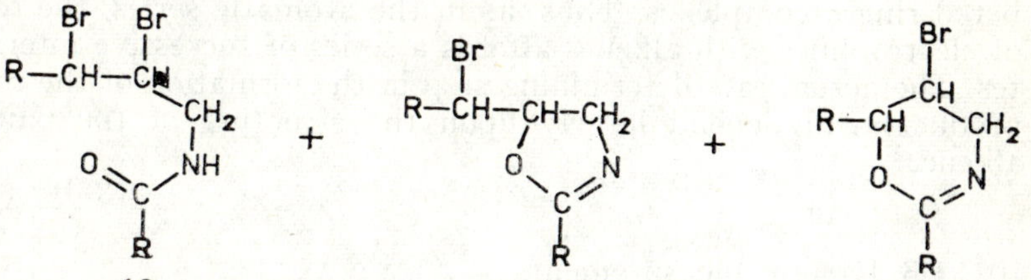

The intervention of the two reaction intermediates has been recently demonstrated by Olah and his co-workers[58]. They have studied the H^1 n.m.r. and C^{13} n.m.r. spectra of some derivatives and complexes of the following alkenes:

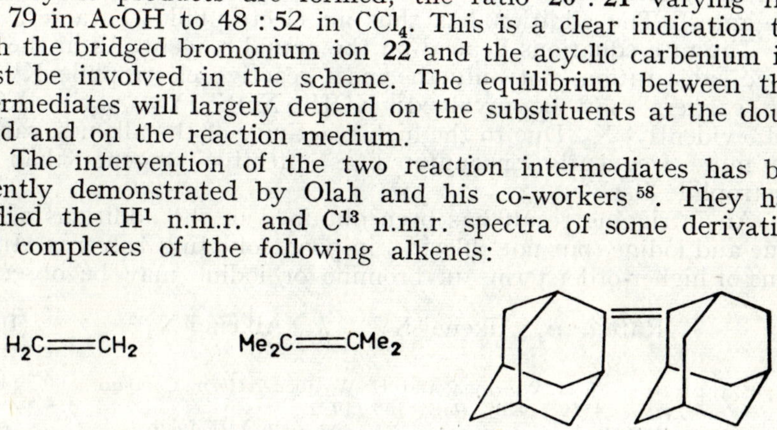

$H_2C{=}CH_2$ $Me_2C{=}CMe_2$ **23**

[56] M. I. Page, *Chem. Soc. Rev.*, 295 (1974).
[57] S. P. McManus and R. A. Hames, *Tetrahedron Letters*, 4549 (1973).
[58] G. A. Olah, P. Schilling, P. W. Westerman and H. C. Lin, *J. Amer. Chem. Soc.*, **96**, 3581 (1974).

The sequence in this series corresponds to increasing steric demands at the reaction centre (double bond). The more crowded of these alkenes is adamantylideneadamantane [59]. Treated with bromine in carbon tetrachloride it readily yields an insoluble complex, stable in neat bromine for at least 60 hrs at 25 °C. In polar solvents it decomposes into components; it may also be decomposed under high vacuum (10^{-7} Torr) or under irradiation. The carbon shifts of the alkenic carbon atoms in the parent alkene, in the bromine complex, in the silver complex (typical π complex) and in the epoxide (typical three-membered ring σ complex) reveal that **23** yields with bromine a loose π complex. A similar comparison between ethylene and its derivatives shows that uncrowded alkenes yield three-membered ring σ complexes. Thus, as in the aromatic series, the reaction of electrophiles with alkenes affords a series of successive intermediates. The actual rate-determining step in the formation of the reaction products will depend largely upon the structure of the particular alkene.

10.2.1.3 Kinetic measurements

The reaction of halogens with alkenes is usually very fast, and difficult to study by conventional kinetic methods. Former works were thus limited to derivatives having at the sp^2 carbon atoms one or more electron-withdrawing groups (COOH, COOR, CN, etc.) [60]. Competitive measurements [40] have also proved to be convenient in establishing relative reactivities. New techniques have recently been developed, notably by Dubois [61–64], allowing the study of simple alkenes. It is now certain that the alkene reactivity increases with its nucleophilicity (electron-repelling groups enhance the reaction rates). The addition has thus an electrophilic character.

The rate equations depend on the actual reagent. As in the aromatic substitution, not only the most reactive electrophile X^+, but also carriers may be involved: XOH, XOY (Y = acyl), XOH_2^+ and, evidently, X_2. Due to the high reactivity of the alkenes, carriers are more frequently encountered as effective reagents than the electrophile itself.

An intriguing result has been obtained in the additions of bromine and iodine (but not chlorine) in poor solvating solvents. Incursions of higher-order terms in bromine or iodine may be observed:

$$\text{Rate} = k_2\,[\text{Alkene}][X_2] + k_3\,[\text{Alkene}][X_2]^2 \qquad (10.18)$$

[59] J. Strating, J. H. Wieringa and H. Wynberg, *Chem. Commun.*, 907 (1969); *Tetrahedron Letters*, 4559 (1970); *ibid.*, 169 (1972).

[60] P. B. D. de la Mare, *Quart. Rev. Chem. Soc.*, 3, 126 (1949).

[61] J. E. Dubois and G. Mouvier, *Tetrahedron Letters*, 1325 (1963).

[62] J. E. Dubois and G. Barbier, *Tetrahedron Letters*, 1217 (1965).

[63] J. E. Dubois and W. V. Wright, *Tetrahedron Letters*, 3101 (1967).

[64] J. E. Dubois, P. Alcais, G. Barmer and R. Uzan, *Bull. Soc. chim. France*, 605, 611, 617 (1968).

The interpretation[60] is that the π complex formed between the alkene and the halogen decomposes rapidly in polar solvents into the halogenium ion and halide, but in the less polar solvents needs a second halogen molecule to afford the halogenium ion.

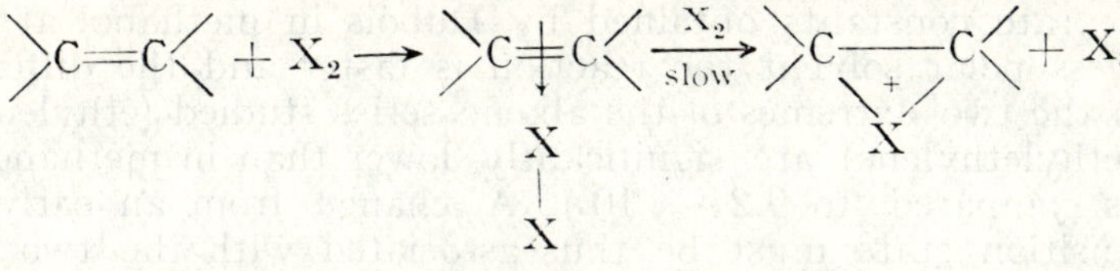

$$\rangle C{=}C\langle + X_2 \rightarrow \rangle C{\#}C\langle \xrightarrow[\text{slow}]{X_2} \rangle C{-}C\langle (X^+) + X_3^- \tag{10.19}$$

(intermediate π complex with X–X: **24**)

Chlorine, which yields an unstable trihalide ion Cl_3^-, is not involved in higher-order rate equations. Either a symmetric (**24**) or an asymmetric (**25**) halogen complex was suggested by other authors[24–26, 65–67]. Poutsma and his co-workers[40, 65] have found that *cis*- and *trans*-2-butene and isobutene are almost equally reactive towards chlorine, whereas large differences are observed in the protonation of these alkenes. Obviously, if an asymmetric complex (**25**) were involved, large differences would be observed in the addition of chlorine.

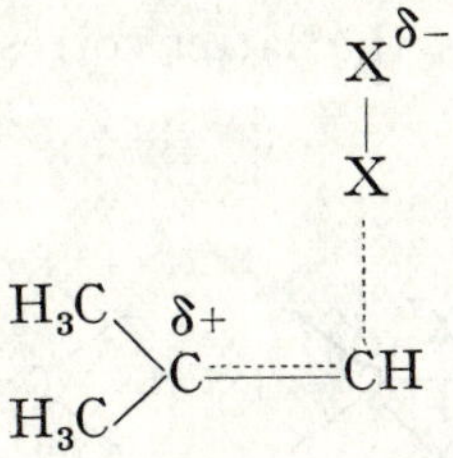

$$(H_3C)_2\overset{\delta+}{C}{\cdots}CH{\cdots}X{-}X^{\delta-} \qquad H_3C{-}\overset{\delta+}{CH}{\cdots}CH(\cdots X{\cdots}X^{\delta-}){-}CH_3$$

25*a* 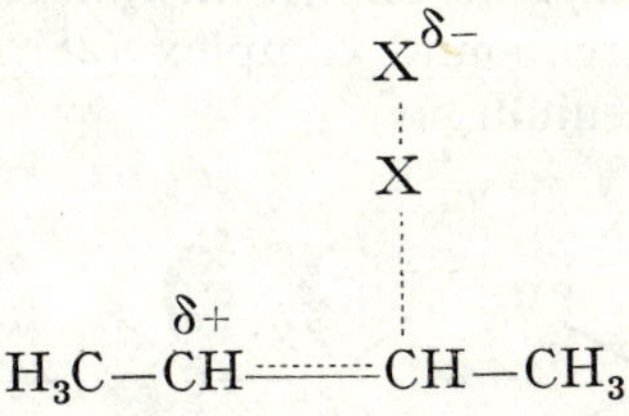 **25*b***

The study of styrenes[66, 67], on the other hand, has indicated asymmetric intermediates. Indeed, an electron-repelling group in α (**26**, $R^1 = Me$; $R^2 = R^3 = H$) brings about a larger rate enhancement than the same group introduced in β ($R^1 = R^2 = H$; $R^3 = Me$).

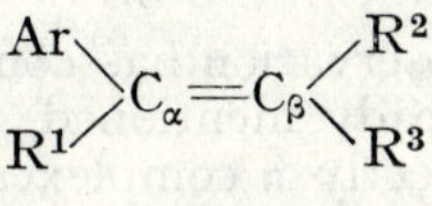

$$(Ar)(R^1)C_\alpha{=}C_\beta(R^2)(R^3) \qquad \rangle C{-}C\langle \text{ bridged by } X^+{-}X$$

26 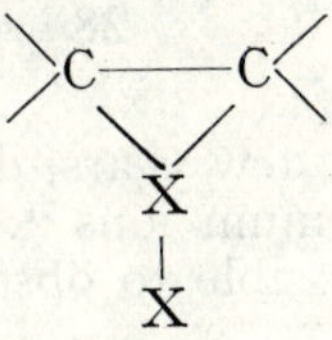 **27**

[65] M. L. Poutsma and J. L. Kartch, *J. Amer. Chem. Soc.*, **89**, 6595 (1967).
[66] R. A. Pincock and K. Yates, *J. Amer. Chem. Soc.*, **90**, 5643 (1968).
[67] J. H. Rolston and K. Yates, *J. Amer. Chem. Soc.*, **91**, 1483 (1969).

This indicates that in the transition state of the rate-determining step a partial positive charge is developed in α.

Olah and his co-workers [68] have determined competitive rate data for the addition of bromine to a series of alkenes in 1,1,2-trichlorotrifluoroethane solution at —35 °C, and compared their data with the rate constants obtained by Dubois in methanol at 25 °C. In the less polar solvent the reaction is faster and the differences between the two extremes of the alkene series studied (ethylene and 2,3-dimethylethylene) are significantly lower than in methanol (5.7×10^3 as compared to 9.24×10^5). A change from an early to a late transition state must be thus associated with the two types of solvents. In solvents with low polarity, π complexes are formed in the rate-determining step.

In the formalism used to designate the intermediates of the halogen addition, there is some confusion, as complexes **24** and **27**, and halogenium ions are rather indiscriminately used in discussions. Clearly, all of them must appear in the various steps of the reaction [69, 70]. Experimental evidence has been found recently [68] which allowed the clear distinction between these species.

According to Olah [68], the significant steps of the halogen addition are the formation of a π complex and the subsequent formation of a σ complex. The former intermediate must be regarded as a three-centre bound complex (**28**), whereas the latter corresponds to the halogenium ion:

$$\text{>C=C<} \longrightarrow \text{>C—C<} \cdots \text{X—X} \;(\mathbf{28}) \longrightarrow \text{>C—C<}\ (\text{bridged } X^{+}, X^{-}) \longrightarrow \text{X—>C—C<—X} \qquad (10.20)$$

In the last few years, direct observation has confirmed the existence of halogenium ions [71]. As already mentioned, in certain cases it was also possible to observe directly π complexes [58].

[68] G. A. Olah and T. R. Hockswender, Jr., *J. Amer. Chem. Soc.*, **96**, 3574 (1974).
[69] M. J. S. Dewar, *The Electronic Theory of Organic Chemistry*, Oxford University Press (1949); *J. Chem. Soc.*, 406 (1946).
[70] D. V. Banthorpe, *Chem. Rev.*, **70**, 295 (1970).
[71] G. A. Olah and J. M. Bollinger, *J. Amer. Chem. Soc.*, **89**, 4744 (1967); **90**, 947, 2587, 6988 (1968).

10.2.1.4 Other data confirming the mechanism

Mechanism (10.20) is also confirmed by the reversibility of iodine[72, 73] and of bromine[54, 74] addition. Thus, cyclohexene and *meso*-methyl dibromosuccinate in DMF yield methyl fumarate beside dibromocyclohexane. The debromination in this solvent is faster in presence of bromide ion. At the same time, bromine addition rates are highly depressed in DMF, despite its high polarity.

Reactions of addition to polyenes confirm the cationic nature of the intermediates, even if it is not quite clear if π complexes and halogenium ions are significantly involved beside open carbenium ions. Indeed, the long-discussed[75] 1,4 addition of halogens to butadiene may either proceed by the delocalised carbocation **29**, or by the five-membered ring cation **30**:

$$>C{=}\underset{|}{\overset{|}{C}}-\underset{|}{\overset{|}{C}}{=}C< \xrightarrow{X_2} >\overbrace{C{\cdots}C{\cdots}C}^{+}-C-X + X^-$$

29 → $X-\overset{|}{\underset{|}{C}}-\overset{|}{C}{=}\overset{|}{C}-\overset{|}{\underset{|}{C}}-X + X-\overset{|}{\underset{|}{C}}-\overset{|}{C}X-\overset{|}{C}{=}C<$

↓

30 (five-membered ring cation with X^+) $\xrightarrow{X^-}$ $HC-CH_2X \; \| \; HC-CH_2X$

If the latter intermediate were to appear in the 1,4-addition, a *cis*-1,4-dihalogenobut-2-ene would result. Mislow and his co-workers have actually shown that both in chlorine[76] and bromine[77] addition, the *trans*-isomer is the single observed 1,4-adduct. Intermediate **30** may be thus completely ruled out.

Large-ring halogenium ions may however be involved in other processes, such as substitution reactions[78]. It has been shown[79] that the five-membered rings are more stable than the three-mem-

[72] N. J. Bythel, P. W Robertson, *J. Chem. Soc.*, 179 (1938).
[73] J. Hine and W. H. Brader, Jr., *J. Amer. Chem. Soc.*, **77**, 361 (1955).
[74] F. Badea, T. Constantinescu, A. Juvara and C. D. Nenitzescu, *Liebigs Ann. Chem.*, **706**, 20 (1967).
[75] J. Thiele, *Liebigs Ann.Chem.*, **306**, 87 (1899).
[76] K. Mislow and H. M. Hellman, *J. Amer. Chem. Soc.*, **73**, 244 (1951).
[77] H. M. Hellman and K. Mislow *J. Amer. Chem. Soc.*, **76**, 1175 (1954).
[78] R. G. Pearson, *Accounts Chem. Res.*, **4**, 152 (1971), and references therein cited.
[79] R. D. Wieting, R. H. Staley and J. L. Beauchamp, *J. Amer. Chem. Soc.*, **96**, 7552 (1974).

bered ring ions, equilibrium (10.21) being completely shifted in favour of the larger ring:

$$H_2C\overset{Br^+}{—}CH_2 + Br(CH_2)_4Br \rightleftharpoons Br(CH_2)_2Br + \text{(five-membered ring } Br^+) \qquad (10.21)$$

With one exception[80], four-membered ring halogenium ions have not been definitely proved. The unfavourable enthalpy (even if lower than for the three-membered ring), coupled with the unfavourable entropy may be an explanation. It has been suggested[81] that in the case of rigid systems, such as **31**, four-membered ring bromonium ions may be involved in the bromine addition.

31 + Br_2 → Br^+ →

Br ⇌ Br^+ → Br, Br

10.2.2 *Other electrophilic additions of non-acid reagents*

A series of other electrophiles, such as sulphenyl chlorides and mercury salts, may add in a similar way to alkenes.

The addition of sulphenyl chloride has been studied by Kharasch[82], product formation being in agreement with the following formulation:

$$>C{=}C< + RSCl \rightarrow >C\!\!-\!\!C< \;(\text{bridged } \overset{+}{S}\text{–R, } \mathbf{32}) \rightarrow >C(SR)\!-\!C\!-\!Cl$$

[80] J. H. Exner, L. D. Kershner and T. E. Evans, *J. Chem. Soc. Chem. Comm.*, 361 (1973).

[81] M. Avram, I. Pogany, F. Badea, I. G. Dinulescu and C. D. Nenitzescu, *Tetrahedron Letters*, 3851 (1969).

[82] See for example N. Kharasch, in *Organic Sulphur Compounds* (N. Kharasch, editor), Pergamon Press, Oxford (1961), p. 375.

The addition is *anti* [83, 84], indicating cyclic episulphonium ions, **32**. The cyclic nature of the intermediate has been demonstrated by Brown[85], by studying the relative rates of the addition to norbornene and to 7,7-dimethylnorbornene ($k_r = k_{norb}/k_{7,7\text{-DM}} = 1800$). The high ratio observed is coupled practically with only one orientation of the reacting species:

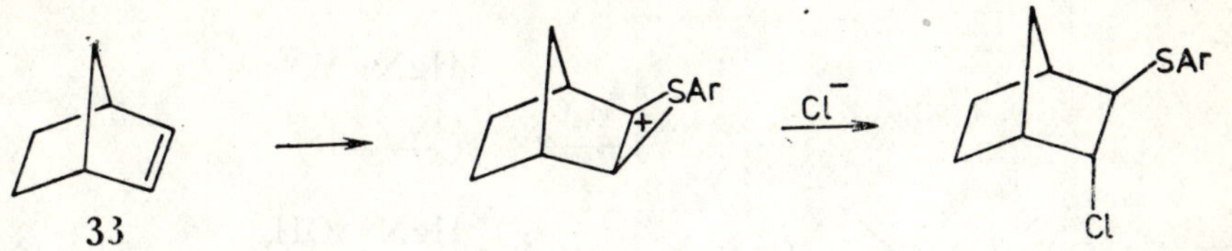

33 → (SAr, +) —Cl⁻→ (SAr, Cl)

(10.22)

Me Me → Me Me (SAr, +) —Cl⁻→ Me Me (SAr, Cl)

34

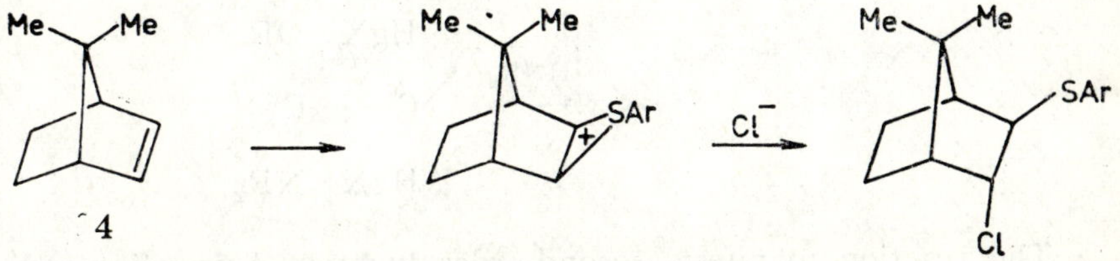

(10.23)

It should be observed that in the photochemical addition of PhSH to these alkenes, the *exo* isomer represents 99,5% from **33** and 95% from **34**, the relative rate, k_r, being only 30. The radical character of the latter reaction (involving thus an open-chain intermediate) clearly indicates that high k_r values must be associated with cyclic intermediates. Mueller [86] made the interesting observation that in the addition of methyl sulphenyl chloride to *cis-trans* cycloocta-1,5-diene, even with a high excess of alkene, a 2:1 addition product is formed in 80—90% yield. This suggests that in the mono-adduct **35** the sulphur atom may activate through space the double bond (transannular activation):

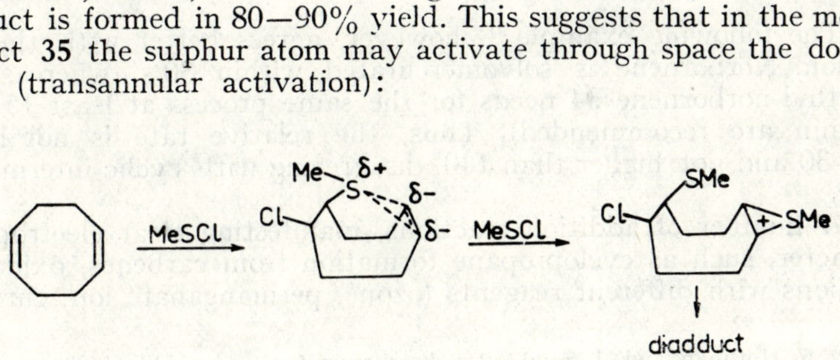

—MeSCl→ Cl, Me–S δ+, δ-, δ- —MeSCl→ Cl, SMe, +SMe → diadduct

[83] N. Kharasch and A. J. Havlik, *J. Amer. Chem. Soc.*, **75**, 3734 (1953).
[84] W. H. Mueller, *Angew. Chem. Intern. Ed. Engl.*, **8**, 482 (1969).
[85] H. C. Brown, J. H. Kawakami and K.-T. Liu, *J. Amer. Chem. Soc.*, **95**, 2209 (1973).
[86] W. H. Mueller, *J. Amer. Chem. Soc.*, **91**, 1223 (1969).

The addition of mercury salts to the double bond has been known since the beginning of the systematic work of Hoffman and Sand [87] în 1900. The electrophilic nature of the reaction is now certain [88], cationic intermediates being trapped by nucleophiles:

$$>C{=}C< + HgX_2 \longrightarrow \begin{cases} \xrightarrow{Y^-} >C(HgX){-}C(Y)< \\ \xrightarrow{H_2O} >C(HgX){-}C(OH)< \\ \xrightarrow{ROH} >C(HgX){-}C(OR)< \\ \xrightarrow{HNR_2} >C(HgX){-}C(NR_2)< \end{cases}$$

The reaction manifests second-order kinetics, being first-order in each reactant. The rate constants increase with increasing nucleophilicity of the alkene, confirming thus the electrophilic character of the reaction. Arguments have been presented either for a cyclic mercurinium ion [89] **36** or for an open-chain intermediate [90] **37**:

$$\underset{\mathbf{36}}{>C{-}C<\ (\text{bridged by } \overset{+}{HgX})} \qquad \underset{\mathbf{37}}{>\overset{+}{C}{-}\overset{|}{\underset{|}{C}}{-}Hg{-}X}$$

The following example [91], however, agrees better with the latter ion. Norbornene is solvomercurated within 30 s, whereas 7,7-dimethyl-norbornene **34** needs for the same process at least 15 min (73 min are recommended). Thus, the relative rate is not lower than 30 and not higher than 140, disagreeing with cyclic intermediates [86].

A number of addition reactions, manifesting clear electrophilic character, such as cyclopropane formation from carbenes, oxidation reactions with different reagents (ozone, permanganate ion, chromyl

[87] K. Hoffman and J. Sand, *Ber. dtsch. chem. Ges.*, **33**, 1340 (1900).

[88] N. S. Zefirov, *Usp. Khim.*, **34**, 1272 (1965); J. Chat, *Chem. Rev.*, **48**, 7 (1951).

[89] H. J. Lucas, F. R. Hepner and S. Winstein, *J. Amer. Chem. Soc.*, **61**, 3102 (1939); T. G. Traylor, *J. Amer. Chem. Soc.*, **86**, 244 (1964).

[90] J. Halpner and H. B. Tinker, *J. Amer. Chem. Soc.*, **89**, 6427 (1967).

[91] H. C. Brown and P. Geoghegan, Jr., *J. Amer. Chem. Soc.*, **89**, 1523 (1967); H. C. Brown, J. H. Kawakami and S. Ikegami, *ibid.*, **89**, 1525 (1967).

chloride, etc.) are better classified as *cycloaddition reactions*[92], as they lead to cyclic adducts. Such reactions will not be treated in this chapter (see Chapter 12). However, it should be noted that the cycloaddition reactions constitute, probably, the most important link between all types of mechanisms implying covalent modifications (heterolytic, homolytic and no-mechanism reactions). The example of permanganate oxidation in alkaline solution[93–94] is illustrative in this respect:

$$\mathrm{{>}C{=}C{<} \;+\; (O{=})_2Mn(O^-)O^- \longrightarrow {>}C(O)\!-\!C(O){<}\,Mn(=O)O^- \longrightarrow {>}C(OH)\!-\!C(OH){<}} \qquad (10.24)$$

According to the mechanism depicted in (10.24), the cyclic adduct (not isolable) is further hydrolysed to the diol. The process resembles Scheme 2 (see p. 486) of the general electrophilic additions. By its stereochemistry (overall *syn*-addition to the double bond) it is related to cycloaddition reactions, being termed 1,3-*anionic cycloaddition*[95].

10.2.3 *Addition of hydrogen halides*

10.2.3.1 Regioselectivity

The regioselectivity of hydrogen halide additions (as well as that of the acid-catalysed hydration of alkenes) has been known for more than a hundred years. Markownikoff[96] thus established that in hydrogen halide additions the halide ion is bonded to the more substituted unsaturated carbon. Considering the electrophilic character of the addition, this rule is easily rationalised:

$$\mathrm{CH_3{\rightarrow}CH{=}CH_2 + H{-}X \longrightarrow CH_3{-}\overset{+}{C}H{-}CH_3 \xrightarrow{X^-}}$$

$$\mathrm{CH_3{-}CHX{-}CH_3} \qquad (10.25)$$

[92] R. Huisgen, *Proc. Chem. Soc.*, **1961**, 357.

[93] A. Lapworth and E. N. Mottram, *J. Chem. Soc.*, **127**, 1628 (1925).

[94] R. Stewart, in *Oxidation in Organic Chemistry*, Part A (K. B. Wiberg, editor), Academic Press, New York (1965); D. G. Lee and J. R. Brownridge, *J. Amer. Chem. Soc.*, **96**, 5517 (1974).

[95] T. Kauffmann, *Angew. Chem.*, **86**, 715 (1974).

[96] W. Markownikoff, *Liebigs Ann. Chem.*, **153**, 256 (1870).

Even Markownikoff was aware of exceptions to his rule. Some of these exceptions are explained by other reaction mechanisms (for example the peroxide-initiated addition of hydrogen bromide). However, anti-Markownikoff additions may be observed even under heterolytic conditions [97–98]:

$$CF_3 \leftarrow CH{=}CH_2 \xrightarrow[AlBr_3]{HBr} CF_3{-}CH_2{-}CH_2Br \qquad (10.26)$$

Here, again, the observed regiospecificity is explained by electronic effects. The electron-withdrawing group CF_3 will deactivate the double bond (as illustrated by the necessity to use a powerful catalyst $AlBr_3$) and will orient the proton to the less substituted carbon atom. The regioselectivity of the reaction clearly indicates the formation of carbocations by protonation of the double bond. An interesting example in that respect is constituted by the additions to t-butylethylene [99]:

$$CH_3{-}\underset{CH_3}{\overset{CH_3}{\underset{|}{\overset{|}{C}}}}{-}CH{=}CH_2 \longrightarrow \underset{\mathbf{38}}{CH_3{-}\underset{CH_3}{\overset{CH_3}{\underset{|}{\overset{|}{C}}}}{-}\overset{+}{C}H{-}CH_3} \longrightarrow$$

$$\underset{\mathbf{39}}{CH_3{-}\underset{CH_3}{\underset{|}{\overset{+}{C}}}{-}\overset{CH_3}{\overset{|}{C}}H{-}CH_3} \longrightarrow \text{products}$$

In order to explain the observed products, the carbocation **38** (resulting by Markownikoff-addition of a proton) must be assumed as the intermediate. A neopentylic rearrangement to the ion **39** is therefore conceivable. Whitmore has thus demonstrated that the regioselectivity of hydrogen halide additions is based upon a cationic mechanism.

10.2.3.2 π- versus σ-intermediates

As in the case of other electrophilic additions, the question of open-chain and bridged ionic intermediates has also arisen. Taft and his co-workers[100] have remarked the dependence of the reaction rates

[97] E. Schmidt, *Liebigs Ann. Chem.*, **267**, 300 (1892).
[98] A. L. Henne and S. Kaye, *J. Amer. Chem. Soc.*, **72**, 3369 (1950).
[99] G. G. Ecke, N. C. Cook and F. C. Whitmore, *J. Amer. Chem. Soc.*, **72**, 1511 (1950).
[100] E. L. Purlee and R. W. Taft, Jr., *J. Amer. Chem. Soc.*, **78**, 5867 (1956).

upon the acidity function h_0, which indicates the formation of a π complex **40**.

$$\rangle C{=}C\langle + HX \longrightarrow \rangle C \overset{\uparrow}{=} C\langle \text{ (}H^+\text{) } \xrightarrow{X^-} \rangle C(H){-}C(X)\langle \qquad (10.27)$$

40

The stereochemistry of the addition to certain olefins, such as 1,2-dimethylcyclopentene[101], is *anti*:

Me, Me (1,2-dimethylcyclopentene) + HCl ⟶ (H, Me; Me, Cl)

in agreement with Taft's formulation. However, Dewar and Fahey[102] demonstrated that a *syn*-addition may be also the predominant course of the reaction, such as in the hydrogen bromide addition to acenaphthylene [102] and indene [103]. Non-stereospecific reactions are also known[104]. According to Dewar and Fahey[103], the strength of the C—E bond formed in the addition of an electrophile E^+ decides the eventual π-complex route.

$$\underset{\mathbf{41}}{\rangle C \overset{E^+}{\overset{\uparrow}{=}} C\langle} \qquad \underset{\mathbf{42}}{\rangle \overset{E}{\overset{|}{C}}{-}\overset{+}{C}\langle}$$

Thus, the tendency to form σ complexes (**42**) must decrease in the order: $H^+ > Cl^+ > Br^+ > I^+$. Iodine will most likely afford a **41** intermediate, whereas proton addition will lead to **42**. The stereochemistry may be explained considering that the addition of hydrogen halides in non-polar solvents manifests third-order kinetics in the acid [105, 106]. This should indicate association of acid molecules, with the probable formation of the ion pair **43***a* and **43***b* [102, 103]. If the initially formed ion pair **43***a* has no time to equilibrate with **43***b*, its collapse will lead to the *syn*-adduct. Non-stereospecific additions

101 G. S. Hammond and C. H. Collins, *J. Amer. Chem. Soc.*, **82**, 4323 (1963).

102 M. J. S. Dewar and R. C. Fahey, *J. Amer. Chem. Soc.*, **84**, 2012 (1962); **85**, 2245 (1963).

103 M. J. S. Dewar and R. C. Fahey, *J. Amer. Chem. Soc.*, **85**, 2248 (1963).

104 C. H. Collins and J. S. Hammond, *J. Org. Chem.*, **25**, 911 (1960).

105 F. R. Mayo and J. J. Katz, *J. Amer. Chem. Soc.*, **69**, 1339 (1947).

106 F. R. Mayo and M. G. Savoy, *J. Amer. Chem. Soc.*, **69**, 1348 (1947).

will be observed if equilibration is rapid and *anti*-adducts will be formed if **43***b* is more stable than **43***a*:

$$\text{>C=C<} + 3HX \longrightarrow \text{>C=C<}\overset{\text{XH}\cdots\text{X}\cdots\text{HX}}{\underset{|}{\text{H}}} \longrightarrow$$

$$\longrightarrow \text{>C(H)}-\overset{+}{\text{C}}\text{<}\cdots\text{X}^{-}(\text{XH}\cdots\text{X}^{-}\cdots\text{HX}) \;\; \rightleftharpoons \;\; \text{>C(H)}-\overset{+}{\text{C}}\text{<}\cdots\text{X}^{-}(\text{XH}\cdots\text{X}^{-}\cdots\text{HX})$$

43*a* → *syn*-adduct (H and X on the same side)

43*b* → *anti*-adduct (H and X on opposite sides)

syn-adduct *anti*-adduct

10.2.3.3 Hydrogen halide additions in acetic acid

The addition of hydrogen chloride to t-butylethylene and to styrene in acetic acid[107] yields both chlorides and acetates:

$$Me_3C{-}CH{=}CH_2 + HCl \xrightarrow{AcOH} \underset{\mathbf{44}}{Me_3C{-}CHCl{-}CH_3} + \underset{\mathbf{45}}{Me_3C{-}CH(OAc){-}CH_3} + \underset{\mathbf{46}}{Me_2CCl{-}CH(CH_3)_2} \quad (10.28)$$

$$PhCH{=}CH_2 + HCl \xrightarrow{AcOH} \underset{\mathbf{47}}{PhCHCl{-}CH_3} + \underset{\mathbf{48}}{PhCH(OAc){-}CH_3} \quad (10.29)$$

[107] R. C. Fahey and C. A. McPherson, *J. Amer. Chem. Soc.*, **91**, 3865 (1969).

In (10.28) the product ratio is **44** : **45** : **46** = 2 : 1 : 2, whereas in (10.29) **47** : **48** is about 13 : 1. The ratio is not affected by adding tetramethylammonium chloride and the reaction rates are proportional to the stoichiometric concentration of hydrogen chloride. The latter result shows that either molecular, or ionised but dissociated hydrogen chloride is involved. The little or negligible effect of added chloride ion indicates that only ion pairs **49** are involved in the process:

$$R{-}CH{=}CH_2 + HCl + AcOH \longrightarrow \left[\begin{matrix} & & Cl^- \\ R{-}\overset{+}{C}H{-}CH_3 \\ AcOH & & \end{matrix}\right] \longrightarrow \text{Products}$$

49

This mechanism has been termed as AdE2. Another mechanism, referred to as Ad3, seems to be involved in both hydrochlorination[108] and hydrobromination[109] of cyclohexene in acetic acid. Suitable deuterium labelling of cyclohexene has enabled the determination of the stereochemistry of hydrogen halide addition. The additions proved to be first-order in cyclohexene and higher-order (1.58) in the acid. The ratio of chloride (chlorocyclohexane) and acetate, C:A, increases from about 0.22 in absence of water to about 1.33 in presence of 1.12 M water. A similar increase of the ratio C:A has been observed by adding tetramethylammonium chloride to the reaction mixture. All these data indicate a dependence on both chloride ion and undissociated hydrogen chloride. Moreover, hydrochlorination and hydrobromination manifest a net preference for the *anti*-addition. The most probable transition state will thus be **50**:

$$\begin{matrix} & & H\cdots\cdots X^{\delta-} \\ & & \vdots \\ R^1{-}CH & \overset{\cdots}{=} & CH{-}R^2 \\ \vdots & & \\ X^{\delta-} & & \end{matrix}$$

50

It can be easily seen that the Ad3 mechanism is the reverse of the E2 elimination with halide ions (see p. 467).

The data obtained in the hydrogen halide addition to cyclohexene in acetic acid, does not rule out, nor does it confirm the π-complex hypothesis of Taft. The study of the $C_2H_5^+$ ion by ion

[108] R. C. Fahey, M. W. Monahan and C. A. McPherson, *J. Amer. Chem. Soc.*, **92**, 2810, 2816 (1970).

[109] R. C. Fahey, C. A. McPherson and R. A. Smith, *J. Amer. Chem. Soc.*, **96**, 4534 (1974).

cyclotron resonance [110], as well as theoretical calculations [111] of the cations involved in a Markownikoff addition, seem also to rule out bridged intermediates in this addition.

10.2.3.4 Hydration of alkenes

The acid-catalysed hydration of alkenes is in many respects similar to acid additions to the double bond. Formation of carbocations by the following mechanism is suggested by the reversibility of the reaction (see the acid-catalysed dehydration):

$$R_2C{=}CR_2 + H_3O^+ \underset{k_{-1}}{\overset{k_1}{\rightleftharpoons}} R_2\overset{+}{C}{-}CHR_2 + H_2O \overset{k_2}{\rightleftharpoons}$$

$$\underset{+OH_2}{R_2C}{-}CHR_2 \rightleftharpoons \underset{OH}{R_2C}{-}CHR_2 + H^+ \qquad (10.30)$$

The involvement of π complexes has been inferred from the linear dependence of the logarithms of the reaction rates upon the $-H_0$ function[112]. This interpretation is also supported by the fact that the hydration of alkenes in deuterated water does not afford deuterated alkene [100]. Levy, Taft and Hammett [113] have also shown that the hydration of 2-methylbut-2-ene and of 2-methylbut-1-ene yields the same alcohol, even if the two alkenes are not isomerised into each other under the reaction conditions. If a carbenium ion were involved in the hydration:

$$CH_3{-}\overset{CH_3}{\overset{|}{C}}{=}CH{-}CH_3 + H^+ \underset{k_1}{\rightleftharpoons} CH_3{-}\overset{CH_3}{\overset{|}{\underset{+}{C}}}{-}CH_2{-}CH_3 \rightleftharpoons$$

$$\downarrow k_2 \quad \text{product}$$

$$\rightleftharpoons CH_2{=}\overset{CH_3}{\overset{|}{C}}{-}CH_2{-}CH_3$$

equilibration of the two alkenes would have been observed. However, considering the similarity between the hydrogen halide addition and the acid-catalysed hydration of alkenes, the carbenium ion mechanism (10.30) seems more probable. The formation of carbenium ions, as opposed to π complexes, is clearly demonstrated by the different relative rates displayed by alkenes in typical π-complex reac-

[110] H. H. Jaffé and S. Billets, *J. Amer. Chem. Soc.*, **94**, 674 (1972).

[111] D. A. Dixon and W. N. Lipscomb, *J. Amer. Chem. Soc.*, **95**, 2853 (1973).

[112] R. H. Boyd, R. W. Taft, Jr., A. P. Wolf and D. R. Christman, *J. Amer. Chem. Soc.*, **82**, 4729 (1960).

[113] J. B. Levy, R. W. Taft, Jr., and L. P. Hammett, *J. Amer. Chem. Soc.*, **75**, 1253 (1953).

tions (halogenations) and in the hydration. Thus, while in the halogenation isobutene and *trans*-but-2-ene are almost equally reactive[65,68], the hydration of isobutene is about 10^3—10^4 faster than the hydration of *trans*-butene[114]. In the hydration of styrenes[115,116] a linear free-energy relationship is found with σ^+ parameters, the slope of the correlation line being $\rho = -3.42$. Both these results suggest classical carbenium ion mechanism (Scheme 1). Direct observation of the protonated species[58,111] also rule out the π complexes as possible intermediates.

10.3 Electrophilic additions to cyclopropanes

The abnormal behaviour of cyclopropane was noticed early. It was attributed by Baeyer[117] to the strain existing in the small rings. As a consequence of this strain, cyclopropane is unstable and undergoes easily ring-opening reactions, leading mostly to 1,3-disubstituted propanes. The observation that cyclopropane may be a reaction product even under the mild conditions of the solvolysis of arenesulphonates[118] (10.31) or in deamination reactions[119] (10.32) of acyclic compounds, has led to the logical proposition that a bridged intermediate, designated as protonated cyclopropane[119,120], is involved either in ring-opening reactions of cyclopropane, or in the cyclopropane ring-closure by a carbocation route.

X → H → + other products (10.31)

$H_3C—CH_2—CH_2—NH_2 \xrightarrow{HNO_2} H_3C—CH_2—\overset{+}{C}H_2 \rightarrow H_3C—CH=CH_2$ 90 %

(10.32)

H^+ → 10 %

[114] R. W. Taft, Jr., cited in P. D. Bartlett and G. D. Sargent, *J. Amer. Chem. Soc.*, **87**, 1297 (1965).

[115] W. M. Schubert, B. Lamm and R. J. Keeffe, *J. Amer. Chem. Soc.*, **86**, 4727 (1966).

[116] W. M. Schubert, *et al.*, *J. Amer. Chem. Soc.*, **94**, 559, 566 (1972).

[117] A. von Baeyer, *Ber. dtsch. chem. Ges.*, **18**, 2277 (1885); see also E. Eliel, *Stereochemistry of Carbon Compounds*, McGraw-Hill, New York (1962), p. 188.

[118] S. Winstein, E. Clippinger, R. Howe and E. Vogelfanger, *J. Amer. Chem. Soc.*, **87**, 377 (1965).

[119] P. S. Skell and I. Starrer, *J. Amer. Chem. Soc.*, **82**, 2971 (1960).

[120] For leading references see: C. J. Collins, *Chem. Rev.*, **69**, 543 (1969); M. Saunders, P. Vogel, E. L. Hagen and J. Rosenfeld, *Accounts Chem. Res.*, **6**, 53 (1973).

Baird and Aboderin[121] reported the first convincing proof concerning the protonated cyclopropane. By treating cyclopropane with deuterosulphuric acid, D_2SO_4, they obtained 1-propanols deuterated in all possible positions. To account for this result, a *corner-*protonated cyclopropane must be admitted as intermediate:

$$\triangle + D^+ \longrightarrow \underset{CH_2\text{-----}CH_2}{\overset{CH_2D}{+}} \longrightarrow CH_2D{-}CH_2{-}CH_2OH$$

$$\swarrow$$

$$\underset{CH_2\text{-----}CH_3}{\overset{CHD}{+}} \begin{array}{l} \nearrow CH_3{-}CHD{-}CH_2OH \\ \searrow CH_3{-}CH_2{-}CHDOH \end{array} \qquad (10.33)$$

Theoretical calculations, as well as experimental data obtained in solvolysis, indicate that among the three possible structures of protonated cyclopropane (Figure 10.3), the corner-protonated species (A) is the more stable[120].

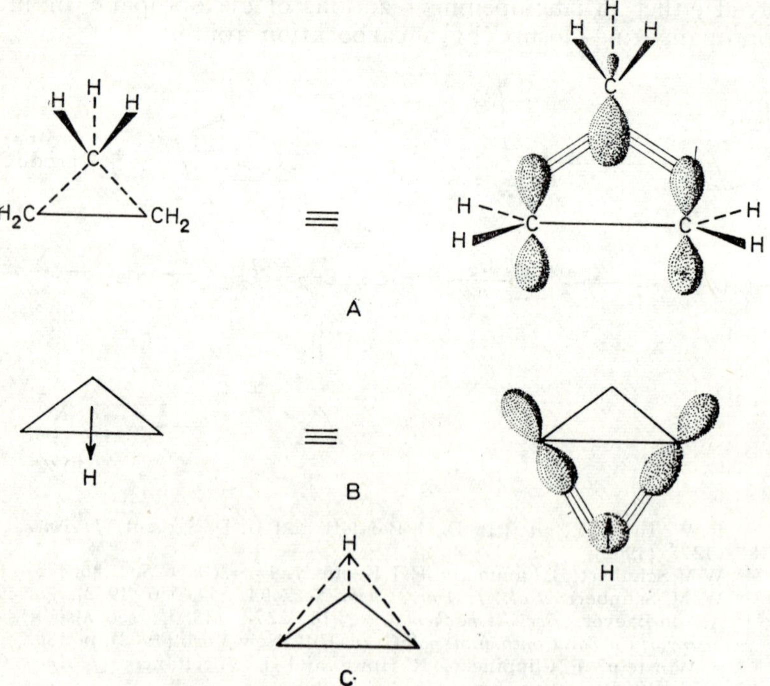

Figure 10.3 Protonated cyclopropane.

[121] R. L. Baird and A. A. Aboderin, *J. Amer. Chem. Soc.*, **86**, 252 (1966).

Corner-protonated species are also likely to appear during the addition of bromine to cyclopropane, in presence of ferric bromide as catalyst:

$$\triangle + Br_2 \xrightarrow{FeBr_3} BrCH_2{-}CH_2{-}CH_2Br + \underset{Br}{CH_2}{-}\underset{Br}{CH}{-}CH_3 + CH_3{-}CH_2{-}CHBr_2 \qquad (10.34)$$

1,1-, 1,2- and 1,3-dibromopropane are formed [122]. Assuming that the catalytic effect of $FeBr_3$ is that of generating the electrophile Br^+, mechanism (10.35) seems plausible. A similar conclusion results from the acetylation of cyclopropane with acetyl chloride, in presence of aluminium chloride [123].

$$FeBr_3 + Br_2 \rightleftharpoons Br^+ + FeBr_4^-$$

$$\triangle \xrightarrow{Br^+} \text{[corner-brominated cyclopropane: } C(H)(H)(Br) \text{ bridging } CH_2{-}CH_2] \xrightarrow{Br^-} \text{1,3 – dibromopropane} \qquad (10.35)$$

$$\downarrow$$

$$\text{[corner-protonated cyclopropane: } C(H)(H)(H) \text{ bridging } CHBr{-}CH_2] \xrightarrow{Br^-} \text{1,1 – and 1,2 – dibromopropane}$$

Considering the Walsh orbitals of cyclopropane [4], edge-protonated cyclopropanes would have seemed more probable in the electrophilic additions to cyclopropanes. The stereochemical study of some electrophilic reactions [124, 125] have indicated that the primary attack of the electrophile is edge-directed, leading in later stages of the reaction to the corner-intermediate.

Theoretical calculations [126] and experiments [127] have demonstrated that among the $C_3H_7^+$ cations, protonated cyclopropane has

[122] N. C. Deno and D. N. Lincoln, *J. Amer. Chem. Soc.*, **88**, 5357 (1966).
[123] H. Hart and R. H. Schlosberg, *J. Amer. Chem. Soc.*, **88**, 5030 (1966).
[124] H. Hart and R. H. Schlosberg, *J. Amer. Chem. Soc.*, **90**, 5189 (1968).
[125] D. H. Rosenblatt, W. H. Denis, Jr., and R. D. Goodin, *J. Amer. Chem. Soc.*, **95**, 2133 (1973).
[126] L. L. Radon, J. A. Pople, V. Buss and P. von R. Schleyer, *J. Amer. Chem. Soc.*, **94**, 311 (1972).
[127] D. J. McAdoo, F. W. McLafferty and P. F. Bente III, *J. Amer. Chem. Soc.*, **94**, 2027 (1972).

an intermediate stability. It is by far more stable than the primary ion, but much less stable than the isopropyl ion:

$$CH_3-CH_2-CH_2^+ \rightleftharpoons \underset{CH_2 \overset{+}{=\!=\!=} CH_2}{\overset{CH_3}{\wedge}} \qquad CH_3-CH_2-CH_2^+ \rightleftharpoons (CH_3)_2CH^+$$

It cannot be decided at present whether the rearrangement of the 1-propyl cation into the secondary ion is a direct process (1,2-hydride shift) or takes place *via* the protonated cyclopropane.

10.4 Nucleophilic additions to the carbonyl group

10.4.1 *Electrophilic character of carbonyl compounds*

Carbonyl compounds may be regarded as carbenium ions, stabilised by α-oxygen:

$$Z(X)C{=}O \longleftrightarrow Z(X)C^+{-}O^- ; \quad Z(X)C{=}\overset{+}{O}H \longleftrightarrow Z(X)C^+{-}OH$$

51*a* **51***b* **52***a* **52***b*

The higher the contribution of the separated-charge structure **51***b* in the resonance hybrid, the higher the electrophilic character of the carbonyl compound will be, and thus its susceptibility to nucleophilic attack. Clearly, the substituents Z and X of the carbon atom can modify the charge density at the carbon atom and therefore the relative contributions of the limit structures. Aldehydes (Z = H, X = alkyl, aryl) will thus exceed in reactivity ketones (Z, X = alkyl, aryl), carboxyl derivatives (Z = alkyl, aryl; X = OR, NR_2, etc.) will be poor electrophiles, the lowest reactivity being displayed by carbonic acid derivatives (X, Z = OR, NR_2, *etc.*). Usually, aldehydes and ketones react with all kind of nucleophiles, either of high reactivity (say carbanions) or of very low nucleophilicity (like the oxygen atom in carboxylic acids, *etc.*), whereas carboxyl reactions involve only strong nucleophiles (carbanions, alkoxides, etc.). Considering the general formulation of a nucleophilic addition to the carbonyl group, the following conclusions are evident:

$$\underset{\mathrm{H}}{\overset{}{\mathrm{Nu}:}} \quad \mathrm{C{=}O} \longrightarrow \underset{\mathrm{H}}{\overset{+}{\mathrm{Nu}}}\mathrm{-C-O^-} \longrightarrow \mathrm{Nu-C-O-H}$$

(a) The nucleophilic addition resembles formally the nucleophilic displacement. In this respect, the reactivity will depend on the reagent and on the reaction medium in the same way as for the nucleophilic substitution.

(b) As in the nucleophilic substitution, the *liate ion*, $\ddot{\mathrm{N}}\mathrm{u}:^-$, will be more reactive than the protonated species :NuH. Considering that the former reagent may be produced by the equilibrium:

$$\ddot{\mathrm{N}}\mathrm{uH} + \mathrm{B}: \rightleftharpoons \ddot{\mathrm{N}}\mathrm{u}:^- + \mathrm{BH}^+ \qquad (10.36)$$

we shall infer that *carbonyl additions can be catalysed by bases.*

(c) The electrophilic character of the carbon atom in the carbonyl group may be significantly enhanced by protonation of the oxygen atom (**52**). Obviously, in the protonated derivative, **52***b* will have a larger contribution than has **51***b* in the free carbonyl compound. *Carbonyl additions will thus be susceptible of acid catalysis.*

Most frequently, carbonyl additions are subject to general catalysis by acids and by bases [128]. However, specific base or acid catalysis may be observed. Thus:

(i) Hydrogen cyanide is only a poor nucleophile. In reactions in which it displays such reactivity, the nucleophilic centre is located at the nitrogen atom. In the hypothetical reaction (10.37), the equilibrium lies toward the side of reactants:

$$\mathrm{HC{\equiv}N}: + \mathrm{C{=}O} \rightleftharpoons \mathrm{H\overset{+}{C}{=}N-C-\ddot{\underset{..}{O}}:^-} \qquad (10.37)$$

However, cyanide ion is a powerful nucleophile. Hydrogen cyanide additions will thus be subject to specific base catalysis, as has been shown by Lapworth [129].

[128] R. P. Bell and B. de B. Darwent, *Trans. Faraday Soc.*, **46**, 34 (1959); R. P. Bell, M. H. Rand and K. M. A. Wynne-Jones, *Trans. Faraday Soc.*, **52**, 1093 (1965).

[129] A. Lapworth, *J. Chem. Soc.*, **83**, 995 (1903); **85**, 1206 (1904).

(ii) Aldehydes may condense with phenols. Evidently, both base catalysis and acid catalysis are possible:

$$C_6H_5O^- + R\text{—}CH\text{=}O \longrightarrow \text{(cyclohexadienone, H, RCH—O}^-\text{)} \longrightarrow p\text{-}^-OC_6H_4\text{—RCH—OH} \quad etc.$$

$$C_6H_5OH + R\text{—}\overset{+}{C}H\text{—}OH \longrightarrow o\text{– } + p\text{ – hydroxymethyl derivatives}$$

However, the acid-catalysed reaction is the single possible process in the case of aromatic hydrocarbons, such as benzene [130].

Condensation reactions of carbonyl compounds have been regarded as nucleophilic additions to carbonyl groups, followed by rapid proton shift and elimination:

$$X\text{—}\ddot{N}H_2 + \overset{|}{\underset{|}{C}}\text{=}O \longrightarrow X\overset{+}{N}H_2\text{—}\overset{|}{\underset{|}{C}}\text{—}O^- \longrightarrow$$

$$XNH\text{—}\overset{|}{\underset{|}{C}}\text{—}OH \xrightarrow{-H_2O} XN\text{=}\overset{|}{\underset{|}{C}}$$

It has been proved that the proton transfer, under certain circumstances, may not be a very rapid process (encounter-controlled reaction). Thus, in the base-catalysed hydrolysis of alkyl benzoates (10.38), Moffat and Hunt [131a] have observed that the rate of the hydrolysis is faster than ^{18}O exchange between the ester and water:

$$R^1\text{—}\overset{*O}{\overset{\|}{C}}\text{—}OR^2 + HO^- \rightleftharpoons R^1\text{—}\overset{*O^-}{\underset{OH}{C}}\text{—}OR^2 \longrightarrow R^1C^*O^*O^- + R^2OH$$

$$\Downarrow \qquad \nearrow \qquad (10.38)$$

$$R^1\text{—}\overset{*O\text{—}H}{\underset{O^-}{C}}\text{—}OR^2 \rightleftharpoons R^1\text{—}COOR^2 + H^*O^-$$

[130] J. E. Hofmann and A. Schriesheim, *Friedel-Crafts and Related Reactions*, vol. 2, Part I (G. A. Olah, editor), Interscience, New York (1964), p. 597.

[131] a. A. Moffat and H. Hunt, *J. Amer. Chem. Soc.*, **81**, 2082 (1959); b. R. E. Barnett, *Accounts Chem. Res.*, **6**, 41 (1973).

Clearly, the equilibrium of proton-transfer must be rate-limiting. Other examples also confirm the hypothesis of a slow proton transfer in carbonyl and acyl-group reactions[131b].

The nucleophilic mechanism of the addition to the carbonyl group is demonstrated clearly by kinetic data.

Thus, it has been shown that many additions to the carbonyl group of aromatic derivatives present a $\rho\sigma$-correlation. Thus, in the formation of semicarbazones of substituted benzaldehydes[132], the correlation line has a slope $\rho = +1.8$. The positive value of the ρ parameter indicates that the carbonyl carbon atom becomes more negative in the transition state, in perfect agreement with the nucleophilic mechanism.

Price and Hammett[133] have found that in the semicarbazone formation, acetone and pinacolone present about the same activation enthalpies ($\Delta H^{\ddagger} = 2.0$ and 1.8 kcal mol^{-1}, respectively). However, the ratio of their rate constants is about 10^3. The difference can be accounted for only by the activation entropies. Indeed, in transition state **53**, the rotation of the bulky group $R = Bu^t$ will be restricted, as compared to the rotation of the less voluminous methyl group in acetone:

$$H_2N{-}CO{-}NH{-}NH_2 + \begin{matrix} CH_3 \\ | \\ C{=}O \\ | \\ R \end{matrix} \longrightarrow \left[H_2N{-}CO{-}NH{-}\overset{\delta+}{NH_2}\cdots\begin{matrix} CH_3 \\ | \\ C \\ | \\ R \end{matrix}\!\!\cdots O^{\delta-} \right]^{\ddagger} \longrightarrow \text{products}$$

53

On the other hand, cyclohexanone reacts nearly 53-fold as fast as cyclopentanone, even though their activation entropies are about the same. The different reaction rates are due to the different activation enthalpies ($\Delta H^{\ddagger} = 1.1$ and 4.0 kcal mol^{-1}, respectively). In the case of cyclopentanone, the rehybridisation of the carbonyl atom from sp^2 to sp^3 will increase the Pitzer strain[134] in the five-membered ring (increase of eclipsing repulsions). Evidently, no eclipsing repulsions will occur in the addition product of cyclohexanone. It must be observed that in carbonyl additions to cyclanones the reactivity is reversed with respect to the reactivity of cyclanyl derivatives in nucleophilic substitution. This is explained by the fact that in substitution reactions an $sp^3 \rightarrow sp^2$ rehybridisation is involved in the rate-determining step, whereas in carbonyl additions the reverse change occurs ($sp^2 \rightarrow sp^3$).

[132] B. M. Anderson and W. P. Jenks, *J. Amer. Chem. Soc.*, **82**, 1773 (1960).
[133] F. P. Price and L. P. Hammett, *J. Amer. Chem. Soc.*, **63**, 2387 (1941).
[134] J. D. Kemp and K. S. Pitzer, *J. Chem. Physics*, **4**, 749 (1936); K. S. Pitzer, *Science*, **101**, 672 (1945).

10.4.2 *Carbanion additions to carbonyl groups*

Carbanion additions to carbonyl groups are involved in a series of reactions of well-known applicability in the laboratory and technological practice. Let us thus mention the condensation reactions of aldehydes and ketones [135], the Claisen condensation of esters [136], Grignard-reagent additions to carbonyl compounds [137, 138], etc.

Condensation reactions (aldolic and crotonic condensations of ketones and aldehydes, Claisen condensations of esters) are closely related by their mechanisms. In all instances, a reactive nucleophile is generated in the first stages of the reaction, which then adds to a carbonyl group. In the reactions of aldehydes and ketones the nucleophile may be generated both under base (10.39) and under acid catalysis (10.40):

$$R^1CH_2{-}COR^2 + B\colon \rightleftharpoons R^1\bar{\ddot{C}}H{-}COR^2 + BH^+ \tag{10.39}$$

$$R^1{-}CH_2{-}C(=O)R^2 + L^+ \rightleftharpoons R^1CH_2{-}\overset{+}{C}(OL)R^2 \rightleftharpoons R^1CH{=}C(OL)R^2 \tag{10.40}$$

Base catalysis is most frequently encountered in ester condensations, even if catalysis by strong Lewis acids [139] is not excluded. This can be explained by the lower reactivity of the carbonyl group in the carboxylic acid derivatives, requiring a nucleophile stronger than the usual ones. The following steps may be involved in aldol condensation:

$$CH_3COR + B\colon \underset{k_{-1}}{\overset{k_1}{\rightleftharpoons}} \underset{\mathbf{54}}{\bar{C}H_2{-}COR} + BH^+ \tag{10.41}$$

$$CH_3{-}C(=O)R + \bar{C}H_2COR \underset{k_{-2}}{\overset{k_2}{\rightleftharpoons}} \underset{\mathbf{55}}{CH_3{-}C(O^-)(R){-}CH_2COR} \tag{10.42}$$

[135] A. T. Nielsen and W. J. Houlihan, *Org. Reactions*, **16**, 1 (1968).

[136] C. R. Hauser and B. E. Hudson, Jr., *Org. Reactions*, **1**, 266 (1942).

[137] E. C. Ashby, *Quart. Rev. Chem. Soc.*, **21**, 259 (1967).

[138] E. C. Ashby, J. Laemmle and H. M. Neumann, *Accounts Chem. Res.*, **7**, 272 (1974).

[139] H. Meerwein, *Ber. dtsch. chem. Ges.*, **66**, 411 (1933); M. J. S. Dewar, *The Electronic Theory of Organic Chemistry*, Oxford University Press (1949), Chapter 7.

$$55 + BH^+ \underset{k_{-3}}{\overset{k_3}{\rightleftharpoons}} CH_3\underset{\underset{R}{|}}{\overset{\overset{OH}{|}}{C}}-CH_2COR + B: \qquad (10.43)$$

56

Assuming that carbanions **54** are the reactive intermediates and that equilibrium (10.43) is rapid, we shall apply the steady-state treatment [140], to find equation (10.44):

$$\text{rate} = -\frac{d[RCOCH_3]}{dt} = \frac{k_1k_2[RCOCH_3]^2\,[B:]}{k_{-1}[BH^+] + k_2[RCOCH_3]} \qquad (10.44)$$

The rate equation (10.44) may be simplified if $k_{-1}[BH^+] \ll k_2[RCOCH_3]$. Thus, if re-protonation of the carbanion is slow as compared to the carbanion addition to the carbonyl double bond, the equation is reduced to:

$$\text{rate} = k_1[RCOCH_3][B:] \qquad (10.45)$$

Second-order kinetics (first-order in aldehyde and in catalyst) have been found in the aldol condensation of acetaldehyde in concentrated solutions. Moreover, the aldol formed has no deuterium if the reaction is performed in deuterated water [141]. Thus, protonation of the carbanion (k_{-1}) must be slower than the addition of the carbanion to the carbonyl group. In more diluted solutions of acetaldehyde, deuterium is incorporated in the molecule of aldol [142], and a more complex rate equation is obeyed.

If protonation is more rapid than the addition of the carbanion to the carbonyl double bond ($k_{-1}[\overset{+}{B}H] > k_2[RCOCH_3]$), equation (10.44) becomes:

$$-\frac{d[RCOCH_3]}{dt} = \frac{k_1k_2}{k_{-1}K_B}[RCOCH_3]^2\,[OH^-] \qquad (10.46)$$

where K_B is the equilibrium constant of the reversible reaction:

$$B: + HOH \rightleftharpoons BH^+ + HO^-$$

The rate equation (10.46) has been found in the case of acetone [135] (and other ketones). The enolate ion of acetone prefers thus to combine with the conjugate acid BH^+, instead of condensing with another

[140] S. W. Benson, *The Foundation of Chemical Kinetics*, McGraw-Hill, New York (1960), p. 573.

[141] K. F. Bonhoeffer and W. D. Walters, *Z. physik. Chem.*, **181A**, 441 (1938).

[142] A. Broche and R. Gibert, *Bull. Soc. chim. France*, 131 (1955).

acetone molecule. This is partly due to the lower reactivity of the carbonyl group, but also to a higher stability of the enolate ion:

$$\underset{\mathbf{57}}{CH_3{\rightarrow}\underset{|}{\overset{}{C}}(O^-){=}CH_2} \quad \text{as compared with} \quad \underset{\mathbf{58}}{CH_2{=}C(O^-){-}H}$$

Being more stable, acetone enolate will be able to discern between different electrophiles.

The condensation of carbonyl compounds in acid medium involves also a nucleophilic addition to the carbonyl group. In this case the reagent is the less nucleophilic enol, but the substrate may be the more reactive protonated carbonyl compound **52.**

The Claisen condensation of esters[136] involves a carbanion attack at an unreactive carbonyl group:

$$B\colon R^1CH_2COOR^2 \underset{k_{-1}}{\overset{k_1}{\rightleftharpoons}} BH^+ + \underset{\mathbf{59}}{\bar{\ddot{C}}H(R^1){-}C(OR^2){=}O \longleftrightarrow CH(R^1){=}C(OR^2){-}O^-} \text{ etc.} \qquad (10.47)$$

$$R^1CH_2{-}C(=O)(OR^2) + \bar{\ddot{C}}HR^1COOR^2 \rightleftharpoons \underset{\mathbf{60}}{R^1CH_2{-}C(O^-)(OR^2){-}CHR^1{-}COOR^2} \qquad (10.48)$$

$$\underset{\mathbf{60}}{R^1CH_2{-}C(O^-)(OR^2){-}CHR^1{-}COOR^2} \rightleftharpoons R^1CH_2COCHR^1COOR^2 + R^2O^- \qquad (10.49)$$

$$R^1CH_2COCHR^1COOR^2 + R^2O^- \rightleftharpoons \underset{\mathbf{61}}{R^1CH_2CO\bar{\ddot{C}}R^1COOR^2} + R^1OH \qquad (10.50)$$

Esters are by far less acidic than ketones[143], and thus stronger bases are necessary in order to obtain sufficient concentrations of the carbanion **59**. Moreover, the addition intermediate **60** has a high

[143] D. J. Cram, *Fundamentals of Carbanion Chemistry*, Academic Press, New York (1965).

tendency to eliminate the carbanion and regenerate the stable carboxyl group; thus, equilibrium (10.48) does not favour the condensation. However, the β-keto ester formed is a stronger carbon acid than is the original ester in equilibrium (10.47). The formation of the delocalised carbanion **61** may shift all the equilibria in the direction of the condensation product. In practice, reaction (10.50) is also shifted in the desired direction by performing the condensation of the ester with sodium metal. Traces of alcohol may be sufficient to start the reaction, forming with sodium the alkoxide. Alcohol is then continuously formed in the process and the equilibrium is shifted by the consumption of the alcohol by the sodium added.

10.4.3 *Additions of Grignard reagents*

The addition of Grignard reagents to carbonyl and carboxyl derivatives is largely used in organic synthesis. Despite its apparent simplicity, it has been recognised[137, 138] that the reaction mechanism is quite complex. A first complication regards the reagent itself. It is well known now that the carbon-magnesium bond in Grignard compounds is highly polar, the alkyl-magnesium halides RMgX differing in their reactivity from the symmetric derivatives RMgR. It has been since long observed, by Schlenk and Schlenk[144], that in ether solution the following species are present in equilibrium ("Schlenk equilibrium"):

$$2\mathrm{RMgX} \rightleftharpoons \mathrm{RMgR} + \mathrm{MgX_2}$$

Both magnesium species, RMgX and RMgR, may attack the carbonyl compound and thus in any mechanism study this equilibrium must be investigated. Actually, the above formulation of the equilibrium is an over-simplification. The monomeric species RMgX and RMgR may equilibrate with associated species. Solvent molecules are also involved, the desolvation and resolvation processes interfering sometimes seriously with the chemical equilibration.

Another complication of the reaction mechanism arises from the fact that the magnesium species (MgX_2 included) may form with the carbonyl derivative various complexes, which may either collapse into the addition product, the magnesium alkoxide, or simply remain as useless side products. Moreover, the complex itself may be a reagent for the carbonyl group[145] (Scheme 5).

$$\text{A. } \mathrm{R^1MgR^2} + \rangle\mathrm{C{=}O} \overset{K}{\rightleftharpoons} \rangle\mathrm{C{=}O}\cdots\mathrm{MgR^1R^2} \xrightarrow{k_1} \text{Products}$$

[144] W. Schlenk and W. Schlenk, *Ber. dtsch. chem. Ges.*, **62**, 920 (1929).
[145] C. G. Swain and H. B. Boyles, *J. Amer. Chem. Soc.*, **73**, 870 (1951).

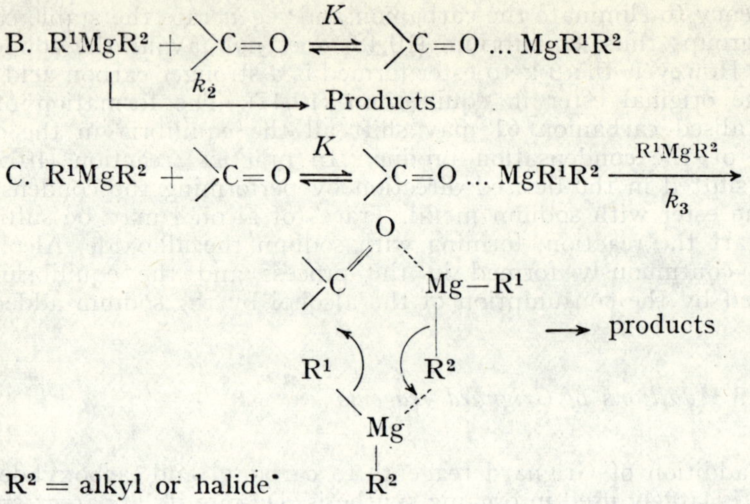

$$\text{B. } R^1MgR^2 + \rangle C{=}O \overset{K}{\rightleftharpoons} \rangle C{=}O \ldots MgR^1R^2$$
$$\xrightarrow{k_2} \text{Products}$$

$$\text{C. } R^1MgR^2 + \rangle C{=}O \overset{K}{\rightleftharpoons} \rangle C{=}O \ldots MgR^1R^2 \xrightarrow[k_3]{R^1MgR^2} \text{products}$$

R^2 = alkyl or halide

Scheme 5

Possible mechanisms in additions of Grignard reagents

The later stages of the reaction contribute in their turn to the complexity of the process. Indeed, the alkoxide formed may associate with other magnesium species, the resulting complexes being in their turn potential reagents. According to Ashby and his co-workers [138, 146], a complete reaction scheme, which neglects however the later stages, is the following:

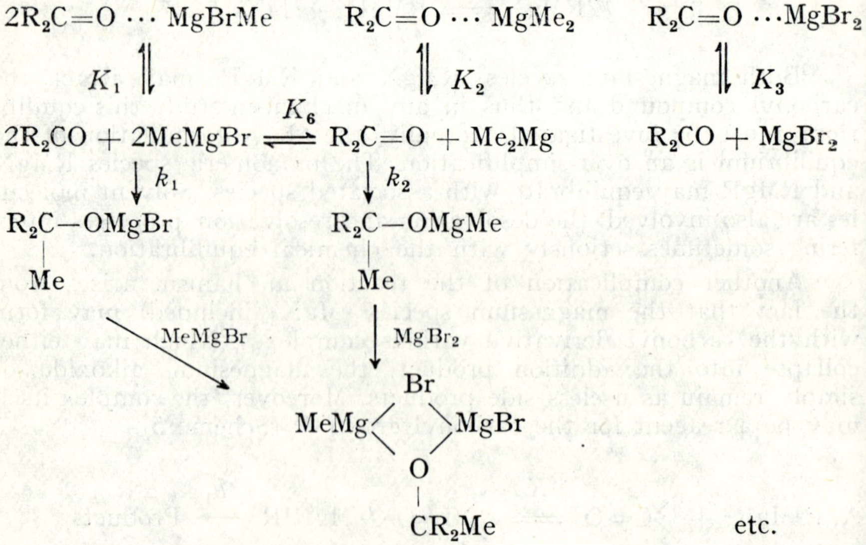

$$2R_2C{=}O \cdots MgBrMe \qquad R_2C{=}O \cdots MgMe_2 \qquad R_2C{=}O \cdots MgBr_2$$
$$K_1 \updownarrow \qquad K_2 \updownarrow \qquad K_3 \updownarrow$$
$$2R_2CO + 2MeMgBr \overset{K_6}{\rightleftharpoons} R_2C{=}O + Me_2Mg \qquad R_2CO + MgBr_2$$
$$\downarrow k_1 \qquad \downarrow k_2$$
$$R_2C(Me){-}OMgBr \qquad R_2C(Me){-}OMgMe$$
$$\xrightarrow{MeMgBr} \qquad \downarrow MgBr_2$$
$$MeMg(\mu\text{-}Br)(\mu\text{-}OCR_2Me)MgBr \qquad \text{etc.}$$

Scheme 6

146 E. C. Ashby, H. M. Newman, F. W. Walker, J. Laemmle and Li-C. Chao, *J. Amer. Chem. Soc.*, **95**, 3330 (1973).

Despite the complexity of the reaction scheme, a kinetic study[146] has confirmed this mechanism. Moreover, spectrometer analysis has allowed the observation of the complexes involved, and their disappearance from the reaction medium has been followed.

10.4.4 *Steric requirements and selectivities in carbanion additions*

It is a well known fact[147] that with highly hindered ketones, the addition of Grignard reagents may not be observed. Both enolisation and reduction of the carbonyl compound may occur instead:

$$R^1_2C{=}O + H{-}CR^2_2{-}CR^3_2{-}MgX \longrightarrow R^1_2CH{-}OMgX + R^2_2C{=}CR^3_2 \qquad (10.51)$$

$$R^1_2CH{-}CO{-}R^2 + R^3MgX \longrightarrow R^1_2C{=}C(OMgX){-}R^2 + R^3H \qquad (10.52)$$

Only these abnormal processes are observed if both ketone and Grignard reagent have large steric demands. Thus, Mosher and his co-workers[147] observed 30% enolisation and 70% reduction, with no traces of addition product, when di-isopropylketone was treated with magnesium isopropyl chloride. Evidently, in the case of the enolisation reaction, the Grignard reagent behaves as a base extracting a proton from the activated position in ketone. The carbanion formed may further undergo other reactions.

Another example of Grignard reactions prevented sterically is found in the cyclobutanone series. Triphenylcyclobutanone **62** yields with phenylmagnesium bromide the acyclic ketone **64**[148]. Proton abstraction leading to carbanion **63** is thus preferred to the nucleophilic addition to the sterically hindered ketone **62**.

62 $\xrightarrow[-RH]{RMgX}$ **63** $\longrightarrow$ PhCH=CH—C(=O)—C̈Ph₂⁻ $\longrightarrow$ PhCH=CH—C(=O)—CHPh₂ **64**

[147] J. Miller, G. Gregoriou and H. S. Mosher, *J. Amer. Chem. Soc.*, **83**, 3966 (1961).
[148] M. W. Rathke and D. F. Sullivan, *J. Amer. Chem. Soc.*, **95**, 3050 (1973).

An interesting example of steric hindrance in a Claisen condensation has been observed recently. Thus, when lithium *N*-isopropylcyclohexylamide (LiICA) is treated with ethyl acetate in hexane at $-78\,°C$, only the lithium salt of ethyl acetoacetate is formed. However, t-butyl acetate affords under the same conditions, in good yields, the lithium salt **65**:

$$2CH_3COOBu^t + 2LiICA \xrightarrow[-78\,°C]{\text{hexane}} \underset{\mathbf{65}\ \text{(insoluble)}}{LiCH_2COOBu^t} +$$

$$\underset{\mathbf{66}\ \text{(soluble)}}{(R_2NH)_2LiCH_2COOBu^t}$$

The removal of the volatile fractions (solvent, di-isopropylamine) leads to 95–100% **65,** which is quite stable and has no tendency to form acetoacetic acid ester. However, with more reactive ketones, **65** reacts readily to form β-hydroxy esters (Reformatsky reaction). The selectivity shown by the lithium derivative in distinguishing between the ester and the ketone carbonyl group is entirely determined by steric effects of the tertiary group. Usually, the highly reactive carbanions **59** are less selective. Carbanions involved in aldol condensation, being more stable, are also more selective. Thus, the treatment of benzophenone with acetaldehyde, in presence of a base catalyst, affords only aldol and never 3,3-diphenyl-3-hydroxy-propanal **67.** The very low reactivity of benzophenone as compared to acetaldehyde determines this clear-cut selectivity of the reaction:

$$Ph_2CO + 2CH_3CHO + B{:} \longrightarrow Ph_2CO + CH_3CHO + BH^+ + \bar{\ddot{C}}H_2CHO$$

$$\underset{\mathbf{67}}{Ph_2C(OH)-CH_2CHO} \not\longleftarrow \quad \longrightarrow CH_3CH(OH)-CH_2CHO$$

It has been found [149] that the aldol synthesis may be directed towards the desired product by appealing to metallated Schiff-bases. The method was discovered in the course of a study of benzyne, generated from fluorobenzene and phenyl-lithium, in presence of

[149] G. Wittig and H. Reiff, *Angew. Chem.*, **80**, 8 (1968).

ethylamine. Beside the expected diethylaniline, α-phenyldiethylamine was also obtained:

$$\text{C}_6\text{H}_5\text{F} \xrightarrow[C_6H_6]{PhLi} \text{C}_6\text{H}_4 \text{ (benzyne)} \xrightarrow{LiNEt_2}$$

$$o\text{-H-C}_6\text{H}_4\text{Li} + \text{N(Et)}{=}\text{CH—Me} \longrightarrow \text{LiN(Et)—CH(Ph)—Me}$$

This reaction clearly involves a reduction, and the Schiff-base was assumed to be the hydride donor [150]. In order to check this hypothesis, benzophenone was treated with lithium diethylamine, in the hope that benzhydrol will be obtained beside the Schiff-base **68.** Actually, benzhydrol resulted, but instead of **68,** adduct **69** was formed:

$$Ph_2C{=}O + LiN(Et)CH_2{-}CH_3 \longrightarrow [Ph_2\overset{+}{C}{-}O{-}Li{-}N(Et){-}CH(H){-}CH_3] \longrightarrow$$

$$\underset{\mathbf{68}}{CH_3{-}CH{=}N{-}Et} + Ph_2CH{-}OLi$$

$$\mathbf{68} \xrightarrow{LiNEt_2} LiCH_2{-}CH{=}N{-}Et \xrightarrow{Ph_2CO} Ph_2C(OLi){-}CH_2{-}CH{=}N{-}Et \text{ (chelate, O–Li–N)} \xrightarrow{H_2O}$$

$$\xrightarrow{H_2O} \underset{\mathbf{69}}{Ph_2C(OH){-}CH_2{-}CH{=}N{-}Et} \text{ (O–H···N)} \longrightarrow Ph_2C(OH){-}CH_2CHO \qquad (10.53)$$

The hydrolysis of this adduct led to aldol **67.** The whole process affords a convenient method for directing the aldol synthesis, consisting in metallating a Schiff-base and treating the resultant reagent

[150] G. Wittig, H.-J. Schmidt and H. Renner, *Chem. Ber.*, **95**, 2377 (1962).

with the carbonyl compound [149]. The method, confirmed in other laboratories [151], makes use of the poorer electrophilic character of Schiff-bases, as compared to the carbonyl derivatives. On the other hand, lithium derivatives of Schiff-bases must be in their turn less nucleophilic than the dissociated carbanions encountered in the usual aldol condensation. In that respect, reaction (10.53) presents higher selectivity than the aldol condensation, the reagent preferring as reaction partner benzophenone (poor electrophile as compared to other carbonyl compounds, but reactive electrophile as compared to Schiff-bases).

10.4.5 *Stereochemistry of carbonyl addition reactions*

The carbonyl group has a symmetry plane and thus may be attacked on either side of it. This is quite clear in the case of monoatomic substituents at the carbonyl group, but is not evident for polyatomic groups. However, considering that most rotations about single bonds have an energy barier of 3—10 kcal mol^{-1} [152], and chemical reactions need higher activation energies, one may consider that any polyatomic group, lacking a chirality element, will adopt the cone symmetry (corresponding to monoatomic ligands), as a result of its rotation. In such cases, the symmetry plane of the carbonyl group is also a symmetry plane for the entire molecule and the synthesis will be symmetric (Figure 10.4.). Nevertheless, if one of the substituents at the carbonyl group has a chirality element, rotations about its bonds will not cancel out the non-equivalence between the two sides of the carbonyl group. Evidently, the point group C_1 must be assigned

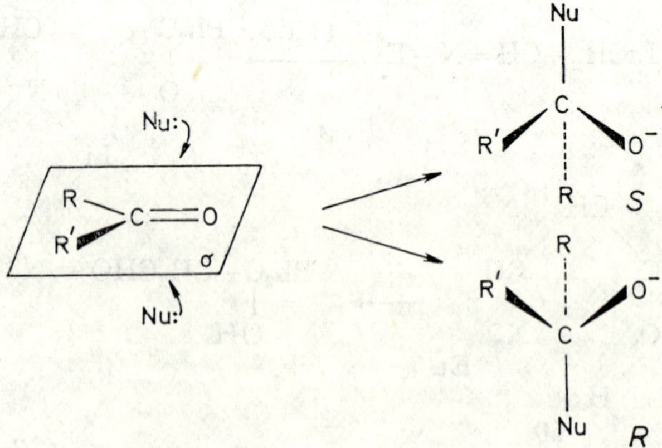

Figure 10.4 Stereochemistry of addition reactions to an achiral carbonyl compound.

[151] H. O. House, D. S. Crumrine, A. Y. Teranishi and H. D. Olmstead, *J. Amer. Chem. Soc.*, **95**, 3310 (1973).

[152] D. J. Cram and F.A. Abdelhafez, *J. Amer. Chem. Soc.*, **74**, 5828, 5846 (1952).

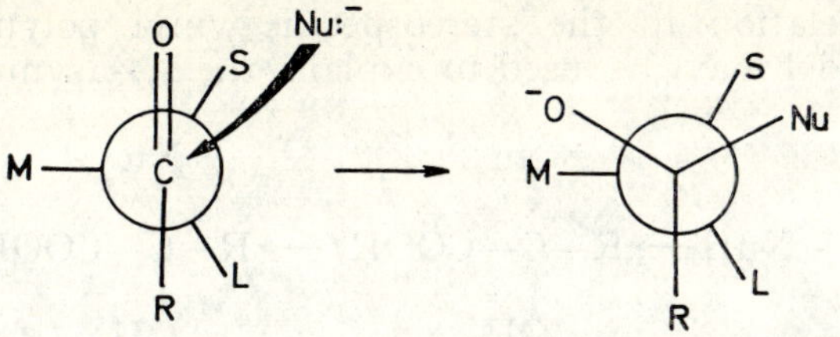

Figure 10.5 Cram's rule; attack on the less hindered side of a carbonyl compound. The designation of the substituents is: L = large, M = medium and S = small.

to such compounds. Non-equivalence of the nucleophilic attack at the carbonyl group should be expected, and this will correspond to a stereospecific addition. This non-equivalence of a local symmetry plane in additions to multiple bonds constitutes the basis of most *asymmetric syntheses*[153]. There are however different models explaining an *asymmetric induction* at a prochiral atom, the first of them being due to Cram and Abdelhafez[152] (*Cram's rule*). This model, later known as *open-chain model*[154], assumes that the stereospecificity of the reaction is determined by the bulk of the substituents in the neighbourhood of the reaction site. The stereochemical rule[153] was formerly based on examples of 1,2-*asymmetric induction* (vicinal chiral centre). According to this model, substituents at a neighbouring centre may be designated as large (L), moderate (M) and small (S). The attack at the prochiral centre occurs from the less hindered side (Figure 10.5), in the most stable conformation of the substrate. Evidently, this model may also apply to carbonyl additions to cyclanones. The open-chain model has been confirmed in numerous works both for 1,2-[155–158] and 1,3-asymmetric induction [154, 159–161]. The 1,3-asymmetric induction (Figure 10.6) is particularly interesting

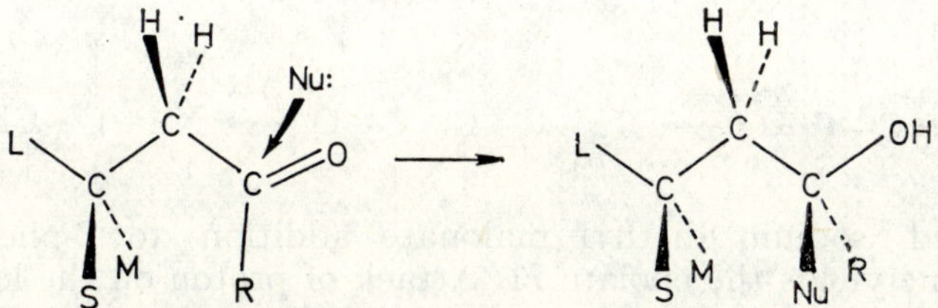

Figure 10.6 The 1,3 asymmetric induction.

[153] P. D. Ritchie, *Asymmetric Synthesis and Asymmetric Induction*, Oxford University Press (1933); J. D. Morrison and H. S. Mosher, *Asymmetric Organic Reactions*, Prentice-Hall, Englewood-Cliffs, New York (1971).

[154] D. J. Cram, *J. Chem. Educ.*, **37**, 317 (1960).

[155] D. J. Cram and J. E. McCarty, *J. Amer. Chem. Soc.*, **76**, 5740 (1954).

[156] D. J. Cram and K. R. Kopecky, *J. Amer. Chem. Soc.*, **81**, 2748 (1959).

[157] J. Mathieu and J. Weill-Raynal, *Bull. Soc. chim. France*, 1211 (1968).

[158] J. Sicher, M. Svoboda, M. Hrda, J. Rudinger and F. Sorm, *Collect. Czech. Chem. Commun.*, **18**, 487 (1953).

[159] J. Furakawa and T. T. Tsuruta, *J. Polymer Sci.*, **46**, 275 (1959).

[160] M. Roha, *Advan. Polymer Sci.*, **4**, 353 (1966).

[161] T. J. Leitereg and D. J. Cram, *J. Amer. Chem. Soc.*, **90**, 4011 (1968).

due to its relation to the stereospecific vinyl polymerisation[162]. The same model may be used to explain the 1,4-asymmetric induction:

$$\mathrm{RCOCOOR^* + NuH \longrightarrow R{-}\underset{OH}{\overset{Nu}{C}}{-}COOR^* \longrightarrow R{-}\underset{OH}{\overset{Nu}{C}}{-}COOH} \qquad (10.54)$$

One chirality element present in the molecule — mixture of diastereomers — excess of one enantiomer (only one chirality centre)

1,4-Asymmetric induction has been used for a long time in asymmetric syntheses, such as (10.54)[163]. Prelog[164] has studied such 1,4-inductions of asymmetry. As in the case of other α-dicarbonyl derivatives, the planar conformation with the carbonyl groups in *anti* is also the most stable conformation of α-ketoesters **70.** The largest group bonded to the C-4 atom determines the direction of the attack from the less hindered side:

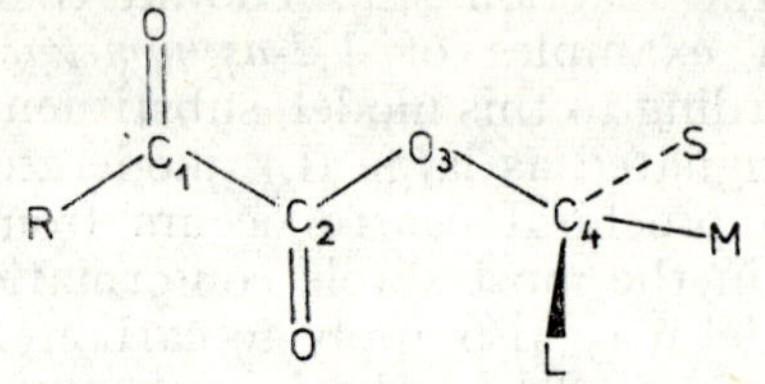

R, C_1, C_2, O_3, C_4, S, M, L, O, O

70

Cram's rule seems also to govern 1,4-additions to α,β-unsaturated carbonyl derivatives, such as Michael condensation[165] (Nu: = carbanion):

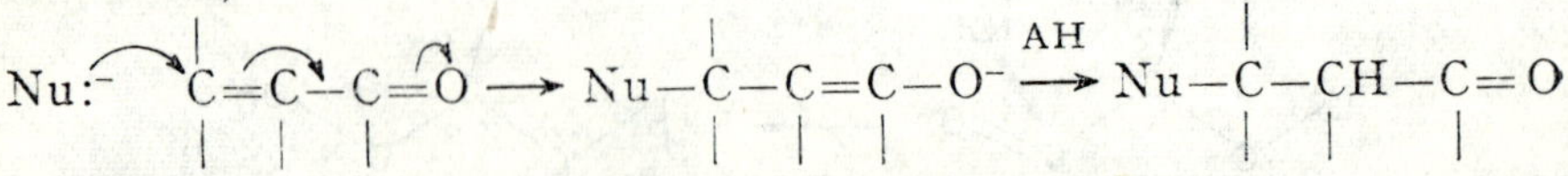

$$\mathrm{Nu{:}^- \; \overset{|}{\underset{|}{C}}{=}\overset{|}{\underset{|}{C}}{-}\underset{|}{C}{=}O \longrightarrow Nu{-}\overset{|}{\underset{|}{C}}{-}\underset{|}{C}{=}\underset{|}{C}{-}O^- \xrightarrow{AH} Nu{-}\overset{|}{\underset{|}{C}}{-}\underset{|}{CH}{-}\underset{|}{C}{=}O}$$

Indeed, sodium diethyl malonate addition to 2-phenylcyclohexen-3-one yields the enolate **71.** Attack of proton on the less hindered side of the plane of the mesomeric ion (remote from the bulky malonate group) affords the *cis*-isomer:

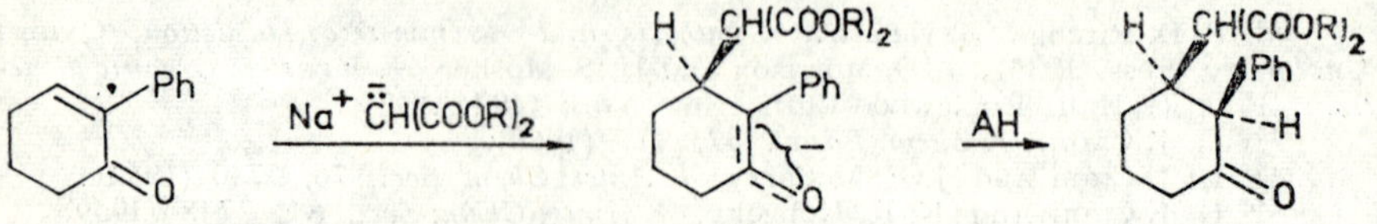

Ph, O — $Na^+\ \bar{\bar{C}}H(COOR)_2$ → H, $CH(COOR)_2$, Ph, O — **71** — AH → H, $CH(COOR)_2$, Ph, H, O

[162] M. Goodman, *Topics in Stereochemistry*, **2**, 73 (1967).
[163] A. McKenzie, *J. Chem. Soc.*, **85**, 1249 (1904).
[164] V. Prelog, *Helv. Chim. Acta*, **36**, 308 (1953).
[165] E. D. Bergmann, D. Ginsburg and R. Pappo, *Org. Reactions*, **10**, 179 (1959).

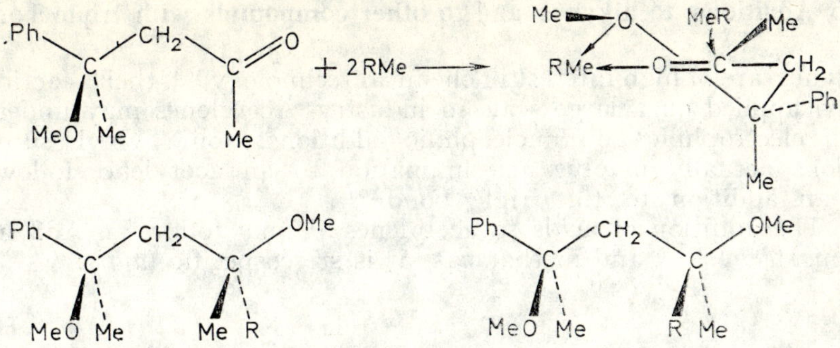

Figure 10.7 The cyclic model in asymmetric induction.

It should be observed that this isomer is not the more stable one. The protonation is thus a kinetically-controlled process.

The cyclic model [154, 166, 167] considers that one isomer may be preferentially formed if it yields stable cyclic chelates under the reaction conditions. This seems to be the case of the organometallic compounds additions to carbonyl derivatives with an OH or an OR substituent in the β-position (Figure 10.7).

It has been shown [166] that other models may also explain the reaction stereochemistry.

The dipolar model [168]. If one of the substituents at the chiral centre is a strongly polar group, the dipolar model explains better the stereochemistry of the reaction than does the open-chain model. The addition to the carbonyl group occurs in the conformation with lower dipole interaction, namely from the less hindered side (Figure 10.8).

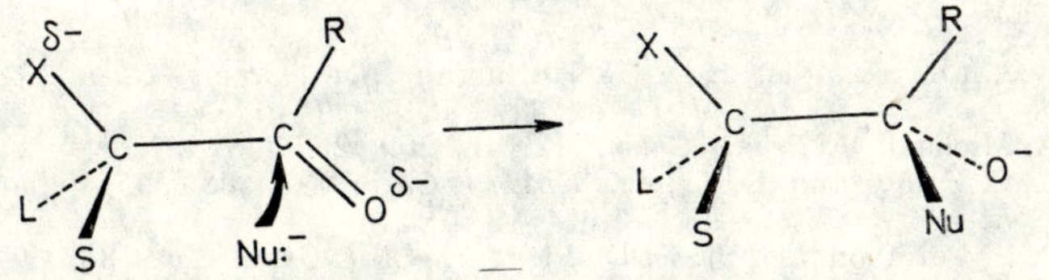

Figure 10.8 The dipolar model in asymmetric induction.

[166] D. J. Cram and D. R. Wilson, *J. Amer. Chem. Soc.*, **85**, 1245 (1963).

[167] T. J. Leiterag and D. J. Cram, *J. Amer. Chem. Soc.*, **90**, 4019 (1968).

[168] J. W. Cornforth, R. H. Cornforth and K. K. Mathews, *J. Chem. Soc.*, 112 (1959).

[169] See for example: J. A. Nieuwland and R. R. Vogt, *The Chemistry of Acetylene*, Reinhold, New York (1945); S. Patai (editor), *The Chemistry of Alkynes*, Interscience, New York.

10.5 Additions to alkynes and to other compounds with triple bonds

Alkynes are of high interest in chemical technology [169], their reactions being applied on a large scale in industry. Acetylenes may undergo both electrophilic and nucleophilic additions. Some vinyl substitutions actually involve an elimination to an acetylene, followed by an addition to the triple bond [170].

The addition of acids to acetylenes [171] may follow an Ad3 mechanism [109, 172, 173] and in that case it is stereospecific (*anti*):

$$-C\equiv C- + 2HX \longrightarrow \underset{\mathbf{72}}{\overset{H\cdots X^{\delta-}}{{}^{\delta-}X\cdots \rangle C\equiv C\langle}} \longrightarrow \overset{\quad\quad H}{\underset{X}{\rangle C=C \langle}}$$

Probably the best known electrophilic addition to acetylene is the hydration in presence of sulphuric acid and mercury sulphate [174]. The mechanism* suggested for this reaction involves an electrophilic attack of mercury:

$$HC\equiv CH + XHg^{+} \longrightarrow \underset{\mathbf{73}}{XHgCH=CH^{+}} \longrightarrow$$

$$XHgCH=CHOH \xrightarrow[H_2O]{H^{+}} CH_3CHO \qquad (10.55)$$

The key intermediate is the vinyl cation **73.** The formation of mercury derivatives was observed even in the earlier works of Kutscherov [174].

Additions of halogens are also found in the industrial practice, as for example the manufacturing of tetrachloroethane and trichloroethylene. However, these processes seem to be homolytic.

Strong nucleophiles, such as alkoxide and thiophenoxide ions [176, 177] may add to the triple bond, however with different

[170] See, for example, H. M. R. Hoffmann, *Ann. Repts. Chem. Soc.*, **648,** 152 (1967).

[171] A. Michael, *J. Prakt. Chem.*, (2) **46**, 209, 289 (1895).

[172] R. C. Fahey and D. J. Lee, *J. Amer. Chem. Soc.*, **88**, 5555 (1966); **89**, 2780 (1967).

[173] P. E. Peterson and J. E. Duddey, *J. Amer. Chem. Soc.*, **88**, 4990 (1966).

[174] M. G. Kutscherov, *Ber. dtsch. chem. Ges.*, **14**, 1532, 1540 (1881); **17**, 13 (1884).

* Mechanism (10.55) seems to have been proposed by Nenitzescu (*Organic Chemistry*, vol. 1, Bucharest, 6th edn. (1966), p. 639) and differs in some respects from other possible schemes (Ref. 175).

[175] See, for example, O. A. Reutov and I. P. Beletskaya, *Reaction Mechanisms of Organometallic Compounds*, North-Holland, Amsterdam (1968), p. 270.

[176] D. E. Jones, R. O. Morris, C. A. Vernon and R. F. M. White, *J. Chem. Soc.*, 2349 (1960).

[177] F. Montari, *Gazz. Chim. Ital.*, **86**, 747 (1956); L. K. Montgomery and L. E. Applegate, *J. Amer. Chem. Soc.*, **89**, 2952 (1967).

stereochemistry[176]. The *anti*-addition, observed in most cases, may again be explained by an Ad3 mechanism.

$$\text{Me}-\text{C}\equiv\text{C}-\text{COOR} + \text{PhS}^- \xrightarrow{(\text{ETOH})} \underset{\textbf{74}}{\overset{\text{ROOC} \qquad \overset{\delta-}{\text{SPh}}}{\underset{\text{EtOH} \qquad \text{Me}}{>\text{C}\equiv\text{C}<}}}$$

$$\longrightarrow \underset{\textit{anti}\text{-adduct}}{\overset{\text{ROOC} \qquad \text{SPh}}{\underset{\text{H} \qquad \text{Me}}{>\text{C}=\text{C}<}}}$$

In their first paper on the Ad3 mechanism, Fahey and his co-workers[108] labelled this mechanism Ad_E3, considering that the electrophilic character is essential in the process. Later[109], however, they recognised that the mechanism is more general, with either the nucleophilic or the electrophilic character more developed. In this respect the works of Martin, Bacaloglu and their co-workers[178–181] are of special interest. These authors have studied the reactions of arylcyanates with amines and phenols and have observed that the addition of amines is catalysed by phenols, whereas the addition of phenols is catalysed by amines:

$$\text{Ar}^1\text{O}-\text{C}\equiv\text{N} + \text{R}_2\text{NH} \xrightarrow{\text{Ar}^2\text{OH}} \text{Ar}^1\text{O}-\underset{\text{NH}}{\underset{\|}{\text{C}}}-\text{NR}_2 \qquad (10.56)$$

$$\text{Ar}^1\text{O}-\text{C}\equiv\text{N} + \text{Ar}^2\text{OH} \xrightarrow{\text{R}_2\text{NH}} \text{Ar}^1\text{O}-\underset{\text{NH}}{\underset{\|}{\text{C}}}-\text{OAr}^2 \qquad (10.57)$$

The reactions were followed either by potentiometric titration or spectrophotometrically. Only very low activation enthalpies were found ($\Delta H^{\ddagger} = 9.6$ kcal mol^{-1} and 8.8 kcal mol^{-1}, for amine and phenol additions, respectively), the reactions being controlled by their high negative activation entropies[178] ($\Delta S^{\ddagger} = -42.5$ and -43.6 e.u.,

[178] D. Martin, *Z. Chem.*, **7**, 123 (1967).

[179] D. Martin, K. Nadolski, R. Bacaloglu and I. Bacaloglu, *J. Prakt. Chem.*, **313**, 58 (1971).

[180] I. Bacaloglu, K. Nadolski, R. Bacaloglu and D. Martin, *J. Prakt. Chem.*, **313**, 839 (1971).

[181] I. Bacaloglu, R. Bacaloglu, D. Martin and K. Nadolski, *J. Prakt. Chem.*, **316**, 529 (1974).

respectively). Cyclic transitions states **75** and **76** were thus suggested:

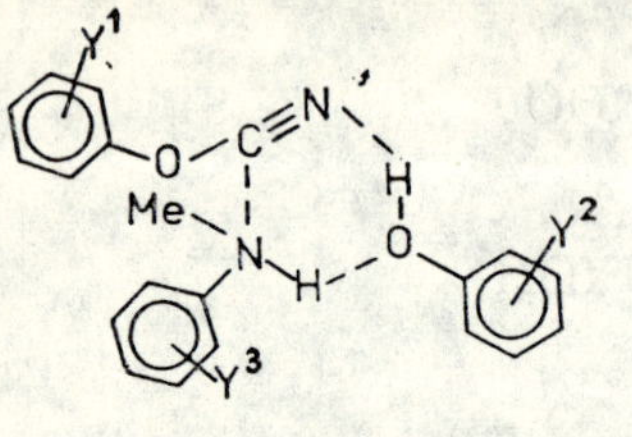

75 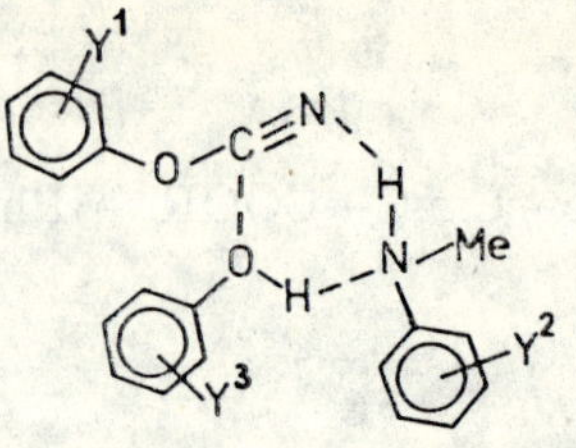 **76**

Later works have demonstrated that the reaction rates are correlated with the Hammett parameters of the substituents either in the aromatic ring of the arylcyanates, or in the phenols, or in the *N*-methylanilines used. The ρ parameters demonstrate both the electrophilic and the nucleophilic character of the reaction, the latter being however more pronounced.

The Ad3 mechanism may thus refer to a very large number of transition states, having as limits the ideal electrophilic and nucleophilic mechanisms. The Ad3 mechanism is probably the most common heterolytic mechanism for addition reactions.

10.6 Polyaddition reactions

Polymer formation involves a polyaddition to multiple bonds. Such processes may be either homolytic or heterolytic. The following heterolytic polymerisation mechanisms are possible:

(a) *Cationic polymerisation.* The intermediates are carbenium ions, which may add to an alkene; the process is repeated then several times to yield the polymer (or oligomer):

Initiation:

Carbenium ion precursor $\longrightarrow R^{1+}$

$$R^{1+} + CH_2{=}CHR^2 \longrightarrow R^1CH_2{-}\underset{\displaystyle R^2}{\underset{|}{CH^+}}$$

Chain propagation:

$$R^1CH_2{-}\underset{\displaystyle R^2}{\underset{|}{CH^+}} +$$

$$CH_2{=}CHR^2 \longrightarrow \ldots \longrightarrow \ldots R^1CH_2\underset{\displaystyle R^2}{\underset{|}{CH}}(CH_2{-}\underset{\displaystyle R^2}{\underset{|}{CH}})_nCH_2{-}\underset{\displaystyle R^2}{\underset{|}{CH^+}}$$

77

Chain termination:

$$\mathbf{77} \xrightarrow{-H^+} \text{Polymerised alkene}$$

Actually, the reaction is more complex than shown above. Wagner-Meerwein rearrangements may accompany the growth of the macromolecular chain. The elimination involved in the chain termination may be accompanied by intermolecular hydride shifts leading to saturated hydrocarbons and polyenes (or to their delocalised cations, respectively). Thus, the polymerisation of isobutylene with sulphuric acid [182] yields a mixture of alkanes (50%, C_{10} to C_{18}) and cations (in the sulphuric layer, 50%). The reaction products are formed by alkylation of the alkenes with the carbenium ions, dealkylation processes, hydride transfer reactions, cyclisation reactions and eliminations. The process may however be controlled at lower temper atures, using other types of catalysts.

(b) *Anionic polymerisation.* It differs from the cationic polymerisation in that the reaction intermediates are anions. Strong nucleophiles (alkoxides, $NaNH_2$) may be used as reaction initiators. The choice monomers for anionic polymerisation are the electron-deficient alkenes such as acrylonitrile. Nucleophilic alkenes, such as styrene [183], yield with alkoxides dimers and polymers. In the case of styrene it has been demonstrated [184] that the dimers result in part by a base catalysis, the products of this reaction being the indene derivative **78** and 1,3-diphenylbutenes (two isomers).

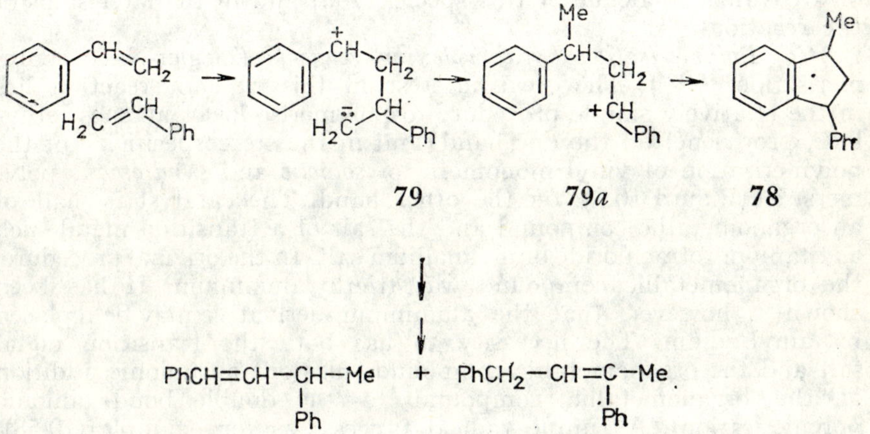

PhCH=CH₂ + H₂C=CH–Ph → PhCH⁺–CH₂–C̈H⁻(Ph) (**79**) → PhCH(Me)–CH₂–C⁺H–Ph (**79a**) → 1-methyl-3-phenylindane (**78**)

↓

PhCH=CH–CH(Ph)–Me + PhCH₂–CH=C(Ph)–Me

[182] N. C. Deno, D. B. Boyd, J. D. Hodge and C. U. Pittman, Jr., *J. Amer. Chem. Soc.*, 86, 1745 (1964).

[183] K. R. Kopecky and S. Evani, *Can. J. Chem.* 47, 4047 (1969).

[184] V. Vîntu, personal communication; M. L. C. F. de Rozas, Ph. D. Thesis, Inst. Petrol, Gaze, Geologie, Bucharest (1973).

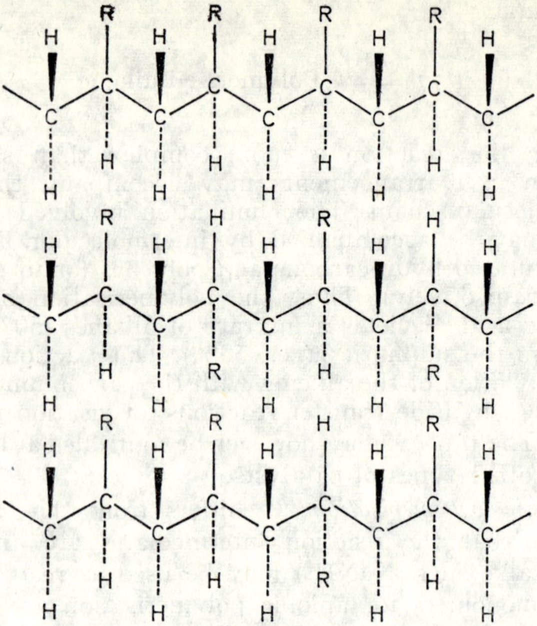

Figure 10.9 The tacticity of a polymer. A. Isotactic; B. syndiotactic; C. atactic.

A head-to-tail process must be assumed in the formation of both derivatives, very usual for the cationic polymerisation, but completely different from the radical pathway [183]. The base catalyst may be involved both in the formation of the intermediate **79** (or of a covalent analogue of this species, **79***a*) and in the later steps of the reaction.

(c) *Polymerisation by complexing reagents* (Ziegler-Natta polymerisation [185, 186]). The great interest in this type of reaction lies in the relatively simple procedure to polymerise lower alkenes (ethylene, propylene) on the one hand, and in the stereospecificity of the polymerisation of vinyl monomers to *isotactic* and *syndiotactic* polymers [162] (Figure 10.9), on the other hand. The catalyst is made of an organometallic compound and the salt of a transition metal such as titanium tetrachloride or a vanadium salt. In the original procedure, the organometallic compound was triethylaluminium. It has been shown[187], however, that the aluminium derivative may be replaced by amylsodium. The necessity to use both the transition metal salt and the organometallic compound rules out an anionic addition of the organometallic compound to the double bond (anionic polymerisation). A simple radical process, as for example (10.58),

[185] K. Ziegler, H. Holzkamp, H. Breil and H. Martin, *Angew. Chem.*, **67**, 541 (1955).

[186] G. Natta, P. Pino, P. Corradini, F. Danusso, E. Mantica, G. Mazzanti and G. Moraglio, *J. Amer. Chem. Soc.*, **77**, 1708 (1955).

[187] C. D. Nenitzescu, C. Huch and A. Huch, *Angew. Chem.*, **68**, 438 (1956).

may be suggested, considering that the nature of the organometallic compound is not essential[187] and that titanium derivatives $RTiCl_3$ have been isolated[188]:

$$RNa + TiCl_4 \longrightarrow RTiCl_3 \longrightarrow R\cdot + TiCl_3 \qquad (10.58)$$

However, the radical mechanism cannot explain the stereospecificity of the polymerisation of vinyl monomers such as propylene. Two further mechanisms have thus been advanced.

The di-metal mechanism[189] considers that two metal centres are necessary to allow the growth of the polymer chain. It is assumed that the active intermediate is formed from the aluminium compound with the titanium halide (**80**), having a structure similar to that of the dimers of organo-boron or -aluminium derivatives (bridged structure). The alkyl group will be attracted by both metal centres:

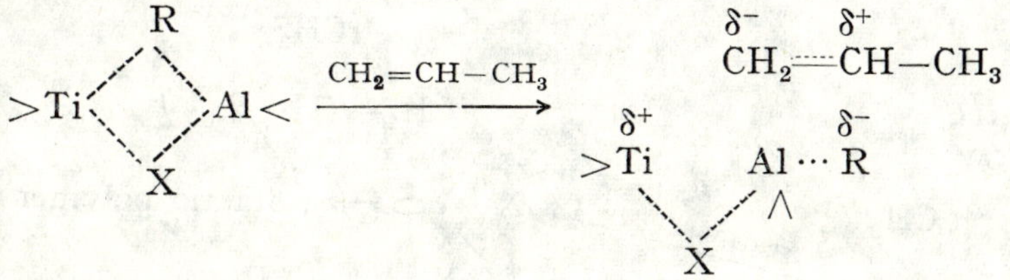

R

>Ti Al< $\xrightarrow{CH_2=CH-CH_3}$

X

δ^- δ^+
CH_2 ⋯ $CH-CH_3$
δ^+ δ^-
>Ti Al ⋯ R

X

80 **81**

CH_3
|
$CH-R$
|
CH_2

>Ti Al<

X

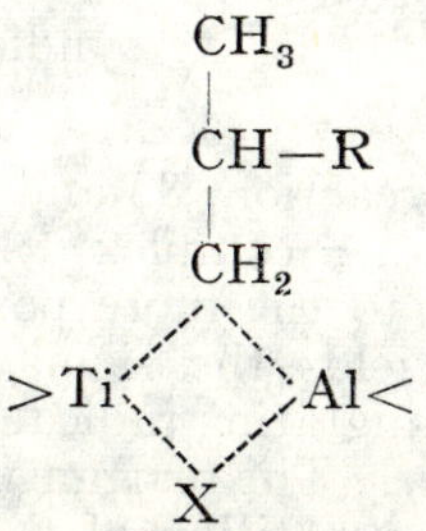

82

The active complex **80** is able to polarise the π bond of the vinyl monomer in the transition state **81**, allowing the insertion of the alkene; this leads to complex **82**, similar to **80**. The process may then be repeated many times. The stereospecificity of the reaction is determined by the substituent of the double bond (in the example chosen methyl), in a 1,3-asymmetric induction.

The mono-metal mechanism[190] assumes that a single metal centre is involved in the transition state, namely the transition metal.

[188] C. Beermann, R. Grof and H. Bestian, *Ger. Pat.* 1023 766 (1958); *Chem. Abstr.*, **54**, 19486 (1960).

[189] G. Natta, *J. Polymer Sci.*, **48**, 219 (1960).

[190] E. Youngman and J. Boor, Jr., *Macromolecular Rev.*, **2**, 33 (1967).

The organometallic compound is only necessary to transfer an alkyl group to the transition metal, which leads to intermediate **83**, the active reagent in the polymerisation process.

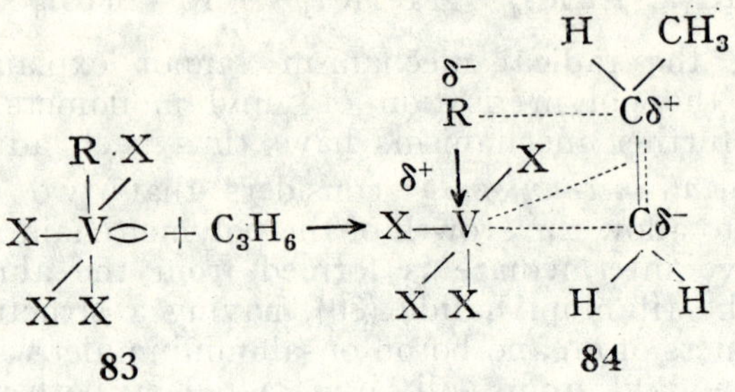

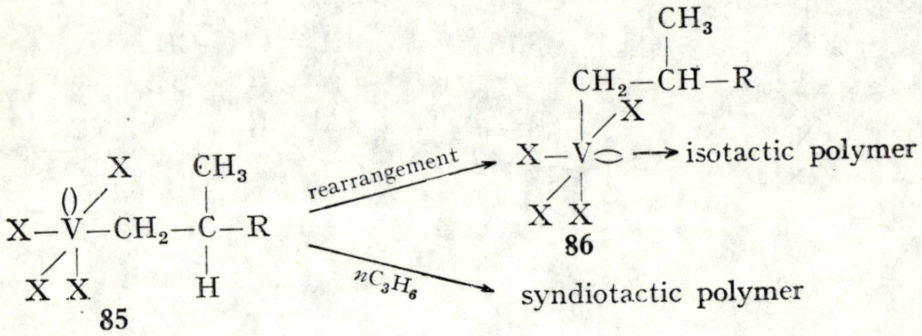

R·X, X—V⊂ (X X) + C_3H_6 ⟶ X—V (X X, X; Rδ^- ⋯ Cδ^+(H)(CH_3); δ^+ ↓; ⋯ Cδ^-(H)(H))

83 **84**

X—V(X)(X X)—CH_2—C(CH_3)(H)—R —rearrangement→ X—V⊂(X)(X X)—CH_2—CH(CH_3)—R ⟶ isotactic polymer

85 **86**

85 —nC_3H_6→ syndiotactic polymer

In the first step of the reaction, **83** and the monomer (propylene in the above example) form a π complex, **84.** The alkyl group R migrates then from the metal to the more positive carbon atom in the complexed alkene, which yields the sigma complex **85.** Repetition of this process, in which the migrating radical will be R—CHMe—CH_2, leads to a syndiotactic isomer. The rearrangement of **85** to **86,** equivalent to **83,** followed by a repetition of these steps, will afford the isotactic polymer. Clearly, equilibration of **85** with **86** must lead to an atactic macromolecular compound.

Obviously, only the mono-metal mechanism accounts for the fact that amylsodium may replace the trialkylaluminium of the original procedure.

10.7 Problems

(1) Consider the frontier orbitals of: (a) ethylene; (b) formaldehyde; (c) crotonaldehyde. Show that the attack of a carbanion (CH_3:$^-$) is not preferred with respect to the reaction of an electrophile (CH_3^+) in case (a), but may be energetically advantageous for (b) and (c).

(2) Demonstrate with experimental and theoretical arguments that all addition mechanisms are closely related to the main reaction mechanisms discussed in the preceding chapters.

(3) Discuss all the differences which may exist between the following intermediates of the electrophilic addition:

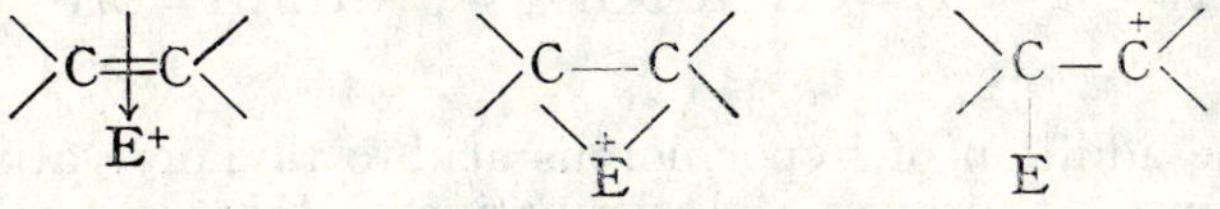

I II III

Suggest reactions in which they would behave differently. What physical method would you use in your study, if the intermediates were observable? What other methods would you apply to distinguish between these reaction intermediates if direct observation is not possible? (Do not neglect the possible equilibration of the species I—III.)

(4) What reaction products would be expected from bromine and norbornadiene, IV, considering the reactivity of norbornene V in addition reactions as well as the reactivity of bicyclo [2.2.1] heptane derivatives in nucleophilic substitutions?

IV V

(5) The hydrogen chloride addition to the deutero-norbornane derivative VI leads to unrearranged (VII) and rearranged (VIII) products:

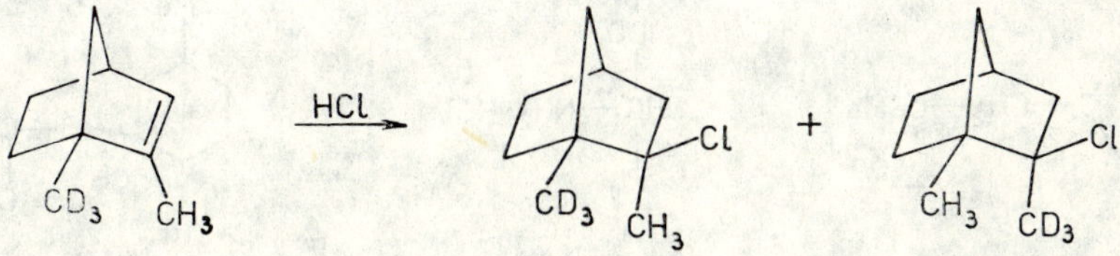

VI $\xrightarrow{HCl}$ VII + VIII

Brown and his co-workers have observed that with short reaction times, the product VII predominates. With longer reaction times, the two products are found in practically equal amounts. What indications do these results offer concerning the reaction mechanism?

(6) How do you account for the fact that the addition of chlorine to t-butylethylene is accompanied by rearrangement, while in the addition of bromine the initial skeleton is conserved?

(7) Indicate which of the following dienes yield high ratios of 1,2-/1,4-addition products in the reaction with halogens:

$CH_2{=}CH{-}CH{=}CH_2$; IX $\qquad$ $PhCH{=}CH{-}CH{=}CHPh$ X

$HOOC{-}CH{=}CH{-}CH{=}CH{-}COOH$; XI $\qquad$ $CH_3CH{=}CH{-}CH{=}CH_2$ XII

(8) The addition of hypochlorous acid to allyl derivatives affords large amounts of propan-1-ol derivative, XIII:

$$CH_2{=}CH{-}CH_2X + HOCl \longrightarrow \underset{\large OH}{CH_2}{-}\underset{\large Cl}{CH}{-}CH_2X + \text{other products}$$

XIII

How do you explain the fact that the amount of XIII increases with the nature of the substituent X, in the order: H < OH < Cl.

(9) The reaction of chlorine with *cis*-1,2-di-t-butylethylene affords only the *anti*-addition product, while the corresponding *trans*-alkene yields, in addition to the *anti*-adduct, a chloroalkene formed by a stereospecific migration of a methyl group:

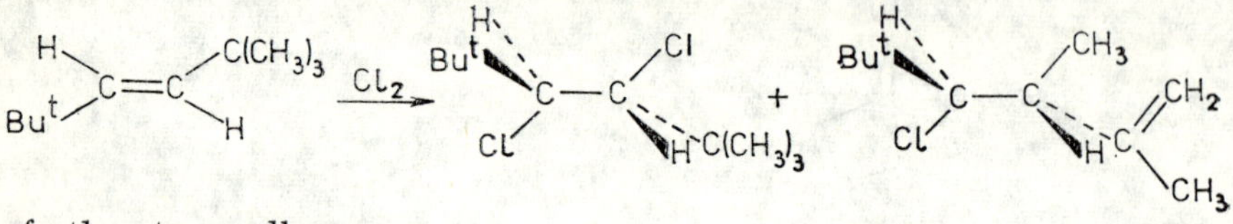

$(H)(Bu^t)C{=}C(C(CH_3)_3)(H) \xrightarrow{Cl_2} (H)(Bu^t)(Cl)C{-}C(Cl)(H)(C(CH_3)_3) + (H)(Bu^t)(Cl)C{-}C(CH_3)(H){-}C(=CH_2)CH_3$

of the two alkenes.

(10) What is the main product expected from the reaction of phenylmagnesium bromide with the following carbonyl compounds:

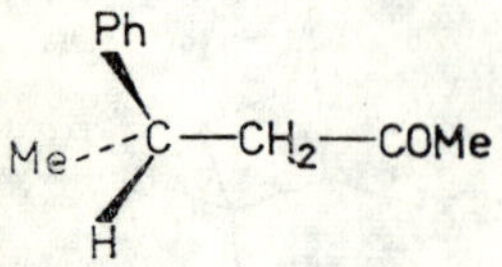

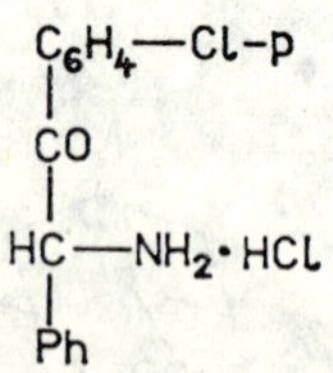

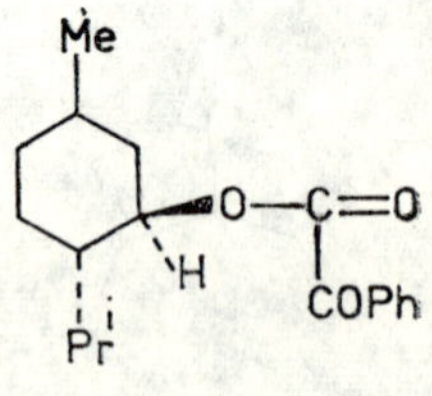

$Me{-}C(Ph)(H){-}CH_2{-}COMe$ $\qquad$ $p\text{-}Cl{-}C_6H_4{-}CO{-}CH(Ph){-}NH_2\cdot HCl$ $\qquad$ (cyclohexyl ester with Me and Pr^i substituents, H) $O{-}C(=O){-}COPh$

(11) Explain the fact that alkynes are more easily attacked by nucleophilic reagents than the alkenes.

(12) Predict the main products of the reduction with $LiAlH_4$ of the following compounds;

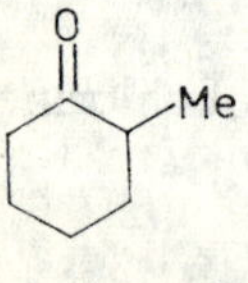

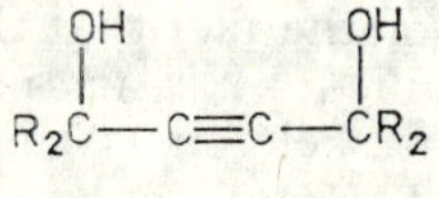

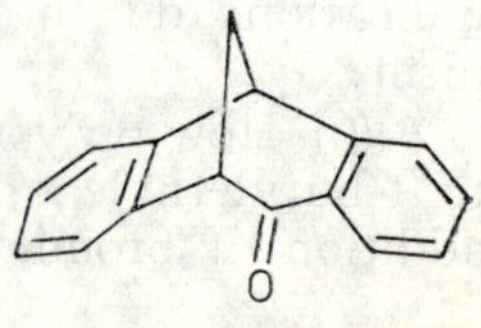

(2-methylcyclohexanone: O, Me) $\qquad$ $R_2C(OH){-}C{\equiv}C{-}CR_2(OH)$ $\qquad$ (bridged dibenzo ketone, O)

PART 4

Homolytic and pericyclic reactions

11
Homolytic reactions

11.1 Free-radical reactions

Reactions involving odd-electron species (reagents or intermediates) are called homolytic reactions. Usually, the reagent is produced in an initial step implying homolysis of a covalent bond and yielding free radicals or free atoms. The free-radical state is then transferred to an organic substrate and the process may repeat itself until the free-radical state is destroyed. As in the heterolytic reactions, the final products may be substitution, addition or elimination compounds. Not infrequently, the same type of reagent can react with the same substrate both in a heterolytic and in a homolytic way. Nevertheless, the reaction products in the two processes are usually quite different.

Staudinger [1] was the first to admit free radicals in polymerisation reactions. A few years later, Taylor [2] suggested the possible involvement of free radicals in other organic reactions. It is now admitted [3] that these reactive intermediates were definitely accepted in 1937, when Kharasch and Mayo[4], on the one hand, and Hey and Waters [5], on the other hand, recognised that the anti-Markownikoff addition of hydrogen bromide [6] is a radical process. In the same year, Flory [7] demonstrated the radical mechanism of vinyl polymerisation.

Radical reactions are involved in many processes of high technological interest such as polymerisation, halogenation, autooxidation, substitution and addition reactions. It is not surprising therefore that such subjects are dealt with in many monographs and reviews [3,8–12].

[1] H. Staudinger, *Ber. dtsch. chem. Ges.*, **53**, 1073 (1920).
[2] H. S. Taylor, *Trans. Faraday Soc.*, **21**, 560 (1925).
[3] C. Walling, *Free Radicals in Solution*, J. Wiley, New York (1957), p. 7.
[4] M. S. Kharasch, H. Engelmann and F. R. Mayo, *J. Org. Chem.*, **2**, 288 (1937).
[5] D. H. Hey and W. A. Waters, *Chem. Rev.*, **21**, 169 (1937).
[6] M. S. Kharasch and F. R. Mayo, *J. Amer. Chem. Soc.*, **55**, 2468 (1933).
[7] P. J. Flory, *J. Amer. Chem. Soc.*, **59**, 241 (1937).
[8] E. W. Steacie, *Atomic and Free Radical Reactions*, Reinhold, New York (1954).
[9] G. H. Williams, *Homolytic Aromatic Substitutions*, Pergamon Press, London (1960).
[10] G. Sosnovsky, *Free Radical Reactions in Preparative Organic Chemistry*, McMillan, New York (1964).
[11] W. A. Pryor, *Free Radicals*, McGraw-Hill, New York (1966).
[12] G. H. Williams (editor), *Advances in Free Radical Chemistry*, Logos Press and Academic Press, vol. 1 (1965); vol. 2 (1967).

11.1.1 *Classification of radical reactions*

According to Walling [3], free-radical reactions fall into four classes:

(a) *Radical-forming reactions.* Free radicals are formed by homolysis of a covalent bond:

$$A-B \longrightarrow A\cdot + B\cdot \tag{11.1}$$

Process (11.1) can be brought about in different ways:

(i) By thermal cleavage of covalent bonds. It is now known that most organic molecules can be split into free radicals at high temperatures. However, certain molecules containing high-energy bonds (see p. 549), can be cleaved at lower temperatures. Examples of such bonds are the peroxide bond, O—O, the C—N bonds in azo derivatives, *etc.*

(ii) By photochemical reactions. Covalent bonds can be more easily broken in excited molecules than in ground state species. Of special interest is the generation of free radicals by the photolysis of ketones [13]:

$$CH_3CO-CH_3 \xrightarrow{h\nu} CH_3CO\cdot + CH_3\cdot \tag{11.2}$$

(iii) By electron-transfer reactions. A labile covalent bond can be broken at low temperatures during electron-transfer reactions. A well-known procedure is the generation of alkoxy radicals by reduction of hydroperoxides with ferrous salts [14]:

$$Fe^{2+} + ROOH \longrightarrow Fe^{3+} + RO\cdot + HO^-$$

The one-electron transfer reaction from metals to cations and other electrophilic substrates has been known for a long time, either in gaseous phase or in solution [15]. More surprising is that even organometallic compounds which are usually nucleophilic can act as electron-donors toward weak electrophiles [16]:

$$C_2H_5Li + C_2H_5I \longrightarrow 2\,C_2H_5\cdot + NaI \tag{11.3}$$

The demonstration of the radical character of reaction (11.3) was only possible by means of modern physical methods (see p. 548).

(b) *Radical transformations.* Free radicals are unstable species, with high tendency to generate more stable compounds. This can be brought about in unimolecular reactions (radical transformations) or in bimolecular processes (radical attack on substrates, radical--destroying reactions, *etc.*).

[13] W. A. Noyes, Jr., L. M. Dorfman, *J. Chem. Phys.*, **16**, 788 (1948).
[14] M. S. Kharasch, A. Fono and W. Nudenberg, *J. Org. Chem.* **15**, 763 (1950).
[15] A. J. Birch, *Quart. Rev. Chem. Soc.*, **4**, 69 (1950).
[16] H. R. Ward, R. G. Lowler and R. A. Cooper, *J. Amer. Chem. Soc.*, **91**, 746 (1969).

The unimolecular reactions are in some respects similar to the unimolecular stabilisation of carbenium ions. Thus the formation of an alkene and a lower radical[17] is similar to the β-fission in the cationic dealkylation (see p. 428).

$$CH_3CH_2CH_2\cdot \longrightarrow CH_3\cdot + CH_2{=}CH_2 \quad (11.4)$$

The formation of a more stable radical, as for example the rearrangement of 2,2,2-triphenylethyl radical to the benzhydryl radical[18] (11.5) resembles the Wagner-Meerwein rearrangement of carbenium ions:

$$Ph_3C{-}CH_2\cdot \longrightarrow Ph_2\dot{C}{-}CH_2Ph \quad (11.5)$$

(c) *Radical attack on substrates* (bimolecular reactions). A free radical can react with stable organic molecules to give another free radical which can be an initiator in addition, polymerisation or substitution reactions. Such reactions are usually *chain reactions* and will be treated in detail in later sections of this chapter.

(d) *Radical-destroying processes.* The free-radical state can be destroyed either by combination of two identical or different radicals:

$$R^1\cdot + R^2\cdot \longrightarrow R^1{-}R^2 \quad (11.6)$$

or by disproportionation:

$$2R{-}CH_2{-}CH_2\cdot \longrightarrow RCH_2CH_3 + RCH{=}CH_2 \quad (11.7)$$

or by electron-transfer:

$$RO\cdot + Fe^{2+} \longrightarrow RO^- + Fe^{3+} \quad (11.8)$$

Both (11.6) and (11.7) are of utmost importance in free-radical processes, and will be treated in the last section of this chapter. Equation (11.8) represents an undesired side reaction.

It must be observed that this type of radical reaction is a bimolecular process. In contrast with other types of bimolecular radical processes, radical-destroying reactions involve the collision of two unstable intermediates, occurring in very small concentrations in the reaction medium. Bimolecular attack on stable substrates will therefore be more probable, even if reactions (11.6) and (11.7) have much lower activation energies. This explains the occurrence of chain processes.

11.1.2 *Characteristics of radical reactions*

As already mentioned, some of the radical reactions involve the same type of reagents and substrates and the same type of reaction (substitution, addition, elimination) as in heterolytic processes.

[17] F. O. Rice, *J. Amer. Chem. Soc.*, 53, 1959 (1931).
[18] D. B. Denney and N. F. Newman, *J. Amer. Chem. Soc.*, 89, 4692 (1967).

However, radical reactions can be distinguished from heterolytic reactions by some characteristics:

(a) Radical reactions are observed both in gaseous and liquid phase. Their rates are only slightly affected by the polarity of the solvent (see however p. 555). In contrast, heterolytic reactions occur only in condensed phase, the rate constants varying sometimes dramatically with the nature and the polarity of the solvent.

(b) Base and acid catalysts (Brönsted or Lewis) have no effect on radical reactions. Homolytic processes are however affected both in rates and products by very small amounts of radical precursors, called *initiators*. Initiators are compounds which are easily decomposed into free radicals by homolysis, under the reaction conditions. Contrary to catalysts, initiators are consumed in the reaction.

(c) Light has a favourable effect on homolytic reactions, due to the photochemical generation of free radicals from the organic molecules. Acceleration by light of a certain process may indicate a radical mechanism.

(d) Free-radical reactions may be influenced by very small amounts of substances called *inhibitors* or *retarders*. Inhibitors have the property of reacting with free radicals to form stable compounds, unable to initiate or continue the homolytic process. Like the initiators, inhibitors are consumed in the reaction. The accidental occurrence of such substances may prevent or delay the starting of the reaction. However, after they have disappeared from the system, the reaction can proceed at the expected rate. Inhibitors can be added on purpose to a chemical system in order to suppress unwanted radical reactions. The time corresponding to the delay of the reaction produced by accidental inhibitors is called the *induction period*.

(e) The effect of initiators, light, inhibitors and the observation of an induction period are characteristic features of a *chain reaction*. In any chain reaction there are three steps: initiation, chain propagation and chain termination.

Initiation. This step generates the initial radical state. Usually, this implies the decomposition of an initiator (In), such as benzoyl peroxide:

$$\mathrm{PhCO{-}O{-}O{-}COPh} \longrightarrow \mathrm{2PhCOO\cdot}$$

or more generally:

$$\mathrm{In} \xrightarrow[\text{or } h\nu]{\Delta} 2\,\mathrm{R\cdot} \tag{11.9}$$

Chain propagation. This step may include two or more elementary reactions. Characteristic of this step is that the radical reagent in the first reaction is a product of the last reaction:

$$\mathrm{R\cdot} + \mathrm{S} \longrightarrow \mathrm{RS\cdot (X\cdot)} \tag{11.10}$$

$$\cdots\cdots\cdots\cdots\cdots\cdots\cdots$$

$$\mathrm{X\cdot} \longrightarrow \text{stable products} + \mathrm{R\cdot} \tag{11.11}$$

Thus, reactions (11.10)—(11.11) may repeat until R· is destroyed.

Chain termination. This step is a radical-destroying reaction, for example combination or dispoportionation. Since collisions of two radical species appearing in very low concentrations are not very frequent, propagation reactions can be repeated a great number of times before the radical state is destroyed. It is thus clear why very small amounts of initiator (of the order of ppm) are sufficient to start a homolytic process. The effect of inhibitors can be explained by comparing reactions (11.10)—(11.11) with (11.12):

$$R\cdot + \text{Inh} \longrightarrow Y\cdot \longrightarrow \text{stable products} \qquad (11.12)$$

If the inhibitor Inh can combine with R· more readily than S (smaller activation energy) to give a reaction product inert toward S, the reaction will be retarded or even stopped. The larger the difference in activation energies between (11.10) and (11.12), the smaller will be the amount of the inhibitor necessary to stop the reaction. Quinones, for example, are efficient inhibitors[19], participating in reactions such as (11.13)—(11.14):

R• + [p-benzoquinone] ⟶ R_0(alkene) + [HO–C₆H₄–O•] etc. (11.13)

R• + [p-benzoquinone] ⟶ [R–O–C₆H₄–O•] etc. (11.14)

Hydroquinones may similarly serve the purpose inasmuch as they are readily converted into the corresponding quinones.

(f) Free radicals display a very low regio- and stereoselectivity in their reactions as compared to the corresponding carbenium ions and carbanions. The low selectivity is determined, on the one hand, by the fact that these neutral and odd-electron species have no special preference for electron-deficient atomic nuclei or electron-rich bonds, as have the nucleophilic and electrophilic reagents, respectively. On the other hand, in their forming reactions (either thermal or photochemical), they result in high energy levels, being therefore more reactive and thus less selective. The reaction products can be controlled by the random encounter of the radical with other molecules, the picture of the products being by far more complex than in a heterolytic reaction.

[19] P. D. Bartlett and H. Kwart, *J. Amer. Chem Soc.*, **72**, 1051 (1950).

Polar effects in the substrate, even if large, may be of no consequence in homolytic reactions. Thus, nitrobenzene is only 2.3 times less reactive than benzene [9] in homolytic aromatic substitution, whereas in the corresponding electrophilic substitution it can be several powers of ten (10^3—10^9) less reactive. At the same time, in a radical substitution all three isomers *(ortho, meta* and *para)* are formed, the *meta* derivative resulting in about the same amount as the *para* isomer.

The kinetics of a homolytic reaction is very complex and only rarely can be represented by a simple equation rate. However, as will be shown later, free-energy relationships are found sometimes, indicating a certain sensibility of the reaction to the electronic characteristics of the substrate. This is usually the case for homolytic reactions involving delocalised radicals [20].

11.1.3 *Cage effects*

A homolytic reaction between a radical R· and a substrate S implies, as other reactions do, a diffusion of the radical reagent to the close neighbourhood of the substrate S. If the diffusion rates are denoted by k_d and k_{-d} for approaching and removing from the substrate, respectively, and the actual chemical rate constant is k_c,

$$R\cdot + S \underset{k_{-d}}{\overset{k_d}{\rightleftharpoons}} R\cdot S \xrightarrow{k_c} \text{products} \tag{11.15}$$

the observed rate will be:

$$k_{obs} = k_c k_d / (k_c + k_{-d}) \tag{11.16}$$

If the diffusion rate constant is high enough, the chemical process will be controlled only by the chemical reactivity (k_c). Otherwise stated, the rate of product formation could be predicted from the rate of the radical-forming reaction. This is generally true for reactions in gaseous phase [21].

In solution, radicals, as very reactive and unselective species, may react with the solvent before colliding with substrate S, yielding thus products derived both from solvent and substrate. If individual rates are known, this cannot be a serious difficulty in appreciating the rates of product formation from the rates of radical formation. However, the generation of radicals in different solvents has shown that their efficiency in initiating radical reactions (for example, polymerisation) is much lower than expected from the rate of their formation. This can be explained if we assume that the radicals are formed in a *solvent cage* [22–24]. In order to start the reaction, the

[20] D. M. Golden and S. W. Benson, *Chem. Rev.*, **69**, 125 (1969).
[21] A. M. North, *Quart. Rev. Chem. Soc.*, **20**, 421 (1966).
[22] J. Frank and E. Rabinovitch, *Trans. Faraday Soc.*, **30**, 120 (1934).
[23] R. M. Noyes, *J. Chem. Phys.*, **18**, 999 (1950).
[24] R. M. Noyes, *J. Amer. Chem. Soc.*, **77**, 2042 (1955); **78**, 5486 (1956).

radicals have to move out of this cage and diffuse into the reaction medium. However, instead of that, they may combine within the solvent cage, yielding stable unreactive molecules. The lowering of radical efficiency in initiating radical reactions is called the *cage effect*. The cage effect can be appreciated from the combination products of the radicals within the solvent cage (for instance dimerisation products), or from the reaction of radicals outside the solvent cage with an efficient radical trapper *(scavenger)*.

Thus, in the thermal decomposition of an equimolar mixture of the azo derivatives **1** and **2**, the products are the symmetrical derivatives **3** and **5**, together with the unsymmetrically deuterated hydrocarbon **4**. In the absence of any cage effect, a ratio 3 : 4 : 5 = 25 : 50 : 25 would be expected:

$$PhCH(CH_3)-N=N-CH(CH_3)Ph\ (\mathbf{1}) + PhCH(CD_3)-N=N-CH(CD_3)Ph\ (\mathbf{2}) \rightarrow PhCH(CH_3)-CH(CH_3)Ph\ (\mathbf{3}) + N_2;\quad PhCH(CH_3)-CH(CD_3)Ph\ (\mathbf{4}) + N_2;\quad PhCH(CD_3)-CH(CD_3)Ph\ (\mathbf{5}) + N_2$$

It has actually been found [25] that **4** represents less than 50% and the symmetrical compounds more than 25% of the combination products. From the observed amounts it has been appreciated that 29.1% of the radicals combine before diffusing outside the solvent cage.

Waits and Hammond [26] have calculated the ratio of the cage reaction from the overall reaction products in the decomposition of the azo derivative **6**, by performing the reaction in presence of very efficient scavengers, such as iodine or 2,2-diphenyl-l-picrylhydrazyl **7**.

6: 1,1'-azobis(cyclohexanecarbonitrile) (two cyclohexane rings joined by N=N, each bearing CN)

7: $Ph_2\ddot{N}-\dot{\ddot{N}}-C_6H_2(NO_2)_3$ (2,4,6-trinitrophenyl)

[25] S. Seltzer and E. J. Hamilton, Jr., *J. Amer. Chem. Soc.*, **88**, 3775 (1966).
[26] H. P. Waits and G. S. Hammond, *J. Amer. Chem. Soc.*, **86**, 1911 (1964).

From the variation of the amount of trapping products with the scavenger (Sc) concentration, they concluded the following reaction scheme:

$$\underset{}{\text{R—N=N—R}} \longrightarrow \underset{\text{solvent cage}}{(\text{R}\cdot + \text{N}_2 + \text{R}\cdot)} \longrightarrow \underset{\text{cage product}}{\text{R—R} + \text{N}_2}$$

$$(\text{R}\cdot + \text{N}_2 + \text{R}\cdot) \longrightarrow \underset{\text{diffused radicals}}{2\text{R}\cdot + \text{N}_2}$$

$$\text{R}\cdot + \text{Sc} \longrightarrow \text{RSc}$$

However, at higher scavenger concentration, the amount of trapping product RSc is greater than expected from the above scheme. The authors consider that under these conditions the scavenger can penetrate the cage, RSc being thus partly a cage product.

It seems that cage effects may be observed, to a smaller extent, even in gaseous phase[27]. In this case, an inert gas, used as diluent, can play the role of the solvent molecules in solution.

Radical reactions in solvent cage can be well observed now by the chemically-induced dynamic nuclear polarisation (CIDNP) method[28, 29]. According to this method, if an n.m.r. spectrum is recorded within the span of time in which the radical reaction takes place, the signals corresponding to the products formed in a solvent cage may suffer the following modifications (Figure 11.1):

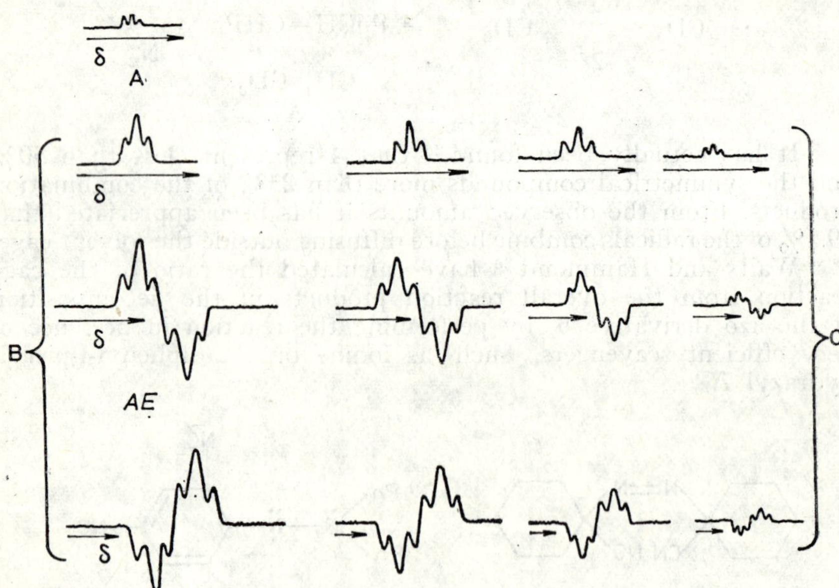

Figure 11.1 Chemically induced nuclear polarisation spectra. A. Normal (unexalted) n.m.r. signals. B. Exalted signals. C. Variation in time of CIDNP signals.

[27] R. K. Lyon, *J. Amer. Chem. Soc.*, **86**, 1907 (1964).
[28] H. R. Ward, *Accounts Chem. Res.*, **5**, 18 (1972).
[29] R. G. Lawler, *Accounts Chem. Res.*, **5**, 25 (1972).

(a) An exaltation of the signals. In the first moments the observed peaks may be 10^2—10^3 more intensive than normal. The exaltation decreases in time, the normal intensity being recorded shortly after the reaction has ended (Figure 11.1.C).

(b) Beside normal absorption signals (A), emission signals coupled with the absorption peaks can appear. They can precede (EA) or follow (AE) the normal absorption peaks. These modifications, in their turn, disappear soon after the end of the reaction.

It is thought that the modifications can be directly related to the reaction mechanism and reaction rates and that they depend upon the nuclear spin states of the molecules undergoing the reaction. Using a radical-pair model the spectra can be quantitatively interpreted [29]. The shape of the CIDNP signals can give valuable information regarding the structure of the free radicals involved, as well as their reactions in the solvent cage. Other applications of this method will be discussed in later sections of this chapter.

The cage effect may be considered to correspond to the internal return (see p. 212) of a tight ion-pair in the heterolytic reactions. It probably occurs in all radical reactions in solution, even if it cannot always be detected experimentally.

11.2. Radical-forming reactions

11.2.1 *Thermal cleavage of covalent bonds*

The ease with which a covalent bond R—X can be broken homolytically depends on the dissociation energy *D* of this bond. Table 11.1 indicates that the most suitable compounds for thermal generation of radicals contain either a carbon-metal bond, or a C—N bond, such as in azo compounds, or an O—O bond such as in peroxides. These derivatives are indeed used most frequently in radical-forming reactions, either in the gaseous or in the liquid phase.

TABLE 11.1

Dissociation energies *D* (kcal mol^{-1}) of some R—X bonds[a]

Bond	*D*	*Bond*	*D*
Me—H	101	MeS—H	80
Et—H	98	MeS—SMe	73
Pr^i—H	89	Me—Br	68
Bu^t—H	85	HO—OH	54
Me—Me	83	MeHg—Me	52
MeCO—Me	72	CH_3—N=N—CH_3	52
MeCO—COMe	60	CBr_3—Br	49
		MeCd—Me	24

[a] After Walling, Reference 3, p. 49.

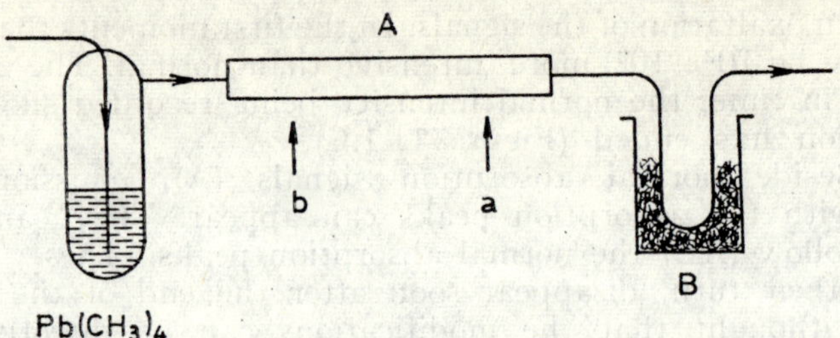

Figure 11.2 Generation of methyl radicals by thermal decomposition of tetramethyllead, $(CH_3)_4Pb$.

(A) *Thermal decomposition of organometallic compounds*

Lower alkyl radicals have been obtained by Panneth [30-32] by heating tetraalkyl lead at low pressure and rapid flow, in an apparatus shown schematically in Figure 11.2.

$$(CH_3)_4Pb \rightleftharpoons 4CH_3\cdot + Pb \tag{11.17}$$

The reaction is reversible. When a carrier gas, such as hydrogen or nitrogen, saturated with tetraalkyl lead, is passed through a quartz tube A, heated strongly at point a, the formation of a lead mirror is observed. In trap B, cooled with liquid nitrogen, no organic product is collected, indicating the complete decomposition by pyrolysis of the organometallic compound. If this experience is repeated with the difference that the tube is heated in another point b, preceding the lead mirror already deposited during the previous experience, one may notice the disappearance of the mirror at point a, whereas in the trap B tetraalkyl lead is collected. This shows that at the pyrolysis temperature equilibrium (11.17) must be completely shifted in the decomposition direction, whereas at lower temperatures the radicals formed can be trapped by metallic mirrors. The trapping of free radicals by metallic mirrors (Pb, Zn, Sb, etc.) is one of the methods used in free radical identifications. From the distance between the metallic mirror and the point of generation of the free radicals, the pressure and the velocity of the gas carrier, as well as from the diameter of the pyrolysis tube, the half-lives of the free radicals can be estimated. For the methyl radical in gaseous phase, a half-life of about 10^{-3} s has been calculated in this manner.

Alkoxides of heavy metals have been successfully used in the generation of free radicals. Usually, the alkoxides are prepared *in situ* [33-35]. The method can be used for the preparation of 1,5

[30] F. A. Panneth and W. Hofeditz, *Ber. dtsch. chem. Ges.*, **62**, 1335 (1929).
[31] F. A. Panneth and W. Lautsch, *Ber. dtsch. chem. Ges.*, **64**, 2702 (1931).
[32] F. A. Panneth and W. Lautsch, *J. Chem. Soc.*, 380 (1935).
[33] E. E. van Tamelen and M. A. Schwartz, *J. Amer. Chem. Soc.*, **87**, 3277 (1965).
[34] K. B. Sharpless, R. P. Hanzlick and E. E. van Tamelen, *J. Amer. Chem. Soc.*, **90**, 209 (1968).
[35] E. E. van Tamelen, B. Akermark and K. B. Sharpless, *J. Amer. Chem. Soc.*, **91**, 1552 (1969).

dienes or dibenzyls, and consists of treating the alcohol with the corresponding amount of titanium (IV) chloride, followed by the reduction with metallic potassium of the alkoxide formed:

$$R^1OH + TiCl_4 \longrightarrow (R^1O)_2TiCl_2 \xrightarrow{K} (R^1O)_2Ti \longrightarrow TiO_2 2R^1\cdot \longrightarrow R^1{-}R^1$$

R^1 being $PhCH_2$ or $R^2_2C{=}CR^2{-}CH_2$

(B) Cleavage of azo compounds

The thermal cleavage of azo-compounds can generate either monoradicals or diradicals:

$$R{-}N{=}N{-}R \longrightarrow 2R\cdot + N_2$$

$$\underset{\mathbf{8}}{\text{(cyclic } N{=}N\text{)}} \longrightarrow \underset{\mathbf{9}}{\dot{C}H_2{-}CH_2{-}\dot{C}H_2} + N_2$$

The energy involved in the cleavage of the C—N bond depends on the structure of R. Thus, while azobenzene, PhN=NPh, can be distilled at atmospheric pressure without decomposition, azotriphenylmethane, $Ph_3CN{=}NCPh_3$, is decomposed even at —40 °C [36]. Clearly, radical stability as expressed by the DE is responsible for this difference in behaviour. The phenyl radical is similar in stability to simple alkyl radicals, whereas the triphenylmethyl radical is one of the most stable radicals due to its high delocalisation energy.

The cleavage of an azo compound may occur either in two steps:

$$R{-}N{=}N{-}R \longrightarrow R{-}N{=}N\cdot + R\cdot \qquad (11.18)$$

$$R{-}N{=}N\cdot \longrightarrow N_2 + R\cdot \qquad (11.19)$$

or in a concerted reaction, in which the two C—N bonds are synchronously broken:

$$R{-}N{=}N{-}R \longrightarrow 2R\cdot + N_2 \qquad (11.20)$$

The stepwise mechanism has been identified in the cleavage of benzylazomethane [37], the more stable benzyl radical being formed in the first step of the reaction.

The concerted cleavage seems to be more frequently encountered [25, 26, 38—40]. By comparing the reactivities of different azo deriva-

36 H. Wieland, H. von Hove and K. Borner, *Liebigs Ann. Chem.*, **456**, 31 (1926).
37 S. Seltzer and F. T. Dune, *J. Amer. Chem. Soc.*, **87**, 2628 (1965).
38 H. C. Ramsperger, *J. Amer. Chem. Soc.*, **49**, 912, 1495 (1927); **51**, 2134 (1929).
39 S. G. Cohen and C. H. Wang, *J. Amer. Chem. Soc.*, **77**, 3628 (1955).
40 C. G. Overberger and A. V. DiGiulio, *J. Amer. Chem. Soc.*, **81**, 2154 (1959).

tives, earlier workers in this field have concluded that the two C—N bonds are broken synchronously. Thus, azoisopropane **10** is less reactive than the compound **11**, a benzyl radical being more stable than an isopropyl radical. If a two-step mechanism were involved, the symmetrical derivative **12** would be at most twice as reactive as **11**, considering the statistical factor. It has actually been found that **12** is decomposed 40 times faster than **11**. Only the concerted mechanism can account for this large difference in reactivity between the two compounds, since the transition state (**13**) must be better stabilised by two benzyl groups than by a single one.

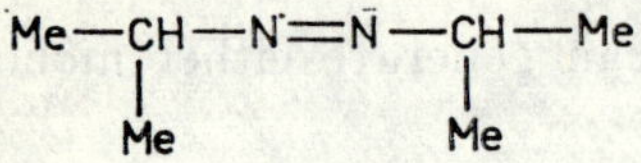

Me—CH(Me)—N=N—CH(Me)—Me **10**

Ph—CH(Me)—N=N—CH(Me)—Me **11**

Ph—CH(Me)—N=N—CH(Me)—Ph **12**

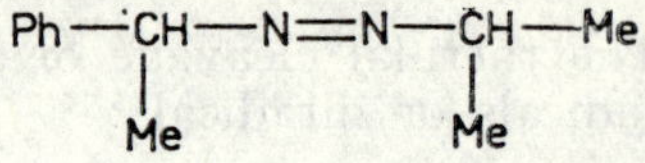

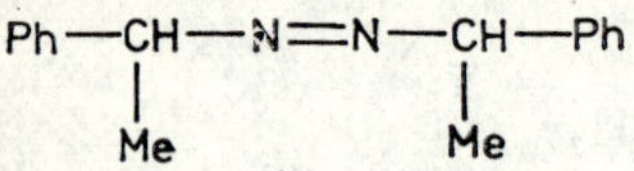

C_6H_5=CH(Me)- -N≡N- -CH(Me)=C_6H_5 **13**

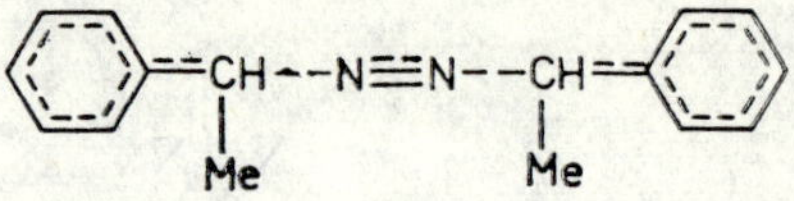

It has been shown that azo compounds such as azoisopropane **10** are thermally isomerised to hydrazones and azines. Thus, at least for **10** and possibly for **11**, the determined rates do not measure the actual rates of C—N cleavage. However, considering that the isomerisation to hydrazone should have a lower activation energy than the bond cleavage, the order of reactivity of derivatives **10**—**12** must be the same, which shows that the conclusion regarding the concertedness of the reaction was not wrong.

More recently, the concerted mechanism was confirmed by careful studies of activation parameters, isotope effects and the stereochemistry of the pyrolysis products of cyclic azo compounds. Thus, Crawford and his co-workers[42–44] have shown that pyrazolines **14**—**16** have the same entropy of activation (~ 10 eu) in pyrolysis, but differ significantly in their activation energies, methylpyrazoline **14** having the highest parameter, whereas the *cis* and *trans* disubstituted derivatives **15** and **16** present nearly the same energy barrier.

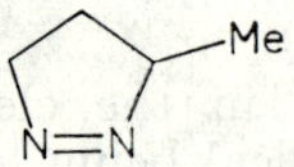

14 (3-methyl-1-pyrazoline: N=N ring, Me) **15** (Me, Me; N=N) **16** (Me, Me; N=N)

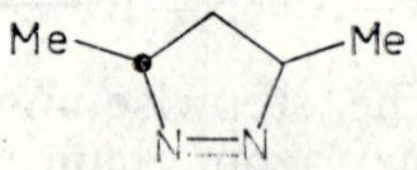

If a stepwise mechanism were involved, this would reflect in the activation entropy, **15** and **16** differing from **14** in $\Delta S^{\ddagger}$ by $R \ln 2$. In the concerted mechanism, the activation entropy is not expected

[41] S.G. Cohen and R. Zand, *J. Amer. Chem. Soc.*, **84**, 586 (1962).
[42] R. G. Crawford and A. Mishra, *J. Amer. Chem. Soc.*, **88**, 3963 (1966).
[43] R. G. Crawford and G. L. Erickson, *J. Amer. Chem. Soc.*, **89**, 3907 (1967).
[44] R. G. Crawford and H. Ali, *J. Amer. Chem. Soc.*, **89**, 3908 (1967).

to vary in this series. However, a lower activation enthalpy is expected for the disubstituted derivatives, as compared to the monomethyl one, due to a better stabilisation of the transition state by the methyl groups. This has actually been confirmed experimentally.

The example given before would indicate a linear relationship in the thermal decomposition of azo derivatives, with the parameters σ_R of all the substituents about the N=N double bond. This has been demonstrated in an elegant study by Nelsen and Bartlett [45], who were able to predict and then verify experimentally the activation parameters of the thermal decomposition of azacumene, using literature data for other azo compounds.

An interesting illustration of the transition state of a concerted cleavage by electron delocalisation is found in the thermal decomposition of polycyclic azo compounds **17**—**20** [46–48]. In all the examples studied, a fused cyclopropane ring near a C—N bond accelerates significantly the decomposition rate. Thus, **18** reacts 10^{17} times faster than **17** [46], and **20** is more reactive than **19** [47–48]. It has been shown [49] that in the transition state the cyclopropane ring contributes two electrons involved in the stabilisation of the incipient diradical (**21**). Thus the easier the breaking of the cyclopropane bond, the higher the reactivity of the azo compound in the thermal cleavage will be.

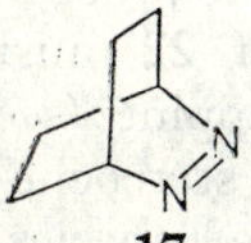

N N N N N N N N N N

17 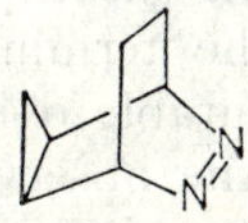 **18** 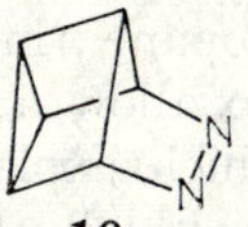 **19** 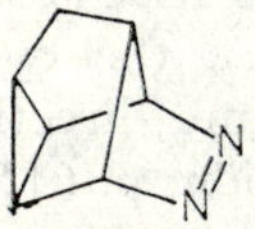 **20** 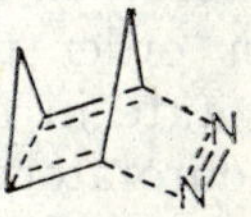 **21**

The thermal decomposition of azo derivatives in solution occurs in a solvent cage [25, 26]. According to Hammond [26], azo alkanes may undergo the following processes during the homolysis:

Cage decomposition: $RN_2R \xrightarrow{k_1} (2\,R\cdot + N_2)$

Cage coupling (primary recombination): $(2\,R\cdot + N_2) \xrightarrow{k_2} R{-}R + N_2$

Diffusion from cage: $(2\,R\cdot + N_2) \xrightarrow{k_3} 2\,R\cdot + N_2$

New cage formation: $2R\cdot \xrightarrow{k_{-3}} (2\,R\cdot)$

[45] S. F. Nelsen and P. D. Bartlett, *J. Amer. Chem. Soc.*, **88**, 137 (1966).
[46] E. L. Allred and K. J. Voorhees, *J. Amer. Chem. Soc.*, **95**, 620 (1973).
[47] E. L. Allred and A. L. Johnson, *J. Amer. Chem. Soc.*, **93**, 1300 (1971).
[48] B. M. Trost and R. M. Cory, *J. Amer. Chem. Soc.*, **93**, 5573 (1971).
[49] H. Schmidt, A. Schweig, B. M. Trost, H. B. Neubold and P. H. Scudder, *J. Amer. Chem. Soc.*, **96**, 622 (1974).

Cage coupling (secondary recombination): $(2\,R\cdot) \xrightarrow{k_4} R{-}R$

Trapping reactions: $R\cdot + S \xrightarrow{k_5} RS$

The higher the concentration of the trapping compound S (substrate or scavenger), the lower the amount of coupling product R—R observed. Nevertheless, the scavenger cannot suppress the formation of the coupling product. It is believed that the scavenger interferes in the process after the initially formed radicals have diffused from the solvent cage. The scavenger can thus suppress the secondary recombination, but not the primary coupling. The effect of a scavenger resembles thus the special salt effect in a heterolysis of a covalent bond (see p. 213).

In the case of cyclic azo compounds, the cage recombination corresponds either to cyclisation of the diradical or to alkene formation. It has been found [42] that the cyclopropane formation in the pyrolysis of pyrazoline has a low stereoselectivity, the *cis*-pyrazoline **15** yielding preferentially 1,2-*trans*-dimethylcyclopropane (66%) whereas the *trans*-pyrazoline **16** affords a higher proportion of *cis* dimethylcyclopropane (72%). This is in agreement with the intermediate formation of a π-cyclopropane having the HOMO of symmetry **22** and the LUMO of symmetry **23**. In the thermal process, in order to close the cyclopropane ring, the termini of **22** must rotate in the same direction to allow a favourable overlapping (see conrotatory motion, p. 615), which explains the observed stereochemistry. If **23** is involved, as would be the case for excited species, the termini must rotate in opposite directions, leading to the reverse product geometry. The small difference in the MO energies of **22** and **23** explains the low selectivity of the reaction [50].

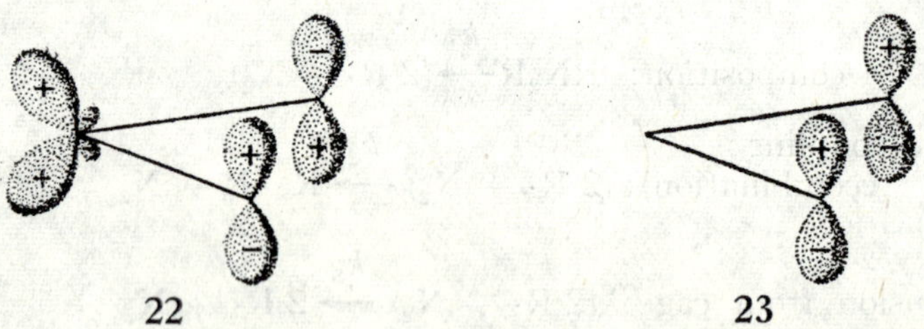

With some exceptions, the diradical intermediates arising in the decomposition of cyclic azo compounds have not been trapped or directly observed. One exception is trimethylenemethane [51–53],

[50] R. Hoffmann, *J. Amer. Chem. Soc.*, **90**, 1475 (1968).
[51] P. Dowd, *J. Amer. Chem. Soc.*, **88**, 2587 (1966).
[52] R. J. Crawford and D. M. Cameron, *J. Amer. Chem. Soc.*, **88**, 2589 (1966).
[53] W. T. Borden, *Tetrahedron Letters*, 259 (1967).

obtained by thermal or photochemical decomposition of the pyrazoline **24**:

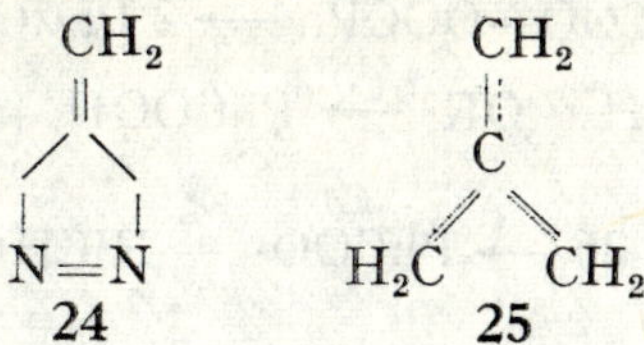

Trimethylenemethane **25** is stabilised by conjugation, its DE calculated by the HMO method being 1.642 β.

(C) Thermal cleavage of peroxide compounds

The generation of free radicals by homolysis of a peroxide bond has many applications in laboratory and industry [54]. As starting material, benzoyl peroxide is largely used. In its thermal decomposition, the homolysis of the O—O bond leads primarily to benzoyl radicals, identified by reactions with efficient scavengers [55].

$$\text{PhCOO—OOCPh} \longrightarrow 2\,\text{PhCOO}\cdot \qquad (11.21)$$

The decomposition of benzoyl peroxide has been recently studied [56] by the CIDNP method. It has thus been confirmed that the benzoyloxy radicals result in a solvent cage. Their decomposition to phenyl radicals by decarboxylation is slow, as compared to the reactions with the solvent. It has been concluded that the lifetime of benzoyloxy radicals is about 10^{-6}—10^{-7} s.

The decomposition rate of benzoyl peroxide is accelerated by the presence of free radicals, such as triphenylmethyl, and inhibited by oxygen [57], which indicates a chain reaction. Unexpectedly for a homolytic reaction, the reaction rates are largely affected by solvent, being slow in carbon tetrachloride and rapid in ethers. It has been shown that in diethyl ether one mole of benzoic acid per mole of peroxide is formed, together with ether **27** [58]. This result, coupled with the kinetics of the thermal cleavage [59], led to the con-

[54] J. O. Edwards (editor), *Peroxide Reaction Mechanism*, Interscience, New York (1962).

[55] G. S. Hammond and L. M. Soffer, *J. Amer. Chem. Soc.*, **72**, 4711 (1950).

[56] R. A. Cooper, R. G. Lawler and H. R. Ward, *J. Amer. Chem. Soc.*, **94**, 552 (1972).

[57] K. Nozaki and P. D. Bartlett, *J. Amer. Chem. Soc.*, **68**, 1686 (1946).

[58] W. E. Cass, *J. Amer. Chem. Soc.*, **69**, 500 (1947).

[59] P. D. Bartlett and K. Nozaki, *J. Amer. Chem. Soc.*, **69**, 2299 (1947).

clusion that the decompositions are induced by solvent, in the following manner:

$$\text{PhCOO—OOCPh} \longrightarrow 2\,\text{PhCOO}\cdot \qquad (11.21)$$

$$\text{PhCOO}\cdot + \text{R}^1\text{CH}_2\text{—OR}^2 \longrightarrow \text{PhCOOH} + \underset{\mathbf{26}}{\text{R}^1\dot{\text{C}}\text{H—OR}^2} \qquad (11.22)$$

$$\text{PhCOO—OOCPh} + \mathbf{26} \longrightarrow \text{PhCOO}\cdot + \underset{\mathbf{27}}{\text{R}^1\text{CH(OR}^2\text{)OCOPh}} \qquad (11.23)$$

The formation of radical **26** explains the induced decomposition. Radical **26** is relatively stable due to heteroatom stabilisation. Other solvent-induced decompositions may occur whenever a stable radical can be formed by reaction with the solvent. Another example is the thermal cleavage of benzoyl peroxide in aromatic hydrocarbons, such as benzene[60, 61]:

$$\text{PhCOO—OOCPh} \longrightarrow 2\,\text{PhCOO}\cdot \longrightarrow 2\,\text{Ph}\cdot + 2\,\text{CO}_2$$

$$\text{Ph}\cdot + \text{C}_6\text{H}_6 \longrightarrow \text{Ph(H)C}_6\text{H}_5\cdot$$

$$\text{PhCOO—OOCPh} + \text{Ph(H)C}_6\text{H}_5\cdot \longrightarrow \text{PhCOOH} + \text{PhCOO}\cdot + \text{Ph—Ph}$$

The rate constants in the decompositions of ring-substituted benzoyl peroxides can be correlated with σ-Hammett parameters[62, 63] The negative value of ρ (—0.38) can be explained[62] by admitting that the peroxide consists of two dipoles, linked at their negative ends (**28**).

$$\underset{\mathbf{28}}{\overset{\text{ArCOO—OOCAr}}{\oplus| \longrightarrow \; \longleftarrow |\oplus}}$$

Electron-repelling substituents will increase the dipole charges, facilitating thus the reaction.

The decomposition of acetyl peroxide yields methyl acetate and ethane[64, 65], beside the products resulted from the reaction of acetyloxy and methyl radicals with the solvent. The former products can be formed by cage recombination of radical pairs. Taylor and Martin[66] have demonstrated by isotopic labelling of the carbonyl

[60] G. B. Gill and G. H. Williams, *J. Chem. Soc.*, 995, 7127 (1965).
[61] G. B. Gill and G. H. Williams, *J. Chem. Soc.*, (B), 880 (1966).
[62] C. G. Swain, W. H. Stockmayer and J. T. Clarke, *J. Amer. Chem. Soc.*, 72, 5426 (1950).
[63] A. T. Blomquist and A. J. Buselli, *J. Amer. Chem. Soc.*, **73**, 3883 (1961).
[64] L. Herk, M. Feld and M. Szwarc, *J. Amer. Chem. Soc.*, **83**, 2998 (1961).
[65] W. Braun, L. Rajbenbach and F. R. Eirich, *J. Phys. Chem.*, **66**, 1591 (1962).
[66] J. W. Taylor and J. C. Martin, *J. Amer. Chem. Soc.*, **88**, 3650 (1966).

oxygen in peroxide that the acetyloxy radicals can couple within the solvent cage before decarboxylation. In isooctane, the cage products represent about 53% of the overall decomposition products:

$$\mathrm{Me{-}\overset{\overset{O^*}{\|}}{C}{-}O{-}O{-}\overset{\overset{O^*}{\|}}{C}{-}Me} \longrightarrow$$

Products from the reaction with the solvent, 46.7%.

$$\begin{array}{ccccc} \uparrow & & \uparrow & & \uparrow \\ \longrightarrow (2\ \mathrm{MeC^*OO\cdot}) & \longrightarrow & (\mathrm{MeC^*OO\cdot + Me\cdot}) & \longrightarrow & (2\ \mathrm{Me\cdot}) \\ \downarrow & & \downarrow & & \downarrow \\ \mathrm{MeC^*O^*O{-}O^*O^*CMe} & & \mathrm{MeC^*O^*OCH_3} & & \mathrm{Me{-}Me} \\ 38\% & & 12.4\% & & 2.9\% \end{array}$$

The mechanism proposed by Taylor and Ward for the decomposition of acetyl peroxide has been confirmed by the CIDNP method [67]. Ward and his co-workers have shown in this way that acyloxy radicals are more readily decarboxylated to alkyl radicals than are the aroyloxy radicals. The thermal decomposition of benzoylpropionyl peroxide [28] is illustrative in this respect. Using CIDNP spectra and product data, the following mechanism was formulated:

$$\mathrm{PhCOOOOCEt} \xrightarrow{\Delta} (\mathrm{PhCOO\cdot\ EtCOO\cdot}) \longrightarrow (\mathrm{PhCOO\cdot\ Et\cdot})$$

$$\swarrow \searrow$$

$$\mathrm{Ph{-}Et} \longleftarrow (\mathrm{Ph\cdot\ Et\cdot}) \qquad \mathrm{PhCOOEt}$$

Even if it does not allow the direct observation of the radicals, the CIDNP method seems to compete successfully with the methods of direct observation [68].

Hydroperoxides and their derivatives are largely used in the generation of alkoxy radicals. Thus, the t-butoxy radical can be formed either from di-t-butylperoxide [69], **29**, or from di-t-butylperoxalate **30**, and other peresters [70, 71], or from t-butyl hypochlorite [72–74], etc. The decomposition of these compounds can be realized either thermally or, better, photochemically. Thus, the half-life of **29** when subjected to thermal decomposition at 45 °C is 10

[67] R. A. Cooper, R. G. Lawler and H. R. Ward, *J. Amer. Chem. Soc.*, **94**, 545 (1972).

[68] H. R. Ward and R. G. Lawler, *J. Amer. Chem. Soc.*, **95**, 946, 1692 (1973).

[69] J. H. Raley, F. F. Rust and W. E. Vaughan, *J. Amer. Chem. Soc.*, **70**, 1336 (1948).

[70] H. Kiefer and T. G. Traylor, *Tetrahedron Letters*, 6163 (1966).

[71] H. Kiefer and T. G. Traylor, *J. Amer. Chem. Soc.*, **89**, 6667 (1967).

[72] C. Walling and B. B. Jackow, *J. Amer. Chem. Soc.*, **82**, 6108, 6113 (1960).

[73] D. L. Carlsson and K. U. Ingold, *J. Amer. Chem. Soc.*, **89**, 4885 (1967).

[74] C. Walling and V. P. Kurow, *J. Amer. Chem. Soc.*, **89**, 4895 (1967).

years[69], whereas under photochemical conditions, at the same temperature, of only 4 hours[71].

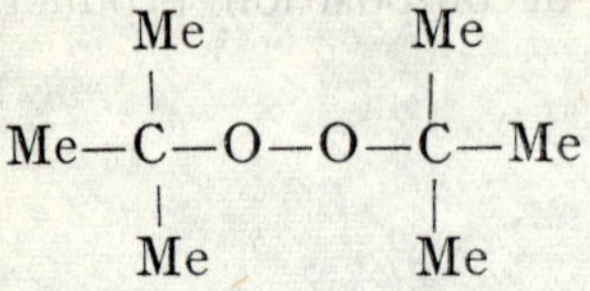

$$\begin{array}{c}\text{Me}\quad\quad\text{Me}\\ |\quad\quad\quad |\\ \text{Me—C—O—O—C—Me}\\ |\quad\quad\quad |\\ \text{Me}\quad\quad\text{Me}\end{array}$$

29

$$\begin{array}{c}\text{Me}\quad\quad\text{O}\;\;\text{O}\quad\quad\text{Me}\\ |\quad\quad\;\;\|\;\;\|\quad\quad |\\ \text{Me—C—O—O—C—C—O—O—C—Me}\\ |\quad\quad\quad\quad\quad\quad |\\ \text{Me}\quad\quad\quad\quad\quad\text{Me}\end{array}$$

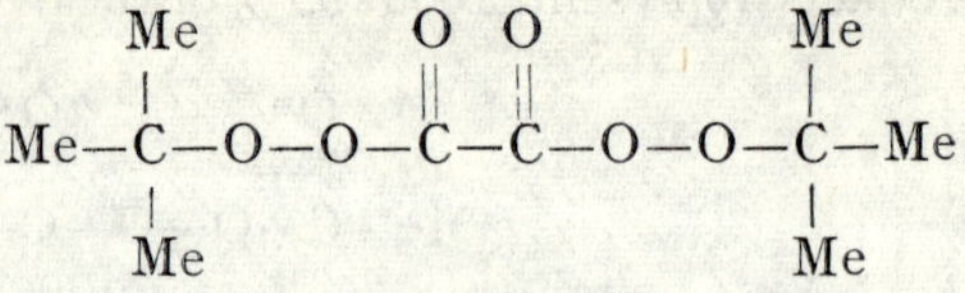

30

Cage effects are very important in these decompositions. Thus, the decomposition rates of **29**—**31** vary significantly with the viscosity of the solvent[71]. It is known that solvent viscosity affects the diffusion rates, so that an increased viscosity leads to more cage-coupling products. On the other hand, a large viscosity will diminish the return of diffused radicals into a new cage. Indeed, in a large variety of solvents of different viscosity, the relative reactivities of **29**—**30** remain the same, indicating that once the radicals have migrated from the cage they do not form another cage, but react with the solvent.

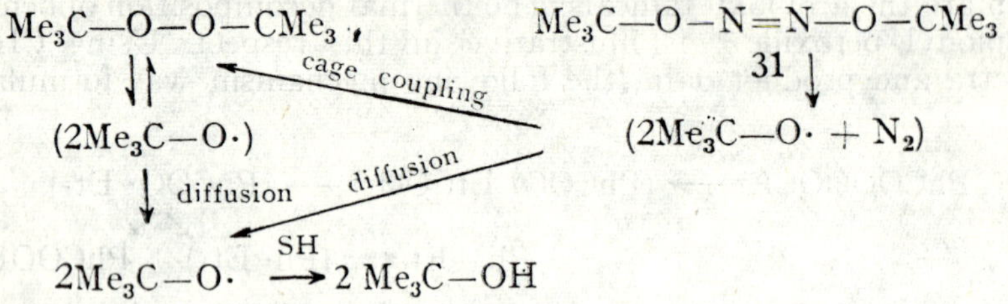

$Me_3C—O—O—CMe_3$ ⇌ $(2Me_3C—O\cdot)$ —(diffusion)→ $2Me_3C—O\cdot$ —(SH)→ $2\,Me_3C—OH$

$Me_3C—O—N{=}N—O—CMe_3$ (**31**) → $(2Me_3C—O\cdot + N_2)$; cage coupling → $Me_3C—O—O—CMe_3$; diffusion → $2Me_3C—O\cdot$

11.2.2 *Photochemical decomposition*

By absorption of a photon, a molecule is excited to a higher energy level. The process consists in promoting an electron from a bonding orbital into an antibonding orbital. The energy difference between the ground state level (E_1) and the excited level (E_2) is given by the well-known relationship:

$$E = E_2 - E_1 = h\nu = hc/\lambda = 286\,000/\lambda \tag{11.24}$$

in which h is Planck's constant, c the velocity of light and λ the wavelength. The energy difference is expressed in kcal mol^{-1} if in equation (11.24) λ is taken in Å. Table 11.2 gives the calculated correspondence between λ and ΔE for λ values in the region of interest for photochemistry.

It can be seen that the energy of a photon in the ultraviolet region is sufficiently high to break a covalent bond (compare the values from Table 11.2 with the values of Table 11.1). However,

TABLE 11.2

Energy-conversion table

Wavelength λ (Å)	*Wave number* $\bar{\nu}$ (cm^{-1})	ΔE kcal mol^{-1}	ΔE eV
2000	50 000	143.0	6.20
2500	40 000	114.4	4.96
3000	33 333	95.3	4.13
3500	28 571	81.7	3.54
4000	25 000	71.5	3.10
4500	22 222	63.5	2.76
5000	20 000	57.2	2.48
6000	16 666	47.7	2.07
8000	12 500	35.7	1.55

bond cleavage can occur only in several cases summarised in Figure 11.3. It has been shown in Chapter 2 that electronic transitions are restricted to cases in which there is no displacement of the atomic nuclei. Another restriction is that the multiplicity of the system (singlet or triplet) must be preserved during the transition. As most of the organic molecules are singlet species in the ground state (S_0), only $S_0 \rightarrow S_1$ transitions will be allowed by the selection rules of photochemical activation. If this electronic transition happens to lead either to a repulsive excited state (Figure 11.3A) or to an ex-

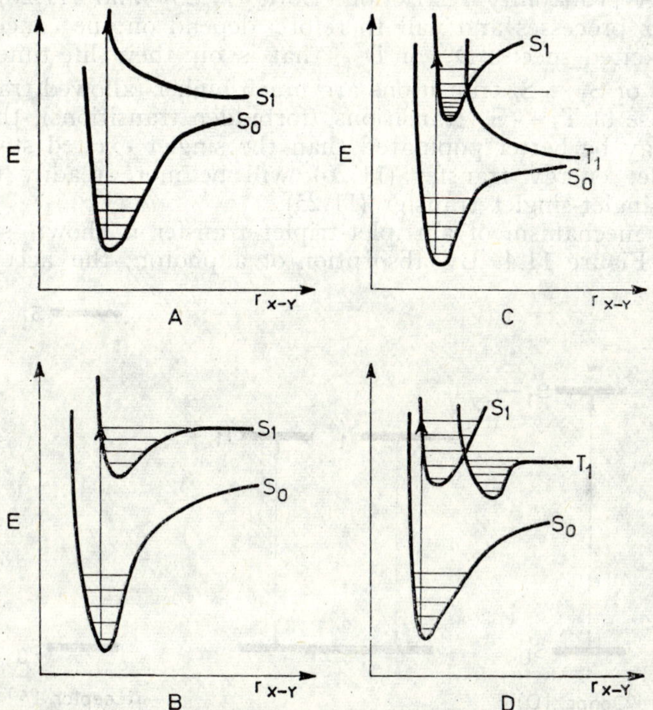

Figure 11.3 Electronic transitions resulting in the dissociation of a diatomic molecule X—Y. A, B from excited singlet states; C, D from excited triplet states.

cited state sufficiently high in energy to permit homolysis (B), the absorption of a photon will result in bond breaking. However, it has been shown that excited states S_1 may undergo intersystem crossing to T_1 states if the energy levels E_{S_1} and E_{T_1} are not separated by a large gap. Intersystem crossing, in its turn, may lead to a triplet repulsive state (C) or to a high vibrational level of T_1 (D), permitting thus bond cleavage. These mechanisms apply strictly to the dissociation of diatomic molecules, but do not differ essentially from the photochemical cleavage of polyatomic molecules.

The excitation of molecules by radiant energy can also be realised by *transfer of electronic energy* [75a] between an excited molecule (energy donor, D) and a ground state molecule (energy acceptor, A):

$$D_{S_1'} + A_{S_0} \longrightarrow D_{S_0'} + A_{S_1} \tag{11.25}$$

$$D_{T_1'} + A_{S_0} \longrightarrow D_{S_0'} + A_{T_1} \tag{11.26}$$

Both processes are allowed by the selection rules of photochemistry, inasmuch as spin is conserved. However, their frequency will also depend on two other factors:

(a) An energy restriction. The energy of the excited state of the donor must be higher than that of the acceptor, $D_{S_1'} > A_{S_1}$ or $D_{T_1'} > A_{T_1}$.

(b) A probability restriction. Both (11.25) and (11.26) are bimolecular processes and will therefore depend on the concentration of the excited species $D_{S_1'}$ or $D_{T_1'}$, that is on their life-times. Since the rates of $S_1 \rightarrow S_0$ transitions are much higher (allowed transitions) than those of $T_1 \rightarrow S_0$ transitions (forbidden transitions), the triplet states may be better populated than the singlet excited states. Triplet-triplet energy transfer (11.26) will be more readily observed than a singlet-singlet transfer (11.25).

The mechanism of a triplet-triplet transfer is shown schematically in Figure 11.4. By absorption of a photon, the acceptor can

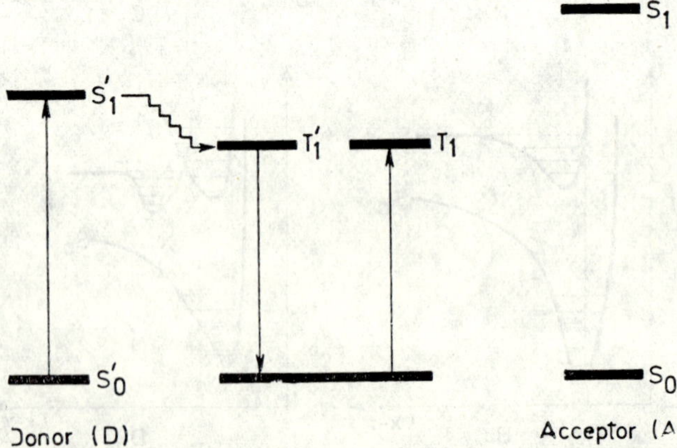

Fig. 11.4. Singlet-triplet transfer of radiant energy (for exemple in benzophenone-photosensitised reactions).

pass into the singlet excited state, S_1. The energy difference between this state and the ground state is very large and thus the molecules in this state will lose readily their activation energy before any chemical transformation occurs. The energy gap between the singlet state S_1 and triplet state T_1 is also too large to permit intersystem crossing. Thus, triplet states cannot be populated in this way and chemical reactions will not be possible by direct irradiation of acceptor A. However, if a suitable molecule (D) can be found, with close S_1' and T_1' levels permitting rapid intersystem crossing and efficient absorption of energy, and if $T_1' > T_1$, the triplet state of the acceptor can be populated and then involved in photochemical transformations. Such processes are called *photosensitised reactions* and the donor enabling such reactions, *photosensitiser*.

A photosensitiser with large applications is benzophenone. The excited singlet state of benzophenone is by 74 kcal mol^{-1} higher than its ground state and corresponds to an $n \rightarrow \pi^*$ transition. It is separated by only 5 kcal mol^{-1} from the triplet excited state (E_T = 69 kcal mol^{-1}). From its spectra it has been determined that each excited molecule S_1 passes over into the triplet state T_1[75b].

In conclusion, photochemical dissociation is possible either by direct irradiation or by indirect energy transfer processes. In the direct irradiation both singlet and triplet states may be involved. In photosensitised reactions, triplet states are the only species involved.

Applying the principle of the spin conservation to an elementary reaction the splitting of an R—X bond can lead either to a singlet cage (11.27) or to a triplet cage (11.28):

$$(\text{R—X})_{S_1} \longrightarrow \underset{\mathbf{32}}{(\text{R}\uparrow + \text{X}\downarrow)} \tag{11.27}$$

$$(\text{R—X})_{T_1} \longrightarrow \underset{\mathbf{33}}{(\text{R}\uparrow + \text{X}\uparrow)} \tag{11.28}$$

If cage effects are small, no difference between the two processes will be observed in the reaction products. However, cage coupling should occur more readily in (11.27) than in (11.28). In triplet splitting, the triplet cage must suffer spin inversion before the coupling of the radicals. Cage effects should therefore be more important in (11.27) than in (11.26). In the photodecomposition of azocumene, Nelson and Bartlett [76] observed no difference in cage effects between

[75] N. J. Turro, *Molecular Photochemistry*, W. A. Benjamin, New York (1965); (a) p. 92; (b) p. 224.

[76] S. F. Nelson and P. D. Bartlett, *J. Amer. Chem. Soc.*, **88**, 141 (1966).

the sensitised and the unsensitised process. They concluded that spin inversion must be more rapid than radical coupling:

$$\begin{array}{c} 2R\cdot \longrightarrow \text{products} \\ \uparrow \\ R{-}N{=}N{-}R \xrightarrow{h\nu} (R\uparrow + N_2 + R\downarrow) \longrightarrow R{-}R + N_2 \\ \uparrow \text{rapid} \\ R{-}N{=}N{-}R \xrightarrow[\text{sensitiser}]{h\nu} (R\uparrow + N_2 + R\uparrow) \\ \downarrow \\ 2R\cdot \longrightarrow \text{products} \end{array}$$

In contrast with this result, Shriner[77] observed large differences between the two process in the photodecomposition of cyclic compounds:

Ph—CH—CH—Me / Ph—N—N=N (triazoline) → 35 (aziridine: H, Ph / C—C / H, Me; N—Ph) + 36 (aziridine: H, Ph / C—C / Me, H; N—Ph)

	34 (*cis* or *trans*)	35	36
(A) direct irradiation			
	cis:	80%	20%
	trans:	24%	76%
(B) photosensitised reaction			
	cis and *trans*:	60%	24%

The lack of stereospecificity in the photosensitised reaction can be explained by the formation of a triplet diradical **37a,** which must undergo spin inversion before cyclisation. Spin inversion must be slower than the molecular rotation about the carbon-carbon bond. The ratio of the isomeric aziridines will reflect their stability. Conversely, in the direct irradiation the singlet diradical **37b** is formed, which undergoes cyclisation more rapidly than molecular rotation. The reaction will thus display a certain stereospecificity.

$$\begin{array}{ccc} Ph{-}CH{-}\dot{C}HMe & & Ph{-}CH{-}\dot{C}HMe \\ |\quad\ \uparrow & ; & |\quad\ \downarrow \\ Ph{-}N\cdot\uparrow & & Ph{-}N\cdot\uparrow \\ \mathbf{37a} & & \mathbf{37b} \end{array}$$

The photochemical decompositions of carbonyl compounds are used to generate alkyl radicals according to the following scheme[13, 78]:

$$RCO{-}R \xrightarrow{h\nu} RCO\cdot + R\cdot$$

$$RCO\cdot \longrightarrow R\cdot + CO$$

[77] P. Shriner, *J. Amer. Chem. Soc.*, **88**, 4759 (1966).

[78] A. S. Kallend and J. N. Pitts, Jr., *J. Amer. Chem. Soc.*, **91**, 1269 (1969).

Cases are known in which the photochemical activation does not lead to the cleavage of the CO—R bond. Thus, chloroacetone [79] is split into a chlorine atom and an acetonyl radical:

$$\mathrm{MeCOCH_2Cl} \xrightarrow{h\nu} \mathrm{MeCOCH_2}\cdot + \mathrm{Cl}\cdot$$

Azo compounds as well as peroxy derivatives can be cleaved either thermally or photochemically. The photochemical decomposition of diacylperoxides in the cavity of an electron spin resonance spectrophotometer can serve for the direct observation of free radicals [80].

It has thus been found that alkyl radicals are the only observable radicals, the acyloxy radicals having a shorter lifetime than the time scale of the electron-resonance phenomenon:

$$\mathrm{(RCOO)_2} \longrightarrow 2\mathrm{R}\cdot + 2\mathrm{CO_2}$$

Photochemical reactions. In photochemical reactions the activation of the chemical process is produced by absorption of a photon. As a result of this absorption, the molecules undergo electronic transitions into excited states differing from the ground state in their energy, electronic configuration and geometry. The reactivity of the excited states will thus be quite different from the reactivity of ground state molecules.

The ratio between the molecules undergoing a particular reaction and the number of molecules which have absorbed a photon of light defines the quantum yield:

$$\Phi = \frac{\text{number of molecules undergoing photochemical reaction}}{\text{number of quanta absorbed by the system}} \tag{11.29}$$

The quantum yields of photochemical reactions are important because they inform us about the ways in which the molecules dispose of the absorbed energy. If each photon absorbed produces an elementary chemical transformation, the quantum yield is equal to unity. For most photochemical reactions the quantum yields are less than unity. However, chain reactions have high quantum yields, $\Phi > 1$.

Photochemical reactions may be either unimolecular or bimolecular processes [75, 81]. The isomerisation of *cis-trans* derivatives (or *syn-anti*) is one of the best known processes involving photochemical activation [82].

$$\mathrm{(H)(R)C{=}C(R)(H)} \underset{}{\overset{h\nu}{\rightleftharpoons}} \mathrm{(H)(R)C{=}C(H)(R)}$$

The mechanism of *cis-trans* isomerisations may be quite complex. For conjugated systems, polyenes or arylalkenes such as stilbenes, a

[79] A. N. Strachan and F. E. Blacet, *J. Amer. Chem. Soc.*, **77**, 5254 (1955).
[80] J. K. Kochi and P. J. Krusic, *J. Amer. Chem. Soc.*, **91**, 3940 (1969).
[81] P. A. Leermakers and G. F. Vesley, *J. Chem. Educ.*, **41**, 535 (1964); J. S. Swenton, *J. Chem. Educ.*, **46**, 7 (1969).
[82] For a review see H. O. Kalinovski and H. Kessler, *Topics in Stereochemistry*, **7**, 295 (1972).

diradical mechanism seems probable. This mechanism implies the promotion of an electron from the highest occupied bonding orbital π into the lowest antibonding orbital π^* by an $S_0 \rightarrow S_1$ transition followed by intersystem crossing to the triplet state T_1. More refined MO calculations have indicated [75] that the most stable configuration for π^* triplet states corresponds to a 90° twisting about the carbon-carbon bond (Figure 11.5). Decay from the perpendicular configuration may lead either to the *cis* or to the *trans* isomer. Delocalisation of the two non-interacting radical systems, as in **38,** may favour this mechanism.

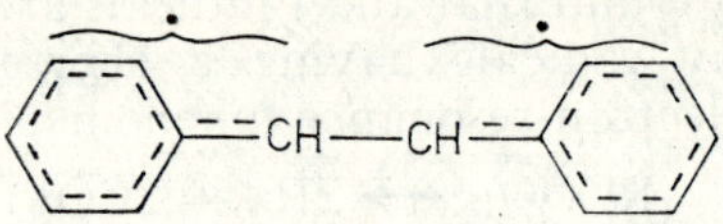

38

For unconjugated alkenes, a stepwise dissociation and recombination mechanism may also compete with the diradical mechanism [82]. In the first step of the reaction (11.30), a vinyl radical is formed in a solvent cage, by photochemical cleavage of the C—H bond. Before cage-coupling between the radical and the hydrogen atom, an inversion at the radical centre may occur by rehybridisation of the sp^2 carbon atom.

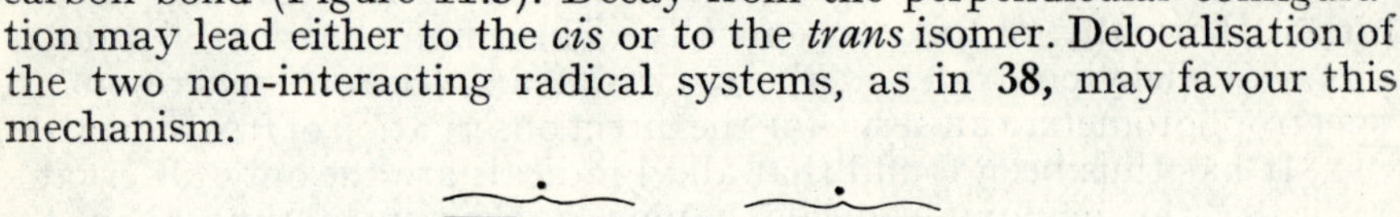

(11.30)

Isomerisation and fragmentation reactions are two other types of unimolecular photochemical reactions. Not very rarely, these reactions may be simultaneously observed, such as in the photochemical transformations of α-alkoxyacetophenones [83, 84]. Again, com-

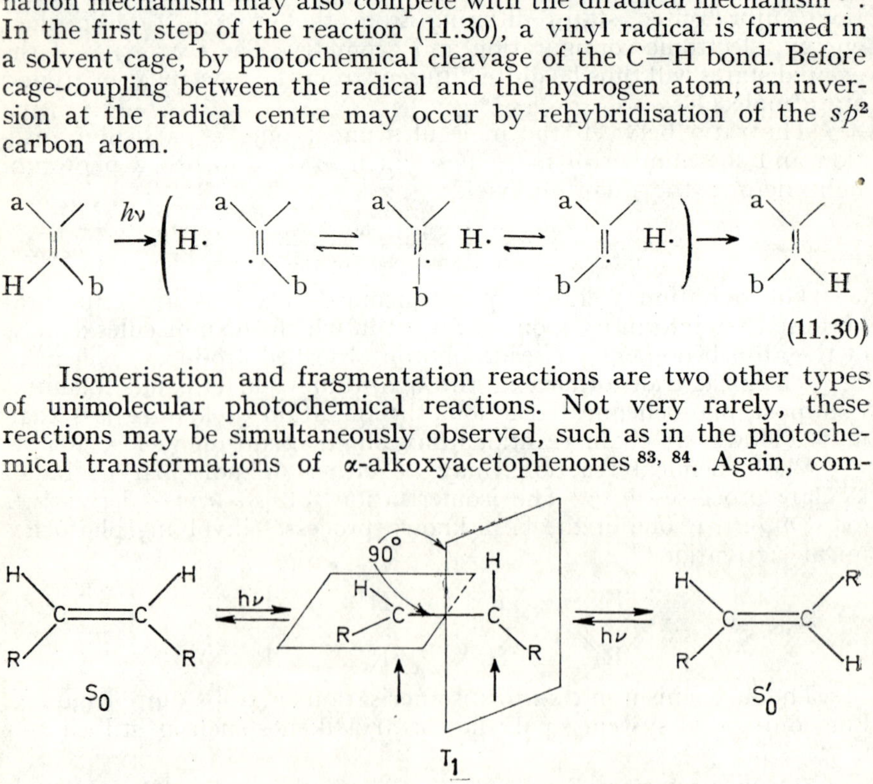

Figure 11.5 Photochemical *cis-trans* isomerisation.

[83] F. D. Lewis and N. J. Turro, *J. Amer. Chem. Soc.,* **92,** 311 (1970).

[84] T. R. Darling, N. J. Turro, R. H. Hirsch and F. D. Lewis, *J. Amer. Chem. Soc.,* **96,** 434 (1974).

mon diradical intermediates may be involved. The formation of these reactive species is favoured by the easy intersystem crossing of the excited carbonyl derivatives.

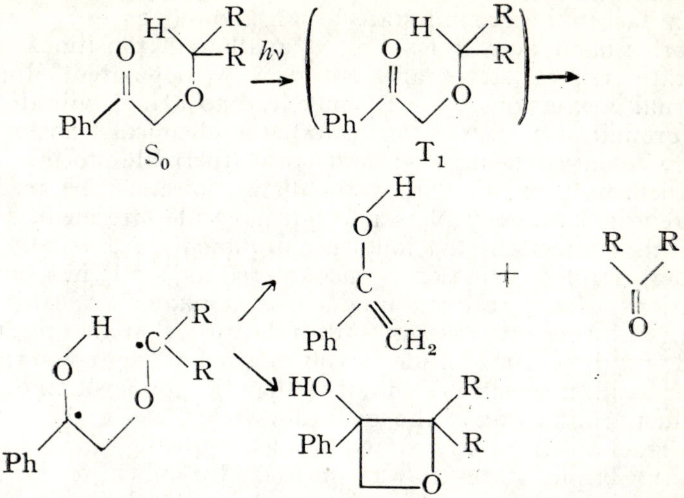

Among isomerisation reactions, valence isomerisation has recently acquired a great importance in the literature. These photochemical reactions are closely related to the thermal isomerisations of unsaturated compounds and will be treated in the last chapter of this book. A similar situation is found for the bimolecular photochemical transformations, such as photochemical cycloadditions:

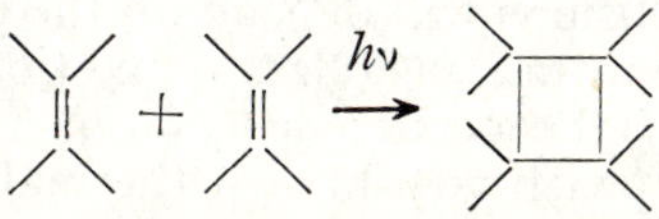

Cycloaddition reactions occur either as thermal or as photochemical processes. Chapter 12 will deal with the mechanism of these reactions.

In any study of a photochemical reaction, the identification of the activated species is one of the most important steps in the elucidation of the mechanism. Reactions from higher excited states ($S_2,S_3, \dots ,T_2,T_3, \dots$) are highly improbable. Indeed, excited molecules can lose their radiant energy either by radiative processes (emission) or by radiationless decay (internal conversions) to the ground state S_0. Both radiative allowed transformations ($S_n \rightarrow S_1$) and internal conversions ($S_n \rightsquigarrow S_1$) from higher excited states to the first excited state occur with high rates, $10^{11}-10^{13}\,s^{-1}$, corresponding to exceedingly short lifetimes. Both radiative and radiationless decay from the first excited states S_1 and T_1 are much slower and may compete with chemical transformations. Thus, the $S_1 \rightarrow S_0$ emissions, called *fluorescence,* are of the order $10^9 - 10^6\,s^{-1}$, while $T_1 \rightarrow S_0$ transitions, called *phosphorescence,* have rates, for the majority of organic molecules, between 10^3 and $10^{-1}\,s^{-1}$. This shows that chemical reactions can be observed both from singlet and from triplet excited states. However, one cannot decide *a priori* which is the most probable excited state that undergoes reaction. Singlet excited states S_1 are directly formed by direct, allowed $S_0 \rightarrow S_1$ transitions, but can decay relatively fast to the ground state by an allowed $S_1 \rightarrow S_0$ transition. On the other hand, even if $T_1 \rightarrow S_0$, as unallowed transitions have a slower rate, triplet states are not directly populated. Moreover, the internal conversions $S_1 \rightarrow S_0$ may lead to a high vibratory level of the ground state S_0 (*hot molecules*), the chemical transformation actually observed being a ground state (thermal) process. Thus, in a photochemical reaction study the distinction must be made between excited singlets and triplets, and hot molecule intermediates. In making this distinction, the following arguments are usually used:

(a) Triplet formation is encountered as a rule in photosensitised reactions. If the reaction products of a photochemical reaction are changed by sensitisers, this is an indication that the reaction performed by direct irradiation involves singlet excited states. On the other hand, if sensitisers do not change the composition of the photoproducts, but increase the quantum yields, it can be assumed that the reaction involves excited triplet states.

(b) Triplet states may be *quenched* by radical or diradical triplet molecules such as oxygen (one of the few molecules having a T_0 ground state configuration). The inhibition of a chemical reaction by triplet quenchers is a strong argument for a photochemical reaction involving the triplet state.

(c) The radiationless decay to higher vibratory levels of the ground state depends on the frequency of the bimolecular collisions between activated molecules and unactivated molecules. Clearly, the frequency of these collisions will depend on the pressure of the system. The study of the quantum yield as a function of the pressure will give an indication of the occurrence of hot-molecule intermediates. Anyway, comparison with well-known thermal processes is necessary in such cases.

(d) Theoretical treatments, based especially on MO theory, may be very useful in deciding the reaction mechanism.

The following example constitutes a good illustration of the application of these principles. Irradiation of cyclohepta-1,3,5-triene in the vapour phase produces toluene as the major product, and the bicyclic compound **39**[85].

(a) The reaction may be photosensitised by mercury[86]. Under these conditions only toluene is formed.

$h\nu$ (Hg) or heat → CH_3

[85] W. G. Dauben and R. L. Cargill, *Tetrahedron*, 12, 186 (1961).
[86] R. Srinivasan, *J. Amer. Chem. Soc.*, **84**, 3432 (1962).

In dilute ether solutions, the major product of irradiation is the bicyclic compound **39** [87]:

$$\text{cycloheptatriene} \xrightarrow[\text{dilute solution}]{h\nu} \underset{\mathbf{39}}{\text{bicyclo[3.2.0]heptadiene}} + \left(\text{toluene } (CH_3)\right)$$

(b) The formation of the photolysis products is not inhibited by oxygen, which indicates the absence of triplet states.

(c) The quantum yields of the bicyclic compound are not affected by pressure, whereas quantum yields of toluene increase with increasing pressure. The only product detected in the pyrolysis of cycloheptatriene at 480 °C is toluene [88].

(d) A MO treatment of the valence isomerisation of cycloheptatriene (see Electrocyclic reactions, Chapter 12) indicates that **39** may be formed in a concerted mechanism from the first excited state of cycloheptatriene.

All these facts show that bicyclo[3.2.0]heptadiene is a product of an excited singlet state S_1, whereas toluene is formed from higher vibratory levels of the ground state S_0 (hot molecule) of cycloheptatriene.

11.3. Radical transformations

The decomposition of a higher free radical into a lower free radical and a stable molecule is quite frequent, as for example:

$$CH_3CH_2CH_2\cdot \longrightarrow CH_3\cdot + CH_2{=}CH_2 \qquad (11.4)$$

$$ArCOO\cdot \longrightarrow Ar\cdot + CO_2 \qquad (11.31)$$

$$RCO\cdot \longrightarrow R\cdot + CO \qquad (11.32)$$

$$(CH_3)_3C{-}O\cdot \longrightarrow CH_3COCH_3 + CH_3\cdot \qquad (11.33)$$

The picture of a homolytic reaction may become very complex due to the occurrence of such decompositions. The lower the reactivity of the substrate towards radical attack, the more important the decomposition will be. In the decomposition of benzoyl peroxide, for example, the benzoyloxy radicals can be trapped by reactive substrates, such as anthracene [88]; with the less reactive benzene, trapping products of both benzoyloxy and phenyl radicals are identified.

[87] W. G. Woods, *J. Org. Chem.*, **23**, 110 (1958).
[88] M. Mingin and K. H. Pausacker, *Austral. J. Chem.*, **18**, 821, 831 (1965).

The t-butoxy radical is relatively stable [69–71]. However, reaction (11.33) may be used for generating methyl radicals from di-t-butyl peroxide [89, 90].

The transformation of acyl radicals according to reaction (11.32) is observed in the photolysis of carbonyl compounds. It may also occur in the decarbonylation of aldehydes. An interesting example in which both a decarbonylation reaction and a radical rearrangement have been observed is the decarbonylation of β-phenylisovaleraldehyde **40** [91]. The initially formed acyl radical is decomposed into the neophyl radical **41,** which partially rearranges to the more stable benzyl radical **42**:

$$\underset{\mathbf{40}}{C_6H_5{-}\overset{\displaystyle CH_3}{\underset{\displaystyle CH_3}{C}}{-}CH_2{-}CH{=}O} \xrightarrow{R^1\cdot} C_6H_5{-}\overset{\displaystyle CH_5}{\underset{\displaystyle CH_2}{C}}{-}CH_2{-}CO\cdot \longrightarrow$$

$$\underset{\mathbf{41}}{C_6H_5{-}\overset{\displaystyle CH_3}{\underset{\displaystyle CH_3}{C}}{-}CH_2\cdot} \longrightarrow \underset{\mathbf{42}}{\cdot\overset{\displaystyle CH_3}{\underset{\displaystyle CH_3}{C}}{-}CH_2{-}C_6H_5}$$

$$\mathbf{41} \xrightarrow{R^2CHO} C_6H_5{-}\overset{\displaystyle CH_3}{\underset{\displaystyle CH_3}{C}}{-}CH_3 \qquad \mathbf{42} \xrightarrow{R^2CHO} C_6H_5{-}CH_2{-}CH(CH_3)_2$$

$R^2CHO = C_6H_5(CH_3)_2CCHO$

The lower the concentration of aldehyde **40,** the higher the ratio of the rearranged hydrocarbon [92]. This shows that the capture of the neophyl radical by R^2CHO competes well with the rearrangement of this radical.

Rearrangement reactions of free radicals, even if possible, are less characteristic of their chemistry than are the rearrangements of the corresponding carbenium ions [93].

[89] D. H. Shaw and H. O. Pritchard, *J. Phys. Chem.*, **70**, 1230 (1966).
[90] D. H. Shaw and H. O. Pritchard, *Can. J. Chem.*, **45**, 2749 (1967).
[91] S. Winstein and F. H. Seubold, Jr., *J. Amer. Chem. Soc.*, **69**, 2916 (1947).
[92] F. H. Seubold, Jr., *J. Amer. Chem. Soc.*, **75**, 2532 (1953).
[93] C. Walling, in *Molecular Rearrangements* (P. de Mayo, editor), Interscience Publishers, New York, vol. 1 (1964), p. 407.

11.4 Radical attack on substrates

11.4.1 *Radical displacements*

A displacement by a free radical X· at a $\gt$C—H bond may in principle occur in two ways:

$$X\cdot + H{-}C\lt \longrightarrow \left[X\cdots H\cdots C\lt\right] \rightarrow XH + \cdot C\lt \qquad (11.34)$$

43

$$X\cdot + \gt C{-}H \longrightarrow \left[X\cdots \overset{|}{\underset{\wedge}{C}}\cdots H\right] \rightarrow X{-}C\lt + H\cdot \qquad (11.35)$$

44

Reaction (11.34) is a homolytic substitution (SH) at hydrogen, whereas (11.35) represents a substitution at the carbon atom. Of the two, reaction (11.34) is preferred. Transition state **43** must be attained more easily than transition state **44,** due to a better compensation of the dissociation energy of the C—H bond by the formation energy of the new bond (generally $E_{XH} > E_{XC}$), a more accessible reaction centre and a higher stability of the resultant radicals than that of the free hydrogen atom.

Numerous examples illustrating equation (11.34) are known. Thus, in the homolytic halogenation of organic molecules [94] a halogen atom is a *chain carrier*, the first reaction of the chain-propagation step corresponding to process (11.34).

$$X_2 \xrightarrow{h\nu} 2X\cdot \qquad (11.36a)$$

$$X\cdot + H{-}R \longrightarrow [X\cdots H\cdots R] \rightarrow HX + R\cdot \qquad (11.36b)$$

$$R\cdot + X_2 \longrightarrow [R\cdots X\cdots X] \rightarrow RX + X\cdot \qquad (11.36c)$$

There are also cases in which the chain carrier in a homolytic halogenation is not a halogen atom. An example is the halogenation with t-butyl hypochlorite [72, 74], in which t-butoxy radicals are the chain carriers:

$$Bu^tOCl \longrightarrow Bu^tO\cdot + Cl\cdot \qquad (11.37a)$$

$$Bu^tO\cdot + H{-}R \longrightarrow Bu^tOH + R\cdot \qquad (11.37b)$$

$$R\cdot + Bu^tOCl \longrightarrow RCl + Bu^tO\cdot \qquad (11.37c)$$

[94] See Reference 3, p. 347.

Sulphuryl chloride is another effective chlorinating agent [95, 96]. The reaction, initiated by peroxides, is generally formulated as:

$$Cl\cdot + RH \longrightarrow HCl + R\cdot \tag{11.38a}$$

$$R\cdot + SO_2Cl_2 \longrightarrow RCl + \cdot SO_2Cl \tag{11.38b}$$

$$\cdot SO_2Cl \longrightarrow SO_2 + Cl\cdot \tag{11.38c}$$

The chain termination reactions, both in the direct halogenation (11.36) and in the indirect halogenations (11.37) and (11.38), are usually combination reactions between free radicals and free atoms.

The homolytic mechanism of the halogenation is demonstrated by numerous experimental data. Thus, the chlorination of ethane can be initiated by compounds like tetraethyl lead [97], which can decompose thermally into free radicals. The halogenation with t-butylhypochlorite may be realised either by photolysis[73] or by a thermal process using as initiator azo-bisisobutyronitrile [74]. The kinetics of the reaction is in agreement with the accepted mechanisms of halogenation. A more convincing argument is based on the stereochemistry of the reaction. Theory predicts that free radicals are planar and thus achiral. If such intermediates are involved, the substitution at a chiral centre will have as a result the complete racemisation of the initial optically active substrate. Brown, Kharasch and Chao [98] have thus observed that the homolytic chlorination of optically active 1-chloro-2-methylbutane yields racemic dichloromethylbutane, which indicates the intervention of a planar radical.

$$\underset{\text{optically active}}{CH_3{-}CH_2{-}\overset{\displaystyle CH_3}{\underset{\displaystyle H}{\overset{|}{\underset{|}{C}}}}{-}CH_2Cl} \xrightarrow{Cl\cdot} CH_3{-}CH_2{-}\dot{C}\begin{matrix} \diagup CH_3 \\ \diagdown CH_2Cl \end{matrix} \xrightarrow{Cl_2}$$

$$\underset{\text{racemic}}{CH_3{-}CH_2{-}CCl(CH_3){-}CH_2Cl}$$

Under certain circumstances, the halogenation may proceed with retention of optical activity. Skell and his co-workers [99] have thus observed that the bromination of optically active 1-bromo-2-methylbutane ($\alpha^{27} = 2.86°$), carried out between 0° and 72 °C, yields an optically active product ($\alpha^{27} = -2.33°$). They have assumed

[95] M. S. Kharasch and H. C. Brown, *J. Amer. Chem. Soc.*, **61**, 2142, 3422 (1939).

[96] For an excellent historical review see H. C. Brown, *Boranes in Organic Chemistry*, Cornell University Press, Ithaca (1972), p. 24.

[97] W. E. Vaughan and F. F. Rust, *J. Org. Chem.*, **5**, 449 (1940).

[98] H. C. Brown, M. S. Kharasch and T. H. Chao, *J. Amer. Chem. Soc.*, **62**, 3435 (1940).

[99] P. S. Skell, D. L. Tuleen and P. D. Readio, *J. Amer. Chem. Soc.*, **85**, 2849 (1963).

that a cyclic radical, similar to a halogenium ion, would be responsible for the stereospecificity of the reaction:

$$CH_3{-}CH_2{-}\underset{H}{\overset{CH_3}{C}}{-}CH_2Br + Br\cdot \longrightarrow CH_3{-}CH_2{-}\underset{\underset{Br}{\vdots}}{\underset{H}{\overset{CH_3\ \ Br}{C}}}{-}{-}CH_2 \longrightarrow$$

45

$$\longrightarrow CH_3{-}CH_2{-}\overset{CH_3\ Br}{C}{-}{-}CH_2 \xrightarrow[-HBr]{Br_2} CH_3{-}CH_2{-}\underset{Br}{\overset{CH_3}{C}}{-}CH_2Br$$

Thus, in the transition state **45** the neighbouring bromine atom assists the cleavage of the C—H bond.

The allylic and benzylic bromination by *N*-bromosuccinimide (NBS)[100] has large applicability in preparative organic chemistry. This is a homolytic chain reaction and it has been believed [101] that succinimidyl radicals are chain carriers:

$$\begin{matrix}CH_2{-}CO\\ |\\ CH_2{-}CO\end{matrix}\!\!>\!N{-}Br \longrightarrow \begin{matrix}CH_2{-}CO\\ |\\ CH_2{-}CO\end{matrix}\!\!>\!N\cdot + Br\cdot \qquad (11.39)$$

46 **47**

$$\mathbf{47} + {-}CH{=}CH{-}CH_2{-} \longrightarrow \begin{matrix}CH_2{-}CO\\ |\\ CH_2{-}CO\end{matrix}\!\!>\!NH + {-}\overset{\cdot}{\overbrace{CH{\equiv}CH{\equiv}CH}}{-}$$

48

$$\mathbf{48} + \mathbf{46} \longrightarrow \mathbf{47} + {-}CH{=}CH{-}CHBr{-} \quad \text{etc.}$$

The assumption was based on the observation that succinimide was quantitatively recovered from the reaction mixture.

Adams, Gosselain and Goldfinger [102] have proposed a different mechanism in which the chain carriers are bromine atoms:

$$\mathbf{46} + HBr \longrightarrow \begin{matrix}CH_2CO\\ |\\ CH_2CO\end{matrix}\!\!>\!NH + Br_2 \qquad (11.40)$$

$$Br_2 \rightleftharpoons 2\,Br\cdot \qquad (11.41)$$

$$Br\cdot + {-}CH{=}CH{-}CH_2{-}\ (\text{or } ArCHR_2) \longrightarrow HBr + \mathbf{48}\ (\text{or } Ar\dot{C}R_2) \qquad (11.42)$$

$$\mathbf{48}\ (\text{or } Ar\dot{C}R_2) + Br_2 \longrightarrow {-}CH{=}CH{-}CHBr{-} \quad (\text{or } ArCBrR_2) + Br\cdot \qquad (11.43)$$

[100] K. Ziegler, A. Spaeta, E. Schaaf, W. Schumann and E. Winkelmann, *Liebigs Ann. Chem.*, **551**, 80 (1942).

[101] G. F. Bloomfield, *J. Chem. Soc.*, 114 (1944).

[102] J. Adams, P. A. Gosselain and P. Goldfinger, *Nature*, **171**, 704 (1953).

This mechanism is supported by experimental data[103–106]. Thus, there is a close parallelism between the bromination by molecular bromine and by NBS. In both reactions substituted toluenes are brominated in the benzylic position, their rates being correlated with σ^+ parameters. The reaction constants ρ are practically identical in both processes[103–104], indicating that the same radical reagent must be involved in the abstraction of the benzylic hydrogen. Clearly, that reagent must be the bromine atom. It is thought that the role of NBS in the reaction consists of maintaining a very low bromine concentration. Indeed, when cyclohexene is treated with molecular bromine, under homolytic conditions, the addition prevails. However, if bromine is added to the system at such a rate that its concentration is kept as low as possible, 3-bromocyclohexene is formed preferentially[104] (allylic substitution).

The transition state in allylic or benzylic substitution seems to have a polar character, as indicated by the negative value of the reaction parameter, $\rho = -1.39$ (80°) *[103]. The origin of the polar interactions during free-radical abstraction is attributed to the influence of dipole-dipole interactions on the energy of the transition state[107]. The transition state in bromination may be represented by the resonance hybrid **49**:

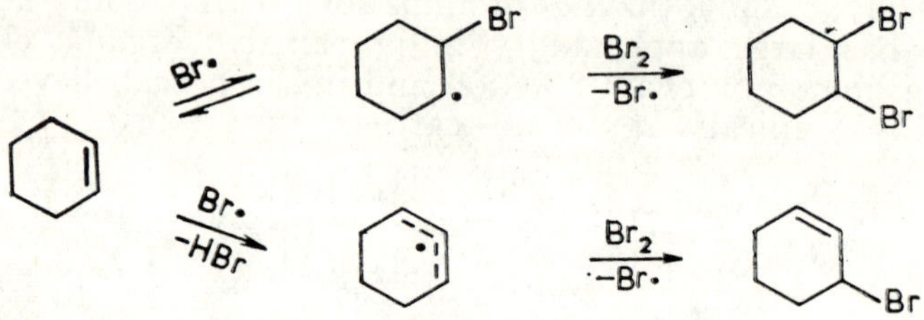

$$\underset{a}{ArCH_2{-}H\ \ Br\cdot} \longleftrightarrow \underset{b}{ArCH_2^{+}H\cdot\ Br^-} \longleftrightarrow \underset{c}{ArCH_2\cdot\ HBr}$$

49

The canonical structure *b* is responsible for the negative value of ρ.

High negative values of ρ have also been observed in other radical abstraction reactions of substituted toluenes[108–110]. The side-chain bromination of alkyl aromatics by polyhalogenomethanes

* Positive ρ values have also been determined recently (W. A. Pryor and W. H. Davis, Jr., *J. Amer. Chem. Soc.*, **96**, 7557 (1974); R. W. Henderson and R. D. Ward, Jr., *ibid*, **96**, 7556 (1974).

[103] R. E. Pearson and J. C. Martin, *J. Amer. Chem. Soc.*, **85**, 354, 3142 (1963); G. A. Russel and K. M. Desmond, *ibid.*, **85**, 3139 (1963).

[104] C. Walling, A. L. Rieger and D. D. Tanner, *J. Amer. Chem. Soc.*, **85**, 3129 (1963).

[105] B. P. McGrath and J. M. Tedder, *Proc. Chem. Soc.*, 80, 199 (1961).

[106] S. S. Friederich, E. C. Friederich, L. J. Andrews and R. M. Keefer, *J. Org. Chem.*, **34**, 900 (1969).

[107] J. M. Tedder, *Quart. Rev. (Chem. Soc.)*, **14**, 336 (1960).

[108] C. Walling and B. Miller, *J. Amer. Chem. Soc.*, **79**, 4181 (1957).

[109] E. S. Huyser, *J. Amer. Chem. Soc.*, **82**, 391, 394 (1960); *J. Org. Chem.*, **26**, 3261 (1961).

[110] A. A. Zavitsas and J. A. Pinto, *J. Amer. Chem Soc,*. **94**, 7390 (1972).

such as bromotrichloromethane. is of special interest. Huyser[109] has proposed a chain-mechanism in which trichloromethyl radicals are chain carriers:

$$\cdot CCl_3 + RH \longrightarrow R\cdot + HCCl_3$$

$$R\cdot + CCl_3Br \longrightarrow RBr + \cdot CCl_3$$

This mechanism is indicated by high yields of bromo-derivative and chloroform and the formation of small amounts of hexachloroethane. However, small amounts of hydrogen bromide are also found, showing that bromine atoms may be in their turn involved in the process[111]. The involvement of free bromine atoms is also indicated by the ρ value (-1.46[103]) which is close to that determined for the bromination by molecular bromine. A recent study of the bromination with bromotrichloromethane[112] has indicated that the ρ parameter decreases from -1.24 to -0.69 if hydrogen bromide is completely scavenged from the system (with powdered potassium carbonate or better by addition of ethylene oxide[113]). Moreover, competitive bromination of toluene and toluene-α-d_3 has indicated deuterium scrambling in the unhalogenated toluene. Therefore, in presence of hydrogen bromide the following mechanism may operate:

$$ArCH_2\cdot + BrCCl_3 \longrightarrow ArCH_2Br + \cdot CCl_3 \qquad (11.44)$$

$$\cdot CCl_3 + HBr \rightleftharpoons CCl_3H + Br\cdot \qquad (11.45)$$

$$ArCH_3 + Br\cdot \rightleftharpoons ArCH_2\cdot + HBr \qquad (11.46)$$

Hydrogen abstraction by bromine atoms is thus reversible and accounts for the deuterium scrambling. Reaction (11.45) is also encountered in the direct bromination by molecular bromine and explains the high negative values of ρ if hydrogen bromide is not scavenged from the system. Hydrogen abstraction reactions are sometimes used in the generation of stable free radicals, such as the cyclohexadienyl radical **50**[114].

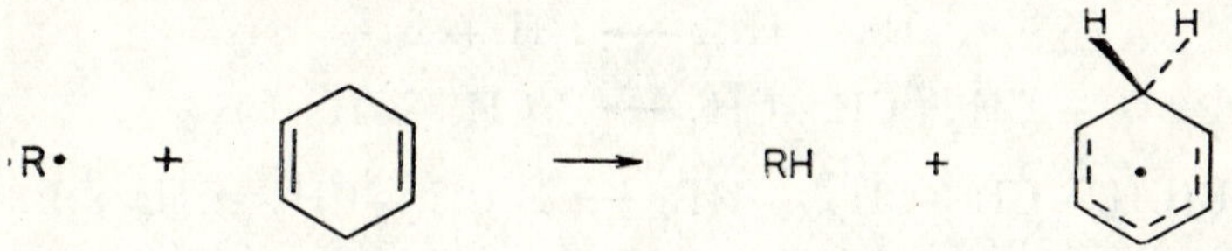

$\cdot R\cdot$ + → RH +

50

Radical displacements at hydrogen according to Scheme (11.34) are the most common radical displacements, but are not limited to that atom. Less frequently observed are radical displacements at the

[111] G. A. Russel and C. De Boer, *J. Amer. Chem. Soc.*, **85**, 3136 (1963).

[112] D. D. Tanner, R. J. Arhart, E. V. Blackburn, N. C. Das and N. Wada, *J. Amer. Chem. Soc.*, **96**, 829 (1974).

[113] E. S. Huyser and R. H. C. Feng, *J. Org. Chem.*, **36**, 731 (1971).

[114] D. G. L. James and R. D. Stuart, *J. Amer. Chem. Soc.*, **86**, 5424 (1964).

carbon atom [3] and even at heteroatoms such as boron and aluminium [115], and phosphorous [116, 117]:

$$R^2O\cdot + BR^1_3 \longrightarrow R^2OBR^1_2 + R^1\cdot$$

$$R^2O\cdot + PR^1_3 \longrightarrow R^2OPR^1_2 + R^1\cdot$$

$$R^2O\cdot + P(OR^1)_3 \longrightarrow R^2OPO(OR^1)_2 + R^1.$$

$$R^2O\cdot + P(OR^1)_3 \longrightarrow (R^1O)_3PO + R^2\cdot \text{ etc.}$$

A generic label for these reactions is S_H2 (substitution, homolytic, bimolecular), in agreement with the symbols used for heterolytic reactions.

11.4.2 *Radical additions*

Radical additions may be represented by the general equations (11.47)–(11.49):

$$X{-}Y \longrightarrow X\cdot \tag{11.47}$$

$$X\cdot + \ \rangle C{=}C\langle \ \longrightarrow X{-}\overset{|}{\underset{|}{C}}{-}\overset{|}{\underset{|}{C}}\cdot \tag{11.48}$$

$$X{-}\overset{|}{\underset{|}{C}}{-}\overset{|}{\underset{|}{C}}\cdot + X{-}Y \longrightarrow X{-}\overset{|}{\underset{|}{C}}{-}\overset{|}{\underset{|}{C}}{-}Y + X\cdot \tag{11.49}$$

A first example in which such a mechanism has been proved is the abnormal addition of hydrogen bromide in the presence of peroxides [4–6]. In this case the chain carriers are bromine atoms.

$$\text{Peroxide} \longrightarrow R\cdot$$

$$R\cdot + HBr \longrightarrow RH + Br\cdot$$

$$Br\cdot + CH_2{=}CH{-}CH_3 \longrightarrow BrCH_2{-}\dot{C}H{-}CH_3$$

$$BrCH_2{-}\dot{C}H{-}CH_3 + HBr \longrightarrow BrCH_2{-}CH_2{-}CH_3 + Br\cdot$$

The bromine atom, like other free atoms and free radicals, attacks the marginal carbon atom of the alkenic linkage. The orientation is partly due to a steric control and partly to a thermodynamic control. Indeed, the addition of free atoms or free radicals to a double bond is reversible and secondary radicals are more stable than primary, due to a stabilisation by hyperconjugation [118].

115 P. J. Krusic and J. K. Kochi, *J. Amer. Chem. Soc.*, **91**, 3942 (1969).
116 C. Walling and M. S. Pearson, *Topics Phosphorus Chem.*, **3**, 1 (1966).
117 J. K. Kochi and P. J. Krusic, *J. Amer. Chem. Soc.*, **91**, 3944 (1969).
118 N. Muller and R. S. Mulliken, *J. Amer. Chem. Soc.*, **80**, 3489 (1958).

Peroxide addition of hydrogen bromide differs strikingly from the heterolytic addition of the same reagent. The heterolytic reaction presents another orientation, is catalysed by Lewis acids and takes place in the dark. The abnormal addition requires an initiation and is accelerated by light. Other additions of X—Y reagents able to add to the double bond either by a homolytic or by a heterolytic mechanism may present some difficulties in ascribing the particular reaction mechanism. One example is the chlorine addition to cyclohexene[119]. Both under homolytic and heterolytic conditions *trans*-1,2-dibromocyclohexane is the main reaction product. The formation of about 4% *cis*-dichlorocyclohexane in the radical reaction cannot be a sound argument in favour of this mechanism, the chlorine addition being less stereospecific even under heterolytic conditions (see p. 493). A clear indication for the radical mechanism is the formation of 4-chlorocyclohexene, which cannot be explained by a heterolytic mechanism, but may be attributed to the highly reactive and unselective chlorine atoms:

+ Cl• ⟶ + (+HCl)

Cl

+

Cl

The homolytic mechanism was further demonstrated by carrying out the addition in presence of cyclohexane; chlorocyclohexane was thus formed, demonstrating unambiguously the involvement of free chlorine atoms.

Rearrangement reactions, if observed, are a clear indication that the homolytic pathway is excluded[120]. In heterolytic additions to bicyclic alkenes, rearrangement is invariably observed, whereas in the corresponding homolytic process the initial framework is preserved[121].

The reactivity of the alkenes with respect to the free radicals may vary with the radical structure. A class of free radicals, called also *nucleophilic free-radicals*, add preferentially to alkenes substituted with electron-attracting groups, whereas another class of radicals, called *electrophilic free-radicals*, react easier with alkenes substituted with electron-repelling groups. This differentiation of the free radicals has bearing in the understanding of the *copolymerisation*

[119] M. L. Poutsma, *J. Amer. Chem. Soc.*, **87**, 2161 (1965).

[120] D. I. Daws and S. J. Cristol, *Advances in Free Radical Chemistry*, Logos Press, London, vol. 1, p. 159 (1965).

[121] T. G. Traylor, *Accounts Chem. Res.*, **2**, 152 (1969).

reaction (see p. 581). In this respect it is interesting to mention the addition of free radical **51** to a mixture of α-methylstyrene and acrylic acid [122], in presence of cuprous salts to inhibit copolymerisation. Radical **51** is a nucleophilic free-radical and will add to acrylic acid to form radical **52.** The latter has an electron-attracting group at the radical centre and is therefore electrophilic, choosing as reaction partner α-methylstyrene. The formation of radicals **53** and **52** can be demonstrated by the reaction products isolated after reduction with cuprous salts.

$$\underset{\mathbf{51}}{CH_3OCO(CH_2)_4CH_2\cdot} + CH_2{=}\underset{|\atop COOH}{CH} \longrightarrow \underset{\mathbf{52}}{CH_3OCO(CH_2)_6{-}\underset{|\atop COOH}{CH}\cdot}$$

$$\mathbf{52} + CH_2{=}\underset{|\atop CH_3}{C}{-}C_6H_5 \longrightarrow \underset{\mathbf{53}}{CH_3OCO(CH_2)_6\underset{|\atop COOH}{CH}{-}CH_2{-}\overset{C_6H_5\atop |}{\underset{|\atop CH_3}{C}}\cdot}$$

The lack of stereospecificity in an addition reaction is an indication (but not a definite proof) for a homolytic process [123]. However, examples of stereospecific homolytic reactions are known, some of them being used in preparative organic chemistry. A well-known stereospecific process is the homolytic addition of hydrogen bromide to compounds like 1-halogenocyclohexenes, yielding *cis*-1,2-dihalogenocyclohexane [124–126]:

X=Br, Cl **54**

A similar mechanism is also assumed for the *trans*-additions of mercaptans to cyclohexenes [127].

If the free radical **54** can explain the high stereospecificity of the addition as a consequence of conformational effects of the cyclohexane ring, the homolytic *trans*-addition of iodine to *cis*- and *trans*-

[122] M. Minisci, M. Cecere, R. Galli and R. Bernardi, *Tetrahedron*, **25**, 2667 (1969).
[123] B. A. Bolm and P. I. Abell, *Chem. Rev.*, **62**, 599 (1962).
[124] H. L. Goering, P. I. Abell and B. F. Aycock, *J. Amer. Chem. Soc.*, **74**, 3588 (1952).
[125] H. L. Goering and L. L. Sims, *J. Amer. Chem. Soc.*, **77**, 3465 (1955).
[126] H. L. Goering and D. W. Larsen, *J. Amer. Chem. Soc.*, **81**, 5937 (1959).
[127] N. A. Le Bel and A. De Boer, *J. Amer. Chem. Soc.*, **89**, 2784 (1967).

but-2-ene is more difficult to interpret. Skell and Pavlis[128] admit cyclic intermediates, similar to halogenium ions:

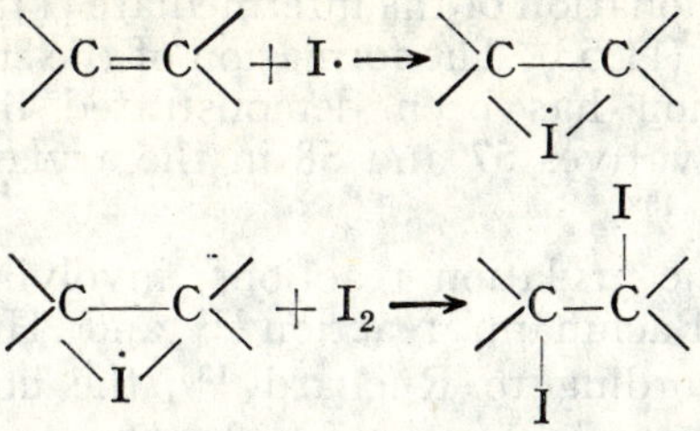

$$>C{=}C< + I\cdot \longrightarrow >C{-}C< \text{ (bridged by I)}$$

$$>C{-}C< \text{ (bridged by I)} + I_2 \longrightarrow >CI{-}CI<$$

The intervention of radical cyclic intermediates is supported by less experimental data than that of the corresponding ionic counterparts. Their postulation in homolytic processes[99, 128–131], even if attractive, must be considered with a certain reserve.

11.4.3 *Homolytic aromatic substitution*

The homolytic aromatic substitution differs essentially from the heterolytic substitutions of aromatic compounds[9], with respect to the reactivity of the aromatic substrate, the orientation and the overall picture of products. As already mentioned, the homolytic aromatic substitution presents a lower substrate selectivity and position selectivity than, for example, the heterolytic electrophilic substitution. Usually, all three possible isomers (*ortho, meta, para*) are formed, the former being in slight excess and the ratio between the latter two oscillating in the neighbourhood of unity. The mechanism for *para*-isomer formation is given below:

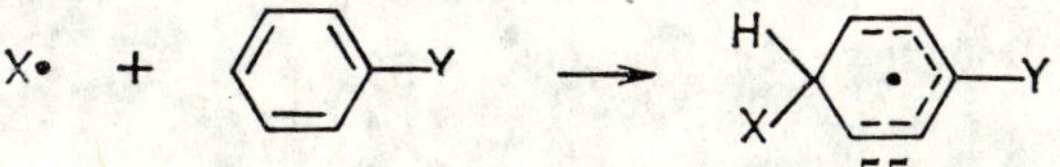

$$X\cdot + C_6H_5Y \longrightarrow \mathbf{55} \quad (11.50)$$

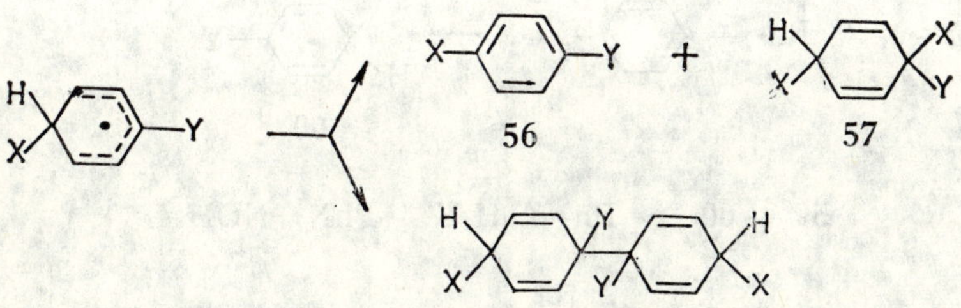

$$\mathbf{55} \longrightarrow \mathbf{56} + \mathbf{57} \quad (11.51)$$

$$\mathbf{55} \longrightarrow \mathbf{58} \quad (11.52)$$

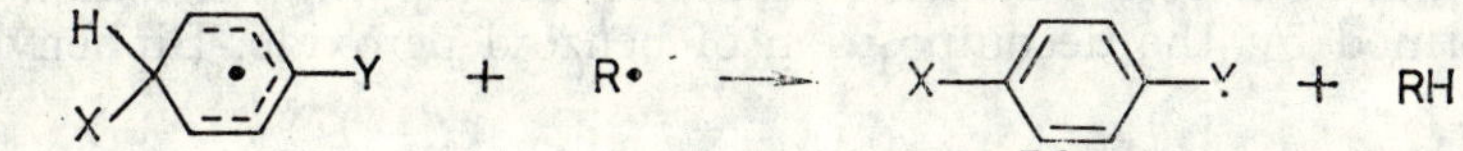

$$\mathbf{55} + R\cdot \longrightarrow \mathbf{56} + RH \quad (11.53)$$

[128] P. S. Skell and R. R. Pavlis, *J. Amer. Chem. Soc.*, **86**, 2956 (1964).
[129] P. D. Readio and P. S. Skell, *J. Org. Chem.*, **31**, 759 (1966).
[130] P. S. Skell and P. D. Readio, *J. Amer. Chem. Soc.*, **86**, 3334 (1964).
[131] J. G. Traynham and W. H. Hines, *J. Amer. Chem. Soc.*, **90**, 5208 (1968).

The intermediates of the reaction are cyclohexadienylic radicals **55** stabilised by delocalisation. The reaction product may result either by disproportionation of the intermediate (11.51) or by hydrogen abstraction from **55** (11.53). The formation of the substitution product by disproportionation has been demonstrated by the isolation of cyclohexadiene derivatives **57** and **58** in the arylation of benzene by diaroylperoxides [132, 133].

Other homolytic arylation reactions, involving diazonium ions, are the Gomberg-Bachmann reaction [134] and Hey reaction [135]. In the latter case, according to Rüchardt [136], the homolytic process is preceded by a series of ionic reactions leading to radical **59** and the phenyl radical. The latter attacks further the aromatic compound forming a cyclohexadienyl intermediate, **60.** Both **59** and **60** are sufficiently stable due to electron delocalisation and cannot lose their radical state before encountering each other.

$$PhN(NO)COMe \xrightarrow{\Delta} PhN{=}NOCOMe$$

$$PhN(NO)COMe \xrightarrow{+MeCOO^-} Ac_2O + PhN{=}NO^-; \qquad PhN{=}NOCOMe \rightarrow Ph\,N{\equiv}N^+\ {}^-OOCMe \text{ (ion pairs)}$$

$$\rightarrow PhN{=}N{-}O{-}N{=}NPh \rightarrow Ph\cdot + \underset{\mathbf{59}}{Ph{-}N{=}N{-}O\cdot} + N_2$$

$$Ph\cdot + C_6H_5{-}Y \longrightarrow \underset{\mathbf{60}}{(Ph)(H)C_6H_5\cdot{-}Y}$$

$$\mathbf{59} + \mathbf{60} \longrightarrow Ph{-}C_6H_4Y + PhN{=}NOH$$

In a preceding section of this chapter it has been shown that benzene is not as effective as other aromatics in trapping benzoyloxy radicals formed by the decomposition of benzoyl peroxide. Biphenyl

[132] D. F. DeTar and R. A. J. Long, *J. Amer. Chem. Soc.*, **80**, 4742 (1958).
[133] E. L. Eliel, S. Meyerson, Z. Welvart and S. H. Wilen, *J. Amer. Chem. Soc.*, **82**, 2936 (1960).
[134] W. E. Bachmann and R. A. Hoffmann, *Org. Reactions*, **2**, 224 (1944).
[135] C. Rüchardt, *Angew. Chem.*, **77**, 974 (1965).
[136] C. Rüchardt and B. Freudenberg, *Tetrahedron Letters*, 3623 (1964).

is the main reaction product. If the decomposition is performed in presence of oxygen[137], phenyl benzoate is formed in 50% yield:

$$(PhCOO)_2 \longrightarrow 2\ PhCOO\cdot$$

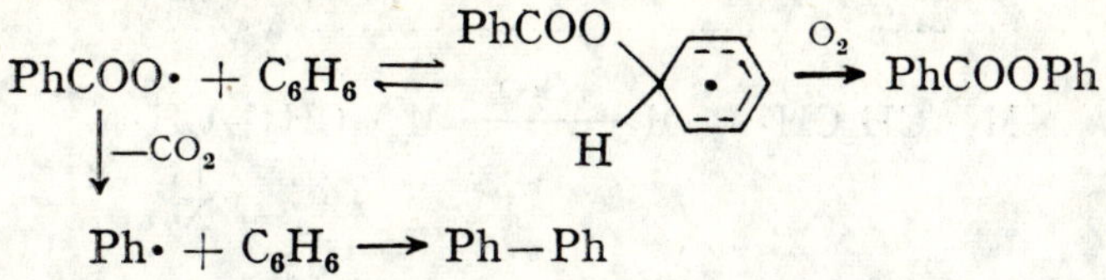

$$PhCOO\cdot + C_6H_6 \rightleftharpoons \text{(PhCOO)(H)}C_6H_5\cdot \xrightarrow{O_2} PhCOOPh$$

$$\downarrow -CO_2$$

$$Ph\cdot + C_6H_6 \longrightarrow Ph-Ph$$

This result seems to indicate that benzoxyl radicals react reversibly with benzene.

11.4.4 *Polymerisation and copolymerisation*

The polymerisation of vinyl compounds is mainly a radical chain reaction[7], its mechanism being now well understood in every detail[138–142]. Scheme 1 represents the reaction mechanism started by an initiator In:

Initiation

$$In \xrightarrow{k_d} 2R\cdot \qquad (11.54)$$

$$\underset{M}{R\cdot + CH_2{=}\underset{|}{CH}X} \xrightarrow{k_i} \underset{M_1\cdot}{R-CH_2-\underset{|}{CH}X\cdot} \qquad (11.55)$$

Propagation

$$R-CH_2-\underset{X}{CH}\cdot + CH_2{=}\underset{X}{CH} \xrightarrow{k_{p,1}} RCH_2\underset{X}{CH}-CH_2-\underset{X}{CH}\cdot \qquad (11.56a)$$

$$M_2\cdot = M_1-CH_2\underset{X}{CH}\cdot$$

[137] T. Nakata, K. Tokumaru and O. Simamura, *Tetrahedron Letters*, 3303 (1967).

[138] C. H. Bamford, W. G. Barb, A. D. Jenkins and P. F. Onion, *The Kinetics of Vinyl Polymerisation by Radical Mechanism*, Academic Press, New York (1958).

[139] T. Alfrey, J. J. Bohr and H. Mark, *Copolymerisation*, Interscience, New York (1952).

[140] B. Vollmert, *Polymer Chemistry*, Springer Verlag, Berlin (1973).

[141] M. Burnett, in *Progr. in Reaction Kinet.* (G. Porter, editor), Pergamon Press, New York (1965), vol. 3, p. 449.

[142] G. E. Ham, *Vinyl Polymerisation*, part I, vol. 1, M. Dekker, New York (1967).

$$\underset{\text{X}}{\text{M}_1\text{CH}_2\overset{|}{\text{CH}}\cdot} + \text{M} \xrightarrow{k_{p,2}} \underset{\text{M}_3\cdot}{\underset{\text{X}}{\text{M}_2\text{CH}_2\overset{|}{\text{CH}}\cdot}} \tag{11.56b}$$

. . . .

$$\underset{\text{X}}{\text{M}_{n-2}\text{CH}_2\overset{|}{\text{CH}}\cdot} + \text{M} \xrightarrow{k_{p,\,n-1}} \underset{\text{M}_n\cdot}{\underset{\text{X}}{\text{M}_{n-1}\text{CH}_2\text{—}\overset{|}{\text{CH}}\cdot}} \tag{11.56c}$$

Chain termination:

$$\underset{\text{X}}{\text{M}_{n-1}\text{CH}_2\text{—}\overset{|}{\text{CH}}\cdot} + \underset{\text{X}}{\text{M}_{p-1}\text{CH}_2\text{CH}\cdot} \xrightarrow{k_t} \underset{\substack{\text{combination}\\\text{product}}}{\text{M}_n\text{—M}_p} + \underset{\substack{\text{disproportionation}\\\text{products}}}{\text{M}_n + \text{M}_{p-1}\text{CH}{=}\underset{\text{X}}{\overset{|}{\text{CH}}}} \tag{11.57}$$

Scheme 1

The above scheme is demonstrated by numerous experimental data. Thus vinyl polymerisation is initiated by compounds which can be thermally decomposed into free radicals, such as benzoyl peroxide [143], triphenylmethylazobenzene [144], azonitriles [145–147], the best-known being azobisisobutyronitrile [148], etc. The radical chain mechanism is also indicated by the photochemical initiation [149] of vinyl polymerisation, in the absence of any initiator.

Fragments of the initiator are incorporated into the molecule of the resulted polymer. Polystyrene and polymethacrylate prepared by initiation with chlorinated [150] or brominated benzoyl peroxides [151] contain halogen. Applying the more sensitive tracer techniques by labelling the initiator with radioactive carbon [152], it has been found that one or two initiator fragments have been incorporated per polymer molecule, in complete agreement with the above scheme.

The kinetic treatment of the polymerisation reactions implies some simplifications of the general mechanism shown in Scheme 1. It can be observed that radicals $M_1\cdot$, $M_2\cdot$, ..., $M_n\cdot$ have the same

143 G. V. Schulz and E. Hussmann, *Z. Physik. Chem.*, **39B**, 246 (1938); G. V. Schulz and F. Blaschke, *Z. Physik. Chem.*, **51B**, 75 (1940).

144 G. V. Schulz, *Naturwissenschaften*, **27**, 659 (1939).

145 K. Ziegler, *Brennstoff. Chem.*, **30**, 181 (1949).

146 F.M. Lewis and M. S. Matheson, *J. Amer. Chem. Soc.*, **71**, 747 (1949).

147 C. G. Overberger, M. T. O'Shaughnessy and H. Shalit, *J. Amer. Chem. Soc.*, **71**, 2661 (1949).

148 L. M. Arnett, *J. Amer. Chem. Soc.*, **74**, 2027 (1952).

149 P. D. Bartlett and C. G. Swain, *J. Amer. Chem. Soc.*, **67**, 2273 (1945).

150 P. D. Bartlett and S. G. Cohen, *J. Amer. Chem. Soc.*, **65**, 543 (1943).

151 C. C. Price, R. W. Kell and E. Krebs, *J. Amer. Chem. Soc.*, **64**, 1103. (1942).

152 J. C. Bevington, H. W. Melville and P. R. Taylor, *J. Polymer Sci.*, **12**, 149 (1954); J. C. Bevington, J. H. Bradbury and G. M. Burnett, *J. Polymer Sci.*, **42**, 469 (1954).

environment at the radical centre and therefore a similar reactivity. It can be thus assumed that $M_1\cdot \sim M_2\cdot \sim ... - M_n\cdot = M\cdot$ and therefrom $k_{p,1} \sim k_{p,2} \sim ... \sim k_{p,n} = k_p$. Scheme 1 then becomes:

Initiation:

$$\text{In} \xrightarrow{k_d} 2\,\text{R}\cdot \qquad (11.54)$$

$$\text{R}\cdot + \text{M} \xrightarrow{k_i} \text{M}\cdot \qquad (11.58)$$

Propagation

$$\text{M}\cdot + \text{M} \xrightarrow{k_p} \text{M}\cdot \qquad (11.59)$$

Chain termination:

$$2\,\text{M}\cdot \xrightarrow{k_t} \text{Products} \qquad (11.60)$$

Scheme 2

A relatively simple equation rate can be derived from this scheme applying the *steady-state* approximation. Thus, considering that the concentrations of the reactive intermediates (R·, M·) are small and do not vary in time, we can write:

$$\frac{d[\text{R}\cdot]}{dt} = 0 = k_d[\text{In}] - k_i[\text{R}\cdot][\text{M}] \qquad (11.61)$$

and

$$\frac{d[\text{M}\cdot]}{dt} = 0 = k_i[\text{R}\cdot][\text{M}] - k_t[\text{M}\cdot]^2 \qquad (11.62)$$

Combining (11.61) with (11.62):

$$[\text{M}\cdot] = (k_d[\text{In}]/k_t)^{1/2} \qquad (11.63)$$

On the other hand, the rate of monomer disappearance in the process is given by:

$$-\frac{d[\text{M}]}{dt} = k_i[\text{R}\cdot][\text{M}] + k_p[\text{M}\cdot][\text{M}]$$

It is reasonable to assume that the first term in the above equation is small as compared to the last one (even if R· and M· are comparable in their reactivity towards the monomer, process (11.58) is repeated a large number of times). Then:

$$-\frac{d[\text{M}]}{dt} = k_p[\text{M}][\text{M}\cdot] = k_p[\text{M}](k_d/k_t)^{1/2}[\text{In}]^{1/2} \qquad (11.64)$$

Equation (11.64) shows that the polymerisation rate must vary with the square root of initiator concentration. Numerous examples are in perfect agreement with this relationship [153]. Moreover, in the photochemical polymerisation, the rates are proportional to the square root of light intensity [149].

The average degree of polymerisation, n, is given by the ratio between the rate of the propagation reaction ($R_p = k_p$ [M·] [M]) and the rate of the termination reaction ($R_t = k_t$ [M·]^2). The higher the frequency of the termination reactions, the lower will be the degree of polymerisation. On the other hand, relation (11.62) shows that the rate of the termination reaction is equal to the rate of initiation ($R_i + k_d$[In]), that is:

$$n = R_p/R_t = R_p/R_i \tag{11.65}$$

Another conclusion is now apparent. The higher the rates of the initiation reactions (therefore the higher the concentration of R·), the lower will be the degree of polymerisation. The decrease of n as a result of the increase of initiation centres is encountered in the polymerisation in reactive solvents, such as carbon tetrachloride:

$$M_i\cdot + CCl_4 \xrightarrow{k_{tr}} M_iCl + \cdot CCl_3 \tag{11.66}$$

$$\cdot CCl_3 + M \longrightarrow M\cdot, \text{ etc.} \tag{11.67}$$

Reaction (11.66) corresponds to a radical displacement reaction. If it can compete with the propagation step, the result will be a low n. On the other hand, reaction (11.67) corresponds to the initiation process. Its occurrence will mean that by the same initiation process (11.58) a greater number of radicals will be produced and even if (11.67) does not change the kinetics, the result will be a decrease of n. Such processes have been called [7] *chain transfer reactions*. Mayo and his co-workers[154] have defined the relation (11.68):

$$\frac{1}{n} = \frac{1}{n_0} + C\frac{[S]}{[M]} \tag{11.68}$$

where n_0 is the average degree of polymerisation which would be observed in the absence of solvent S, and C is the *transfer constant* of the solvent, $C = k_{tr}/k_p$. By plotting the measurable degree of polymerisation *vs*. the ratio [S]/[M], C can be determined from the slope of the resultant line. Table 11.3 gives some typical values of C.

[153] K. Nozaki and P. D. Bartlett, *J. Amer. Chem. Soc.*, **68**, 2377 (1946).
[154] F. R. Mayo *et al.*, *J. Amer. Chem. Soc.*, **65**, 2324 (1948); **70**, 2373 (1948); **75**, 3530 (1953); *Discussions Faraday Soc.*, **2**, 328 (1947).

TABLE 11.3

Typical transfer constants for styrene polymerisation in different solvents, at 60° and 100 °C [a]

Solvent	10^4C_{60}	10^5C_{100}
Cyclohexane	0.024	0.16
Benzene	0.018	0.184
Toluene	0.125	0.64
Isopropylbenzene	0.82	2.0
$ClCH_2CH_2Cl$	0.32	—
$BrCH_2CH_2Br$	—	6.6
Carbon tetrachloride	90	180

[a] Values cited in Reference 3, p. 152

By comparing Table 11.1 with Table 11.3 it can be seen that the higher the dissociation energy of a C—X bond corresponding to the solvent molecule, the lower the value of the transfer constant. The formation of stable radicals in reactions of type (11.66) increases the value of C. An important result is that transfer reactions are favoured by increasing temperature.

Chain-transfer reactions may also involve monomer molecules and even polymer molecules. Probably due to this and other factors, relation (11.68) is verified only at very low monomer conversions.

It was recently shown [155] that if all transfer reactions are included in the termination reactions and differential equations of type (11.65) are integrated over all conversions of the monomer, a more general equation may be obtained relating the average degree of polymerisation to transfer parameters. Such equations are verified in the full range of monomer conversions.

The treatment applied to polymerisation reactions is similar to the treatment of copolymerisation of two vinyl monomers, M_1 and M_2 (Scheme 3).

Initiation:

$$\mathrm{In} \longrightarrow 2\,\mathrm{R}\cdot \tag{11.69}$$

$$\mathrm{R}\cdot + \underset{\mathbf{M_1}}{\mathrm{CH_2{=}\underset{\displaystyle X}{\underset{|}{C}H}}} \longrightarrow \underset{\mathbf{61}a}{\mathrm{R{-}CH_2\underset{\displaystyle X}{\underset{|}{C}H}\cdot}} = \mathrm{M_1}\cdot \tag{11.70}$$

$$\mathrm{R}\cdot + \underset{\mathbf{M_2}}{\mathrm{CH_2{=}\underset{\displaystyle Y}{\underset{|}{C}H}}} \longrightarrow \underset{\mathbf{62}a}{\mathrm{R{-}CH_2\underset{\displaystyle Y}{\underset{|}{C}H}\cdot}} = \mathrm{M_1}\cdot \tag{11.71}$$

[155] M. Dimonie, *Rev. Roumaine Chim.*, **21**, 763 (1976).

Propagation:

$$M_1\cdot \xrightarrow[k_{11}]{M_1} M_1CH_2{-}\underset{\underset{\textstyle\mathbf{61}b}{X}}{CH}\cdot \simeq M_1\cdot \tag{11.72}$$

$$M_1\cdot \xrightarrow[k_{12}]{M_2} M_1CH_2{-}\underset{\underset{\textstyle\mathbf{62}b}{Y}}{CH}\cdot \simeq M_2\cdot \tag{11.73}$$

$$M_2\cdot \xrightarrow[k_{21}]{M_1} M_2CH_2{-}\underset{\underset{\textstyle\mathbf{61}c}{X}}{CH}\cdot \simeq M_1\cdot \tag{11.74}$$

$$M_2\cdot \xrightarrow[k_{22}]{M_2} M_2CH_2{-}\underset{\underset{\textstyle\mathbf{62}c}{Y}}{CH}\cdot \simeq M_2\cdot \tag{11.75}$$

Considering the arguments discussed before (see p. 574), it can be inferred that the reactivity of intermediates **61**–**62** will depend only on the electronic effects of groups X and Y and will not be affected by the more remote groups. It can be thus assumed that **61***a* $\simeq$ **61***b* $\simeq$ **61***c* $= M_1\cdot$ and **62***a* $\simeq$ **62***b* $\simeq$ **62***c* $= M_2\cdot$. Applying these simplifications, the rates of monomer disappearance will be given by:

$$-\frac{d[M_1]}{dt} = k_{11}[M_1\cdot][M_1] + k_{21}[M_2\cdot][M_1] \tag{11.76}$$

$$-\frac{d[M_2]}{dt} = k_{12}[M_1\cdot][M_2] + k_{22}[M_2\cdot][M_2] \tag{11.77}$$

Applying the steady-state treatment and assuming that the rates of the conversion of the radicals into one another are equal, $k_{12}[M_1\cdot][M_2] = k_{21}[M_2\cdot][M_1]$, the following relation is found:

$$\frac{d[M_1]}{d[M_2]} = \frac{[M_1]}{[M_2]}\,\frac{r_1[M_1] + [M_2]}{r_2[M_2] + [M_1]} \tag{11.78}$$

where $r_1 = k_{11}/k_{12}$ and $r_2 = k_{22}/k_{21}$. These two parameters are called the *monomer reactivity ratios* and are characteristic of the particular monomer pair. They measure the ease of combination of an intermediate radical with its monomer (for example $M_1\cdot$ with M_1) as compared with the other monomer.

In order to obtain a perfect alternation of monomers in the polymer chain, the ratios r_1 and r_2 must be both less than unity and as much as possible of the same magnitude. The fulfilment of these conditions depends largely on the electronic effect of the substituents X and Y in monomers (see Scheme 3). Indeed, if X is an electron-attracting group and Y an electron-repelling group, it results that $M_1\cdot$ will be an electrophilic radical which will react more readily with the nucleophilic monomer M_2. The reverse will be true for radical $M_2\cdot$. An example of a good alternation of the monomer units in the macromolecular chain is the styrene-methyl methacrylate copolymer for which the reactivity ratios were found to be $r_{St} = 0.52$ and $r_M = 0.46$ [156].

The stereochemistry of the polymerisation reaction presents analogies with the other radical additions. It is now known [157] that even radical polymerisations can lead to stereospecific products (isotactic and syndiotactic polymers). Thus, the polymerisation of acrylic and methacrylic esters at low temperatures [158, 159] leads to syndiotactic polymers. The reaction stereospecificity can be explained by a 1,3 induction (see p. 527), in which the substituent X from the radical **63** determines the orientation of the monomer in the transition state:

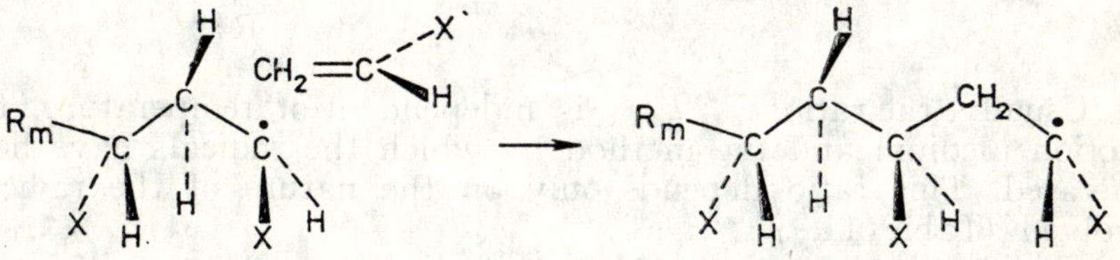

63 → **64** (syndiotactic)

At higher temperatures, rotations about the C—C bonds in **63** and **64** are favoured, determining the loss of the stereospecificity.

An oriented transition state in the radical polymerisation is also demonstrated by the formation of an optically active polymer in the copolymerisation of maleic anhydride and methacrylic ester of the optically active α-methylbenzyl alcohol [160]. The optical activity is preserved even after the removal by hydrolysis of the optically active alcohol from the resultant copolymer.

[156] F. R. Mayo and M. M. Lewis, *J. Amer. Chem. Soc.*, **66**, 1594 (1944).

[157] M. Goodman, *Topics in Stereochemistry*, vol. 2 (N. L. Allinger and E. L. Eliel, editors), Interscience, New York (1967), p. 73.

[158] B. S. Garrett, W. E. Goode, S. Gratch, J. F. Kincaid, C. L. Leoesque, A. Spell, J. D. Stroupe and W. A. Watanabe, *J. Amer. Chem. Soc.*, **81**, 1007 (1959).

[159] T. G. Fox, B. S. Garret, W. E. Goode, S. Gratch, J. F. Kincaid, A. Spell and J. D. Stroupe, *J. Amer. Chem. Soc.*, **80**, 1768 (1958).

[160] N. Beredjick and C. Schuerch, *J. Amer. Chem. Soc.*, **80**, 1933 (1958).

11.5 Combination and disproportionation reactions

The combination and disproportionation reactions between identical or different free radicals are the most frequently encountered reactions in which radical state in destroyed:

$$\begin{matrix} R^1 \\ R^1 \end{matrix}\rangle CH-\dot{C}R^1_2 + \begin{matrix} R^2 \\ R^2 \end{matrix}\rangle CH-\dot{C}R^2_2 \xrightarrow{k_{disp}} \begin{matrix} R^1 \\ R^1 \end{matrix}\rangle C=CR^1_2 + \begin{matrix} R^2 \\ R^2 \end{matrix}\rangle CH-CHR^2_2 \qquad (11.79)$$

$$\begin{matrix} R^1 \\ R^1 \end{matrix}\rangle CH-\dot{C}R^1_2 + \begin{matrix} R^2 \\ R^2 \end{matrix}\rangle CH-\dot{C}R^2_2 \xrightarrow{k_{comb}} \xrightarrow{k_{comb}} \begin{matrix} R^1 \\ R^1 \end{matrix}\rangle CH-CR^1_2-CR^2_2-CH\langle \begin{matrix} R^2 \\ R^2 \end{matrix} \qquad (11.80)$$

Usually, the ratio k_{disp}/k_{comb} is independent of temperature, the reaction medium and the method by which the radicals have been generated. This ratio depends only on the nature of the radicals involved (Table 11.4).

The invariance of the ratio k_{disp}/k_{comb} with temperature is considered to indicate either that the activation energies of the two processes are equal ($E_{disp} - E_{comb} = 0$) or that both processes have an activation energy about zero. According to certain authors this is better explained by admitting that there is a common transition state **65** for both disproportionation and combination of the radicals [161–168].

$$CH_2\langle \begin{matrix} CH_2 \\ H \end{matrix} \rangle CH_2-CH_3 \; \mathbf{65} \longrightarrow \begin{cases} \xrightarrow{k_{disp}} C_2H_6 + C_2H_4 \\ \xrightarrow{k_{comb}} C_4H_{10} \end{cases}$$

[161] J. M. Hay, *J. Chem. Soc.*, *B* 1175 (1967).
[162] J. A. Kerr and A. F. Trotman-Dickenson, *Progr. Chem. Kinet.*, **1**, 105 (1961).
[163] P.S. Dixon, A.P. Stefani and M.Szwarc, *J. Amer. Chem. Soc.*, **85**, 2551 (1963).
[164] J. W. Kraus and J. G. Calvert, *J. Amer. Chem. Soc.*, **79**, 5921 (1957).
[165] C. A. Heller and A. S. Gordon, *J. Phys. Chem.*, **60**, 1315 (1965).
[166] D.H. Slater, S.S. Colier and J.G. Calvert, *J. Amer. Chem. Soc.*, **90**, 268 (1968).
[167] S. W. Benson and R. Shaw, *J. Amer. Chem. Soc.*, **89**, 5331 (1967).
[168] J. N. Bradley, *J. Chem. Phys.*, **35**, 748 (1961).

TABLE 11.4

The ratio k_{disp}/k_{comb} in the disproportionation and combination of two radicals, $R^1\cdot$ and $R^2\cdot$

R^1	R^2	k_{disp}/k_{comb}	*Reference*
Me	Et	0.04	161, 162
Et	Et	0.14–0.26	163
Pr^i	Pr^i	0.41; 0.61; 0.60	114, 164, 165
Bu^i	Bu^i	0.075	166
Bu^s	Bu^s	2.3	164
Bu^t	Bu^t	4.6	164
cyclohexyl	cyclohexyl	0.5	162
cyclohexadienyl	cyclohexadienyl	0.5	167
cyclohexadienyl	Pr^i	0.52	114
dimethylbenzyl	dimethylbenzyl	0.04–0.07	45

The data of Table 11.4 may indicate that the ratio k_{disp}/k_{comb} is determined by the substituents in α and to a lesser extent (if at all) by the bulkiness of the radical. However, for radical **66** the ratio is much smaller [45] (0.04–0.07) than for the t-butyl radical **67** (4.6 [164]). As the transition state must have about the same structure as the free radical, it is concluded that the way in which the free radical is stabilised must be decisive in determining the pathway adopted in the collapse of the transition state (either disproportionation or combination). Hyperconjugation will stabilise the t-butyl radical. The limit structures of **67** suggest that disproportionation will be favoured over combination for this radical. Dimethylbenzyl radical, on the other hand, must be stabilised by electron delocalisation; dimerisation will thus outweigh other processes.

$C_6H_5{=}C(CH_3)_2$ (**66**) $\qquad H\cdot\ H_2C{=}C(CH_3)_2 \longleftrightarrow H_3C{-}\dot{C}(CH_3)_2$ (**67**)

Table 11.4 indicates that the combination of two radicals is preferred in most cases. Similar observations have been made in other radical reactions involving stable molecules having either radical or diradical structures.

Thus, in the reaction of alkoxy radicals with nitrogen oxide (a molecule with an unpaired electron), the ratio k_{disp}/k_{comb} is fairly constant (0.2—0.27) with temperature, the structure of the alkoxy radical and the reaction conditions[169–171]:

$$\begin{matrix}CH_3\\ \\ R\end{matrix}\!\!\!\!\!\!>\!CH{-}O\cdot + NO \xrightarrow{k_{disp}} CH_3COR + (HNO \rightarrow 1/2\,N_2O + 1/2\,H_2O)$$

$$\begin{matrix}CH_3\\ \\ R\end{matrix}\!\!\!\!\!\!>\!CH{-}O\cdot + NO \xrightarrow{k_{comb}} \begin{matrix}CH_3\\ \\ R\end{matrix}\!\!\!\!\!\!>\!CH{-}ONO \quad R = H, Me, Et$$

Similar to the combination of free radicals with nitrogen oxide is the reaction of free radicals with molecular oxygen. Such processes are involved in autooxidation reactions[172]. The molecule of oxygen is a triplet in the ground state and will thus resemble diradicals. By combination with free radicals another radical is produced:

$$R\cdot + O_2 \rightarrow R{-}O{-}O\cdot$$

The reaction of free radicals with molecular oxygen has been used in the determination of the structure of free radicals resulting from the earlier stages of a decomposition. Bartlett and his co-workers[173] have thus observed that by the decomposition of peresters **68***a* and **68***b* in presence of oxygen, at atmospheric pressure, a mixture of stereoisometric hydroperoxides is formed, indicating the planar structure of the radical intermediates. If the decomposition of the peresters is performed at high oxygen pressures, the two isomeric peresters may yield different mixtures of hydroperoxides, the reaction being stereospecific. This result may indicate that the initially formed radical has a pyramidal structure.

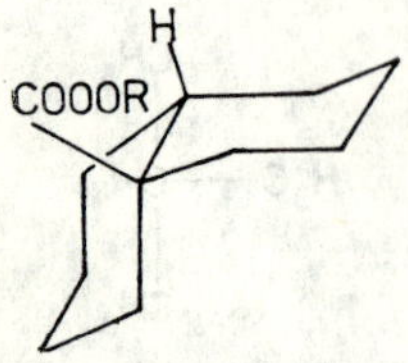

H
COOOR
↓ **68***a*

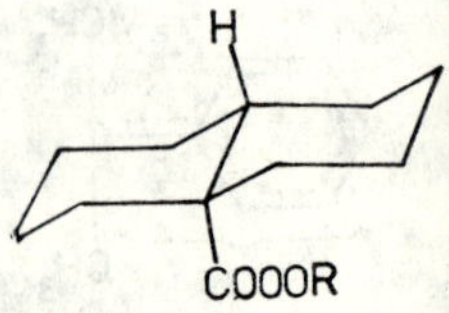

H
COOOR
↓ **68***b*

[169] G. A. Hughes and L. Phillips, *J. Chem. Soc. A*, 894 (1967).
[170] R. L. East, J. R. Gilbert and. L. Phillips, *J. Chem. Soc. A*, 1673 (1968).
[171] P. F. Walker and L. Phillips, *J. Chem. Soc. A*, 2103 (1968).
[172] J. L. Bolland, *Quart. Rev. Chem. Soc.*, **2**, 1 (1949); L. Bolleman, *ibid.*, **8**, 147 (1954).
[173] P. D. Bartlett, R. E. Pincock, J. H. Rolston, W. G. Schindel and L. A. Singer, *J. Amer. Chem. Soc.*, **87**, 2590 (1965).

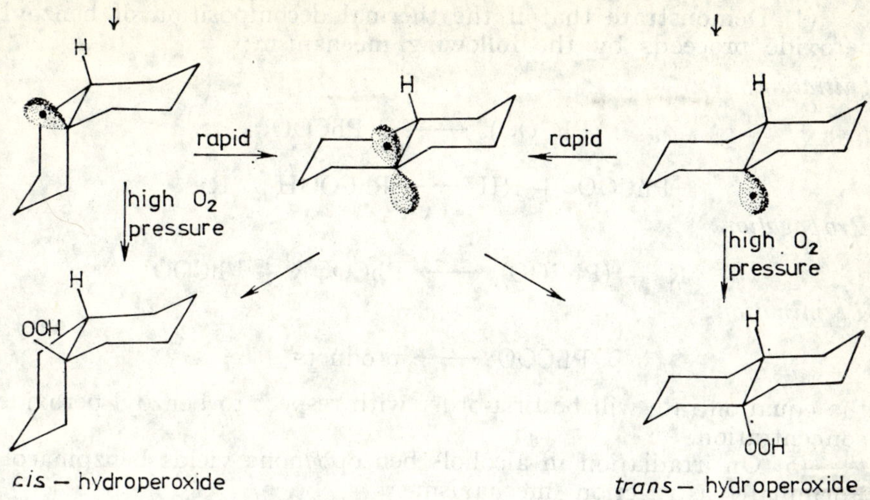

The autooxidation of organometallic compounds of Group III metals is also considered to involve hydroperoxide radicals in a chain reaction [174, 175].

$$ROO\cdot + AlR_3 \rightarrow ROOAlR_2 + R\cdot$$
$$R\cdot + O_2 \rightarrow ROO\cdot, \text{ etc.}$$

1.1.6 Problems

(1) The decomposition of an equimolecular mixture of azocompounds **1** and **2** leads to a mixture of hydrocarbons **3**—**5** (see p. 545). Considering that the cage-coupling products represent 29.1%, calculate the ratio **3** : **4** : **5** in the following hypotheses:

(a) the reactions outside the cage are only radical combinations;

(b) the reactions with the solvent outside the cage represent 50%.

(2) Determine the dissociation energy D of the C—H bond in methane, knowing the following data:

$$CH_4 + Br\cdot \rightarrow CH_3\cdot + HBr \quad E = 17.6 \text{ kcal mol}^{-1}$$
$$CH_3\cdot + HBr \rightarrow CH_4 + Br\cdot \quad E = 2.0 \text{ kcal mol}^{-1}$$
$$H\cdot + Br\cdot \rightleftharpoons HBr \quad \Delta H = -\ 87.0 \text{ kcal mol}^{-1}$$

(3) Explain why azobenzene cannot be thermally decomposed into free radicals.

[174] A. G. Davies and B. P. Roberts, *J. Chem. Soc. B*, 1074 (1968).
[175] P. F. Allies and P. B. Brindley, *Chem. and Ind.*, 319 (1967).

(4) Demonstrate that if the thermal decomposition of benzoyl peroxide proceeds by the following mechanism:

Initiation:

$$(PhCOO)_2 \xrightarrow{k_1} 2\ PhCOO\cdot$$

$$PhCOO\cdot + RH \xrightarrow{k_2} PhCOOH + R\cdot$$

Propagation:

$$R\cdot + (PhCOO)_2 \xrightarrow{k_{ex}} PhCOOR + PhCOO\cdot$$

Termination:

$$PhCOO\cdot \xrightarrow{k_t} \text{products}$$

the equation rate will be first-order with respect to benzoyl peroxide concentration.

(5) On irradiation in alcohols benzophenone yields benzpinacol. Formulate a reaction mechanism.

(6) A photochemical reaction of a compound A is sensitised by benzophenone ($E_T = 69$ kcal mol^{-1}) and fluorene ($E_T = 68$ kcal mol^{-1}). With lower quantum yields it may also occur with phenanthrene ($E_T = 62$ kcal mol^{-1}) or naphthalene ($E_T = 61$ kcal mol^{-1}). No reaction is observed if nitrobenzene ($E_T = 60$ kcal mol^{-1}) is used as sensitiser. How can these results be interpreted?

(7) In contrast with the homolytic chlorination of cyclohexene, the homolytic bromination yields only the addition product beside 3-bromocyclohexene (no 4-halogenocyclohexene is formed). What can be concluded from this result?

(8) The decomposition of the peresters derived from *cis* and *trans* α-bromocinnamic acids yields bromostyrenes in a stereospecific reaction. The similar decomposition of the peresters corresponding to α-methyl and α-phenylcinnamic acids is not stereospecific.

(a) Formulate the reactions.

(b) Give an interpretation of the results.

(9) The experimental DE of the cycloheptatrienyl radical was found to be 31 kcal mol^{-1}. Show whether this value agrees with the DE predicted by the simple HMO theory.

(10) The reaction of acrylic acid with α-methylstyrene in presence of initiators and cuprous salts yields lactone I. In methanol, beside I, ether II is also formed. Formulate a mechanism for the generation of these two products.

```
                          Me                                    Me
                          |                                     |
R—CH2—CH—CH2—C—Ph          R—CH2—CH—CH2—C—Ph
           |             |                        |             |
          CO ———— O                      COOH         OMe
                 I                                     II
```

Explain why the cuprous salt interacts with radical **53** and not with **51** and **52** (see p. 574).

12
Thermal and photochemical transformations of unsaturated compounds

12.1 Introduction

12.1.1 *History*

It has long been known that unsaturated compounds, containing a delocalised π-system, can undergo rearrangements into cyclic compounds and other π-systems, or can add on other π-systems, leading to cyclic compounds, in uncatalysed thermal processes [1]. Such transformations were only incidentally studied until about 1930. Diels and his co-workers [2] discovered in 1925 the addition of dienes to diazodicarboxylic ester; shortly afterwards Diels and Alder [3] published their first paper on the *diene synthesis (Diels-Alder reaction)*. Subsequent work of Alder and his co-workers [4, 5] proved the generality of the reaction, its high regio- and stereo-selectivity. Thus, the interest in the thermal transformations of unsaturated compounds started.

In 1940 Cope and Hardy [6] discovered a rearrangement of unconjugated dienes presenting the same experimental characteristics as both the diene synthesis (12.1) and the much-earlier known Claisen rearrangement (12.2) [7]. At that time, the experimental data seemed puzzling. Indeed, the reactions occurred in a thermal uncatalysed process, with equal ease in gaseous or condensed phase. The polarity of the solvent had only a minor effect on the reaction rates, while Lewis and Brönsted acids did not affect the course of the reactions. Those characteristics were reminiscent of homolytic reactions, which were acquiring at that time a growing importance from the point of view of organic mechanisms. It is thus not surprising that a two-step mechanism, involving diradicals, was first proposed for the diene synthesis [8]. However, the reactions displayed also the main features of heterolytic processes. They were not affected by promoters and

[1] See, for example, S. Patai (editor), *The Chemistry of Alkenes*, Interscience, New York (1964).
[2] O. Diels, J. Hansen Blom and W. Koll, *Liebigs Ann. Chem.*, **443**, 242 (1925).
[3] O. Diels and K. Alder, *Liebigs Ann. Chem.*, **460**, 98 (1928).
[4] K. Alder and G. Stein, *Liebigs Ann. Chem.*, **504**, 222 (1933).
[5] K. Alder and G. Stein, *Angew. Chem.*, **50**, 514 (1937).
[6] A. C. Cope and E. M. Hardy, *J. Amer. Chem. Soc.*, **62**, 441 (1940).
[7] L. Claisen, *Ber. dtsch. chem. Ges.*, **45**, 3157 (1912).
[8] G. B. Kistiakowsky and W. W. Ransom, *J. Chem. Phys.*, **7**, 725 (1938); see also S. W. Benson, *J. Chem. Phys.*, **46**, 4920 (1967).

inhibitors, while polar substituents were found to change dramatically the reaction rates. Polar intermediates were therefore considered as an alternative[9]. Neither the radical pathway nor the polar mechanism accounted in a satisfactory manner for the observed regio- and stereo-selectivity of the reactions. Evans[10] was the first to propose a concerted mechanism, involving a cyclic planar transition state resembling a six-electron aromatic compound. The concerted mechanism was later proved to be correct, both for the diene synthesis and for the Claisen rearrangement. Similar transition states were also proposed[11], differing from that of Evans in that the reaction centres did not lie in the same plane.

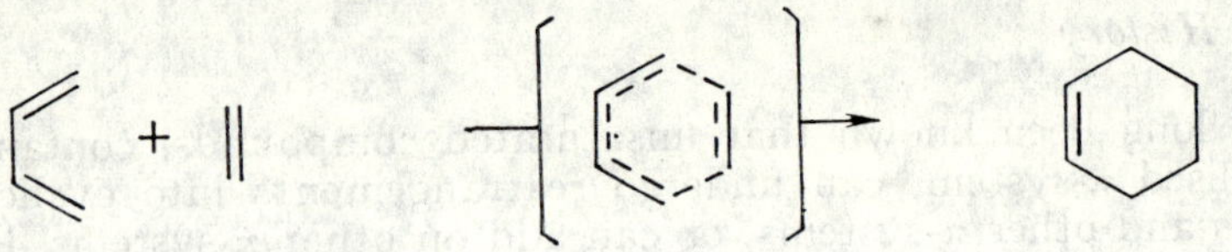

(12.1)

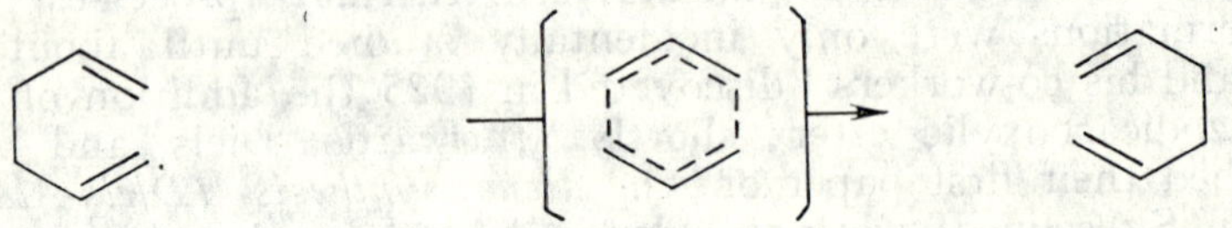

(12.2)

Considering that in these regiospecific reactions bonds are formed and broken between four centres, Bartlett[12] called them *four-centre reactions*. In order to stress the experimental ambiguity of the reaction mechanisms, Doering[13] introduced the term of *no mechanism reactions*. Evidently, the reacting system passing through a certain transition state does also possess a reaction mechanism; here the proposed term merely suggests that both characteristics of the common homolytic and heterolytic reactions are observed experimentally.

The preparation of heterocyclic compounds by addition to double and triple bonds of certain reagents like diazoalkanes has been used in organic chemistry since the end of the last century. The reaction mechanisms were obscure and did not receive much attention in the early period. In the late 1950s, Huisgen's work in the field was outstanding. He showed[14, 15] that

[9] See for example R. Gompper, *Angew. Chem.*, **81**, 348 (1969) and the literature cited therein.

[10] M. G. Evans and E. Warhurst, *Trans. Faraday Soc.*, **34**, 614 (1938); M. G. Evans, *ibid.*, **35**, 824 (1939).

[11] S. J. Rhoads, in *Molecular Rearrangements* (F. de Mayo, editor), Interscience, New York, vol. 1 (1963), p. 655.

[12] P. D. Bartlett, in *Organic Chemistry*, vol. 3 (H. Gilman, editor), J. Wiley, New York (1953), p. 4.

[13] W. von E. Doering, quoted by S. J. Rhoads [11].

[14] R. Huisgen, *Proc. Chem. Soc.*, 357 (1961); R. Huisgen, R. Grashey and J. Sauer, in Ref. 1, p. 739.

[15] R. Huisgen, *Angew. Chem.*, **75**, 604 (1963); *Helv. Chim. Acta*, **50**, 2421 (1967).

the reagents can be formally considered as 1,3-dipoles (**1**) which may add to bonds in a concerted way (12.3) to yields cyclic adducts. He extended the reaction by finding numerous other examples of reagents, establishing their kinetics, stereochemistry and applicability.

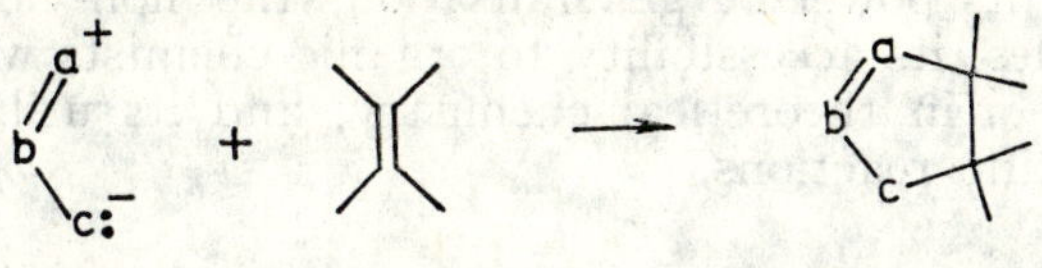

(12.3)

1

Huisgen has thus established the similarity between 1,3-*dipolar additions* and diene syntheses. For these reactions and for other similar additions (for example cyclobutane formation (12.4) from alkene) he introduced the general term *cycloaddition reactions*.

(12.4)

The enormous work of Huisgen and his co-workers in cycloaddition reactions has revived the interest in such processes. It was soon recognised that some of them may occur with high regio- and stereo-selectivity, in a concerted way (diene synthesis, dipolar additions), whereas others, such as cyclobutane formation [16], are stepwise reactions involving diradical intermediates.

Parallelling the reactions mentioned above, which are all thermal processes, another field developed, initially touching only the chemistry of steroids, namely the photochemical transformations of unsaturated compounds [17]. In the case of ergosterol, for example, Havinga [18] demonstrated that the reactions are stereospecific and differ from the thermal transformations of the compound.

Photochemical, as well as thermal reactions of the polyenes, may involve in some instances the synchronous formation and breaking of bonds. It has been pointed out by Oosterhoff [19] that the symmetry of the molecular orbitals may be responsible for the observed results. By this time the data concerning both thermal and photochemical transformations of π-systems were sufficient to allow a theoretical generalisation. This was actually done in 1965 in three famous papers by Woodward and Hoffmann [20–22], devel-

[16] J. D. Roberts and C. M. Sharts, *Org. Reactions*, **12**, 1 (1962).
[17] See for example N. J. Turro, *Molecular Photochemistry*, W. A. Benjamin, New York (1965), p. 162.
[18] E. Havinga and J. L. M. A. Schlatmann, *Tetrahedron*, **16**, 151 (1961).
[19] L. J. Oosterhoff, quoted by Havinga [18].
[20] R. B. Woodward and R. Hoffmann, *J. Amer. Chem. Soc.*, **87**, 395 (1965).
[21] R. Hoffmann and R. B. Woodward, *J. Amer. Chem. Soc.*, **87**, 2046 (1965).
[22] R. B. Woodward and R. Hoffmann, *J. Amer. Chem. Soc.*, **87**, 2611 (1965).

oped later by the same authors [23, 24]. It can be said that the Woodward-Hoffmann theory represented a revolution in organic chemistry. Never has a theoretical treatment provoked such an enormous explosion in the chemical literature as has their theory of symmetry control in concerted reactions. The success of their treatment is due in part to its power of generalisation, the non-violation of the established rules, its accessibility to organic chemists working either in preparative or in theoretical chemistry, and its utility in predicting stereospecific reactions.

12.1.2 *Pericyclic reactions*

Woodward and Hoffmann [24] have introduced the term *pericyclic reactions* for clearly differentiable reaction types, like (12.1)—(12.4), which have in common a cyclic transition state, are highly stereospecific, their stereochemistry depending on the total number of π- and σ-electrons essentially involved in the formation or breaking of bonds. Table 12.1 lists the most common pericyclic reactions, indicating the admitted transition state and the number of electrons delocalised in the transition state.

TABLE 12.1
Pericyclic reactions

No.	*Reaction type and examples*	*Transition state (TS)*	*Number of delocalised electrons in TS*
	(A) *Unimolecular transformations*		
(1)	Sigmatropic reactions		
(1a)	Cope rearrangement (thermal)		6
(1b)	Claisen rearrangement (thermal)		6

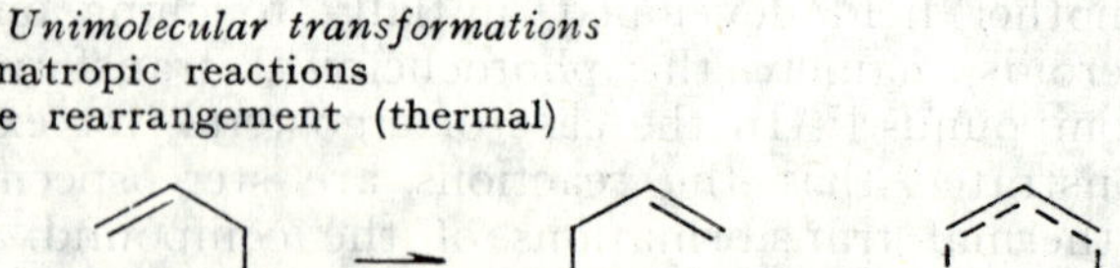

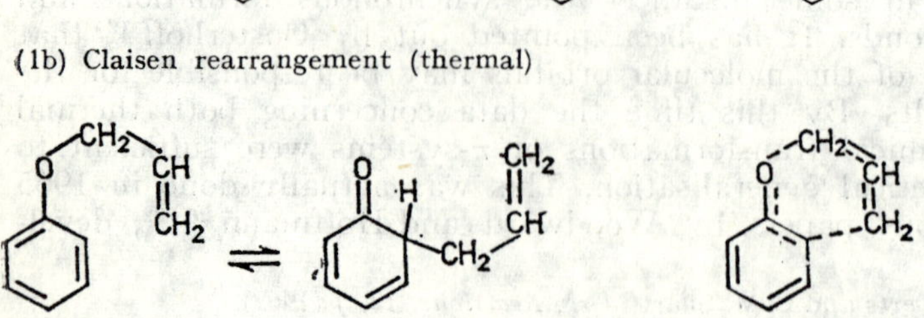

[23] R. B. Woodward, *Aromaticity*, Special Publication no. 21, The Chemical Society London (1967), p. 211; R. Hoffmann and R. B. Woodward, *Accounts Chem. Res.*, **1**, 17 (1968).

[24] R. B. Woodward and R. Hoffmann, *The Conservation of Orbital Symmetry*, Verlag Chemie, Weinheim (1970); *Angew. Chem.*, **81**, 797 (1969).

TABLE 12.1 — continued

(1c) Hydrogen shifts (thermal or photochemical)

$R^1_2C{=}CH-(CH{=}CH)_k-CHR^2_2 \rightleftarrows$ $R^1_2C\text{⋯}CH\text{⋯}(CH\text{⋯}CH)_k\text{⋯}CR^2_2$ (bridged by H) 4q+2 or 4q

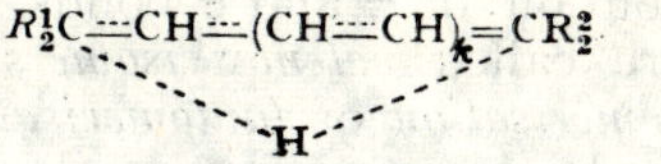

$\rightleftarrows R^1_2CH-(CH{=}CH)_k-CH{=}CR^2_2$

(2) Electrocyclic reactions (thermal or photochemical)

(2a) 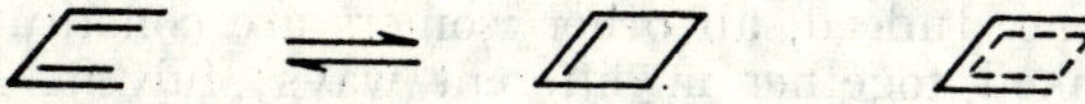 4

(2b) 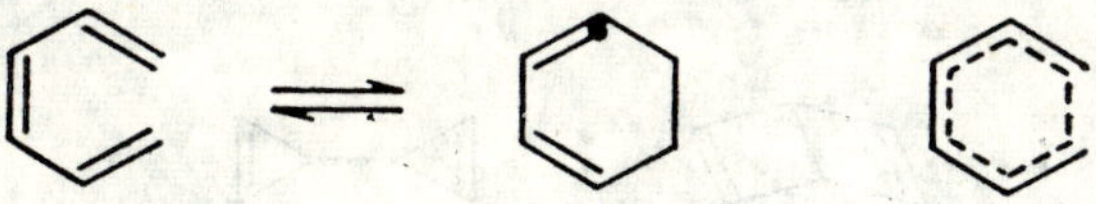 6

(2c)  + ◁ ⟶ 2

(B) *Bimolecular transformations*

(3) Cycloaddition reactions

(3a) Diene synthesis

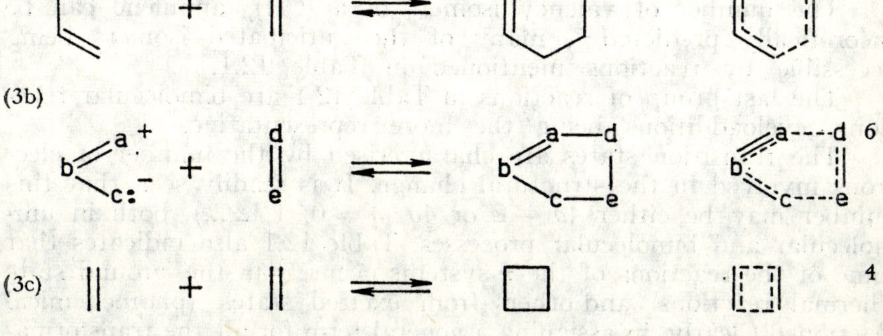

(3d)  $+ \ :CR_2 \longrightarrow$ CR_2 CR_2 2

(4) Intermolecular hydrogen shifts

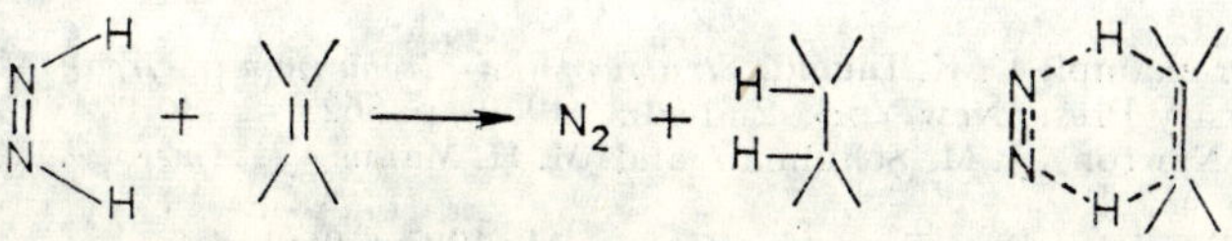

N_2 + 6

A first group of pericyclic reactions (A) comprises reversible unimolecular rearrangements of unsaturated compounds. If the equilibrating isomers consist of identical groups, but have a different distribution of π and σ bonds, as is the case in Table 12.1, the isomers are called *valency isomers* or *tautomers* and the reaction *valency isomerisation* or *tautomerisation* [25]. For a clearer distinction between valency isomerism and other types of isomerism, the following example is useful. From the seven isomers with the molecular formula C_6H_6 [26] (2*a*—2*g*), only fulvene 2*g* is not a valency isomer of benzene. Indeed, all other isomers are constituted only of CH groups, linked together in different ways; fulvene has a quaternary carbon and a CH_2 group.

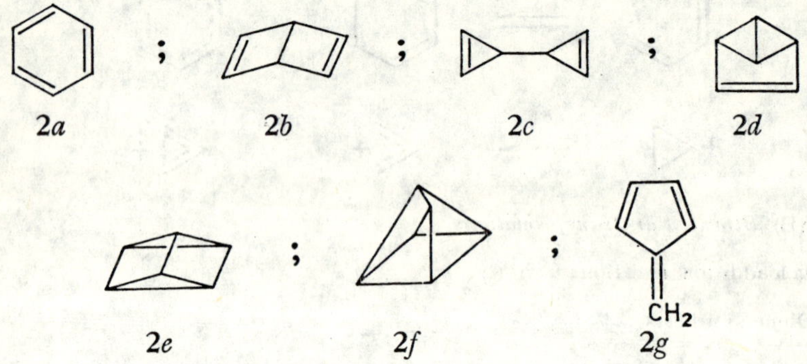

The number of valency isomers of a $(CH)_n$ annulene can be theoretically predicted [27], many of the anticipated isomers being accessible by reactions mentioned in Table 12.1.

The last group of reactions in Table 12.1 are bimolecular reactions, cycloadditions being the more representative.

The transition states are characterised by the number of electrons involved in the structural change. It is readily seen that this number may be either $4q + 2$ or $4q$ ($q = 0, 1, 2, \ldots$), both in unimolecular and bimolecular processes. Table 12.1 also indicates that some of the reactions of the π-systems occur from the ground state (thermal reactions) and others from excited states (photochemical reactions). Clearly, in assigning a general term for all the transformations included in the table, the term of *no mechanism reactions* would be unsuitable. Indeed, this name was coined only for ground state reactions in which both heterolytic and homolytic features were observed. Bartlett's proposal could hardly be used, some of the reactions being either three- (2c, 3d) or six-centre (4) processes. The term pericyclic reactions seems therefore more suitable, in-

[25] See for example C. K. Ingold, *Structure and Mechanism in Organic Chemistry*, Cornell University Press, New York, 2nd edn. (1969), p. 862.

[26] M. D. Newton, J. M. Schulman and M. H. Manus, *J. Amer. Chem. Soc.*, **96**, 17 (1974).

[27] A. T. Balaban, *Rev. Roumaine Chim.*, **11**, 1097 (1966).

cluding even ionic reactions (2c, 3b) besides the photochemical and thermal processes of non-ionic reagents.

Another possible designation is that of $n\pi \rightarrow n\sigma$ transformations. This term was initially used by Hoffmann and Woodward[21] in order to classify cycloaddition reactions and indicates the number (n) of π orbitals which are converted in the process into σ orbitals. Actually, if n is 0, 1, 2..., all reactions shown in Table 12.1 can be designated in a similar manner. Thus, in a sigmatropic reaction (A.1) the number of π and σ bonds remains unchanged. A sigmatropic reaction will be thus a $0\pi \rightarrow 0\sigma$ transformation. An electrocyclic reaction (A.2) can be named a $\pi \rightarrow \sigma$ transformation, only one π bond being converted to a σ bond. In the following discussions we shall use this classification since it seems to designate more precisely the actual electronic changes.

12.1.3 *Conservation of orbital symmetry*

It was shown in Chapter 2 that the molecular orbitals can be regarded as resulting from overlap interactions of atomic orbitals. In any such interaction, two atomic orbitals give rise to two molecular orbitals, one bonding, of low energy, and the other antibonding, of high energy (Figure 12.1).

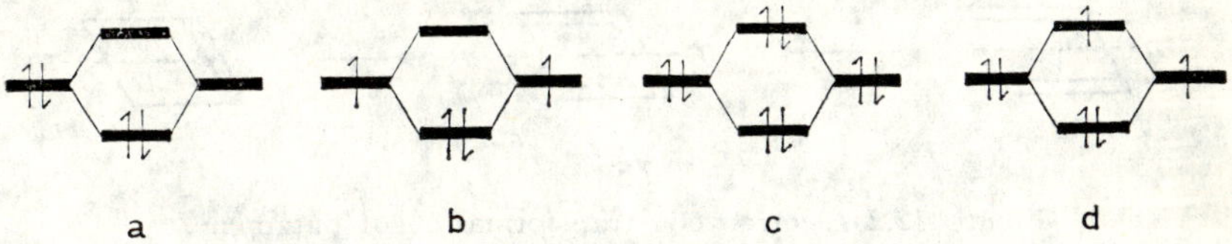

a b c d

Figure 12.1 Overlap interactions of atomic orbitals.

The overall electronic energy of the resulting orbitals depends on how the interacting orbitals are occupied and on the energy difference between them. The smaller this difference, the lower the energy of the bonding orbital will be. The electronic energy will thus be low in case *a* of Figure 12.1, it will not differ from that of the atomic orbitals in *c* and will be high in *d*. We have assumed a similar interaction between molecular orbitals (see p. 83) in order to simplify the calculations for more complex molecules. As Figure 12.1 clearly shows, for those cases only the frontier orbitals will be relevant for the electronic energy of the resultant perturbed orbitals (highest occupied and lowest vacant). We have further shown (p. 85) that the same procedure (interaction of molecular orbitals) can be applied in the theoretical treatment of reaction rate and reactivity, if the initial state is considered to be characterised by the relevant unperturbed orbitals of the reagents and the final state by the perturbed orbitals of the transition state. Occasionally, we have stressed the importance of orbital symmetry in such treatments. Invariably, a symmetric orientation with respect to a reflection plane bisecting the interacting nuclei will generate a bonding orbital, an antisymmetric orientation will correspond to an antibonding orbital in the final state (see Figure 2.6). Symmetry is thus conserved in orbital interactions and this is the basic idea of the Woodward-Hoffmann theory.

Let us consider the simple case of the reversible conversion of butadiene to cyclobutene (Figure 12.2) and split this equilibrium into two opposite reactions. The molecular orbitals of the cisoid butadiene will mix together to generate the molecular orbitals of the transition state. We do not know the geometry of the transition state, nor the form of its molecular orbitals. We know however that symmetry must be preserved in some way. The same must be true for the reverse reaction.

Now, since the reaction is reversible, we may assume that an identical transition state is reached starting from either direction. That means that the molecular orbitals of butadiene and those of cyclobutene must be related by a common symmetry element. Otherwise stated, the symmetry of molecular orbitals is preserved in pericyclic reactions. A bonding orbital in reagents will generate a bonding orbital in products.

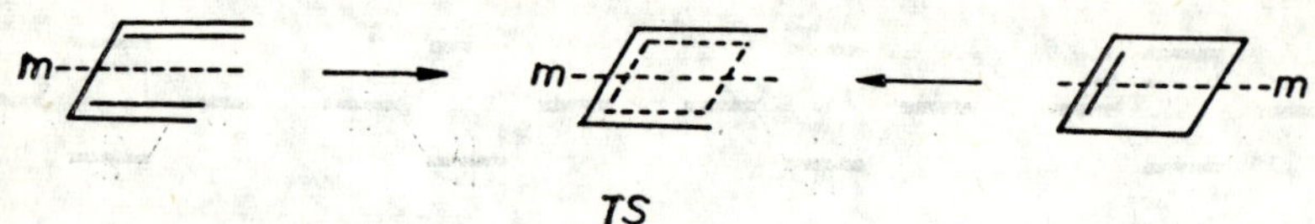

Figure 12.2 Electrocyclic transformation of butadiene.

If that is not the case, the reaction is said to be *symmetry-forbidden;* otherwise, the reaction is *symmetry-allowed.*

The Woodward-Hoffmann principle can be applied either by considering the symmetry of the frontier orbitals (see Sections 12.2 and 12.3) or, in a more quantitative way, by constructing correlation diagrams (see Section 12.4). In both methods the following rules must be observed:

(a) In order to get a better insight into the reaction, the interacting system must be reduced to a more simple model (experimentally possible or impossible), with the highest symmetry with respect to both initial and final state, but involving the same bond modifications as the actual reaction. The reader must keep in mind that substituents rarely change the course of a reaction, but do drammatically affect the symmetry of a compound. Thus, the diene synthesis starting from isoprene and methyl-maleic anhydride (12.5), as well as other similar examples in which no symmetry element is found in the transition state, can be reduced to the hypothetical reaction of butadiene and ethylene (12.6), with the same bond modifications, but with the highest possible symmetry in the transition state:

$$\text{Me-substituted butadiene} + \text{Me(H)C=C anhydride} \longrightarrow \text{cyclohexene anhydride (Me, Me)} \tag{12.5}$$

$$\text{butadiene} + \text{ethylene} \longrightarrow \text{cyclohexene} \tag{12.6}$$

(b) Only the symmetries of the relevant molecular orbitals involved in the reactions are considered. Thus, in reaction (12.6) only six π orbitals in the reagents (4 π orbitals for butadiene, Ψ_1, Ψ_2, Ψ_3, Ψ_4, and 2 π orbitals in ethylene, π and π^*) are essential in the transformation. Other orbitals, for example the σ orbitals corresponding to C—H bonds, even if they suffer modifications (rehybridisations), are not affected by the symmetry operations and may be neglected. Similarly, for the reverse reaction, there are six essential molecular orbitals of cyclohexene, four sigma, σ_1, σ_2, σ_1^* and σ_2^* and two π orbitals, π and π^*.

(c) If the formal MOs, predicted by the simple theory, do not share a symmetry element with the transition state, the semi-localised orbitals (between two nuclei) will be replaced by suitable delocalised orbitals, applying the general method of combination of AOs or MOs, accounting for the overall symmetry of the molecule. Thus, the eight orbitals of the C—C bonds in cyclobutane (σ_1, σ_2, σ_3, σ_4, and the corresponding antibonding orbitals) do not account for the symmetry of the molecule (D_{4h}). The delocalised orbitals formed by linear combinations of the semi-localised orbitals will reflect the symmetry of the molecule. However, in cycloaddition (12.4) it is sufficient to consider the delocalised orbitals formed by two pairs of the semi-localised MOs (say $\sigma_1\sigma_1^*$ and $\sigma_2\sigma_2^*$), which are symmetrically common in the transformation. Such procedures are possible considering the vectorial properties of AOs and MOs.

(d) Throughout this chapter the molecular orbitals will be represented by the atomic orbitals displaying the same symmetry as the MOs, with no consideration of the actual value of the coefficients involved in the linear combination (12.7):

$$\Psi_i = \sum_{j=1}^{j=n} c_{ij}\varphi_j \tag{12.7}$$

This simplification can be applied whenever only qualitative considerations are made. For quantitative treatments, the value of the coefficients c_{ij} must be known. For a polyenic system, the sign and the value of the coefficients are given by the relation (12.8):

$$c_{ij} = \sqrt{\frac{2}{n+1}} \sin \frac{\pi i j}{n+1} \tag{12.8}$$

where n is the number of carbon atoms in the chain, i is the index of the molecular orbital considered and j the label of the carbon atom in the system.

Quantitative treatments will not be made in this chapter. However, relation (12.8) can be used for the determination of the sign of a particular atomic orbital in the MOs of the reacting molecules.

12.2 Sigmatropic reactions

12.2.1 *The electronic mechanism of sigmatropic reactions*

A sigmatropic reaction of the order $[i, j]$ is defined by Woodward and Hoffmann [22] as a migration of a sigma bond flanked by one or more π-electron systems, to a new position whose termini are $i-1$ and $j-1$ atoms removed from the original bonding loci, in an uncatalysed intramolecular process. The well-known Claisen and Cope rearrangements are sigmatropic changes of the order [3,3].

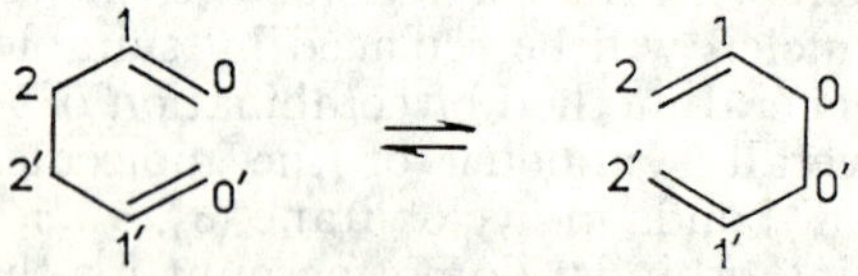

$$i-1 = j-1 = 2$$
$$i = j = 3$$

The most simple sigmatropic reaction is the migration of a hydrogen atom within a polyenic system (12.9). In this case, the migrating σ bond (C—H) has only one terminus and therefore $i-1 = 0$ and $i = 1$. The order of these reactions will be $[1, j]$.

$$\underset{1\quad 2\,.\,.\,.\,.\,.\,.\,.\,.\,j}{R_2C{=}CH{-}(CH{=}CH)_k{-}CHR_2} \rightarrow \underset{1\,.\,.\,.\,.\,.\,.\,.\,.\,.\,.\,j}{R_2CH{-}(CH{=}CH)_k{-}CH{=}CR_2}$$

$$\mathbf{3}a \qquad\qquad \mathbf{3}b \tag{12.9}$$

The electronic mechanism of a sigmatropic transformation has been discussed by Woodward and Hoffmann [22, 28], by applying simple molecular orbital considerations of the frontier orbitals involved in the reaction. Let us first take the example of the simple

[28] R. Hoffmann and R. B. Woodward, *J. Amer. Chem. Soc.*, **87**, 4389 (1965).

reaction (12.9). We shall first assume that the reaction proceeds stepwisely, radical **4** and a hydrogen atom being the intermediates. Their combination may either give the initial polyene, **3***a*, or the rearranged hydrocarbon **3***b*:

$$3a \rightleftharpoons R_2C{=}CH{-}(CH{=}CH)_k{-}\dot{C}R_2 \rightleftharpoons 3b$$

4

In the concerted migration of hydrogen, in order to avoid a high-energy orbital in the transition state, a positive overlap between the atomic orbital 1*s* of hydrogen and the lobes of the *p* orbitals of C_1 and C_j, in the corresponding molecular orbitals of **4**, must be maintained. These MOs will be the highest occupied (HO) for a ground state (thermal) reaction and the lowest unoccupied (LU) for an excited state (photochemical) reaction. In the radical **4**, the HOMO is a non-bonding orbital, **5**, whose symmetry is readily ascertained. It can be seen that the symmetry of the marginal atomic orbitals in the non-bonding orbital will depend upon the number of carbon atoms in the conjugated chain:

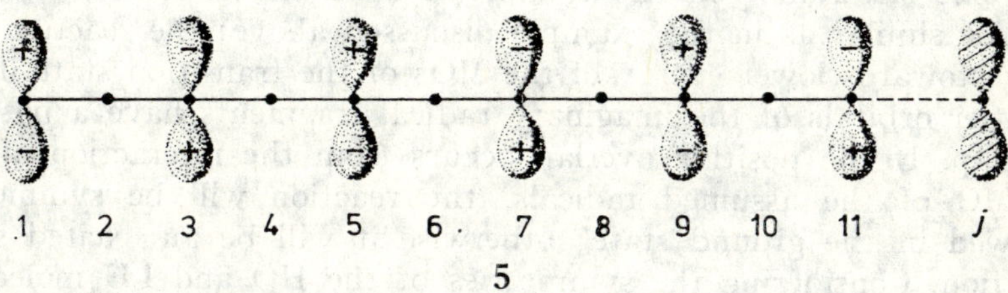

5

The positive lobes of the AOs of C_1 and C_j may be situated either on the same side of the polyenic chain if *k* in **3** is odd, or on opposite sides if *k* is even. It results that there will be two different ways in which a hydrogen atom may migrate [22] in a concerted reaction. In one of them (**6**), called the *suprafacial shift*, the hydrogen transferred is associated with the same face

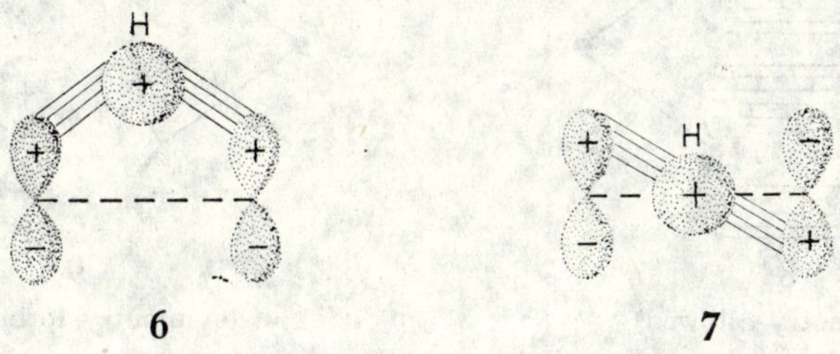

6 **7**

of the π-system. In the other (**7**), called the *antarafacial shift*, the migrating hydrogen atom passes from the top face of a carbon atom (C_j) to the bottom face of the other (C_1). From these simple MO considerations, Woodward and Hoffmann [22] have derived the following rule: a sigmatropic change of order [1, *j*] will be in the ground state a suprafacial change if in polyene **3** *k* is odd (1, 3, 5...) and antarafacial if *k* is even (0, 2, 4...). Considering that in the LUMO of **4** the symmetries of the AOs of C_1 and C_j are reversed, the rule for the excited states will be: a sigmatropic change of order [1, *j*] will occur in the excited state as a suprafacial migration if *k* in **3** is even (0, 2, 4...) and as a antarafacial shift if *k* is odd (1, 3, 5...).

For sigmatropic changes [*i*, *j*] in which both *i* and *j* are greater than 1, we may adopt a similar treatment. Let us consider the Cope rearrangement (12.2) and assume that the reaction occurs stepwisely, involving allyl radicals as intermediates:

(12.10)

The electronic conditions imposed upon the concerted reaction will be similar as in the example discussed above: the reaction will move towards lower energy of the MOs of the transition state if the frontier orbitals of the imaginary radical fragments have a positive overlap. If the positive overlap occurs from the interaction of the HOMO of the assumed radicals, the reaction will be symmetry-allowed in the ground state; otherwise it will be an excited state reaction. Considering the symmetries of the HO and LU molecular orbitals of the allyl cations in (12.10), it can be concluded that the Cope rearrangement is a ground state sigmatropic rearrangement:

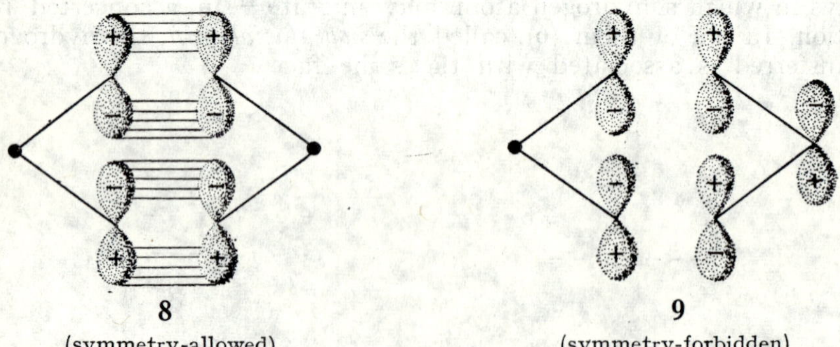

8 (symmetry-allowed)

9 (symmetry-forbidden)

If the symmetries of the allyl and pentadienyl radicals are considered, we can conclude that a sigmatropic change of the order [3,5] will be symmetry-forbidden in the ground state, but symmetry-allowed in the excited state:

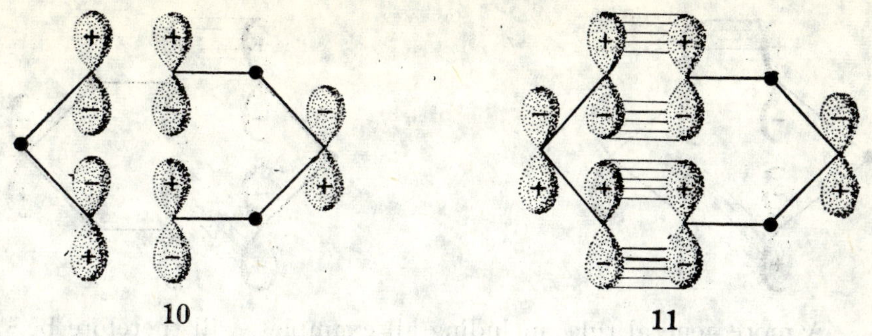

10 11

These two examples are sufficient to illustrate the following rule: a sigmatropic change of order $[i, j]$ $(i, j > 1)$ will be symmetry-allowed in the ground state and symmetry-forbidden in the excited state if $i + j = 4q + 2$ $(q = 1, 2, 3...)$. If $i + j = 4q$, the rule is reversed.

The procedure used above to set up the last rule is a simplification of the treatment adopted by Woodward and Hoffmann [22, 24]. It has the advantage of a rapid decision, as the imaginary interacting radicals can easily be found and the symmetry of the relevant AO in the frontier MOs readily established without any calculation. This procedure also allows a rapid control of the initially assumed order of the sigmatropic change. Indeed, the order of a sigmatropic change is also determined by the numbers i and j corresponding to the numbers of atoms in the hypothetical interacting radicals in the transition state. The procedure is also suitable for the demonstration of a more general rule, stated below, regarding sigmatropic reactions.

In **8**—**11** we have assumed a *supra-supra* interaction of the radical fragments, which is equivalent (**12**) to an *antara-antara* interaction:

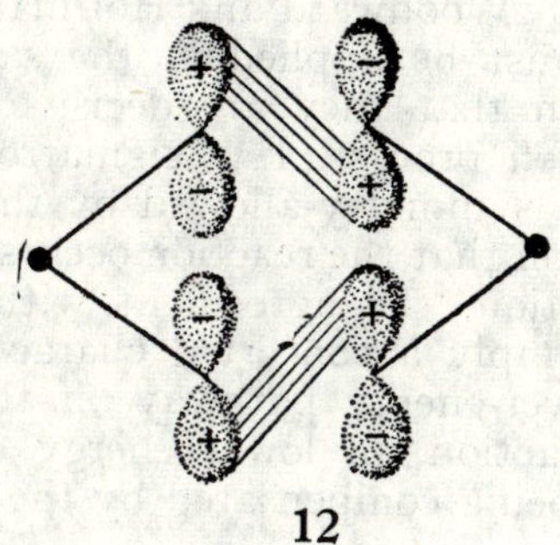

12

However, there are two other possibilities of interaction between these π-systems, namely *antara-supra* and *supra-antara* (equivalent or non-equivalent, depending on the reacting system):

A more general rule, including all examples, will therefore be [24]: a sigmatropic change of order $[i, j]$ will be *supra-supra* or *antara-antara* in the ground state if $i + j = 4q + 2$, and in the excited state, respectively, if $i + j = 4q$ ($q = 0, 1, 2...$). For *antara-supra* processes the rule is reversed: *antara-supra* and *supra-antara* changes are allowed in the ground state if $i + j = 4q$ and in the excited state if $i + j = 4q + 2$.

As already stated, the theory of Woodward and Hoffmann is one of the first generally used theories of organic reactivity based on MO concepts. Previous treatments, like Dewar's perturbation theory, even if more precise and correct in the derived relations, have acquired less interest in the literature due to the more complex mathematical apparatus and to a less suggestive chemical representation of the phenomena involved. Woodward and Hoffmann's concept of orbital symmetry conservation has the advantage, for organic chemists, of being more easily understood in its graphical representations than are other MO theories. Other, more complex treatments of pericyclic reactions [24, 29–31] are in perfect agreement with the rules established by Woodward and Hoffmann.

12.2.2 *Sigmatropic changes of order* [1, *j*]

The rules established by Woodward and Hoffmann for these and other pericyclic reactions must be applied in the same way as the laws of thermodynamics, in that they are decisive when indicating the impossibility of a given process. If a sigmatropic change, observed experimentally, is not symmetry-allowed by the selection rules, this will be a clear indication that the reaction occurs stepwisely. However, for an observed reaction, the agreement with the selection rules does not necessarily imply a concerted character. The reaction will always follow the lower-energy pathway; if that pathway happens to be a concerted reaction (of lower energy demand, the energies of the bonds broken being compensated by the energies of the bonds

[29] N. T. Anh, *Les Règles de Woodward-Hoffmann*, Ediscience, Paris (1970).

[30] M. J. S. Dewar, *Tetrahedron Suppl.*, **8**, Part I, 85 (1956); *idem, Aromaticity*, Chem. Soc. Spec. Publ., **21**, 177 (1967); H. Zimmermann, *J. Amer. Chem. Soc.*, **88**, 1564, 1566 (1966).

[31] N. D. Epiotis, *J. Amer. Chem. Soc.*, **94**, 1924 (1972), and the following papers.

formed), the process will be stereospecific. A multistep reaction, favoured by electronic stabilisation of the intermediates or imposed by geometrical requirements, will be unstereospecific. The difference between the two possible mechanisms will also be reflected in the activation parameters. A concerted reaction will have lower activation enthalpy (or activation energy) and lower activation entropy than a stepwise reaction.

Woodward-Hoffmann's theory predicts that a [1,3] hydrogen shift will be in the ground state an antarafacial process. However, even if allowed by the rules, the reaction is impossible for geometrical reasons. Thus, triene **13** is thermally stable[32], even if in acid catalysis readily rearranges to the more stable aromatic compound.

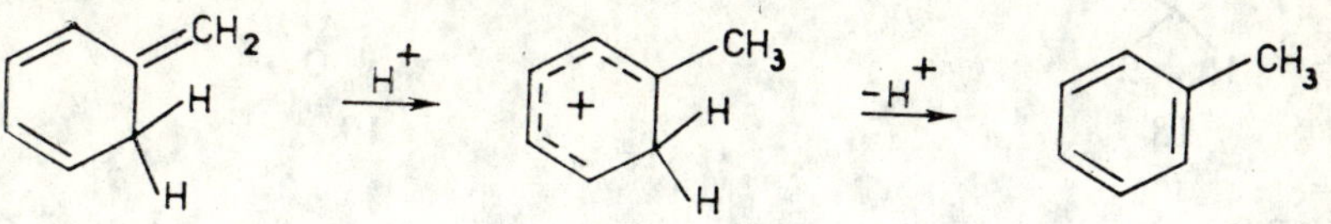

CH_2 H H **13** $\xrightarrow{H^+}$ CH_3 + H H $\xrightarrow{-H^+}$ CH_3

The photochemical [1,3] shift, allowed by symmetry orbital considerations, is a geometrically possible suprafacial migration. Such excited state reactions have actually been observed[33].

An interesting example of concerted hydrogen shifts in a polyenic system, either under thermal or photochemical conditions, is found in the isomerisation reactions of 1,4-di-(7,7′-cycloheptatrienyl) benzene **14**[34]. On irradiation ($\lambda = 2537$ Å), this compound affords in a 60% yield the 1,1-isomer, **15**, whereas on heating (165 °C) it yields (70%) the 3,3′-isomer **16**. The latter is converted on irradiation ($\lambda = 3500$ Å) into the 2,2′-isomer, **17**. The changes in the polyenic moieties can be regarded either as thermally-allowed [1,5] sigmatropic changes or as photochemically-allowed [1,7] hydrogen migrations:

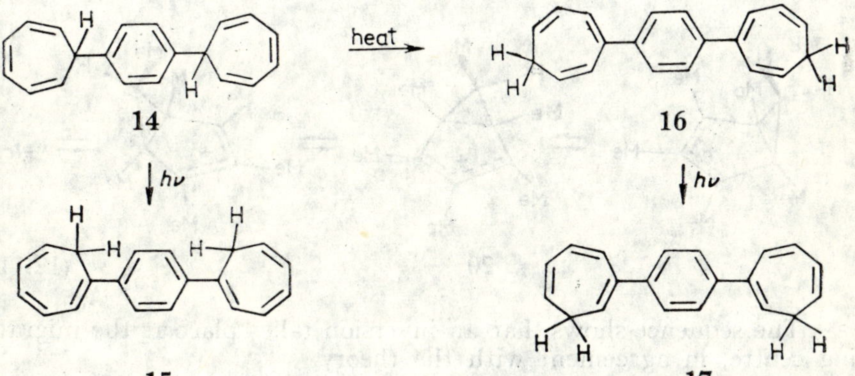

14 $\xrightarrow{\text{heat}}$ **16**

14 $\xrightarrow{h\nu}$ **15**

16 $\xrightarrow{h\nu}$ **17**

[32] W. J. Bailey and R. A. Baylouny, *J. Org. Chem.*, **27**, 3476 (1962).
[33] W. G. Dauben and W. T. Wipke, *Pure Appl. Chem.*, **9**, 539 (1964).
[34] R. W. Murray and M. L. Kaplan, *J. Amer. Chem. Soc.*, **88**, 3527 (1966).

The sigmatropic $[i, j]$ change is not confined to hydrogen shifts. Other bonds, like C—C bonds, can also be involved in the process. Obviously, for such examples the general rule will be applied, the reaction being associated either with overall retention or inversion at the reaction centres if $i + j = 4q + 2$, or with one inversion if $i + j = 4q$ ($q = 0, 1, 2, \ldots$). The thermal isomerisation of **18** into **19**[35] is a convincing example of a *supra-antara* transformation in a sigmatropic shift of order [1,3]:

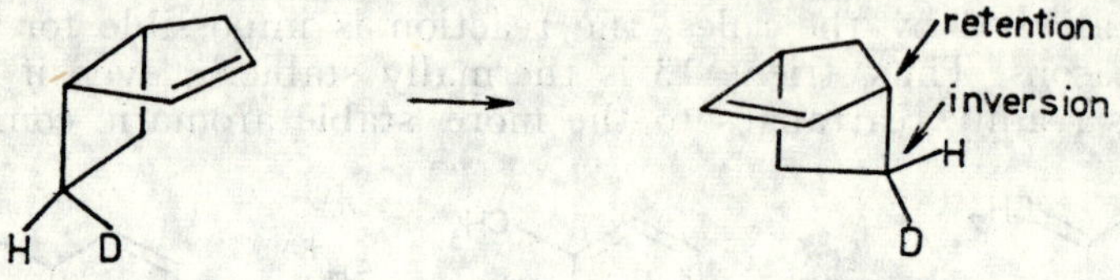

retention
inversion

18 **19**

Sigmatropic changes are also possible in ionic systems. The well-known Wagner-Meerwein rearrangement is a ground state allowed suprafacial sigmatropic shift of order [1,2].

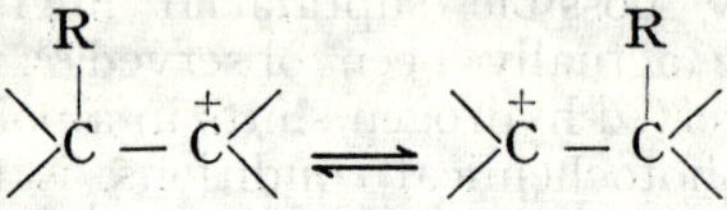

R = H, alkyl, etc.

A more interesting example of an ionic sigmatropic change of order [1,4] was discussed by Childs and Winstein[36]. Hexamethyl-bicyclo[3,1,0]hexenyl ion **20** presents different signals for the methyl groups in the n.m.r. spectra at low temperature. At higher temperatures, except for the methyl group bonded to the cyclopropyl ring, the signals of the other methyls coalesce into a single peak, indicating the equilibrium (12.11):

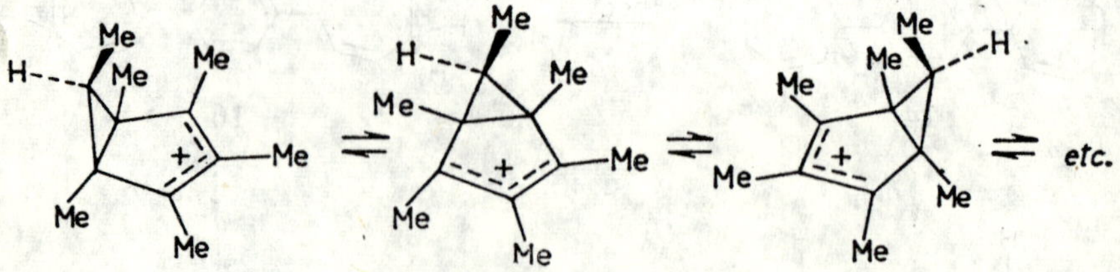

20 (12.11)

The sequence shows that an inversion takes place at the migrating centre, in agreement with the theory.

[35] J. A. Berson and G. L. Nelson, *J. Amer. Chem. Soc.*, **89**, 5303 (1967).

[36] R. F. Childs and S. Winstein, *J. Amer. Chem. Soc.*, **90**, 7146 (1968); see also: R. F. Childs, M. Sakai, B. D. Barrington and S. Winstein, *ibid.*, **96**, 6403, 6409 (1974); R. F. Childs and M. Zega, *ibid.*, **96**, 6418 (1974).

In ground state, hydrogen shifts are not always concerted processes. Stepwise reactions involving radical intermediates are frequently encountered. Such a process, which deserves mention, is the isomerisation of **21** [37]. The products are best explained as arising from a 1,4-dehydrobenzene, **22**.

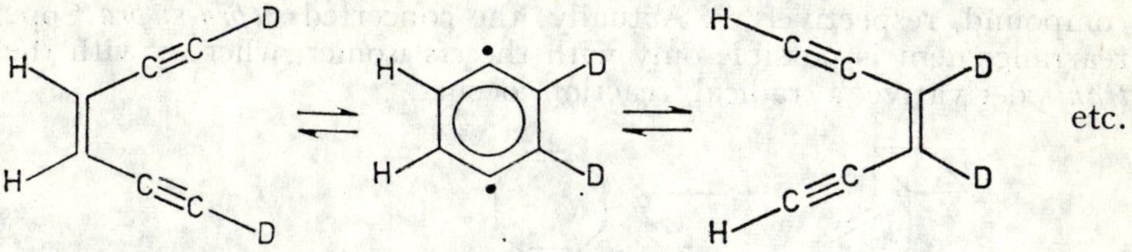

21 ⇌ **22** ⇌ **23** etc.

The intermediate **22** must be planar. Even if not an isomer of benzene, it resembles more the structure of benzene proposed in 1886 by Dewar [38], than does the unplanar valence isomer *2b*, improperly called Dewar benzene.

12.2.3 *Cope rearrangement*

Cope and his co-workers [6, 39, 40] have observed that substituted diallyl derivatives undergo skeleton modification on heating in the absence of any catalyst or radical promoter.

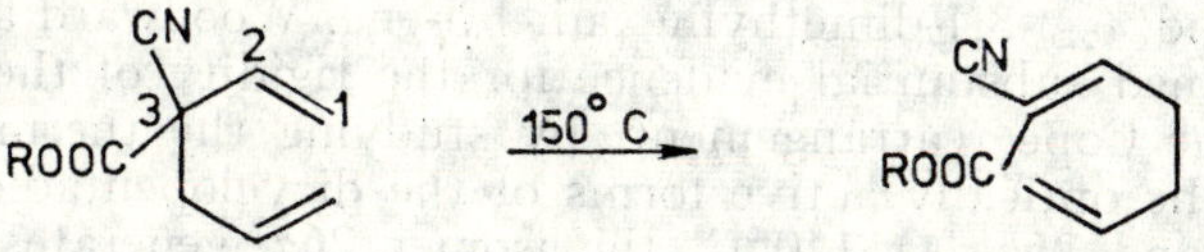

$$\xrightarrow{150^\circ C} \qquad (12.12)$$

$$\xrightarrow{300^\circ C} \qquad (12.13)$$

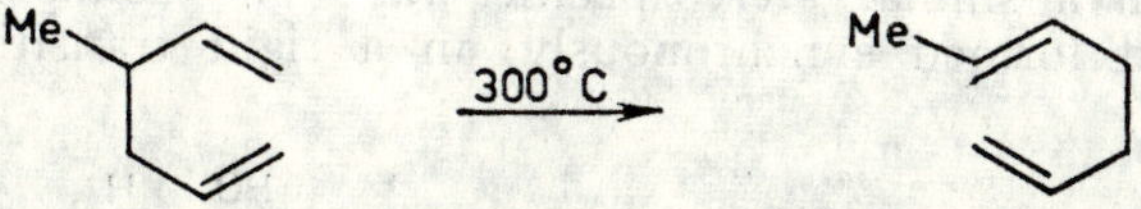

The reaction kinetics are first-order [40, 41], the rate constants varying only slightly with the medium. Nevertheless, the polarity of substituents in position 3 may affect the reaction rates. Electron-withdrawing substituents in that position favour the reaction, whereas electron-repelling groups have the opposite effect [40] (compare equation (12.12) with equation (12.13)).

[37] R. R. Jones and R. G. Bergman, *J. Amer. Chem. Soc.*, **94**, 660 (1972).
[38] J. Dewar, *Proc. Roy. Soc. (Edinburgh)*, 84 (1866).
[39] H. Levy and A. C. Cope, *J. Amer. Chem. Soc.*, **66**, 1684 (1944).
[40] E. G. Foster, A. C. Cope and F. Daniels, *J. Amer. Chem. Soc.*, **69**, 1893 (1947).
[41] G. R. Aldridge and G. W. Murphy, *J. Amer. Chem. Soc.*, **73**, 1158 (1951).

The activation entropies have large negative values, indicating a rigid transition state, in agreement with a concerted mechanism. Lower absolute values of this parameter may suggest a stepwise radical reaction. Thus, the thermal isomerisation of the two divinylcyclobutanes (**24**) into cyclooctadiene presents an entropy of activation of −11.7 e.u. for the *cis* isomer and of −1.2 e.u. for the *trans* compound, respectively [42]. Actually, the concerted *supra-supra* Cope rearrangement is possible only with the *cis* isomer, whereas with the *trans* derivative a radical reaction occurs.

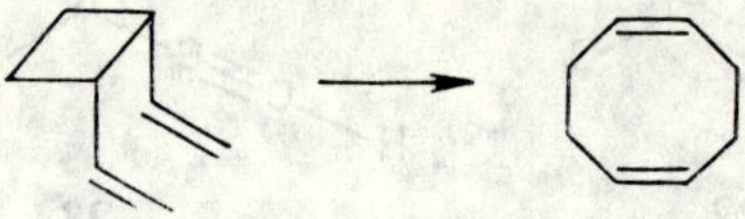

24*a* **25**

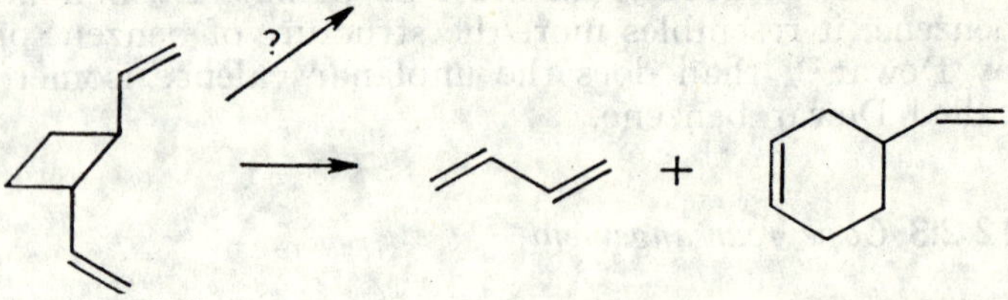

?

24*b*

The four-centre character of the reaction has been demonstrated by Doering and Roth [43] in an elegant study of the isomerisation of *meso*- and (±)-3,4-dimethylhexadi-1,5-ene. Woodward and Katz [44] have obtained substantial evidence for the rigidity of the transition state in the Cope rearrangement, by studying the thermal isomerisation of the optically active forms of the dicyclopentadiene derivatives **26*a*** and **26*b***. At 140 °C, the isomer **26*a*** generates only **27*a***, with complete retention of the optical activity, whereas **26*b*** yields only **27*b*** in a similar stereospecific way. The bonds are therefore broken and formed synchronously, in a rigid transition state **28**.

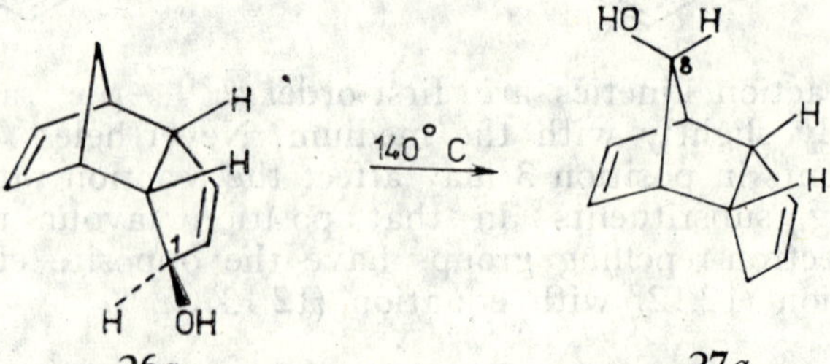

H H 1 H OH 140° C HO H 8 H H

26*a* **27*a***

[42] G. S. Hammond and C. D. De Boer, *J. Amer. Chem. Soc.*, **86**, 899 (1964).
[43] W. von E. Doering and W. R. Roth, *Tetrahedron*, **18**, 67 (1962).
[44] R. B. Woodward and T. J. Katz, *Tetrahedron*, **5**, 70 (1959).

H OH 8 H H 140° C H H 1 OH H bond broken bond formed

26*b* 27*b* 28

The known experimental data regarding the Cope rearrangement [11, 45] are in complete agreement with a symmetry-allowed *supra-supra* process of order [3,3]. However, one should not think that the unconcerted pathway is encountered only in the case of geometrically unfavourable systems, such as 24*b*. A diradical pathway, as indicated by Doering and his co-workers [46], may well compete with the concerted mechanism, electronic factors deciding the ultimate result of this competition. In the diallyl derivatives **29**, for instance, a phenyl group in position 3 (**29***b*) should only slightly affect a diradical mechanism, but will be expected to accelerate a concerted reaction. At the same time, phenyl groups in position 2 and 5 (**29***c*, **29***d*) must have a stabilising effect on the diradical **30**. Dewar and Wade [47] have actually found that the diphenyl derivative **29***d* reacts 2000 times faster than the unsubstituted diallyl (**29***a*), whereas the monophenyl derivatives **29***b* and **29***c* rearrange only 18- and 41-fold, respectively, as fast as the parent compound.

R^1 R^2 R^3 → R^1 R^2 R^3 → R^1 R^2 R^3

29*a* $R^1 = R^2 = R^3 = H$
b $R^1 = Ph;\ R^2 = R^3 = H$
c $R^1 = R^3 = H;\ R^2 = Ph$
d $R^1 = H;\ R^2 = R^3 = Ph$

[45] A. Jefferson and F. Scheinmann, *Quart. Rev.*, **22**, 391 (1968).
[46] W. von E. Doering, V. G. Toscano and G. H. Beasley, *Tetrahedron*, **27**, 5299 (1971).
[47] M. J. S. Dewar and L. E. Wade, *J. Amer. Chem. Soc.*, **95**, 290 (1973).

Obviously, the competition between the concerted mechanism and the diradical pathway is definitely won by the latter in the case of the diphenyl derivative, whereas for the monosubstituted derivative **29*b*** the concerted mechanism prevails, the reaction always choosing the lowest energy-demanding route.

12.2.4 *Claisen rearrangement*

Claisen [7] has found that the allyl ethers of phenols are converted by heating into allyl substituted phenols. According to the position in which the allyl moiety is found in the product, *ortho* and *para* Claisen rearrangements are encountered [11, 48]. The latter type is illustrated by allyl phenolic ethers having both *ortho* positions occupied by alkyl groups.

O–CH$_2$–CH=CH$_2$ (R, R) **30**; R=H → **31** (H, CH$_2$–CH=CH$_2$) → rapid → **32** (OH, CH$_2$–CH=CH$_2$) (12.14); R=alkyl → (R, R, H, CH$_2$–CH=CH$_2$) → rapid → (OH, R, R, CH$_2$–CH=CH$_2$) (12.15)

Claisen rearrangements are not confined to phenolic derivatives. Vinyl ethers [49a] as well as unsaturated sulphur compounds [49b] readily undergo similar transformations. The following examples show clearly that the Cope and the Claisen rearrangements are

[48] D. S. Tarbell, *Org. Reactions*, **2**, 1 (1944).

[49] (a) E. Vogel, *Angew. Chem.*, **74**, 829 (1962); (b) H. Takahashi, K. Oshima, H. Yamamoto and H. Nozaki, *J. Amer. Chem. Soc.*, **95**, 5803 (1973).

similar reactions. The latter differs from the former by the heteroatoms of the diallyl system.

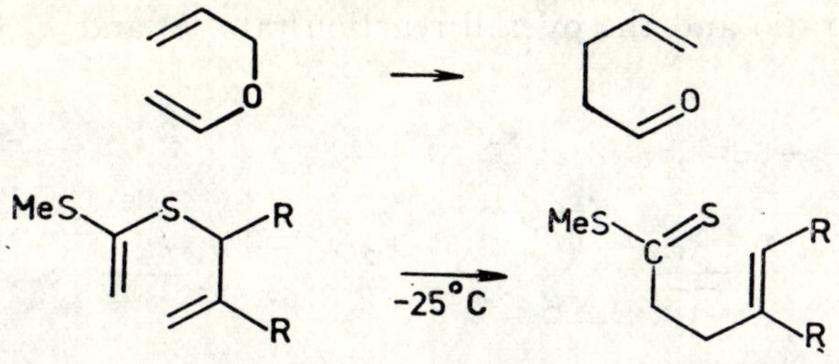

Actually, the primary product of a Claisen rearrangement **31**, is not stable and isomerises rapidly into a phenol **32**. The prototropic change of **31** can be regarded as a sigmatropic process of order [1,3]. However, as for the polyene **13**, the allowed concerted process is not possible for geometrical reasons (see p. 603). Nevertheless, the reaction product **32** may function as an acid catalyst, determining the rapid isomerisation of **31**.

The *ortho* Claisen rearrangement has first-order kinetics [50–52], and the reaction rate is not affected by the reaction medium. It has been demonstrated, by substitution with alkyl groups in the allyl system [53] and by carbon labelling in the migrating moiety [54], that during the reaction the positions in the allyl system are reversed. The allyl rearrangement can be best explained by a cyclic transition state **33**.

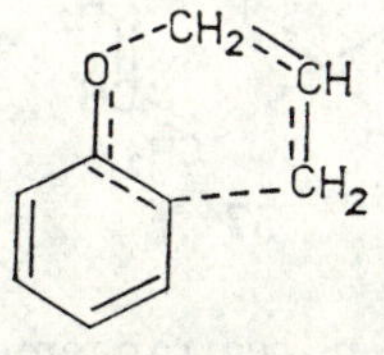

33

The *para* Claisen rearrangement has similar kinetics [55, 56] to the *ortho* process. In contrast with the latter, the migrating allyl groups does not change its initial geometry [54, 56, 57]. It has been demonstrated that cyclohexadienones (**31**) are intermediates in the *para* Claisen

[50] D. S. Tarbell and J. F. Kincaid, *J. Amer. Chem. Soc.*, **62**, 728 (1940).
[51] W. N. White, D. Gwynn, R. Schlett, C. Girard and W. Fife, *J. Amer. Chem. Soc.*, **80**, 3271 (1958).
[52] H. L. Goering and R. R. Jacobson, *J. Amer. Chem. Soc.*, **80**, 3277 (1958).
[53] W. M. Lauer and R. M. Leekley, *J. Amer. Chem. Soc.*, **61**, 3043 (1939).
[54] W. Haegele and H. Schmid, *Helv. Chim. Acta*, **41**, 657 (1958).
[55] J. F. Kincaid and D. S. Tarbell, *J. Amer. Chem. Soc.*, **61**, 3085 (1939).
[56] S. J. Rhoads and R. L. Crecelius, *J. Amer. Chem. Soc.*, **77**, 5057 (1965).
[57] S. J. Rhoads, R. Raulins and R. Reynolds, *J. Amer. Chem. Soc.*, **75**, 2531 (1953).

rearrangement. Thus, Curtin and Crawford[58] have prepared the unsaturated ketone **34** and submitted it to thermal isomerisation. Both the phenyl-ether **35** and the *para*-allyl phenol **36** were formed. From their ratio and the overall reaction rate, k_1 and k_2 have been calculated.

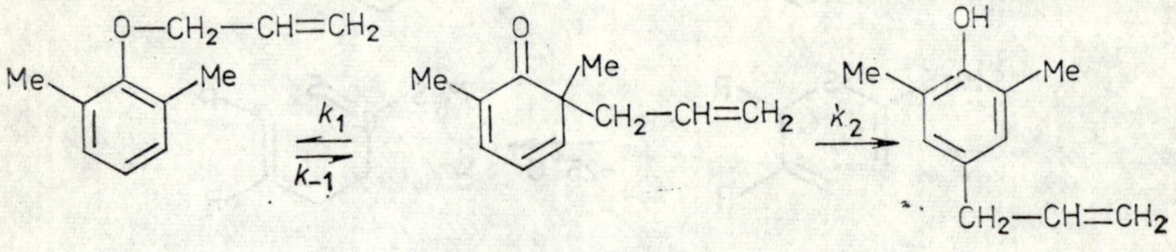

O—CH₂—CH=CH₂ Me Me $\underset{k_{-1}}{\overset{k_1}{\rightleftharpoons}}$ O Me Me CH₂—CH=CH₂ $\xrightarrow{k_2}$ OH Me Me CH₂—CH=CH₂

35 **34** **36**

These results show that the *para* transformation is preceded by an *ortho* Claisen reaction, which must be reversible in its first step. Moreover, as the ratio k_1/k_2 is independent of the reaction temperature, similar transition states must be involved in both isomerisations of **34**. Since in the overall process **35** → **36** the atoms of the allyl system do not change their initial positions, but do rearrange in the first step **35** → **34**, it follows that the transition state for the *para* migration of the allyl group must be **37**:

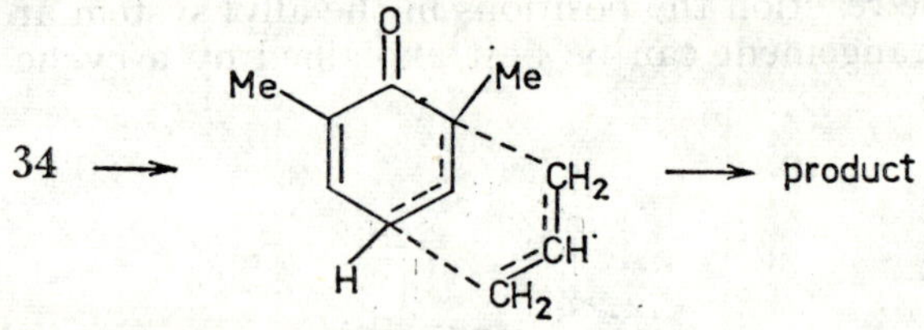

34 ⟶ O Me Me H CH₂ CH CH₂ ⟶ product

37

Thus, the *para* Claisen rearrangement is a succession of two sigmatropic changes of order [3,3], the first step corresponding to an *ortho* Claisen rearrangement and the second to a true Cope rearrangement. The last step, obviously, will be a two-step proton migration.

12.2.5 *Sigmatropic changes in polycyclic polyenes*

The increasing use of n.m.r. spectroscopy in chemical research has revealed fascinating new reactions of organic compounds, otherwise impossible to observe. An example is constituted by those rearrangement reactions in which both substrate and product have the same structure. These reactions have been called *degenerate*

[58] D. Y. Curtin and R. J. Crawford, *J. Amer. Chem. Soc.*, 79, 3156 (1957).

isomerisations and the molecules liable to such transformations *flipping molecules* [59]. A simple degenerate Cope transformation undergoes bicyclo[5,1,0] octadiene **38**. The hydrogen atoms H^1, H^2, H^3, H^4 and H^5 are clearly identified in the low-temperature spectrum. As the temperature is increased, the peaks corresponding to protons H^1 and H^5, on one hand, and to H^2 and H^4, on the other, coalesce indicating equilibrium (12.16) [60]:

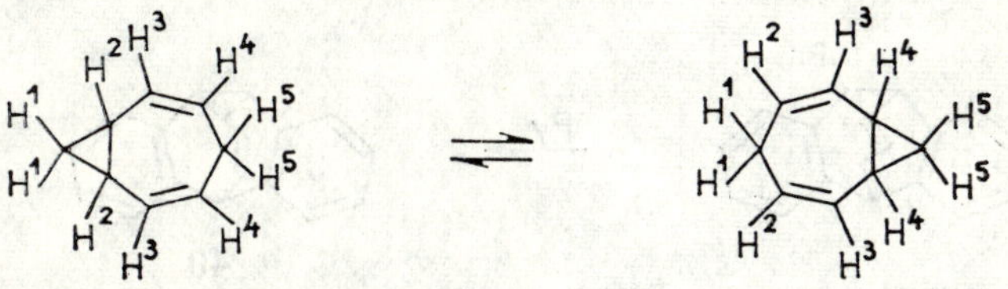

(12.16)

38

A more involved example was predicted by Doering [59] for bullvalene **39**, first prepared by Schröder [61], and later, by a different method, by Doering and Rosenthal [62]. At low temperatures, this substance gives different signals in the n.m.r. spectra, in agreement with its C_{3v} symmetry. However, at 120 °C a single peak is observed, indicating that the initial hydrocarbon has either rearranged into the highest symmetrical hydrocarbon $C_{10}H_{10}$ possessing $D_{10}\,h$ symmetry, or that **39** undergoes degenerate transformations, as shown in equation (12.17), where all CH groups equilibrate:

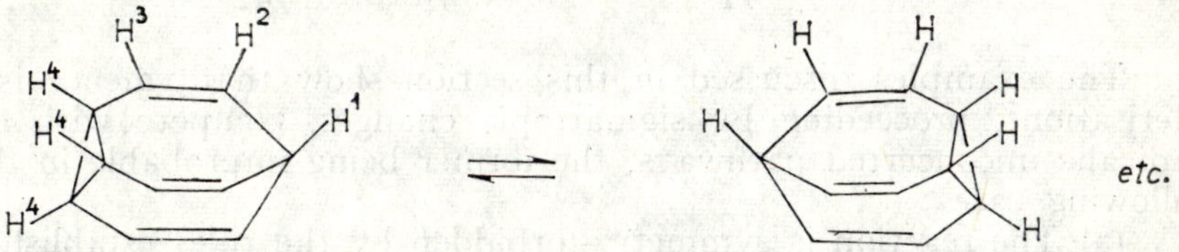

etc.

39

(12.17)

Bullvalene is recovered unchanged after heating at 120 °C and thus equilibria (12.17) must account for the observed spectra. For a perfect equilibration of all CH groups $10!/3 = 1\,309\,800$ Cope degenerate transformations are necessary. Actually, no classical bond should exist in bullvalene at 120 °C.

By annelation of the double bonds in bullvalene with aromatic rings, the reactivity of the parent compound is dramatically changed. Thus, dibenzobullvalene, prepared by sequence (12.18) [63], has no

[59] W. von E. Doering and W. R. Roth, *Angew. Chem.*, **75**, 27 (1963).
[60] W. von E. Doering and W. R. Roth, *Tetrahedron*, **19**, 715 (1963).
[61] G. Schröder, *Chem. Ber.*, **97**, 3140 (1964); R. Merényi, J. F. M. Ott and G. Schröder, *ibid.*, **97**, 3150 (1964); *Angew. Chem.*, **79**, 458 (1967).
[62] W. von E. Doering and J. W. Rosenthal, *J. Amer. Chem. Soc.*, **88**, 2078 (1966).
[63] M. Banciu, M. Elian, E. Ciorănescu and C. D. Nenitzescu, *Tetrahedron Letters*, 2573 (1972).

tendency to rearrange at 200 °C [64]. At 400 °C it is transformed, in 60% yield, into **41**, beside 10% of **42** [64]. The formation of **41**

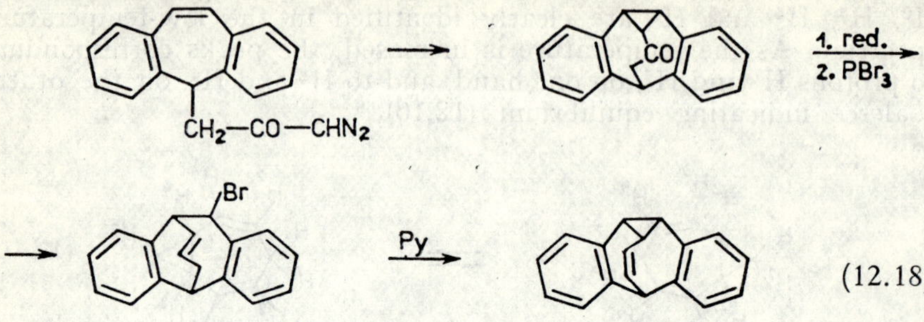

$$\text{CH}_2\text{—CO—CHN}_2 \longrightarrow \xrightarrow[\text{2. PBr}_3]{\text{1. red.}} \longrightarrow \xrightarrow{\text{Py}} \textbf{40} \qquad (12.18)$$

from **40** can be regarded as a 1,3-shift in a vinylcyclopropane system [65]. Such rearrangements are predicted to be *supra-antara* transformations, impossible to realise in **40** for geometrical reasons. The reaction must therefore be radicalic:

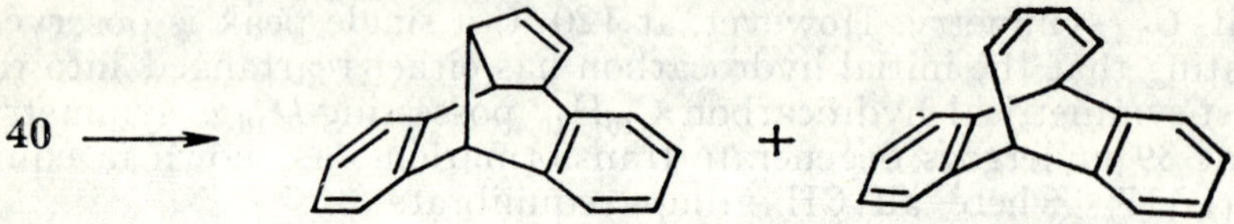

$$\textbf{40} \longrightarrow \textbf{41} + \textbf{42}$$

The examples discussed in this section show that valence isomerisations, proceeding by sigmatropic changes, compete with diradicalic unconcerted pathways, the former being improbable in the following cases:

(a) The reaction is symmetry-forbidden by the rules established by Woodward and Hoffmann.

(b) The reaction is impossible for geometrical reasons, even if allowed by the selection rules.

(c) Alternative unconcerted pathways, with lower activation energies, are possible, either for geometrical or for electronic reasons.

The same conclusions are true for other transformations discussed in this chapter and will not be repeated in the following sections.

12.3 Electrocyclic trasformations

An electrocyclic reaction is defined as a concerted closure of a ring, starting from a conjugated open system in which a π orbital is con-

[64] M. Elian, Ph. D. thesis, Bucharest (1973).
[65] M. C. Flowers and H. M. Frey, *J. Chem. Soc.*, 3547 (1961).

verted into a σ orbital. The stereochemistry of such intramolecular transformations is imposed by two different motions of the termini of the acyclic carbon chain. In one of them, called *disrotatory rotation* (12.19), the two termini of the polyenic chain rotate in different directions, whereas in the *conrotatory* process (12.20) the marginal atoms are rotating in the same direction:

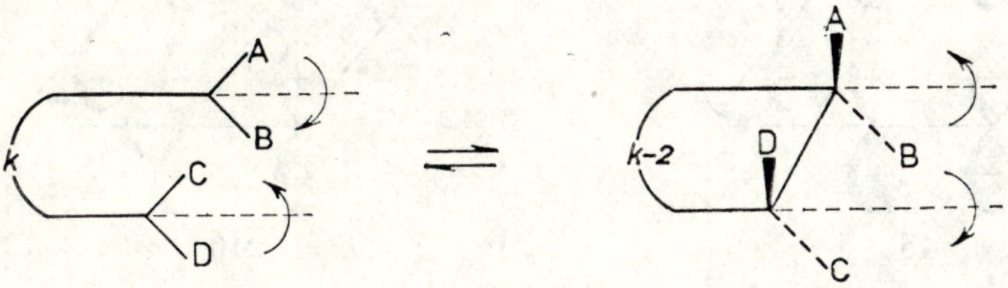

(12.19)

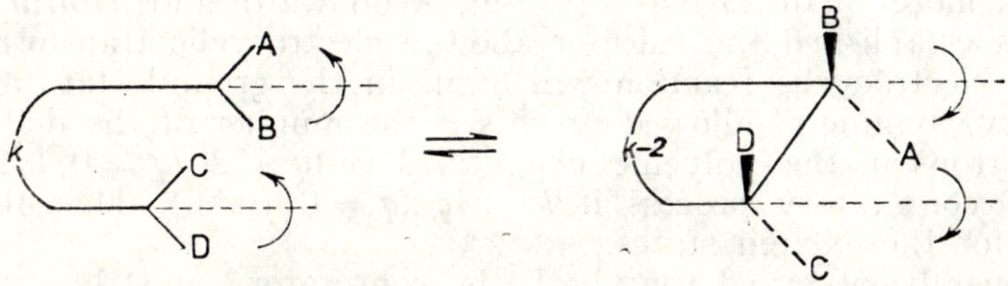

(12.20)

The stereochemistry actually observed in such valence isomerisations is determined by the symmetry of the relevant MOs of the polyenes involved [20, 24].

Let us first mention the example of the reversible transformation of butadiene to cyclobutene (Figure 12.2). In order to form the new σ bond of cyclobutene, the lobes of the AOs in the corresponding π orbitals of butadiene must rotate in the direction which allows a positive overlap at the termini of the polyenic chain. The consideration of the frontier orbitals in diene may be sufficient for deciding the way in which the polyenic chain will rotate. It is readily seen that the HOMO in butadiene (43) must move in a conrotatory way, whereas the LUMO will produce the σ bond by a disrotatory motion.

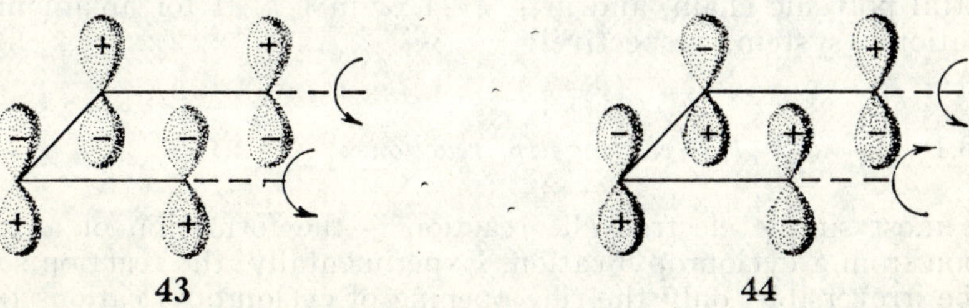

By contrast, the cyclisation of hexa-1,3,5-triene will be possible either by applying a disrotatory motion of the HOMO (**45**) or by imposing a conrotatory process to the LUMO.

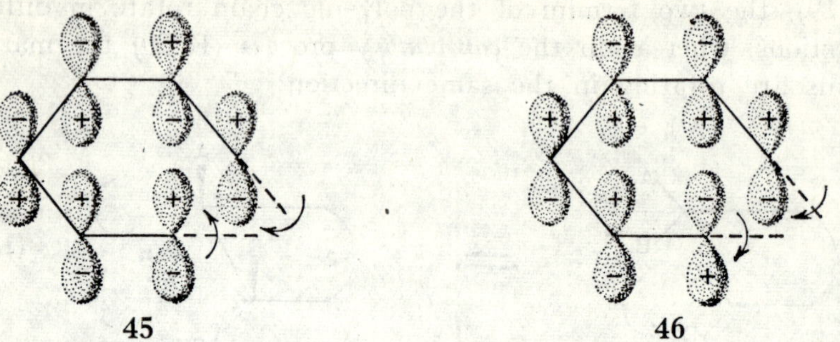

In a more quantitative way[20, 24], Woodward and Hoffmann have thus established the rule for allowed electrocyclic transformations: an electrocyclic reaction will occur in the ground state as a disrotatory, symmetry-allowed process if the number of the delocalised electrons in the polyenic chain is $k = 4q + 2$ $(q = 0, 1, 2...)$ and as a conrotatory process if $k = 4q$ $(q = 0, 1, 2...)$. The rule is reversed for the excited states.

It must be observed that both the conrotatory and the disrotatory motions can occur in two different ways. The disrotatory process, for instance, may proceed either by an *inward* or by an *outward* rotation. The reverse reaction, the electrocyclic opening of a cyclic polyene, must be subject to the same selection rules as the direct cyclisation process, that is an allowed disrotatory electrocyclic cyclisation will be coupled with an allowed disrotatory opening of the corresponding ring. However, in order to preserve the same stereochemistry, an inward disrotatory motion in the direct process must correspond to an outward rotation in the reverse reaction.

When discussing the typical examples of electrocyclic reactions, we shall arbitrarily classify them with respect to the relative position of the newly formed σ bond.

Table 12.2 shows some other aspects of the electrocyclic reactions: electrocyclic processes are also observed with ionic systems, both cationic and anionic; decisive for the stereochemistry is the number of electrons in the delocalised polyenic system; for a given number of electrons k, the new bond can be formed in positions 1, k of a neutral polyenic chain, and in 1, $k-1$ or in 1, $k+1$ for an anionic or a cationic system, respectively.

12.3.1 *[1,3] and [1,5] electrocyclic reactions*

The most simple electrocyclic reaction is the formation of an allyl cation from a cyclopropyl cation. Experimentally, the reaction seems to be irreversible, only the ring opening of cyclopropyl cations being

TABLE 12.2

Electrocyclic reactions

Number of electrons in the relevant MOs of the polyene	*Reaction type*	*Examples*	*Ground state*[a)]	*Excited state*[a)]
2	[1,3]	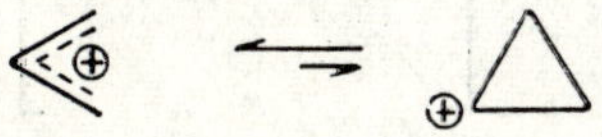 allyl cation ⇌ cyclopropyl cation	D	C
4	[1,3]	 allyl anion ⇌ cyclopropyl anion	C	D
4	[1,4]	 butadiene ⇌ cyclobutene	C	D
4	[1,5]	 pentadienyl cation ⇌ cyclopentenyl cation	C	D
6	[1,5]	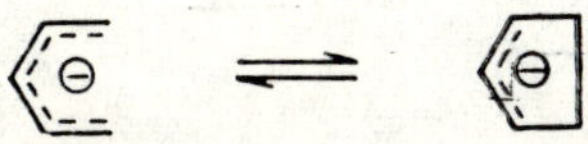 pentadienyl anion ⇌ cyclopentenyl anion	D	C
6	[1,6]	 hexatriene ⇌ cyclohexadiene	D	C
6	[1,7]	heptatrienyl cation ⇌ cycloheptadienyl cation	D	C

a) C = conrotatory; D = disrotatory.

observed. Even if both reactions are allowed by orbital-symmetry considerations, the cyclopropyl cation must be disfavoured thermodynamically with respect to the allyl cation (because of the ring-strain in the former system).

Cyclopropyl cations are reactive intermediates, and are generally formed from stable covalent compounds **47.** However, theory and experiment agree [20, 24, 66–68] that ionisation and ring opening could be synchronous, the electrons of the breaking σ bond in cyclopropane assisting the departure of the leaving group, X. It can be seen that groups situated on the same side of the ring as the leaving group

[66] C. H. DePuy, L. G. Schnack and J. W. Hauser, *J. Amer. Chem. Soc.*, **88**, 3343 (1966); see also Ref. 24, p. 46.

[67] C. H. DePuy, L. G. Schnack, J. W. Hauser and W. Wiedemann, *J. Amer. Chem. Soc.*, **87**, 4006 (1965).

[68] S. J. Cristol, R. M. Sequeira and C. H. DePuy, *J. Amer. Chem. Soc.*, **87**, 4007 (1965).

will move inwardly; the same process will impose an outward disrotatory motion to the groups situated in *anti*:

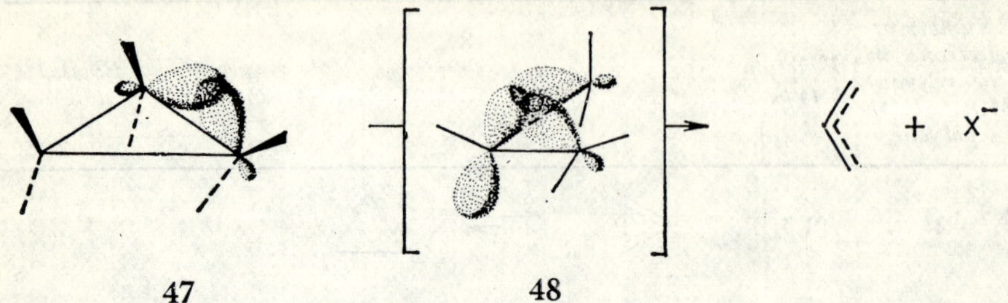

It follows that two different stereochemistries can be observed in the ring opening of cyclopropyl derivatives. *All-cis* derivatives **49** will generate in process (12.21) an *all-cis* allyl cation. Conversely, the *anti* derivative **50** will form in (12.22) the *all-trans* cation:

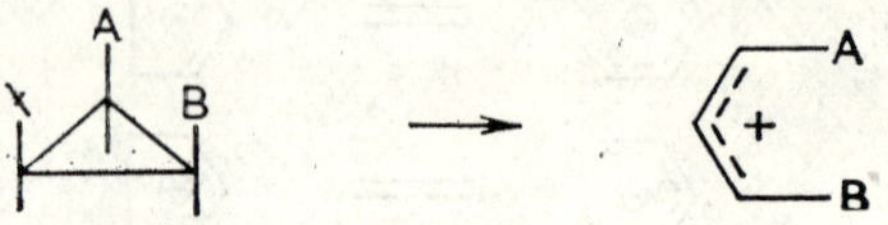

A, X, B → A, +, B (12.21)

49

A, B, X → A, +, B (12.22)

50

It can be thus anticipated that the outward disrotatory reaction (12.22) will be favoured over the inward process (12.21) if A and B are bulky groups. Schleyer *et al.* [69] have confirmed this experimentally, showing that the *anti* derivatives **50** solvolyse faster than the corresponding *syn* derivatives **49**. Conversely, the outward motion will be disfavoured over the inward process if A and B form a medium-sized ring [68].

Table 12.2 shows that the [1,5] electrocyclic reaction of a cation must be conrotatory in the ground state and disrotatory in the excited state. Cyclic cations with a medium ring can only undergo disrotatory closure, the conrotatory motion leading to a geometrically impossible *trans*-fused ring. In agreement with the theory, Doering and his co-workers[70] have shown that heptamethylcyclohexadienyl cation is relatively stable on heating (see p. 349), whereas Winstein and

[69] P. von R. Schleyer, G. W. van Dine, U. Schöllkopf and J. Paust, *J. Amer. Chem. Soc.*, **88**, 2868 (1966).

[70] W. von E. Doering *et al.*, *Tetrahedron*, **4**, 178 (1958).

Childs[36, 71] demonstrated the easy ring closure of such cations to bicyclo[3,1,0]hexenyl ions under photochemical conditions:

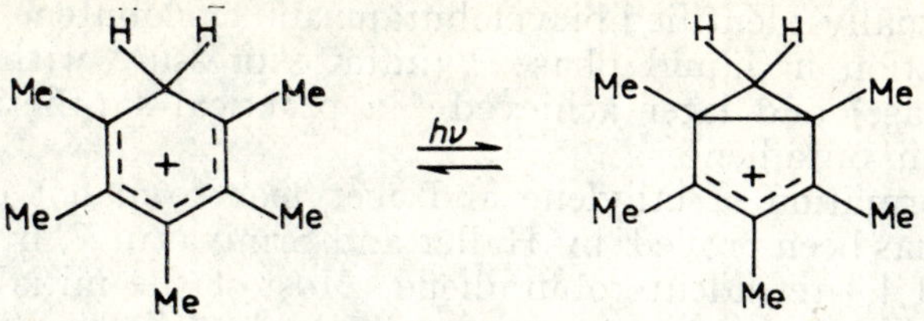

The field of ionic electrocyclic reactions, even if new, fits well into the general pattern of pericyclic reactions.

12.3.2 [*1,4*] *Electrocyclic reactions*

Dienes and cyclobutenes are related by a reversible [1,4] electrocyclic transformation. However, the two reactions can be separately carried out. Indeed, cyclobutene must be thermodynamically unstable with respect to the diene. Therefore, in a thermal process only ring opening of the cyclobutene is observed[72]. However, under photochemical conditions dienes are more efficient in absorbing radiant energy than are simple alkenes. Consequently, ring closure of dienes can be carried out in high yields photochemically.

Actually, the cyclisation of butadiene, the most simple example of a direct [1,4] electrocyclic reaction, is not easily achieved experimentally. It has been known that on irradiation in gaseous phase butadiene yields a complex mixture of ethylene, acetylene, but-1-ene, methylallene, hydrogen and polymers[73, 74]. The quantum yields of acetylene and ethylene formation increase with the pressure[75]. It has been concluded that the products result as hot molecules, liable to

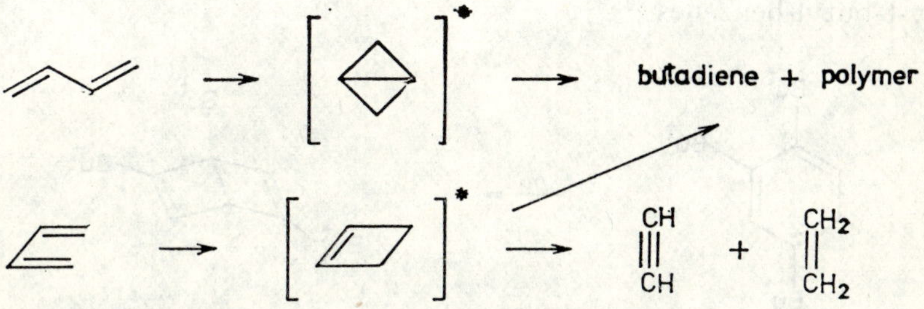

butadiene + polymer

CH ≡ CH + CH_2 = CH_2

[71] F. R. Childs, M. Sakai and S. Winstein, *J. Amer. Chem. Soc.*, **90**, 7144 (1968).

[72] R. Willstaetter and W. von R. Schmaedel, *Ber. dtsch. chem. Ges.*, **38**, 1992 (1905); see also M. Avram, C. D. Nenitzescu and E. Marica, *Chem. Ber.*, **90**, 1857 (1957); F. Badea, S. Roșca, I. G. Dinulescu, M. Avram and C. D. Nenitzescu, *Rev. Roumaine Chim.*, **10**, 1201 (1965).

[73] J. Collin and F. P. Lossing, *Can. J. Chem.*, **35**, 778 (1957).

[74] R. Srinivasan, *J. Amer. Chem. Soc.*, **82**, 5063 (1960).

[75] R. Srinivasan, *J. Amer. Chem. Soc.*, **85**, 4045 (1963).

undergo further transformations. Srinivasan [75, 76] considered that such high-energy intermediates could be isolated by increasing the reaction pressure.

He actually identified bicyclobutane and cyclobutene as products of the reaction in liquid phase [75] (infinite pressure with respect to gaseous phase) and later achieved [76] a practical synthesis of cyclobutene from butadiene.

The formation of ethylene and acetylene from hot cyclobutene molecules has been proved by Haller and Srinivasan [77], by irradiating gaseous 1,1,4,4-tetradeuterobutadiene. Most of the initial deuterium was found in ethylene, in agreement with the following mechanism:

CD_2 CD_2 → D_2 D_2 → CH≡CH + CD_2=CD_2

Cyclic dienes may react in a similar way. Typical examples have been studied by van Tamelen *et al.* [78-80]. Sequence (12.23) represents the most accessible route in the synthesis of the Dewar benzene **2***b*.

51 —$h\nu$→ —$Pb(OAc)_4$→ **2***b* (12.23)

The direct synthesis of the Dewar benzene, starting from benzene, by a similar electrocyclic process, must be disfavoured by the high energy gap between the stable aromatic compound and the strained bicyclic derivative. In the benzenes *ortho*-disubstituted with bulky groups, the energy difference of about 60 kcal mol^{-1} [81] between the two valence isomers should be partially compensated. The first recorded electrocyclic transformations of benzenoid compounds involved poly-t-butyl-benzenes [80, 82]:

Bu^t Bu^t Bu^t —$h\nu$→ Bu^t Bu^t Bu^t

[76] R. Srinivasan and K. A. Hill, *J. Amer. Chem. Soc.*, **88**, 3765 (1966).
[77] J. Haller and R. Srinivasan, *J. Amer. Chem. Soc.*, **88**, 3694 (1966).
[78] E. E. van Tamelen and S. P. Pappas, *J. Amer. Chem. Soc.*, **85**, 3297 (1963).
[79] E. E. van Tamelen and S. P. Pappas, *J. Amer. Chem. Soc.*, **84**, 3789 (1962).
[80] E. E. van Tamelen, S. P. Pappas and K. L. Kirk, *J. Amer. Chem. Soc.*, **93**, 6092 (1971).
[81] W. Schäfer, *Angew. Chem.*, **78**, 716 (1966).
[82] K. E. Wilzbach and L. Kaplan, *J. Amer. Chem. Soc.*, **87**, 4004 (1965).

The formation of Dewar benzene in the above example reduces the non-bonding interactions between the bulky t-butyl groups. It is now known [83] that even benzene can be converted photochemically to its valence isomers such as *2b*. One must however explain why such highly unstable isomers do not rearrange instantly into the more stable aromatic compounds. The most satisfactory answer is given by the theory of conservation of orbital symmetry. The symmetry-allowed concerted opening of bicyclo[2.2,0]hexadienes is a conrotatory process, impossible for geometrical reasons, as it would lead to a *trans*-cyclohexatriene. In order to avoid this forbidden pathway, the reaction moves toward higher activation parameters. Activation enthalpies between 20 and 30 kcal mol $^{-1}$ have actually been found [84–87], confirming that the energy barrier is important enough to prevent even a highly exothermic reaction when it is not allowed by the symmetry rules.

Cyclobutenes are thermally converted into butadienes [72, 88–91]. It has been demonstrated [88, 89] that the ring opening corresponds to a conrotatory process:

ROOC COOR —130° C→ H COOR H COOR *(cis-trans)*

Fused-ring cyclobutenes are converted to monocyclic dienes under more drastic conditions [90, 91]:

H H —400° C→ H H

necessary to remove the geometrical interdiction of the symmetry-allowed reaction.

[83] For recent reviews see L. T. Scott and M. Jones, Jr., *Chem. Rev.*, **72**, 181 (1972); E. E. van Tamelen, *Accounts Chem. Res.*, **5**, 186 (1972).
[84] J. E. Baldwin, A. H. Andrist and R. K. Pinschmidt, *Accounts Chem. Res.*, **5**, 402 (1972).
[85] J. F. M. Oth, *Angew. Chem.*, **80**, 633 (1968).
[86] R. Breslow, J. Napierski and A. H. Schmidt, *J. Amer. Chem. Soc.*, **94**, 5906 (1972).
[87] P. Lechtken, R. Breslow, A. H. Schmidt and N. J. Turro, *J. Amer. Chem. Soc.*, **95**, 3025 (1973).
[88] E. Vogel, *Angew. Chem.*, **68**, 189 (1956).
[89] E. Vogel, *Liebigs Ann. Chem.*, **615**, 14 (1958).
[90] W. G. Dauben and R. L. Cargill, *Tetrahedron*, **12**, 186 (1961).
[91] R. Criegee and H. Furrer, *Chem. Ber.*, **97**, 2949 (1964).

A few spectacular examples of [1,4] electrocyclic reactions have been described by Chapman and his co-workers[92–95]. Thus, the irradiation of α-pyrones[93–95] in an argon matrix at 8 K affords a β-lactone **52** which further yields the transient cyclobutadiene[96]. Cyclobutadiene is stable enough under these conditions to be trapped, or analysed by i.r. spectrometry.

$$\text{α-pyrone} \xrightarrow[8\ \text{K}]{h\nu} \mathbf{52} \xrightarrow[8\ \text{K}]{h\nu} \text{cyclobutadiene} + CO_2$$

The intermediate formation of the lactone **54** during the generation of 1,2-dehydrobenzene (benzyne) has been proved by a similar procedure[96].

$$\text{phthaloyl peroxide} \xrightarrow{h\nu} \mathbf{53} \underset{}{\overset{h\nu}{\rightleftharpoons}} \mathbf{54} \longrightarrow \text{benzyne} + CO_2$$

Some [1,4] electrocyclic transformations of natural compounds, competing with other electrocyclic processes, will be mentioned in the next section.

12.3.3 [*1,6*] *Electrocyclic reactions*

The equilibration of a hexa-1,3,5-triene with a cyclohexa-1,3-diene (12.24) differs in some respects from the electrocyclic processes discussed above.

$$\text{hexa-1,3,5-triene} \rightleftharpoons \text{cyclohexa-1,3-diene} \qquad (12.24)$$

One aspect concerns the geometry of the acylcic triene. Cyclisation in the ground state is possible only from the *cis* isomer. In the excited states, however, the geometry of the initial triene is no

92 O. L. Chapman and C. L. McIntosh, *J. Amer. Chem. Soc.*, **92**, 7005 (1970).
93 O. L. Chapman *et al.*, *J. Amer. Chem. Soc.*, **95**, 244, 247, 614 (1973).
94 O. L. Chapman, C. L. McIntosh, J. Pacansky, C. V. Calder and G. Orr, *J. Amer. Chem. Soc.*, **95**, 4061 (1973).
95 O. L. Chapman, D. De La Cruz, R. Roth and J. Pacansky, *J. Amer. Chem. Soc.*, **95**, 1338 (1973).
96 C. D. Nenitzescu, *Angew. Chem.*, **73**, 300 (1961).

longer important, since the *trans* and *cis* isomers equilibrate rapidly under photochemical conditions.

trans $\overset{h\nu}{\rightleftharpoons}$ *cis* $\rightleftharpoons$ $\overset{h\nu}{\rightleftharpoons}$

There seems to be no great difference in free energy between the acyclic triene and the cyclic diene, due to a smaller strain in the six-membered ring, as compared with the four-membered system. On the other hand, in a [1,6] electrocyclic process the equilibrating polyenes absorb efficiently in near regions of the u.v. spectrum. Thus, another difference between the [1,4] and [1,6] electrocyclic reactions is that the latter are reversible under both thermal and photochemical conditions. Side reactions are more frequent in the reversible transformations of trienes than in the reactions of dienes. One of them is a [1,4] electrocyclic reaction, the triene behaving as a substituted diene. On the other hand, the resulting cyclic diene may undergo ring closure to bicyclic compounds. The dehydrogenation of cyclic dienes to aromatic compounds may also occur in the thermal processes.

The simplest triene, hexa-1,3,5-triene, was obtained as a *cis-trans* mixture (1:7) by the thermal decomposition at 100 °C of the quaternary ammonium hydroxide **55** [97]. The dehydration of alcohol **56** at 325 °C affords only the *trans* isomer, beside cyclohexadiene [98].

$$CH_2{=}CH{-}CH_2{-}\underset{\underset{+NMe_3}{|}}{CH}{-}CH{=}CH_2$$

55

$$CH_2{=}CH{-}CH{=}CH{-}\underset{\underset{OH}{|}}{CH}{-}CH_3$$

56

Thus, *cis*-hexa-1,3,5-triene is thermally stable at 100 °C, but undergoes electrocyclic closure at higher temperatures. On heating, cyclohexadiene yields benzene and hydrogen [99]. The electrocyclic ring opening is prevented by the energetically more favourable process, the aromatisation.

[97] J. C. H. Hiva, P. L. di Benneville and H. J. Simms, *J. Amer. Chem. Soc.*, **82**, 2537 (1960).

[98] K. Alder and H. von Brochel, *Liebigs Ann. Chem.*, **608**, 195 (1957).

[99] F. O. Rice and H. L. Stallbaumer, *J. Amer. Chem. Soc.*, **64**, 1527 (1942).

Hexa-1,3,5-triene gives on irradiation benzene, hydrogen and hexa-1,2,4-triene [100]. The photochemical process must be the intermediate formation of cyclohexadiene. Benzene and hydrogen probably result from higher vibrational levels of the latter. The irradiation of the higher polyene, octa-2,4,6-triene, is outstanding with respect to the selection rules. An equilibrium is reached between this triene (90—95%) and *trans*-1,2-dimethyl-cyclohexa-3,5-diene [101]:

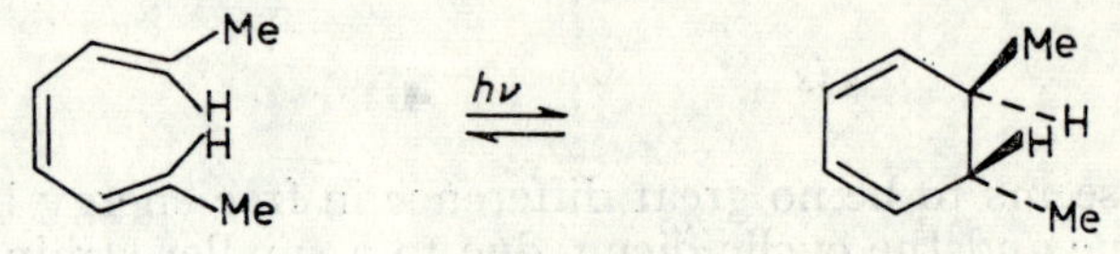

The process is conrotatory in both directions.

Cyclic trienes behave similarly. Cope and his co-workers [102] have shown that 1,3,5-cyclooctatriene equilibrates at 80 °C with the bicyclic compound **57**; the isomer **58** has not been observed:

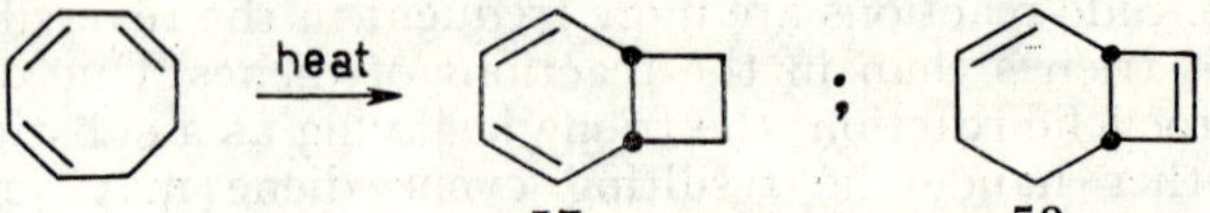

Indeed, in a medium ring system only a disrotatory process, leading to a *cis*-fused bicyclic compound is possible. The formation of **57** corresponds to an allowed [1,6] disrotatory ring closure of cyclooctatriene, whereas **58** would be the result of a [1,4] electrocyclic process forbidden in the ground state.

Steroid chemistry has amply exemplified the selection rules, providing, as mentioned before, the first hint concerning the electronic factors which govern the stereochemistry of pericyclic reactions. Thus, ergosterol, a steroid of the anti series (9αH; 10βMe), irradiated with u.v. light equilibrates with lumisterol (L), precalciferol (P), tachisterol (T) and vitamin D_2 [103]. Using a labelling technique with radioactive carbon, Havinga *et al.*[104] demonstrated that precalciferol must be the initial product in the photochemical reaction of ergosterol (E). From the variation of the quantum yields with the reaction conditions they have concluded [104–106] that both lumisterol and tachisterol

[100] R. Srinivasan, *J. Amer. Chem. Soc.*, **84**, 3982 (1962).

[101] J. Meinwald and P. H. Mazzochi, *J. Amer. Chem. Soc.*, **88**, 2850 (1966).

[102] A. C. Cope, A. C. Haven, Jr., F. L. Ramp and E. R. Trumball, *J. Amer. Chem. Soc.*, **74**, 4867 (1952).

[103] A. Windhaus, K. Dimroth and W. Bregwisch, *Liebigs Ann. Chem.*, **543**, 240 (1940).

[104] E. Havinga, A. Verloop, and A. L. Koevoet, *Rec. Trav. Chim.*, **75**, 371 (1956).

[105] M. P. Rappold, P. Westerhof, K. H. Hannewald and J. A. Keverling-Buisman, *Rec. Trav. Chim.*, **77**, 241 (1958).

[106] M. P. Rappold and E. Havinga, *Rec. Trav. Chim.*, **79**, 369 (1964).

are not essential for the formation of vitamin D_2, the latter resulting from precalciferol by a thermal process. The following reaction scheme [18] accounts for all experimental data:

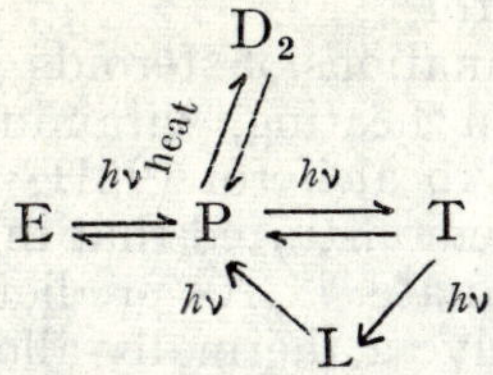

$$\begin{array}{ccccc} & & D_2 & & \\ & & \uparrow\downarrow \; {\scriptstyle h\nu,\ heat} & & \\ E & \rightleftharpoons & P & \underset{h\nu}{\rightleftharpoons} & T \\ & {\scriptstyle h\nu}\nwarrow & & \swarrow{\scriptstyle h\nu} & \\ & & L & & \end{array} \tag{12.25}$$

Precalciferol and tachisterol are trienes related by *cis-trans* isomerism, whereas lumisterol belongs to the *anti* series (9βH; 10αMe), being the epimer of ergosterol. Precalciferol results by photochemical ring-opening of the *anti* compounds E and L, the processes corresponding to a conrotatory motion, in agreement with theory. Moreover, the specific formation of ergosterol from precalciferol, in low quantum yields, and of lumisterol from tachisterol, are well accounted for by the detailed mechanism represented in (12.26):

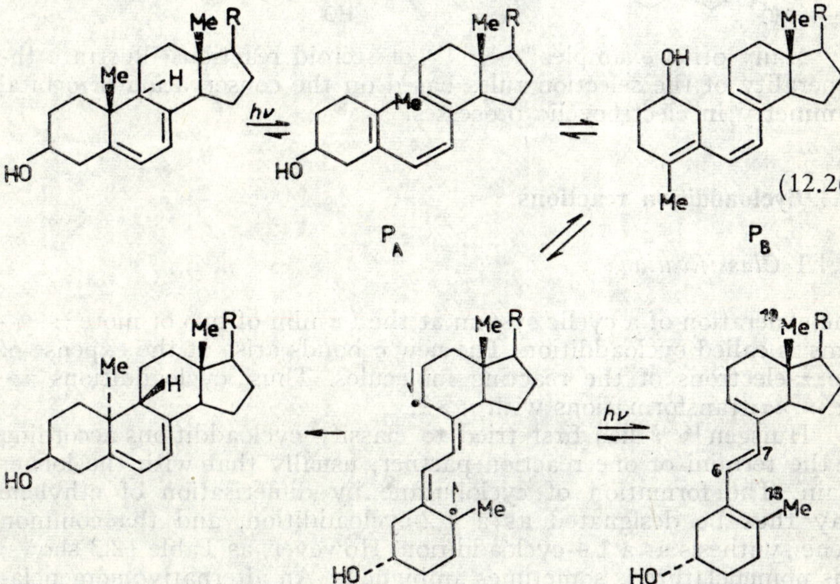

(12.26)

From the two conformers of precalciferol, only the less stable P_A can undergo cyclisation. Consequently, the quantum yields are expected to be low, as it is actually observed [104]. By absorption of light, the *trans*-triene T is excited to the singlet species T*. The latter can decay either back to T or to P (*cis-trans* equilibration) or may yield cyclic products by rotation about the C-6—C-7 axis. In principle, both E and L can form, depending on the sense of rotation. However, the upward rotation of the cyclohexenol moiety, leading to E, is sterically prohibited by the large groups 18-Me and 19-Me which may interfere. Cyclisation occurs therefore in the sense corresponding to less steric interactions, lumisterol being thus the single product of tachisterol.

The thermal transformations of steroids differ from their photochemical reactions [107]. On heating, vitamin D_2 yields pyrocalciferol (9αH, 10αMe), and isopyrocalciferol (9βH, 10βMe), both belonging to the *syn* series. It was demonstrated that in this process precalciferol is again the key intermediate [108]. As predicted by the theory, these cyclisation reactions involve a thermally-allowed disrotatory motion. The conrotatory opening is geometrically impossible for the steroids with a syn geometry at C-9 and C-10. Consequently, one can only expect a [1,4] electrocyclic transformation leading to cyclobutenes [109–111]:

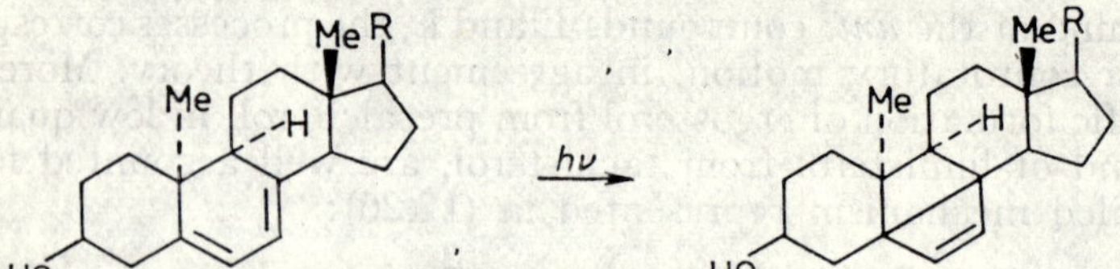

Many other examples [107, 112, 113] of steroid reactions illustrate the generality of the selection rules based on the conservation of orbital symmetry in electrocyclic processes.

12.4 Cycloaddition reactions

12.4.1 *Classification*

The generation of a cyclic system at the termini of two or more π-systems is called cycloaddition. The new σ bonds arise at the expense of the π-electrons of the reacting molecules. Thus, cycloadditions are $n\pi \rightarrow n\sigma$ transformations with $n \geqslant 2$.

Huisgen [14, 15] has first tried to classify cycloadditions according to the termini of one reaction partner, usually that with the longer chain. The formation of cyclobutane by dimerisation of ethylene may thus be designated as a 1,2-cycloaddition, and the common diene synthesis as a 1,4-cycloaddition. However, as Table 12.3 shows, this nomenclature is sometimes ambiguous. An alternative nomenclature (last column of the table), generally adopted in recent papers,

[107] For reviews see: L. Fieser and M. Fieser, *Steroids*, Reinhold, New York (1959), p. 137; N. L. Vendel, in *Molecular Rearrangements* (P. de Mayo, editor), Interscience, New York (1963), vol. 2, p. 1020.

[108] A. Verloop, A. L. Koevoet and E. Havinga, *Rec. Trav. Chim.*, **76**, 689 (1957).

[109] A. Windaus and K. Dimroth, *Ber. dtsch. chem. Ges.*, **70**, 1631 (1937).

[110] W. G. Dauben and G. J. Fonken, *J. Amer. Chem. Soc.*, **79**, 2971 (1957).

[111] W. G. Dauben and G. J. Fonken, *J. Amer. Chem. Soc.*, **81**, 4060 (1959).

[112] R. L. Autrey, D. H. R. Barton and W. H. Rensch, *Proc. Chem. Soc.*, 55 (1959).

[113] W. H. Schuller, R. N. More, J. E. Howkins and R. W. Lawrence, *J. Org. Chem.*, **27**, 1178 (1962).

is based on the number of atoms of the two systems involved. The Diels-Alder reaction will thus be a 4+2 cycloaddition, easily distinguished from other reactions involving dienes as the longer chain partner (for example 4+1, 4+4 cycloadditions, see Table 12.3).

TABLE 12.3
Classification of cycloaddition and fragmentation reactions

Type of transformation	*Type of ring involved*	*Examples*	*Name of the reaction* a)	b)
(1) $2\pi-2\sigma$ cycloaddition				
(a)		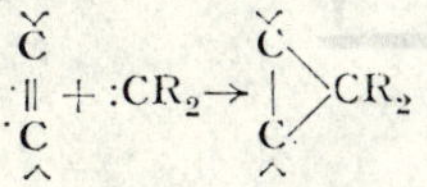 C=C + :CR_2 → C–C >CR_2	1,1-cycloaddition	2+1 cycloaddition
(b)		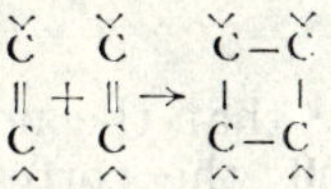 C=C + C=C → C–C / C–C	1,2-cycloaddition	2+2 cycloaddition
(c)		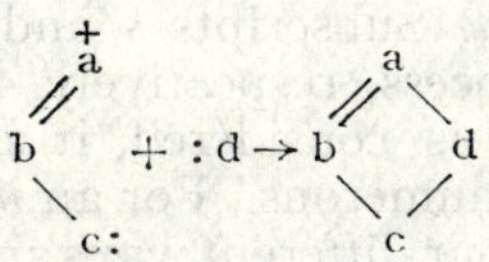 $\overset{+}{a}$=b–c: + :d → a=b–c–d	1,3- or 1,1-cycloaddition	3+1 cycloaddition
(d)	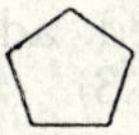	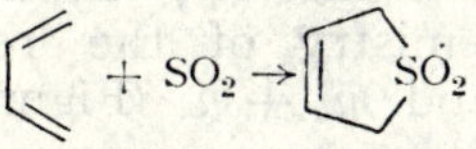 diene + SO_2 → SO_2	1,4- or 1,1-cycloaddition	4+1 cycloaddition
(e)		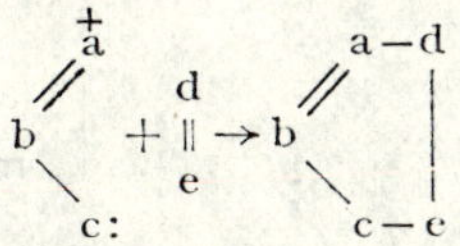 $\overset{+}{a}$=b–c: + d=e → a–d, b, c–e	1,3-dipolar addition or 1,2-cycloaddition	3+2 cycloaddition
(f)		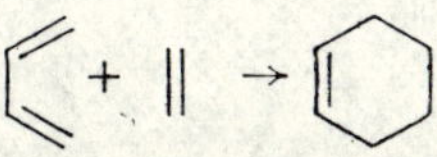 diene + ‖ → cyclohexene	1,4- or 1,2-cycloaddition	4+2 cycloaddition
(2) $2\sigma-2\pi$ fragmentation	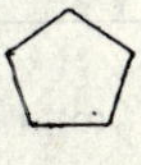	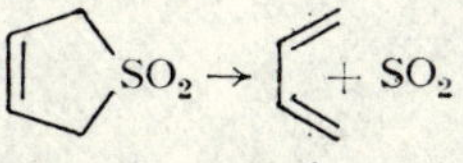 SO_2 → diene + SO_2	—	4+1 fragmentation
	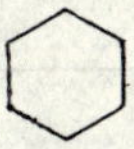	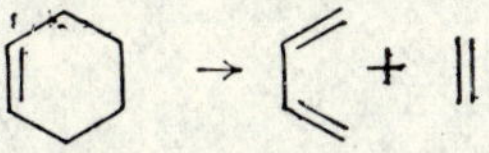 cyclohexene → diene + ‖	—	4+2 fragmentation
(3) $3\pi-3\sigma$ cycloaddition	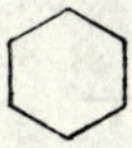	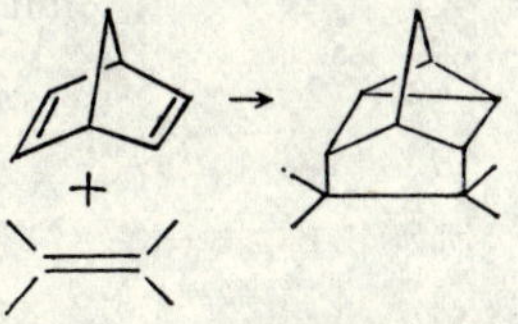 norbornadiene + alkene →	—	2+2+2 cycloaddition

a) According to the positions in the reacting molecule at which the new σ bonds develop.
b) According to the number of atoms which generate the cyclic system.

Cycloadditions are usually reversible, the reverse reaction being called *fragmentation* or *cycloreversion*. The latter can be designated in the same manner as for cycloadditions.

There are two different ways in which the new bond can develop at the termini of a π system. First, the bonds may be formed on the same face of the reacting system, in a *suprafacial cycloaddition* **59**.

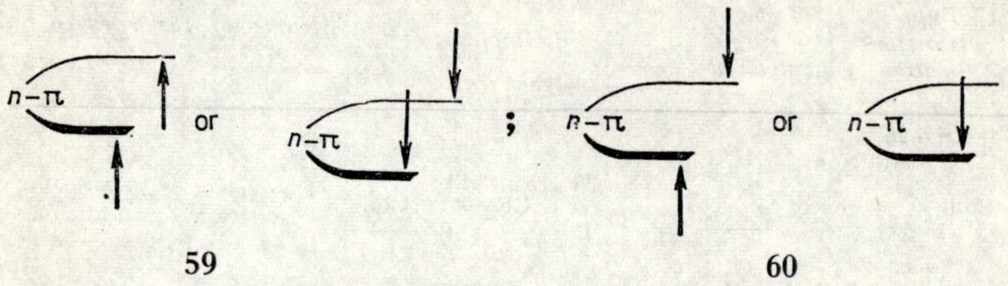

It is also possible[23, 24] that the new bonds develop on opposite sides of the π-system, **60**, the corresponding process being called *antarafacial cycloaddition*. Subscripts *s* and *a* are used to indicate a *supra* and an *antara* process, respectively. If the number of components in a cycloaddition is considered, it appears that the possible pathways are still more numerous. For an $m+n$ cycloaddition, where m and n are different, four different ways are conceivable, distinguishable by the stereochemistry of the resulted adducts: $m_s + n_s$, $m_a + n_a$ and $m_s + n_a$ and $m_a + n_s$ (Figure 12.3).

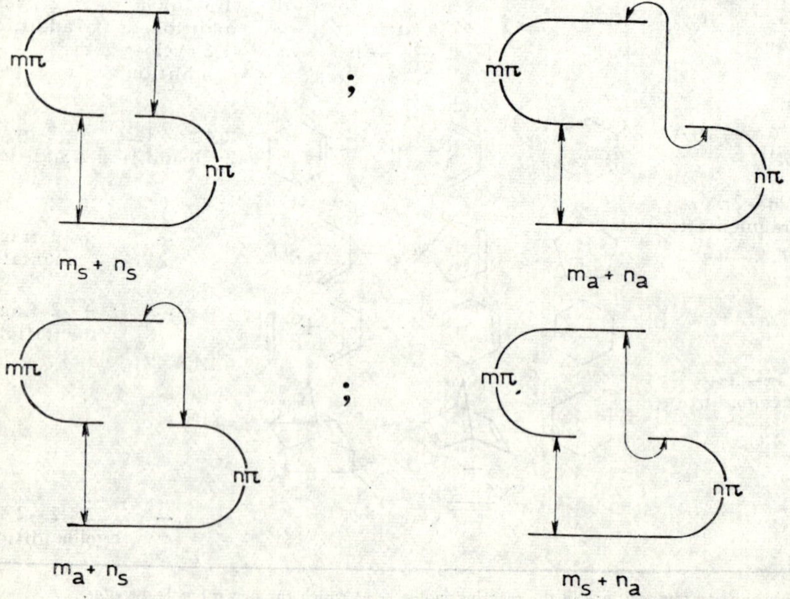

Figure 12.3 Possible interactions in an $m + n$ cycloaddition.

The principle of orbital-symmetry conservation in concerted processes decides the actual course of a cycloaddition. It must be remembered that even if the reaction is symmetry-allowed, the system may adopt a stepwise mechanism if the latter involves a lower activation energy than the concerted pathway.

The orientation of the components in the transition state is closely related to the conservation of orbital symmetry.

12.4.2 *Parallel orientation of components in the transition state (the supra-supra process)*

The *supra-supra* process may illustrate the applications of correlation diagrams in establishing the rules of orbital-symmetry conservation in pericyclic reactions [21, 24]. We shall consider two typical examples of $2\pi \rightarrow 2\sigma$ transformations, namely a 2+2 and a 2+4 cycloaddition. We shall try to gain a maximum insight into the problem by choosing the most simple molecules, leading to high-symmetry transition states, even if the actual reactions cannot be achieved experimentally.

The dimerisation of ethylene to cyclobutane may occur *via* a transition state in which the two molecules are parallel (Figure 12.4). The initial state A has a lower symmetry (D_{2h}) than the final state (D_{4h}) C. However, two reflection planes (**1** and **2**) will be common to both A and C and thus to the transition state, B, too.

It is thus possible to classify the essential orbitals, suffering more drastic modifications in the transition state, with respect to these elements, and then to analyse how their symmetry is conserved. (We therefore assume that the reaction is reversible.) We shall denote the symmetry of an orbital by two letters, the first referring to reflection plane **2**, the second to plane **1**; if the orbital is symmetric, the designation is *S*, if antisymmetric, *A*.

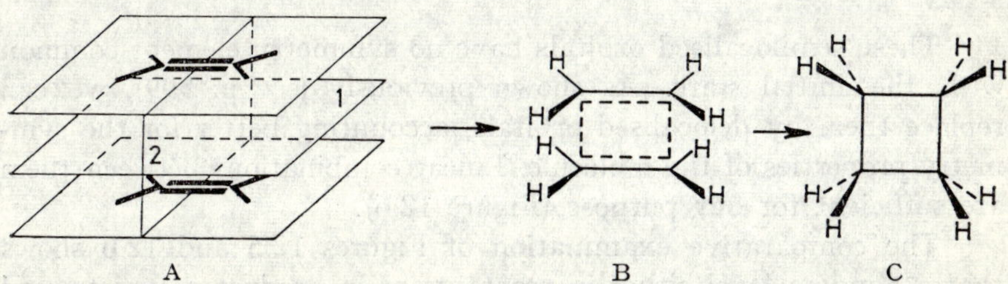

Figure 12.4 Possible mechanism for the dimerisation of ethylene (2+2 cycloaddition).

When the reaction starts from ethylene, the essential orbitals are the two bonding π orbitals of the two molecules, π_1 and π_2, and two antibonding orbitals, π_1^* and π_2^*. The symmetric orientation of these orbitals with respect to the reflection plane 2 corresponds to a bonding orbital in reactants. Thus, SX ($X = S$ or A) will have lower energy than AX. If the two molecules are beyond the interacting distance, the symmetry with respect to plane 1 will not affect the energies of the corresponding orbitals. If interactions are possible, an XS orbital will be lower in energy than XA. Now we are able to arrange the four orbitals in an increasing order of energy (Figure 12.5), as a function of their symmetry:

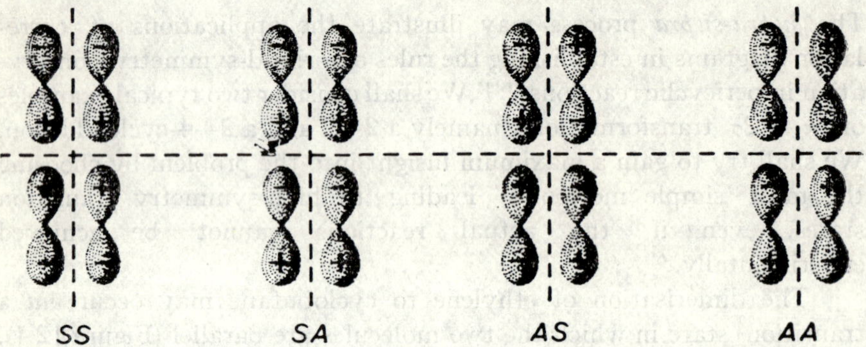

Figure 12.5 The symmetry of the π orbitals of two interacting ethylene molecules with respect to two reflection planes.

For the transition state of the fragmentation of cyclobutane, the essential orbitals to be examined are σ_1, σ_2, σ_1^* and σ_2^*:

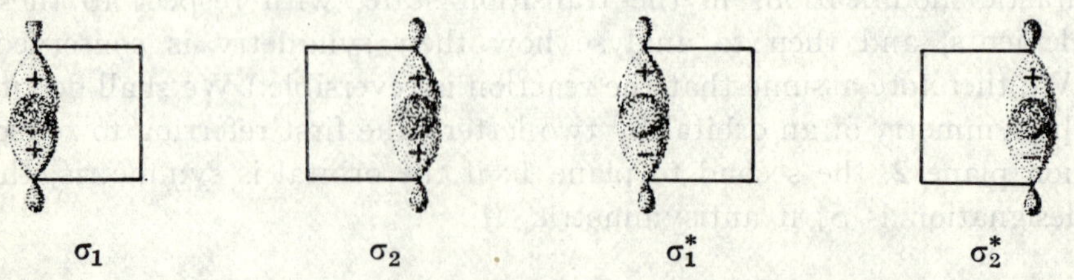

These semilocalised orbitals have no symmetry element common with the initial state. As shown previously (see p. 599), we can replace them by delocalised orbitals, accounting better for the symmetry properties of the molecule. Linear combinations between them are sufficient for our purpose (Figure 12.6).

The comparative examination of Figures 12.5 and 12.6 shows that a bonding level either in reactants or in product moves toward a higher energy level (antibonding) in the transition state. This is

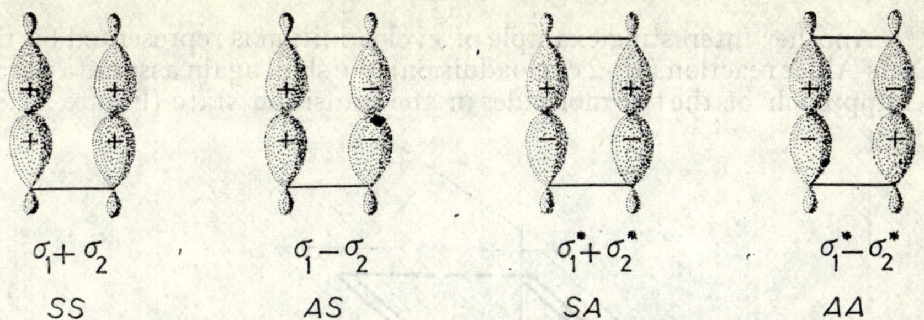

Figure 12.6 The symmetry of the σ orbitals of cyclobutane in the transition state of fragmentation, with respect to two reflection planes.

better shown in the correlation diagram (Figure 12.7). The correlation diagram is constructed by placing on the left side the energy levels of the orbitals of the reactants and on the right side the corresponding levels of the product, with indication of their symmetry. Levels of like symmetry are then connected. We may notice that a bonding reacting level in reactant is correlated with an antibonding level in the product:

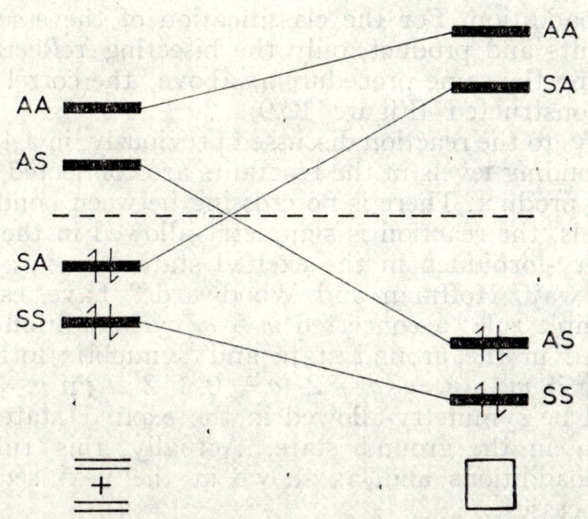

Figure 12.7 Correlation diagram for a 2+2 cycloaddition.

Clearly, if orbital symmetry is to be conserved, two ground state ethylene molecules cannot combine concertedly to give cyclobutane. In other words, the reaction is *symmetry-forbidden* in the ground state. However, if electrons are promoted to the antibonding orbital, as it happens in a photochemical reaction, the process may become *symmetry-allowed* in the excited state.

Such conclusions can be reached qualitatively, without detailed calculations; for precise estimations of the energies involved, as well as for more complex cases, elaborate calculations are necessary [24].

Another interesting example of cycloaddition is represented by the Diels-Alder reaction (4+2 cycloaddition). We shall again assume a parallel approach of the two molecules in the transition state (Figure 12.8).

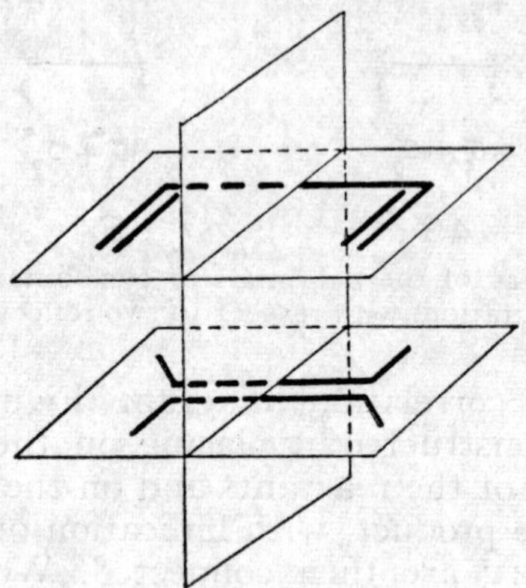

Figure 12.8 Parallel orientation of butadiene and ethylene.

Clearly, in this case the transition state is less symmetric than in a 2+2 cycloaddition. For the classification of the essential orbitals of the reagents and product, only the bisecting reflection plane can be used. Using the same procedure as above, the correlation diagram is readily constructed (Figure 12.9).

Contrary to the reaction discussed previously, in a 4 + 2 cycloaddition the bonding levels in the reactants are connected with bonding levels in the product. There is no crossing between bonding and antibonding levels, the reaction is symmetry-allowed in the ground state and symmetry-forbidden in the excited state.

In this way, Hoffman and Woodward [21] have established the following simple rule: a concerted $m + n$ cycloaddition will be symmetry-allowed in the ground state and symmetry-forbidden in the excited state if $m + n = 4q + 2$ ($q = 0, 1, 2 \ldots$); if $m + n = 4q$, the reaction will be symmetry-allowed in the excited state and symmetry-forbidden in the ground state. Actually, this rule applies to $m_s + n_s$ cycloadditions and, as shown in the next section, also to $m_a + n_a$ processes.

12.4.3 *Orthogonal orientation of components in the transition state (supra-antara processes)*

Careful examination of Figure 12.9 shows that the delocalised orbital $\sigma_1 + \sigma_2$ in cyclohexene can be regarded as resulting from the positive overlap of the LUMO of butadiene (Ψ_3) with the HOMO of ethylene (π), and $\sigma_1 - \sigma_2$ by the positive overlap of the HOMO of butadiene (Ψ_2) with the lowest vacant MO of ethylene (π^*). As the overlap is possible, the reaction will be symmetry-allowed in the ground state. A similar consideration of Figure 12.7 leads to the conclusion that the

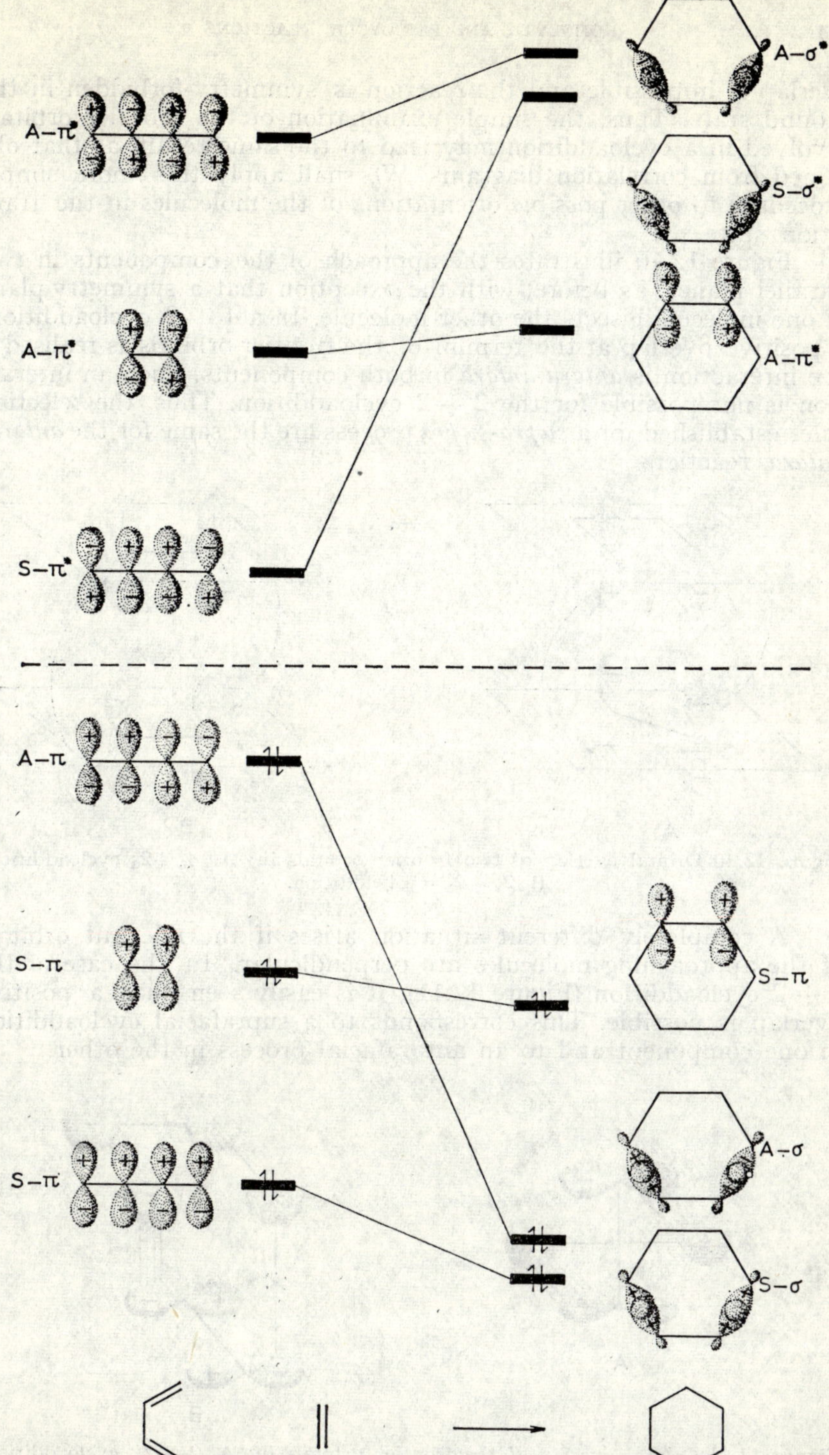

Figure 12.9 Correlation diagram for 4+2 cycloaddition.

overlap is impossible and the reaction is symmetry-forbidden in the ground state. Thus, the simple examination of the frontier orbitals involved in a cycloaddition may lead to the same result as that obtained from correlation diagrams. We shall apply this more simple procedure to other possible orientations of the molecules in the transition state.

Figure 12.10 illustrates the approach of the components in two parallel planes, as before, with the exception that a symmetry plane of one molecule bisects the other molecule. In a $4 + 2$ cycloaddition, a positive overlap at the termini of the frontier orbitals is realised if the interaction is *antara-antara* in both components. Such an interaction is not possible for the $2 + 2$ cycloaddition. Thus, the selection rules established for a *supra-supra* process are the same for the *antara-antara* reactions.

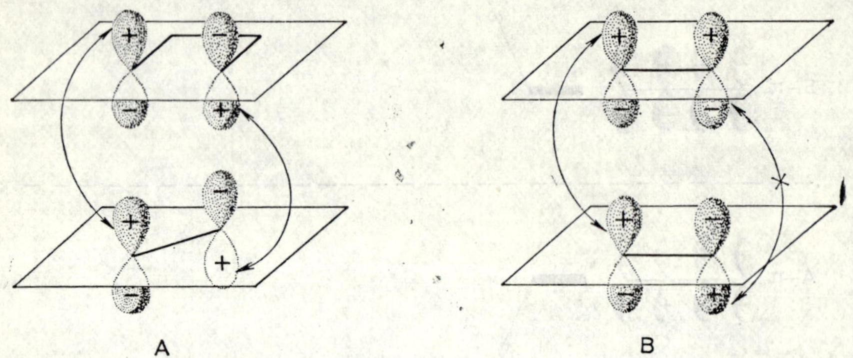

Figure 12.10 Orbital overlap of two frontier orbitals in: A. $4_a + 2_a$ cycloaddition; B. $2_a + 2_a$ cycloaddition.

A completely different situation arises if the relevant orbitals of the approaching molecules are perpendicular. In the case of the $2 + 2$ cycloaddition (Figure 12.11), it is easily seen that a positive overlap is possible. This corresponds to a suprafacial cycloaddition in one component and to an antarafacial process in the other.

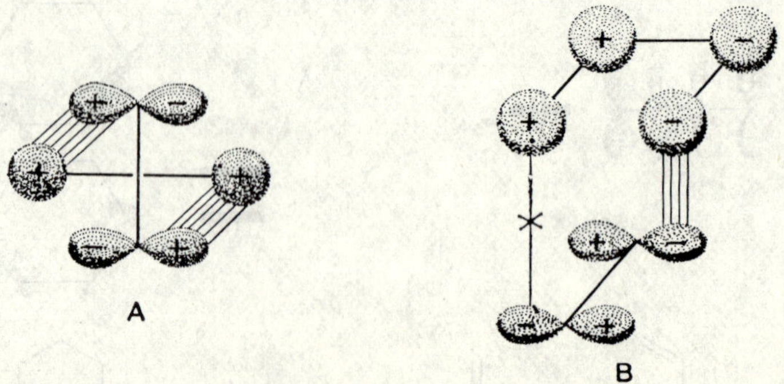

Figure 12.11 Orbital overlap of the frontier orbitals in: A. $2_a + 2_s$ cycloaddition; B. $4_s + 2_a$ cycloaddition.

Clearly, the $4_s + 2_a$ cycloaddition cannot be a ground state concerted reaction. Thus, the selection rules for the *supra-antara* (and *antara-supra*) processes must be reversed with respect to the rules established for *supra-supra* cycloadditions. Table 12.4 summarises all the rules for cycloaddition reactions.

TABLE 12.4

Generalised rules for cycloaddition reactions

$m + n$	*Allowed in the ground state* *Forbidden in the excited state*	*Allowed in the excited state* *Forbidden in the ground state*
$4q$	$m_s + n_a$	$m_s + n_s$
	$m_a + n_s$	$m_a + n_a$
$4q + 2$	$m_s + n_s$	$m_s + n_a$
	$m_a + n_a$	$m_a + n_s$

The *antara-supra* processes, even if less studied than the more common *supra-supra* reactions, are frequently encountered in recent works on cycloaddition reactions and confirm in their turn the utility of orbital-symmetry considerations.

12.4.4 σ *Bonds involved in cycloaddition reactions*

The theory of concerted cycloadditions may be extended to other reactions involving σ bonds. Let us consider for instance the cyclobutane fragmentation into two molecules of ethylene (Figure 12.12).

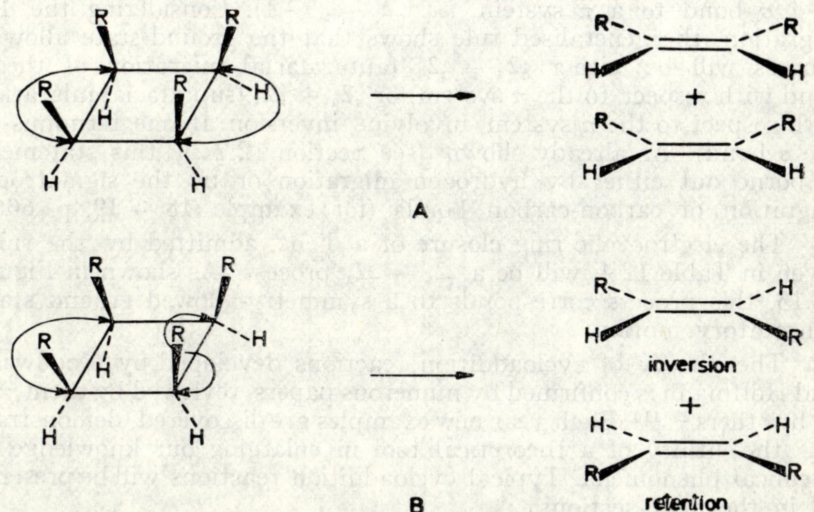

Figure 12.12 Concerted fragmentation of substituted cyclobutenes into ethylenes. A. $\sigma 2_s + \sigma 2_s$; B. $\sigma 2_s + \sigma 2_a$.

The process can be regarded as an addition of the breaking σ bonds to the other σ bonds, either in a *supra-supra* (A) or in a *supra-antara* (B) manner. According to the generalised rules summarised in Table 12.4, A will be symmetry-forbidden, whereas B will be symmetry-allowed.

The other pericyclic reactions discussed in this chapter can be treated in the same way. In this connection it is necessary to introduce the following definitions and rules:

(1) A σ bond is considered to be involved in a cycloaddition in a *supra* manner if configuration is either retained or inverted at both of its termini in the course of the reaction. The retention of configuration at one terminus and the inversion at the other will indicate an *antara* addition.

In the example given in Figure 12.12 A, the *all-cis* cyclobutane yields only a *cis*-alkene. Configuration must be retained (or inverted) at both termini of the two σ bonds involved, the reaction being a *supra-supra* process. In B, a *cis*-alkene and a *trans*-alkene are formed. Thus, at both termini of one σ bond the configuration must be retained, whereas the configuration is retained only at one terminus of the other bond. The process will be *supra-antara.*

(2) In the designation of a cycloaddition involving σ bonds (addition of σ bonds to σ bonds, addition of σ bonds to π bonds), using the same symbolism as for typical cycloadditions (addition of π bonds to π bonds), a subscript (π or σ) preceding the number of electrons in the relevant orbital is added, for example ${}_{\pi}2_s$; ${}_{\sigma}2_a$; ${}_{\sigma}2_s$, etc.

The cycloregression of cyclobutane, admitted by the selection rules, will be ${}_{\sigma}2_s + {}_{\sigma}2_a$.

At this point we may extend the generalised rules of cycloaddition reactions to the electrocyclic and sigmatropic reactions.

A sigmatropic change of the order [1, *j*] will be a cycloaddition of a σ bond to a π system, *i.e.* ${}_{\sigma}2 + {}_{\pi}(j-1)$. Considering the 1,3 migration, the generalised rule shows that the ground-state allowed process will be either ${}_{\sigma}2_s + {}_{\pi}2_a$ (antarafacial migration of the σ bond with respect to the π system) or ${}_{\sigma}2_a + {}_{\pi}2_s$ (suprafacial migration with respect to the π system, involving inversion at one terminus of the σ bond). As already shown (see Section 12.2.2), this statement is borne out either by hydrogen migration or by the sigmatropic migration of carbon-carbon bonds (for example **18** $\rightarrow$ **19**, p. 606).

The electrocyclic ring closure of a diene, admitted by the rules given in Table 12.4, will be a ${}_{\pi}2_s + {}_{\pi}2_a$ process. As shown in Figure 12.13, this process corresponds to a symmetry-allowed ground state conrotatory motion.

The theory of cycloaddition reactions developed by Woodward and Hoffmann is confirmed by numerous papers, reviewed by them [23,24] or by others [29,114]. Each year new examples are discovered, demonstrating the utility of a theoretical tool in enlarging our knowledge of chemical phenomena. Typical cycloaddition reactions will be presented in the next sections.

[114] D. Seebach, *Fortschr. Chem. Forsch.*, **11**, 177 (1969).

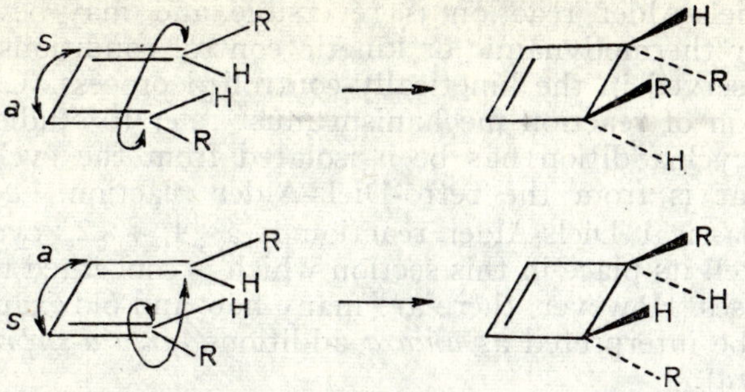

Figure 12.13 Electrocyclic ring closure of a 1,4-*trans-trans*-disubstituted diene ($_{\pi}2_s + _{\pi}2_a$).

12.5 Diels-Alder reaction (4+2 cycloaddition)

The typical example of a no mechanism reaction was considered to be the diene synthesis[11] (the Diels-Alder reaction). The reaction is carried out simply by heating a diene or another conjugated π system with a reactive unsaturated compound (*dienophile*) and it leads to a six-membered ring derivative. Usually, the reaction is not sensible to catalysts, and light does not affect its course. Depending on the specific components, either carbocyclic[115–117] or heterocyclic[118, 119] products can be obtained.

The great advantage of the Diels-Alder reaction in organic synthesis lies in its high regio- and stereo-selectivity[120]. Thus, in the reactions of a substituted diene **61** with an alkene **62** if the possible orientations (**63** and **64**) as well as the number of chiral centres (four) developed in the adduct are considered, 32 isomers would be expected to result:

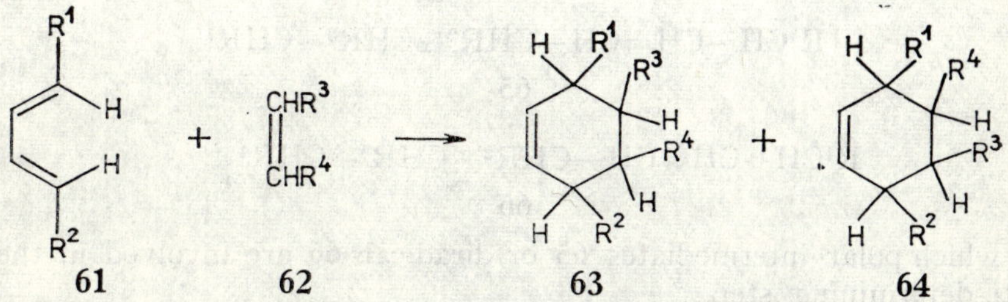

61 + **62** ⟶ **63** + **64**

Actually, one orientation predominates (high regioselectivity) and only one diastereoisomer is produced (high stereoselectivity).

[115] M. C. Kloetzel, *Org. Reactions*, **4**, 1 (1948); H. L. Holmes, *Org. Reactions*, **4**, 60 (1948).
[116] J. Sauer, *Angew. Chem.*, **78**, 233 (1967).
[117] J. Sauer, *Angew. Chem.*, **79**, 76 (1967).
[118] S. B. Needleman and M. C. Kuo, *Chem. Rev.*, **62**, 405 (1962).
[119] J. Harner, *1,4-Cycloaddition Reactions*, Academic Press, New York (1967).
[120] J. G. Martin and R. K. Hill, *Chem. Rev.*, **61**, 537 (1961).

The Diels-Alder reaction is reversible and may be carried out either under thermodynamic or kinetic control conditions. Its selectivity is observed in the kinetically-controlled process. Clearly, any interpretation of reaction mechanism must refer to studies in which the direct cycloaddition has been isolated from the cycloregression process, that is from the retro-Diels-Alder reaction.

The classical Diels-Alder reaction is a ${}_{\pi}4_s + {}_{\pi}2_s$ cycloaddition, and finds well its place in this section which is concerned mainly with such processes. However, there are many new and old examples [121, 122] which can be interpreted as *antara* additions (*antara-supra* and even *antara-antara*).

12.5.1 *Kinetic measurements*

The Diels-Alder reaction is a second-order reaction [123–126], first-order in each component:

$$\text{rate} = k_2[\text{diene}]\,[\text{dienophile}] \tag{12.27}$$

The equation rate does not change when the reacting system is transferred from liquid phase to gaseous phase and the rate constants are only slightly affected by the polarity of the solvent. However, the presence of Lewis acids, even if not essential, may accelerate the reaction rates and affect the regioselectivity [127–129].

The reverse reaction, the *retro*-Diels-Alder process, is a clean first-order reaction [128, 129], both in the gaseous and condensed phase. As for the direct reaction, rate constants are not significantly modified by the medium.

$$\text{rate} = k_1\,[\text{adduct}] \tag{12.28}$$

Rate equations (12.27) and (12.28) are in agreement with a concerted reaction. However, they do not exclude a stepwise reaction

$$R^1\overset{+}{C}H-CH=CH-CHR^2-CHR^3-\overset{-}{C}HR^4$$

65

$$R^1\dot{C}H-CH=CH-CHR^2-CHR^3-\dot{C}HR^4$$

66

in which polar intermediates **65** or diradicals **66** are involved in the rate-determining step.

[121] J. Meinwald, A. Eckel and K. L. Erickson, *J. Amer. Chem. Soc.*, **87**, 3532 (1965); see also Reference 101.
[122] W. Merk and R. Petit, *J. Amer. Chem. Soc.*, **89**, 4787 (1967).
[123] A. Wasserman *et al.*, *J. Chem. Soc.*, 1511 (1935); 432, 1028 (1936); 362, 371, 381, 870 (1939); 612 (1942); 979 (1953).
[124] J. Sauer, H. Wiest and A. Mielert, *Z. Naturforsch.*, **17b**, 203 (1962).
[125] J. Sauer, D. Lang and H. Wiest, *Z. Naturforsch.*, **17b** 206 (1962).
[126] G. B. Kistiakowsky and J. R. Lacher, *J. Amer. Chem. Soc.*, **58**, 123 (1936).
[127] G. N. Schrouzer and P. Glockner, *Chem. Ber.*, **97**, 2451 (1964).
[128] J. Sauer, D. Lang and H. Wiest, *Chem. Ber.*, **97**, 3208 (1964).
[129] K. N. Houk and R. W. Strozier, *J. Amer. Chem. Soc.*, **95**, 4094 (1973).

Intermediates have occasionally been identified in the Diels-Alder reaction [132]. Their occurrence explains also the observed orientation and side products. However, many other experimental data agree better with the concerted mechanism. Thus, the activation parameters of the direct reaction or of the cycloreversion reaction indicate a highly oriented transition state, such as would correspond to a synchronous formation (or breaking) of the bonds. Indeed, most of the observed activation entropies of the diene addition (see Table 12.5) have large negative values [123, 128], whereas the reverse reaction has activation entropies of about zero [130, 131], suggesting that the geometry of the transition state resembles the geometry of the adduct.

12.5.2 *Orientation* (*regioselectivity*)

Two different orientations are possible in a diene addition involving substituted reaction components. Thus, a 1-substituted diene with a monosubstituted dienophile may form an *ortho* and a *meta* adduct; the same dienophile with a 2-substituted diene can yield a *meta* and a *para* adduct. Actually, it has been observed that 1-substituted dienes form preferentially *ortho* adducts, whereas with 2-substituted derivatives *para* adducts are favoured [133–135].

These results could be regarded as being consistent with a mechanism involving a polar intermediate in which the substituents may conjugate with the adjacent charged atoms:

Ph COOR → Ph O C OR → Ph COOR

However, the regioselectivity of a diene addition may be better explained by a concerted mechanism. The transition state of a diene addition may be regarded as a six π-electron system, similar to that of benzene [10], except for the fact that beside *p*-π overlap, *p*-σ overlap must also be considered:

Ph CH=O → Ph CH=O → Ph CH=O

67

[130] B. S. Khombota and A. Wassermann, *J. Chem Soc.*, 375 (1939).

[131] W. C. Herndon, C. R. Grayson, and J. M. Manion, *J. Org. Chem.*, **32**, 526 (1967).

[132] S. W. Benson, *J. Chem. Phys.* **46**, 4920 (1967).

[133] I. N. Nazarov, Yu. A. Titov, and A. I. Kuznetsova, *Izv. Akad. Nauk SSSR, Otdel Khim. Nauk*, 1270 (1959).

[134] Yu. A. Titov, *Usp. Khim.*, **31**, 529 (1962); *Russ. Chem. Rev.*, **31**, 267 (1962).

[135] A. A. Petrov and A. P. Sopov, *Zhur. Obshch. Khim.*, **27**, 1795 (1957).

As in benzene, substituents in *ortho* and *para* will be better conjugated than those in *meta*. These qualitative aspects regarding the regioselectivity of the reaction have been treated more quantitatively by quantum mechanical methods[136–139]. The generalised frontier orbital method[138, 139], based on empirical and calculated energies of frontier orbitals as well as on the coefficients of the HO and LUMOs, may well account for the observed regioselectivity.

A more complicated problem is the orientation in the catalysed Diels-Alder reaction. Lewis acids may have a very favourable effect on the orientation, increasing for example the *para*/*meta* ratio of a 2-substituted diene about 15-fold[140]. Applying the frontier orbital method to the acrolein dimerisation in which one molecule (the dienophile) is activated by protonation, Houk and Strozier[129] have demonstrated that the catalysed reaction may also be concerted, the diene-HO-dienophile-LU interactions accounting both for the increased reaction rate and the increased selectivity. Secondary orbital interactions (see p. 646) can be more important in the catalysed process than they are in the uncatalysed reaction.

12.5.3 *Diene reactivity*

In order to participate in a synchronous Diels-Alder reaction, dienes must adopt a cisoid geometry about the bond (*s-cis**), which for an acyclic diene is less stable than the *s-trans* conformation. The observed rate constant of reaction (12.29) will thus depend on the equilibrium constant K:

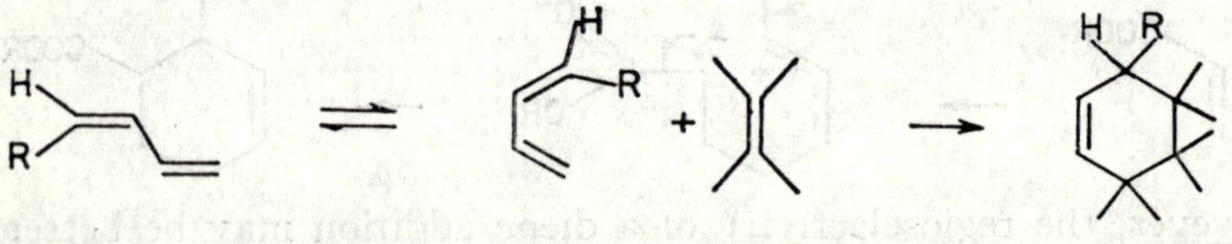

(12.29)

In *cis*-1-substituted dienes, the *s-cis* conformation must be disfavoured with respect to the *s-cis* form of the corresponding *trans* isomer, due to non-bonding interactions between the substituent and the 4-hydrogen. In a concerted reaction, the *trans* isomer should be therefore more reactive than the *cis* derivative. No large differences in reactivity may be expected in a non-synchronous reaction. Actually, the experimental data are in agreement with the first alternative. Indeed, *cis*-1-phenylbutadiene[142] with maleic anhydride, under the conditions

* The symbol *s* (*s* = single) has been introduced by Mulliken[141] in order to distinguish different planar conformations of dienes, resulted by rotation about single bonds.

136 J. Feuer, W. C. Herndon and L. H. Hall, *Tetrahedron*, **24**, 2575 (1968).

137 R. Sustman, A. Ansmann and F. Vahrenholt, *J. Amer. Chem. Soc.*, **94**, 8099 (1972).

138 K. N. Houk, *J. Amer. Chem. Soc.*, **94**, 8953 (1972).

139 K. N. Houk, *J. Amer. Chem. Soc.*, **95**, 4092 (1973).

140 T. Inukai and T. Kojima, *J. Org. Chem.*, **32**, 872 (1967) and references quoted therein.

141 R. S. Mulliken, *Rev. Mod. Phys.*, **14**, 265 (1942).

142 O. Grummit and F. J. Christoph, *J. Amer. Chem. Soc.*, **73**, 3479 (1951).

under which the *trans* isomer is converted quantitatively into the reaction product, forms an adduct only in 5% yield, indicating that the *trans* isomer is about 10^2—10^3-fold more reactive than the *cis*.

It is now well known that beside the planar *s-trans* and *s-cis* conformations *, dienes may also exist in non-planar *skew* or *gauche* forms, usually possessing higher stability than the *s-cis* conformer[143–145]. It is believed [146] that acyclic dienes, even unsubstituted, react in this conformation in a stepwise or quasi-synchronous reaction, in order to avoid the unstable *s-cis* conformation.

The conformation of cyclic and polycyclic dienes is frozen either in the cisoid (**68**) or in the transoid (**69**, **70**) form. Irrespective of the mechanism (concerted or stepwise), the reaction is only possible from cisoid forms.

The formation of Diels-Alder adducts is currently used in the determination of the position of double bonds in polycyclic hydrocarbons [107].

When polar effects are considered, distinction must be made between diene addition with normal electron demand and inverse electron demand [147–149]. In the former reaction, more frequently encountered, the diene component is electron-rich, whereas the dienophile is electron-deficient. Electron-repellent substituents in the diene will favour the reaction, while electron-withdrawing groups will decrease the rate. Indeed, in the reaction with maleic anhydride, in dioxane at 30 °C, 2-methylbutadiene [150] is twice as reactive as butadiene, whereas the latter, under similar conditions, reacts 10 times faster than 2-chlorobutadiene. In the Diels-Alder reaction with inverse electron demand, the diene is the electron-deficient component. Electron-withdrawing substituents, such as in hexachlorocyclopentadiene, will determine the electron-acceptor character of the diene component.

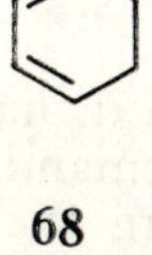

; ; ; Ph O Ph

68 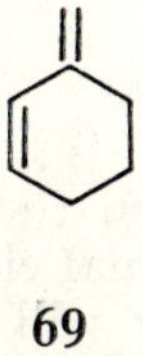 **69** 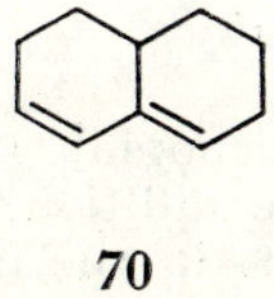 **70** 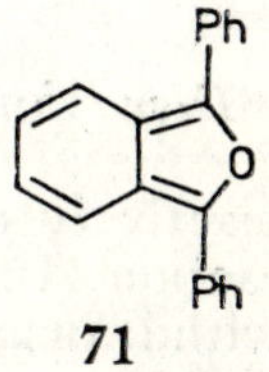 **71**

* Conjugation is maximum in the planar conformation ($\tau = 0°$) and is annihilated if the torsional angle τ formed by the planes of the two bonds is 90°. Non-bonding interactions are minimum in the latter geometry. In order to preserve a part of the conjugation energy and at the same time diminish the non-bonding interactions, a torsional angle of about 30—60° is actually adopted.

143 L. Pauling, *The Nature of the Chemical Bond*, 3rd edn., Cornell University Press, Ithaca, New York (1960), p. 130—142.

144 L. A. Radon and J. A. Pople, *J. Amer. Chem. Soc.*, **92**, 4786 (1970).

145 K. Hogen and K. Hedberg, *J. Amer. Chem. Soc.*, **95**, 1103 (1973).

146 W. von E. Doering, personal communication.

147 W. E. Bachmann and N. C. Deno, *J. Amer. Chem. Soc.*, **71**, 3062 (1949).

148 J. Sauer and H. Wiest, *Angew. Chem.*, **74**, 353 (1962).

149 J. Sauer and J. Heinrichs, *Tetrahedron Letters*, 4979 (1966).

150 J. Sauer, D. Lang and A. Mielert, *Angew. Chem.*, **74**, 352 (1962).

Thermodynamically unstable compounds are the most reactive dienes. *ortho*-Quinodimethanes, such as **71**, are thus frequently used as trapping agents[116, 151] for alkenes.

Cyclobutadiene deserves a special mention. Later works[96] have shown that cyclobutadiene is extremely unstable, the classical methods used failing to generate this compound. Nenitzescu, Avram and their co-workers[152] have shown however that cyclobutadiene may form stable π complexes with metal derivatives, their decompositions serving as models for the reactions of the ligand. Pettit and his co-workers[153, 154] have demonstrated that cyclobutadiene may play the role either of diene or of dienophile in cycloaddition reactions. (It seems that cyclobutadiene is thermodynamically stable, although highly reactive *. Simple HMO calculations (see p. 72) indicate a diradical structure, in agreement with its reactivity.)

The reaction between an electron-rich diene and an electron-deficient dienophile can be well interpreted by quantum mechanical methods[31, 129, 139], assuming a concerted process. The lower the difference between the interacting frontier orbitals and the higher the coefficients of the termini atoms in these orbitals, the higher the reactivity of the diene in the cycloaddition will be.

Indeed, considering the perturbation theory (see p. 83) applied to the frontier orbital method, the stabilisation energy resulted by the interaction of two molecular obitals is given by relation (12.30)[155]:

$$E = \frac{\sum (a_{i,R}\, b_{j,S}\, \beta)^2}{E_R - E_S} \tag{12.30}$$

where E_R, E_S are the energies of the frontier orbitals in components R and S, respectively, while $a_{i,R}$ and $b_{j,S}$ are the coefficients of the termini atoms in the frontier orbitals i and j of the two components. A high reactivity is either determined by a small difference $E_R - E_S$ or by large values of the coefficients $a_{i,R}$, $b_{j,S}$.

12.5.4 *Dienophile reactivity*

The reactivity of the dienophile depends on the electron demand of the reaction. In a diene addition with normal electron demand, electron-withdrawing groups in the dienophile will accelerate the reac-

* Although often assumed, there is no direct relation between thermodynamic stability and chemical reactivity. Molecules which are stable in the absence of other species can be highly reactive in the presence of suitable reagents.

151 I. G. Dinulescu, M. Avram and C. D. Nenitzescu, *Stud. Cerc. Chim.*, **11**, 419 (1963).

152 M. Avram, E. Marica and C. D. Nenitzescu, *Chem. Ber.*, **92**, 1088 (1959); M. Avram, G. D. Mateescu, I. G. Dinulescu, E. Marica and C. D. Nenitzescu, *Tetrahedron Letters*, 21 (1961).

153 L. Watts, J. D. Fitzpatrick and R. Pettit, *J. Amer. Chem. Soc.*, **88**, 623, 1328 (1966).

154 P. Reeves, J. Henery and R. Pettit, *J. Amer. Chem. Soc.*, **91**, 5888, 5890 (1969).

155 M. J. S. Dewar, *The Molecular Orbital Theory of Organic Chemistry*, McGraw-Hill, New York (1969), p. 205.

tion rates (Table 12.5). Thus, tetracyanoethylene, a dienophile with typical electron-withdrawing groups, is 4.6×10^8-fold as reactive as cyclopentadiene, an electron-rich dienophile. For simple alkenes, this difference must be still larger *. Table 12.5 shows that the activation entropies are fairly constant and therefore the relative rates are mainly determined by the activation enthalpies. This is an indication that the rate constants must satisfy a linear free-energy relationship in which the sum of the substituent parameters (σ_R^*) at the double bond must be considered. Such estimations have been made by Dewar [156] which has shown that a disubstituted derivative, R—CH=CH—R, should react twice as fast as the monosubstituted compound, R—CH=CH_2 (due to the statistical factor), if a stepwise mechanism is involved. If the reaction proceeds through a cyclic transition state **72** and **73** respectively, two substituents should be more efficient in stabilising it than a single substituent. Earlier works, as well as recent studies of Sauer *et al.* [124, 125, 128, 157, 158], have shown that substituted ethylenes (*trans* isomers) react at least 50—100-fold as fast as the corresponding monosubstituted derivatives (Table 12.5), which indicates a concerted mechanism.

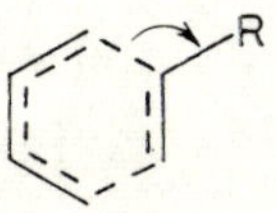

72 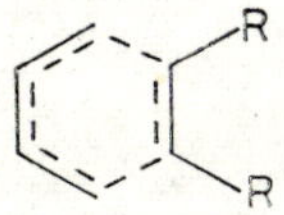 **73**

Transient intermediates like benzyne (see p. 418) or cyclobutadiene (see p. 622), are extremely reactive dienophiles. Their reactions have been used either for synthetic purposes [159, 160] or as trapping reactions, such as (12.31) [161], in demonstrating the occurrence of other highly reactive intermediates:

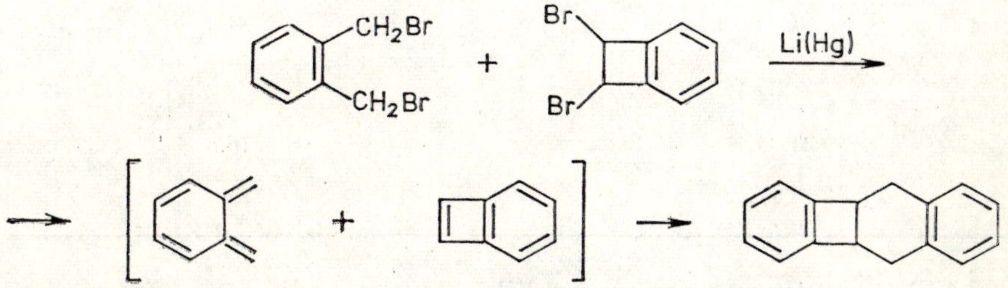

CH_2Br CH_2Br + Br Br Li(Hg)

(12.31)

* Simple alkenes do not react with electron-rich dienes.

156 M. J. S. Dewar, *Tetrahedron Letters*, (4), 16 (1959).

157 J. Sauer, H. Wiest and A. Mielert, *Chem. Ber.*, **97**, 3183 (1964).

158 J. Sauer, D. Lang and H. Wiest, *Chem. Ber.*, **97**, 3208 (1964).

159 M. Stiles and R. G. Miller, *J. Amer. Chem. Soc.*, **82**, 3802 (1960).

160 M. Stiles, R.G. Miller and U. Burkhardt, *J. Amer. Chem. Soc.*, **85**, 1792 (1963).

161 C. D. Nenitzescu, M. Avram and D. Dinu, *Chem. Ber.*, **90**, 2541 (1957); M. Avram, I. G. Dinulescu, D. Dinu, G. D. Mateescu and C. D. Nenitzescu, *Tetrahedron*, **19**, 309 (1963).

TABLE 12.5

Reactivity of dienophiles towards cyclopentadiene, in dioxan, at 20 °C [157]

No.	*Dienophile*	$10^6 k_2$ *l mol* s^{-1}	$\Delta H^{\ddagger}$ *kcal mol*$^{-1}$	$\Delta S^{\ddagger}$ *e.u.*
1	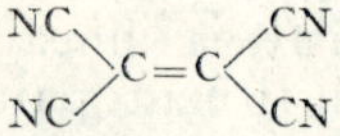 $(NC)_2C{=}C(CN)_2$	430 000 000	—	—
2	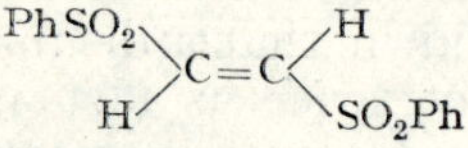 $PhSO_2(H)C{=}C(H)SO_2Ph$ (trans)	115 000	7.8	−36
3	$HC(CO)_2NPh$ (maleimide, $\|$—CO—NPh—CO—)	70 500	7.5	−38
4	$\|$—CO—O—CO— (maleic anhydride)	55 600	8.3	−36
5	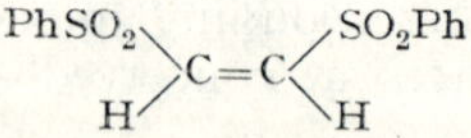 $PhSO_2(H)C{=}C(H)SO_2Ph$ (cis)	1 940	10.8	−34
6	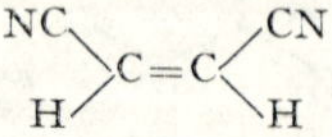 $NC(H)C{=}C(H)CN$ (cis)	910	10.8	−35
7	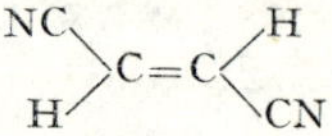 $NC(H)C{=}C(H)CN$ (trans)	806	11.6	−33
8	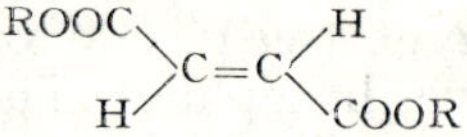 $ROOC(H)C{=}C(H)COOR$ (trans)	742	11.2	−35
9	$CH_2{=}CH{-}COOR$	11.8	14.3	−32
10	$CH_2{=}CH{-}CN$	10.4	—	—
11	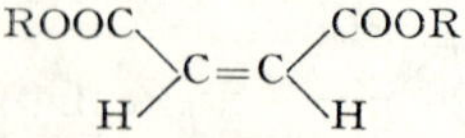 $ROOC(H)C{=}C(H)COOR$ (cis)	6.28	14.1	−34
12	$CH{-}CH$ / $\|\ \ \|$ / $CH\ \ CH$ / CH_2 (cyclopentadiene)	0.94	16.6	−29

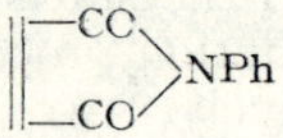

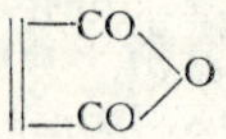

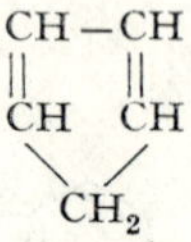

The reactivity of a dienophile can be well accounted for by the frontier orbital theory [129, 139], considering a concerted reaction. However, these and other similar treatments cannot exclude the non-synchronous mechanism, at least for certain components.

In the diene addition with inverse electron demand simple alkenes are the most reactive. These reactions have been used for the identification of simple alkenes [162].

[162] M. Avram, I. Dinulescu, E. Marica and C. D. Nenitzescu, *Chem. Ber.*, **95**, 2248 (1962).

12.5.5 *Stereochemistry*

The structure of a Diels-Alder adduct can be sometimes accurately predicted considering polar effects in the components (for the regiospecificity) as well as the rules established by Alder and Stein[4,5] (for the stereospecificity).

The rule of cis addition. The relative orientation of substituents in both components of a diene addition is preserved in the resulted adduct. Thus, a 1,4-*trans*-disubstituted diene, **74**, will form a cyclohexene derivative **75** having the two substituents in equatorial-equatorial positions (*trans*). A diene, for example cyclopentadiene, will react with the two isomers of a 1,2-disubstituted dienophile yielding a *cis* and a *trans* adduct, respectively:

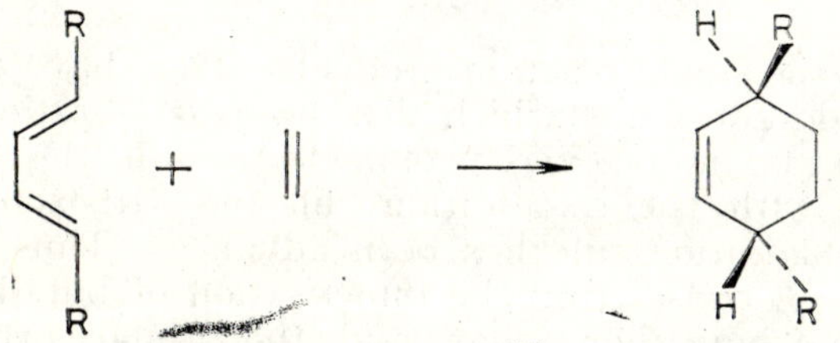

trans-trans
74

equatorial-equatorial (*trans*)
75

fumarate diester

maleate diester

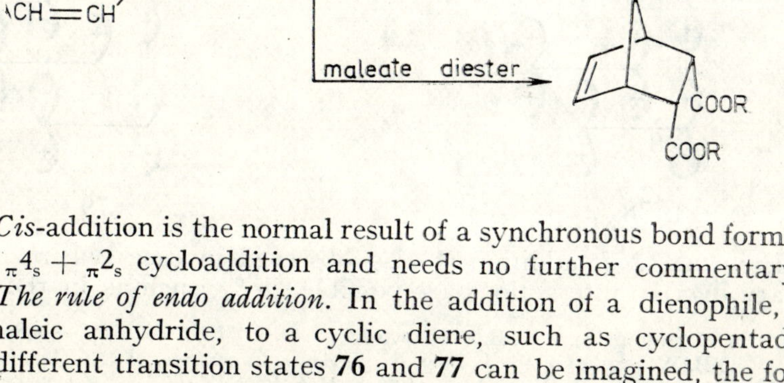

Cis-addition is the normal result of a synchronous bond formation in a $_{\pi}4_s + {}_{\pi}2_s$ cycloaddition and needs no further commentary.

The rule of endo addition. In the addition of a dienophile, such as maleic anhydride, to a cyclic diene, such as cyclopentadiene, two different transition states **76** and **77** can be imagined, the former leading to an *endo* adduct and the latter to the *exo* isomer.

Alder and his co-workers [4] have found that only the *endo* product forms, if there is no equilibration of the initially formed adduct. Otherwise, the more stable *exo* derivative may also result

CH=CH CH CH₂ CH + CH–CO CH O CO → [CO O CO 76 → *endo*; CO O CO 77 → *exo*]

or even be the single reaction product. They have thus established the *endo*-addition rule which has been later rationalised [5, 115] as the rule of maximum overlap of π-electrons in the transition state.

Unlike with the *cis*-addition rule, no satisfactory explanation of the *endo*-addition rule has been advanced. Thus, Hoffmann and Woodward [163], considering the dimerisation of butadiene in a Diels-Alder scheme, have shown that beside the primary orbital interactions between C-1 and C-4 of the diene component and C-1 and C-2 of the dienophile, there are also secondary interactions between C-2 of the former component and C-3 of the latter. Such interactions are only possible in the *endo*-orientation (**79**), stabilising the transition state, and are impossible in the *exo*-orientation (**78**). The same must be true in other diene additions involving delocalised π-systems in the dienophile:

78 **79**

However, *endo*-addition is also observed in cases where no secondary overlap interactions are possible [164, 165], such as in the diene

[163] R. Hoffmann and R. B. Woodward, *J. Amer. Chem. Soc.*, **87**, 4388 (1965).
[164] K. B. Wiberg and W. J. Bartley, *J. Amer. Chem. Soc.*, **82**, 6375 (1960).
[165] S. J. Cristol, W. K. Seifert and S. B. Soloway, *J. Amer. Chem. Soc.*, **82**, 2351 (1960).

addition involving cyclopropane as dienophile. Herndon and Hall[166] have demonstrated that in such cases primary interactions, assuming the geometry of a concerted transition state, are sufficient to account for the *endo*-addition. The secondary interactions may play an important role in some cycloadditions involving highly electron-deficient dienophiles[139].

Anyway, whatever the explanation of the actual stereochemistry, the *endo*-orientation is consistent with a concerted mechanism.

The *endo*-addition rule is not as universal as the *cis*-addition rule. It can be annihilated by the more general rule which states that the reagents approach each other from the less hindered side[120]. If that side happens to correspond to an *exo*-orientation, *exo*-adducts will be observed both under kinetic and thermodynamic conditions.

An interesting example has been studied by Jacobson[167] who, in order to verify the importance of secondary overlap interactions, has subjected the naphthalene derivatives **80***a*—**80***c* to a diene addition using dimethyl fumarate and 4-phenyl-1,2,4-triazoline-3,5-dione, **81**, as dienophiles:

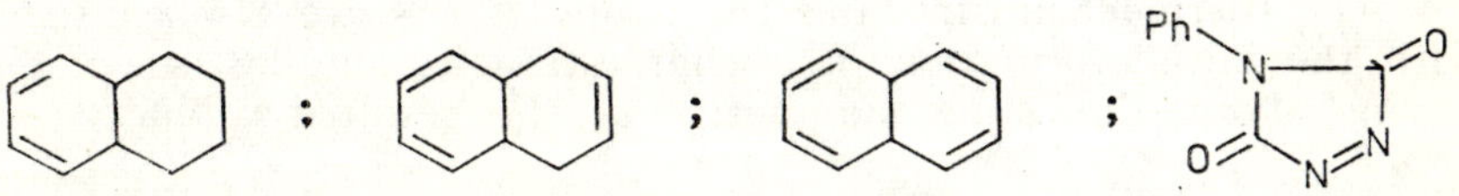

Ph N O O N N

80*a* ; **80***b* ; **80***c* ; **81**

Attack on the less hindered side (*exo*-orientation) was preferred in all cases, the ratio *exo*/*endo* varying from 19 (for **80***a*) to 4 (for **80***c*). The increase in the weight of *endo*-orientation for **80***c* cannot be explained by secondary overlap interactions, but are well accounted for by the decrease of geometrical constraints. Indeed, the steric requirements of the saturated cyclohexane ring in **80***a* are higher than the steric requirements of the sp^2 carbon atoms in **80***c*.

12.5.6 *Thermochemistry. Isotope effects*

The Diels-Alder reaction is highly exothermic. According to Hammond's postulate[168], the transition state should then be similar to the initial system formed by the diene and dienophile, the direct reaction presenting thus an early transition state, whereas the *retro*-Diels-Alder reaction a late transition state (see also p. 137). These predictions are confirmed by secondary isotope effects.

[166] W. C. Herndon and L. H. Hall, *Tetrahedron Letters*, 3095 (1967).
[167] B. M. Jacobson, *J. Amer. Chem. Soc.*, **95**, 2579 (1973).
[168] G. S. Hammond, *J. Amer. Chem. Soc.*, **77**, 334 (1955).

Indeed, it is well known that in both heterolytic and homolytic reactions the rehybridisation of the reaction centre from sp^3 to sp^2 is accompanied by a deuterium kinetic isotope effect, corresponding either to an α-hydrogen [169, 170] (bonded to the reaction centre) or to a β-hydrogen [171, 172]. The β-isotope effect is believed to be the result of a better hyperconjugation of a H—C bond than of a D—C bond. The α-isotope effect is fairly constant and small, $k_H/k_D = 1.12$. In a Diels-Alder reaction the reacting centres suffer a reverse rehybridisation, $sp^2 \rightarrow sp^3$. Thus the substitution by deuterium of an α-hydrogen is expected to be accompanied by an increase of the reaction rate, showing an isotope effect $k_D/k_H = 1.12$ if a late transition state is involved (as in the endothermic heterolysis or homolysis). For an early transition state, practically no isotope effects should be observed. Conversely, in a *retro*-Diels-Alder reaction, the rehybridisation of the reaction centres is $sp^3 \rightarrow sp^2$, this process occurring in an endothermic reaction; an isotope effect $k_H/k_D \cong 1.12$ is to be expected. Actually, an isotope effect of 1.01 —1.07 has been observed in the diene addition, whereas in the cycloreversion k_H/k_D agrees well with the predicted value [117]. The combination of α- and β-isotope effects in the same type of Diels-Alder processes confirms in its turn the concerted mechanism [173].

The thermochemistry and the istope effects are also consistent with the initial formation of a charge-transfer complex (see p. 350), having about the same symmetry as the resulted adduct [174, 175].

12.5.7 *Unconcerted* 4 + 2 *cycloadditions*

The evidence for the concerted Diels-Alder reaction presented in the preceding sections is sometimes regarded as consistent with an unconcerted reaction, involving as intermediate either a polar species, **65**, or a diradical, **66** (p. 638). Indeed, let us suppose that the reaction proceeds through **65** and that the process **65** → adduct has a higher activation energy than the formation of **65**. The polarity of the solvent will then oppose the disappearance of **65**, an overall decrease of the reaction rate with increasing polarity of the solvent being observed. The reverse must be true if the first step is rate-determining. However, if the two reaction steps have nearly the same activation energies, the solvent effects will compensate mutually, the reaction appearing little affected by the change in the solvent polarity, as should also be expected in a concerted reaction. In the same manner it can be shown [9] that all the experi-

[169] A. Streitwieser, Jr., R. H. Jagow, R. C. Fahey and S. Suzuki, *J. Amer. Chem. Soc.*, **80**, 2326 (1958).
[170] S. Seltzer, *J. Amer. Chem. Soc.*, **83**, 2625 (1961).
[171] E. S. Lewis, *Tetrahedron*, **5**, 143 (1959).
[172] V. J. Shiner, Jr., *Tetrahedron*, **5**, 243 (1959).
[173] S. Seltzer, *J. Amer. Chem. Soc.*, 1534 (1965).
[174] R. B. Woodward, *J. Amer. Chem. Soc.*, **64**, 3058 (1942).
[175] L. J. Andrews and R. M. Keefer, *J. Amer. Chem. Soc.*, **77**, 6284 (1955).

mental arguments (regarding kinetics, reactivity, orientation, stereochemistry, thermodynamics) are not convincing evidence for the concerted mechanism. According to Gompper [9], only the stepwise reaction can be experimentally proven with certainty, by identifying the reaction intermediates.

As already shown, one way of identifying a reactive intermediate is by trapping. It must be observed that both **65** and **66** can form a 4 + 2 and a 2 + 2 adduct:

$R^1CH{=}CH{=}CH{-}CHR^2{-}CHR^3{-}CHR^4 \longrightarrow$ (4 + 2 adduct) + (2 + 2 adduct)

65 or **66**

The formation of the two adducts in the same rate-determining step is convincing evidence for the stepwise reaction.

The 4 + 2 cycloadditions between polar dienes and polar dienophiles are usually two-step reactions, involving dipolar intermediates. As an example we shall mention the reaction of an enamine with a heteroatomic diene [176]. The intermediate **82** has been trapped with tetracyanoethylene:

$Me_2C{=}CHNMe_2 + H_2C{=}CH{-}C(Me){=}O \longrightarrow$ **82** (TCNE)

A reaction involving intermediate **83**, similar to **82**, is the Nenitzescu synthesis of hydroxyindoles [177, 178]. The mechanism, proposed by Răileanu and Nenitzescu [179], involves the addition of an enamine to quinone with formation of the zwitter-ion **83**. The cyclisation of the latter is possible only in one way, namely leading to an indole derivative:

176 I. Fleming and M. H. Karger, *J. Chem. Soc.* C, 226 (1967).
177 C. D. Nenitzescu, *Ber. dtsch. chem. Ges.*, **58**, 1063 (1925).
178 G. Domschke, *Zeitschr. für Chemie*, **6**, 41 (1966).
179 D. Răileanu and C. D. Nenitzescu, *Rev. Roumaine Chim.*, **10**, 339 (1965); D. Răileanu, M. Palăghiță and C. D. Nenitzescu, *Tetrahedron*, **27**, 5031 (1971).

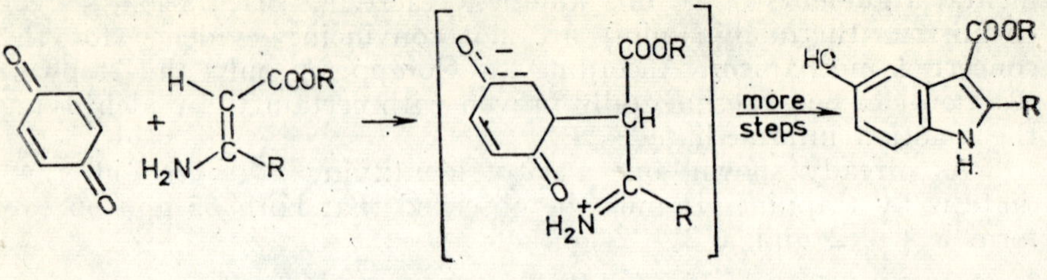

83

Later works[179–181] with conjugated enamines have confirmed the mechanism of Nenitzescu's indole synthesis and at the same time the stepwise reaction in the cycloaddition of such components.

In the gaseous phase, non-polar dienes and dienophiles are expected to form diradicals **66** if they do not react in a concerted way. An example extensively studied is the dimerisation of butadiene[132]. Intermediate **84** can lead to the cyclic dimers **24**, **25** and **85**. The thermal decomposition of these dimers as well as the kinetics of their reaction and of the dimerisation process agree well with a common intermediate **84**:

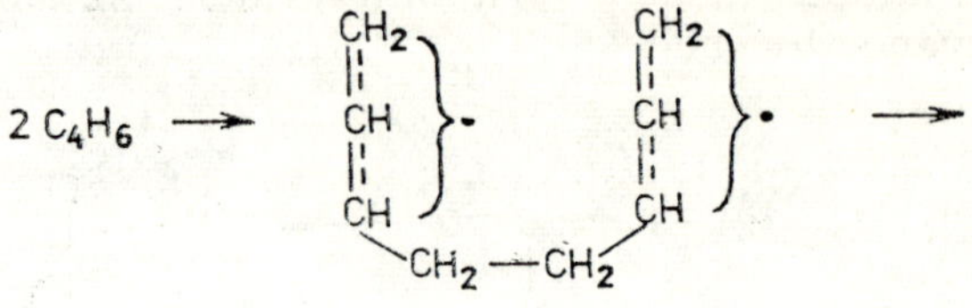

$2\,C_4H_6 \longrightarrow$

84

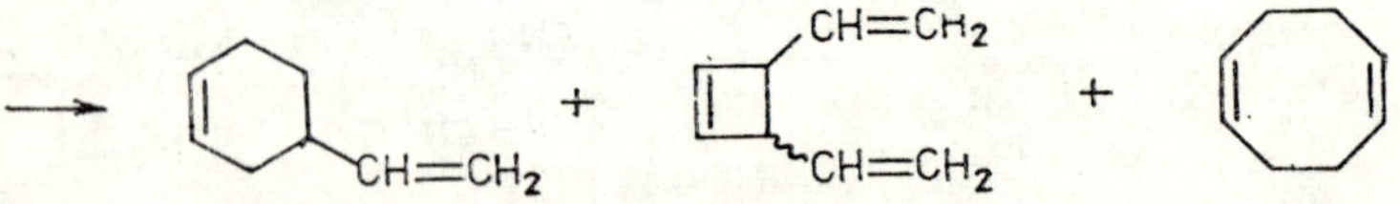

85 24 25

The rapid transformation of the diradical intermediate can determine an overall stereospecific reaction, despite the non-synchronous formation of the bonds[146]. We may say that **66** can memorise the initial conformation of the substrate. Benson[132] has estimated the lifetime of **84** at 10^{-11}—10^{-12}, in agreement with the above supposition.

[180] R. Huisgen, G. Szeimes and L. Mobius, *Chem. Ber.*, **100**, 2494 (1967); see also G. Cook (editor), *Enamines, synthesis, structure and reactions*, Marcel Dekker, New York (1969).

[181] N. Anghelide, C. Drăghici and D. Răileanu, *Tetrahedron*, **30**, 623 (1974).

Applying a perturbation method, Herndon and Feuer[182] were able to calculate the ratio of the cyclohexenic and cyclobutanic adducts in reactions (12.32)–(12.34), in perfect agreement with the experimental data[183–185], assuming a diradical intermediate **66**.

$$\text{Me}_2\text{C=CH–CH=CH}_2 + (\text{NC})_2\text{C=C(CN)}_2 \xrightarrow[2^\circ\text{C}]{\text{THF}} \text{cyclohexene adduct (12\%)} + \text{cyclobutane adduct (88\%)} \tag{12.32}$$

$$\text{CH}_2\text{=CH–CH=CH}_2 + \text{CH}_2\text{=C(CN)OAc} \xrightarrow[127^\circ\text{C}]{\text{hexane}} \text{cyclohexene adduct (82\%)} + \text{cyclobutane adduct (18\%)} \tag{12.33}$$

$$\text{Ph}_2\text{C=CH–CH=CH}_2 + (\text{NC})_2\text{C=C(CN)}_2 \longrightarrow \text{cyclohexene adduct (6\%)} + \text{cyclobutane adduct (66\%)} \tag{12.34}$$

Such examples show that the unconcerted 4 + 2 reaction is by no means rare and that the synchronous process is not always proved with certainty. However, theoretical arguments coupled with experimental evidence are reliable enough to decide that the concerted reaction is not an exception. Only when the activation energy of the unconcerted reaction is expected to be lower than the activation energy of the synchronous one, will a thermal $4_s + 2_s$ process choose the two-step pathway. These data are thus in perfect agreement with the theory of Woodward and Hoffmann.

182 W. C. Herndon, and J. Feuer, *J. Org. Chem.*, **33**, 417 (1968).
183 C. A. Stewart, Jr., *J. Amer. Chem. Soc.*, **84**, 117 (1962).
184 J. C. Little, *J. Amer. Chem. Soc.*, **87**, 4020 (1965).
185 J. K. Eisch and G. R. Henk, *J. Org. Chem.*, **31**, 589 (1966).

12.6 Other examples of cycloadditions

12.6.1 *2 + 1 Cycloadditions*

The 2 + 1 cycloadditions lead to three-membered rings. Most commonly encountered is the carbene addition to double bonds (see also p. 430):

$$>C=C< \quad + \quad :CR_2 \longrightarrow >C\underset{CR_2}{—}C< \qquad (12.35)$$

As has been shown in Chapter 9, carbenes may have either a singlet or a triplet structure. According to Skell[186], the addition of singlet carbenes is similar to the formation of halogenium ions:

$$>C=C< \quad + \quad Br^+ \longrightarrow >C\underset{\overset{+}{Br}}{—}C< \qquad (12.36)$$

and is therefore stereospecific. The examples known confirm amply this supposition[187]. The addition of a triplet carbene should be similar to the addition of atomic oxygen[188]:

$$>C=C< \quad + \quad :\ddot{\underset{\cdot}{O}}: \longrightarrow >C\underset{O}{—}C< \qquad (12.37)$$

and as such should be unstereospecific. Thus, the stereochemistry of the products can be used to decide whether the intermediate carbene had a singlet or a triplet structure[186].

A quantum mechanical treatment[189] of the cycloaddition of methylene, $:CH_2$, to ethylene has indicated that the components can be correlated in the ground state with ground state trimethylene **86**, which in its turn can be correlated with ground state cyclopropane. The reaction is thus symmetry-allowed in the ground state. In the transition state, the reagents should approach each other in a non-symmetric way **87** in agreement with the initial mechanism proposed by Skell[186].

[186] P. S. Skell and A. Y. Garner, *J. Amer. Chem. Soc.*, **78**, 5430 (1956).
[187] W. Kirmse, *Carbene Chemistry*, Academic Press, New York (1964); *Carbene, Carbenoide und Carbenanaloge*, Verlag Chemie (1969).
[188] R. J. Cvetanovič, *Can. J. Chim.*, **36**, 623 (1958).
[189] R. Hoffmann, *J. Amer. Chem. Soc.*, **90**, 1475 (1968).

Even if the occurrence of carbenes and their structure is exactly proven [190], it is now generally believed [187] that in cyclopropane synthesis carbenoid intermediates are involved rather than carbenes (see p. 431). It is interesting that whenever carbenoid derivatives of carbenes :CHX are involved, the reaction is *endo*-specific. Thus, if ArCH: is generated from $ArCHN_2$ (which is assumed to yield a true carbene) the addition is *cis*, affording however equal amounts

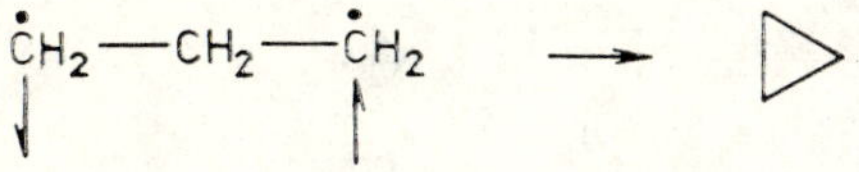

86 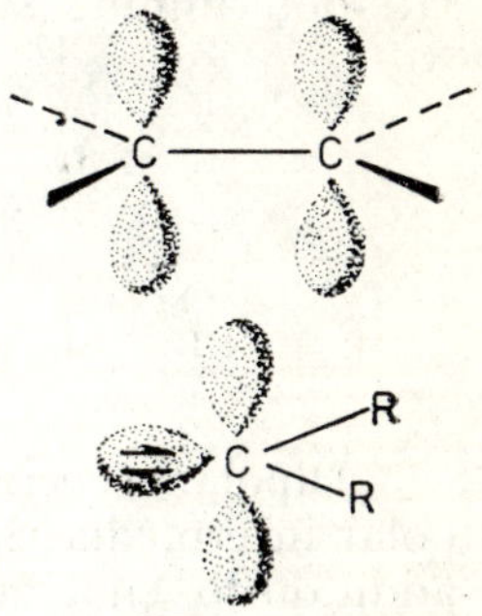 **87**

of the *endo* and *exo* products **88** and **89**. If the carbene is generated by α-elimination from $ArCHBr_2$ (assumed to yield a carbenoid species), the *endo*-adduct is preferentially formed [191]:

88 **89**

There is so far no satisfactory explanation for the *endo*-addition.

Similar to carbene additions to double bonds are the additions of nitrenes, :NR, yielding aziridines [192]:

[190] W. B. DeMore and S. W. Benson, in *Advances in Photochemistry* (W. A. Noyes, G. S. Hammond and J. N. Pitts, editors), Interscience, New York, vol. 2 (1964), p. 219.

[191] C. L. Closs and R. A. Moss, *J. Amer. Chem. Soc.*, 84, 4042 (1964).

[192] L. Lee Muller and J. Hammer, *1,2-Cycloaddition reactions. The Formation of three- and four-membered heterocycles*, Interscience, New York (1967).

12.6.2 *2 + 2 Cycloadditions*

The cyclodimerisation of fluorinated alkenes and the formation of cyclobutane from alkenes activated by electron-withdrawing groups has found applications in laboratory and industry relatively recently[193]. Simple heating is usually sufficient to carry out the reaction. One may assume that these thermal reactions are stepwise, the $_{\pi}2_s +$ $+ _{\pi}2_s$ cycloaddition being symmetry-forbidden in the ground state. Either dipolar intermediate **90** or diradicals **91** may result in the rate-determining step:

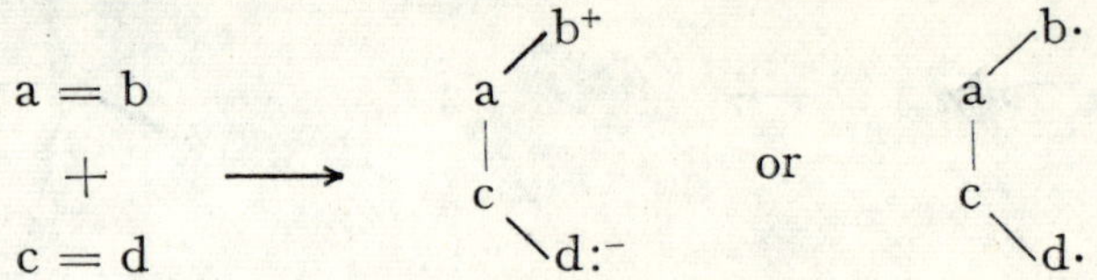

90 **91**

Dipolar species have been evidenced in the cycloaddition of polar intermediates[180, 181, 194]. In a recent study of tetracyanoethylene addition to enol ethers, Huisgen and Steiner[195] have demonstrated the formation of a zwitter-ion **92** which can rotate about the newly formed bond before the cyclisation, yielding both a *cis* and a *trans*-adduct:

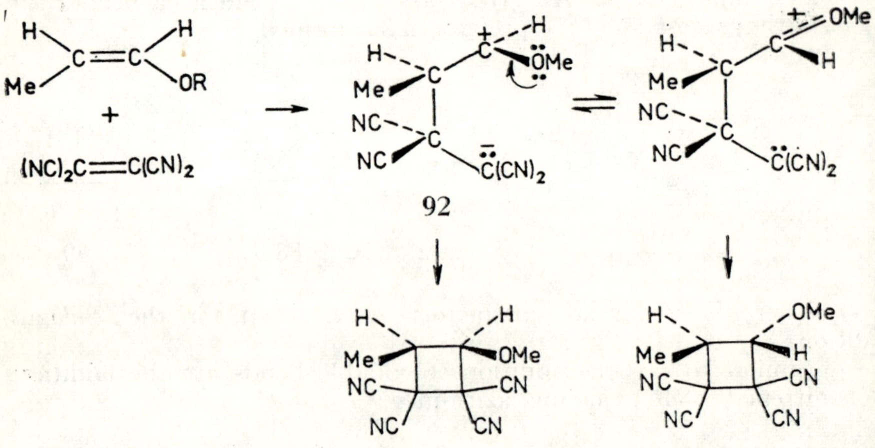

92

The non-stereospecific reaction has a greater weight in polar solvents (acetonitrile) than it has in non-polar solvents (benzene). Indeed, it can be assumed that the solvation of the two charged centres in **92** enables the rotation, the polar solvent acting better in that respect than the non-polar benzene.

[193] J. D. Roberts and C. M. Sharts, *Org. Reactions*, **12**, 1 (1962).

[194] P. D. Bartlett, L. K. Montgomery and B. Seidel, *J. Amer. Chem. Soc.*, **86**, 616 (1964).

[195] R. Huisgen and G. Steiner, *J. Amer. Chem. Soc.*, **95**, 5044, 5055, 5056 (1973).

Diradical intermediates are found more commonly in thermal $2+2$ cycloadditions. Thus, Bartlett and his co-workers [196, 197] have demonstrated that the cycloaddition of 1,1-dichloro-2,2-difluoroethylene to stereoisomeric hexa-2,4-dienes yields a mixture of isomeric cyclobutanes in a non-stereospecific way, the orientation corresponding to the formation of the most stable diradical:

$$CCl_2{=}CF_2 + CH_3{-}CH{=}CH{-}CH{=}CH{-}CH_3 \longrightarrow$$

$$\longrightarrow \begin{array}{c} CF_2{-}\dot{C}Cl_2 \\ | \\ CH \\ H_3C \quad CH\overset{\cdot}{\overline{=\!=}}CH\overline{=\!=}CH{-}CH_3 \end{array} \longrightarrow \text{products}$$

A comparison between the relative rates in typical diene additions and the relative rates of dichlorodifluoroethylene cycloadditions [198] has also indicated an unconcerted pathway.

Thermal concerted $2+2$ reactions are predicted to occur in an *antara-supra* process. Clearly, a ${}_{\pi}2_s + {}_{\pi}2_a$ reaction involves serious steric restrictions and will be observed only in special cases, such as the ketene addition to alkenes [199, 200]:

$$R^1CH{=}CH_2 + R^2CH{=}CO \longrightarrow \begin{array}{l} R^1CH{-}CH_2 \\ \;\;|\qquad\;\; | \\ R^2CH{-}CO \end{array}$$

It has been shown that the reaction is *supra* in the alkene component and must therefore be *antara* in the ketene component. The process must be favoured by the electrophilic character of ketenes [201].

$$\text{[orbital diagram: } {}^{-}O{-} \ldots R^1, H, R^2 \text{]} \longrightarrow \ddot{O}{=}C{-}\ldots{-}C\begin{smallmatrix}H\\R^2\end{smallmatrix} \equiv \text{cyclobutanone with } R^1, H, R^2, H, H, H$$

In the t-butylketene addition to styrene, the reaction has been demonstrated to be both stereospecific and stereoselective [202]. It is

[196] P. D. Bartlett and L. K. Montgomery, *J. Amer. Chem. Soc.*, **86**, 628 (1964).
[197] R. Gompper and W. Elser, *Tetrahedron Letters*, 1971 (1964).
[198] L. K. Montgomery, K. Schueller and P. D. Bartlett, *J. Amer. Chem. Soc.*, **86**, 622 (1964).
[199] R. Huisgen, L. Feiler and G. Binisch, *Angew. Chem. Intern. Ed. Engl.*, **3**, 753 (1964).
[200] R. Huisgen and P. Otto, *J. Amer. Chem. Soc.*, **90**, 5342 (1968).
[201] See for example Reference 29, p. 101.
[202] M. Gheorghiu, F. Kerek and M. Avram, *Rev. Roumaine Chim.*, **20**, 75 (1975).

believed that in the transition state the bulky t-butyl group chooses the most remote position from the phenyl group, so that the reaction leads to a *cis* adduct. This unfavourable stereochemistry of the adduct indicates an early transition state.

Photochemical cycloadditions $2_s + 2_s$ are quite frequent [203], even if they can compete with diradical mechanisms. If the two double bonds are in the same molecule, the concerted reaction may have high yields, as in the following examples:

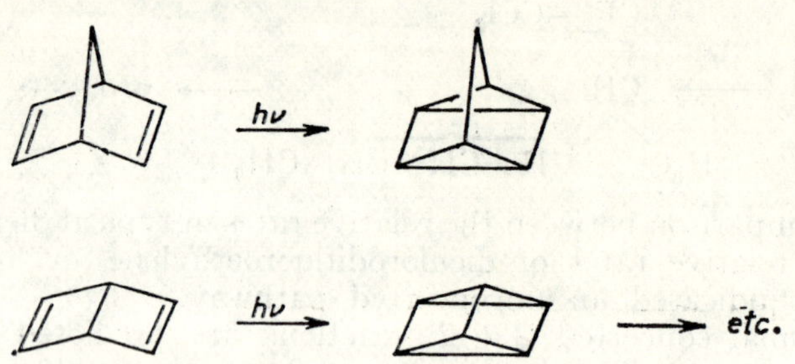

The 2 + 2 cycloadditions confirm in their turn the generality of the selection rules of Woodward and Hoffmann.

12.6.3 *2 + 3 Cycloadditions* (*1,3-dipolar cycloadditions*)

The 2 +3 cycloaddition implies as reagents a 1,3-dipole and an unsaturated compound, called dipolarophile. The product of the reaction is a five-membered ring heterocycle. The theory of the 1,3-dipolar addition and its applications in chemistry have been developed by Huisgen [14, 15].

Some of the 1,3-dipoles are stable compounds, such as diazoalkanes, azides, ozone, etc. However, the majority of them are unstable derivatives and must be generated *in situ*. A 1,3-dipole is a mesomeric compound, its dipolar character being only recognised by writing down all possible mesomeric structures. According to Huisgen [14, 15] there are four classes of dipoles, distinguished by their bonds and the way in which they are stabilised by conjugation (see Table 12.6).

TABLE 12.6

1,3-Dipolar systems

1,3-Dipole		*Examples*
$\overset{+}{a}=\ddot{b}-\overset{-}{c}:$	Nitrile ylides	$-\overset{+}{C}=\ddot{N}-\bar{\ddot{C}}<\longleftrightarrow-C\equiv\overset{+}{N}-\bar{\ddot{C}}<$
$\updownarrow$	Nitrile imines	$-\overset{+}{C}=\ddot{N}-\underset{..}{\bar{\ddot{N}}}-\longleftrightarrow-C\equiv\overset{+}{N}-\underset{..}{\bar{\ddot{N}}}-$
$a\equiv\overset{+}{b}-\overset{-}{c}:$	Nitrile oxides	$-\overset{+}{C}=\ddot{N}-\underset{..}{\ddot{O}}:^{-}\longleftrightarrow-C\equiv\overset{+}{N}-\ddot{O}:^{-}$

203 W. L. Dilling, *Chem. Rev.*, **66**, 373 (1966).

TABLE 12.6. (continued)

1,3-Dipole		*Examples*
	Diazoalkanes	$:\overset{+}{N}=\ddot{N}-\overset{-}{\ddot{C}}< \longleftrightarrow :N\equiv\overset{+}{N}-\ddot{C}<^{-}$
	Azides	$:\overset{+}{N}=\ddot{N}-\overset{-}{\ddot{N}}-R \longleftrightarrow :N\equiv\overset{+}{N}-\overset{-}{\ddot{N}}-R$
	Nitrogen monoxide	$:\overset{+}{N}=N-\ddot{O}:^{-} \longleftrightarrow :N\equiv\overset{+}{N}-\ddot{O}:^{-}$
$\ddot{a}-b=c$	Vinyl carbenes	$-\ddot{C}-C=C< \longleftrightarrow -\overset{+}{C}=C-\overset{-}{\ddot{C}}<$
$\updownarrow$		
$\overset{+}{a}=b-\overset{-}{c}:$	Ketocarbenes	$-\ddot{C}-C=\ddot{O}: \longleftrightarrow -\overset{+}{C}=C-\ddot{O}:^{-}$
	Vinyl imines	$:\ddot{N}-C=C< \longleftrightarrow :\overset{+}{N}=C-\ddot{C}<^{-}$
	Ketoimines	$:\ddot{N}-C=\ddot{O}: \longleftrightarrow :\overset{+}{N}=C-\ddot{O}:^{-}$
$\overset{+}{a}-\ddot{b}-\overset{-}{c}:$	Azomethine ylides	$>\overset{+}{C}-\ddot{N}-\overset{-}{\ddot{C}}< \longleftrightarrow >C=\overset{+}{N}-\overset{-}{\ddot{C}}<$
$\updownarrow$	Azomethine imines	$>\overset{+}{C}-\ddot{N}-\overset{-}{\ddot{N}}- \longleftrightarrow >C=\overset{+}{N}-\overset{-}{\ddot{N}}-$
$a=\overset{+}{b}-\overset{-}{c}:$	Nitrones	$>\overset{+}{C}-\ddot{N}-\ddot{O}:^{-} \longleftrightarrow >C=\overset{+}{N}-\ddot{O}:^{-}$
	Azoxy compounds	$-\overset{+}{N}-\ddot{N}-\ddot{O}:^{-} \longleftrightarrow -\ddot{N}=\overset{+}{N}-\ddot{O}:^{-}$
	Nitro compounds	$:\overset{+}{\ddot{O}}-\ddot{N}-\ddot{O}:^{-} \longleftrightarrow :\ddot{O}=\overset{+}{N}-\ddot{O}:^{-}$
	Carbonyl oxides	$>\overset{+}{C}-\ddot{O}-\ddot{O}:^{-} \longleftrightarrow >C=\overset{+}{O}-\ddot{O}:^{-}$
	Ozone	$:\overset{+}{O}-\ddot{O}-\ddot{O}:^{-} \longleftrightarrow :O=\overset{+}{O}-\ddot{O}:^{-}$
$\dot{a}-b-\dot{c}$	Trimethylene	$\dot{C}H_2CH_2\dot{C}H_2 \longleftrightarrow \overset{+}{C}H_2CH_2\overset{-}{\ddot{C}}H_2$
$\updownarrow$		
$\overset{+}{a}-b-\overset{-}{c}:$		
etc.		etc.

(a) Dipoles with double bond in which the electron-deficient atom acquires the octet by conjugation with unshared electrons from the adjacent atom:

$$\overset{+}{a}=\ddot{b}-c:^{-} \longrightarrow a\equiv\overset{+}{b}-c:^{-}$$

The central atom must have a lone pair of electrons and must be able to develop four bonds. It can therefore be only a nitrogen atom. Atoms a and c can however be carbon, nitrogen or oxygen.

(b) Dipoles with double bond and without octet stabilisation. Such dipoles are similar to vinylcarbenes [204]:

$$\ddot{a} - b = c \longleftrightarrow \overset{+}{a} = b - c{:}^-$$

(c) Dipoles without double bond and with octet stabilisation. The electron-deficient atom is stabilised by conjugation with the unshared electrons of the adjacent atom:

$$\overset{+}{a} - \ddot{b} - c{:}^- \longleftrightarrow a = \overset{+}{b} - c{:}^-$$

The central atom can be either nitrogen or oxygen, whereas the marginal atom may be carbon, nitrogen or oxygen.

(d) Dipoles without double bond and without octet stabilisation. A representative example of this class is trimethylene **86**.

1.3-Dipolar cycloadditions display a striking similarity with diene additions. If the 1,3-dipole is a stable compound, the reaction is simply carried out by mixing the components with or without heating, in the absence of any catalysts or light. Kinetic measurements have shown that the reaction is second-order (first-order in each component), the rate constants being only slightly affected by the polarity of the solvent. Moreover, most of the 1,3-dipoles behave as electron-donors and react with the same kind of electron-deficient components as do dienes in the Diels-Alder reaction. Reactive dienophiles are also reactive dipolarophiles [14, 15,205, 206]. The activation entropies have large negative values [206], as has been observed in the Diels-Alder reaction. The following example, illustrating also the generation of an unstable 1,3-dipole *in situ*, shows that 1,3-cycloadditions are highly stereospecific:

$$\text{PhXC=N—NHPh} \xrightarrow{Et_3N} \text{Ph}\overset{+}{\text{C}}\text{=}\ddot{\text{N}}\text{—}\overset{-}{\ddot{\text{N}}}\text{Ph} \longleftrightarrow \text{PhC}\equiv\overset{+}{\text{N}}\text{—}\overset{-}{\text{N}}\text{Ph}$$

not isolated

93

93 —fumaric diester→ PhC=N–NPh ring; H, ROOC on one C; COOR, H on the other C (trans)

93 —maleic diester→ PhC=N–NPh ring; H, ROOC on one C; H, COOR on the other C (cis)

[204] R. Huisgen, *Angew. Chem.*, **73**, 368 (1962).

[205] R. Huisgen, H. J. Sturm and H. Wagenhoffer, *Z. Naturforsch.*, **17b**, 202 (1962).

[206] A. Eckel, R. Huisgen, R. Sustmann, G. Wallbillich, D. Grashby and E. Spindler, *Chem. Ber.*, **100**, 2192 (1967).

The concertedness of the reaction is demonstrated also by the orientation of the reagents in the transition state. Very frequently the observed orientation does not correspond to the charge distribution in the dipole and the electron effects in the dipolarophile [14, 15]:

$$\mathrm{PhClC{=}N{-}CH_2{-}C_6H_4{-}NO_2} \xrightarrow{Et_3N} \mathrm{Ph\overset{+}{C}{=}N{-}\overset{..}{\bar{C}}H{-}C_6H_4{-}NO_2}$$

94

$$\mathbf{94} + \mathrm{PhCH{=}O} \longrightarrow \mathbf{95}$$

95

The formation of the adduct **95** can be explained by the steric demands in the two components. As in the diene addition, the reagents must approach each other from the least hindered side. The steric demands of the transition state are observed even with reagents with very small steric requirements.

The similarity between the 1,3-dipolar addition and the Diels-Alder reactions results from the structure of the dipoles. It has been shown [14, 15, 24, 29, 207] that dipoles can be treated as four-electron systems and thus the reaction will be a ground-state allowed process similar to the diene addition.

12.7 Problems

(1) 2,3-Dimethylbutadiene suffers on irradiation an electrocyclic cyclisation to 1,2-dimethylcyclobutene, the quantum yield of this process being $\Phi = 0.12$. How can you explain the fact that in the similar reaction of 1-methylbutadiene (piperylene) the electrocyclic transformation has a quantum yield of only 0.03 ?

(2) Indicate the products which may be obtained by the irradiation of cycloheptatriene and cycloocta-1,3-diene. What process do you think would involve the higher quantum yield? Justify your answer.

(3) Tarbell and Kincaid [50, 55] have studied the kinetics of the Claisen rearrangement under a variety of conditions. Their rate data in neat liquid satisfy better a zero-order kinetics (the line of the plot conversion *versus* time does not pass through the origin) instead of the asumed first-order kinetic law. Find an interpretation for this result.

[207] R. Huisgen, *Angew. Chem.*, **80**, 329 (1968).

(4) In the pyrolysis of compound I, the polycyclic hydrocarbon II is obtained:

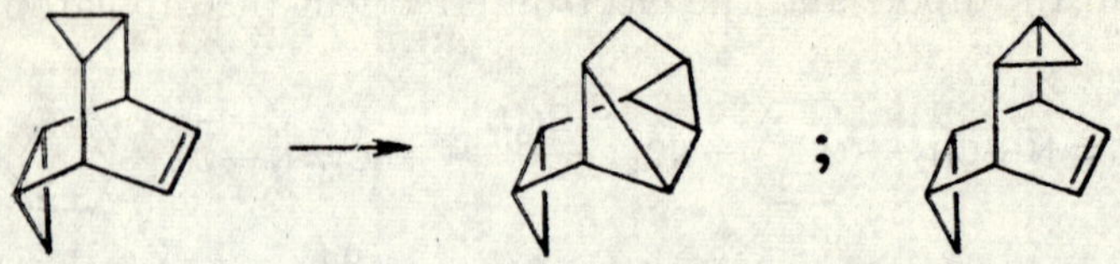

I II ; III

(D. Kaufmann and A. D. Meijere, *Tetrahedron Letters*, 3831 (1974)).

If this process is concerted, what type of pericyclic mechanism would you ascribe to the pyrolysis? Propose a mechanism corresponding to an unconcerted reaction. Of the two isomers, I and III, which is expected to have a higher tendency to react concertedly?

(5) The treatment of the tropylium ion (symmetry D_{7h}) with cyclopropylene, in water solution, leads to two alcohols, which on oxidation yield the ketones IV and V. Write a mechanism for these processes. The alcohols involved are products of concerted or unconcerted reactions?

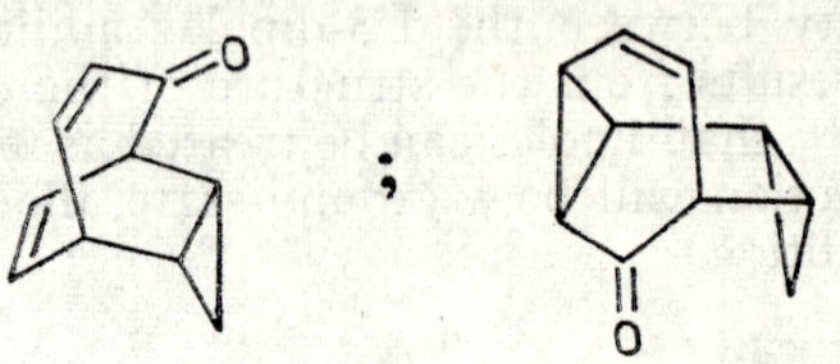

IV ; V

What products are expected from the reaction of tropylium ion with acenaphthylene?

Indication: See S. Itô, I. Itoh, I. Saito and A. Mori, *Tetrahedron Letters*, 3887 (1974).

(6) *cis*-Divinyl compounds VI may undergo a thermal isomerisation to VII:

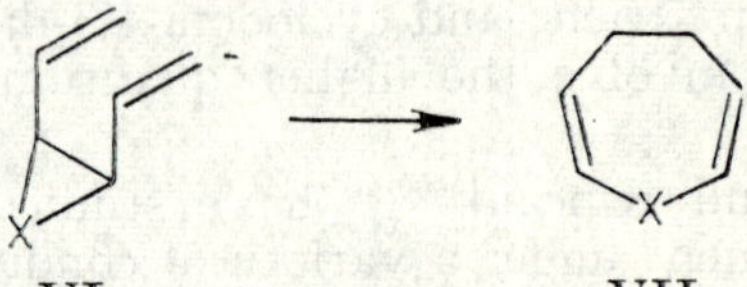

VI → VII (a)

(a) Show to what type of pericyclic transformation corresponds reaction (a).

(b) What variations of the rate constants would you predict for reaction (a) if X is, in turn, CH_2, O, NR?

(c) Indicate the products formed in reactions similar to (a) starting from the following derivatives:

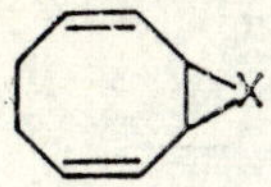

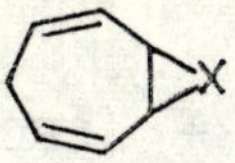

VIII ; IX

Indication: See J. C. Pommelet and J. Chuche, *Tetrahedron Letters*, 3897 (1974).

(7) The solvolysis of compound X yields small amounts of naphthalene. The naphthalene ratio in the reaction products does not increase if compound X is pyrolysed over 300 °C. However, high yields of naphthalene are obtained if the solvolysis of X is carried out in ultraviolet light.

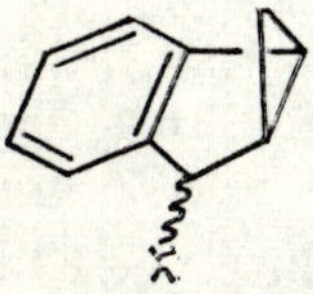

X

Explain these results.

Indication: See V. Wertheimer, A. Răzuş, F. Badea and A.-M. Glatz, *Tetrahedron Letters*, 1799 (1975).

(8) The treatment of cyclopentene with bromochlorocarbene, :CClBr, yields two isomers, α and β. The α isomer loses chloride ion at the same rate as the adduct obtained from cyclopentene and dichlorocarbene. The β isomer reacts in a similar way as the adduct obtained from cyclopentene and dibromocarbene. Indicate the structure of the two isomers, α and β.

(9) What is the reaction product expected from diene XI and maleic anhydride:

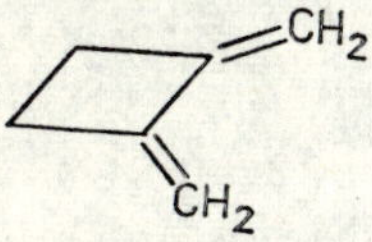

CH_2

CH_2

XI

under the following conditions: (a) equimolar amounts of reagents and temperature below 100 °C; (b) excess of maleic anhydride and temperature over 100 °C.

(10) In the 2 + 3 cycloaddition:

$$\text{(norbornadiene with 7-X)} + RCH{=}N{=}N \longrightarrow \text{(adduct)}$$

only the *endo-anti* isomer is formed. Explain this result.

Indication: See M. Frank-Neumann and M. Sedrati, *Angew. Chem.*, **86**, 673 (1974).

Subject Index

B

E

G

I

N

O

Q

R

S

T

U

V

W

X

Y

Z